T0210526

# THE ROUTLEDGE HANDBOOK OF DISABILITY AND SEXUALITY

This handbook provides a much-needed holistic overview of disability and sexuality research and scholarship. With authors from a wide range of disciplines and representing a diversity of nationalities, it provides a multi-perspectival view that fully captures the diversity of issues and outlooks.

Organised into six parts, the contributors explore long-standing issues such as the psychological, interpersonal, social, political and cultural barriers to sexual access that disabled people face and their struggle for sexual rights and participation. The volume also engages issues that have been on the periphery of the discourse, such as sexual accommodations and support aimed at facilitating disabled people's sexual well-being; the socio-sexual tensions confronting disabled people with intersecting stigmatised identities such as LGBTI or asexual; and the sexual concerns of disabled people in the Global South. It interrogates disability and sexuality from diverse perspectives, from more traditional psychological and sociological models, to various subversive and post-theoretical perspectives and queer theory. This handbook examines the cutting-edge, and sometimes ethically contentious, concerns that have been repressed in the field.

With current, international and comprehensive content, this book is essential reading for students, academics and researchers in the areas of disability, gender and sexuality, as well as applied disciplines such as healthcare practitioners, counsellors, psychology trainees and social workers.

**Russell Shuttleworth** is Senior Lecturer in Social Work at Deakin University, Australia. An anthropologist, social worker and disability support worker/personal assistant by training, his major research areas include the social and cultural construction of disability and impairment, disability and sexuality and disability and masculinities.

**Linda R. Mona** has worked as a clinical psychologist for the VA Long Beach Healthcare System in California, USA, for the past 19 years. As the clinical director of spinal cord injury/disorder psychology, she has prioritised disability affirmative psychological services and sexual health assessment and treatment for people with disabilities.

# THE ROUTLEDGE HANDBOOK OF DISABILITY AND SEXUALITY

*Edited by*
*Russell Shuttleworth and Linda R. Mona*

Routledge
Taylor & Francis Group

LONDON AND NEW YORK

First published 2021
by Routledge
2 Park Square, Milton Park, Abingdon, Oxon OX14 4RN

and by Routledge
52 Vanderbilt Avenue, New York, NY 10017

*Routledge is an imprint of the Taylor & Francis Group, an informa business*

*British Library Cataloguing-in-Publication Data*
A catalogue record for this book is available from the British Library

*Library of Congress Cataloging-in-Publication Data*
Names: Shuttleworth, Russell, 1953- editor. | Mona, Linda, editor.
Title: The Routledge handbook of disability and sexuality/edited by
Russell Shuttleworth and Linda Mona.
Description: Milton Park, Abingdon, Oxon; New York, NY:
Routledge, 2021. | Includes bibliographical references and index.
Identifiers: LCCN 2020034304 (print) | LCCN 2020034305
(ebook) | ISBN 9781138593237 (hardback) |
ISBN 9780429489570 (ebook)
Subjects: LCSH: People with disabilities–Sexual behavior.
Classification: LCC HQ30.5 .R68 20221 (print) |
LCC HQ30.5 (ebook) | DDC 306.7087–dc23
LC record available at https://lccn.loc.gov/2020034304
LC ebook record available at https://lccn.loc.gov/2020034305

ISBN: 978-1-138-59323-7 (hbk)
ISBN: 978-0-429-48957-0 (ebk)

Typeset in Bembo
by Deanta Global Publishing Services, Chennai, India

*Dedicated to Paul Chappell, a gifted researcher and advocate for diverse voices of disabled sexualities...(1972–2018)*

# CONTENTS

Contents

Contents

# ILLUSTRATIONS

## Figures

## Tables

# CONTRIBUTORS

**David Abbott** is Professor of Social Policy at the Norah Fry Centre for Disability Studies at the University of Bristol, UK. He completed a collaborative study about the lives of LGBT+ people with intellectual disabilities in 2005 and has an interest in the intersection of disability and sexuality.

**Courtney Andree** earned a doctorate in English Literature and Film and Media Studies from Washington University in St. Louis in 2015. Her recent work has appeared in *Film History*, *Disability Studies Quarterly*, *The Journal of Literary and Cultural Disability Studies*, *Literature Compass*, and *Modern Fiction Studies*, and she has pieces forthcoming in *An Interdisciplinary Companion to Slapstick Cultures*.

**Erin E. Andrews** is the Psychology Program Manager for the nationally designated Veterans Affairs TeleMental Health Clinical Resource Hub. She is a board-certified rehabilitation psychologist. Dr. Andrews has published and presented extensively in the areas of disability identity, inclusion in psychology training and cultural competence and disability.

**Julia Bahner** is postdoc at the School of Social Work, Lund University, Sweden. Her current research concerns the impact of inaccessibility on disabled people's opportunities for dating, socialising and sexual expression. Previous work includes an international project on sexual support policies and practices, and sex education in special schools for pupils with mobility impairments in Sweden.

**Pamela Block** is Professor of Anthropology at Western University, a past president of the Society for Disability Studies and a fellow of the Society for Applied Anthropology. Her research focuses on disability culture and cultural perceptions of disability in the United States, Brazil and Canada. She is particularly interested in multiple marginalisation and the intersections of gender, race, poverty and disability in movements for disability liberation (justice and rights) and disability oppression (eugenics, sterilisation, mass-incarceration and killing in Brazil, the United States and Canada). Finally, she engages actively in discussions of the emergence of neurodivergence and disability studies in Brazil and other Global South countries.

**Inge Griet Emy Blockmans** has a background in linguistics, literature and theatre (MA, Antwerp University, Belgium) and social psychology (MSc, University of Surrey, UK). As a joint PhD candidate at the Department of Special Needs Education (Faculty of Psychology and Pedagogical Sciences, Ghent University, Belgium) and the Institute for Family and Sexuality Studies (Faculty of Medicine, KU Leuven, Belgium), with the support of an FWO-Flanders PhD fellowship, she focused on the subjective experience of their bodies and intimate relationships of women with spinal cord injuries. Throughout her work with auto-ethnography, life stories, collective biography and ethnographic fieldwork, she aims to find ways to study embodiment away from objectification and immobilisation of people and their life stories.

**Alexander A. Boni-Saenz** is an Associate Professor of Law at Chicago-Kent College of Law, USA. Professor Boni-Saenz holds a J.D. magna cum laude from Harvard Law School, a M.Sc. with distinction in Social Policy from the London School of Economics, and an A.B. magna cum laude with highest honors in Psychology and Government from Harvard College.

**Christine Borst** is a licensed marriage and family therapist. She received her PhD from East Carolina University in Medical Family Therapy, and proceeded to work in medical practice transformation for several years, integrating behavioral health and medicine. She is an adjunct professor in the College of Health Solutions at Arizona State University, an artist, and children's book author/illustrator.

**Stine Hellum Braathen** is a research manager at SINTEF Digital, Department of Health Research, Norway. She has a background in anthropology, and also holds a PhD in Psychology from Stellenbosch University and an M.Phil. in international community health from the University of Oslo, Norway. She has been doing research in the field of disability studies since 2004, with a main focus on various aspects of inclusion and participation, including in the field of sexual and reproductive health.

**Sarah S. Brindle** received her Ph.D. in Counseling Psychology from the University of Iowa and has practiced in California as a licensed psychologist for over 14 years. She has provided services to veterans at the Long Beach VA Healthcare System in the Spinal Cord Injury Service in Long Beach, California, working with individuals with physical disabilities and chronic health problems. Her published work and professional presentations have focused on topics related to women's health issues, positive sexual expression across the lifespan and for people with disabilities, various psychological issues affecting people with disabilities, and spirituality among people with disabilities. She is also a certified yoga instructor with expertise in adaptive yoga for people with physical disabilities.

**Mark T. Carew** is a Senior Researcher at Leonard Cheshire and an Honorary Research Associate at University College London (UK). A social psychologist by training, his experience and research interests lie in tackling problematic issues and barriers that affect people with disabilities globally, including exclusion from sexual health.

**Mussa Chiwaula** is the Director-General of the Southern Africa Federation of the Disabled (SAFOD), based in Botswana, where he provides leadership in coordinating the activities of Disability Umbrella Organisations that work to promote the rights of persons with disabilities in 10 countries in Southern Africa. Also, a visual artist and a well-travelled disability activist for over twenty years, he has a wealth of work experience in disability rights and policy advocacy.

**Colleen Clemency Cordes** is a Clinical Professor and the Assistant Dean for Non-Tenure-Eligible Faculty in the College of Health Solutions at Arizona State University. A licensed psychologist, her work focuses on developing clinically and culturally competent behavioural health providers for the integrated primary care workforce. She has a particular interest in how behavioural health providers in primary care can promote the sexual health and well-being of people with disabilities, and other diverse populations.

**Frédérique Courtois** is Full Professor and Director of the Sexology Department at the University of Quebec at Montreal. Her research focuses on the sexual function of men and women with spinal cord injury, and she has been working as a clinician for the past 25 years at Montreal's and Quebec's rehabilitation centres in Canada. She has authored (and co-authored) over 100 peer-reviewed articles, chapters, and books on sexual medicine and rehabilitation, and has given more than 100 conference presentations on her work across Europe and North America.

**Karen Cuthbert** is currently a Research Associate in Sociology at the University of Sheffield, UK, after completing her PhD at the University of Glasgow. Her research centres around "non'" sexualities, such as asexuality, abstinence and celibacy, paying particular attention to how these intersect (both discursively and experientially) with social relations of gender, disability, "race," class and age. She has also published on the experiences of queer religious young people.

**Elisabeth De Schauwer** has a background in Educational Studies. She is working as a Guest Professor at the Department of Special Needs Education (Faculty of Psychology and Pedagogical Sciences), Ghent University, Belgium). As a researcher, her field of interest is situated in Disability Studies. She is intrigued by the role of difference/disability in (pedagogical) relations.

**Geoffrey Edwards** is a senior researcher whose research has spanned many different disciplines, ranging from engineering and physics, to health science and the arts. His recent work has focused on disability studies and questions around human vulnerability. He is also working on a series of novels, the first of which should appear in 2020.

**Ethan Eisen** is a Licensed Clinical Psychologist who received his PhD in clinical psychology from The George Washington University, as well as rabbinic ordination from Yeshiva University's Rabbi Isaac Elchanan Theological Seminary (RIETS). Ethan is currently a clinical supervisor at a community mental health agency in Israel. Ethan writes and lectures on the overlap between Jewish Law/Thought and contemporary issues in psychology, and he has developed a specific interest in issues of disability in the Jewish community.

**Kathryn Ellis** is the CEO and founder of the Institute for Sex, Intimacy and Occupational Therapy, which offers continuing education services for occupational therapy professionals and client services for sexuality counseling. Kathryn is a Doctor of Occupational Therapy and an AASECT-certified sexuality counselor. Her mission is to dismantle restrictive norms related to sexuality and intimacy, which limit occupational therapists and clients so that all individuals can thrive as sexual beings.

**Paul Enzlin** has a background in Educational Studies and Sexology and is full professor at the Institute for Family and Sexuality Studies (Faculty of Medicine, KU Leuven, Belgium), Programme Director of the Master of Human Sexuality Studies (KU Leuven) and Head of the Centre for Clinical Sexology and Sex Therapy (CeKSS, UPC KU Leuven). As a researcher, his field of interest is related to the connections between (chronic) illness and disability and sexual functioning and sexual experience, sexual distress, asexuality and sex education.

**Loree Erickson** is currently Ethel Louise Armstrong postdoctoral fellow at Ryerson School of Disability Studies, USA. She is a sessional instructor, teaching classes on sexuality, transformative justice, gender studies, disability justice and pop culture. Her research and activism draw on her experience as an award-winning queercrip porn maker and a forerunner in theorising and thriving through collective care.

**Michael Feely** is an Assistant Professor in the School of Social Work and Social Policy at Trinity College Dublin, Ireland. He is interested in theories relating to disability as well as the mind and body more generally. He is also involved in supporting self-advocacy by people with intellectual disabilities.

**Anjali J. Forber-Pratt** is an Assistant Professor in the Department of Human and Organizational Development at Vanderbilt University in Nashville, Tennessee. Her work adopts a social-ecological framework, looking at issues related to identity development, inclusion and school climate, with a large focus on disability. She is the author of over 30 peer-reviewed journal articles and numerous book chapters. As a wheelchair-user herself for over 30 years, and a two-time Paralympian and medalist, Dr. Forber-Pratt is also nationally and internationally recognized as a disability leader and mentor.

**Patsie Frawley** is an Associate Professor of Disability and Inclusion Studies at the University of Waikato, New Zealand. Her research focuses on sexuality and relationships rights of people with intellectual disabilities. She developed the peer-led program Sexual Lives and Respectful Relationships in Australia and through this promotes inclusive research and program development in sexuality, relationships and violence and abuse prevention with people with disabilities.

**Véronique Gauthier** is a CRSH Fellow and is affiliated with three research centres: Centre interdisciplinaire de recherche en réadaptation et intégration sociale (CIRRIS); Centre de recherche interdisciplinaire sur la violence familiale et la violence faite aux femmes (CRI-VIFF); and Groupe de recherche interdisciplinaire sur les maladies neuromusculaires (GRIMN). Since 2010, she has worked as a research professional in research projects on the following themes: disability, sexuality, empowerment, prevention, gerontology, violence, intervention in crisis situations, patient-oriented research, patient-partners, feminist research, qualitative and mixed research.

**Marivete Gesser** is a Professor in the Post-Graduate Programme in Psychology at the Federal University of Santa Catarina, Brazil (UFSC), Coordinator of the Centre for Disability Studies (NED) at the same university, and a member of the Gender Studies Institute (IEG). She has experience as a psychologist in the field of sexual and reproductive rights involving women with disabilities, and also as a training facilitator in this area. Dr. Gesser's studies focus on the following topics: intersections of gender, sexuality, race, poverty and disability in Brazil; psychology and disability studies; public policy and human rights for disabled people; disability, social justice and inclusive education.

**Anita Ghai** is Professor in the School of Human Studies, Ambedkar University, Delhi and was formerly President of the Indian Association for Women's Studies. Her academic interests include disability studies, sexuality, and psychology and gender. As a former Fellow at the Nehru Memorial Museum Library, Teen MurtiBhawan, Anita employed a feminist and disability theory frame in research on care of disabled women recipients (i.e., their daughters and providers of care).

**Gerard Goggin** is Wee Kim Wee Professor of Communication Studies, Nanyang Technological University, Singapore. He is also Professor of Media and Communications at the University of Sydney. Gerard has a longstanding interest in media, disability and technology, with books including *Routledge Companion to Disability and Media* (2020; with Katie Ellis and Beth Haller);

*Normality and Disability: Intersections Among Norms, Laws and Culture* (2018; with Linda Stella and Jess Cadwallader), and *Disability and the Media* (2015; with Katie Ellis).

**Antoine Guérette** is an architecture graduate with master's degrees in both professional architecture and architectural sciences. His research interests focus on a wide range of aspects concerning user-centred architecture and evidence-based design, such as inclusive architecture and universal design. His recent work has studied the effects of population ageing on primary healthcare clinics in order to adapt these buildings to the growing number of frail elderly patients.

**Alisa Grigorovich** is a critical gerontologist and a Postdoctoral Fellow at KITE-Toronto Rehabilitation Institute, University Health Network. Her research focuses on the intersection of stigma, aging, and disability in institutional and home care settings, with particular attention to persons living with dementia.

**Sylette Henry-Buckmire** is a Fulbright Scholar and adjunct faculty in the Occupational Therapy Department at the University of the Southern Caribbean (USC) in the twin island Republic of Trinidad and Tobago. Her research interests lie at the intersection of disability, mobility and occupational justice and health promotion, addressing health disparities within marginalised groups across varying axes of power. Her recent publication is the chapter 'Mobility and Occupation of Physically Disabled People in Trinidad and Tobago. A form of non-confrontational activism?' In the edited book *The New Disability Activism: Current Trends, Shifting Priorities and (Uncertain) Future Directions*, a Routledge Handbook.

**Sally J. Hodges** is the Lead Therapist at the Kansas City Relationship Institute, a private practice she helped establish in the Kansas City Metro that specialises in couples therapy. Additionally, she teaches Mental Health and Aging at Kansas State University.

**Xanthe Hunt** is a researcher at the Institute for Life Course Health Research in the Department of Global Health at Stellenbosch University (South Africa). Xanthe's work in the field of disability studies centres on women's health and sexual and reproductive health. She has a keen interest in public health approaches to improving access to sexual and reproductive health for people with disabilities in low-and middle-income countries, as well as participatory methods.

**Angela Ingram** works in international special education leadership. She graduated from the University of Oregon with a PhD in special education. Her experience includes teaching a wide range of diverse students with disabilities in the US and internationally in China, Qatar and Zambia. Grounded in Disability Studies in Education, her research focuses on equity, secondary transition, and understanding intersectional barriers faced by youth with disabilities.

**Michelle Jarman** is Associate Professor of Disability Studies at the Wyoming Institute for Disabilities at the University of Wyoming, where she directs the undergraduate Minor in Disability Studies. Dr. Jarman received her Ph.D. in English from the University of Illinois at Chicago. Her broad research interests focus upon intersecting literary cultural representations of disability, gender and ethnicity. She is co-editor of *Barriers and Belonging: Narratives of Disability*, published in 2017 by Temple University Press. Jarman's essays have appeared in academic journals such as *Disability Studies Quarterly*, the *Journal of Literary and Cultural Disability Studies*, *Feminist Formations* and the *Journal of American Culture*, as well as in literary and disability studies anthologies.

**Debra Keenahan** is a multi-disciplinary artist, academic and author at the University of Western Sydney whose work focuses upon the personal and social impacts of disability.

Having achondroplasia dwarfism, she brings a personal insight to understanding the dynamics of interpersonal interactions and social structures that include/exclude the visibly different from equitable social relations. Through her employment and personal experience, Debra has learnt the power of the arts in communicating with and engaging people on highly political and difficult issues.

**Pia Kontos** is a senior scientist at KITE-Toronto Rehabilitation Institute, University Health Network and is Associate Professor in the Dalla Lana School of Public Health, University of Toronto. She is a critical scholar committed to the transformation of long-term dementia care so that it is more humanistic and socially just. She draws on the arts to enrich the lives of people living with dementia. She also creates research-based dramas to challenge structural violence in dementia care settings and to foster relational caring. She has presented and published across multiple disciplines on embodiment, relationality, ethics and dementia.

**Omolola A. Ladele** is currently a Senior Lecturer in the Department of English, Lagos State University (LASU), Nigeria, where she teaches several graduate and undergraduate level courses including African literature, Postcolonial literature, Women and Gender Studies, Feminist literature and Literary theory. She has presented academic papers at international and local conferences in North America, Europe, the Middle East, and Nigeria. Many of these have appeared in peer-reviewed local and international journals. She is manuscript reviewer of local and international journals and she is, currently, co-editor of *LAPES –the LASU Papers in English Journal*.

**Alicia Lamontagne** is an architecture student at Université Laval who has focused on 3-D printing and virtual reality.

**Sally Lee** is a Lecturer in Social Work at Bournemouth University, splitting her time between teaching and undertaking research on a variety of projects, including adult safeguarding and sexual well-being. She completed her doctoral research in 2016, exploring social work practice, physical disability and sexual well-being, and she is currently developing learning materials (including a short film) around sexual well-being for social care and health practitioners. Sally brings to her academic role extensive social work practice experience built up during more than 25 years of working in diverse practice settings and services. Her research interests focus on often-marginalised populations, including investigating the experience of financial abuse and the detriment to individuals and society over and above financial loss.

**Kirsty Liddiard** is currently a Research Fellow in the School of Education and a co-leader of the Institute for the Study of the Human (iHuman) at the University of Sheffield. Prior to this post, she was the inaugural Ethel Louise Armstrong Postdoctoral Fellow at the School of Disability Studies, Ryerson University, Toronto, Canada. A disabled feminist and public sociologist, Kirsty's research explores dis/ableism in the everyday lives of disabled people. She is the author of *The Intimate Lives of Disabled People* (Routledge, 2018). You can learn more about Kirsty's research and activism at https://kirstyliddiard.wordpress.com and follow her on Twitter at @kirstyliddiard1.

**Gwynnyth Llewellyn** is Professor of Family and Disability Studies and Co-director, NHMRC Centre for Research Excellence in Disability and Health at the University of Sydney, Australia. Her work addresses disability and inequity across the lifespan, and in the context of family and parenting, discrimination and personal safety. Her work on children and young people with disability led to her appointment as Expert Advisor on Disability to the Royal Commission on Institutional Responses to Child Sexual Abuse in Australia 2013–2017.

**Emily M. Lund** is an Assistant Professor of rehabilitation counselling and counsellor education in the Department of Educational Studies in Psychology, Research Methodology and Counseling at the University of Alabama. Her research interests include interpersonal violence, trauma, suicide, and non-suicidal self-injury among people with disabilities, LGBTQ+ issues and graduate education and training in psychology and counselling. She has authored over 80 peer-reviewed articles and several book chapters.

**France Maphosa** is a Professor of Sociology and Anthropology at the University of Botswana. Maphosa's research interests include the sociology of organisations, participation, disability, rural livelihoods, labour studies and alternative dispute resolution (ADR). He has been awarded several research grants including the ORREA Senior Scholars Research Grant, CODESRIA Advanced Research Fellowship Grant and the CODESRIA National Working Group Research Grant. He has more than 15 years' experience of University teaching. He has taught at both undergraduate graduate levels. Maphosa has written extensively. He has a number of books and papers to his name.

**Alan Santinele Martino** is a PhD candidate in the Department of Sociology at McMaster University, Ontario, Canada. His primary areas of interest include the sociology of sexualities, sociology of gender, and critical disability studies, as well as their intersections. His dissertation looks at the romantic and sexual experiences of adults with intellectual disabilities in Ontario, Canada.

**Helen Meekosha** is Honorary Associate Professor in Social Sciences at UNSW. Her work in Critical Disability Studies broke new ground in setting disability in a context of neo-liberalism and globalisation, arguing the case for an examination of Global North/South relations that affect the incidence and production of disability. Helen has been involved in Women with Disabilities Australia (WWDA) since its inception and, as President of WWDA, she accepted the prestigious National Human Rights Award in 2001. Helen has undertaken collaborative research with colleagues in Japan, the US, UK and Canada; as a Visiting Scholar at Ruskin College, Oxford; Noted Scholar in feminist disability studies at the University of British Columbia, Vancouver; and as Visiting Scholar for the Ed Roberts Program at UC Berkeley. Her book with Karen Soldatic, *The Global Politics of Impairment and Disability: Processes and Embodiments* (2014), sought to address the processes of scholarly colonialism in disability.

**Linda R. Mona** has worked as a clinical psychologist for the VA Long Beach Healthcare System for the past 19 years. As the clinical director of spinal cord injury/disorder psychology, she has prioritised disability-affirmative psychological services and sexual health assessment and treatment for people with disabilities. Linda is actively involved in VA psychology training, both locally and nationally, to ensure that disability diversity and identity is part of the work of future generations of psychologists. She has facilitated over 100 trainings and has over 50 publications, focusing on disability, sexuality and disability diversity. Linda's work has been recognized in mainstream media including Canadian Discovery Health, PBS, NBC's *Today* show and *Self Magazine*.

**Ernesto Morales** is an architect with a master's degree in industrial design. He earned his Ph.D. in Design at the University of Montreal and then completed a postdoctoral position in "Environmental Gerontology" and then went on to finish a second postdoctoral position on "Healing Environments" in the Netherlands. He is currently Associate Professor at Université

Laval in Quebec City and Researcher at the Centre interdisciplinaire de recherche en réadaptation et intégration sociale.

**Amie O'Shea** is a Lecturer and researcher in the School of Health & Social Development at Deakin University. Amie's research and teaching takes a poststructuralist approach to sexuality and gender for people with intellectual disability, and continues the search for meaningful and collaborative research methodologies. As the National Co-coordinator for *Sexual Lives & Respectful Relationships* (SL&RR), a peer education programme for people with intellectual disability and acquired brain injury, Amie is committed to sexuality rights and the prevention of violence against people with intellectual disability.

**Christina Pierpaoli Parker** is a doctoral candidate in Clinical Psychology and resident specialising in adult behavioural medicine. Her research and clinical interests explore the intersection of older adults' physical and psychological health, focusing on the adjustment to and behavioural management of chronic health challenges (e.g., metabolic syndrome, osteoarthritis, HIV). Ongoing programmatic work involves developing applied behavioural health programmes to ease or prevent those challenges, including the Senior Sex Education Experience (SEXEE) study, a psychoeducational intervention for improving late-life sexual functioning.

**Dian Ramawati** is a doctoral candidate at SUNY at Stony Brook, New York, concentrating on Disability Studies. She is from Central Java, Indonesia, where she works as a Paediatric Nursing professor for undergraduate nursing students and focuses her research on children with disabilities and their families. Dian developed a keen interest in sexuality and disability and plans to continue conducting research in this area following the completion of her doctoral dissertation. She incorporates health sciences, nursing and disability studies into her research collaborations across multiple disciplines.

**Stacy Reger** currently works as a clinical geropsychologist in the Spinal Cord Injury/Disorder (SCI/D) Service at the VA Long Beach Healthcare System. She specialises in adjustment to disability and chronic health concerns, end-of-life, and neuropsychological and capacity testing. She is the director of SCI/D Neuropsychology testing clinic as well as being active in the Los Angeles County Psychological Association.

**Poul Rohleder** is a clinical psychologist and Senior Lecturer at the Department of Psychosocial and Psychoanalytic Studies at the University of Essex (UK). He completed his PhD on HIV and disability at Stellenbosch University. His research interests focus on sexuality, sexual health and disability. He is author (with Stine Hellum Braathen and Mark Carew) of *Disability and Sexual Health: A Critical Exploration of Key Issues* (Routledge, 2018)

**Tafadzwa Rugoho** is a lecturer at Great Zimbabwe University in the Department of Development Studies. He has authored a variety of book chapters and journal papers on disability issues as well as presenting papers at research conferences in this area over the past five years. Tafadzwa coedited two books, titled *Sexual and Reproductive Health of Adolescents with Disabilities* and *Disability and Media – African Perspectives*. He has worked for a variety of disability organisations for more than 15 years. Tafadzwa is disabled and he is a disability activist.

**Dikaios Sakellariou** is a Reader at the School of Healthcare Sciences, Cardiff University, UK. His research focuses on health inequalities, experiences of disability and disablement, and the intersubjective nature of care practices. His most recent book is *Disability, Normalcy, and the Everyday* (2018), co-edited with Gareth Thomas.

**Ann Fudge Schormans** is an Associate Professor in the School of Social Work at McMaster University, Ontario, Canada. Employing inclusive, co-researcher methodologies and knowledge production, arts-informed methods and writing with disabled co-researchers, her current research projects include: developing curriculum materials with survivors of institutions; disabled masculinity; experiences of parenting for people labelled/with intellectual disabilities.

**Samantha Sharp** completed her Master's degree in Clinical Psychology at California State University, Northridge, and is currently completing her Psy.D. in Clinical Psychology at Pepperdine University. Her clinical work has oriented around multicultural and cognitive behavioural psychotherapies, particularly in their application and evidence base for trauma populations. Her published and professional works have focused on topics such as spirituality and religiosity in post-trauma populations, multiculturally informed interventions for sexual assault, and community-based treatments of trauma.

**Russell Shuttleworth** is currently Senior Lecturer in Social Work at Deakin University, Australia. An anthropologist, social worker and disability support worker/personal assistant by training, his major research areas include the social and cultural construction of disability and impairment, disability and sexuality and disability and masculinities . He has published in a wide variety of journals including *Sexuality and Disability, Disability Studies Quarterly, Intellectual and Developmental Disabilities, Technology and Disability, Men and Masculinities, Critical Sociology* and *Medical Anthropology Quarterly*.

**Leslie Swartz** is a clinical psychologist and a Distinguished Professor in Psychology at Stellenbosch University (South Africa). He is an internationally-rated researcher in South Africa, who has published widely in disability studies and mental health. He has headed a number of research projects and training programmes in these areas. He is founding editor-in-chief of the *African Journal of Disability*.

**Maggie Syme** is an Assistant Professor and Associate Director of Research for the Center on Aging at Kansas State University. She received specialised training in ageing, capacity assessment and dementia and currently focuses her research on biopsychosocial and cultural aspects of sexuality in later life.

**Geert Van Hove** has a background in Educational Studies. He is Head of Department of Special Needs Education (Faculty of Psychology and Pedagogical Sciences, Ghent University) and Endowed Chair Disability Studies (DSIN) (Free University Amsterdam). As a researcher, his field of interest encompasses disability studies and inclusive education.

**Jess Whatcott** completed her doctorate at the University of California, Santa Cruz and is currently an Assistant Professor in Women's Studies at San Diego State University (USA). Dr Whatcott studies formations of gender, sexuality, race and disability in United States history. Dr. Whatcott's current research uses queer theory, crip theory, and political economy to examine eugenics policies in early twentieth century California.

**Rachel Wotton** is a PhD candidate at Western Sydney University and a migrant sex worker who has lived and worked throughout Australia and overseas for over two decades. She has enjoyed many opportunities to present sex worker issues at local, national and international conferences. Rachel is a founding member of Touching Base (www.touchingbase.org), which brings people with disability and sex workers together to advocate for the rights for both marginalised communities and to decrease stigma and discrimination.

# Introduction

# INTRODUCTION

## Contextualising disability and sexuality studies

*Russell Shuttleworth, Julia Bahner and Linda R. Mona*

Disability and sexuality is a diverse and multi-disciplinary area of study. There is increasing recognition cross-culturally that a focus on this intersection is important for critical scholarship, research and applied practice. It is a far cry from when the two editors (RS and LRM) first taught a disability and sexuality studies class in 2001 at San Francisco State University. That class was one of the first to address disability and sexuality from a holistic perspective, incorporating physiological, psychological, social, political and cultural perspectives into what was, at the time, still often taught as a primarily medical model unit. There has since been a steady expansion in disability studies (both in its critical and applied versions), sexuality studies and diversity studies' programmes, which include some variation of disability and sexuality as a unit or module of study. Accordingly, the purpose of this handbook is not to provide an in-depth view of any particular issue, sub-population, disciplinary or theoretical approach—it is to capture a sense of the diversity of issues and perspectives within a selective overview of the state of research and scholarship in this vibrant subject area. In this way, we hope to engage a broad readership including students, academic scholars, researchers and practitioners from a range of disciplines who are interested in state-of-the-art research that connect their own interests to the larger multi-disciplinary field of study.

In this introductory chapter to the handbook, the authors have several aims. First, we think it is necessary to provide a historical and political contextualisation for the current diversity of work in disability and sexuality studies. Today's multi-disciplinary and inter-disciplinary theoretical approaches did not emerge ex nihilo. Mapping the historical and political contours and development of certain applied and academic perspectives on relevant issues within disability and sexuality is important to understanding what is happening currently. We recognise that our historical and political mapping is largely situated within the handbook editors' Anglo-centric background, and, as a corrective, we also include sections on the Nordic experience of disability and sexuality issues and the recent work on the sexuality and sexual lives of people within the Global South, much of it by indigenous researchers and scholars. We do not, however, claim to provide a complete history, conceptualisation and contextualisation of all aspects of the field in all local contexts. We then discuss several areas of entrenched and emerging focus within critical disability and sexuality studies. We further outline the structure of the book, presenting specific introductions to each of the volume's thematic sections before synopses of individual chapters and finish with some concluding thoughts.

## Medical dominance

Early research in disability and sexuality developed from clinical medical practice which narrowly focused on sexual function and rehabilitation, most notably of men with traumatic injury acquired during wartime. The disability and sexuality research followed suit through much of the 1970s and was primarily generated in response to a loss of sexual function and/or sensation experienced by heterosexual men in early to mid-adulthood following a major physical trauma, often a spinal cord injury. Emphasizing function over a holistic sense of the person, it is tempting to read this obsession with sexual and reproductive function as cultural anxiety over loss of penile potency, perceived as a diminishment of masculinity. This early research largely ignored women, and completely ignored racial and sexual minorities, reflecting the prioritised white, heterosexual masculinity (Shuttleworth 2003).

While a primary Post World War II medical focus was clearly about restoration of masculine sexual and reproductive function for those who had experienced traumatic injury, for those person with what is termed a "developmental disability," medicalisation of their sexuality has a more sinister past.[1] Medical experts have often routinely sterilised the latter population, especially those with intellectual disability, going back to the embryonic period of the eugenics movement up through the present (Dunn 2018; Kempton and Kahn 1991; Trent 1994). Involuntary sterilisation is much less frequent today as an official policy in western countries, albeit not completely absent (Elliott 2017). More common are the less physically invasive but no less dehumanising "risk-prevention" practices such as being prescribed long-term contraception without consent, or pressured to abort the foetus (Björnsdóttir et al. 2017; Tilley et al. 2012).

Indeed, the eugenics policies of the past now take more subtle forms like the inherent biases against impairment in genetic and abortion counselling of expectant parents (Saxton 2010). Much of the history of sterilisation for people with an early onset impairment through today's supposedly more "benign" practices has still yet to be written. The obsession to restore the sexual function of individuals who have experienced traumatic injury and the efforts to exclude people with a "developmental disability" from having sexual and reproductive lives both stem from the same normative impetus overseen by medical authority. Medical model research which focuses exclusively on sexual function goes hand in hand with the power and authority medicine wields in disabled people's everyday sexual lives (Joseph et al. 2018; Shuttleworth 2000).

Normative understandings of how the body is supposed to function underlie medical diagnoses, treatments and rehabilitation, and any deviations are targeted for correction or exclusion. In the case of disease, the medical model is appropriate in order for the patient to "get better." It is true that an impairment can accompany some chronic health conditions and some impairments can make one more susceptible to bodily discomfort. However, the long and often permanent duration of many impairments may be better framed as bodily variations rather than "deviations." Much research has shown how many people with disability reach maturity with poor body images and low sexual self-esteem based on their comparison to normative models of functioning (Liddiard 2018; Mona et al. 1994; Taleporos and McCabe 2001a, 2001b; Wiseman 2014).[2] Many have had a history of medical procedures, surgeries, and rehabilitation therapies which were aimed at correcting and normalising their bodies, often to no avail (Shuttleworth 2000). Coupled with a lack of (adequate) sex education (Bahner 2018; Bollinger and Cook 2019), this medical framing of their bodies as deviant not only negatively influences their sense of their own bodies but also negatively affects their sexual selves (Bahner 2012; Helmius 1999; Liddiard 2018; Shuttleworth 2000; Wiseman 2014). This negative medical framing of disabled people's bodies and their sexuality has further consequences, such as the medical profession's avoidance of discussing sexual health issues with their disabled clients during physical and other medical exams (O'Dea et al. 2012; Shah 2019).

## The politics of sex and disability

In the US, the emergence of the Disability Rights and Independent Living Movements in the 1970s was a major impetus for challenging the medical model's dominance in understanding disability. This movement was contemporaneous with similar developments in the UK, which led to the conceptualisation of the social model of disability. This critique of medicalisation of disability was only weakly extended in the case of the approach to disability and sexuality. It has often been observed that the disability rights movement and academic disability studies chose to focus on rights to access basic needs, such as adequate sustenance, the built environment, housing, education and employment (Liddiard 2018; Shakespeare et al. 1996; Shuttleworth 2000; Waxman and Finger 1989). This eschewing of issues pertaining to disabled people's sex lives because they were not essential for survival is often assumed to be a relatively calculated move. And the argument was often presented that it was more important to keep the eye on the ball of getting more basic rights covered. One of the editors (Russell) recalls gauging the interest in research on facilitated sex and disability at several Independent Living Centres in the US at the turn of the millennium and being told precisely this. But this hierarchy of "needs" narrative deployed by the movement also reflects a larger cultural assumption that, at a basic level, sexuality and access to sexual well-being is less important than other domains of life. Since sex is "obviously" not considered essential for survival, not focusing on it politically, and for governments, not funding it within the spectrum of social services is considered par for the course in most cultural contexts (Bahner 2020; Sakairi 2020; Shapiro 2002; Shildrick 2007).

During the late 1970s through the 1980s, there were voices raising the visibility and political viability of a focus on disability and sexuality. A more holistic understanding of disability and sexuality seemed to be taking hold, as people with disability, disability advocates and rehabilitation professionals strove to put this issue on the education, health and social service agenda (Jacobson 2000; Kempton and Kahn 1991). But as the 1980s wore on, this became no easy feat given the conservative political climate in countries like the US and the UK (Jacobson 2000; Shuttleworth 2003). Other historical factors also came into play that affected the dynamic. For example, the impetus for professionals working with people with intellectual disability to provide sexual awareness and social skills training was given a boost during the HIV/AIDS crisis, as fears about this population's vulnerability to being sexually exploited, as well as increasing concerns about the number of sex offenders who had an intellectual disability were voiced. As Kempton and Kahn note "it was recommended that education about AIDS should be integrated into all special education curricula" (Kempton and Kahn 1991, p. 107). Whereas ostensibly this focus was on protection, it did provide these professionals with a firmer foothold to develop a human rights–based agenda with people with intellectual disability (Kempton and Kahn 1991).

For some disabled people and their allies involved in disability rights and disability studies, the connection between the personal and political was never more apparent than in sex and disability. Several of these voices tried to put sexuality on the political and research agendas of the Disability Rights Movement. As early as 1981, Harlan Hahn, a disabled political scientist, railed against the medicalisation of disability and sexuality and a pronounced lack of social and political analysis of disability and sexuality issues (Hahn 1981). In another important article, Hahn (1988a) asks questions such as, "Can Disability Be Beautiful"? He explored many of the themes that have been picked more recently in critical and aesthetic approaches to disability and sexuality. Certainly his discussion of disability as evoking two kinds of anxiety, aesthetic and existential (Hahn 1988b), can be considered a forerunner of such approaches as Shildrick's (2009) post-conventional perspective where she interrogates some of the psychic discomfort many non-disabled people have toward disabled bodies in their eschewing of disabled people as sexual possibilities (cf. Malmberg 2012).

Barbara Waxman-Fidducia and Anne Finger wrote a seminal, critical article in the pages of Disability Studies Quarterly (DSQ), calling for politicisation of many disability and sexual issues, "The Politics of Sex and Disability" (1989). This was an important manifesto for disabled people to accept and embrace their different bodies and to politicise the many barriers to their sexual expression, core ideas of which remain very relevant in the current environment. It was in this manifesto that the first version of the infamous quote often singularly attributed to Finger (1991) was born, in which they proclaimed that people with disability were "more concerned with being loved and finding sexual fulfilment than getting on a bus" (1989, p. 1). In the very same issue of DSQ, Carol Gill (1989) provided an updated critique of how the medical model bias had imbued the majority of research on disability and sexuality up through that time.

Echoing the arguments over medicalisation of disability and sexuality and the lack of attention from disability studies, advocates and scholars, Tom Shakespeare, Kath Gillespie-Sells and Dominic Davies published "The Sexual Politics of Disability: Untold Desires" in 1996. The book was based on research in the United Kingdom on disability and sexuality, and it was the first systematic research on this issue from a predominantly socio-political perspective. In their work, Shakespeare et al. (1996) prioritised the relatively simple goal of documenting in disabled people's own voices what they think and feel about their sexuality and sexual lives. These researchers captured much of the sexual oppression encountered by disabled people, as well as their more positive sexual experiences. Importantly, Shakespeare et al. introduced Ken Plummer's notion of sexual citizenship into the disability rights mix. As they argue, "If disability politics is centrally about civil rights and citizenship demands, its failure to campaign in the area of sexual citizenship is a major omission" (1996, p. 207).

With increasing calls for sexual inclusion, the disability and sexuality focus was beginning to expand. More research was being conducted with disabled women, and with people with a wider range of different impairments, while psychological and social perspectives were also gathering more momentum in the field. By the turn of the millennium, many disabled people and allies, their advocates and some professionals working with them were calling for more participatory disability and sexuality research and sexual rights (e.g., Frawley et al. 2003, Johnson et al. 2001a; McCarthy 1999; Shakespeare 2000; Shuttleworth and Mona 2002). Beginning at this time, the negative medical framing of disabled people's sexual lives also led some scholars to question the inadequate attention given to disabled people's capacity for sexual pleasure by both medical practitioners and social researchers (e.g., Gill 2015; Loeser et al. 2018; Tepper 2000).

The sexual access approach to disability and sexuality was founded in the socio-political perspectives that developed from the 1970s through the 1990s, especially the civil rights approach embodied in the Independent Living perspective in the US and the social model perspective in the UK. Yet, the politically expedient view among many in the disability rights movement was that, since sex was not a need that would enable one to survive and flourish in the public sphere, it should remain a background issue until substantial social access gains had been made. With the increasing political concern with the barriers to sexual participation and the sexual rights of people with disabilty at the turn of the millennium, maintaining this strict barrier between public and private forms of access became less and less rationally defensible. At the time, use of the term "sexual access" seemed to register the required degree of starkness necessary to bring the stakes into full view.

Use of the notion in a disability and sexuality political sense was meant to subvert its narrow bio-evolutionary meaning as strictly about the sexual act and to imply the entire sociosexual imaginary in the nurturance of sexual well-being. Rather than focusing on one barrier to sexual participation at a time, it strove to show the interconnectedness of the range of physical, existential, psychological, social, political, symbolic and materialist factors in creating (using the language of the time) "the myth of asexuality of disabled people." With the recent claiming of asexual identity, use of the term asexual has been problematised, and using non-sexual might currently be more neutral and appropriate. Suffice to say, using the concept of sexual access was an attempt to more radically politicise disability and sexuality issues and put them squarely on the disability rights and sexual rights' agendas, which continues to be a struggle up through the present (e.g., Bahner 2020; De Boer 2015; Duncan and Goggin 2002; Fischel and O'Connell 2019; Fritsch et al. 2016; Kanguade 2009; Liddiard 2018; O'Toole 2002; Shuttleworth 2007; Shuttleworth and Mona 2002; Wade 2002).

## The queering of disability and sexuality

Accompanying the more explicit politicisation of disability and sexuality in the new millennium were important intersectional shifts in disability and sexuality research. By the turn of the new millennium, the issues of disabled LGBTQ people were slowly being addressed (O'Toole 2000; Shakespeare et al. 1996). Because of the low priority of intersectional and especially queer perspectives from much of the history of disability and sexuality literature, we present here a short discussion of some relevant issues and research that has emerged in recent years, as well as a few theoretical frameworks that have emerged from intersectional confluences in the current conceptual terrain of the field. Again, this is simply a selection of several intersections and intersectional issues that occupy researchers in disability and sexuality.

While gendered experiences of sexuality had been common themes, they were often limited to (heterosexual) women and men (Kristiansen and Traustadóttir 2004; Meekosha and Dowse 1997; Morris 1996; Smith and Hutchison 2004; Thomas 2006). Nevertheless, numerous studies have offered important insights that complicate otherwise one-dimensional understandings of "disabled" sexual experience. The ways in which sexuality becomes one of the main aspects described in performing femininity or masculinity and becoming a "real" woman or man has been emphasised (Abbott et al. 2019; Helmius 1999; Slater et al. 2018). It is interesting to note that these descriptions of the development of disabled gender identity were sometimes outlined as problematically conforming to traditional gender roles, owing to the often routine de-gendering of disabled women and men (Apelmo 2012; Malmberg 2012).

Recent explorations in disability, gender and sexual identity show that struggles with heteronormativity and ableism can be used productively as a community-building strategy through identity politics and arts: crip-queer alliances (García-Santesmases Fernández et al. 2017; Reynolds 2007; Sandahl 2003). This is understandable as disabled LGBTQ people will have different experiences compared to heterosexuals in certain respects (Mona et al. 2007). Firstly, the increased risk of discrimination on the grounds of intersecting identities can pose considerable challenges (Duke 2011; McCann et al. 2016; Toft et al. 2019). For example, the "fetishisation of mobility" is an oft-experienced barrier to accessing and feeling welcome in queer spaces (Drummond and Brotman 2014, p. 541). Similarly, ideals of intellectual ability can lead to fewer opportunities to develop one's sexual identity through socialisation in bars and clubs, where it is common to meet potential sexual partners (Blyth 2012). Within service contexts, LGBTQ people with intellectual disability can experience a higher rate of discrimination than either heterosexual people with disability or LGBTQ people without disability (Leonard and Mann 2018).

A general lack of support for the sexuality of LGBTQ people can also extend to their sexual expression (Abbott and Howarth 2007). As Wilson et al. (2018, p. 191) argue in their narrative literature review of people with ID who identify as LGBTIQ, there is a need for "recognition of [their]…marginalisation…and the personal and systemic barriers they face…awareness raising and advocacy; and…targeted education, support and information services."

With regards to gender identity, other challenges include transformations in one's identity as a result of changes in body, health, impairment and the use of aids. For example, an addition of an ileostomy to a disabled queer woman's body "prompted another shift in her aesthetic, in particular a progression from butch to femme," due to choice of clothes in relation to practicality (Drummond and Brotman 2014, p. 544). On the other hand, some studies suggest that it is more common for disabled people than non-disabled people to identify as sexual minorities, although reasons may differ across impairment groups (Bollier et al. 2019; Pecora et al. 2016). A possible reason may be that confidence in one's sexual identity could make it easier to challenge the prevailing body normativity in gay and queer circles (Abbott et al. 2019, p. 694).

Much of the recent research on disabled LGBTQ people and sexual identity has been influenced by queer theory, which began to be felt at the turn of the millennium. Shelley Tremain (2000) published an important article that argued that disability and sexuality scholarship and research had an inherently heteronormative bias and challenged the field to incorporate queer perspectives. Queer theory interrogates the heteronormative assumptions that imbue gender and sexual norms and, in fact, the assumptions that regulate deviancy of any kind of sexual activity or identity (Johnson 2014). Robert McRuer and Abby Wilkerson (2003) edited a special disability issue of GLQ that queered a range of issues, including intersections of queer and crip identities in performance (Sandahl 2003) and the tensions within and limitations of "coming out" discourse for someone who is both queer and disabled (Samuels 2003). Mark Sherry (2004) also published a paper in Disability and Society that discussed how queer theory might benefit critical disability analysis, and vice versa. Sherry discussed how both disciplines have been inspired by the feminist deconstruction of the public/private divide, the social constructionist understanding of sex/gender and impairment/disability, and the rejection of fixed identities and ideologies of normalcy. He rejected an "additive" approach of identities and instead argued for an analysis that could illuminate how "queerness is implicated in the construction of disability, and vice versa" (ibid., p. 770).

Margrit Shildrick began publishing a series of provocative texts, employing a range of post-conventional theories, including queer theory, on the intersection of disability and sexuality (see, for example, Shildrick 2004; 2007; 2009). Eschewing conventional approaches and queering disability and sexuality, Shildrick provides positive valence for the flexible and fluid connections that are perhaps most evident in disabled people's embodiment and everyday lives but which are inherent in post-conventional understanding of embodied self-hood in general. Reflecting on one of the primary concepts orienting queer theory, Adrienne Richs' "compulsory heterosexuality," Robert McRuer puts forward a theory of what he calls "compulsory able-bodiedness." "Like compulsory heterosexuality, then, compulsory able-bodiedness functions by covering over, with the appearance of choice, a system in which there actually is no choice" (2006, p. 8). As with compulsory heterosexuality then, compulsory able-bodiedness is a disciplinary formation "similarly emanat[ing] from everywhere and nowhere" (2006, p. 8).

Queer theory and the more recently formulated crip theory are meant to unsettle and disrupt the normative and compulsory assumptions inherent in sexuality, gender and disability discourses (Santinele Martino 2017). Crip theory is an interdisciplinary critical disability theory that strongly aligns and allies with queer theory's interrogation of the categorisation of "normal" and abject forms of embodiment, identities and pleasures (McRuer 2006). Indeed, key sites of contestation in the cultural imaginary, such as neoliberalism will implicate the ascendency of both heterosexuality and able-bodiedness and call for critical interrogation from both theoretical perspectives (Kafer 2003; McRuer 2006). Whereas disability studies was often critiqued for its lack of attention to diversity and intersectional analysis (e.g., Corker and Shakespeare 2002; Meekosha 2006), crip theory celebrates moving across boundaries and the diversity of subject locations within the disability community. As Hall (2019) notes, "crip theory is highly inclusive and invites coalitions among pathologised persons." This coalition making can be seen in the struggle for trans disabled people's rights and recognition. Jen Slater and Kirsty Liddiard (2018) recently argued that there are many commonalities between trans and critical disability studies, and that collaborations in research, theorisation as well as activism would bring about fruitful developments for both areas of scholarship.

McRuer has been especially influential in advancing the trajectory of crip theory in the critical disability studies literature. In relation to disabled sex, McRuer and Mollow (2012, p. 32) explain how they think that disability can "transform sex, creating confusions about what and who is sexy" and "what counts as sex." A recent expansion of this theme relates to the common histories of pathologisation and stigmatisation of disabled people's sexualities and those who perform bondage, discipline, dominance, submission, and sadomasochism (BDSM) practices (Goldberg 2018; Tellier 2017). Emma Sheppard (2018; 2019) writes about how disabled people living with chronic pain conditions can find BDSM useful in exploring pain, sexuality and emotion in non-normative ways.

But crip theory has itself also been consistently critiqued for not encompassing a sufficiently wide array of disability experiences (Jenks 2019; Löfgren-Mårtenson 2012). Not all disabled people identify with a politicised disability identity in the first place (Bone 2017), and some may not be comfortable with the often provocative and sexually explicit crip performance style (García-Santesmases Fernández et al. 2017). Elisabet Apelmo (2012, p. 37f) makes the important point that far from all people with disability enjoy the cultural and economic capital necessary for the privilege of "noncompliance and moving across boundaries as a political strategy for social change."

## Disability affirmative clinical practice

Missing from our discussion so far has been the relevance of psychology, allied health and sexual health services for people with disability. This does not fit easily into a general historical trajectory of disability and sexuality from a predominantly socio-political perspective. But it is an important area that needs to be included for a holistic understanding of the topic, and which can nevertheless have socio-political implications. In fact, discussion of psychology, allied health and sexual health services for people with disability has been mostly absent from the discourse of sexuality and disability scholarship, with the exception of heeding caution about the ways in which mental health providers lack a full understanding of the disability experience. This is likely attributed to concerns that disabled persons' experiences might be over-pathologised and inherently misinterpreted by psychologists and other mental health providers. However, over the past 15 years, psychologists with and without disability have been advocating for disability culturally inclusive psychological assessment and intervention for the disability community (Andrews et al. 2019; Clemency Cordes et al. 2016; Dunn and Andrews 2015; Mona et al. 2006; Mona et al. 2019; Olkin 1999) and specifically within the context of sexual expression (Carcieri and Mona 2017; Clemency Cordes et al. 2013; Eisenberg et al. 2015; Mona et al. 2013). In an effort to better educate and train psychologists about disability-affirmative sexual health, attention has been directed towards increasing the understanding of disability identity, culture and ideas about navigating sexual activity with different bodies, sensory abilities and thoughts/emotions.

Historically, the medicalisation of disabled bodies and minds has been reflected in the clinical practice of healthcare providers. The facilitation of sexual health services from both medical and psychological providers has varied significantly across the years and across regions. The range of services have included little to no discussion of sexual health and disability to a gradual change with the incorporation of these topics into clinical practice. The avoidance of these discussions in healthcare probably stems from the lack of understanding of the lived experience of disability and embracing the belief that people with disability have the opportunity to experience vibrant sexual lives, although these behaviours may appear different from traditional sexual activity. According to Mona et al. (2017 p. 1005), "It is important to understand that disability is unique from other cultural experiences and identities in that it represents a difference in the 'body' inclusive of non-normative physical, sensory, cognitive, and/or psychological representations, in addition to shared social, political, and identity experiences. This embodied difference, coupled with sexual activity, with its salient physical dimensions, creates a clinical challenge for most psychologists who begin to assess and treat people with disability for sexual health."

As definitions of health have shifted from a focus on medical conditions toward quality of life, researchers have documented the need for providers to serve in advocacy roles in order to promote whole person healthcare (Krahn and Campbell 2011). This shift has opened the door for room to discuss disability and sexuality in a variety of healthcare settings. Where do disabled people receive sexual health guidance? How do they learn strategies to best navigate sexual activity given certain limitations? In the past, this information has been circulated among the disability community through personal narratives, social media and online disability specific articles. Directing the dissemination of information to mental health providers has not always been seen as a preferred vehicle to exchange this information. Both strategies have an important role. Disability peer-to-peer discussions have a unique level of trust coming from someone with lived experience. And, having access to and trust in providers who can support and give resources to people with disabilitiy, is equally as valuable, given the amount of time people with disability interact with healthcare systems. But this process can only be successful with incorporating the view of disability as a diverse life experience inclusive of the ways in which sexual self-identity and sexual expression can be affected by living with disability.

Conversations about sexuality and intimate relationships are particularly important for people with disability because sexuality may be overlooked due to the belief that disability precludes or de-prioritises one's sexual self (Mona et al. 2014), or because other health needs are considered more salient. Sexual relationships and expression are an important component of well-being, and obstacles faced with sexual functioning and adaptation can challenge a person's sense of self and hope for the future (ibid.). Interestingly, positive sexual expression and sexual connection in a relationship have been found to serve as psychological buffers that help maintain positive mood and emotional wellness (Sánchez-Fuentes et al. 2014). Thus, clinical providers are called to prioritise sexuality during all psychological assessments of people with disability and are prepared to engage in services that are inclusive of disability-specific knowledge.

An emerging area in which there has been minimal research is the role of sexuality for people who are faced with a "life limiting" condition (Lion 1990; Redelman 2008), especially the meanings that sex has for young people with "life-limiting" (LL) or "life-threatening" impairments (LTIs) (Liddiard et al. forthcoming). In research on young people with muscular dystrophy, Abbott et al. (2019, p. 687) emphasise the need to acknowledge "the challenges of planning for an 'unanticipated' adult life" in which development of gender and sexual identity has often been neglected. Other recent research suggests that the personal experiences of young people with LL or LTIs are rarely heard and there appears to be a specific gap in the dissemination of sexuality and intimacy information within the transition from childhood to adulthood (Blackburn 2019). Hordon and Currow (2003) offered practical suggestions to healthcare providers in terms of assessing and providing support to people living with LL or LTIs, but they did not provide a specific focus on younger adults. This issue cries out for a broader clinical perspective on sexuality and a disability-affirmative approach in practice. As disability and sexuality are explored historically and across disciplines, we are called to have a larger voice about the life span developmental experiences inclusive of mortality, death and dying at younger ages.

In addition to the field of psychology, occupational therapy has also moved towards a more systematic approach to sexual health and assessment of disabled people. Within occupational therapy, activities of daily living are defined as behaviours that are aimed towards caring for one's own body (American Occupational Therapy Association 2014, p. 620). Thus, sexual expression per se can be conceptualised as falling under that definition. Occupational therapists are trained in assessing cognitive, emotional and physical components of any activity as well as the needs for barriers of the environment to support the success of activity. This, combined with the field's attention to the evaluation of cultural and social norms, places occupational therapy in an ideal place to work alone or with other healthcare disciplines in the facilitation of sexual relationships and activity among people with disability.

Similarly, social workers have been involved in addressing sexuality of disabled people for decades, inclusive of people with intellectual disability (Ballan 2008; Linton et al. 2017). However, specialised education for disability is often lacking within social work programmes (Linton et al. 2017; Moyle 2016; Soldatic and Meekosha 2012), as well as sustained social work research on disability issues (see Bigby et al. 2018, for a review of research in Australia). Given this history, it can be surmised that disability and sexuality issues are likewise rarely included in social work education and research. Social workers learn on the job, doing their best with the guidance of generalist social work principles to ethically address disabled people's sexual health and well-being needs within what is most often a minimal context of governmental and organisational policy and practice (Linton et al. 2017). Nevertheless, over the past several years, there have been efforts to raise the visibility of social work as an important player in supporting, advocating and allying with people with disability for their sexual rights and sexual well-being needs (e.g. Lee et al. 2019; Linton et al. 2017). Social workers, if not social work as a discipline, have ramped up both assessment and treatment of disability and sexuality issues.

Given that rapport and trust between individuals and allied health providers can develop between and among any of the disciplines, interdisciplinary care relating to disability and sexuality issues is becoming highly relevant for people with disability (Clemency Cordes et al. 2016). Another notable development is the introduction by Mona et al. (2017) of the Disability and Sexuality Healthcare Competency Model (DASH-CM) to address the need for comprehensive assessment of disability and sexuality health and wellness of people with disability. Specific conceptual and behavioural sexuality and disability competencies that correspond to the provision and facilitation of these services are delineated in this model. The DASH-CM is offered as a bidirectional tool that can be used for provider self-assessment, in addition to developing clinical expertise in this area of practice. Engaging in a dual process of psychologist self-assessment and the exploration of patients' beliefs and values fosters the delivery of culturally competent healthcare. Prioritising sexual health and wellness assessment of people with disability in clinical practice, engaging in self-assessment related to disability and sexuality and incorporating disability-inclusive material in training programmes will help to transform psychological healthcare toward a truly disability- affirmative ideal. "The DASH-CM goes beyond typical models of cultural competence in that it calls for providers to understand disability life experiences that may be fairly inaccessible to providers and hard to conceptualise, in addition to radically recon-ceptualising sexual expression" (Mona et al. 2017, p. 1009).

One way that clinical providers can adapt their therapeutic approach is to explore the crea-tion of various disability-specific sexual activity scenarios. Brainstorming sexual desires within the context of impairment-specific limitations assists with preparation for either partnered or solo sexual activity, such as timing sexual activity to follow regular bowel and bladder care, making potential modifications in sexual positioning for catheters or other assistive equipment and/or setting reminders to cue for memory enhancement of planned sexual activity. Also, for individuals whose daily pain levels may vary, scheduling sexual acts during times that are associated with less pain can be helpful, but may require a new way of thinking about how to incorporate sexuality and intimacy into a daily routine. Experimenting with a range of sexual behaviours can help individuals develop a repertoire of options for sex play such as kissing and skin-to-skin contact that may be less likely to result in pain. Assisting people with disability by talking through the sexual scripts, planning for potential challenges and emphasising preparation of the sexual environment simultaneously addresses the specific sexual needs of disabled people.

Overall, the logistical and practical components of assessing and assisting people with disabil-ity with improving their sexual expression have been under-explored. Disability studies, medical anthropology, sociology and other disciplines have spent years documenting the diverse sexual life experiences of disabled people from historical, theoretical and social lenses (e.g., Bahner 2012; 2020; Liddiard 2018; Kulick and Rydström 2015; Shakespeare et al. 1996; Shuttleworth 2000). Among these conversations, facilitation of sexual activity has been broached through discussion of sexual surrogacy (Shapiro 2002; 2017), sex work services (Sanders 2007; 2010; Wotton and Isbister 2010; Wotton 2016; Fritsch et al. 2016), facilitated sex provided by sup-port workers (e.g., Earle 1999; Bahner 2012; 2013; 2015a; 2015b; Kulick and Rydström 2015; Shuttleworth and Taleporos forthcoming), sexual assistants (Sakairi 2020),[3] and adaptive sexual activity (Morales et al. 2018).

However, incorporating psychology within these sexual enhancement discussions has been mostly absent (see, however, Limoncin et al. 2014; Mannino et al. 2017). With the recent schol-arship emphasizing disability-inclusive clinical services, there seems to be a trend towards more comprehensive clinical assessment of these issues. In addition, those healthcare systems operating on integrated care models have an increased chance of providing cross-discipline person-centred care. The DASH-CM (Mona et al. 2017) makes the shift from "person-centred sexual health-care" to "disabled person-centred sexual healthcare" and provides mental health providers with the tools necessary to assist people with disability with their sexual expression process inclusive of their disability identity. Using this model across interdisciplinary health care assessment and intervention would serve as a comprehensive approach to systematically address the breadth of issues and barriers affecting the sexual expression of people with disability.

## The Nordic experience

The history of disability and sexuality research and also the provision of services is often written in an Anglo-centric way, which assumes that countries besides the United States, the United Kingdom and other predominantly English-speaking societies, do not have a past that speaks to this important subject. The Nordic countries show how histories of this topic can be divergent. Indeed, the Nordic countries themselves are often lumped together in discussions about welfare state regime type in general, and access to disability services in particular—however, there are significant differences under the surface. Focusing specifically on sexuality issues, recent research from Sweden has illuminated differences between Denmark and Sweden with regards to disability advocacy, sexual cultures and access to sexual support (Bahner 2019; Kulick and Rydström 2015; Rydström 2019). Whereas, in both countries, there have historically been activist groups led by disabled people grounded in a "fully-fledged historical-materialist critique of capitalist economy" (Rydström 2019, p. 1653), their approaches to sexual rights differed markedly. In Sweden the most influential activist, Vilhelm Ekensteen, expressed wariness about directing too much attention towards sexuality issues and it did not become highlighted in activism or advocacy. Instead, issues around sexuality and relationships were discussed more informally in the members' journal of the Swedish National Union of the Handicapped (De Handikappades Riksförbund [DHR]) (ibid.). On the policy and practice level the discourse was individualised and largely medicalised, with reports and suggestions to do with sexual rehabilitation and access to sex aids (Kulick and Rydström 2015).

In contrast, the Danish disability activists in the influential Handi-Kamp group published special issues on sexuality on an almost yearly basis (Rydström 2019). Another contrast was the framing of sexual issues on the basis of Handi-Kamp's socialist ideology, namely as situated in an economic structure. To this also came an understanding of the need to abolish segregation and institutionalisation, as well as to counter commercial beauty ideals that excluded disabled bodies. A further difference compared with Sweden was the influence from gay activists, and specifically the wish to develop positive, proud identities. Gender issues were also frequently discussed (ibid.).

In other words, sexuality was prioritised in the Danish disability advocacy and seen as important and on par with other, more public issues on the agenda. This has influenced policy and practice as well. Whereas there are no national policies or regulations in Sweden around sexual support or even sexual health and rights for people with disability, the Danish National Board of Health and Welfare (*Socialstyrelsen*) has issued a dedicated handbook (Kulick and Rydström 2015). It details how staff in disability services are expected to work with ensuring service users' sexual health and support needs. The initial progress came along with de-institutionalisation and the so-called "normalisation principle," with focus on people with intellectual disability, as developed by Niels Erik Bank-Mikkelsen at the National Board of Health and Welfare. The various principles for a "normal" life course and activities of daily living included a principle on the need for support around sexuality. Since then, both the disability movement and professional actors have regarded sexual support as a natural part of service provision (ibid.).

In Sweden, there is instead great insecurity among service providers as well as service users about whether or not sexual expression is even legitimate to support (see Bahner, Chapter 11 in this volume). This insecurity is fuelled by the lack of disability policy focusing on sexuality issues, as well as missing disability perspectives in sexual health policies. However, following the recent population study on sexual health by the Public Health Agency, which did not include measures on disability, a pilot study was initiated about people with intellectual disability. It was based on previous research on the lack of sexual access and rights (Löfgren-Mårtenson 2009; 2011; Löfgren-Mårtenson et al. 2018). In general, Swedish research and practice has focused more on issues to do with intellectual disability. A new strand of research that has been developing in recent years within this field concerns multicultural experiences in general and the honour context in particular (Löfgren-Mårtenson and Ouis 2018).

Another reason for the insecurity around sexual access and support has to do with the ongoing silence in the broader disability movement on sexual rights, with only a few youth organisations being active on the subject in recent years by partaking in temporary projects. In fact, most work around sexuality and disability has been driven by sexual rights organisations in externally funded projects, rather than by disabled people's organisations (Bahner 2019). In an ongoing doctoral study, disabled people's organisations that have been involved with sexuality-related project are interviewed about their engagement and views on the status of sexuality issues in the broader disability community (Karlsson 2020).

Looking at the neighbouring countries Finland, Norway and Iceland, there seems to be less research published (at least in English). A Norwegian study explored the experiences of disabled LGBTQ people (Haualand and Grønningsæter 2013). From Finland, an ethnographic study of the living conditions of people with profound impairments found sexuality to be a neglected but not unimportant topic, not least from a gender perspective (Vehmas 2019). Similar insights are detailed in an Icelandic paper (Björnsdóttir et al. 2017). In all countries, there are examples of activist groups and education projects working on sexual rights, for example the Icelandic feminist disability activist group Tabú that engages vigorously in sexual rights issues (see Slater 2015).

Denmark stands out again, although their research is also primarily published in Danish. Most notably the journal of Handicap History (*Handicaphistorisk Tidsskrift*) published a special issue on sexuality in 2003, edited by the well-known special education researcher and advocate Birgit Kirkebæk. The special issue showcases Danish research in the field, often with a social constructivist perspective and with an analysis of how power operates in the sexual lives of disabled people. Kirkebæk herself has written an article in English on her archival research of the construction of "feebleminded" girls who were institutionalised at the island of Sprogø (Kirkebæk 2005). She discusses how they were described homogeneously and stereotypically with emphasis on their "sexual appetite" (see also Engwall 2004 from Sweden). More recently, a study of the experiences of sex life from the onset of a spinal cord injury to 10 years after was published in *Sexuality and Disability* (Angel and Kroll 2020). There are also examples of projects and publications in the disability movement, most recently by the Association of Youth with Handicap (*Sammenslutningen af Unge Med Handicap*). They conducted development work around sexuality and sexual support, which resulted in the publication "Sexual well-being for people with multiple impairments. A study in Danish housing and respite care services" (Stefánsdóttir and Rasmussen 2017). In other words, sexuality is continuously researched and worked on within Denmark, in contrast with the other Scandinavian countries.

## Disability and sexuality in a globalised world

For many years research on disability and sexuality in the Global North has tended to dominate the literature (Chappell 2019; Shuttleworth 2010). Recently, critical disability scholars have decried a hegemonic Global North focus in the discipline as a whole (Meekosha 2011; Meekosha and Soldatic 2011), which is prompting more research on a range of issues in the Global South, including disability, sexuality and reproductive concerns (e.g., Chappell and de Beer 2019; Addlakha et al. 2017). In the Global South, where disability support systems are not widely available and more conservative norms around gender and sexuality prevail, disabled people may experience additional struggles, often in connection with poverty and lacking accessibility in general. This is succinctly demonstrated in the recent seminal publication "Disabled Sexualities in the Global South" (Chappell and de Beer 2019). Issues around (post) colonialism, race, ethnicity and indigenous populations may further exacerbate disability prejudice and sexual stereotypes (Meekosha, 2006). Recent research from various Global South contexts is deepening our understanding of the complex, situated and intersectional nature of sexuality and disability experience. Although 80% of disabled people in the world live in the Global South (World Health Organization 2011), their representation in research is lacking, with research and theorisation developing from Global North perspectives (Erevelles 2011). Since disability and sexuality research in the Global South represents a relatively emergent area of study, it is important to provide a brief overview of some of the research which is occurring.

A Japanese study of disabled women's autobiographies showed that, instead of conforming to the expected gender roles, their own unique sexual pathways were forged and gender identities became important strategies (Yasuda and Hamilton 2013, p. 50f). For some, a strategy included becoming sex workers, which provided an opportunity to challenge normative assumptions about their presumed asexuality, unattractiveness and disability status. However, such strategies seem most available to those who have stronger social positions in society. In contrast, the interviewed women in Naoko Kawaguchi's (2019) study barely said the words "sex" and "sexuality" and were largely isolated from sexual opportunities. This led to participants being more apt to talk about issues around marriage and reproduction and several had experienced matchmaking meetings, which are common in the general culture, usually arranged by the family. However, disabled women often felt disempowered as the meetings seemed to be tailored to disabled men's needs, such as their wishes for women to be good at housework and caring. There were also experiences of potential partners' family members objecting to the prospect of marriage with a disabled woman.

In Zimbabwe where heterosexuality is the norm and other sexualities prohibited by law, disabled women may use an active sex life as a way to refute myths of being passive and asexual (Peta et al. 2016). This could be the case, even though sexual relations were abusive (see also Bartley (2019), reporting on similar issues in Jamaica). Conforming to traditional cultural practices such as elongating one's labia and inserting various vaginal products to attract and preserve a sexual relationship with a male partner were ways of being seen as any woman in the community. Paul Chappell (2014) found similar patterns of conforming to gender roles and local traditions instead of necessarily focusing on disability identity among a group of disabled youth in South Africa. The heteronormative discourse was evident in the young women's legitimisation of cheating and "intimate partner violence as an acceptable component of married life" (Chappell 2017, p. 595). The young men, in turn, held conservative views of what would be considered acceptable clothing, behaviour and recreational activities for young women.

In Cambodia, agreement from family is needed for marriage and the initiative could only be taken by a man. The prospective wife's ability to run the household, reproduce and take care of the family were important criteria that many did not believe disabled women were capable of (Gartrell et al. 2017). Several of the interviewed women had internalised beliefs of these perceived inabilities as inherent to their impairments and of being "a burden" (see also Ghosh 2019 for a similar example from India, and Tefera et al. 2017 from Ethiopia). They were afraid of not being able to sustain a marriage in the long term and therefore avoided the prospect of marriage altogether. Others were happily married to husbands who respected them and complimented their abilities in the household. In one of the few studies where men with disability were also included, from Sri Lanka, similar experiences of not being seen as capable enough to be a husband and father were common (Hettiarachchi and Attanayake 2019).

Other factors that can affect sexual access include lack of adequate infrastructure in health services, poverty among disabled populations and widespread prevalence of gender-based violence (Rugoho and Maphosa 2017; Tefera et al. 2017). In China, "marrying a man without disabilities from the countryside is often considered an option also by many low-income families of girls with disabilities living in cities" (Aresu and Mac-Seing 2019, p. 290). Here we see how disability prejudice, gender, class and rural-urban relations intersect. Similarly, in Zimbabwe, some disabled women saw no other option for livelihood than becoming sex workers, resulting in precarious positions due to both the illegality of the occupation and increased risk of violence and abuse resulting from disabling barriers as well as unequal gendered relations (Rugoho 2017).

But the often difficult circumstances need not be the end of the story. Devine and colleagues (2017) report from a three-year programme of participatory action research with disabled women in the Philippines which aimed to explore barriers to accessing sexual and reproductive health (SRHR) services, including violence response, and to suggest improvements. For many of the participating women, this was the first time that they had discussed sexual rights and the project greatly improved their knowledge about rights and services. It was also the first time that most of them had met with other women with disability, resulting in enhanced self-confidence and increased capacity to cope with discrimination. Some of them went on to teach other women and to form self-advocacy and peer support groups. Learning about SRHR also led to wider impacts for some as they gained confidence to seek paid employment, re-negotiate destructive relationship patterns with partners and become more independent.

In summary, the discussed examples of barriers in gaining access to sexual opportunities, sexual activity and SRHR services in the Global South show the wide variety of experiences. Although some experiences may be specific to the circumstances in the country under study, they are often based on familiar structures discussed in previous sections. Societal norms around gender, health and sexuality exist to much the same extent, however, with additional hardships and complexities resulting from poverty and lack of basic services and social security systems.

## A diversity of current issues

The history of scholarship and research on disability and sexuality issues was often reflected during the late 1970s through the 1990s in the pages of the journal Sexuality and Disability. In the latter part of these years, the focus on medical model research gradually began making room for more psychological and social perspectives. Since the turn of the millennium, special issues of journals and book-length monographs focusing on disability and sexuality have increasingly been published and they similarly chronicle the introduction of key ideas and themes in scholarly work on the subject: e.g., critical perspectives and applied issues (Mona and Shuttleworth 2000), sexual access as a socio-political issue (Shuttleworth and Mona 2002); a dialogue between disability studies and queer theory (McRuer and Wilkerson 2003); the link between research, advocacy and policy (Shuttleworth 2007); politics, access and identity issues (Shuttleworth and Sanders 2010); cultural studies approaches (McRuer and Mollow 2012); international perspectives on sexual health/sexuality for people with intellectual disability (ID) (McCarthy and Wilson 2014); diverse theoretical approaches to disabled people's embodiment (Loeser et al. 2018); and disability and sexuality in the Global South (Chappell and de Beer 2019). This list of edited books and special editions of journals on disability and sexuality is by no means exhaustive.

The growth of the field reflects an embracing of disability and sexuality as a significant area of multidisciplinary and interdisciplinary research and analysis. Although it is impossible to cover all the current areas of disability and sexuality research and scholarship in this introduction, there are several issues we want to especially highlight as examples of current research before providing an overview of this handbook.

## *Sexual expression in institutional/quasi-institutional settings and the issue of sexual consent*

Nowhere are sexual boundaries felt to be more restrictive for people with disability than in institutional and quasi-institutional settings. These kinds of contexts are often designed to be both physically and socially risk averse, which can drastically minimise disabled people's sexual opportunities. For disabled people living in group homes, attending sheltered workshops or similar segregationist spaces, "much of the discourse around sexuality and intellectual disability can be classified as 'crisis responsive' or 'harm reducing'" (Gill 2015, p. 7). Service users' sexuality is often regarded not only as risky but even dangerous (either to themselves or to others, including staff) and therefore needs to be restricted, according to this approach (Hollomotz 2011). The organisation of daily living in residential homes, special schools and activity centres often works to discourage sexual and gender expression (Björnsdóttir et al. 2017; Feely 2016; Löfgren-Mårtenson 2004; O'Shea and Frawley 2019; Peuravaara 2013; Shuttleworth 2007). Privacy and necessary support are also commonly denied or simply not prioritised (Brown and McCann 2018; Hamilton 2009; Kelly et al. 2009).

These practices are still common despite the decades of rights-based work and advocacy; progressive approaches can be diffused before becoming fully implemented in policy and practice. The sexual opportunities for individuals with intellectual disability are still often mediated by protective and risk-averse family members, disability policies, support services and staff (Wilson and Frawley 2016). In the introduction to an important special issue of the Journal of Intellectual and Developmental Disability (JIDD), McCarthy and Wilson (2014) argue that, although there has been some changes to policy within the intellectual and disability rehabilitation field, "support for sexuality primarily remains reactive where regressive, punitive, and sometimes bizarre responses to the sexual expression of people with intellectual and developmental disability are still evident" (McCarthy and Wilson 2014, p. 123). But countless studies have shown how such approaches may in fact have the opposite effect, namely to "disable individuals from becoming competent social and sexual actors and from accessing information and services that have the potential to reduce sexual 'vulnerability'" (Hollomotz 2011, p. 11). Although it is true that disabled people, especially women, are more at risk of being sexually abused and assaulted, silence, lack of accessible information and inadequate support is not the answer (United Nations Population Fund 2018).

Issues around consent, which are often at the centre of discussions in this area, need to be approached with dignity and respect for every individual's agency and capacity for pleasurable sexual experiences, whatever form they may take (Gill 2015). While restrictions in laws on consent and mental capacity vary across countries and states, the Convention on the Rights of Persons with Disabilities (CRPD) demands a reconsideration of the attitudes underlying oppressive legal frameworks (Arstein-Kerslake 2015; Perlin and Lynch 2015). Namely, it is necessary to challenge the individualised understanding of autonomy and capacity to consent, which obscures many of the situational and relational dynamics of sexual relationships for persons with intellectual disability and also others with a diagnosis of major neurocognitive impairment, such as some form of dementia (Boni-Saenz 2015; Clough 2014). Indeed, when there is adequate support, sexual expression and even parenthood is possible for persons with intellectual disability (Johnson et al. 2001a, 2001b). An excellent example of the kind of support necessary is provided by the Sexual Lives and Respectful Relationships (SL&RR) approach, which has been successful in linking individuals with intellectual disability with professionals and service organisations to work together on their sexual rights, the core of which is a peer education programme run by people with intellectual disability (Frawley et al. 2017).

## *Ageing, disability and sexuality*

There is a vast archive of literature that highlights the barriers to sexual access and well-being for older adults, including those with dementia, both within and outside institutional residences. However, until recently, this work had been minimally framed in terms of disability terminology and theory. In fact, disability studies have yet to engage more widely with critical ageing studies, which similarly problematises norms around ability, normalcy and oppressive support systems (Grigorovich and Kontos 2018; Kontos et al. 2016; Sandberg and King 2019; Syme et al. 2018). While intersectionality scholars have studied age as a social category in relation to ageism as a dimension of oppression (Hulko 2009), such work has not often been used within disability studies (Kahana and Kahana 2017; Priestley 2001; 2003). However, a recent study by Shakespeare and Richardson (2018), "The Sexual Politics of Disability, Twenty Years On," takes stock of ageing disabled people's experiences of their sexual identities and lives. In this qualitative study, one participant reflected that, having lived with disability had helped them manage sexuality as they aged, compared to others who may need to adjust to new age-related bodily changes. Another interviewee said that compared to their youth they now felt "more like everyone else" since most people's bodies were changing with age (ibid., p. 84). Several others mentioned that they were also more comfortable and confident in their bodies.

Similarly, research on the sexual lives of older people, who experience an impairment in their later years, has not been much theorised by notions relevant to disability studies – until very recently. Comparing sexual discourse around people with cognitive disability and older individuals with dementia, Sanberg et al. (2020) illustrate "the production of sexual normates" and "risky sexual subjects." The circumstances for persons with dementia are especially complex, whose gender and sexuality are often seen as less relevant as cognitive functioning progressively declines (Sandberg 2018)—much like the experiences of being deemed child-like and therefore asexual for people of all ages with an intellectual disability. Unsurprisingly, for older disabled adults living in assisted accommodation, including those with dementia, sexual expression can be especially circumscribed by staff's attitudes or organisational regulations (Frankowski and Clark, 2009; Simpson et al. 2017). Pia Kontos, Alisa Grigorovich and others have also recently published several papers relevant for critical disability studies, arguing for the usefulness of the notion of relational citizenship for the sexuality of persons with dementia (e.g., Grigorvich and Kontos 2018; Kontos et al 2016).

## Overview of the book

Current research and scholarship in disability and sexuality studies is challenging the traditional boundaries of the sexual imaginary. The authors who contributed chapters to this volume encompass a wide range of issues and perspectives across disciplines on culture, history, aesthetics, embodiment, discourse, representation, performance, social media, identity, inclusiveness, progressive social policy, support and professional practice, sexual rights and sexual well-being. Several authors address issues that are still controversial in many countries in the worlds, such as the various kinds of facilitated sex; other authors present scholarship on emerging issues such as the socio-sexual tensions and dilemmas that confront disabled people with intersecting stigmatised identities like transgender or asexual, and the sexual concerns of disabled people in the Global South. Taken together, the chapters in this handbook fittingly provide a selective overview of the breadth, diversity and dynamism that currently exists in disability and sexuality studies.

## *Part I: Theoretical frames and intersections*

Theoretical development is important to any field of study. Importantly, in critical disability studies, including critical approaches to disability and sexuality, it is not just theory for its own sake that is worshipped but theoretical frames that provide innovative modes of critiquing how disabled people's sexual lives are generally understood and of envisioning their transformation. Conceptual work in the field currently interrogates disability and sexuality from diverse points of view, ranging from more traditional psychological and sociological models of, for example, body image and stigma (e.g., Kedde and Berlo 2006; Taleporos and McCabe 2002, Ünal 2018), to post, queer and crip approaches (e.g., García-Santesmases Fernández et al. 2017; Goodley and Lawthom 2011; Liddiard 2018; McRuer 2006; 2011; McRuer and Wilkerson 2003; Sheppard 2018; 2019; Shildrick 2009). The chapters in this section of the handbook often provide various frames that not only critique barriers to sexual participation but also point toward a transformation in ways of theorising a range of issues within the purview of critical disability and sexuality studies.

The first chapter by Kirsty Liddiard provides an overview of current perspectives in social and cultural theory relevant to theorising the sexual lives of disabled people. Using sexual stories of disabled people from her own research, Liddiard exemplifies a range of ways to theorise the issues they raise. Situating her general approach within Critical Disability Studies, she provides a strong critique of biological essentialism as inherently "othering" disabled people's sexual lives, bodies and identities. Liddiard presents a case of "imagining otherwise" via employing queer theory and crip theory, but, despite showing their usefulness, she also challenges whether they are radical enough to theorise the sexual politics of disabled people's everyday lives. She then considers some of the recent critical theorising posited by Goodley and others employing post-human approaches and Dis/Human studies, again showing their value, but similarly questioning their explanatory power in being able to adequately conceptualise the sexual, intimate and erotic lives of disabled people.

Whereas the intersection of queer theory and crip theory are discussed as part of Liddiard's overview, Alan Santinele Martino and Ann Fudge Schormans focus specifically on the rich exchange occurring between these two critical approaches. For Martino and Fudge Schormans, questioning normalcy is central to both approaches as, by extension, are compulsory heterosexuality and/or compulsory ablebodiedness/mindedness. The authors then delve deeper into the usefulness of crip theory, specifically in challenging the hegemonic perceptions of disability and sexuality, presenting a range of theoretical tools to understand how people with disability experience and negotiate their sexual lives. They illustrate many of these concepts by showing how they play out in the sexual lives of people labelled/with intellectual disability. An overarching theme of their chapter is that work in crip theory can be mutually beneficial for theoretical perspectives in both disability studies and sexuality studies.

In contrast to the first two chapters, Michael Feely applies a specific approach, a Deleuzian theoretical perspective, which can provide innovative insight into the ways bodies are situated in material–discursive contexts in which some capacities may be enabled and others disabled. Altering the context in some way will thus affect the ascribing of these capacities. Feely illustrates this by employing the case of the reproductive capacity of men with Down syndrome. From a biological point of view, these men are almost universally perceived as infertile. However, a more expansive Deleuzian sense would view their reproductive situation as being potentially alterable via the interaction of non-static, dynamic factors and processes such as biological, economic, technology and legislative. Feely argues that men with Down syndrome could in fact become fathers, if societies *focused* on enabling this capacity.

There has been minimal attention in critical disability studies to the sociology and phenomenology of persons with dementia (however, see Shakespeare et al. 2017), let alone the intersection of dementia and sexuality. Alisa Grigorovich and Pia Kontos' chapter is a welcome addition to the limited literature on the latter. These authors offer a critical analysis of literature on disability, dementia and sexuality. Their goal is to develop an approach to sexual rights-oriented development in long-term residential care homes that focuses on the structural barriers and facilitators. They problematise common trends and concepts within the Critical Disability Studies and Critical Dementia Studies literature and point toward new directions in future scholarship and research. A productive coalition of these two critical foci is sketched and the authors' recently developed schema notion of relational citizenship for the sexuality of persons with dementia is productively outlined.

In the next chapter, Emily Lund, Anjali Forber-Pratt and Erin Andrews look at sexual identity development among people with disability within theoretical and historical contexts and outline new perspectives on the impact of ableism and heterosexism within this identity development process. The authors argue that people with disabilitiy have historically been denied the opportunity to develop and express their sexual identities despite having the same spectrum of sexual and romantic wants and needs as their non-disabled peers. They add that, as the psychology field has recognized the benefits and importance of positive, integrated disability identity for well-being, more attention has been paid to sexual identity development of people with disability. Lund, Forber-Pratt and Andrews explore these fascinating constructs by highlighting the importance of historical, disability-affirmative and intersectional lenses when exploring disability sexual identity development among people with disabilitiy.

Finally, Marivete Gesser's chapter starts with the understanding that sexual and reproductive rights for disabled people are a human rights and social justice issue. Building from this notion and other feminist understandings, the author sets forth to examine disability as a category of analysis to explore the sexual and reproductive rights for disabled people within Brazil. Gesser begins this discussion by highlighting some research results from Brazil that outline the main barriers that disabled people face daily in practising their sexuality. These barriers are further examined within the biomedical model of disability that is embraced by some in Brazil. Lastly, Gesser discusses fundamental theoretical assumptions to be incorporated into professional practices committed to ensuring the sexual and reproductive rights of disabled people.

## Part II: Subjugated histories and negotiating conventional discourses

Scholarship on the subjugated histories of disabled people both within various modern contexts such as institutions and in terms of various disability movements, policies and disability support practices has been highlighted in the literature (e.g., Tilley et al. 2012; Shuttleworth and Meekosha 2017). The case can be forcefully made for also interrogating the sexual histories of disabled people (e.g., Tilley et al. 2012). Several of these sexual histories are highlighted in this section of the handbook, dealing with a range of historical themes that shed light on present-day policies and practices. Disability histories can show outward changes over time such as deinstitutionalisation and the guise of progressive policy thinking, but nevertheless these policies might still evidence current structural restrictions and inequalities. Disability histories often intersect with the history and politics of other divisive marginalised groups where coalitions between marginalised discourses might be fruitful. Disability histories are also complexly interwoven into the cultural traditions and values of a culture or religion that, while in certain ways recognising disabled personhood, can also restrict to varying degrees access to physical and social contexts in the sexual lives of people with disability.

Jess Whatcott's chapter maps eugenic discourse and its material consequences for certain "defective classes" in the beginning of the 20th century in California. In order to protect the "health of the normative body politic" during this time period, there was legally sanctioned institutionalisation and sterilisation, which caused trauma and abuse on a large scale. By reviewing archival evidence, Whatcott shows how gendered categories of disability, madness and sexual deviance were constructed through institutionalisation and sterilisation. Sexual deviance was perceived as a kind of physical and mental disability. In addition, the "disabilities" of women and girls were constructed as inherently sexually deviant. Although there are inherent risks in viewing this era as a co-history of sorts, Whatcott nevertheless argues, similarly to Jarman, that it opens up the possibility of creating an important political coalition of shared campaigning.

Employing perspectives from feminist disability studies and reproductive justice, Michelle Jarman's chapter investigates shared historical legacies of racial and disability discrimination in reproductive domains aimed at people of colour and people with disability in the United States, from the reproductive control during slavery to eugenics programmes in the 20th century. These histories inform current approaches to abortion debates. Jarman's argument is nuanced around recent conservative activist appropriation of disability perspectives in their targeted termination of Down syndrome foetuses, as part of their larger agenda to ban abortions and limit access to birth control and reproductive health care. Jarman's primary point is to show how recognition of these interwoven histories provides a frame for an intersectional politics of reproductive justice that can critically interrogate the new eugenics, without condemning personal reproductive decisions. She concludes with an approach to reproductive freedom that is shared by scholars and activists in both feminist disability studies and reproductive justice.

In her chapter, Gwynnyth Llewellyn traces the historical roots of today's policy responses to sexual abuse towards disabled young people and children in Australia. The chapter begins with the question of why these children, who are under increased risk of sexual abuse, are so absent from the discourse? Llewellyn shows how practices of severe institutionalisation in the 1800s continue to be reproduced in the twenty-first century, albeit under a different guise. Although modern-day disability policies are more inclusive and based on human rights, they nevertheless frame children as non-autonomous dependents of their families, rather than as agents in their own right. A structural condition that further exacerbates vulnerability to abuse is the organisation of support services, which are mainly segregated and thus "out of the public eye." Llewellyn argues that this, together with denial of agency and personhood, contributes to ongoing abuse and neglect—which to this day is left without consistent recording in Australia.

In chapter 10, Ethan Eisen offers a complex exploration of sexuality and people with disabilitiy, who identify and live as Orthodox Jews. This author points out that the formal framework of Jewish ritual law (*halakha*) provides unique benefits and challenges with regard to sexuality and people with disability. Eisen describes that, according to the *halakha* framework, the main principle of valuing each human life in Orthodox Judaism necessitates participation in intimacy and sexuality inclusive of rituals associated with upholding this value. Furthermore, the author notes that the ability to access and participate in rituals may vary depending upon the need for assistance and environmental accessibility. Lastly, Eisen calls for an increase in efforts to include people with disability into traditional Orthodox Jewish cultural practices, as well as the need for Rabbinic leaders to commit focus and scholarship related towards sexuality and disabilitiy.

## Part III: Politics, policies and legal frames across the world

In this section, we take a wide-angle focus on political process and sociosexual change, from the dissensus of politics through the encoding of policies to the interpretation and supplementing of legal doctrines. Disability and sexuality politics is often about advocating for the recognition of disabled people's diverse sexualities and their rights to a meaningful sexual life. Depending on the particular local context, there has generally been a history of effort to highlight the lack of policies about various forms of sexual and reproductive support. In most Global South countries, however, there is even less policy development to support disabled people's sexual lives than in the Global North and minimal acknowledgement of their sexualities. The majority of chapters in this section focus on sexual support as a political and policy issue. The last chapter argues that certain supplements to the legal doctrine of consent and its reconceptualisation would result in the more inclusive sexual participation of people with cognitive impairments.

In her chapter, Julia Bahner reports on a Swedish study of sexual facilitation in personal assistance services (PAS) to people with mobility impairments, that is, the support needed for some to achieve one's desired sexual practice. In a policy context where disabled people's sexual rights are rarely highlighted, disability services are reluctant to take responsibility. This results in disabled service users being unsure of their rights as well as being afraid of repercussions. Bahner uses the concept of sexual citizenship to explore the political, organizational and social power dimensions surrounding claims for sexual support.

Tafadzwa Rugoho and France Maphosa explore barriers to accessing sexual and reproductive health (SRHR) among disabled women in Zimbabwe. They start off by setting the policy and cultural scene, namely the lack of disability rights policy implementation, the widespread medical and charitable perspective on disabled people and their lives, and disabled people's organisations' missing knowledge and interest in sexuality issues. The particular barriers encountered by disabled women when visiting SRHR clinics comprise negative attitudes by staff, inaccessible buildings and information, and lack of confidentiality. Rugoho and Maphoso importantly also discuss how most of the barriers are rooted in a life in poverty.

David Abbott's chapter details the lived experiences of people with learning disability who are part of dedicated LGBT support groups. The groups offer social support, a safe place, a place to meet other people, to have fun—and with the potential to change lives, not least for people whose support workers exhibit prejudicial attitudes towards LGBT identities. On the other hand, staff "with their radar out" who are often LGBT too, provide important ways of opening up spaces and conversations. However, there is a risk that facilitation of sexual access is taken away amidst current cuts to social services. Abbott ends the chapter with a timely problematisation of the place of research in policy and practice development.

In their chapter, Dian Rahmawati and Pam Block explore the opportunities for sex education for young adults with intellectual disability in Central Java, Indonesia. They conduct a review of a number of intensively reported court cases detailing sexual violence and abuse directed at young people with disability in special schools. Their analysis shows that vulnerability to abuse is paramount, even in relation to adults in educational and caring contexts. Against this background, the research empirically investigates how parents and educators introduce and talk about sexuality with disabled young people, with the goal to improve the quality of education and health services.

In the final chapter in this section, Alex Boni-Saenz first explains how certain legal doctrines regulate and control sexual access for some individuals with cognitive impairments, such as dementia and Down syndrome, in the United States. Under the current legal regime, an individual with a cognitive impairment may in fact lack the legal capacity to provide consent for sex. Legal doctrines governing capacity in turn influence legal liability and social norms. Boni-Saenz outlines two possible supplements to the current legal framework for capacity to consent. For those individuals with acquired cognitive impairments such as dementia, "advance consent" means that an individual develops a sexual advance directive for periods of cognitive impairment while residing in long-term care institutions. For individuals with lifelong cognitive impairments, such as Down syndrome, "network consent" would rely on the assistance of a supportive network to realise legal capacity to consent to sex.

## Part IV: Representation, performance and media

The cultural representation of disabled people is an important area of critical disability studies scholarship (e.g., Darke 2004; Mitchell and Snyder 2000). This work is generating critiques of the representation of disabled sex in a range of popular media (e.g., Stevens 2010; Malinowska 2018). Cultural representation of disabled sex is, as Malinowska (2017, p. 364) puts it, under-represented and "mediated by normative representations of sexual performance and popular images of disability". This critique can extend to more mundane representations of people with disability as sexual beings in the news media (Houston 2019). Works of literature can also reveal the intersectional issues and oppressive conditions that can affect disabled people's sexual lives. On the other hand, current disability art is celebrating disabled people as erotic, desiring and desirable in a myriad of ways. For example, in performances that challenge bodily and sexual norms (e.g., Berne 2013; García-Santesmases Fernández et al. 2017; Quippings: Disability Unleashed 2020; Sins Invalid 2019) and through historical narratives of transgressive and transformative sexual performance and lifestyle (Moore and Shuttleworth 2019). Other current work is more explicitly connected to the various intellectual arguments around disability aesthetics (Shildrick 2009; Siebers 2010). In this section of the book, authors take up these representational issues across a range of media.

In chapter 16, Debra Keenahan provides a reflective analysis of what a disability and sexuality aesthetic might look like. Keenahan notes that, within the history of art, social discourse on sexuality has predominantly excluded disability. She first analyses three portraits of dwarfs from the photographic exhibition *Intimate Encounters*. Then she compares the process and images produced by the photographer of this exhibition with self-representations by herself, an artist and woman with dwarfism. The comparison reveals and challenges the discourse and pejorative gaze of the viewer, with a shift from stereotypical exclusion to representation of the subjects as sexual beings in dynamic intimate relationships. Keenahan cogently critiques the concept of disability aesthetics discussed by Siebers (2010) as being founded upon a restrictive understanding, and she extends this disability aesthetic sense into a dynamic and interactive process that more adequately represents people with disability as sexual beings.

In her chapter, Courtney Andre sets out to explore the rich and complicated history of disability and sexuality on screen. From more recent movies and television shows, e.g., *The Sessions* (2012) and *The Theory of Everything* (2014) to older films containing themes about returning war veterans which sought to convince women to marry and care for disabled ex-servicemen. In addition, the author takes a deeper dive into the history of disability and sexuality within film by exploring eugenics films of the 1930s, which posed that the desires and sexual activities of people with disability imperilled the nation in addition to women's pictures from the mid-century which fetishised the sexual violation of women with disability and connected love to cure for disability. Andre connects her analysis to recent scholarship on disability and film and further explores considerations of film industry practices and narrative, and representational strategies.

Omolola Ladele argues that, in many postcolonial African societies, disabled people's sexuality and sexual lives remain a marginalised discourse. Their sexualities are often obscured in the larger society and their sexual identities can be rendered invisible. Ladele notes the proliferation and range of contemporary African literature that problematises the corporeality and oppressive conditions that constitute the everyday lived experience of disabled sexualities. She analyses two African texts that show the complex issues that variously disabled people confront. In these powerful fictional narratives, she reconstructs the sexualities of disabled, poor people and renders visible how disability intersects with gender. Throughout her chapter, she grapples with the question: what are the implications from this literature for the sexual rights of people with disability in Africa?

In her chapter, Loree Erickson discusses how a queercrip porn community challenges normative ideals of undesirability and produces new knowledge and manifestations of queercrip sexual embodiment and desire. The research explored Erickson's queercrip porn collaborators' experiences of living in an ableist and disabling culture which left them with feelings of in/ visibility, shame and exclusion. Queercrip porn instead offers a space for imagining new ways of being and living, based on access, empowerment and solidarity, and where cultivating sexual agency and bodily autonomy are made possible.

In Goggin and Meekosha's chapter, they show how the news media represents disabled people via desexualising images, especially those whose bodies, communicative styles and sexualities are non-normative. They illustrate this by analysing the case of Anna Stubblefield, a former Professor of Philosophy at Rutgers University in the US who, in late 2015, was convicted of two counts of aggravated sexual assault of Derrick Johnson, a 29-year-old man with cerebral palsy. Desexualised and infantilising images of Derrick Johnson, such as being a disabled man who wears diapers, were often employed when describing him in reportage of the trial. The authors argue that these typical and oppressive ways of representing disabled people in the media as non-sexual are highly influential in public culture. The authors also discuss several political challenges posed by the Stubblefield case and the representation of Derrick Johnson in the media, including issues of sexual consent and communication.

## Part V: Sexual narratives and (inter)personal perspectives

The chapters in this section focus on disabled people's (inter)personal perspectives on sexuality and narratives of their sexual lives. In these often complex accounts, the ways in which people with disability experience and interpret their own bodies and identities within the context of sexuality are shown to be diversely constructed by their own sense of the other's gaze, but also implicitly and sometimes explicitly revealing a resistant and/or transformative agency. In these accounts, disabled people's choices can seem complicated and may involve negotiating strictly within the parameters of a culture's sexual imaginary or trying to reformulate and imagine disabled sexuality differently. Various of these chapters also show that in the study of disabled (a)sexualities, traditional human science research approaches can be meaningfully augmented by autoethnographic lenses and innovative theoretical and methodological frameworks.

Angela Ingram uses qualitative data and photography to explore the lived experiences of disabled transgendered youth in high school. She begins with emphasising research data, suggesting that, compared with heterosexual disabled youth, LGBTQ disabled adolescents are more likely to face discrimination and negative experiences. Then, by utilising in-depth phenomenological interviews and research participant's own pictures, Ingram analyses the ways in which the complexities of "transgender" and "disability" identities impact high school experiences. The participants in this study speak to the ways in which hegemonic domains of power have functioned to oppress them, and their photographs illustrate these themes clearly. The intersections of ableism and cisgenderism that are embedded in the everyday fabrics of society are discussed within the context of transgender youth. The quest to find and embrace sexuality within this trans disability identity is also explored.

In their chapter, Inge Blockmans, Elisabeth De Schauwer, Paul Enzlin and Geert Van Hove explore the desire for sexual pleasure among several women with spinal cord injury within a wider search for the intimate (re)exploration of their changed and vulnerable bodies. A primary question the authors ask is, what is the potential space and scope for manoeuvres these women experience in their rehabilitation journey? Another crucial task the authors set themselves is to reformulate how the material-discursive practices around sexuality, touch and (health)care, and the women's own construction of the meaning of their sexual pleasure and their bodies, contribute to the degree of pleasure experienced (or not). Blockmans, De Schauwer, Enzlin and

Hove employ a highly experimental style of writing to report on their qualitative research that holds their findings and participants' voices in a lived tension that remains non-finalised.

Set against the backdrop of the Trinidad and Tobago Carnival, Sylette Henry-Buckmire explores narratives of disruptions and interpersonal manoeuvrings of persons often excluded from sexual participation, most significantly people with physical disability. Employing a broadly occupational science conceptual framework, Henry-Buckmire presents the stories of four women with physical disability, who are in sexual relationships. She relies on several feminist ethnographic techniques, including face-to-face interviewing, travel diaries and accompanying participants, who are highly mobile in the community, to capture multiple facets of their accounts. Henry-Buckmire argues that these women's sexual narratives show how current social practices around disability and sex are framed and controlled both formally and informally by historical discourses. In doing so, she reveals the everyday implications of sexual deprivation, alienation, marginalisation and imbalance.

In her chapter, Anita Ghai challenges readers about the unique issues that affect the sexuality of people with disability in South Asia. The salient issues regarding sexuality and disability in India are addressed within the context of existing literature and Ghai's own personal perspective as a disabled woman. The missing discourse of sexual desire in studies of disability and sexuality in India are highlighted within the framework of theoretical influences and debates informing prevailing assumptions about disability and sexuality in this region of the world. An emphasis is placed on the sexuality of disabled women and their conceptualization as "asexual" or "hypersexual" via medical interpretations. Furthermore, the ways in which the intersections among sexuality, able-bodiedness, moral panic, access and desire, intimacy and sexuality for people with a disability are explored. Lastly, the significance of sex education for people with disability in India is discussed.

Drawing on qualitative research with disabled asexual people, in her chapter, Karen Cuthbert explores the "intersectional black hole" of disabled asexual subjectivity. While many disabled people have struggled to be acknowledged as sexual, as most chapters in this book also show, there is often a lack of understanding of those who have an asexual orientation or identity. Discussing the personal accounts and issues of her participants, Cuthbert offers important insights into asexual disabled experience and gives suggestions for how disability studies can be more accommodating of non-normative asexual identities, and likewise, how asexual communities can be more inclusive of disabled people without fear of pathologising asexuality.

In the next chapter, Poul Rohleder, Mussa Chiwaula, Leslie Schwartz, Stine Hellum Braathen, Mark Carew and Xanthe Hunt describe a participatory action research project in South Africa meant to challenge the dominant myths about disability and sexuality and the sexual lives of people with disability, presenting their social and interpersonal experiences from their own perspectives. A highpoint of the chapter is the methodological creativity of the researchers, as they employ a diverse mix of research methods to capture both social attitudes and subjective lived experience including media and narrative methods. The use of collaborative and innovative media methods provides participants with powerful modes to convey their personal sexual experiences and challenge the hegemonic and stigmatising images of the sexual and gendered lives of disabled people.

In Chapter 27, Amie O'Shea and Patsie Frawley present a sex education project by, for and with people with intellectual disability. Through co-development of sexual stories over the past two decades, their inclusive research approach allows for a sex education programme that is based on the needs and wants of those concerned. The co-researchers have been involved fully in collecting data, telling stories, and developing training and resources. Furthermore, the project has contributed to previously largely unheard narratives from the sexual lives of people with intellectual disability, including those who identify as LGBTQ. Rather than lack of "capacity," the stories show that it is ableist and disabling barriers that hinder many people with intellectual disability from realising their aspirations regarding love and sexuality.

## Part VI: Accommodation, support and sexual well-being

There are many other chapters in the book that could have been included in this section. Indeed, there are several chapters in the politics and policy section that are specifically focused on forms of sexual support. However, what has guided the editors in those cases is their overtly controversial and political nature: the fact that progressive sexual policies around explicit forms of sexual support are resisted in so many local contexts. Similarly, several chapters in the previous section also dealt with promoting sexual well-being but focused on the importance of giving voice to disabled people about their sexuality and sexual experiences. Although no less implicitly political or narrative focused, the chapters in this section take up more explicitly the facilitation of sexual well-being in disabled people's lives. Taking into consideration all of the theory, policy, and historical discussions on disability and sexuality, we now turn to the experience of sexual expression. Integrating theory, politics, and psychology, the following chapters focus on the practical domain of sexual expression. How are people with disability having sex, what obstacles do they face and what are the suggested ways to improve access to and pleasure with sexual activity? Both logistical and social barriers are explored by authors who address adaptive sexual practices and disability-inclusive healthcare and counselling practises. As several of these chapters show, immersing sexual health clinical approaches within the context of disability culture proves a foundation for respect and rapport building within a traditionally focused medical model system and allows individuals to access the information that they require to express their sexuality. Several chapters point the way for disability studies to join forces with psychology, social work, and occupational therapy, which will enhance our research knowledge of the intersection of theory and clinical provision.

Maggie Syme, Stacy Reger, Christina Pierpaoli Parker, and Sally Hodges set forth their chapter with the underlying philosophy that sexuality is an integral part of well-being across the life course. This chapter examines sexual citizenship for older people with disability (PWDs) via a life course perspective, while considering the biological, psychological, social, and cultural influences on life-long sexual trajectories. For older PWDs, age and disability status intersect to create unique sexual experiences and needs that have long been stigmatised from both disability and ageing perspectives. The authors focus on the socio-cultural realities of ageism and ableism that pose specific challenges to the pursuit of sexual wellness and serve to invalidate and/or inhibit access to a satisfying intimate life.

Ernesto Morales and colleagues, many with training in architectural design, address access to pleasure for people with disability within the context of socially constructed ideas about sexuality, disability and health. They begin by recognising that many adults with disability utilise sexual self-stimulation as their primary means of sexual activity and that access to this activity may not always be an option because some may not be capable of engaging in this activity independently. Using a co-design methodology, participants in this study included individuals with upper limb mobility limitations. These participants were engaged throughout the research process to develop design solutions that would lead towards sexual autonomy and well-being in their masturbation practices. Research results reveal interesting data on current masturbation practices, barriers experienced by participants and the testing of product design solutions.

In the next chapter, Rachel Wotton brings us an important perspective on the availability and use of sex work, sexual surrogacy, sexual assistants, and "compassionate masturbation services" among people with disability. This chapter examines a variety of socially structured frameworks people with disability are offered to explore their sexual expression, outside of the usual dating and relationship paradigms. Wotton carefully delineates this discussion by including a focus on these different sexual services via mapping their geographical locations and discussing specific issues with these services across each modality. Wotton advocates for positive legislative changes that would allow for these sexual practices to be legally recognised occupations in all countries.

In their chapter, Colleen Clemency Cordes and Christine Borst discuss family-centred integrated behavioural health practices as one strategy to promote comprehensive sexual healthcare for women with disability (WWD). This clinical approach involves behavioural health professionals working alongside other medical professionals as part of an interdisciplinary team. This practice has been identified in the US as a viable approach for culturally inclusive practice for people with disability. Given that multiple healthcare disciplines may be required to assess any given person, having a team approach often ensures that improved care is delivered. In this chapter, the sexuality of WWD is addressed and the role of behavioural health providers in providing disability culturally responsive assessment and treatment to promote sexual health and well-being is examined. Finally, relevant clinical concerns that may emerge in the context of family-centred integrated behavioural health care are described followed by the illustration of a clinical case highlighting these concepts in a healthcare setting.

Occupational therapists (OTs) Dikaios Sakellariou and Kathryn Ellis argue on the basis of their clinical experience and research evidence that sexuality should be seen as an integral part of OT practice, namely among the other taken-for-granted "activities of daily living." Firstly, Sakellariou and Ellis discuss the need to incorporate sexual knowledge within OT education, preferably using situated and experiential learning, including that by people with disabilitiy themselves. They then go on to propose an occupational empowerment framework and rights-based approach which addresses issues beyond physical impairment, namely discussions related to relaxation techniques, self-esteem, identity and unlearning harmful messages and beliefs.

In Chapter 33, Sally Lee develops an approach in social work practice that employs multiple models, including human rights, the social and affirmative models of disability and several models of sexual and social health practice to develop a multifaceted perspective in working with supporting disabled people's sexual well-being. Based on her own practice as well as qualitative research with disabled people and practitioners, Lee urges social workers to further use an exploratory, collaborative approach, where the insecurities of both parties can be handled productively in an effort to challenge discrimination and promote sexual inclusion.

In our last chapter, Sarah Brindle and Samantha Sharp explore the intersectionality of disability, sexuality and spirituality within the context of psychological assessment and treatment. They argue that healthcare providers across the globe receive little or no training in the assessment and treatment of people with disability who identify religion and/or spirituality as important to their identity and world view. The authors note that the strengths and advantages of acknowledging spirituality and religion are important for some disabled people and that particular attention paid to mind-body practices (e.g., adapted yoga, meditation and prayer) that focus on addressing disability and sexuality from a broader perspective is crucial. Brindle and Sharp call for mental health providers' self-assessment on the complex interplay among disability, sexuality and religion/spirituality.

## Concluding thoughts

Our primary goals in soliciting chapters from an international and diverse range of authors for this handbook were to explore disability and sexuality conceptually, historically, globally, from multiple disciplines, and to explore the intersectional nature of scholarship and practice. Similarly, we also thought it important to highlight the variations within the disability community—variations encompassing differences in impairment and social categories such as gender and sexuality, as well as cultural variations within societies and across the world, often intersecting in multiple ways within disabled people's sexual lives. We feel we were variously successful on these fronts, some more than others. In compiling this volume, we also wanted to recognise that, despite progress in many areas, there are still many barriers faced by people with disability in terms of fighting for their sexual rights, justice, identity, sexual expression and access to pleasure and sexual well-being. Struggles include the right to recognition in sexual health policy and population studies of the diverse sexual and reproductive health needs of disabled people across the world, recognition of sexuality as an important and legitimate area in disability services and education of allied health and social care professionals, to explicitly name just a few. Indeed, in many countries, the resistance to the development of pro-sexuality disability services and support policy and practice still demands both activism and scholarship. Finally, we wanted to highlight scholarship that shows not only how many disabled people are in fact exploring and living meaningful sexual lives in both mundane and imaginative ways, but that, in so doing, how they implicitly work against the forms of sexual regulation (e.g., heteronormativity, body normativity, behavioural normativity) and towards transforming the parameters of sexual possibility in the 21st century.

## Notes

1  The term "intellectual disability" is used to refer to a disability characterised by significant limitations in both intellectual functioning and in adaptive behaviour, which covers many everyday social and daily living skills. This disability originates before the age of 18. In contrast, the term "developmental disability" is a broader characterisation that can refer to one of multiple types of chronic conditions and/or impairments in physical, learning, language or behaviour areas (see Schalock et al. 2010).
2  In this introductory chapter, we use the terms "people with disability" and "disabled people" interchangeably. We mean to conceptually highlight "disability" as the social structural, disabling process that engulfs many people whose function is impaired in some way. We do not employ the phrase "people with disabilities," as this seems to prioritise functional impairment over a more relational process. There is no terminological standardisation across the chapters of this handbook. Each author employs the disability terminology they are most comfortable with.
3  For a detailed description of most of these types of facilitated sex, see Wotton's chapter in the current volume.

## References

Abbott, D., & Howarth, J. (2007) Still off-limits? Staff views on supporting gay, lesbian and bisexual people with intellectual disabilities to develop sexual and intimate relationships. *Journal of Applied Research in Intellectual Disabilities* 20(2), pp. 116–126.

Abbott, D., Carpenter, J., Gibson, B.E., Hastie, J., Jepson, M., & Smith, B. (2019) Disabled men with muscular dystrophy negotiate gender. *Disability & Society* 34(5), pp. 683–703.

Addlakha, R., Price, J., & Heidari, S. (2017) Disability and sexuality: Claiming sexual and reproductive rights. *Reproductive Health Matters* 25(50), pp. 4–9.

American Occupational Therapy Association. (2014) Occupational therapy practice framework: Domain and process. *American Journal of Occupational Therapy* 68(Suppl. 1), pp. 1–48.

Andrews, A.E., Forber-Pratt, A.J., Mona, L.R., Lund, E.M., Pilarski, C.R., & Balter, R. (2019) #SaytheWord: A disability culture commentary on the erasure of 'disability'. *Rehabilitation Psychology* 64(2), pp. 111–118.

Angel, S., & Kroll, T. (2020) Sex life during the first 10 years after spinal cord injury: A qualitative exploration. *Sexuality and Disability* 38(1), pp. 107–121.

Apelmo, E. (2012) Crip heroes and social change. *Lambda Nordica* 2012(1–2), pp. 27–52.

Aresu, A., & Mac-Seing, M. (2019) When sexuality meets disability: Experiences, attitudes and practices from China. In P. Chappell & M. de Beer (eds.), *Diverse voices of disabled sexualities in the Global South*. Cham, Switzerland: Palgrave Macmillan, pp. 277–299.

Arstein-Kerslake, A. (2015) Understanding sex: The right to legal capacity to consent to sex. *Disability & Society* 30(10), pp. 1459–1473.

Bahner, J. (2012) Legal rights or simply wishes? The struggle for sexual recognition of people with physical disabilities using personal assistance in Sweden. *Sexuality and Disability* 30(3), pp. 337–356.

Bahner, J. (2013) The power of discretion and the discretion of power. Personal assistants and sexual facilitation in Swedish Disability Services. *Vulnerable Groups and Inclusion* 4(1). https://www.tandfonline.com/doi/full/10.3402/vgi.v4i0.20673

Bahner, J. (2015a) Risky business? Organizing sexual facilitation in Swedish personal assistance services. *Scandinavian Journal of Disability Research* 18(2), pp. 164–175.

Bahner, J. (2015b) Sexual professionalism: For whom? The case of sexual facilitation in Swedish personal assistance services. *Disability & Society* 30(5), pp. 788–801.

Bahner, J. (2018) Cripping sex education: Lessons learned from a programme aimed at young people with mobility impairments. *Sex Education* 18(6), pp. 640–654.

Bahner, J. (2019) Mapping the terrain of sexuality and disability: From policy to practice. *Ars Vivendi Journal* 11(March), pp. 27–47.

Bahner, J. (2020) *Sexual citizenship and disability: Understanding sexual support in policy, practice and theory*. London & New York: Routledge.

Ballan, M. (2008) Disability and sexuality within social work education in the USA and Canada: The social model of disability as a lens for practice. *Social Work Education* 27, pp. 194–202.

Bartley, M. (2019) Navigating dating and relationships with a disability in Jamaica. In: P. Chappell & M. de Beer (eds.), *Diverse voices of disabled sexualities in the Global South*. Cham, Switzerland: Palgrave Macmillan, pp. 109–126.

Berne, P. (2013) Sins invalid: An unshamed claim to beauty in the face of invisibility. Harriman, New York, New Day Films.

Bigby, C. (2018) Social Work Research in the Field of Disability in Australia: A Scoping Review. *Australian Social Work* 71(1), pp. 18–31.

Björnsdóttir, K., Stefánsdóttir, Á., & Stefánsdóttir, G.V. (2017) People with intellectual disabilities negotiate autonomy, gender and sexuality. *Sexuality and Disability* 35(3), pp. 295–311.

Blackburn, M.C. (2019) *Sexuality, relationships and reproductive choices in young adults with life-limiting and/or life-threatening conditions*. Unpublished PhD thesis, United Kingdom: The Open University.

Blyth, C. (2012) Members only: The use of gay space(s) by gay disabled med. In R. Shuttleworth & T. Sanders (eds.), *Sex & disability: Politics, identity and access*. Leeds: The Disability Press, pp. 41–58.

Bollier, A.M., King, T., Austin, S.B., Shakespeare, T., Spittal, M., & Kavanagh, A. (2019) Does sexual orientation vary between disabled and non-disabled men? Findings from a population-based study of men in Australia. *Disability & Society*. Published online: 3 Dec 2019. doi: 10.1080/09687599.2019.1689925.

Bollinger, H., & Cook, H. (2019) After the social model: Young physically disabled people, sexuality education and sexual experience. *Journal of Youth Studies* 23(7), pp. 837–852.

Bone, K.M. (2017) Trapped behind the glass: Crip theory and disability identity. *Disability & Society* 32(9), pp. 1297–1314.

Boni-Saenz, A. (2015) Sexuality and incapacity. *Ohio State Law Journal* 76, pp. 1201–1253.

Brown, M. & McCann, E. (2018) Sexuality issues and the voices of adults with intellectual disabilities: A systematic review of the literature. *Research in Developmental Disabilities* 74, pp. 124–138.

Carcieri, E.M., & Mona, L.R. (2017) Assessment and treatment of sexual health issues in rehabilitation: A patient-centered approach. In M.A. Budd, S. Hough, S.T. Wegener, & W. Stiers (eds.), *Practical psychology in medical rehabilitation*. Springer International Publishing, pp. 287–294.

Chappell, P. (2014) How Zulu-speaking youth with physical and visual disabilities understand love and relationships in constructing their sexual identities. *Culture, Health & Sexuality* 16(9), pp. 1156–1168.

Chappell, P. (2017) Dangerous girls and cheating boys: Zulu-speaking disabled young peoples' constructs of heterosexual relationships in Kwazulu-Natal, South Africa. *Culture, Health & Sexuality* 19(5), pp. 587–600.

Chappell, P. (2019) Situating disabled sexual voices in the Global South. In P. Chappell & M. de Beer (eds.), *Diverse voices of disabled sexualities in the Global South*. Cham, Switzerland: Palgrave Macmillan, pp. 1–25.

Chappell, P., & de Beer, M. (eds.) (2019) *Diverse voices of disabled sexualities in the Global South*. Cham, Switzerland: Palgrave Macmillan.

Clemency Cordes, C., Cameron, R.P., Mona, L.R., Syme, M.L., & Coble-Temple, A. (2016) Perspectives on disability within integrated healthcare. In: L. Suzuki, M. Casas, C. Alexander, & M. Jackson (eds.), *Handbook of multicultural counseling* (4th ed). Thousand Oaks, CA: Sage Publications, pp. 401–410.

Clough, B. (2014) Vulnerability and capacity to consent to sex - asking the right questions? *Child and Family Law Quarterly* 26(4), pp. 371–396.

Corker, M., & Shakespeare (2002) Mapping the terrain. In M. Corker & T. Shakespeare (eds.), *Disability/postmodernity: Embodying disability theory*. London: Continuum, pp. 1–17.

Darke P. (2004) The changing face of representations of disability in the media. In J. Swain, et al. (eds.), *Disabling barriers, enabling environments*. London: Sage, pp. 100–105.

De Boer, T. (2015) Disability and sexual inclusion. *Hypatia* 30(1), pp. 66–81.

Devine, A., Ignacio, R., Prenter, K., Temminghoff, L., Gill-Atkinson, L., Zayas, J., & Vaughan, C. (2017) "Freedom to go where I want." Improving access to sexual and reproductive health for women with disabilities in the Philippines. *Reproductive Health Matters* 25(50), pp. 55–65.

Drummond, J.D., & Brotman, S. (2014) Intersecting and embodied identities: A queer woman's experience of disability and sexuality. *Sexuality and Disability* 32(4), pp. 533–549.

Duke, T.S. (2011) Lesbian, gay, bisexual, and transgender youth with disabilities: A meta-synthesis. *Journal of LGBT Youth* 8(1), pp. 1–52.

Duncan, K., & Goggin, G. (2002) "Something in your belly": Fantasy, disability and desire in my one legged dream lover. *Disability Studies Quarterly* 22(4), pp. 127–144.

Dunn, C. (2018) *Sterilisation of girls with disability: The state responsibility to protect human rights*. Hobart, Tasmania: Women with Disabilities Australia (WWDA).

Dunn, D.S., & Andrews, E.E. (2015) Person-first and identity-first language: Developing psychologists' cultural competence using disability language. *American Psychologist* 70, pp. 255–264.

Earle, S. (1999) Facilitated sex and the concept of sexual need: Disabled students and their personal assistants. *Disability & Society* 14(3), pp. 309–323.

Eisenberg, N., Andreski, S., & Mona, L.R. (2015) Sexuality and physical disability: A disability-affirmative approach to assessment and intervention within health care. *Current Sexual Health Reports* 7(1), pp. 19–29.

Elliott, L. (2017) Victims of violence: The forced sterilisation of women and girls with disabilities in Australia. *Laws* 6(3), pp. 1–19.

Engwall, K. (2004) Sinnesslöa kvinnor och sexualitet i ett historiskt perspektiv [Feebleminded women and sexuality in an historical perspective]. In K. Barron (ed.), *Genus och funktionshinder [Gender and disability]*. Lund: Studentlitteratur, pp. 53–82.

Erevelles, N. (2011) *Disability and difference in global contexts: Enabling a transformative body politic*. New York: Palgrave Macmillan.

Feely, M. (2016) Sexual surveillance and control in a community-based intellectual disability service. *Sexualities* 19(5–6), pp. 725–750.

Finger, A. 1991 Forbidden fruit. *New Internationalist* 233, pp. 8–10.

Fischel, J., & O'Connell, H. (2019) Cripping consent: Autonomy and access. In Fischel, J. (ed.), *Screw consent: A better politics of sexual justice*. Oakland: University of California Press, pp. 135–171.

Frankowski, A.C., & Clark, L.J. (2009) Sexuality and intimacy in assisted living: Residents' perspectives and experiences. *Sexuality Research and Social Policy Journal of NSRC* 6(4), pp. 25–37.

Frawley, P., Johnson, K., Hiller, L., & Harrison, L. (2003) *Living safer sexual lives: A training and resource pack for people with learning disabilities and those who support them*. Brighton: Pavilion Publishing Ltd.

Frawley, P., O'Shea, A., Stokoe, L., Cini, V., Davie, R., & Wellington, M. (2017) *Sexual lives and respectful relationships training manual*. Geelong: Deakin University.

Fritsch, K., Heynen, R., Ross, A.N., & van der Meulen, E. (2016) Disability and sex work: Developing affinities through decriminalization. *Disability & Society* 31(1), pp. 84–99.

García-Santesmases Fernández, A., Vergés Bosch, N., & Almeda Samaranch, E. (2017) 'From alliance to trust': Constructing crip-queer intimacies. *Journal of Gender Studies* 26(3), pp. 269–281.

Gartrell, A., Baesel, K., & Becker, C. (2017) "We do not dare to love": Women with disabilities' sexual and reproductive health and rights in rural Cambodia. *Reproductive Health Matters* 25(50), pp. 31–42.

Ghosh, N. (2019) Fashioning selves: Femininity, sexuality and disabled women in India. In P. Chappell & M. de Beer (eds.), *Diverse voices of disabled sexualities in the Global South*. Cham, Switzerland: Palgrave Macmillan, pp. 55–74.

Gill, C. (1989) Sexuality and disability research: Suffering from a case of the medical model. *Disability Studies Quarterly* 9(3), pp. 12–15.

Gill, M. (2015) *Already doing it: Intellectual disability and sexual agency*. Minneapolis: University of Minnesota Press.

Goldberg, C.E. (2018) Fucking with notions of disability (in)justice: Exploring BDSM, sexuality, consent, and Canadian law. *Canadian Journal of Disability Studies* 7(2), pp. 123–160.

Goodley, D., & Lawthom, R. (2011) Disability, Deleuze and sex. In F. Beckman (ed.), *Deleuze and sex*. Edinburgh: Edinburgh University Press, pp. 89–105.

Grigorovich, A., & Kontos, P. (2018) Advancing an ethic of embodied relational sexuality to guide decision-making in dementia care. *The Gerontologist* 58(2), pp. 219–225.

Hahn, H. (1981) The social component of sexuality and disability. *Sexuality and Disability* 4(4), pp. 220–233.

Hahn, H. (1988a) Can disability be beautiful? *Social Policy* 18, pp. 26–31.

Hahn, H. (1988b) The politics of physical differences: Disability and discrimination. *Social Issues* 44(1), pp. 39–47.

Hall, Melinda C. (2019) Critical disability theory. *The Stanford Encyclopedia of Philosophy* (Winter Edition), Edward N. Zalta (ed.). https://plato.stanford.edu/archives/win2019/entries/disability-critical/

Hamilton, C. (2009) "Now I'd like to sleep with Rachael" – researching sexuality support in a service agency group home. *Disability & Society* 24(3), pp. 303–315.

Haualand, H.M., & Grønningsæter, A.B. (2013) Dobbelt diskriminert? LHBT-personer med nedsatt funksjonsevne [Doubly discriminated? LGBT persons with disability]. In A.B. Grønningsæter, B.R. Lescher-Nuland, & H.W. Kristiansen (eds.), *Holdninger, levkeår og livsløp - forskning om lesbiske, homofile og bifile [Attitudes, living conditions and life courses – research on lesbians, gays and bisexuals]*. Oslo: Universitetsforlaget, pp. 77–100.

Helmius, G. (1999) Disability, sexuality and sociosexual relationships in women's everyday life. *Scandinavian Journal of Disability Research* 1(1), pp. 50–63.

Hettiarachchi, S., & Attanayake, S. (2019) Candid conversations: Narratives of young adults with disabilities in Sri Lanka on intimate partner relationships. In P. Chappell & M. de Beer (eds.), *Diverse voices of disabled sexualities in the Global South*. Cham, Switzerland: Palgrave Macmillan, pp. 131–150.

Hollomotz, A. (2011) *Learning difficulties and sexual vulnerability: A social approach*. London: Jessica Kingsley Publishers.

Hordern, A.J., & Currow, C.D. (2003) A patient-centred approach to sexuality in the face of life-limiting illness. *The Medical Journal of Australia* 179(6), pp. 8–11.

Houston, E. (2019) 'Risky' representation: The portrayal of women with mobility impairment in twenty-first-century advertising. *Disability & Society* 34(5), pp. 704–725.

Hulko, W. (2009) The time- and context-contingent nature of intersectionality and interlocking oppressions. *Affilia* 24(1), pp. 44–55.

Jacobson, D. (2000) The sexuality and disability unit: Applications for group training. *Sexuality and Disability* 18, pp. 175–177.

Jenks, A. (2019) Crip theory and the disabled identity: Why disability politics needs impairment. *Disability & Society* 34(3), pp. 449–469.

Johnson, K. (2014) Queer theory. In T. Teo (ed.), *Encyclopedia of critical psychology*. New York: Springer, pp. 1618–1624.

Johnson, K., Hillier, L., Harrison, L., & Frawley, P. (2001a) *Living safer sexual lives final report*. Australian Research Centre in Sex, Health and Society La Trobe University.

Johnson, K., Traustadóttir, R., Harrison, L., Hillier, L., & Sigurjónsdóttir, H.B. (2001b) The possibility of choice: Women with intellectual disabilities talk about having children. In: M. Priestley (ed.), *Disability and the life course: Global perspectives*. Cambridge: Cambridge University Press, pp. 206–218.

Joseph, M., Saravanabavan, S., & Nisker, J. (2018) 'Physicians' perceptions of barriers to equal access to reproductive health promotion for women with mobility impairment. *Canadian Journal of Disability Studies* 7(1), pp. 62–100.

Kafer, A. (2003) Compulsory bodies: Reflections on heterosexuality and able-bodiedness. *Journal of Women's History* 15(3), pp. 77–89.

Kahana, J.S., & Kahana, E. (2017) *Disability and aging: Learning from both to empower the lives of older adults*. Boulder, Colorado: Lynne Rienner Publishers.

Kanguade, G. (2009) Disability, the stigma of asexuality and sexual health: A sexual rights perspective. *Review of Disability Studies: An International Journal* 5(4), pp. 22–36.

Karlsson, M.M. (2020) *"Gå eller rulla – alla vill knulla." Om Cripaktivism i nyliberal tid ["Walking or rolling – Everybody wants to fuck." Crip acitivism in neoliberal times]*. Lund: Arkiv.

Kawaguchi, N. (2019) Difficulties disabled women in Japan face with regard to love, marriage, and reproduction. *Ars Vivendi Journal* 11(March), pp. 48–60.

Kedde, H., & van Berlo, W. (2006) Sexual satisfaction and sexual self images of people with physical disabilities in the Netherlands. *Sexuality and Disability* 24(1), pp. 53–68.

Kelly, G., Crowley, H., & Hamilton, C. (2009) Rights, sexuality and relationships in Ireland: "It'd be nice to be kind of trusted". *British Journal of Learning Disabilities* 37(4), pp. 308–315.

Kempton, W., & Kahn, E. (1991) Sexuality and people with intellectual disabilities: A historical perspective. *Sexuality and Disability* 9(2), pp. 93–111.

Kirkebæk, B. (2005) Sexuality as disability: The women on Sprogø and Danish society. *Scandinavian Journal of Disability Research* 7(3–4), pp. 194–205.

Kontos, P., Grigorovich, A., Kontos, A.P., & Miller, K.L. (2016) Citizenship, human rights, and dementia: Towards a new embodied relational ethic of sexuality. *Dementia: The International Journal of Social Research and Practice* 15(3), pp. 315–329.

Krahn, G., & Campbell, V.A. (2011) Evolving views of disability and public health: The roles of advocacy and public health. *Disability and Health Journal* 4, pp. 12–18.

Kristiansen, K., & Traustadóttir, R. (2004) *Gender and disability research in the Nordic countries*. Lund: Studentlitteratur.

Kulick, D., & Rydström, J. (2015) *Loneliness and its opposite. Sex, disability, and the ethics of engagement*. Durham: Duke University Press.

Lee, S., Fenge, L., & Collins, B. (2019) Disabled people's voices on sexual well-being. *Disability & Society* 35(2), pp. 303–325.

Leonard, W., & Mann, R. (2018) The everyday experience of lesbian, gay, bisexual, transgender and intersex (LGBTI) people living with disablity, No.111 Gay & Lesbian Health Victoria, Australia Research Council in Sex, Health and Society, La Trobe University, Melbourne: La Trobe University.

Liddiard, K. (2018) *The intimate lives of disabled people*. London: Routledge.

Liddiard, K., Whitney, S., Watts, L., Evans, K., Vogelmann, E., Spurr, R., Goodley, D. (forthcoming) *Living life to the fullest: Disability, youth and voice*. Bingley: Emerald Points.

Limoncin, E., Galli, D., Ciocca, G., Gravina, G.L., Carosa, E., Mollaioli, D., Lenzi, A., & Jannini, E.A. (2014) The psychosexual profile of sexual assistants: An internet based explorative study. *PLOS One*. https://doi.org/10.1371/journal.pone.0098413

Linton, K.F., Rueda, H.A., & Williams, L.R. (2017) *Disability, intimacy and sexual health: A social work perspective*. Washington, DC: National Association of Social Workers.

Lion, E. (1990) *Sexuality of the dying: What dying participants, their spouses and their caregivers teach us about the sexuality of the dying*. Unpublished Ed.D. thesis, Indiana University.

Loeser, C., Pini, B., & Crowley, V. (2018) Disability and sexuality: Desires and pleasures. *Sexualities* 21(3), pp. 255–270.

Löfgren-Mårtenson, L. (2004). "May I?" About sexuality and love in the new generation with intellectual disabilities. *Sexuality and Disability* 22(3), pp. 197–207.

Löfgren-Mårtenson, L. (2009) The invisibility of young homosexual women and men with intellectual disabilities. *Sexuality and Disability* 27(1), pp. 21–26.

Löfgren-Mårtenson, L. (2011) "I want to do it right!" A pilot study of Swedish sex education and young people with intellectual disabilities. *Sexuality and Disability* 30(2), pp. 209–225.

Löfgren-Mårtenson, L. (2012) "Hip to be crip?" Om cripteori, sexualitet och personer med intellektuell funktionsnedsättning. *Lambda Nordica* (1–2), pp. 53–76.

Löfgren-Mårtenson, L., Molin, M., & Sorbring, E. (2018) H@ssles and hopes on the internet: What professionals have to encounter in dealing with internet use and sexuality among youths with intellectual disabilities. *International Papers of Social Pedagogy* 8(1), pp. 14–35.

Löfgren-Mårtenson, L., & Ouis, P. (2018) "It's their way of protecting them": Between care and control in an honor context for youths with intellectual disabilities. *International Papers of Social Pedagogy* 8(1), pp. 14–35.

Malinowska, A. (2018) Lost in representation: Disabled sex and the aesthetics of the 'norm'. *Sexualities* 21(3), pp. 364–378.

Malmberg, D. (2012) "To be cocky is to challenge norms." The impact of bodynormativity on bodily and sexual attraction in relation to being a cripple. *Lambda Nordica* 1–2, pp. 194–216.

Mannino, G., Giunta, S., & La Fiura, G. (2017) Psychodynamics of the sexual assistance for individuals with disability. *Sexuality and Disability* 35(4), pp. 495–506.

McCann, E., Lee, R., & Brown, M. (2016) The experiences and support needs of people with intellectual disabilities who identify as LGBT: A review of the literature. *Research in Developmental Disabilities* 57, pp. 39–53.

McCarthy, M. (1999) *Sexuality and women with learning disabilities*. London: Jessica Kingsley Publishers.

McCarthy, M. (2014) Women with intellectual disability: Their sexual lives in the 21st century. *Journal of Intellectual & Developmental Disability* 39(2), pp. 124–131.

McCarthy, M., & Wilson, N. (2014) Introduction to the special series of papers on sexual health and sexuality. *Journal of Intellectual and Developmental Disability* 39(2), pp. 123–124.

McRuer, R. (2006) *Crip theory: Cultural signs of queerness and disability*. New York, NY: New York University Press.

McRuer, R. (2011) Disabling sex: Notes for a crip theory of sexuality. *GLQ: A Journal of Lesbian and Gay Studies* 17(1), pp. 107–117.

McRuer, R., & Mollow, A. (2012) Introduction. In R. McRuer & A. Mollow (eds.), *Sex and disability*. Durham and London: Duke University Press, pp. 1–34.

McRuer, R., & Wilkerson, A. (eds.) (2003) Desiring disability: Queer theory meets disability studies. Special issue of *GLQ: A Journal of Lesbian and Gay Studies* 9(1–2).

Meekosha, H. (2006) What the hell are you? An intercategorical analysis of race, ethnicity, gender and disability in the Australian body politic. *Scandinavian Journal of Disability Research* 8(2–3), pp. 161–176.

Meekosha, H. (2011) Decolonising disability: Thinking and acting globally. *Disability and Society* 26, pp. 667–682.

Meekosha, H., & Dowse, L. (1997) Enabling citizenship: Gender, disability and citizenship in Australia. *Feminist Review* 57, pp. 49–72.

Meekosha, H., & Soldatic, K. (2011) Human rights and the Global South: The case of disability. *Third World Quarterly: Journal of Emerging Areas* 32, pp. 1383–1397.

Milbrodt, T. (2019) Sexy like us. Expanding notions of disability and sexuality through burlesque performance. *Journal of Literary & Cultural Disability Studies* 13, pp. 377–392.

Mitchell, D.T., & Snyder S.L, (2000) Representation and its discontents: The uneasy home of disability in literature and film. In Mitchell D.T. & S.L. Snyder (eds.), *Narrative prosthesis: Disability and the dependencies of discourse*. Ann Arbor: University of Michigan Press, pp. 15–46.

Mona, L.R., Cameron, R.P., & Clemency Cordes, C. (2017) Disability culturally competent sexual healthcare. *American Psychologist* 72(9), pp. 1000–1010.

Mona, L.R., Cameron, R.P., & Fuentes, A.J. (2006) Broadening paradigms of disability research to clinical practice: Implications for conceptualization and application. In K.J. Hagglund & A. Heinemann (eds.), *Handbook of applied disability and rehabilitation research*. New York, NY: Springer Publications, pp. 75–102.

Mona, L.R., Gardos, P.S., & Brown, R.C. (1994) Sexual self-views of women with disabilities: The relationship among age-of-onset, nature of disability, and sexual self-esteem. *Sexuality and Disability* 12(4), pp. 261–277.

Mona, L.R., Hayward, H., & Cameron, R.P. (2019) Cognitive behavior therapy and people with disabilities. In G.Y. Iwamasa & P.A. Hays (eds.), *Culturally responsive cognitive behavior therapy: Practice and supervision*. Washington, DC: American Psychological Association, pp. 257–285.

Mona, L., & Shuttleworth, R. (2000) Introduction to the special issue. *Sexuality and Disability* 18(3), pp. 155–158.

Mona, L.R., Syme, M.L., Cameron, R.P., Clemency Cordes, C., Fraley, S.S., Baggett, L.R., Roma, V.G. (2013) Sexuality and disability: A disability-affirmative approach to sex therapy. In Y.M. Binik & K. Hall (eds.), *Principles and practices of sex therapy*. (5th Ed.). New York: Guilford, pp. 457–481.

Moore, F., & Shuttleworth, R. (2019) *How to handle an anthropologist*. Berkeley: Inter-Relations.

Morales, E., Gauthier, V., Edwards, G., & Courtois, F. (2018) Co-designing sex toys for adults with motor disabilities. *Sexuality and Disability* 36, pp. 47–68.

Morris, J. (ed.) (1996) *Encounters with strangers: Feminism and disability*. London: Women's Press.

Moyle, J. (2016) Including disability in the social work core curriculum: A compelling argument. *Australian Social Work* 69(4), pp. 495–502.

O'Dea, S.M., Shuttleworth, R.P., & Wedgwood, N. (2012) Disability, doctors and sexuality: Do healthcare providers influence the sexual wellbeing of people living with a neuromuscular disorder? *Sexuality and Disability* 30(2), pp. 171–185.

Olkin, R. (1999) *What psychotherapists should know about disability*. New York, NY: The Guilford Press.

O'Shea, A., & Frawley, P. (2019) Gender, sexuality and relationships for young Australian women with intellectual disability. *Disability & Society* 35(4), pp. 654–675.

O'Toole, C. (2000) The view from below: Developing a knowledge base about an unknown population. *Sexuality and Disability* 18(3), pp. 207–224.

O'Toole, C. (2002) Sex, disability and motherhood: Access to sexuality for disabled mothers. *Disability Studies Quarterly* 22(4), pp. 81–101.

Pecora, L.A., Mesibov, G.B., & Stokes, M.A. (2016) Sexuality in high-functioning Autism: A systematic review and meta-analysis. *Journal of Autism & Developmental Disorders* 46(11), pp. 3519–3556.

Perlin, M.L., & Lynch, A.J. (2015) "Love is just a four-letter word": Sexuality, international human rights, and therapeutic jurisprudence. *Canadian Journal of Comparative and Contemporary Law* 1(1), pp. 9–48.

Peta, C., McKenzie, J., Kathard, H., & Africa, A. (2016) We are not asexual beings: Disabled women in Zimbabwe talk about their active sexuality. *Sexuality Research and Social Policy* 14, pp. 410–424.

Peuravaara, K. (2013) Theorizing the body: Conceptions of disability, gender and normality. *Disability & Society* 28(3), pp. 408–417.

Priestley, M. (ed.) (2001) *Disability and the life course: Global perspectives*. Cambridge: Cambridge University Press.

Priestley, M. (2003) *Disability: A life course approach*. Oxford: Polity Press.

Quippings: Disability Unleashed. https://www.facebook.com/quippings/ accessed 5 January, 2020.

Redelman, M. (2008) Is there a place for sexuality in the holistic care of patients in the palliative care phase of life? *American Journal of Hospice and Palliative Medicine* 25(5), pp. 366–371.

Reynolds, D. (2007) Disability and BDSM: Bob Flanagan and the case for sexual rights. *Sexuality Research and Social Policy* 4(1), pp. 40–52.

Rugoho, T. (2017) Fishing in deep waters: Sex workers with disabilities in Harare, Zimbabwe. *International Journal of Gender Studies in Developing Societies* 2(3), pp. 227–240.

Rugoho, T., & Maphosa, F. (2017) Challenges faced by women with disabilities in accessing sexual and reproductive health in Zimbabwe: The case of Chitungwiza town. *African Journal of Disability* 26(6), pp. 2223–9170.

Rydström, J. (2019) Disability, socialism and autonomy in the 1970s: Case studies from Denmark, Sweden and the United Kingdom. *Disability & Society* 34(9–10), pp. 1637–1659.

Sakairi, E. (2020) *Sexuality: the experiences of people with physical disabilities in contemporary Japan. Untold stories and voices silenced in the name of taboo*. Unpublished PhD thesis, University of Auckland.

Samuels, E. (2003) My body, my closet: Invisible disability and the limits of coming-out discourse. *GLQ: A Journal of Lesbian and Gay Studies* 9(1–2), pp. 233–255.

Sánchez-Fuentes, M., Santos-Iglesias, P., & Sierra, J.C. (2014) A systematic review of sexual satisfaction. *International Journal of Clinical and Health Psychology* 14, pp. 67–75.

Sandahl, C. (2003) Queering the crip or cripping the queer?: Intersections of queer and crip identities in solo autobiographical performance. *GLQ: A Journal of Lesbian and Gay Studies* 9(1–2), pp. 25–56.

Sandberg, L.J. (2018) Dementia and the gender trouble? Theorising dementia, gendered subjectivity and embodiment. *Journal of Aging Studies* 45, pp. 25–31.

Sandberg, L.J., Bertilsdotter Rosqvist, H., & Grigorovich, A. (2020) Regulating, fostering and preserving: the production of sexual normates through cognitive ableism and cognitive othering. *Culture, Health and Sexuality*. Published online 10 Aug 2020. doi: 10.1080/13691058.2020.1787519

Sandberg, L.J., & King, A. (2019) Queering gerontology. In: D. Gu & M.E. Dupre (eds.), *Encyclopedia of gerontology and population aging*. Cham: Springer, pp. 1–7.

Sanders, T. (2007) The politics of sexual citizenship: commercial sex and disability. *Disability & Society* 22(5), pp. 439–455.

Sanders, T. (2010) Sexual citizenship, sexual facilitation and the right to pleasure. In R. Shuttleworth & T. Sanders (eds.), *Sex and disability: Politics, identity and access*. Leeds: The Disability Press, pp. 139–154.

Santinele Martino, A. (2017) Cripping sexualities: An analytic review of theoretical and empirical writing on the intersection of disabilities and sexualities. *Sociology Compass* 11(5). https://doi.org/10.1111/soc4.12471

Saxton, M. (2010) Disability rights and selective abortion (3rd ed.). In L. Davis (ed.), *The disability studies reader*. New York: Routledge, pp. 120–132.

Schalock, Robert L., Borthwick-Duffy, Sharon A., Bradley, Valerie J., Buntinx, Wil H.E., Coulter, David L., Craig, Ellis M., Gomez, Sharon C., Lachapelle, Yves, Luckasson, Ruth, Reeve, Alya, Shogren, Karrie A., Snell, Martha E., Spreat, Scott, Tasse, Marc J., Thompson, James R., Verdugo-Alonso, Miguel A.,

Wehmeyer, Michael L., & Yeager, Mark H. (2010) *Intellectual disability: Definition, classification, and systems of supports* (11th ed.). Washington, DC: American Association on Intellectual and Developmental Disabilities.

Shah, S. (2019) 'Forgotten as adults': A digital ethnography of ageing and access to healthcare for women with Cerebral Palsy. In *Nordic Network on Disability Research 15th conference: Inclusion and Exclusion in the Welfare Society*. Copenhagen 8–10 May 2018.

Shakespeare, T. (2000) Disabled sexuality: Toward rights and recognition. *Sexuality and Disability* 18(3), pp. 159–166.

Shakespeare, T., Gillespie-Sells, K., & Davies, D. (1996) *The sexual politics of disability: Untold desires*. London: Casell.

Shakespeare, T., & Richardson, S. (2018) The sexual politics of disability, twenty years on. *Scandinavian Journal of Disability Research* 20(1), pp. 82–91.

Shakespeare, T., Zeilig, H., & Mittler, P. (2017) Rights in mind: Thinking differently about dementia and disability. *Dementia* 18(3), pp. 1075–1088.

Shapiro, L. (2002) Incorporating sexual surrogacy into the Ontario Direct Funding Program. *Disability Studies Quarterly* 22(4), pp. 72–81.

Shapiro, L. (2017) The disabled sexual surrogate. *Reproductive Health Matters* 25(50), pp. 134–137.

Sheppard, E. (2018) Using pain, living with pain. *Feminist Review* 120(1), pp. 54–69.

Sheppard, E. (2019) Chronic pain as emotion. *Journal of Literary & Cultural Disability Studies* 14(1), pp. 5–20.

Sherry, M. (2004) Overlaps and contradictions between queer theory and disability studies. *Disability & Society* 19(7), pp. 769–783.

Shildrick, M. (2004) Queering performativity: Disability after Deleuze. *Scan: Journal of Media Arts Culture* 1(3). http://scan.net.au/scan/journal/display.php?journalid=36

Shildrick, M. (2007) Contested pleasures: The sociopolitical economy of disability and sexuality. *Sexuality Research & Social Policy* 4(1), pp. 53–66.

Shildrick, M. (2009) *Dangerous discourses of disability, subjectivity and sexuality*. Basingstoke: Palgrave Macmillan.

Shuttleworth, R. (2000) The search for sexual intimacy for men with cerebral palsy. *Sexuality and Disability* 18(4), pp. 263–282.

Shuttleworth, R. (2003) Disability and sexuality: From medical model to sexual rights. *American Sexuality* 1(1). http://nsrc.sfsu.edu/

Shuttleworth, R. (2007) Disability and sexuality: Toward a constructionist approach to disability and sexuality and the inclusion of disabled people in the sexual rights movement. In N. Teunis & G.H. Herdt (eds.), *Sexual inequalities and social justice*. Berkeley and Los Angeles, California: University of California Press, pp. 174–207.

Shuttleworth, R. (2010) Towards an inclusive sexuality and disability research agenda. In R. Shuttleworth & T. Sanders (eds.), *Sex and disability: Politics, identity and access*. Leeds: The Disability Press, pp. 1–20.

Shuttleworth, R., & Meekosha, H. (2017) Accommodating critical disability studies in bioarchaeology. In J. Byrnes & J. Mueller (eds.), *Bioarchaeology of impairment and disability: Theoretical, ethnohistorical and methodological perspectives*. Cham, Switzerland: Springer, pp. 19–38.

Shuttleworth, R., & Mona, L. (2002) Disability and sexuality: Towards a focus on sexual access. *Disability Studies Quarterly* 22(4), pp. 2–9.

Shuttleworth, R., & Sanders, T. (eds.) (2010) *Sex and disability: Politics, identity and access*. Leeds: The Disability Press.

Shuttleworth, R., & Taleporos, G. (forthcoming) Disability, facilitated sex and sexual autonomy. *Sexuality and Disability*.

Siebers, T (2010) *Disability aesthetics*. Ann Arbor: Michigan University Press.

Simpson, P., Wilson, C.B., Brown, L.J.E., Dickinson, T., & Horne, M. (2017) "We've had our sex life way back": Older care home residents, sexuality and intimacy. *Ageing and Society* 38(7), pp. 1478–1501.

Sins Invalid. (2019) *Skin, tooth, and bone: The basis of movement is our people* (2nd ed.). Berkeley, CA.

Slater, J. (2015) *Youth and disability: A challenge to Mr Reasonable*. London: Ashgate.

Slater, J., Ágústsdóttir, E., & Haraldsdóttir, F. (2018) Becoming intelligible woman: Gender, disability and resistance at the border zone of youth. *Feminism & Psychology* 28(3), pp. 409–426.

Slater, J., & Liddiard, K. (2018) Why disability studies scholars must challenge transmisogyny and transphobia. *Canadian Journal of Disability Studies* 7(2), pp. 83–93.

Smith, B., & Hutchison, B. (eds.) (2004) *Gendering disability*. New Brunswick, NJ: Rutgers University Press.

Soldatic, K., & Meekosha, H. (2012) Moving the boundaries of feminist social work education with disabled people in the neoliberal era. *Social Work Education* 31(2), pp. 246–252.

Stevens, B. (2010) Crip sexuality: Sk(r)ewed media representation. In R. Shuttleworth & T. Sanders (eds.), *Sex and disability: Politics, identity and access.* Leeds: The Disability Press, pp. 59–78.

Syme, M., Cohn, T., Stoffregen, S., Kaempfe, H., & Schippers, D. (2018) "At my age ..." Defining sexual wellness in mid- and later life. *The Journal of Sex Research* 56(7), pp. 832–842.

Taleporos, G., & McCabe, M.P. (2001a) Physical disability and sexual esteem. *Sexuality and Disability* 19(2), pp. 131–148.

Taleporos, G., & McCabe, M.P. (2001b) The impact of physical disability on body esteem. *Sexuality and Disability* 19(4), pp. 293–308.

Taleporos, G., & McCabe, M.P. (2002) Body image and physical disability – personal perspectives. *Social Science & Medicine* 54(6), pp. 971–980.

Tefera, B., Van Engen, M., Van der Klink, J., & Schippers, A. (2017) The grace of motherhood: disabled women contending with societal denial of intimacy, pregnancy, and motherhood in Ethiopia. *Disability & Society* 32(10), pp. 1510–1533.

Tellier, S. (2017) Advancing the discourse: Disability and BDSM. *Sexuality and Disability* 35, pp. 485–493.

Tepper, M.S. (2000) Sexuality and disability: The missing discourse of pleasure. *Sexuality and Disability* 18(4), pp. 283–290.

Thomas, C. (2006) Disability and gender: Reflections on theory and research. *Scandinavian Journal of Disability Research* 8(2–3), pp. 177–185.

Tilley, E., Walmsley, J., Earle, S., & Atkinson, D. (2012) 'The silence is roaring': Sterilization, reproductive rights and women with intellectual disabilities. *Disability & Society* 27(3), pp. 413–426.

Toft, A., Franklin, A., & Langley, E. (2019) "You're not sure that you are gay yet": The perpetuation of the 'phase' in the lives of young disabled LGBT + people. *Sexualities* 23(4) pp. 516–529.

Tremain, S. (2000) Queering disabled sexuality studies. *Sexuality and Disability* 18(3), pp. 291–299.

Trent, J.W. (1994) *Inventing the feeble mind. A history of mental retardation in the United States.* Berkeley: University of California Press.

Ünal, B. (2018) *The development of disability pride through challenging internalized idealist and ableist norms in Turkish society: a grounded theory study.* Unpublished PhD thesis, Middle East Technical University.

United Nations Population Fund. (2018) *Young persons with disabilities: Global study on ending gender-based violence and realizing sexual and reproductive health and rights.* New York, NY: UNFPA.

Vehmas, S. (2019) Persons with profound intellectual disability and their right to sex. *Disability & Society* 34(4), pp. 519–539.

Wade, H. (2002) Discrimination, sexuality and people with significant disabilities: Issues of access and the right to sexual expression in the United States. *Disability Studies Quarterly* 22(3), pp. 9–30.

Waxman, B., & Finger, A. (1989) The politics of sex and disability. *Disability Studies Quarterly* 9(3), pp. 1–5.

Wilson, N.J., & Frawley P. (2016) Transition staff discuss sex education and support for young men and women with intellectual and developmental disability. *Journal of Intellectual and Developmental Disability* 41(3), pp. 209–221.

Wilson, N., Macdonald, J., Hayman, B. Bright, A.M., Frawley, P., & Gallego, G. (2018) A narrative review of the literature about people with intellectual disability who identify as lesbian, gay, bisexual, transgender, intersex or questioning. *Journal of Intellectual Disabilities* 22(2), pp. 171–196.

Wiseman, P. (2014) *Reconciling the 'private' and 'public': Disabled young people's experiences of everyday embodied citizenship.* Unpublished PhD thesis, University of Glasgow.

World Health Organization. (2011) *World report on disability.* Geneva: World Health Organization (WHO).

Wotton, R. (2016) *Sex workers who provide services to clients with disability in New South Wales, Australia.* Unpublished MA thesis, University of Sydney.

Wotton, R., & Isbister, S. (2010). A sex worker perspective on working with clients with a disability and the development of Touching Base Inc. In R. Shuttleworth & T. Sanders (eds.), *Sex and disability: Politics, identity and access.* Leeds: The Disability Press, pp. 155–178.

Yasuda, H., & Hamilton, C. (2013) Investigating the sexuality of disabled Japanese women: Six autobiographical accounts. *Women's Studies Journal* 27(2), pp. 44–53.

Þórný Stefánsdóttir, N., & Rasmussen, K.S. (2017) *Seksuel trivsel for personer med multiple funktionsnedsættelser. En undersøgelse blandt danske bo- og aflastningstilbud [Sexual well-being for people with multiple impairments. A study in Danish housing and respite care services].* Taastrup: Sammenslutningen af Unge Med Handicap (SUMH).

# PART I

# Theoretical frames and intersections

# 1

# THEORISING DISABLED PEOPLE'S SEXUAL, INTIMATE, AND EROTIC LIVES

## Current theories for disability and sexuality

*Kirsty Liddiard*

## Introduction

Whereas there is now considerable literature and research on disability and sexuality, there remain fewer explorations of disabled people's intimate and interpersonal relationships with sexual partners, and experiences of and engagements with intimacy and love (Ignagni et al. 2016; Liddiard 2018). Culturally, disabled people are routinely assumed to lack the capabilities and capacities to embody and experience sexuality and desire, as well as the agency to love and be loved by others and to build their own families, if they so choose. In the past, much of this sexual oppression and exclusion was rendered unimportant by academics and activists, in comparison with other areas of life, and, as such, was deprioritised within both academic and activist contexts in favour of a focus on disabled people's social and political histories (Finger 1992; Shakespeare et al. 1996). This omission was felt strongly within disability rights movements, although it was seldom publicly acknowledged—as the words of activist Anne Finger (1992, p. 8), in an early edition of the *New Internationalist*, show:

> Sexuality is often the source of our deepest oppression; it is also often the source of our deepest pain. It's easier for us to talk about—and formulate strategies for changing—discrimination in employment, education and housing than to talk about our exclusion from sexuality and reproduction.

Finger's words are reproduced in many scholarly writings on disability and sexuality—you'll likely find them throughout this important book. This is no coincidence: Her words reached the very heart of key problems within the disability rights and justice movements of the era because they highlight the marked exclusion of sexuality. In later years, however, following the publication of the text by Shakespeare et al. (1996, p. 1), *Untold Desires: The Sexual Politics of Disability*, there have been myriad attempts to commence a dialogue about the sexualities of disabled people across multiple disciplines and activist movements: Sexology, Rehabilitation

39

Studies, Sociology, Cultural Studies, the Disability Rights Movement, Crip Activism, Critical Sexuality Studies and, of course, Disability Studies.

In this chapter, I contemplate a smörgåsbord of approaches to theorising the sexual, intimate and erotic lives of disabled people. Throughout, I purposefully centre the sexual stories of disabled people co-constructed through my own sociological research (Liddiard 2018). As both a disability studies scholar and a disabled woman, it is crucial for me to continually locate the lived and material realities of disability life into my theoretical reworkings. As such, the politics of location are vital to my work (see Liddiard 2018). Through this positionality, I show how we often live within and through the theories we construct; how theory isn't as detached from the mundanity of everyday life as we typically assume; and how theory gets under our skin and just won't leave us be (see Goodley 2016). I begin, then, by situating Critical Disability Studies as a politicised, ethical and transdisciplinary foundation from which to explore disabled sexualities. Next, I consider the material makings of sexual bodies and selves, drawing upon the impacts of biological essentialism as a reductionism that routinely Others disabled people's sexual lives, bodies and identities. Following this, I imagine otherwise through queer theory and crip theory, questioning their ability to forge a more radical sexual politics of disability that speaks to the everyday lived lives of disabled people. In the final section, I draw upon more recent developments to disability theory, namely Posthuman Disability Studies and DisHuman Studies (Goodley et al. 2014b; see also dishuman.com), which seek "more expanded, crip [and] relational forms of the human" (Goodley et al. 2017). I explore the extent to which these emerging perspectives can pave new ways of imagining and advocating for disabled people's sexual and intimate futures. I conclude this chapter by questioning whether critical social theories, as currently constructed, can adequately theorise the lived, embodied and material realities of disabled people's sexual, intimate and erotic lives.

## Theorising with Critical Disability Studies

As I state in my recently authored book, *The Intimate Lives of Disabled People* (Liddiard 2018, p. 15), "theory and theorising, quite rightly, can be messy jobs." To understand and theorise the sexual, intimate and erotic lives of disabled people means applying myriad critical theories, such as feminism, interactionism, phenomenology and poststructuralism, and crip, queer, postmodern and psychoanalytic approaches to disability, the body, gender, sexuality, identity, embodiment and subjectivity. Critical Disability Studies is the vehicle through which I do so in my own explorations of disabled sexualities. Critical Disability Studies, to me, is a politicised, ethical and transdisciplinary space—crucially, one that "connects the aspirations and ambitions of disabled people with the transformative agendas of class, feminist, queer, and postcolonial studies" (Goodley 2011a, p. 174). Critical Disability Studies enables a clear focus on, and validation of, the intersections of disability life, connecting disability with the politics of class, race, ethnicity, gender, sexuality and nation. As Meekosha and Shuttleworth (2009, p. 50) suggest, in doing so, Critical Disability Studies emphasises that that the struggle for social justice and diversity subsists "on another plane of development—one that is not simply social, economic, and political, but also psychological, cultural, discursive and carnal." Such an intersectional understanding of disability markedly builds upon the predominantly Marxist and materialist-orientated approaches to disability promulgated through the social model. The social model of disability laid "the blame for disabled people's oppression clearly at the feet of economic relations in capitalistic society" (Meekosha and Shuttleworth 2009, p. 55). With its unrelenting focus on civil rights and structural disablism, the social model simultaneously omitted equal political focus towards the private and intimate lives of disabled people (see Liddiard 2018 for a longer history; Shakespeare et al. 1996).

Key to Critical Disability Studies is its commitment to destabilising and contesting disablism and ableism. For clarity, Campbell (2009, p. 44) defines ableism as a "network of beliefs, processes and practices that produces a particular kind of self and body (the corporeal standard) that is projected as the perfect, species-typical and therefore essential and fully human." Disablism, on the other hand, is the resultant oppressive treatment of disabled people. Where I use the term "dis/ableism" I refer to the iterative processes of ableism and disablism that "casts [disabled people] as a diminished state of being human." Significantly, Critical Disability Studies acknowledges that studies of disability and ability are routinely political and politicised. Theory, as currently constructed, is not separate to nor detached from the everyday lived realities of living with and through disability, but begins there. Therefore, quite often I follow others (Campbell 2001; 2009; Davis 2002; Goodley 2014; Rose 2001) in turning my attention away from disability and onto ableist hegemony or "ableist-normativity" (Campbell 2008, p. 1)—something now known as Critical Ableism Studies or Ability Studies (Wolbring 2008). A clear focus on ableism, then, disrupts "those normative homelands that all of us are forced to populate" (Goodley 2014, p. 194). Elsewhere, I (and others—see Gill 2015) have called this, "sexual ableism": an understanding that the Othering of disabled sexualities emerges through the dominance of the deeply ableist (and hetero/sexist) institutions of heteronormativity and heterosexuality (discussed later). Thus, sexual ableism—to me, as someone who has explored the complexities and intricacies of disabled people's own sexual stories through research (see Liddiard 2018)—is the impossibility of sexual normalcy for myriad bodies and minds. As Lucille, a heterosexual married 36-year-old cisgender woman with an acquired spinal injury, said in my research (Liddiard 2018, p. 103):

> Why would you want to have sex if you couldn't feel anything other than a weird nerve pain and why would someone want sex with a girl who couldn't orgasm?

Lucille's words emphasise the ways that people with acquired impairment can feel desexualised following the transition to a disabled identity, but also that her newly queered body (which no longer achieves pleasure in normative ways) is uncomfortable, because it challenges culturally dominant preconceptions of what (and where) pleasure and erogenous sensation should take place. Thus, the impossibility of meeting the demands of sexual normalcy can be a root cause of the (sexual) oppression of disabled people, and many other sexual minorities that don't *fit*. Mike Gill (2015, p. 151) defines sexual ableism as "the system of imbuing sexuality with determinations of qualification to be sexual based on criteria of ability, morality, physicality, appearance, age, race, social acceptability and gender conformity."

Critically, the sexual stories of disabled people that underpin my book emphasised sexual ableism and the privileging of sexual normalcy as significantly impactful to individual, lived and psycho-emotional experiences of sexual opportunities, identities and intimate relationships with others (Liddiard 2018). In this framing, disability can only ever be lived and experienced as being troublesome towards meeting heteronormative ideals—a product of sexual ableism. For example, informants' sexual stories (Liddiard 2018, p. 166):

> Privileged normative sexuality as a central theme: normative gender categories and heterosexuality were upheld and privileged as given, natural and fixed. Intimate relationships and coupledom were strongly desired and seen to affirm a sense of intimate citizenship (Plummer 2003), as well as to confirm worth and desirability. Sexual expression and gratification were understood as inherently natural (particularly in the context of male bodies) and obtaining pleasure served to proffer social value, evidence one's humanness and 'constitute full subjectivity' (Shuttleworth 2000, p. 280).

Normative bodily aesthetics were revered and strived for (most notably by women), and the prescribed phallocentric and orgasmic "mechanics" of heteronormative practice remained the immovable embodied norms from which other alternative sexual methods, interactions and pleasures were judged.

This analysis shows the ways in which heteronormativity, and its bedfellow, sexual ableism, continue to shape the sexual subjectivities of, and have psycho-emotional consequences for, those who are excluded from its narrow boundaries. Sexual ableism propagates a normalcy, a perfection of body, self and pleasure, which no one, in reality, can ever achieve (see McRuer 2006a). Sexual ableism, then, merely serves to marginalise a diversity of bodies and minds, and guard the boundaries of, as Margrit Shildrick (2007, p. 221; original emphasis) puts it, "the contested question of who is to count as a sexual subject." In the remaining sections of this chapter, I explore multiple theoretical perspectives and understandings of disability, sexuality and the body to consider what they offer towards theorising the lived, embodied and material realities of disabled people's sexual, intimate and erotic lives.

## Theorising sexual bodies and material selves

Early constructionism, which was embedded in phenomenological and interactionist sociology, redefined the scholarly field of sexualities from the 1960s onwards (see Gagnon and Simon 1973). Constructionism enabled a refusal of essentialist "pre-social" notions of sexuality towards an understanding of sexuality as socially, politically and culturally shaped and produced (Jackson and Scott 2010). Or, as Villanueva (1997, p. 18) suggests, sexualities are "shaped through a system of social, cultural, and interpersonal processes." In short, essentialist perspectives of sexuality root sexual expression and desire as being purely biological in nature, which arguably denies human agency and autonomy (Jackson 1999). Sexuality emerges, then, as "ethological fallacy" (Gagnon and Simon 1973, p. 3) that pays little attention to humans as "complex, arbitrary and changeable creatures" (Weeks 1986, p. 46). In this section, I consider the material makings of sexual bodies and selves, drawing upon the impacts of biological essentialism as a form of reductionism that routinely Others disabled sexualities.

Biological conceptualisations of sexuality sustain and reproduce a heteronormative agenda based upon normative binary modes of gender, from which disabled and myriad Othered bodies are routinely excluded (Liddiard 2014). For example, the "male" is reduced to being only dominant, animalistic and powerful: A reductionist phallocentrism that locates stamina, performance and bodily function as being integral to maleness, which some have argued functions to emasculate disabled men and obstruct or deny disabled masculinities (Drench 1992; Murphy 1990; Shakespeare 1996). Inexorably, essentialist approaches centre reproductive function at the heart of sexuality. For biological determinists, "sexuality is both definable and explicable in terms of a reproductive imperative" (Jackson 1999, p. 5). Thus, in an ableist cultural imaginary, where rights to reproduction are awarded only to those deemed the fittest and most *able* (see Tepper 2000), a mode of sexuality so heavily defined by reproduction is deeply problematic in contexts of disability. As Waxman Fiduccia (2000, p. 169) suggests, "sexual rights have always and only been awarded to those who are proclaimed to deliver quality offspring"; thus, biomedicine seeks to dominate and regulate the fertility of the dangerous disabled female. By regulating female sexuality and reproduction, as suggested by Waxman Fiduccia (2000) and others (e.g. Anderson and Kitchen 2000; Kent 2002; Lee and Heykyung 2005), the disabled female body is denied reproductive justice and freedom (Waxman and Finger 1991).

Biological essentialism is also central to the discipline of Sexology—an "empiricist approach that focuses on the study and classification of sexual behaviours, identities and relations" (Bland and Doan 1998, p. 1). Sexology is argued to be the science that made sex an "object of study" for the first time (Hawkes 1996, p. 56). For example, the works of Kinsey (1953), Chesser (1950) and Masters and Johnson (1966) advocated biomedical sexological knowledge of human sexuality and desire, that made measurable for the first time what came to be known as the key stages of the standard human sexual experience. Such physiological norms and boundaries, established by sexual medicine, are problematic for a majority of bodies, but particularly for those labelled impaired (as shown through Lucille's words above). To offer another example, sexological discourses of physical sexuality led Phillip, who became disabled through a motorbike accident at age 35 (just three years before our interview took place), and other disabled men in my research, to describe their physical sexual experiences in terms of their difference:

> It's very hard to describe actually, but you get … obviously you've lost outer sensation and the ability to climax, but it's amazing how strong the mind is and the enjoyment you get from, you know, the act, if you will, of sex. So … that has diminished … it's diminished the kind of … I guess the, it's not enjoyment as such because I love having sex, but it's the … there's … I could say there's something missing in it, actually.

Phillip's assertion of something being "missing" (to quote his words) supports Tepper's (2000, p. 289) research with men with Spinal Cord Injury (SCI), which found that most described post-injury pleasure as "not the same." The homogeny of the charted experience of (masculine) sexual pleasure doesn't make space for Phillip to understand his experiences as "different", only "missing". Such experiences are undoubtedly shaped by sexual normalcy as constructed through Masters and Johnson's (1966) sexual response cycle—which became the benchmark for new scientific and cultural understandings of human sexuality—which quantified and charted the physiological aspects of sexuality as a normative, linear trajectory of pleasure: Attraction, arousal, climax, orgasm and resolution. This firmly established a "physiological norm" of the sexually able body, legitimised through expert clinical and scientific knowledge. Such homogenising material norms firmly rooted the ability and necessity to "achieve", reach and strive for orgasm (Tepper 2002) as central to sexuality in ways that rendered alternative experiences of pleasure as dysfunctional, inadequate and in need of treatment (Bullough 1994; Hawkes 1996). Phillip's words above show him trying to make sense of his new experiences; notably, there is little accessible lexicon for him to describe his new engagements with pleasure outside of standardised physiologies of pleasure.

The scientific construction of a functionally normative sexual body—re/produced through what Tepper (2000, p. 288) calls "a genitally focused and performance-orientated conception of sexuality"—diminishes the potentialities of impairment and bodily difference for sexual pleasure, expression and desire. Moreover, a persistent emphasis on bodily *function* (and *lack* of it) ensures that sexuality falls under the responsibility of health professionals who become "gatekeepers" within sexual and intimate life (Shakespeare et al. 1996). For disabled people in particular, it places rightful access to sexual expression and an erotic life at the mercy of a paradigm that routinely devalues the possibilities of their bodies (Anderson and Kitchen 2000; Hahn 1981; Milligan and Neufeldt 2001; Tepper 1999; 2000). This serves to (further) medicalise their lives and bodies and constructs disability (and sexual 'failure') only on individual terms. Inevitably, medicine's propagation of the physiological norm of the sexually able body, Shakespeare et al. (1996, p. 66) argue, invites a biomedical gaze which both underpins and advocates the need for (sexual) treatments and therapies, and intensifies the medical voyeurism of disabled people as "subjects and fetishised objects" (Shakespeare et al. 1996, p. 3; see also Solvang 2007; Waxman Fiduccia 1999). In the following section, I explore queer and crip as the means to contest these essentialist and biomedical understandings of disability, desire and the body, which are routinely embedded in a sexual ableism that inherently denies "an understanding of disability as a valuable difference that yields unique perspectives of personhood, competence, sexuality, agency and ability" (Gill 2015, p. 3).

## Thinking through queer and crip

As I stated at the outset, Critical Disability Studies interrogates and problematises hegemonic normalcy, its politics, its power, its language and its identity (Wilchins 2004)—in ways similar to queer theory's political project to de-essentialise identity (see Halperin 1995). Disability scholars have long acknowledged queer theory's contribution to a more radical Disability Studies agenda (Breckenridge and Vogler 2001; Corbett 1994; McRuer 2006a; Sherry 2004; Sinecka 2008), acknowledging "a synthesis of queer and disability theories" (Goodley 2011a, p. 41). Within this section of the chapter, I outline my current understandings of queer and crip, exploring where queer ends and crip begins. Whether queer and crip even have distinct boundaries remains in flux (see McRuer 2006a); as Carrie Sandahl (2003, p. 27) states, "the fluidity of both terms makes it likely that their boundaries will dissolve." Furthermore, it's pertinent to remember that queer theory is by no means a "unified perspective" (Jackson and Scott 2010, p. 19). It is also valid to contemplate the countless critiques of queer theory—who and what it includes and excludes. For example, queer theory has been criticised for overlooking the specifically gendered experiences of sexual dissidents; neglecting the material conditions of women's lives through ignoring the structural inequalities routine within in their lives (Jackson 1999); "discriminating against the interests of lesbians" (Jeffreys 1994, p. 459); acting as masculinist theory in costume (Smyth 1992); reducing gender to lexicon and overlooking embodiment (Bordo 1993); and distancing the category of woman from everyday lived reality (Fraser 1999).

Despite these significant critiques, however, considering queer offers a radical conversation about bodies, boundaries and binaries that cannot be ignored when theorising disabled sexualities. Like crip (as I will argue later), queer challenges normalcy across multiple registers: "Corporeal, mental, sexual, social, cultural, subcultural" (Sandahl 2003, p. 26). In terms of sexuality, this materialises in its attempts to both refuse and mediate the strict hegemonies of pleasure, desire and eroticism and its commitment to disband the naturalised and normalised binaries of sex and gender—both of which alienate and Other disabled people with impaired bodies (amongst many other kinds of people and bodies) (Liddiard 2018). Thus, thinking *with* and *through* queer as a lens facilitates new conceptualisations of disability, sex and gender in terms of their revolutionary potential, which is lost through other approaches, such as materialist understandings of disability (Rembis 2010). Therefore, making use of a queered Critical Disability Studies enables a noteworthy contestation of the assumed, taken-for-granted hegemonies of sex, gender, disability and the body. This contestation enables new conceptualisations to flourish, positioning disability as a productive and vital threat to—or emancipation from—the heterosexual matrix: "The grid of cultural intelligibility through which bodies, genders and desires are naturalised" (Butler 1990, p. 5).

Importantly, the critical and transgressive politics enabled by queer theory makes space to disrupt the constructed categories of "able," "sexual" and "normal" reified through biological essentialism (discussed earlier). As Goodley et al. (2015, p. 53) suggest, "queer contests the able individual, disputes the psychological, geographical, and cultural normative centre and breaks fixed binaries." But queer remains "contested terrain, with theorists and activists continuing to debate what (and whom) the term encompasses or excludes" (Kafer 2013, p. 16). For McRuer (2002, p. 222), queer theory emerges as "a diverse array of projects that explores the construction and shifting contemporary meanings of sexuality." Fittingly, Critical Disability Studies interrogates the dis/ableist institutions and practices that reproduce the normalcy and naturalness of the "able" body and which contribute to thanatopolitics, defined by Rose (2001) as the increasingly ableist-obsessed nature of everyday life. It is through the construction of the disabled body as Other, grotesque, and monstrous (see Goodley et al. 2015) that hegemonic normalcy is upheld (Davis 1995). Thus, not only does normalcy, as currently constructed, make possible disability, race, class and gender, the construction of these Othered states legitimises and provides authority to notions of normalcy. Other theorists (see Michalko 2002) have highlighted the fragility of the non-impaired body by using the term TAB, or Temporarily Able Bodied, to purposefully destabilise the boundaries of dis/abled and normal/other. McRuer (2006a, p. 2) speaks to a similar politic through his concept of "compulsory able-bodiedness" based upon feminist/queer notions of "compulsory heterosexuality" (Rich 1989):

> The system of compulsory able-bodiedness, which, in a sense, produces disability, is thoroughly interwoven with the system of compulsory heterosexuality that produces queerness: that, in fact, compulsory heterosexuality is contingent on compulsory able-bodiedness and vice versa.

Thus, a body deemed, and labelled, as "able" is not a queer one, and a queered body is one that is "disabled". In much the same way that Butler (1990) posits heterosexual hegemony as maintained through repetitive performances of heterosexuality and heteronormativity, McRuer (2006a, p. 9) contends that "institutions in our culture are showcases for able-bodied performance." He maintains:

> The culture asking such questions assumes in advance that we all agree: able-bodied identities, able-bodied perspectives are preferable and that we all, collectively, are aiming for. A system of compulsory able-bodiedness repeatedly demands that people with disabilities embody for others an affirmative answer to the unspoken question, 'Yes, but in the end, wouldn't you rather be more like me?'
>
> *(McRuer 2006a, p. 9)*

More significant, as McRuer (McRuer 2006a, p. 9) suggests, is that despite the routine repetition of the heterosexual and able-bodied identity, both will always fail: "They are incomprehensible in that each is an identity that is simultaneously the ground on which all identities supposedly rest and an impressive achievement that is always deferred and thus never really guaranteed". Therefore, based upon what he labels "ability trouble" (extended from Butler's concept of "gender trouble"), McRuer (2006a) proposes that, despite its compulsory nature, able-bodiedness is an impossibility, therefore making everyone "virtually disabled" (Goodley 2011a, p. 41).

To draw in *crip* more explicitly, McRuer (2006a, p. 35) demarcates crip theory as that which "questions—or takes a sledgehammer to—that which has been concretised." Crip doesn't just look inwards, but outwards: speaking back to non-disabled and disabled liberalism, and non-disabled and disabled neoliberalism: "Crip experiences and epistemologies should be central to our efforts to counter neoliberalism and access alternative ways of being" (McRuer 2006a, p. 41). Embracing crip means thinking about impairment as more than just an "unwelcome presence" (Shildrick 2009, p. 32), a bodily abnormality to be accommodated, apologised for, and explained. Crip shatters pathological and essentialist discourses of disability that render bodies only as unintelligible and undesirable, and makes the necessary space to contemplate and re-imagine the pleasurable and erotic possibilities of the (crip) body. For example, crip repositions impairment as the basis for alternative forms of pleasure and fulfilment (see Shuttleworth 2010), to potentially "open up new (sexual) horizons" (Shildrick 2009, p. 36) in ways that exceed normative expectations and boundaries of what we have come to know as "the body", "sex" and the "sexy body." The experiences of a heterosexual couple, Shaun and Hannah, interviewed within my research (Liddiard 2018) show this explicitly:

*Shaun:* I have very sensitive areas on my shoulders and … 'cos that's where I was injured so that's kind of a natural thing … so it's nice just for the touching side of things, really.
*Hannah:* Yeah, I remember the first time, because I didn't know that about spinal injury and I was stroking Shaun's shoulder and he was like 'wow!' [Collective laughs] I was like, 'What?!' I think I must have stroked it for an hour!
*Shaun:* She gets bored after a couple of minutes now! [Laughs]
*Hannah:* So that was an eye opener, that wow, so … I think you could get to the stage of having an orgasm through touching above the injury, which is amazing really.

The ability to orgasm through one's shoulder undoubtedly crips the sexually embodied norms of the conventional erotic body that dictates that orgasms are bound only to genitals (Ostrander 2009). Philosopher Abby Wilkerson (2002, p. 48) emphasises the promises of the crip body: "Diffused sensuality, including orgasms centred in earlobes, nipples, sensitive areas of the neck and else." Notably, Shaun and Hannah's experiences show the possibilities of pleasure, through exploration, that impairment can generate (Parker and Yau 2012). The wondrousness of this displaced erogenous zone—a product of Shaun's impaired body—reinscribes their heterosexual sex with new meanings for them as people engaged in an intimate relationship. Furthermore, it speaks to a deeply erotic body—a body whereby pleasure isn't bound to mapped convention—a body that materialises as a space of vivacity and production (Overboe 2007), and which can "expand and envelop in exciting ways" (Goodley 2011a, p. 158).

Within the politics of crip, assimilation is never the goal; 'passing'—performing normalcy—is counterintuitive. Crip doesn't seek to normalise or individualise disability or desire, but draws upon and centres its very queerness as a moment of revision and reflection—making possible, as McRuer (2006a, p. 71) puts it, "that a disabled world is possible and desirable." Thus, crip bodies (seen through Shaun's embodied experiences above) are dynamic in their non-normativity: Abandoning medically imposed notions of deficit to reimagine bodies, that may be ill, sick or impaired, as transgressive and vital. This "new mode of representation" (Siebers 2008, p. 54) of the crip sexual body has, in part, been understood through a hybrid of disabled/crip queer activism (McRuer 2006a). For example, sentiments such as: "Trached dykes eat pussy all night without coming up for air" (O'Toole 2000, p. 212) exemplify the productive realities of crip bodies within the confines of pleasure and sexuality.

Such redefinition of the naturalised body, then, is powerfully transformative and can be richly applied to the project of exploring and theorising disabled sexualities and constructing an embodied, intersectional and eroticised sexual politics of disability. However, it seems necessary at this juncture to ask some critical questions: Do crip and queer give enough consideration to the social and institutional conditions under which the majority of disabled people live? Who is included and excluded? How might we use crip theory to politicise the lives of people with the label of learning disability—or others—whose lived experiences of disability are seldom considered within its boundaries? We might also question the relevance or proximity of crip and queer: The ways in which such perspectives remain largely out of reach of disabled people outside of the academy and beyond spaces of queer and/or radical politics (see Liddiard and Slater 2017; see also Bone 2017). We might also ask, for example, whether destabilising sex and gender is an altogether helpful project when theorising the sexual and intimate lives of disabled people—many of whom are, as their stories in my book emphasised (Liddiard 2018)— striving to be included in the seemingly 'fixed' and 'stable' normative sex and gender categories that dis/ableist cultures routinely deny them. Ultimately, what are the costs of such radical and transgressive sexual politics in the contexts of disability?

Whereas asking these questions undoubtedly risk me sounding polemical, or worse, erotophobic (Wilkerson 2002), as I outlined at the beginning of the chapter, it is pertinent to me, as both a disabled woman and a researcher, to remain faithful to the material realities in which the majority of disabled people live, many of whom are routinely pathologised, infantilised, institutionalised, segregated and criminalised, particularly in contexts of sexuality. It is critical not to overlook the fact that many disabled people are merely surviving in these deeply precarious times (as are many marginalised others), where they are fighting for rights to life, to care and to love, all the while being re/cast as a burden, problem and excess (Ignagni et al. 2016). I now move on to address some of these questions, drawing upon recent posthuman influences to critical disability theory to trouble some of the tensions identified in this chapter so far.

## DisHuman: Which way to the future?

In this final section, I draw upon Posthuman Disability Studies, and their contributions to the emerging theoretical modes of the DisHuman and DisSexual (Liddiard 2018), which work towards, I suggest, paving new ways of imagining and advocating for more emancipatory sexual and intimate futures in contexts of disability. Importantly, I suggest that a DisHuman lens offers new ways to understand such tensions, and enables alternative ways forward when critically imagining disabled people's sexual and intimate futures. Importantly, through the DisHuman position, it becomes possible to "recognise the norm, the pragmatic, and the political value of claiming the norm" (Goodley and Runswick-Cole 2014, p. 5) while always seeking to disrupt and contest it. The prefix of "dis-" is used here to denote the "necessity to critically question and—at certain junctures—disrespect and critique that which it precedes" (Goodley and Runswick-Cole 2014, p. 5). To offer an example, speaking about the lives of those with the label of intellectual impairment, Goodley and Runswick-Cole (2014, p. 2) clarify: "Intellectual disability is always profound because it enlarges, pauses, questions, and clarifies what it means to be human. Intellectual disability 'disses' (or disrespects) the human but it also desires the human." Therefore, as Goodley and Runswick-Cole (2014, pp. 3–4) suggest, a DisHuman analysis "allows us to claim (normative) citizenship (associated with choice, a sense of autonomy, being part of a loving family, the chance to labour, love and consume), while simultaneously drawing on disability to trouble, re-shape and re-fashion liberal citizenship," ultimately, then, to invoke alternative kinds of citizenship and ways of being in the world. Relatedly, the disabled sexualities that unfolded in my book through informants' sexual stories (Liddiard 2018)—while striving for normalcy—also showed themselves to be unquestionably non-normative, queered, cripped, and, at points, interdependent and radically relational—entities "entangled in multiple assemblages" (Fritsch 2010, p. 3)—connected both inter-corporeally and sometimes literally with multiple technologies, bodies and services. Therefore, rather than abject, aberrant and asexual—as informants' own storytelling largely maintained (an understandable psycho-emotional response to the impacts of sexual ableism)— disability showed itself to be productive, radical and vital, because of the ways it already surpasses and subverts the strict confines of the human and, by extension, human sexuality (Liddiard 2014).

To be clear, a DisHuman analysis at once appreciates and comprehends disabled desires for sexual normalcy, comprehending the power of such a category, yet, at the same time, it acknowledges a space for seeing otherwise, to appreciate the lived and everyday on another register. It seeks "more expanded, crip [and] relational forms of the human" (Goodley et al. 2015), which, I want to suggest, pave new ways of imagining and advocating for disabled people's sexual and intimate futures. In short, thinking through the DisHuman illuminates the presence and power of disability to exceed and expand the confines of heteronormative sexuality and pleasure (which limit all sexual subjects) in ways that shift sexual desire and practice for dynamic and provocative effect.

To re/encounter informants' sexual stories through a DisHuman framework reveals a multitude of transgressions: Actively displacing, decentring or demoting the orgasm as a marker of pleasure, dislodging it from heteronormative scripts as the "natural outcome" (Cacchioni 2007, p. 306); queering heterosexual practices through exploration that facilitated discovery of a multitude of pleasures (tongues and fingers as cocks); expanding normative modes of pleasure through incorporating the visual, staring, fantasy, verbal sex and (sexual) technologies and enhancements, uncoupling human sexuality from its "long-presumed biological essence" (Plummer 2015, p. 44); the discovery of new erogenous zones—orgasmic pleasures through stroking arms, shoulders and necks—queering the embodied norms of the conventional erotic body (Ostrander 2009; see also Liddiard and Goodley in press); and, through an acknowledgement that disability can disrupt the autonomous, independent, self-governing human subject as crip sexual bodies, come to be collectively and collaboratively maintained by chosen others, such as partners, family, carers, sex workers and care professionals (see Earle 1999; see also Fritsch 2010). Without doubt, then, disability, by its very nature, offers possibilities for opening up new ontologies of pleasure and alternative economies of desire (Liddiard 2018).

Echoing DisHuman politics, the DisSexual is a mode through which disabled people can claim their humanness through conventional modes of sexuality and gender if they so choose, yet simultaneously defy, crip and exceed such boundaries (Liddiard 2018). Its bifurcated position provides ways to negotiate tensions mentioned in earlier sections—the simultaneity and connection in binaries produced through sexual ableism: Normalised/Other, crip/queer, rejected/desired, natural/technologised, autonomous/collaborative, intimate/commodified, private/exposed and orgasmic/non-orgasmic, as well as the liminal spaces between. This is important, because, whereas the majority of disabled people in my research expressed their desires for sex and intimacy in ways that might be deemed as illustrative of a typically sexually functioning and normatively gendered human subject, the input of personal assistants, technologies, and, sometimes, sex workers, as well as the opening up of the totality of the body and impairment as sites of multifarious pleasure, reconfigures how we think of and enact sex, desire and intimacy.

To end, then, what I've tried to do in this final section is to ponder the possibilities of a posthuman disabled sexual subject, as well as to locate these in the context of the everyday lived lives of disabled people. The DisSexual is, I argue, a means through which to do so theoretically–but also ethically, faithfully and honourably—in ways that stay true to informants' own constructed sexual subjective realities. Thus, the beauty of a DisHuman analysis is that it reframes what would otherwise be a problematic desire for the norm—a yearning to fit into binaries, a longing for sameness and a will to be included into the category of the human. At the same time, it encourages thinking *with* crip (McRuer 2006b), even if the majority of disabled people's lived and conscious realities remain far removed from these understandings—largely because sexual ableism suppresses access to crip and therefore sexual liberation for disabled people (and many others). Rather than call, then, for the total splintering of the human, of binaries and of normalcy—risky projects in times where disabled people remain marginalised, impoverished and excluded—the DisHuman and DisSexual enable access to the power, relative security and social (and sexual) capital of the human, while always holding these categories to account.

## Drawing some conclusions

Through this chapter, I have sketched out various theoretical, intellectual and political foundations of disabled sexualities and the intersections of disability and desire. The (purposefully) messy way in which I have made use of theory, drawing upon critical explorations of sexuality, pleasure, gender, the body and embodiment in contexts of dis/ability and dis/ableism, supports Shakespeare and Corker's (2002, p. 15) assertion that "the global experience of disabled people is too complex to be rendered within one unitary model or set of ideas." In opening up this theoretical space, then, I have outlined the contributions of a range of theoretical and political perspectives and have highlighted the tensions and synergies between them, at the same time as considering their usefulness and applicability towards understanding the sexual, intimate and erotic lives of disabled people. Importantly to me, as a disabled woman scholar, making use of a range of critical social theories in this way ensures that I remain politically and ethically aligned to Critical Disability Studies, a framing that embraces the multiple dimensions of disability: Social, political and economic, psychic, cultural, discursive and carnal (Meekosha and Shuttleworth 2009). It also safeguards that disabled people's own lived experiences—their own testimonies and stories of sexuality, intimacy and love—are rendered most important of all. I maintain, then, that it is critical to not privilege theory, intellectualism and radicalism over the everyday, the lived and the mundane—the marked realities of disability life in the current violent and oppressive contexts of disablism and ableism. Whereas Goodley (2016, p. viii) reminds us that, as scholars and activists, we should never apologise for our use of theory, on the grounds that "contemporary times demand contemporaneous theories," as people interested and invested in disability, it is our political, ethical and moral responsibility to remain faithful to the realities of disabled people's everyday lives.

## References

Anderson, P. and Kitchen, R. (2000) Disability, space and sexuality: Access to family planning services. *Social Science & Medicine* 51, pp. 1163–1173.

Bland, L. and Doan, L. (1998) *Sexology uncensored: The documents of sexual science*. Cambridge: Polity Press.

Bone, K.M. (2017) Trapped behind the glass: Crip theory and disability identity. *Disability & Society*. doi:10.1080/09687599.2017.1313722.

Bordo, S. (1993) *Unbearable weight: Feminism, western culture, and the body*. Berkeley: University of California Press.

Breckenridge, C.A. and Vogler, C. (2001) The critical limits of embodiment: Disability's criticism. *Public Culture* 13(1), pp. 349–357.

Bullough, V. (1994) *Science in the bedroom*. New York: Basic Books.

Butler, J. (1990) *Gender trouble*. London: Routledge.

Cacchioni, T. (2007) Heterosexuality and "the labour of love": A contribution to recent debates on female sexual dysfunction *Sexualities* 10(3), pp. 299–320.

Campbell, F.K. (2001) Inciting legal fictions: Disability's date with ontology and the ableist body of the law. *Griffith Law Review* 10, pp. 42–62.

Campbell, F.K. (2008) Refusing able(ness): A preliminary conversation about ableism. *M/C Journal*. Online. Available from: http://journal.media-culture.org.au/index.php/mcjournal/article/view/46 [accessed: 6/7/2017].

Campbell, F.K. (2009) *Contours of ableism: Territories, objects, disability and desire*. London: Palgrave Macmillan.

Chesser, E. (1950) *Sexual behaviour: Normal and abnormal.* London: Hutchinson.

Corbett, J. (1994) A proud label: Exploring the relationship between disability politics and gay pride. *Disability & Society* 9(3), pp. 343–357.

Davis, L.J. (1995) *Enforcing normalcy: Disability, deafness, and the body.* New York: Verso.

Davis, L.J. (2002) *Bending over backwards: Disability, dismodernism, and other difficult positions.* New York: New York University Press.

Drench, M. (1992) Impact of altered sexual function in spinal cord injury *Sexuality and Disability* 10, pp. 3–14.

Earle, S. (1999) Facilitated sex and the concept of sexual need: Disabled students and their personal assistants. *Disability & Society* 14(3), pp. 309–323.

Finger, A. (1992) Forbidden fruit. *New Internationalist* 233, pp. 8–10.

Fraser, M. (1999) Classing queer: Politics in competition. *Theory, Culture and Society* 16(2), pp. 107–131.

Fritsch, K. (2010) Intimate assemblages: Disability, intercorporeality, and the labour of attendant care. *Critical Disability Discourse* 2, pp. 1–14.

Gagnon, J. and Simon, W. (1973) *Sexual conduct: The social sources of human sexuality.* Piscataway, NJ: Transaction.

Gill, M. (2015) *Already doing it: Intellectual disability and sexual agency.* Minneapolis and London: University of Minnesota Press.

Goodley, D. (2011) *Disability studies: An interdisciplinary introduction.* London: Sage Publications.

Goodley, D. (2014) *Dis/ability studies: Theorising disablism and ableism.* London: Routledge.

Goodley, D. (2016) *Disability studies: An interdisciplinary introduction.* 2nd ed. London: Sage.

Goodley, D., Lawthom, R. and Runswick-Cole, K. (2014) Posthuman disability studies. *Subjectivity* 7(4), pp. 342–361. doi:10.1057/sub.2014.15.

Goodley, D. and Runswick Cole, K. (2014) Becoming DisHuman: Thinking about the human through dis/ability. *Discourse: Cultural Politics of Education* 36(1), pp. 1–15. doi:1080/01596306.2014.930021.

Goodley, D., Runswick-Cole, K. and Liddiard, K. (2015) The DisHuman child. *Discourse: Studies in the Cultural Politics of Education, Special Issue Fabulous Monsters: Alternative Discourses of Childhood in Education* 37(5). doi:10.1080/01596306. 2015.1075731.

Hahn, H. (1981) The social component of sexuality and disability: Some problems and proposals *Sexuality and Disability* 4(4), pp. 220–233.

Halperin, D. (1995) *Saint Foucault: Towards a gay hagiography.* Oxford: Oxford University Press.

Hawkes, G. (1996) *A sociology of sex and sexuality.* Buckingham: Open University Press.

Ignagni, E., Fudge-Schormans, A., Liddiard, K. and Runswick-Cole, K. (2016) Some people aren't allowed to love: Intimate citizenship in the lives of people labelled with intellectual disabilities. *Disability & Society.* doi:10.1080/09687599.2015. 1136148.

Jackson, S. (1999) *Heterosexuality in question.* London: Sage.

Jackson, S. and Scott, S. (2010) *Theorizing sexuality.* Berkshire: Open University Press.

Jeffreys, J. (1994) The queer disappearance of lesbians: Sexuality in the academy. *Women's Studies International Forum* 17(5), pp. 459–472.

Kafer, A. (2013) *Feminist queer crip.* Bloomington and Indianapolis: Indiana University Press.

Kent, D. (2002) Beyond expectations: Being blind and becoming a mother. *Sexuality and Disability* 20(1), pp. 81–88.

Kinsey, A.C. (1953) *Sexual behavior in the human female.* London: W. B. Saunders.

Lee, O.E.K. and Heykyung, O. (2005) A wise wife and good mother: Reproductive health and maternity among women with disability in South Korea. *Sexuality and Disability* 23(3), pp. 121–144.

Liddiard, K. (2014) The work of disabled identities in intimate relationships. *Disability & Society* 29(1), pp. 115–128. doi:10.1080/09687599.2013.776486.

Liddiard, K. (2018) *The intimate lives of disabled people.* London: Routledge.

Liddiard, K. and Goodley, D. (in press) Disability and impairment. In B. Turner, et al. (eds.) *Encyclopedia of social theory.* London: Wiley.

Liddiard, K. and Slater, J. (2017) "Like, pissing yourself is not a particularly attractive quality, let's be honest": Learning to contain through youth, adulthood, disability and sexuality. *Sexualities, special issue, Disability and Sexual Corporeality* 21(3), pp. 319–333.

Masters, W.H. and Johnson, V.E. (1966) *Human sexual response.* Boston, MA: Little, Brown.

McRuer, R. (2002) Critical investments: AIDS, Christopher Reeve, and queer/disability studies. *Journal of Medical Humanities* 23(3), pp. 221–237.

McRuer, R. (2006a) *Crip theory: Cultural signs of queerness and disability.* New York and London: New York University Press.

McRuer, R. (2006b) We were never identified: Feminism, queer theory, and a disabled world. *Radical History Review* 94, pp. 148–154.

Meekosha, H. and Shuttleworth, R. (2009) What's so "critical" about critical disability studies? *Australian Journal of Human Rights* 15(1), pp. 47–76.

Michalko, R. (2002) *The difference that disability makes.* Philadelphia, PA: Temple University Press.

Milligan, M. and Neufeldt, A. (2001) The myth of asexuality: A survey of social and empirical evidence. *Sexuality and Disability* 19(2), pp. 91–109.

Murphy, R. (1990) *The body silent.* New York: W.W. Norton.

Ostrander, N. (2009) Sexual pursuits of pleasure among men and women with spinal cord injuries. *Sexuality and Disability* 27, pp. 11–19.

OToole, C.J. (2000) The view from below: Developing a knowledge base about an unknown population. *Sexuality and Disability* 18(3), pp. 207–224.

Overboe, J. (2007) Disability and genetics: Affirming the bare life (the state of exception) *Canadian Review of Sociology and Anthropology* 44(2), pp. 220–235.

Parker, M.G. and Yau, M.K. (2012) Sexuality, identity and women with spinal cord injury. *Sexuality and Disability* 30(1), pp. 15–27.

Plummer, K. (2003) *Intimate citizenship: Private decision and public dialogues.* Seattle and London: University of Washington Press.

Plummer, K. (2015) *Cosmopolitan sexualities: Hope and the humanist imagination.* Cambridge: Polity Press.

Rembis, M.A. (2010) Beyond the binary: Rethinking the social model of disabled sexuality. *Sexuality and Disability* 28, pp. 51–60.

Rich, A. (1989) Compulsory heterosexuality and lesbian existence. In L. Richardson and V. Taylor (eds.) *Feminist frontiers II: Rethinking sex, gender and society.* New York: Random House.

Rose, N. (2001) The politics of life itself. *Theory, Culture and Society* 18(6), pp. 1–30.

Sandahl, C. (2003) Queering the crip or cripping the queer? Intersections of queer and crip identities in solo autobiographical performance. *GLQ: A Journal of Lesbian and Gay Studies* 9(1), pp. 25–56.

Shakespeare, T. (1996) Power and prejudice: Issues of gender, sexuality and disability. In L. Barton (ed.) *Disability and society: Emerging issues and insights.* Harlow: Longman, pp. 191–214.

Shakespeare, T. and Corker, M. (2002) *Disability/post-modernity: Embodying disability theory.* London: Continuum.

Shakespeare, T., Gillespie-Sells, K. and Davies, D. (1996) *Untold desires: The sexual politics of disability.* London and New York: Cassell.

Sherry, M. (2004) Overlaps and contradictions between queer theory and disability studies. *Disability & Society* 19(7), pp. 769–783.

Shildrick, M. (2007) Dangerous discourse: Anxiety, desire and disability. *Studies in Gender and Sexuality* 8(3), pp. 221–244.

Shildrick, M. (2009) *Dangerous discourse of disability, subjectivity and sexuality.* New York: Palgrave Macmillan.

Shuttleworth, R. (2000) The search for sexual intimacy for men with cerebral palsy. *Sexuality and Disability* 18(4), pp. 263–282.

Shuttleworth, R. (2010) Towards an inclusive disability and sexuality research agenda. In R. Shuttleworth and T. Sanders (eds.) *Sex and disability: Politics, identity, and access.* Leeds: The Disability Press, pp. 191–214.

Siebers, T. (2008) *Disability theory.* Ann Arbor: University of Michigan Press.

Sinecka, J. (2008) "I am bodied. I am sexual. I am human." Experiencing deafness and gayness: A story of a young man. *Disability & Society* 23(5), pp. 475–484.

Smyth, C. (1992) *Lesbians talk queer notions.* London: Scarlett Press.

Solvang, P. (2007) The amputee body desired: Beauty destabilized? Disability re-valued? *Sexuality and Disability* 25, pp. 51–64.

Tepper, M. (1999) Letting go of restricted notions of manhood: Male sexuality, disability and chronic illness. *Sexuality and Disability* 17(1), pp. 37–52.

Tepper, M. (2000) Sexuality and disability: The missing discourse of pleasure. *Sexuality and Disability* 18(4), pp. 283–290.

Tepper, M. (2002) Forbidden wedding: Movie review. *Disability Studies Quarterly* 22(4), pp. 162–164.

Villanueva, M.I.M. (1997) *The social construction of sexuality: Personal meanings, perceptions of sexual experience, and females.* Sexuality in Puerto Rico. Unpublished book, Faculty of the Virginia Polytechnic Institute and State University.

Waxman, B. and Finger, A. (1991) The politics of sexuality, reproduction and disability. *Sexuality Update, National Task Force on Sexuality and Disability* 4(1), pp. 1–3.

Waxman Fiduccia, B. (1999) Sexual imagery of physically disabled women: Erotic? Perverse? Sexist? *Sexuality and Disability* 17(3), pp. 277–282.

Waxman Fiduccia, B. (2000) Current issues in sexuality and the disability movement. *Sexuality and Disability* 18(3), pp. 167–174.

Weeks, J. (1986) *Sexuality*. Chichester: Ellis Horwood/Tavistock Publications.

Wilchins, R. (2004) *Queer theory, gender theory*. Los Angeles: Alyson Books.

Wilkerson, A. (2002) Disability, sex radicalism and political agency. *NSWA Journal* 14(3), pp. 33–57.

Wolbring, G. (2008) The politics of ableism. *Development* 51, pp. 252–258.

# 2

# THEORETICAL DEVELOPMENTS

## Queer theory meets crip theory

*Alan Santinele Martino and Ann Fudge Schormans*

## Introduction

The intersection of disabilities and sexualities remains under researched and under theorised in both sociology of sexualities and critical disability studies, resulting in significant gaps in our understanding of the sexual and intimate lived experiences of disabled people (Kattari 2015; Liddiard 2013; McRuer and Mollow 2012). This gap, and the potential of each field to contribute to our understanding, guides our exploration here. The sociology of sexualities and disability studies scholarship have only recently been put into dialogue, the overlap of these two fields remaining "sparse" (Jungels and Bender 2015, p. 78). Nevertheless, as Jungels and Bender (2015) have noted, and we agree:

> There is a great potential and urgent need for disability and sexuality scholars to bring their fields together to more fully understand the sexual lives and needs of people with disabilities, especially given the ability to create positive changes in the lived experiences of disabled individuals.
>
> *(p. 179)*

The sociology of sexualities is a burgeoning field of sociology (Bernstein 2013; Seidman 2015), emerging in response to common, limited and limiting understandings of sexuality as an individual, biological and psychological issue, a matter of natural drives that are steeped in human nature (DeLamater and Plante 2015; Seidman 2015). This essentialist approach demonstrates a belief in an underlying "essence" behind behaviours that appear to be shared by different people (Vance 2006). Sexualities scholars have challenged these understandings, noting that sexualities are too complex to be understood solely on biological and psychological terms (Rubin 1984), providing a distinct set of theories to engage in the major project of "denaturalizing" how we understand sexualities (Seidman 2015). Drawing on social constructionism, sociologists have proposed that sexualities are socially produced, transformed and institutionalised through multiple social institutions and discourses (Foucault 1978; Vance 2006). Sexualities studies have, however, rarely attended to the sexualities and experiences of disabled people (Kattari 2015).

The field of critical disability studies has similarly been engaged in problematising essentialist understandings of disability, shifting our attention from understandings based solely on medical, charity and individualised approaches, first to a more structural approach, and increasingly to more complex theorising that acknowledges the social, structural, historical and material factors involved in understandings and experiences of disability (Feely 2016; Shakespeare 2014). Disability studies have done the radical work of placing disability at the centre of analysis (Shakespeare 2014). Yet, for a long time, disability studies and activism treated sexualities as apolitical or a personal matter (Finger 1992; Shakespeare et al. 1996; Shuttleworth and Mona 2002). This has changed (Santinele Martino and Campbell 2019). Some critical disability studies scholars have more recently invigorated the study of sexualities among disabled people (see, for example, Gill 2015; Liddiard 2018; McRuer and Mollow 2012). Critiquing normalcy and ablebodiedness/mindedness as they factor into understanding and experience of disabled sexualities, these scholars are challenging conventional thinking and expanding the conversation to attend, for example, to desire, romance and eroticism. They are generating an exciting body of empirical literature and theoretical perspectives for examining the intersections of disabilities and sexualities.

Rich theoretical dialogues between disability studies and other emancipatory theories, including queer and feminist theories, have also emerged (Kafer 2013; McRuer and Mollow 2012; Sandahl 2003). For example, feminist perspectives have played an important role in the development of crip theory ("crip" being shorthand for "cripple"), primarily through their critique of a lack of a complex understanding of gender identities and an intersectional lens, as well as the historical pathologisation of women's bodies (Egner 2016). Queer theory also engages with questions of identity, social binaries and intersectionality, paying particular attention to unsettling heterosexuality as a norm. As Johnson (2014) explains, "'to queer' means to disrupt or make something 'strange,' twisting or unsettling meanings, pushing the invisible into the spotlight" (p. 1618). Sharing similar goals, these theoretical perspectives have informed and transformed each other through questioning normalcy and compulsory heterosexuality and ablebodiedness/mindedness (Santinele Martino 2017). Crip theory has benefited greatly from this intersectional theorising. Crip theorists have articulated how, "rather than seeking assimilation and acceptance, our understandings of disability and queerness need to radically change," as do our understandings of sexualities (Santinele Martino 2017, p. 7).

In this chapter, we suggest that crip theory is a productive way of bridging critical disability studies and the sociology of sexualities. We concur with Goodley's (2014) claim that "queer and disability studies have the potential to unsettle one another and find shared vocabularies" (p. 38). Drawing on some of the major works and implications of crip theory, we suggest directions for its further development and demonstrate how this theoretical approach is uniquely positioned to contribute to both critical disability and sexualities studies. On one hand, crip theory sheds light on theoretical tools that disability studies scholars can use to understand the ways disabled people experience and negotiate their intimate lives through a multi-level analytic approach. On the other hand, it suggests disability studies insights that can contribute to theoretical perspectives in sexualities studies. To do so, we contextualise this discussion in relation to the sexualities of people labelled/with intellectual disabilities (ID),[1] a disability label that appears to complicate questions of sexualities.

## Intellectual disability and sexualities

We have chosen to centre our discussion of crip theory's usefulness for theorising the intersection of disability and sexualities on "intellectual disability" for multiple reasons. Importantly, intellectual disability rarely figures into these conversations and theorising in critical disability studies, queer, or sexualities fields (Abbott 2015; Löfgren-Mårtenson 2013). Where there is an agreement within disability communities that people labelled/with ID should have access to intimate citizenship, they are often treated as "special exceptions" requiring a different approach due to their disability label (Gill 2010, p. 210). Furthermore, as Santinele Martino and Fudge Schormans (2018) have noted, people labelled/with ID have "historically been afforded few opportunities to communicate their experiences and desires concerning their personal and intimate lives, including their friendships, parenting, family relationships or romantic and sexual experiences" (p. 2).

At this moment, attention to intellectual disability is largely absent in crip theorising. Its absence from queer theorising can perhaps be understood in light of broader critiques of queer theorists' failure to consider disability in their theorising (Egner 2016). This has resulted in the dialogue between crip and queer theorists being "rather one-sided, with crip theorists making overtures and remaining politely unsatisfied with queer theorists' openness to the relationship" (Johnson 2015, p. 252). Nonetheless, queer theory offers ways to explore the intersections between disability and sexualities, which we illustrate in this chapter. Acknowledging the exclusions existing in crip theory itself, McRuer has noted, "my relationship to crip theory is often concerned with who is at the margins, and with paths to those margins, with who is not necessarily a candidate for the face of the movement" (as cited in Peers et al. 2012, p. 151). As Löfgren-Mårtenson (2013) articulates, crip theory often "proceeds from people with physical disabilities, i.e., individuals who have a voice, who can write about their situation and organise dissent, and who are often found in the international disability rights movement" (p. 420). People with intellectual and less visible impairments are thus excluded (Bone 2017; Goodley 2011). Crip theory "carries with it the embodied history and a preference for the visibly disabled body" (Bone 2017, p. 1306). Our intention in this chapter is not to further well-known "hierarchies of disability" (Deal 2003), but to propose fruitful directions for theorising the particular experiences of people labelled/with ID.

A possible argument for the failure to attend to intellectual disability in queer, crip and disability theorising of sexualities may be that people labelled/with ID trouble normative assumptions about sexualities. Tenacious infantilising assumptions of (in)ability have led to their being desexualised—neither understanding nor desiring sexualities and intimacy, and unable to consent and make decisions about sexual identity or their intimate lives because of the disability label (Gill 2010; Runswick-Cole and Goodley 2015). Very few researchers focus on the romantic and sexual experiences of people labelled/with ID, particularly from a critical lens, and even more so for people labelled/with "profound" intellectual disability (Gerschick 2000). The common de-sexualisation of people labelled/with ID promotes the notion that intellectual disability and sexuality are simply incompatible (Emmens 2009). As a group, people with this label have historically experienced high rates of multiple forms of sexual abuse, violence and exploitation; rooted in social devaluation, dehumanisation and ableism, this violence continues and may be increasing (Fudge Schormans and Sobsey, 2020). Being identified as always and only vulnerable often leads to restrictions on their rights to, and opportunities, for sexual expression.

Working in opposition and parallel to de-sexualisation and vulnerability is the disturbing and stubborn notion of sexual deviance, which marks individuals labelled/with ID as dangerous, as a threat (Löfgren-Mårtenson 2013). Deemed unable to "control" their "excessive," "bad," "deviant," "unacceptable" or "unnatural" sexualities (Albertz and Lewiecki-Wilson 2008; Noonan and Gomez 2011), they are understood to pose a risk to themselves, and (of perhaps even greater concern) to non-disabled others (Santinele Martino and Fudge Schormans 2018).

A third way people labelled/with ID (women in particular) trouble normative assumptions concerns reproduction. The question of people labelled/with ID becoming parents brings together discourses of vulnerability, deviance and risk. It is commonly understood (and feared) that disabled people may pass on their "disorders," "contaminating" future generations of "healthy people" (Grekul 2002). People labelled/with ID challenge the "good parent" ideal and related understandings and expectations of reproduction and its relationship to the future. Centred on the figure of the child as the future, and thus concerned with who is deemed (in) capable of reproducing and raising this child of the future, people labelled/with ID are rarely considered as potential parents (Ignagni and Fudge Schormans 2016). At the same time, conse-quent to largely unquestioned attitudes and entangled with reproduction and futurity, having a child with an ID is often seen as the "worst thing" that could possibly happen for expectant non-disabled parents (Fudge Schormans and Sobsey, 2020).

The power of these discourses is such that, whether people labelled/with ID are perceived to be potential victims, expected to express unacceptable sexual behaviours, and/or to pose a threat to the normative future, their sexual rights are commonly restricted. They are always understood to be in contrast to "normate" sexualities, that occupy a privileged position within Rubin's (1984) charmed circle of sexuality, always falling outside of normalcy. As Rubin (1984) has artic-ulated, "sexuality is political. It is organised into systems of power, which reward and encourage some individuals and activities, while punishing and suppressing others" (p. 171). For people labelled/with ID, whose sexualities are often understood to be "bad" sexualities, forms of sexual expression and intimacy are often controlled, contained and punished. Their intimate lives have been marked by a history of violence, protectionism and infantilisation (Winges-Yanez 2014), including forced sterilisation and institutionalisation (Malacrida 2015). The entanglement of this history of eugenics, alongside contemporary newgenics practices, tenacious medical and charity understandings of disability, and neoliberal constructions of the "ideal citizen," continue to shape the intimate lives of people labelled/with ID (Ignagni and Fudge Schormans 2016).

On the whole, empirical literature exploring the sexual and intimate lives of people labelled/with ID has taken a quite narrow focus (Gill 2015). Studies have focused on questions of sexual abuse of people labelled/with ID, as well as on labelled sexual offenders and their potential risk to other community members (Dickman and Roux 2005; Steptoe et al. 2006). Studies have also surveyed the perceptions and attitudes of non-disabled social actors regarding the sexualities of people labelled/with ID, including parents and family members, support work-ers and teachers (e.g., Dupras and Dionne 2014; Meaney-Tavares and Gavidia-Payne 2012). Very few have sought the perceptions and experiences of people labelled/with ID themselves (Ignagni and Fudge Schormans 2016; Santinele Martino and Fudge Schormans 2018). The emphasis on vulnerability, risk and "appropriate" sexual behaviours "tends to deflect attention from [their] wider sexuality and sexual needs" (Cambridge and Mellan 2000, p. 294), while simultaneously hiding the rich array of sexual practices and identities among people labelled/with ID (Santinele Martino 2017). Whereas much of the public discourse around sexuality and disability has focused on "deviance and inappropriate behaviour, abuse and victimisation, asexuality, gender and orientation with regard to women, and reproductive issues in women and men," Tepper (2000) has identified a "missing discourse of pleasure" (p. 283). An emerging body of literature examines sexual pleasure and desires in the lives of disabled people; most of these studies focus on experiences of people with physical impairments (e.g., Connell et al. 2014; Tepper et al. 2001), with far less attention being paid to the experiences of people labelled/with ID (Turner and Crane 2016).

People labelled/with ID have romantic and sexual feelings, needs and desires, and therefore, should be entitled to the same sexual rights, protections, opportunities and choices as non-disabled people (Dupras 2015). Yet, they experience innumerable barriers to remaining sexual and exercising choice in terms of identity, sexual partnerships and intimacy (Gill 2015; Kulick and Rydström 2015). In what follows, we begin to explore the ways that crip theory might be used to inform our understanding of the sexualities of people labelled/with ID.

## Introducing crip theory

Building on queer theory's insights, crip theory is a young and growing, theoretical perspective within critical disability studies (Egner 2016; Goodley 2014): Put simply, it is a multifaceted analytic approach, focusing on disabled embodiment and lived experiences (McRuer 2006). Emerging out of this conjunction of insights from critical disabilities studies, feminist theory and queer studies, crip theory problematises the "naturalness" of compulsory heterosexuality and able-bodiedness/able-mindedness (McRuer 2006). Queer theory, with its attention to non-binary thinking and interrogation of the construction of normalcy, including the ways that the world is divided into "heterosexual" and "homosexual," and the consequent lack of acknowledgement of and attention to diversity, played an important role in setting the groundwork for crip theory to emerge (Egner 2016; McRuer 2006). Sexualities scholars have challenged "compulsory heterosexuality" by turning our attention to the history, social construction and institutionalization of heterosexuality (Rich 1980; Seidman 2015). A particularly important insight is that heterosexuality requires complex social processes to shape people into exclusive heterosexuality. The social processes that maintain heterosexuality as compulsory, however, take place under the guise of ideologies that make heterosexuality seem both natural and fixed (Rich 1980). It is the intention of queer theory to highlight the instability of gender and sexual categories, as well as the rich variety of sexualities and sexual desires that exist (Johnson 2014). Also, as Johnson (2014) articulates, queer theory has "evolved by arguing for the legitimacy of a range of marginalised subjects and practices" and "paying increasing attention to the way other categories of difference [...] intersect with gender and sexuality" (p. 1618). Crip theorists have done the important work of introducing disability and dis/ableism to this theorising.

Pointing to the limitations of binaries to represent and describe the variable ways we embody, experience and perform our social practices, crip theory makes space for experiences that do not fit into existing social categories and binaries. In a manner similar to queer theorists' critique of compulsory heterosexuality, crip theorists critique compulsory ablebodiedness/mindedness. Like compulsory heterosexuality, compulsory able-bodiedness/mindedness works to "contain" and "manage" disability (McRuer as cited Peers et al. 2012, pp. 148–149). In this way, both queer and crip theorising offer a lens for understanding how social categories, such as "normal" and "abnormal," and "disabled" and "non-disabled," are co-constitutive. In other words, the very categories of heterosexual/homosexual and non-disabled/able-bodiedness/mindedness can only exist and have meaning in relation with their "opposite." These theoretical perspectives turn our attention to the "invisible centre" that compulsory heterosexuality and compulsory able-bodiedness/mindedness inhabit. Both raise a series of important questions: What are "normal" (and thus, acceptable) sexualities? Why are some bodies/minds seen as more (or less) desirable than others? And who gets to decide? (see for example Kafer 2013; McRuer 2006; Samuels 2003).

Disabled and queer people also share a long and difficult history of pathologisation and oppression (McRuer 2002; Sandahl 2003)— this includes the ways that non-normative sexualities and gender identities have been pathologised and constructed as disabilities (Egner 2016; Santinele Martino 2017). For instance, some asexual activists have actively maintained a distance from disability as a response to pathologising constructions of asexuality as a sign of "unhealthiness" (Kim 2011). Disabled sexualities have been similarly pathologised and denied (Egner 2016). Historically, both terms, queer and crip, have been used in derogatory ways to label particular social groups deemed to be "deviant" and "dangerous" (Löfgren-Mårtenson 2013; Rydström 2012). These terms, however, have recently been reclaimed by some crip and queer scholars/ activists as a way to flag pride and to mobilise (Löfgren-Mårtenson 2013; Sandahl 2003). Warner (1999) has articulated the need for a "stigmaphile" lifestyle, one in which queer subjects contest dominant culture by displaying their "deviance" (p. 43). In a similar manner, crip theory opens up opportunities for disabled people to command social interactions by "embracing the stigma" (Rydström 2012, p. 14). By re-appropriating the term, and calling oneself "crip," the intention is to take control over representation and language to challenge dominant understandings of what it means to be human or "normal." It is a form of defiance and resistance, a kind of empower- ment and a marker of disability pride (Schalk 2013) that is meant to be "provocative" (Löfgren- Mårtenson 2013, p. 414).

"Queer and crip sexual practices," Cohen (2015) argues, "deeply challenge heterosexual norms of position, limit, seclusion, independence and reproduction" (p. 156), thus reconstructing sex and sexualities beyond normative forms (Fritsch et al. 2016; McRuer and Mollow 2012). By "cripping" sexual practices, disability expands conceptions of what it means to be sexual, highlighting the ways that sexual norms are limited and limiting and continue to shape our understanding of, for example, how sex happens and what it means to be a man or a woman. The inability to perform normative sexual and gender scripts marks people as different, as lesser, and sometimes leads disabled people to feel like a "sexual failure" (Liddiard 2018, p. 7). Siebers (2008) suggests that we need to understand disability "not as a defect that needs to be overcome to have sex but as a complex embodiment that enhances sexual activities and pleasure" (p. 148). Siebers proposes that disabled individuals should work towards the construction of a "sexual culture" that would challenge dominant, ableist approaches to pleasure, desire and sexual prac- tices. Thus, rather than trying to assimilate and come as close as possible to normative forms of sexualities, it is important to introduce new forms of understanding about what it means to be sexual. Take, for instance, the experiences of disabled people who develop and bring into view new erogenous zones, or who contest fixed lines between public and private when engaging in assisted sex (Siebers 2008).

## People labelled/with ID already "cripping" sexualities

As we have noted, crip theory is not without critique regarding its suitability as an informa- tive theoretical tool (Bahner 2018). For example, some disability studies scholars are explor- ing how applicable crip theory can be to the context of the lives of people labelled/with ID (e.g., Löfgren-Mårtenson 2013). Others have pointed to the experiences of people with chronic illness who may not identify as disabled (Bone 2017), or disabled people who disagree with reclaiming such a derogatory and historically charged term as crip (García-Santesmases Fernández et al. 2017). The accounts of some people labelled/with ID who we have interviewed and worked with across multiple projects have likewise revealed tensions. There is a similar reluctance or refusal on the part of some to use the language of crip, notably by people who strongly identify as self-advocates. Others choose not to identify as disabled. Other accounts, however, illustrate how people labelled/with ID—despite significant constraints—are "crip- ping" sexualities, re-constructing ways of thinking about sexualities, gender and parenting. Here, we share a few examples to demonstrate how people labelled/with ID are questioning and de- centring dominant understandings of sexualities and relationships.

The first example emerged from a study conducted by the first author focused on the inti- mate lives of people labelled/with ID and revealed how participants thought about family and parental responsibilities in rather expansive ways. William,[2] a young man labelled/with ID, spoke about a previous relationship and his desire to have children:

> We wanted children, I mean, anybody wants children, I mean, you might have a bit of a tougher time supporting it but *there is family, friends, people who will support you*—so I think that was our dream and it didn't last long enough ... I honestly think there really is a lot of drawbacks for people with disabilities, people think "oh, you can't have children."
>
> *(Santinele Martino 2019a, emphasis added)*

Evident in this man's account is the active discouragement from parenting that many people labelled/with ID face due to expectations of deficit marking them unable to be good parents—an experience also noted by participants in two projects conducted by the second author (Fudge Schormans unpublished work; Ignagni and Fudge Schormans 2016). In rebuttal, William "crips" the very notion of family beyond the dominant idea of two parents solely responsible for raising their children. Rather, he suggests that care work could be a more shared, community effort, one that includes "family, friends, people who will support you." As participants made plain in a study by the second author, such care work can be further "cripped" when it is provided *by* people labelled/with ID (Fudge Schormans 2015).

A second study, conducted by the first author, revealed participants' efforts to crip dominant sexual scripts. For example, a young woman labelled/with an ID, Andrea, reflected on the dominant sexual scripts of sex:

> I don't pick up on the weird like secret messages of dating [...] I feel like when you are having sex it has to be like a certain behaviour, like not making jokes, or being goofy, or anything [...] I always think it's weird, you need to be a certain way during [sex].

Sexual scripts refer to norms regarding sexuality that sexual actors internalise and embrace through socialisation (Simon and Gagnon 1986). Scripts are used by sexual actors to guide their practices, including how they engage and participate in sexual situations. In her statement, Andrea makes reference to the requirement to perform sexually in a "certain," normative way that draws boundaries around how sex should happen. By referring to dominant sexual scripts, in which, for example, sexual partners do not make jokes during sex, Andrea de-centres this script by naming it as "weird."

Attempts to subvert dominant sexual scripts, however, can come at a cost. The constraints and surveillance experienced by people labelled/with ID can be so significant that finding opportunities for any sexual expression can be challenging for them. When provided with information and opportunities, people labelled/with ID are typically offered a "limited menu of options" for sexual expression which, for example, rarely includes possibilities for queer identities and experiences (Santinele Martino 2019b). They are commonly seen as lacking the "ability" or maturity to be sexual and exercise intimate citizenship. Who gets to define "normal" and "appropriate" sexualities and sexual practices for people labelled/with ID? As consistently shown in the literature, parents, caregivers, service providers and restrictive policies are typically imbued with the power to define the parameters regarding the sexualities of people labelled/with ID (e.g., Santinele Martino 2019a; Wilton and Fudge Schormans, 2020).

An example of the power caregivers hold over acceptable sexualities for people labelled/with ID is drawn from a study conducted by the second author on how people labelled/with ID make use of city space. This example involves a man living in a group home that afforded him very limited opportunities for any kind of independence, solitary time or socialising beyond the confines of the group home walls and/or other residents and staff. In his interview, this man talked of having met a woman at (a similarly highly supervised) summer camp programme several years previously and that they have shared a romantic relationship since that meeting. No supports or opportunities are provided for them to physically meet outside of the summer camp, so they maintain their relationship via regular Skype meetings (set up by staff) and the annual summer camp. The degree of surveillance is such that sex is not possible, yet they identify as romantic partners and are committed to each other. The question is whether this is a subversion of normative conventions, a reworking of such or a compromise formation. It must be noted that this relationship has lasted for some time. This man has repeatedly professed the happiness he derives from this Skype relationship—a relationship he understands to be special, and to be his right. At the same time, he also reports a repeatedly frustrated desire for opportunities to meet in person—without supervision—outside of summer camp.

Whether living with families, relatives, caregivers or in institutionalised group homes, or participating in organised programmes like summer camps, the possibilities for normative performances, let alone to "play" with performances of sexualities, sex and gender are often curtailed for many people labelled/with ID due to protectionism and social control. Forms of sexual expression, that are usually accepted, tolerated or excused for most non-disabled people, are often punished if exhibited by people labelled/with ID (Löfgren-Mårtenson 2013). It is important to acknowledge, in instances where they exercise subversion, that the potential consequences of such can be quite serious. In our research endeavours about disability and sexuality, we have heard multiple instances of people labelled/with ID experiencing different types of punishment for expressing different forms of sexualities; even "harmless" forms of sexual expression, such as holding hands or a kiss on the cheek, can be reason for punishment and greater surveillance.

## What crip theory brings to sexualities

By sharing how disabled people have shown both agency and creativity by reconstructing sex and sexualities in ways that go beyond normative sexualities (Fritsch et al. 2016; McRuer and Mollow 2012), crip scholars both affirm how disabled people are both "subjects and objects of a multiplicity of erotic desires and practices" (McRuer 2011, pp. 107–108), and theorise how disability can "transform sex, creating confusions about 'what and who is sexy' and 'what counts as sex'" (McRuer and Mollow 2012, p. 32). Critical disabilities studies scholars have drawn on the work of sociologist Ken Plummer to articulate the importance of intimate citizenship in the lives of people labelled/with ID (Ignagni et al. 2016; Liddiard 2018). Intimate citizenship, Plummer (1995) explains, refers to people's social and political rights "to choose what we do with our bodies, our feelings, our identities, our relationships, our genders, our eroticisms and our representations" (p. 17). Despite the promise of these protections, intimate citizenship is often elusive for people labelled/with ID (Ignagni et al. 2016). Liddiard (2018) proffers that "one way to characterise the sexual oppression experienced by disabled people is through the absence of […] 'intimate citizenship'" (p. 56), and points to how the "lack of intimate citizenship and lack of a *claim for rights* to intimate citizenship is embedded in these spaces of asexualisation, desexualisation and fetishisation" (p. 75). This is attributed, in part, to the failure of researchers to adequately attend to the intersection of intellectual disability, sexuality, intimacy and other relationships, reproduction and parenting or pleasure (Rogers 2016).

Scholars in the sociology of sexualities have paid great attention to intersectionality, exploring how sexualities are intertwined with other stratification systems (Gamson and Moon 2004; Plummer 2012). Yet, despite this awareness that social categories intersect to shape sexual experiences and sexualities of individuals and social groups, sexualities researchers have paid little attention to the experiences of disabled people (Kattari 2015). Disability studies scholars, like Shuttleworth and Mona (2002), have critiqued the sexualities literature for not considering the importance of "sexual access" in shaping the sexualities of marginalised groups. Although Shuttleworth's own work is more focused on the experiences of people with physical impairments, the notion of sexual access is useful for theorising the experiences of people labelled/with ID. What sexualities scholars sometimes take for granted is access to what Green (2014), using Bourdieu, calls sexual fields—meaning competitive sexual arenas that bring together sexual actors in their pursuit of romantic and sexual partnership. This is a significant gap considering that not every sexual actor can enter and participate in many, if any, sexual fields due to structural and other constraints (Santinele Martino 2019b). For example, we still know little about the spaces that people labelled/with ID are able to claim and participate in as sexual actors (Ignagni et al. 2016; Wilton and Fudge Schormans 2020). There is, nonetheless, an understanding that people labelled/with ID face a lack of access to spaces where sexual exploration can take place—in the home, online and public spaces (Wilton and Fudge Schormans 2020). Access to spaces is only one aspect of sexual access. The notion of sexual access involves also "access to the psychological, social and cultural contexts and supports that acknowledge, nurture and promote sexuality in general or disabled people's sexuality specifically" (Shuttleworth and Mona 2002, para. 4).

The Western social world reproduces a "culture of undesirability," as Erickson (2016) puts it, which constructs disabled sexualities as being non-existent (p. 15). This invisibility of disabled sexualities is also shaped by shame, which can serve as a tool for social control, individualising the "problem" of disabled sexualities. Shame can have a deep impact as it enters the psyche and keeps some disabled people from accessing sexuality (Erickson 2016). Disabled people experience everyday ableism by not being asked on a date, by being responded to as not sexual, through infantilisation. This is a culture in which non-disabled people are sometimes shocked to learn that people labelled/with ID desire romance and sexual pleasure. Moreover, stories of multifaceted and fulfilling disabled sexualities remain limited.

> It is made up of isolation from another night at home while everyone else goes to the party. The fear of being left by the people you love and who are supposed to love you. The pain of staring or passing, the sting of disappointment, the exhaustion of having the same conversations over and over again.
>
> *(Erickson 2016, p. 14)*

In a "culture of undesirability," marginalised sexualities and social groups often experience punishment, containment and surveillance (Erickson 2016; see also Rubin 1984). Studies have consistently shown that the sexualities of people labelled/with ID are also controlled and restricted through state practices, such as funding programes and, for example, housing policies that discourage disabled people from getting married, as the act of getting married may lead to reduction in funding or loss of housing. Other policies discourage people labelled/with ID from having children because funding is solely allocated to supporting disabled people without considering supports to the family more broadly.

A reframing of sexualities, in which disabled people are subjects of desire who find ways to express their sexualities, has occurred other than at a theoretical level (Santinele Martino and

Campbell 2019). As an example, initiated by a self-described "queer crip," Andrew Gurza, the hashtag #DisabledPeopleAreHot has received hundreds of (re)tweets (Gurza 2019). Disabled people from around the world have shared photos of themselves along with empowering messages about how disabled people are beautiful and desirable. Most of the activism in this area, however, has been led by people with physical impairments. Mainstream media has largely ignored activism by people labelled/with ID, especially regarding sexualities. One example of activism is a campaign entitled "It's My Right" that features a series of posters with people labelled/with ID, often nude and embracing a partner, with the following text: "People with learning disabilities enjoy sex. It's a fact of life" (Family Planning Association 2019). There is an urgent need for more spaces for people labelled/with ID to articulate their own claims regarding their sexualities.

In addition to expanding spaces for people labelled/with ID to share their own understanding of sexualities, it is also invaluable to further refine our theoretical tools. One way of doing that is by continuing to draw on sexualities literature for theoretical inspiration and experimentation. For instance, we would like to suggest that disability studies scholars can benefit from tools from sexualities scholarship, that can help us develop other ways of engaging in a multi-level analytic approach. As Shuttleworth and Mona (2002) have appropriately noted, the "individualising focus" on disabilities and sexualities:

> Has tended to draw attention away from the socio-structural relations between disabled and non-disabled people, the symbolic meanings of disability and desirability in the larger culture and the psychological implications of experiencing multiple barriers to sexual expression and establishing sexual relationships.
>
> *(p. 2)*

As a potential way forward, we propose that the sexual fields framework (Green 2014) is one rich, theoretical approach to the question of sexual desire (on both a micro- and a macro-level), that allows for an analysis of the sexual lives of people with ID from a multi-level analytic approach (Santinele Martino 2017). Previous studies using the sexual fields framework have made a significant contribution to the sociology of sexualities by demonstrating, empirically, how desire and desirability are collectively produced (Green 2014). The sexual fields theoretical framework, including the concepts of sexual fields, sexual capital and erotic habitus, provide analytical tools to systematically explore the non-random organization of desire through a rich set of concepts (Green 2014), which have not yet been applied in the context of disabled people's sexual lives, and can help us better understand the sexual experiences of people labelled/with ID.

The current sexual fields literature, however, also has limitations, especially when we consider its application to the experiences of people labelled/with ID (Santinele Martino 2017). While the literature reveals some consideration of the experiences of sexual actors who, in some ways, "opt out" from sexual fields, often because they lack the sexual capital necessary to be competitive "players," it is assumed that they are adults with some level of knowledge and autonomy that allow them to participate in sexual fields, if they choose. Yet, the study by Santinele Martino (2019b) on the romantic and sexual experiences of adults labelled/with ID has revealed that many participants are *kept out* of sexual fields through a series of disabling social processes (e.g., infantilisation, de-sexualisation), which have not been considered within sexual fields literature. This is one example of how disability and sexualities studies can inform and enrich one other. Some seminal texts in disability studies, as Sherry (2004) notes, "make little more than passing reference to sexuality and unproblematically rely on binary gender categories" (p. 7699). This is a significant gap, firstly, because there are disabled people—including people labelled/with ID—who identify as queer, trans or non-binary, and disability studies scholars have a responsibility to challenge forms of inequality and exclusion, including homophobia and transphobia (Sherry 2004; Slater and Liddiard 2018). Also, it is not just about queer disabled people, but how insights from queer theory and disability studies can produce an understanding of how they intersect. Finally, much crip theorising has come from scholars in the humanities and literary studies fields (Egner 2016), providing a myriad of opportunities for social scientists to use crip theory as a guide for empirical exploration.

## Conclusion

With this nascent interdisciplinary, theoretical conversation, we hope we have provided some new theoretical tools for disability studies that will prove useful in understanding how individuals labelled/with intellectual disabilities (might better) negotiate sexual desire and experience. We also hope we have contributed to the sociology of sexualities by examining new ways in which to apply its theories in studying the experiences of people labelled/with ID. Broadening the rather narrow understanding of sexualities is good for everyone, not just disabled people (Santinele Martino and Campbell 2019), and there is much that can be learned from this blended approach. To begin with, by exploring how people labelled/with ID deconstruct normative expectations of sexualities via subversive practices, this dialogue may provide "alternative narratives" to sexualities (Egner 2016, p. 182), narratives that might shine light on aspects often taken for granted by sexualities scholars, including, for example, sexual access (Shuttleworth and Mona 2012). Aligned with the interest in the sexualities literature in examining questions of desire and desirability (Green 2014), crip theory also raises important questions, such as: What do we desire? Who do we desire? (Goodley 2014). More importantly, it pushes forward a project of subverting taboo and turning attention to *desiring disability* (e.g., Samuels 2003; Fudge Schormans 2015). Crip theory is about non-compliance; it is about understanding sexualities differently, creatively and expansively. Crip theory can be further developed through the inclusion of the experiences and desires of people labelled/with ID. Sexual stories (Plummer 1995), especially for marginalised sexualities, can serve as a space "to enter into community, to cultivate and circulate possibility and tactics and to foster community building" (Erickson 2016, p. 18), allowing us to envision a "better future" that finds "disability desirable" (Kafer 2013, p. 14), rather than one perpetuating deficit-based understandings of disability. Despite the many challenges to be and remain sexual, people labelled/with ID have agency and engage in subversive acts to become and be sexual (Gill 2015; Noonan and Gomez 2011). What is needed is for people labelled/with ID to, in the words of Canadian activist Andrew Gurza, show the world what it means to be "deliciously disabled."

## Notes

1 In this chapter, we use the terms "people labelled/with ID" and "disabled people" to acknowledge critique that people-first language risks depoliticising disability, and perpetuates a focus on impairment (Titchkosky 2001), while also recognising that the term "people with ID" is preferred by many self-advocacy groups in Canada. The political terms, "disabled" and "labelled" people, make plain that labels have been imposed and are "not always owned by the individual with regard to whom it is used" (McClelland et al. 2012, p. 809), acknowledging too the harmful consequences of being labelled (Spagnuolo 2016). The addition of the slash—labelled/with—demonstrates the heterogeneity amongst people identified/identifying with "intellectual disability".

2 Pseudonyms have been used.

# References

Abbott, D. (2015) Love in a cold climate: Changes in the fortunes of LGBT men and women with learning disabilities? *British Journal of Learning Disabilities* 43(2), pp. 100–105.

Albertz, M., & Lewiecki-Wilson, C. (2008) Let's talk about sex… and disability baby! *Disability Studies Quarterly* 28(4). https://doi.org/10.18061/dsq.v28i4.151

Bahner, J. (2018) Cripping sex education: Lessons learned from a programme aimed at young people with mobility impairments. *Sex Education* 18(6), pp. 640–654.

Bernstein, M. (2013) The sociology of sexualities: Taking stock of the field. *Contemporary Sociology* 42(1), pp. 22–31.

Bone, K.M. (2017) Trapped behind the glass: Crip theory and disability identity. *Disability and Society* 32(9), pp. 1297–1314.

Cambridge, P., & Mellan, B. (2000) Reconstructing the sexuality of men with learning disabilities: Empirical evidence and theoretical interpretations of need. *Disability & Society* 15(2), pp. 293–311.

Cohen, J.J. (2015) Queer crip sex and critical mattering. *GLQ: A Journal of Lesbian and Gay Studies* 21(1), pp. 153–162.

Connell, K.M., Coates, R., & Wood, F.M. (2014) Sexuality following trauma injury: A literature review. *Burns & Trauma* 2(2), pp. 61–70.

Deal, M. (2003) Disabled peoples attitudes toward other impairment groups: A hierarchy of impairment. *Disability & Society* 18(7), pp. 897–910.

DeLamater, J., & Plante, R.F. (2015) *Handbook of the sociology of sexualities.* Dordrecht, NL: Springer.

Dickman, B.J., & Jane Roux, A. (2005) Complainants with learning disabilities in sexual abuse cases: A 10-year review of a psycho-legal project in cape town, South Africa. *British Journal of Learning Disabilities* 33(3), pp. 138–144.

Dupras, A. (2015) Sexual rights of disabled persons: Between uniformity and diversity. *Sexualité, Santé, Droits de lHomme/Sexuality, Health and Human Rights* 24(3), e59–e63.

Dupras, A., & Dionne, H. (2014) The concern of parents regarding the sexuality of their child with a mild intellectual disability *Sexologies* 23(4), pp. 79–83.

Egner, J. (2016) A messy trajectory: From medical sociology to crip theory. In *Sociology looking at disability: What did we know and when did we know it* (Research in Social Science and Disability, Vol. 9). Bingley, UK: Emerald Group Publishing Limited, pp. 159–192.

Emmens, E.F. (2009) Intimate discrimination: The states role in the accidents of sex and love. *Harvard Law Review* 122, pp. 1307–1402.

Erickson, L. (2016) Transforming cultures of (un)desirability: Creating cultures of resistance *Graduate Journal of Social Science* 12(1), pp. 11–22.

Family Planning Association. (2019) "It's my right." Retrieved April 10, 2019. https://www.fpa.org.uk/sexual-health-week/its-my-right

Feely, M. (2016) Disability studies after the ontological turn: A return to the material world and material bodies without a return to essentialism. *Disability & Society* 31(7), pp. 863–883.

Finger, A. (1992) Forbidden fruit. *New Internationalist* 233, pp. 8–10.

Foucault, M. (1978) *The history of sexuality,* Vol. 1. New York: Random House.

Fritsch, K., Heynen, R., Ross, A.N., & van der Meulen, E. (2016) Disability and sex work: Developing affinities through decriminalization. *Disability & Society* 31(1), pp. 84–99.

Fudge Schormans, A. (2015) People with ID (visually) re-imagine care. In M. Barnes, N. Ward, L. Ward, & T. Brannelly (Eds.), *Renewing care: Critical international perspectives on the ethics of care.* Bristol: Bristol University Press, pp. 179–193.

Fudge Schormans, A., & Sobsey, D. (2020) Maltreatment of children with developmental disabilities. In Ivan Brown & Maire Percy (Eds.), *Developmental disabilities in Ontario* (4th edn). Toronto, ON: Delphi Graphic Communications.

Gamson, J., & Moon, D. (2004) The sociology of sexualities: Queer and beyond. *Annual Review of Sociology* 30, pp. 47–64.

García-Santesmases Fernández, A., Vergés Bosch, N., & Almeda Samaranch, E. (2017) From alliance to trust: Constructing crip-queer intimacies. *Journal of Gender Studies* 26(3), pp. 269–281.

Gerschick, T.J. (2000) Toward a theory of disability and gender. *Signs: Journal of Women in Culture and Society* 25(4), pp. 1263–1268.

Gill, M. (2010) Rethinking sexual abuse, questions of consent, and intellectual disability. *Sexuality Research and Social Policy* 7(3), pp. 201–213.

Gill, M. (2015) *Already doing it: Intellectual disability and sexual agency*. Minneapolis, MN: University of Minnesota Press.

Goodley, D. (2011) *Disability studies: An interdisciplinary introduction*. London: Sage.

Goodley, D. (2014) *Dis/ability studies: Theorising disablism and ableism*. London: Routledge.

Green, A.I. (2014) *Sexual fields: Toward a sociology of collective sexual life*. Chicago: University of Chicago Press.

Grekul, J.M. (2002) *The social construction of the feebleminded threat: Implementation of the Sexual Sterilization Act in Alberta 1929–1972*. Ph.D. Thesis, University of Alberta.

Gurza, A. [theandrewgurza]. (2019, February 17) I am not afraid of your body, so why are you afraid of mine? #DisabledPeopleAreHot [Tweet]. Retrieved from https://twitter.com/theandrewgurza/status/1097184353900261377

Ignagni, E., & Fudge Schormans, A. (2016) Reimagining parenting possibilities: Towards intimate justice. *Studies in Social Justice* 10(2), pp. 238–260.

Ignagni, E., Fudge Schormans, A., Liddiard, K., & Runswick-Cole, K. (2016) "Some people are not allowed to love": Intimate citizenship in the lives of people labelled with ID. *Disability & Society* 31(1), pp. 131–135.

Johnson, K. (2014) Queer theory. In Thomas Teo (Ed.), *Encyclopedia of critical psychology*. New York: Springer, pp. 1618–1624.

Johnson, M.L. (2015) Bad romance: A crip feminist critique of queer failure. *Hypatia* 30(1), pp. 251–267.

Jungels, A.M., & Bender, A.A. (2015) Missing intersections: Contemporary examinations of sexuality and disability. In *Handbook of the sociology of sexualities*. Cham: Springer International Publishing, pp. 169–180.

Kafer, A. (2013) *Feminist, crip, and queer*. Bloomington, IN: Indiana University Press.

Kattari, S. (2015) "Getting it": Identity and sexual communication for sexual and gender minorities with physical disabilities. *Sexuality & Culture* 19(4), pp. 882–899.

Kim, E. (2011) Asexuality in disability narratives. *Sexualities* 14(4), pp. 479–493.

Kulick, D., & Rydström, J. (2015) *Loneliness and its opposite: Sex, disability, and the ethics of engagement*. London, UK: Duke University Press.

Liddiard, K. (2013) (S)exploring disability: Intimacies, sexualities, and disabilities. A research summary. Retrieved from https://warwick.ac.uk/fac/soc/sociology/staff/sypgbj/accessible_research_summary_second_draft_repaired.pdf

Liddiard, K. (2018) *The intimate lives of disabled people*. New York: Routledge.

Löfgren-Mårtenson, L. (2013) "Hip to be crip?" About crip theory, sexuality and people with ID. *Sexuality & Disability* 31(4), pp. 413–424.

Malacrida, C. (2015) *A special hell: Institutional life in Albertas eugenic years*. Toronto: University of Toronto Press.

McClelland, A., Flicker, S., Nepveux, D., Nixon, S., Vo, T., Wilson, C., et al. (2012) Seeking safer sexual spaces: Queer and trans young people labelled with ID and the paradoxical risks of restriction. *Journal of Homosexuality* 59(6), pp. 808–819.

McRuer, R. (2002) Compulsory able-bodiedness and queer/disabled existence. In Sharon L. Snyder, Brenda Jo Brueggemann, & Rosemarie Garland-Thomson (Eds.), *Disability studies: Enabling the humanities*. New York, NY: Modern Language Association, pp. 88–99.

McRuer, R. (2006) *Crip theory: Cultural signs of queerness and disability*. New York: New York University Press.

McRuer, R. (2011) Disabling sex: Notes for a crip theory of sexuality. *GLQ: A Journal of Lesbian & Gay Studies* 17(1), pp. 107–117.

McRuer, R., & Mollow, A. (2012) *Sex and disability*. Durham: Duke University Press.

Meaney-Tavares, R., & Gavidia-Payne, S. (2012) Staff characteristics and attitudes towards the sexuality of people with intellectual disability. *Journal of Intellectual and Developmental Disability* 37(3), pp. 269–273.

Noonan, A., & Gomez, M.T. (2011) Who's missing? Awareness of lesbian, gay, bisexual and transgender people with intellectual disability. *Sexuality and Disability* 29(2), pp. 175–180.

Peers, D., Brittain, M., & McRuer, R. (2012) Crip excess, art, and politics: A conversation with Robert McRuer. *Review of Education, Pedagogy, and Cultural Studies* 34(3–4), pp. 148–155.

Plummer, K. (1995) *Telling sexual stories: Power, change and social worlds*. London: Routledge.

Plummer, K. (2012) Critical sexualities studies. In G. Ritzer (Ed.), *The Wiley-Blackwell companion to sociology*. Malden, MA: John Wiley, pp. 243–269.

Rich, A. (1980) Compulsory heterosexuality and lesbian existence. *Signs* 5(4), pp. 631–660.

Rogers, C. (2016) Introduction—Intellectual disability and sexuality: On the Agenda? *Sexualities* 19(5–6), pp. 617–622.

Rubin, G. (1984) Thinking sex. In Carole Vance (Ed.), *Pleasure and danger*. New York: Routledge & Kegan Paul.

Runswick-Cole, K., & Goodley, D. (2015) *Learning disabled children*. Manchester: Manchester Metropolitan University.

Rydström, J. (2012) Introduction: Crip theory in Scandinavia. *Lambda Nordica* 17(1–2), pp. 7–20.

Samuels, E. (2003) My body, my closet: Invisible disability and the limits of coming out discourse. *GLQ: A Journal of Lesbian and Gay Studies* 9(1–2), pp. 233–255.

Sandahl, C. (2003) Queering the crip or cripping the queer? Intersections of queer and crip identities in solo autobiographical performance. *GLQ: A Journal of Gay and Lesbian Studies* 9(1/2), pp. 25–56.

Santinele Martino, A. (2017) Cripping sexualities: An analytic review of theoretical and empirical writing on the intersection of disabilities and sexualities. *Sociology Compass* 11(5), e12471.

Santinele Martino, A. (2019a) Struggles over the sexualities of individuals with ID in Alberta, Canada. In *Dis/consent: Perspectives on sexual consent and sexual violence*. Winnipeg, MB: Fernwood, pp. 98–107.

Santinele Martino, A. (2019b) Being kept out of sexual fields: The intimate lives of adults with ID in Ontario, Canada. Paper presented at the American Sociological Association Meeting, New York City.

Santinele Martino, A., & Campbell, M. (2019) Exercising intimate citizenship rights and (re)constructing sexualities: The new place of sexuality in disability activism. In Maria Berghs, Tsitsi Chataika, Yahya El-Lahib, & Kudakwashe Dube (Eds.), *The Routledge handbook of disability activism*. New York: Routledge, pp. 97–109.

Santinele Martino, A., & Fudge Schormans, A. (2018) When good intentions backfire: University Research Ethics Review and the intimate lives of people labelled with ID. *Forum: Qualitative Social Research* 19(3), Art. 9.

Schalk, S. (2013) Coming to claim crip: Disidentification with/in disability studies. *Disability Studies Quarterly* 33(2), n.p. Available at http://dx.doi.org/10.18061/dsq.v33i2.3705 (Accessed Sep 7, 2020).

Seidman, S. (2015) *The social construction of sexuality*. New York and London: W. W. Norton & Company.

Shakespeare, T. (2014) *Disability rights and wrongs revisited*. New York: Routledge.

Shakespeare, T., Gillespie-Sells, K., & Davies, D. (1996) *The sexual politics of disability: Untold desires*. London and New York: Cassell.

Sherry, M. (2004) Overlaps and contradictions between queer theory and disability studies. *Disability & Society* 19(7), pp. 769–783.

Shuttleworth, R., & Mona, L. (2002) Disability and sexuality: Toward a focus on sexual access. *Disability Studies Quarterly* 22(4), pp. 2–9.

Siebers, T. (2008) *Disability theory*. Ann Arbor: University of Michigan Press.

Simon, W., & Gagnon, J.H. (1986) Sexual scripts: Permanence and change. *Archives of Sexual Behavior* 15(2), pp. 97–120.

Slater, J., & Liddiard, K. (2018) Why disability studies scholars must challenge transmisogyny and transphobia. *Canadian Journal of Disability Studies* 7(2), pp. 83–93.

Spagnuolo, N. (2016) Building backwards in a post institutional era: Hospital confinement, group home eviction, and Ontarios treatment of people labelled with ID. *Disability Studies Quarterly* 36. https://doi.org/10.18061/dsq.v36i4.5279

Steptoe, L., William, R.L., Forrest, D., & Power, M. (2006) Quality of life and relationships in sex offenders with intellectual disability. *Journal of Intellectual & Developmental Disability* 31(1), pp. 13–19.

Tepper, M.S. (2000) Sexuality and disability: The missing discourse of pleasure. *Sexuality and Disability* 18(4), pp. 283–290.

Tepper, M.S., Whipple, B., Richards, E., & Komisaruk, B.R. (2001) Women with complete spinal cord injury: A phenomenological study of sexual experiences. *Journal of Sex and Marital Therapy* 27(5), pp. 615–623.

Titchkosky, T. (2001) Disability: A rose by any other name?—"People First" language in Canadian society. *Canadian Review of Sociology and Anthropology* 38(2), 25–40.

Turner, G., & Crane, B. (2016) Pleasure is paramount: Adults with ID discuss sensual intimacy. *Sexualities* 19(56), pp. 677–697.

Vance, C.S. (2006) Social construction theory: Problems in the history of sexuality. In I. Grewal & C. Kaplan (Eds.), *An introduction to womens studies: Gender in a transnational world*. New York: McGraw Hill, pp. 29–32.

Warner, M. (1999) *The trouble with normal: Sex, politics and the ethics of queer life*. Cambridge, MA: Harvard University Press.

Wilton, R., & Fudge Schormans, A. (2020) "I think they're treating me like a kid": intellectual disability, masculinity and place in Toronto, Canada. *Gender, Place and Culture* 27(3), pp. 429–451.

Winges-Yanez, N. (2014) Why all the talk about sex? An authoethnography identifying the troubling discourse of sexuality and intellectual disability. *Sexuality and Disability* 32(1), pp. 107–116.

# 3

# THINKING DIFFERENTLY WITH DELEUZE ABOUT THE SEXUAL CAPACITIES OF BODIES AND THE CASE OF INFERTILITY AMONGST MEN WITH DOWN SYNDROME

*Michael Feely*

In this chapter, I wish to elucidate a theoretical framework—borrowed from Gilles Deleuze—for thinking differently and creatively about the sexual capacities of bodies labelled as impaired. Adopting a Deleuzian perspective towards bodies means avoiding the temptation to ascribe them fixed identities or inherent and context-independent characteristics (for example, "an impaired body" or "an infertile body"). Instead, this approach asks us to think of bodies as always existing within specific material-discursive contexts, which will enable some of their capacities while restricting other potential capacities. It then encourages us to think about the capacities these bodies could gain if their material-discursive contexts were altered or changed. The approach can be utilised to think about the sexual capacities of any body but, in this chapter, it shall be explicated, using the example of the reproductive capacities of male Down syndrome bodies.

The chapter will begin with a discussion of infertility amongst men with Down syndrome, an issue I became aware of during a research project about the treatment of sexuality within a service for adults with intellectual disabilities. It shall then introduce a Deleuzian theoretical framework. After this, it will present a Deleuzian analysis of infertility amongst men with Down syndrome as a concrete example of how the framework can be used to explore sexual capacities. It will then conclude by suggesting the framework, or aspects of it, could be taken up to think differently about a wide range of sexual problems.

## Infertility amongst men with Down syndrome

In 2013, I was researching the treatment of sexuality within an Irish intellectual disability service. I had collected hundreds of stories, told by service providers and service users, about sexual incidents within the service and had begun the process of analysing these. One recurring theme in these stories pertained to hopes and fears regarding reproduction. Many of the service users I spoke to—including a man with Down syndrome—told me that they hoped for a future involving marriage and children. Service providers, by contrast, tended to fear pregnancies amongst service users because they suspected they would be held accountable by their employers and/or service users' family carers for allowing these to occur. Service providers shared multiple stories about having to prevent service users—including men with Down syndrome—from having sex, within the service.

I had been researching literature on the theme of reproduction for some time when I made what was, for me, a startling discovery: Men with Down syndrome are almost universally infertile. Exceptions to this rule are exceedingly rare. So rare, in fact, that there were *no* recorded cases of men with Down syndrome fathering pregnancies until 1989 (Sheridan et al. 1989). This discovery raised multiple issues and questions. To begin, it seemed to invalidate some of the service providers' fears around reproduction. I also began to wonder whether my participant with Down syndrome, who planned to start a family, realised that he was most probably infertile. I suspected that he, and the many other men I knew with Down syndrome, did not. Finally, the discovery got me thinking about the question of whether men with Down syndrome were unalterably infertile or if there were reproductive technologies available that would allow them to have children. I began a provisional search for literature around this. What I discovered was a huge amount of research documenting efforts to aid infertile men with genetic conditions, such as Klinefelter's and Kartagener's syndromes (which are associated with physical rather than intellectual impairments), coupled with a complete silence on research documenting efforts to help men with Down syndrome. Nobody, it seemed, was interested in finding out whether men with Down syndrome could, with the relevant supports, have children.

I eventually dropped the topic of Down syndrome and infertility from my analysis but, in the years since then, I have sometimes found myself thinking about this. When asked to contribute a chapter to this volume, I decided to return to the topic of infertility amongst men with Down syndrome, to find out whether there had been any technological developments, and to provide a concrete example of a novel theoretical framework for thinking about the sexual capacities of all bodies. It is this framework that we will now explore.

## Thinking about bodies: Essentialism and its discontents.

Philosophical essentialism holds that every type or kind of entity (e.g., the human body) has a set of defining characteristics relating to its form and function. Individual entities (e.g., a human body) must possess these defining characteristics in order to secure their identity and membership of their type or kind. Put differently, an entity (e.g., a human body) must have a particular set of characteristics relating to form and function (e.g., a particular number and configuration of limbs and digits, an ability to walk upright, an ability to reason and so on) in order to be accepted as a full exemplar of the human species. These characteristics are assumed to be timeless, inherent and context independent.

This way of thinking isn't new, but it does tend to take different forms in different eras. Aristotle postulated that all species have a timeless essence or ideal form that individual members approximate to, to a greater or lesser extent, without ever fully embodying it (Futuyma 2009). Davis (2013) argues that, during the nineteenth century, the classical notion of the ideal human body was superseded by the neo-essentialist notion of the normal body. During this period, eugenic scientists calculated statistical norms and split populations into normal and abnormal groups according to a range of embodied and cognitive traits. Statistical norms soon became moral norms, and subnormal outliers—those who fell to the left of the bell curve's arch—were understood to be defectives, who should be eliminated through eugenic measures. For example, in many western nations, those with subnormal IQs were institutionalised and involuntarily sterilised to prevent them from reproducing.

It is perhaps unsurprising that disability scholars have devoted a tremendous amount of energy to challenging embodied and cognitive norms and contesting essentialist understandings of anomaly as pathology, or difference as defect. One popular and powerful way of doing this has been to embrace social constructionist epistemologies and methodologies, such as Foucault's genealogy. Genealogy, when applied to the phenomenon of disability, focuses on how the linguistic categorisation of our morphologically and cognitively diverse human species changes over time, and explores the social consequences of these categories. This reveals that:

> Socially constructed definitions of disability have varied throughout history. They have been both the cause for celebration of physical difference, and the cause of discrimination, loathing and pity.
>
> *(Brzuzy 1997, p. 82)*

If normal/impaired distinctions are not essential (i.e., transhistorical and transcultural) but, rather, socially constructed in language, then it follows that society could, in theory, discard them and find less oppressive and more inclusive ways of describing human diversity and treating those with anomalous minds. In disability studies, it is this understanding of the mutability of categorical distinctions that contributed to the emergence of the "crip theory" and its attempts to deconstruct, undermine and ultimately collapse normal/impaired distinctions (see McRuer 2006).

Social constructionism has proved a powerful tool for disability studies but has some major limitations. Critics argue that, in focusing on language and meaning, it fails to attend to the importance of materiality and of material embodiment, including embodied experiences. For many people with disabilities, overlooking inaccessible material environments, unpleasant embodied experiences, including pain, and the real functional limitations of certain bodies, are serious shortcomings (see Feely 2016b).

In response to the perceived restrictiveness of social constructionism epistemologies, recent years have seen the emergence of new-materialist ontologies and methodologies. Latour's actor network theory, Barad's agential realism, and Deleuzian assemblage theory can all be understood as anti-essentialist, material-semiotic approaches that attempt to supplement social constructionism's focus on language and meaning, with an increased attentiveness to materiality. While these approaches could all provide productive ways of attending to a phenomenon like infertility, here we will adopt a Deleuzian framework. The Deleuzian materialist ontology differs from social constructionist epistemologies in two important respects. Firstly, social constructionists struggle to discuss material reality or material bodies because they assume that efforts to describe material reality or the material body will inevitably serve to construct the meaning of this supposed "reality." In Butler's (1990, p. 10) words: "There is no reference to a pure body which is not, at the same time, a further formation of that body." Consequently, they tend to limit their focus to exploring how discourse shapes the meaning of a relatively passive material world.

## The Deleuzian materialist ontology

Deleuze, by contrast, embraces reality, which he suggests encompasses both the discursive and the material. Deleuze agrees with social constructionists that discourse is important and affects material entities, but, at the same time, he insists that material entities are also important and that they affect discourse. In short, Deleuze collapses the discourse/matter distinction and proposes that both discursive statements and material entities are equally real and are mutually affecting.

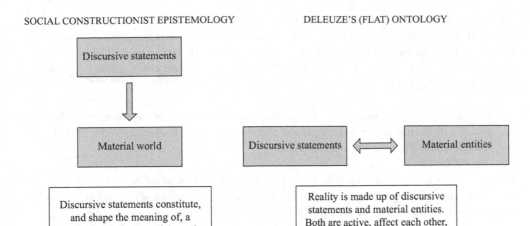

*Figure 3.1* The Deleuzian ontology.

Put differently, material entities (e.g., a river) and semiotic entities (e.g., a nation's constitution) have the same ontological status. Hence, reality, for Deleuze, can be imagined as a flat plane of different, but mutually affecting, material and semiotic entities. (Figure 3.1)

Deleuze also differs from social constructionists in his understanding of change. Social constructionists have demonstrated that linguistic categories are continually changing, rather than static. Again, Deleuze agrees, but he extends this observation to the material aspects of existence, too. Nothing in the Deleuzian universe is static. Everything is in motion, morphing, mutating, continually *becoming* at different rates of speed (an apple falling from a tree, becoming a pig's breakfast, becoming masticated mush in the mouth, becoming chyme in the belly, becoming nutrients and energy, becoming faeces, becoming pig slurry on a farm, becoming fertiliser spread across a field…) and slowness (unicellular organisms, becoming Animalia, becoming chordates, becoming tetrapods, becoming mammals, becoming primates, becoming the Hominidae, becoming *Homo* sp., becoming *Homo sapiens*, becoming men who access assistive reproductive technologies…). Because everything in the Deleuzian universe is in motion, there can be no eternal essences. Indeed, the belief that entities have eternal essences can be understood as arising from "a kind of 'optical illusion' produced by relatively slow rates of change" (DeLanda 2006, p. 49).

The Deleuzian ontology, then, suggests that we inhabit a world of equally real material and semiotic entities that are always in process, rather than static. To make this ontology work, however, requires a new way of discussing these real entities (rather than representations of them) that avoids essentialism.

## Affect

Deleuze's solution is to reject the essentialist assumption that entities (e.g., a human body) have fixed identities and to discuss them instead in terms of their *affects*:

> In the same way that we avoided defining a body by its organs and functions, we will avoid defining it by species or genus characteristics; instead, we will seek to count its affects.
>
> *(Deleuze and Guattari 2004, p. 283)*

To avoid confusion, we need to be clear about what Deleuze means by "affect." Affect, in this formulation, does not pertain to human emotions. Rather, Deleuze is using the term as short-hand for something like a body's capacities or the things it can do (it can sing, it can be tattooed, it can write a mystery novel...). Importantly, affects or capacities are *always* context dependent. In Srnicek's (2007) words:

> Concrete individuals are never ontologically isolatable from their environment ... In fact, the division between individuals and their environment is, in many ways, already an abstraction from the real situation.

Because a body (or any entity) is always embedded in a specific material-semiotic context, and because contexts are open to endless change, its list of affects or capacities will necessarily be ongoing. For example, the things a particular body can do will vary radically, depending on whether it is on land or underwater, on Earth or on Mars, online or offline, rich or poor, at room temperature or cryogenically frozen and so on. Any body (regardless of whether it has been labelled as "impaired" or "normal") will face real restrictions or things that it cannot do in certain material-semiotic contexts (e.g., a body may be unable to fly) but these restrictions remain context specific and are never essentialised as inherent characteristics of the body (i.e., the body might be able to fly if it had access to a plane).

This way of thinking also applies to sexual capacities. Imagine a man called Jack with Down syndrome. Jack's current material-semiotic context will enable a certain set of sexual affects or capacities (sexual things he can do and have done to him). For example, Jack might be able to meet his girlfriend for a date in a restaurant and, afterwards, kiss her goodnight at his bus-stop. At the same time, his material-semiotic context will also constrain and limit the sexual things he can do and have done to him. In his current context, for instance, Jack may be unable to invite his girlfriend back to his residential service to sleep with him because the service has no double bedrooms, its policies forbid residents from having visitors after 9 p.m. and there is a staff member "on nights" to enforce the rules. Similarly, Jack and his girlfriend may be unable to have a child because Jack's specific testicular phenotype makes a spontaneous conception highly unlikely. These are very real limitations. Nonetheless, we need to avoid the temptation to essentialise them as immutable and context-independent facts about Jack. *It is possible* to imagine a world where Jack could sleep with his girlfriend and access relevant reproductive technologies. Moreover, while we don't know for sure, *it might be possible* to create this world.

To allow us to think about what a body can presently do, and what it might be able to do, the Deleuzian framework offers a distinction between a body's *actual* and *virtual* capacities to affect and be affected. Actual affects refer to capacities that are already realised or, put differently, things the body can do in its current material-discursive context. Virtual affects, by contrast, refer to potential but currently non-actualised affects or to the infinite number of other things the body *could* do if its context was altered or changed. Each and every body will have an ongoing list of virtual capacities that *could be* materialised by altering its material-semiotic context.

All of this means that, for Deleuze, the pertinent question to ask about a body (or any entity) is not "What is a body?" (which suggests a list of essential characteristics) but rather questions like "What can a body do?" (which relates to the things it can currently do) and "What else can a body do?" (which relates to the things it could do in other contexts). The only way to answer these questions is to think creatively about and experiment with all the things a body can do in its current context or could do if its context was changed. In the case of infertility amongst men with Down syndrome, exploring whether fertility is a virtual affect or capacity that could be actualised would involve thinking about the material-semiotic context that men with Down syndrome inhabit and how this could be changed.

## Assemblages

So far, we have discussed how the capacities of an entity (e.g., a human body) to affect and be affected are dependent on material-semiotic context. This is a convenient way to introduce the concept of affect but risks oversimplification. What we really mean is that an entity's capacities will depend on the range of other entities that it has relationships with or the other entities that it affects and that affect it. Returning to the example of Jack, instead of talking about Jack's material-semiotic context, it is perhaps more productive to say that his sexual capacities to affect and be affected are determined by his relationships with other entities (for example, his relationships with his girlfriend, his parents, his disability service and his absence of relationships with reproductive technologies and so on). In order to think about networks of relations like this, Deleuze proposes the concept of the *assemblage*.

An assemblage is a temporary conglomeration or network of heterogeneous components belonging to aspects of existence often treated as discreet or separate (e.g., the architectural, the biological and the discursive). The unity of these components comes purely from the fact that they work together to produce something. For example, we might think of an *in vitro* fertilisation (IVF) assemblage as consisting of components including a reproductive clinic, two prospective parents, a reproductive endocrinologist, sperm and ova, petri dishes, the capital that pays for the procedure and the policies and legislation that enable the procedure, all working together, to produce a pregnancy.

To think about the relationships between heterogeneous components or forces that make up an assemblage, Deleuze and Guattari (2004) offer the metaphor of the *rhizome*. A rhizome is an underground stem which produces a subterranean tangle of roots which spreads horizontally in all directions and has no centre. All its parts are intricately interconnected and affect each other. Thus, thinking rhizomatically about an assemblage means thinking in a nonlinear fashion, producing an analysis that follows multi-directional connections between the heterogeneous components or forces that make up the assemblage:

> A rhizome ceaselessly establishes connections between semiotic chains, organizations of power and circumstances relative to the arts, sciences and social struggles.
>
> *(Deleuze and Guattari 2004, p. 7)*

More specifically, thinking rhizomatically about an IVF assemblage would involve a nonlinear exploration of how various architectural, human, technological, economic and legislative forces work together to produce a pregnancy (Figure 3.2).

An assemblage is not a fixed entity. Rather, its components (buildings, bodies, technologies, policies) came together over time; will work together for a time to produce something (a pregnancy); and, in time, will fall apart. In the words of Jackson and Mazzei (2012, p. 1): "An assemblage isn't a thing—it's the *process* of making and unmaking the thing." As such, research questions about an assemblage should not relate to the identity of the assemblage (i.e., what is an IVF assemblage?). Rather, they should relate, firstly, to what it does or how its components work together to produce the phenomenon in question (i.e., a pregnancy), and, secondly, what else it could do if its components were altered (i.e., could the IVF assemblage be changed to include men with Down syndrome?). With this theoretical framework established, we can now move on to a concrete case study, which will demonstrate how it might allow us to explore sexual capacities.

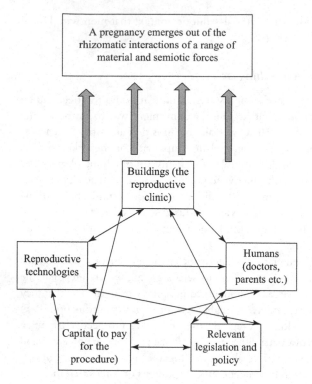

*Figure 3.2* An IVF assemblage produces a pregnancy.

## Case study

This section will offer an assemblage analysis of the production of infertility amongst men with Down syndrome, with a focus on Ireland. We will begin by discarding the assumption that infertility is an inherent characteristic of Down syndrome bodies. Instead, we will assume that infertility might be something that emerges, or is produced, in the relationships between bodies with Down syndrome and a range of other entities or forces (biological, technological, economic, legislative, attitudinal and so on). Consequently, our first research question is:

• How is infertility produced (and maintained) amongst Irish men with Down syndrome?

To answer this question, we must think rhizomatically about how infertility emerges through the complex interactions of material and semiotic forces. This will involve jumping between various orders of existence and analysing how a range of relevant forces interact to produce and maintain infertility. Where we begin our analysis is somewhat arbitrary but, in this case, let's start in the material realm by looking at biological bodies and—more specifically—at literature about the biological bodies of men with Down syndrome.

## *Biology*

The first thing that becomes apparent, when one begins researching the reproductive biology of men with Down syndrome, is the paucity of literature and the gaps in the scientific knowledge concerning this topic. Of the literature that does exist, the majority relates to causes of infertility and rare exceptions to this rule.

Pradhan et al. (2006) draw on the available literature to suggest possible biological causes of infertility amongst men with Down syndrome, including: Undescended testes; small testicles and penis; decreased spermatogenesis; sexual impotence; unusual hormone levels; and progressive gonadal failure. The review of the literature by Stefanidis et al. (2011) also points to a number of possible biological causes, relating to hormonal, morphological and cytogenetic issues. Ultimately, though, they conclude that the biological causes of infertility remain contested and somewhat of a medical mystery.

There is also a small body of literature documenting rare exceptions to the rule of infertility. Sheridan et al. (1989) present the first documented case of fertility in a male with non-mosaic Down syndrome. The man in question had a relationship with a housemate in his residential service culminating in a pregnancy, which ended in miscarriage. Interestingly, their discussion of the significance of this anomalous case does not frame it as significant or of interest to other men with Down syndrome who may wish to have children. Rather, Sheridan et al. (1989, p. 297) offer it as evidence for the need for societal vigilance around contraception in the post-institutional era:

> This observation emphasises the need to maintain adequate contraceptive cover, especially as more mentally handicapped adults are removed from supervised institutions and encouraged to live within the community.

Five years later, Zuhlke et al. (1994) report a similar case involving a 29-year-old man and a female housemate, this time culminating in the birth of a chromosomally typical child. As with the previous example, they do not frame this case as significant for men with Down syndrome who want children but, this time, the case is presented as cautionary tale for parents of children with Down syndrome, rather than society at large: "Parents of children with trisomy 21 should be carefully counselled regarding fertility" (Zuhlke et al. 1994, p. 326).

More than a decade later, Pradhan et al. (2006) announced the third case of fertility in a man with Down syndrome. In this instance, the man fathered a healthy boy with his wife, who is described as "normal." As before, the authors see the case as suggesting a need for contraception. However, in contrast to the previous cases, Pradhan et al. (2006, p. 1765.e1) understand this information to be of relevance to men with Down syndrome rather than society at large or their parents: "It is important to advise post-pubertal Down syndrome males on contraceptive measures."

In light of all of this, we might imagine relevant biological forces in producing infertility as encompassing the anomalous reproductive biology of men with Down syndrome, a lack of scientific knowledge about (and interest in) the precise biological origins of this infertility and a tendency amongst scientists to identify rare instances of fertility as a threat to society or parents, rather than promising news for men with Down syndrome.

## Legislation

The tendency to construct the sexuality and fertility of people with intellectual disabilities as a threat is not confined to the biological literature. If we follow a rhizomatic link from the material to the semiotic—and, more specifically, from biology to legislation—we find more forces that serve to limit rather than enable the sexual capacities of intellectually disabled bodies. Most importantly, in the Irish context, is the fact that that, until 2017, consensual sex with a person with an intellectual impairment (or between people with intellectual impairments) was illegal (see Law Reform Commission 2013). For men with Down syndrome, this meant that trying to impregnate a partner naturally (i.e., without reproductive technologies) was illegal. Moreover, if such a man did manage the extremely rare feat of impregnating his partner, this could, theoretically, serve as incriminating evidence against him. Interestingly, under this law, attempting to have a child through IVF would not have been illegal.

This legislation was finally repealed and replaced in 2017 by the Criminal Law (Sexual Offences) Act 2017, which allows people with intellectual disabilities to have consensual sex relations. These developments mean there no longer seems to be legal prohibitions on either "natural" or assisted reproduction amongst adults with intellectual disabilities. In addition to this, in 2018, Ireland belatedly ratified United Nations' Convention on the Rights of Persons with Disabilities (UN 2006, p. 15), which includes explicit acceptance of "the rights of persons with disabilities to decide freely and responsibly on the number and spacing of their children."

In short, pertinent legislative forces in the production and maintenance of infertility include Ireland's criminalisation of sexual relationships up until 2017. This said, since 2017, less restrictive legislation has been introduced and the reproductive rights of people with disabilities have been recognised. Moreover, there is now no explicit legislative barrier to men with Down syndrome accessing reproductive technologies.

## Reproductive technologies

Having established that there is no legal prohibition, we might follow another rhizomatic link from the legislative to the technological, to explore whether there are technologies that *could* facilitate reproduction. Moving then on to research with respect to reproductive technologies, one is struck by the large amount of literature in this area and the major technological developments that have occurred over recent decades, developments that have allowed millions of male bodies that were previously thought infertile to reproduce. Also notable is the amount of literature on efforts to help men with a range of genetic conditions that affect physical, rather than intellectual, functioning. This is coupled with one very significant exception, with a silence around assisting men with Down syndrome to reproduce.

Fauser and Edwards (2005) offer a brief overview of developments in reproductive technologies over the twentieth century. They suggest that research on the *in vitro* fertilization of mammalian eggs began as early as the 1930s. In 1977, research involving humans culminated in the birth of the first IVF child in the United Kingdom. The classical IVF technologies developed during this time—which involve extracting ova from the ovaries, allowing sperm to fertilise the ova in a liquid in the laboratory, and then implanting a fertilised egg in the woman's uterus—spread rapidly and have allowed millions of bodies, that were once thought to be infertile, to gain the capacity to reproduce.

Classical IVF technologies were helpful for both women and some men with fertility issues. However, in the early 1990s, the arrival of intracytoplasmic sperm injection (ICSI)— which involves a single sperm being injected directly into the oocyte cytoplasm—allowed a great many more men who produce very few, unusually shaped or relatively immobile sperm to reproduce (see Boulet et al. 2015). ICSI proved beneficial to men with a variety of genetic anomalies that affect fertility and physical (rather than intellectual) functioning. For example, men with Kartagener's syndrome quickly gained access to ICSI technology and, with it, the capacity to reproduce (Zumbusch et al. 1998). The technique also proved extremely successful for men with Klinefelter's syndrome (Dávila Garza and Patrizio 2013). It is interesting to note that papers which announce the efficacy of ICSI for men with physical disabilities tend to frame these scientific developments as positive breakthroughs, albeit tempered by sober discussions regarding the possibility of disability amongst offspring.

Turning to Down syndrome, which is associated with intellectual as well as physical disabilities, a literature search returned very little research on using reproductive technologies to enable men with Down syndrome to have children, coupled with a lot of literature on technologies which seek to prevent "normal" couples, who access reproductive technologies having children with Down syndrome. These technologies include preimplantation genetic screening and preimplantation genetic diagnosis, which Braude et al. (2002, p. 941) suggest, offer "a practical alternative to prenatal diagnosis and termination of pregnancy."

When I first consulted the literature in 2013, there was simply no research on the use of reproductive technology to enable men with non-mosaic Down syndrome to have children. However, when I returned to the literature again, I found a breakthrough had occurred. In 2015, Aghajanova et al. (2015, p. 1409) announced the birth of a "healthy infant born after intracytoplasmic sperm injection-*in vitro* fertilization (ICSI-IVF) with preimplantation genetic screening (PGS), using sperm from a man with non-mosaic trisomy 21." The man in question, and his chromosomally typical wife, had strong extended family support for their desire to start a family. They accessed a fertility treatment in California through a reduced-fee scheme, which eventually led to the birth of a chromosomally typical child. Aghajanova et al. (2015, p. 1413) end their case report with a consideration of the ethical implications of their breakthrough and suggest some preconditions for treating other men with Down syndrome:

> At a minimum, one must take into account the couple's ability to parent, the potential impact on offspring of having a cognitively impaired parent, as well as the availability of family, medical and community support systems.

To sum up, we might see relevant technological forces as including a longstanding lack of scientific interest in, and research into, helping males with Down syndrome to have children, using technologies that have allowed a huge number of men with other genetic conditions (that do not affect intellectual functioning) to reproduce. This said, in the recent past, there has been a breakthrough and it seems that fathering children is no longer something men with Down syndrome *might* be able to do. Rather, it appears reproduction is something many men with Down syndrome *could do* if they were able to access the relevant technologies.

## *Economic forces*

Whereas technological advances may make fertility a virtual capacity for many men with Down syndrome, this capacity will not be actualised unless they can access these technologies, and there are reasons to believe that this may remain exceedingly difficult. Following a rhizomatic link from reproductive technology to economics, we find that technologies, such as IVF, are expensive and which sections of the population can and cannot afford to access them depends on several factors, including government policies and private wealth. In Ireland, IVF is not currently provided by the public health service and is only available on a private basis (HSE 2019). Costs vary but, in 2014, were estimated to be €4000 to €4500 per round of IVF and €5000 to €5500 for ICSI. Multiple rounds of treatment may be required, and couples can expect to incur a range of additional costs from consultant's fees to additional tests or procedures. Couples may end up paying in the region of €15,000 with no guarantee of a successful outcome (see McBride 2014). In short, accessing reproductive technologies requires large amounts of capital.

Most Irish men with Down syndrome live on a means-tested government payment called the Disability Allowance. The maximum amount payable is €198 per week. Receiving additional income from work or by having savings, investments or property can lead to this payment being reduced or withheld (Citizens' Information 2019). Thus, the process of saving the tens of thousands needed for IVF might, in fact, reduce the amount of Disability Allowance one receives, making further savings impossible. These welfare policies, and the poverty traps they create, mean that, for many men with Down syndrome, reproductive technologies remain prohibitively expensive.

The low income of men with Down syndrome may cause problems in other ways, too. It may act as a deterrent to having children and, additionally, as an impediment to successfully raising children. In reviewing the literature on parents with intellectual disabilities, Sheerin (1998, p. 132) concludes that "in the absence of proper social supports, including parenting education, there may be problems in the adequacy of parenting provided by some learning-disabled people." His contention is that parenting problems are not inevitable; many people with intellectual disabilities *could* be adequate parents but they *would* need extra supports. Extra supports, of course, cost extra money and, in the absence of personal or familial wealth, parents with intellectual disabilities are reliant on social care services. If, as Begley et al. (2009) suggest, such supports are often not available, it seems reasonable to assume that some fathers with Down syndrome might struggle to raise children, conceived either naturally or through reproductive technologies.

To summarise, relevant economic forces in producing and maintaining infertility in mean with Down syndrome relate to the high costs of accessing reproductive technologies. Men with Down syndrome tend to have low incomes, and this makes accessing privately owned reproductive clinics very unlikely. Low income, combined with a lack of publicly funded supports for parents with intellectual disabilities, also creates barriers for men with Down syndrome from successfully raising children.

## *Sex education and policies around sexuality*

A degree of affluence—which many men with Down syndrome don't possess—is not the only thing one needs in order to access reproductive technologies. For men with fertility issues, being cognisant of these issues may be another precondition for seeking and gaining access to reproductive technologies. Therefore, at this point, we might follow another rhizomatic connection, this time from the field of economics to the field of sex education and service policies around sexuality.

Whereas literature on sex education is scarce, the study by Allen and Seery (2007, p. 6) of disability services around Ireland found that:

> There is no national structured approach to the provision of RSE (relationship and sexuality education) to people with intellectual disability. Whatever RSE is provided is piecemeal and many people with an intellectual disability are not receiving it.

Small-scale qualitative sexuality research involving Irish adults with intellectual disabilities seems to corroborate this. Kelly et al. (2009), for example, found that a large proportion of her participants who were in a relationship desired but had not received sex education, and their knowledge about sex—including biological knowledge about their own bodies—was limited.

All of this raises an interesting question: "Do Irish men with Down syndrome realise that they are infertile?" Unfortunately, there appears to be no empirical research—Irish or international—on this topic. Lapsing into the anecdotal, in the thirteen years I have worked and conducted inclusive research with men with Down syndrome, I have encountered a great many men who hoped, and firmly believed, they would have children, but not a single individual that mentioned their infertility as a potential barrier to realising their dreams.

In addition to a lack of sex education, Irish disability services often have restrictive policies relating to sexuality in general, which may make finding a partner with whom to start a family exceptionally difficult. Inclusion Ireland (2017) suggests that the pre-2017 legislation contributed to extremely conservative policies around sexuality in services, which made it extremely difficult for people with intellectual disabilities to access privacy and supports. Empirical studies, including Feely (2016a), have also reported a high level of sexual surveillance within disability services, prohibitions around exhibiting affection physically and a complete lack of private spaces. There is, as yet, no research regarding if and how legislative change has affected these policies.

To conclude, relevant forces related to sex education and sexuality policies include a lack of sex education and sexual knowledge amongst men with Down syndrome. As long as men with Down syndrome men remain unaware of their infertility, it is highly improbable that they will—collectively or individually—demand access to relevant reproductive technologies. Meanwhile, restrictive policies around sexuality in disability services may also make it extremely difficult for men with Down syndrome to find partners with whom to start families.

## Attitudes

If the infertility of men with Down syndrome is not inevitable but rather sustained by potentially alterable conditions, including current economic relations and a lack of sex education, infertility becomes a political rather than simply a biological issue. With this in mind, we might follow one last rhizomatic connection, from the field of sex education to the question of attitudes, in order to gauge what, if any, support there is available to help men with Down syndrome become fathers. Here, we find reason to believe that, whereas support for this proposition has increased rapidly, the sexual and reproductive rights of people with intellectual disabilities remain controversial.

At a national level, in 2011, the National Disability Authority's (2011) survey of public attitudes revealed that a minority of 37% of respondents agreed that adults with intellectual disabilities or autism should have children if they wished to. By the time of the 2017 survey, however, a majority of 62% of respondents agreed that adults with intellectual disabilities should have children if they wished (National Disability Authority, 2017). On a smaller scale, attitudes of family carers and disability service staff may also be significant. The available research on these attitudes is extremely limited but, again, the existing studies suggest that the prospect of adults with intellectual disabilities becoming parents remains a highly contentious issue amongst family carers and disability service staff (Begley et al. 2009; Drummond 2006; Evans et al. 2009). The final relevant force in producing and maintaining infertility, then, relates to reasonably widespread (albeit diminishing) hostility amongst the public, disability service providers, and family carers, to the idea that men with Down syndrome should have children if they wish.

## What it all means

In presenting an assemblage analysis of the production of Down syndrome amongst Irish men with intellectual disabilities, we started by jettisoning the assumption that infertility was an inherent characteristic of these men. Instead, an assemblage approach suggests that infertility is something that emerges or is produced and maintained by a range of mutually affecting material and discursive forces. In elucidating these forces, we followed a series of rhizomatic connections between different orders of existence. Starting with biological bodies we found a lack of scientific knowledge about, or interest in, the reproductive biology of men with Down syndrome and a tendency to construct rare instances of fertility in terms of threat. Jumping from the biological to the legislative we found the sexuality of adults with intellectual disabilities constructed again as a threat and we explored longstanding (but recently repealed) legislation that criminalised sex with, or between, adults with intellectual disabilities.

Moving then to the technological realm, we uncovered a long-standing lack of interest in attempting to use existing reproductive technologies and procedures to help men with Down syndrome have children. However, one recent case gave us reason to suspect that fertility is a virtual capacity of many men with Down syndrome that could be actualised with current reproductive technologies. We then followed another rhizomatic link to the realm of economics and found that the high costs associated with accessing reproductive technologies and raising children and the relatively low income of men with Down syndrome may form a significant barrier to these men's capacity for reproduction being actualised. Leaping from economics to education, we found that a lack of sex education and a consequent lack of biological knowledge amongst men with Down syndrome may be another force that prevents the actualisation of fertility. Having established that infertility was probably not inevitable but potentially alterable, we took one final rhizomatic jump to examine public, familial and staff attitudes in order to gauge the appetite for change. Here we found widespread (but rapidly decreasing) hostility to the idea of men with intellectual disabilities becoming fathers (Figure 3.3).

As with deciding where to begin assemblage analysis, the decision to end a rhizomatic journey through an assemblage is somewhat arbitrary; it is always possible to follow more links and explore more material-semiotic forces. But, in offering this particular assemblage analysis, we are suggesting that the production and maintenance of infertility cannot be reduced to any one of the heterogeneous forces explored but, rather, is an emergent property of complex interactions amongst all of them (and, no doubt, a myriad more unexamined forces).

## What else can an infertility assemblage do?

An assemblage analysis of a social or sexual problem, like infertility amongst men with Down syndrome, can allow us to draw a detailed, but never complete, map of how the problem is produced. But the point of this type of analysis is not simply to create a rich description. Rather, understanding how a range of material-semiotic forces are working together to produce something in the present allows us to think about how we could (and, perhaps, *if* we should) intervene and attempt to change the workings of the assemblage. In Nail's (2017, p. 37) words:

> Once we understand how the assemblage functions, we will be in a better position to perform diagnosis: To direct or shape the assemblage toward increasingly revolutionary aims.

This means that the second pertinent question for our analysis is:

- Could the infertility assemblage be altered to produce something else (i.e., fertility)?

There are some reasons to believe that the infertility assemblage is highly *territorialised* (a Deleuzian term for an assemblage that is highly regulated and rigid) and may be difficult to

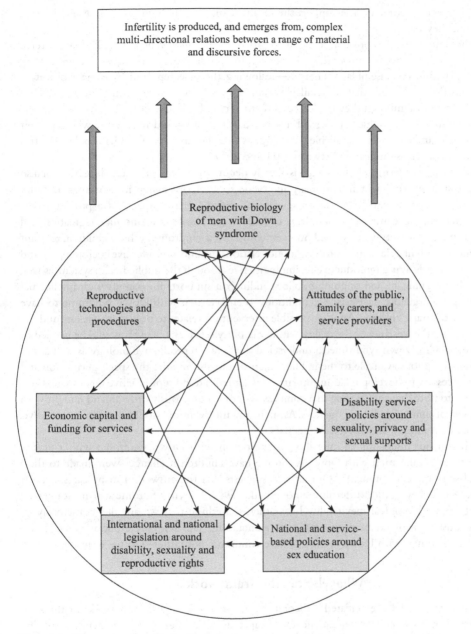

Infertility is produced, and emerges from, complex multi-directional relations between a range of material and discursive forces.

Reproductive biology of men with Down syndrome

Reproductive technologies and procedures

Attitudes of the public, family carers, and service providers

Economic capital and funding for services

Disability service policies around sexuality, privacy and sexual supports

International and national legislation around disability, sexuality and reproductive rights

National and service-based policies around sex education

*Figure 3.3* The production of infertility.

change. There are, at present, no recorded cases of Irish men with Down syndrome fathering children, either with or without technological assistance; the means of artificial reproduction in Ireland remains privately owned and hence prohibitively expensive for most men with Down syndrome; the resources and supports necessary to aid parents with intellectual disabilities to raise happy and healthy children are often not available; service-level policies around sexuality mean that many service users have been denied sex education and basic knowledge about their bodies; and many people do not support the idea that men with Down syndrome should have children if they wish.

At the same time, it is possible to point to processes of *deterritorialisation* (a Deleuzian term for processes that allow for movement, change and the production of the new), that are already occurring within the assemblage: There is—following the case reported by Aghajanova et al. (2015)—reason to believe that the available technologies could enable Irish men with Down syndrome to start families if they could access them; recent legislative changes have decriminal-ised consensual sex amongst people with intellectual disabilities and have acknowledged their right to start families; and the available data suggests that the support of the Irish public for this right increased very significantly between 2011 and 2017.

In addition to exploring how change is already occurring within an assemblage, we can also think about how we might intentionally intervene to alter or change its workings. Because assemblage analysis is equally attentive to material and discursive aspects of a problem, sug-gested interventions can relate to any level of analysis (i.e., biology, economics, legislation and so on). If we, as a society, wanted to allow men with Down syndrome to have children, relevant interventions might relate, firstly, to economic relations around reproductive technologies and raising children. Making reproductive technologies free through the public health system, as well as providing significant and ongoing public spending on supports for parents with intellectual disabilities, are arguably necessary preconditions to enable men with Down syndrome to have and adequately raise children. Another possible intervention relates to policy changes around sex education and the treatment of sexuality within disability services. Changes would be needed to enable men with Down syndrome to understand their own reproductive biology as well as the reproductive options available to them. These men would also require the space, permission and privacy necessary to develop relationships, have sex and/or attend reproductive services. It is also important to note that intervening in a complex system is notoriously difficult and fraught with the danger of unintended consequences. As such, any intervention would need to be tentative, continually monitored and altered as necessary.

We began our discussion of infertility amongst men with Down syndrome with a practi-cal question: "Could men with Down syndrome have children if changes were made to their material-semiotic environment?" Our analysis suggests that the answer, in many cases, is most probably "Yes, they could." Infertility amongst men with Down syndrome can no longer be thought of as a biological inevitability but, rather, something that we are all—consciously or unconsciously—involved in producing and maintaining. The relevant question, then, is no longer "Could men with Down syndrome have children?" but rather "Do we want them to?"

## What else can this framework do?

In this chapter, we have elucidated a theoretical and methodological framework for thinking creatively about the context-dependent sexual capacities of bodies with reference to a specific case study. But exploring infertility amongst men with Down syndrome is not the only thing this framework can do. Whether one is interested in bodies that have been defined as sexually impaired, impotent, insentient, incapable of giving informed consent and so on, a Deleuzian approach can open up new ways of thinking about, and responding to, the problem. The basic approach to exploring each of these problems, as outlined in this chapter, would be: To reject essentialist notion that the limitation in question is an inherent and immutable characteristic of the body; to assume, instead, that the limitation is contextual and produced by an assemblage or network of material-semiotic forces all working together; to identify what these heterogeneous material-semiotic forces are and how they work together to produce the problem; and, finally, to think about, and experiment with, ways the assemblage could be altered or changed to ame-liorate the problem. These aforesaid Deleuzian approaches are never prescriptive, and readers of this chapter are invited to adopt and adapt the concepts and procedures outlined here in creative and unfaithful ways in their own projects and for their own purposes.

# References

Aghajanova, L., Popwell, J., Chetkowski, R., & Herndon, C. (2015) Birth of a healthy child after preimplan-tation genetic screening of embryos from sperm of a man with non-mosaic Down syndrome. *Journal of Assisted Reproduction and Genetics* 32(9), pp. 1409–1413.

Allen, M., & Seery, D. (2007) *The current status of sex education practice for people with an intellectual disability in Ireland*. Cork: The Sex Education Centre.

Begley, C., Higgins, A., & Lalor, J. (2009) *Women with disabilities: Barriers and facilitators to accessing services dur-ing pregnancy, childbirth and early motherhood*. Dublin: School of Nursing and Midwifery, Trinity College Dublin.

Boulet, S.L., Mehta, A., Kissin, D.M., Warner, L., Kawwass, J.F., & Jamieson, D.J. (2015) Trends in use of and reproductive outcomes associated with intracytoplasmic sperm injection. *JAMA* 313(3), pp. 255–263.

Braude, P., Pickering, S., Flinter, F., & Mackie Ogilvie, C. (2002) Preimplantation genetic diagnosis. *Nature Reviews Genetics* 3, pp. 941–955.

Brzuzy, S. (1997) Deconstructing disability. *Journal of Poverty* 1(1), pp. 81–91.

Butler, J. (1990) *Gender trouble*. New York: Routledge.

Citizens Information (2019) Disability allowance. *Citizens Information Website*. http://www.citizensinform ation.ie/en/social_welfare/social_welfare_payments/disability_and_illness/disability_allowance.html. Accessed 15th July 2019.

Dávila Garza, S.A., & Patrizio, P. (2013) Reproductive outcomes in patients with male infertility because of Klinefelters syndrome, Kartageners syndrome, round-head sperm, dysplasia fibrous sheath, and stump tail sperm: An updated literature review. *Current Opinion in Obstetrics and Gynecology* 25(3), pp. 229–246.

Davis, L. (2013) *The disability studies reader*. New York: Routledge.

DeLanda, M. (2006) *A new philosophy of society: Assemblage theory and social complexity*. New York: Continuum.

Deleuze, G., & Guattari, F. (2004) *A thousand plateaus: Capitalism and schizophrenia*. New York: Continuum.

Drummond, E. (2006) Attitudes towards sexuality: A pilot study in Ireland. *Learning Disability Practice* 9(4), pp. 28–34.

Evans, D.S., McGuire, B.E., Healy, E., & Carley, S.N. (2009) Sexuality and personal relationships for people with an intellectual disability. Part II: Staff and family carer perspectives. *Journal of Intellectual Disability Research* 53(11), pp. 913–921.

Faucer, B., & Edwards, R. (2005) The early days of IVF. *Human Reproduction Update* 11(5), pp. 237–238.

Feely, M. (2016a) Sexual surveillance and control in a community-based intellectual disability service. *Sexualities* 19(5–6), pp. 725–747.

Feely, M. (2016b) Disability studies after the ontological turn: A return to the real world and material bod-ies without a return to essentialism. *Disability & Society* 31(7), pp. 863–883.

Futuyma, D. (2009) *Evolution*. Massachusetts: Sinauer Associates.

HSE (2019) IVF. *HSE Website*. https://www.hse.ie/eng/health/az/i/ivf/. Accessed 15th July 2019.

Inclusion Ireland (2017) *Sexual offences against vulnerable persons*. Inclusion Ireland: Dublin.

Jackson, A., & Mazzei, L. (2012) *Thinking with theory in qualitative research*. New York: Routledge.

Kelly, G., Crowley, H., & Hamilton, C. (2009) Rights, sexuality and relationships in Ireland: "It'd be nice to be kind of trusted". *British Journal of Learning Disabilities* 37(4), pp. 308–315.

Law Reform Commission (2013) *Report: Sexual offences and capacity to consent*. Dublin: Law Reform Commission.

McBride, L. (2014) How to control costs when going down the IVF route. Irish Independent, March 23rd. https://www.independent.ie/business/personal-finance/how-to-control-costs-when-going-down-the-ivf-route-30116444.html. Accessed 15th July 2019.

McRuer, R. (2006) *Crip theory: Cultural signs of queerness and disability*. New York: New York University Press.

Nail, T. (2017) What is an assemblage? *SubStance* 46(1), pp. 21–37.

National Disability Authority (2011) *National survey of public attitudes to disability in Ireland 2011*. Dublin: National Disability Authority.

National Disability Authority (2017) *National survey of public attitudes to disability in Ireland 2017*. Dublin: National Disability Authority.

Pradhan, M., Dalal, A., Khan, F., & Agrawal, S. (2006) Fertility in men with Down syndrome: A case report. *Fertility and Sterility* 86(6), pp. 1765e1–1765e3.

Sheerin, F. (1998) Parents with learning disabilities: A review of the literature. *Journal of Advanced Nursing* 28(1), pp. 126–133.

Sheridan, R., Lerena, J., Matkins, S., Debenham, P., Cawood, A., & Bobrow, M. (1989) Fertility in a male with trisomy 21. *Journal of Medical Genetics* 26, pp. 294–298.

Srnicek, N. (2007) *Assemblage theory, complexity and contentious politics*. Unpublished thesis. University of Western Ontario.

Stefanidis, K., Belitsos, P., Fotinos, A., Makris, N., Loutradis, D., & Antsaklis, A. (2011) Causes of infertility in men with Down syndrome. *Andrologia* 43, pp. 353–357.

United Nations (2006) *Convention on the rights of persons with disabilities*. New York: United Nations.

Von Zumbusch, A., Fiedler, K., Mayerhofer, A., Jeßberger, B., Ring, J., & Vogt, H.J. (1998) Birth of healthy children after intracytoplasmic sperm injection in two couples with male Kartageners syndrome. *Fertility and Sterility* 70(4), pp. 43–646.

Zuhlke, C., Thies, U., Braulke, I., Reis, A., & Schirren, C. (1994) Down syndrome and male fertility: PCR-derived fingerprinting, serological and andrological investigations. *Clinical Genetic* 46, pp. 324–326.

# 4

# A CRITICAL RETHINKING OF SEXUALITY AND DEMENTIA

## A prolegomenon to future work in critical dementia studies and critical disability studies

*Alisa Grigorovich and Pia Kontos*

## Introduction

Critical disability scholars and activists have made important contributions to broadening the discourse on sexuality and to more fully accommodating rights related to sexuality (Shakespeare 2000; Shildrick 2007; McRuer and Mollow 2012; Gill 2015; Liddiard 2017). However, this scholarship has concentrated primarily on the sexuality and intimacy-related needs of younger people and, in particular, people living with physical disabilities. Although there has been some attention to the sexualities of younger adults with intellectual disabilities (Gill 2015; McRuer and Mollow 2012), for the most part, there has been little attention paid to the intersection of ageing, cognitive disability, and sexuality. In consequence, the sexual rights of older adults living with dementia continue to be neglected. This neglect can be traced to the fact that, within critical disability studies, dementia itself has not been treated as a "disability." Furthermore, it is only recently, with the increased emphasis on citizenship and human rights (Bartlett and O'Connor 2007; Shakespeare et al. 2019), that the "rights" of persons living with dementia have been considered in critical dementia studies. Despite that, there has been little engagement between critical disability and critical dementia studies. As we will demonstrate in this chapter, such engagement can yield significant insights regarding sexuality.

There is growing attention to sexuality and sexual rights across ageing studies literature, and this scholarship has largely been dominated by two inter-related discourses on dementia that are contributing to misunderstandings and prejudice regarding the sexualities of persons living with dementia: the "cerebralisation" of selfhood (Vidal 2009; Kontos 2012a; Katz 2013), and the biomedical discourse of dementia (Herskovitz 1995; Volicer and Hurley 2003). Together, these discourses perpetuate a cultural imaginary (Gatens 1996) or a collective representation of dementia as a total erasure of self, which is part of a larger stigmatising "decline narrative" of ageing (Gullette 2004). As a consequence, persons living with dementia are largely constructed as "unagentic" and "failed" ageing subjects.

This cultural imaginary is evident in the dominance of metaphors and images in public policy and popular culture that represent dementia as "the funeral without end," "the loss of self," or "a living death" (Behuniak 2011; Kontos 2012a; Mitchell et al. 2013). In the context of sexuality, this is demonstrated by the recent emergence of "inappropriate sexual behaviour" (ISB) and the proliferation of professional and empirical literature focused on the management of sexual dysfunction of persons living with dementia and the prevention of sexual harm in long-term residential care homes, also known as nursing homes (Kontos et al. 2016; Grigorovich and Kontos 2018). This leaves unaddressed the ethical imperative of balancing the right of persons living with dementia to protection from harm with their "right to experience a pleasurable sexuality" (HERA, quoted in Dixon-Mueller et al. 2009, p. 12).

Our interest here is to offer a critical analysis of literature on disability, dementia, and sexuality, and to inform the development of a more just approach to support sexual rights in long-term residential care homes that focuses on structural barriers and facilitators. As a prolegomenon, this chapter is necessarily a partial and fragmentary theoretical account that is not intended to address all of the existing literature on sexuality and dementia. Instead, it is an attempt to problematise common trends and conceptual elements across the literature, and to signal critical directions for future inquiry.

## Scholarship on sexual rights

Citizenship and human rights activists and scholars have contributed to the recognition of individual rights to sexual expression and definition. Generally, sexual rights include the right to have control of one's sexuality, to be free from sexual violence and coercion, and to have access to appropriate sexual and reproductive health care. Sexual citizenship emerged as a distinct sub-field of study, with the recognition that conceptualisations of citizenship needed to be broadened to accommodate not only class and race, but also gender and sexuality (Cossman 2007; Richardson 2000). Sexual citizenship has been used to articulate and support rights claims (economic, legal and social) based on sexual conduct, identity and relationships (Richardson 1998; 2000). An alternative stream of sexual rights discourse, based on principles of human rights, emerged around the same time as sexual citizenship (Lottes 2013), focusing on ensuring global access to sexual and reproductive health services (e.g., abortion), protection of individuals from sexual and gender-based violence and challenging criminalisation of homosexuality (Dixon-Mueller et al. 2009; Lottes 2013).

Collectively, this scholarship and activism has been pivotal to more fully recognising and accommodating rights related to sexuality across the globe. However, for the most part, scholarship and activism regarding sexual rights has concentrated primarily on the sexuality and intimacy-related needs of younger people (e.g., ensuring access to reproductive services and access to marriage) without disabilities (Grigorovich and Kontos 2018; Kontos et al. 2016). This is concerning as the sexual rights of people living with disabilities continue to be systematically restricted. Sexual rights violations include lack of access to knowledge about sexuality, restricted access to sexual services and high rates of sexual violence. Moreover, the overriding focus of this theoretical scholarship is on the sexual rights of the "choosing subject," who is capable of self-definition, choice and autonomy, which has contributed to the neglect of the sexual needs and rights of individuals for whom these dimensions are considered to be impaired (Richardson 2015; Shildrick 2007), such as in the case of persons living with dementia.

## Abjection, disability and sexuality

Critical disability studies literature has been central to the recognition of the socio-cultural discourses that are implicated in the symbolic and material sexual exclusion of persons with disabilities. A key theoretical contribution of this literature has been the broadening of the construct of "abjection" in relation to disability and sexuality (Butler 2000; Garland-Thompson 2006; McRuer 2006; Shildrick 2009). Whereas the concept of abjection has been applied differently across philosophical and theoretical literature in different fields (McClintock 2013; Tyler 2013), it is generally used to explore the iterative and performative processes of identification and exclusion through which the borders of the self/other are produced. Central to the concept of abjection is the idea that "other" and "self" are co-constructed so the "abject is everything that the [self] seeks to expunge...it is also a symptom of the failure of this ambition" (McClintock 2013, p. 129). This makes the concept of abjection a productive resource for exploring and challenging how "difference" becomes signified as "deviance" to justify the oppression of persons living with disabilities.

Broadening earlier feminist and queer theorising, critical disability scholars have demonstrated that the boundaries of "appropriate sexuality" and "appropriate embodiment" are simultaneously co-constructed. More specifically, disabled bodies serve a normative function in that their discursive repression and exclusion produces and reinforces the ideal sexual subject and its boundaries. The cultural idealisation of bounded, independent, and economically productive bodies serves as a prescriptive "standard of bodylines" (Shildrick 1999), a baseline against which sexual norms are defined. McRuer (2006) terms this "compulsory able-bodiedness," which he argues is intimately intertwined with compulsory heterosexuality, and, more broadly, heteronormativity (Butler 2000), in that the most successful heterosexual subject is one who is able-bodied. According to Shildrick (2009), the sexualities of persons with disabilities stir cultural anxiety and repulsion as they exacerbate the vulnerability and danger elicited by the intercorporeality of sexual relations. She adds that, given that the sexualities of persons with disabilities rarely follow the heteronormative script (e.g. dominant and penetrative sexuality, reproductive, private), they cannot readily be recuperated through established regulatory practices and norms that seek to mitigate the perceived risk to corporeal integrity that sexuality poses. Consequently, the sexualities of persons with disabilities are "always already deviant" (Kafer 2003, p. 82) and are largely represented as either asexual or as excessively sexual. This discursive exclusion of the sexualities of persons with disabilities, in turn, enables material forms of discrimination and abuse, including sexual violence, professional and familial surveillance of sexualities and barriers to accessing sexual and reproductive services (Kafer 2003; Shildrick 2005b).

Critical disability scholarship has been pivotal to broadening thinking about the sexualities of persons living with disabilities, yet, as Kafer (2003, 2013) has argued, this field continues to focus on "able-bodiedness" as the central problematic, implicitly neglecting how "able-mindedness" operates to marginalise persons with cognitive or intellectual disabilities. For example, scholarship on sexuality and disability often centres on the "rights" of persons with disabilities to express their sexualities without interference, without challenging the foundational reliance of liberal rights-based discourse on "compulsory able-mindedness" (Carlson 2003; Kittay and Carlson 2010). The lack of attention to "compulsory able-mindedness" in the context of sexuality is problematic as the sexualities of persons with disabilities are often restricted, based on the assumption that their dependence on others denotes a lack of cognitive capacity with respect to sexuality. While the assumption that sexuality requires a particular "threshold" of cognitive capacity is problematic for all persons with disabilities, it is especially so for persons living with dementia.

As critical dementia studies scholarship makes clear, in the context of dementia, abjection is also sustained by the "ontological pre-eminence" (Ortega 2009) of the "cerebralisation" of selfhood that has its philosophical roots in the rise of the "modern self" in the 17th century and the socially constructed brain-self consubstantiality (Katz 2012; Kontos 2014). As Vidal has argued, this ideology of the self—the "the individualism characteristic of western and westernised societies, the supreme value given to the individual as autonomous agent of choice and initiative, and the corresponding emphasis on interiority" (2009, p. 7)—treats the brain as the organ responsible for the functions with which the self is identified. The naturalisation of brain-self coupling is also foundational to the biomedical discourse on dementia (Innes and Manthorpe 2012; Volicer and Hurley 2003), which largely constructs persons living with dementia as "non-persons" or as burdensome bodies to be managed by individuals and society. The actions, needs and movements of persons living with dementia are thus often dismissed as aimless and are perceived as disease-driven "symptoms" or "responsive behaviours," rather than meaningful attempts to communicate or to engage with others (Dupuis et al. 2012; Grigorovich et al. 2019; Kontos et al. 2018).

## Sexuality and dementia

The biomedical discourse and the naturalisation of the brain–self coupling are implicated in how the sexualities of persons living with dementia are conceptualised in professional and empirical literature on dementia. In particular, through the recent establishment of the concept of "inappropriate sexual behaviour" (ISB), the sexualities of persons living with dementia have become constructed as mental pathologies that require medical assessment and intervention (Guay 2008; Stubbs 2011; Mendez and Shapira 2013; Cipriani et al. 2016). The biomedicalisation of sexualities of persons living with dementia is part of a broader history of medicalisation of sexuality that has legitimised the bioauthority of psychiatry over the treatment of mental illness through diagnostic categories that establish "sex as a cause, symptom and form of madness, particularly when sexualities [contest] heteronormative and patriarchal discourses" (Tosh 2011, p. 3). Scholarship on ISB is far from uniform; there is a lack of consensus on both terminology and classification criteria, and there is no universal agreement as to why the sexualities of persons living with dementia are "inappropriate" and thus in need of professional management (Gibson et al. 1999; Tsatali et al. 2011). Definitions of "inappropriateness" across the empirical and theoretical literature are at once expansive and circular, and examples of ISB have included everything from handholding to physical assault, masturbation, the use of pornography, and a "change" in sexual preference (Black et al. 2005; Guay 2008; Cipriani et al. 2016). For example, one review of literature on ISB (Mahieu et al. 2011) described it as *uninhibited* sexual behaviour or hypersexuality [that is] … *beyond the person's control* … and mostly arises *due to dementia*. It may include touching intimate body parts of care providers and bystanders, kissing and hugging that exceeds plain affection, disrobing oneself and others, using sexually suggestive language, attempting intercourse and compulsive masturbation in both public and personal areas" (Mahieu et al. 2011, p. 1141; added emphasis). As this description demonstrates, the sexualities of persons living with dementia are constructed as "inappropriate" largely because they are deemed to be "unagentic" symptoms of dementia-related neurodegeneration. Other descriptions of ISB similarly emphasise the lack of sexual agency of persons living with dementia by describing their sexualities as "repetitive" and/or "inconsistent" with their sexual history pre-diagnosis of dementia (Cipriani et al. 2016; Kaplan and Krueger 2010; Torrisi et al. 2016).

The boundaries of "appropriate sexuality" are largely left undefined within literature on ISB, suggesting that it is not particular sexual expressions per se that are inappropriate, but rather it is the subjective perceptions of others that these are "unacceptable" within the context in which they are expressed (De Giorgi and Series 2016). While this suggests that sexual expression may be deemed acceptable in some social contexts, regardless of context and because of cognitive impairment, in practice the sexual expressions of persons living with dementia are never deemed to be acceptable. This unacceptability is most prominent in long-term care homes where every aspect of life is carried out in the immediate view of, and under the control and scrutiny of, care providers. Given that care providers have been shown to have negative attitudes towards later-life sexualities and to perceive support of sexuality as being less important than that of other aspects of care (Gott *et al.* 2004; Yelland and Hosier 2017), the indeterminacy of ISB creates the possibility that care providers' personal moral values and preferences prompt them to unfairly restrict the sexualities of persons living with dementia. Although literature on ISB acknowledges that the sexualities of persons living with dementia may also cause distress or some form of "harm" to the person living with dementia (Mendez and Shapira 2013), the overwhelming emphasis is on the distress or disruption that these sexualities pose for informal caregivers and care providers (Black *et al.* 2005; De Giorgi and Series 2016; Mendez and Shapira 2013). The sexualities of persons living with dementia are thus restricted primarily because they "interfere" with the work and life of others (Black *et al.* 2005; Mendez and Shapira 2013). Indeed, professional literature on ISB explicitly recommends treating (e.g. suppressing) sexual desires of persons living with dementia for the purpose of decreasing caregiver "burden" and/ or delaying or preventing transition into institutional care (Cipriani et al. 2016; Tucker 2010).

The influence of the biomedical discourse and the "cerebralisation" of selfhood is further evidenced by the predominance of the four moral principles approach of bioethics as the framework for decision making with respect to sexuality in professional literature and in guidelines for long-term residential care (Grigorovich and Kontos 2018). This approach supports intervening in the sexual freedom of persons living with dementia when their sexual expressions are judged as posing a risk of harm to the self or to others. While in principle, this approach aims to achieve a balance between respecting individuals' right to autonomy with their right to protection from harm (Kuczewski 1999), in practice, such an approach leans towards paternalism and risk-averse decision-making. In part, this is because this approach privileges cognition and the "qualities of mind [such] as rationality, self-sovereignty and impartiality" (Shildrick 2005a) as the central basis for decision making. This, in turn, leads to a narrow and overdetermined focus on "informed consent" as the surrogate measure of what is ethically acceptable. In the context of sexuality, this is demonstrated by the preoccupation in the literature with formally establishing cognitive capacity for sexual decision-making as the criterion for interfering in the sexual freedom of persons living with dementia. Although there is no consensus on how to assess or determine sexual capacity, generally, it is required that persons living with dementia verbally articulate their expectations regarding desired sexual intimacy, as well as demonstrate awareness of potential sexual "risks" and the ability to avoid exploitation, before they are permitted to engage in sex (Wilkins 2015). The adoption of an informed consent model for sex based on this ethical approach implicitly denies persons living with dementia the "dignity of risk" and holds them to a higher standard of sexual decision-making than persons without cognitive disabilities, who are generally allowed to make sexual decisions without first "weighing the pros and cons or the biological implications of their decisions" and without interference (Lindsay 2010, p. 314). Furthermore, given that persons living with dementia experience progressive deterioration in cognitive reasoning capacity, such an approach may give rise to an absolutisation of rationality and cognition, and set the bar for undue interference significantly, necessitating erring on the side of the duty to intervene, rather than on allowing freedom of self-expression (Everett 2007; Holm 2001).

Empirical literature on sexuality and dementia demonstrates an emphasis on risk management and neglect for the support of sexual expression of persons living with dementia (Miles and Parker 1999; Parker 2007). For example, research on providers' reactions and perceptions has found that they tend to respond to the sexualities of persons living with dementia with expressions of disgust, denial and discouragement (Roach 2004; Ward et al. 2005; Cornelison and Doll

2013; Syme 2014). Care providers have also been found to hold negative or patronising attitudes toward the sexualities of persons living with dementia, and the sexualities of men, in particular, are perceived as threatening to both providers and other residents (Parker 2007; Roach 2004; Ward et al. 2005;Yelland and Hosier 2017).When confronted with overt sexual desires or activity, many providers report feeling shame and discomfort, as well as the need to intervene in the sexual expression (Roach 2004; Shuttleworth et al. 2010;Villar et al. 2015).This is most apparent in the context of long-term care homes where the sexuality of persons living with dementia becomes a matter of public concern due to the congregate nature of the living environment (Benbow and Beeston 2012;Taylor and Gosney 2011).

Few long-term care homes have policies that specifically address sexual expression and which protect residents' rights to sexual privacy ( Cornelison and Doll 2013; Hill 2014; Shuttleworth et al. 2010). Information about residents' sexuality is routinely shared with their spouses and children, without staff considering this to be a violation of residents' right to sexual privacy (Bauer 1999). Homes rarely have policies that explicitly permit residents to display and use sexually themed materials in their private rooms, or offer private spaces for sexual activity (Hill 2014; Shuttleworth et al. 2010). In the absence of clear policy directives, providers tend to rely on their personal values and judgement, as well as the expressed wishes of residents' families, which may conflict with the wishes of residents and reflect ageist and heteronormative sexual norms (Archibald 1998; Bauer et al. 2014; Simpson et al. 2015;Ward et al. 2005).This is not an unfounded concern given that a recent case report on the pharmacological restriction of same-sex sexual expression noted that the same-sex sexual expression was judged to be inappropriate, not only because it was directed at female care providers, but because the family believed the sexual expression was inconsistent with their mother's heterosexuality (Sarikaya and Sarikaya 2018).As a consequence, the sexual expressions of persons living with dementia are at best discouraged, and at worst actively restricted through a combination of pharmacological (e.g., anti-depressants, hormones) and non-pharmacological interventions (e.g., isolation, environmental restraint, behavioural modification).

## Relational and embodied ethic of sexuality

To redress the lack of support for the sexual expressions of persons living with dementia requires that we adopt an ethical framework for sexuality that does not rest on cognitive capacity as the criterion for the moral permissibility of sexual expression.To confine sexuality to the order of cognitive understanding reflects a hypercognitivism (Post 1995) that holds persons living with dementia to a higher standard of decision-making regarding sexuality than is typically demanded of persons living without cognitive disabilities. Some scholars have argued that this is particularly problematic as decision-making regarding sexual expression could, in most cases, be considered to "lie closer to the decision about ice cream [preferences] than to decisions about major surgery" (Richardson and Lazur 1995, p. 123). Moreover, sexual decision-making is relational and dynamic in nature and can change with the type of relationship and the context in which it is located.Thus, although we do not deny that persons living with dementia may be vulnerable to sexual abuse, and thus require protection, their cognitive disability should not be used to deny them the opportunity to have a sexual life, should that be their desire.

Elsewhere, we have elaborated a new ethic of sexuality that is grounded in the relational model of citizenship (Grigorovich and Kontos 2018; Kontos et al. 2016) and that can inform the development of such an approach. The central contribution of this model to sexuality in the context of dementia is that, contrary to the construction of persons living with dementia as "unagentic," it recognises the agential status of embodied self-expression as well as its inherent relationality, and it upholds sexuality as fundamental to self-expression and human flourishing. More specifically, this model is premised on two tenets—"embodied selfhood" and "relationality." Embodied selfhood takes its theoretical bearings from Merleau-Ponty's (1962) understanding of non-representational intentionality and the primordial body, and Bourdieu's (1990) sociological theory of the logic of practice, focusing on his concept of habitus, which links bodily dispositions to structures of the social world. With the novel integration of these ideas, Kontos has argued that pre-reflective intentionality is a fundamental source of selfhood (Kontos 2006; 2012a; 2012b; Kontos and Martin 2013), a source defined as "the existential capacity of the body to engage with the world, a capacity that is manifest in the body's natural (pre-social) relationship with the world by way of the body's power of natural expression, as well as in the ongoing socio-cultural relationships between the pre-reflective body and the world (history, culture, power and discourse)" (Grigorovich et al. 2019, p. 177). Further, embodied selfhood is itself relational, in that the body is always intertwined with a shared world.

These tenets of the relational model of citizenship foreground the agentic capacity of the body to perceive and experience the world (Kontos 2006; 2012a; 2012b; Kontos and Martin 2013), and thus importantly challenge assumptions of loss of agency with dementia. Given that embodied self-expression is the primary means of engagement for persons living with dementia (Kontos 2012b; Kontos and Grigorovich 2018a; 2018b), this model brings a new and critical dimension to the discourse on sexual rights, which has the potential to more fully support the realisation of sexual rights of persons living with dementia through institutional policies, structures and practices. In particular, grounding the ethic of sexuality in this model of citizenship underscores that the support of sexual rights is a matter of social justice and that states have responsibility to support these rights through regulation and redistribution of social and economic resources.

## Discussion

To restrict the support of sexual rights for persons living with dementia to the prevention of harm is to disregard the importance of sexuality for self-expression, self-empowerment and human flourishing. Sexuality is critical for affection and intimacy (Bauer *et al.* 2013; Gott and Hinchliff 2003; Ní Lochlainn and Kenny 2013), and sexual expression has beneficial effects for well-being; in the context of older people, sexual expression has been found to be associated with decreased pain sensitivity, increased relaxation and lower levels of depression (Ní Lochlainn and Kenny 2013; Syme 2014). A more just approach to the recognition of and support for the sexual rights of persons living with dementia thus requires that we focus not only on protection, but also on the enablement of access to sexual pleasure, and the formation of intimate and sexual relationships.

To support the sexual rights of persons living with dementia demands reform of the structural conditions implicated in both access to sexual pleasure and protection from sexual harm. For example, residents' lack of access to the outdoors, as well as their lack of autonomy and privacy, have all been implicated in aggression (Duxbury and Whittington 2005; Tufford 2017). There may also be a link to poor work organisation and understaffing (Banerjee et al. 2012; Lachs et al. 2016). Finally, sexual aggression may also be associated in these settings with existing prohibitions on autonomous sexual expression and the use of sexual materials or services (Miles and Parker 1999; Kontos et al. 2016; Grigorovich and Kontos 2018). Addressing these factors through structural and organisational changes would improve efforts to prevent harm against individuals living with dementia without excessively restricting individual liberty.

These reforms further require that socio-cultural, political and organisational barriers to sexual expression be addressed for persons living with dementia (Appel 2010; Shildrick 2007). At the socio-cultural level, this would require the development of public health, professional training, and policy initiatives to reduce the stigma associated with later-life sexuality and dementia, and to redress the discriminatory and marginalising attitudes and practices of care providers and others. Enhancing the ability of providers to determine mutual assent (e.g., interest, voluntariness and enjoyment) in the context of sexual expression, and to recognise the importance of such expressions for well-being and quality of life, will be crucial to enhancing their capacity to recognise and support the sexual rights of persons living with dementia. Residential long-term care homes will also need to develop organisational policies and practices that explicitly acknowledge respect for sexual rights by developing sexual expression policies and correcting oppressive practices. Examples could include the removal of implicit and explicit prohibitions against the use of sexual materials or engagement in sexual activity, sensitising care providers so that they respect the sexual privacy of residents (e.g., knocking before entering residents' private rooms) and the provision of private spaces in homes for sexual expression for residents who are living in shared accommodation. The development of such policies should be based on consultations with care providers, families and people living with dementia.

Finally, to support rights to sexuality for persons living with dementia will additionally require facilitation to ensure that they not only have access to opportunities for sexual expression, but also the support needed to realise their rights. Here we concur with Earle (2001) that while facilitation encompasses the active involvement of care providers, the scope of their role should be thought of as existing on a continuum from provision of information and advice to more hands-on support. Furthermore, facilitation should encompass support of opportunities for autonomous (e.g., masturbation) and shared (e.g., with another person) sexual expressions. Notable examples of facilitation of sexual rights in long-term care homes to date have included: a UK home that hosts "bar nights," involving the creation of a bar in the main area of the home where residents may socialise with each other and purchase drinks (Archibald 2002); the development and hosting of diverse social events and romantic outings for residents (Kamel and Hajjar 2004); and enabling residents to engage providers with particular expertise in sexuality and disability to provide the residents with sexual education and counselling, or sexual surrogacy (Kulick and Rydstrom 2015). There have also been examples of homes where providers facilitated residents' sexual expression by assisting residents to purchase commercial sexual services (Gardner 2013; Obermueller 2011).

## Conclusion

What we have offered here is a prolegomenon to future work in dementia studies and critical disability studies. The approach that we suggest, based on our embodied and relational ethic of sexuality, sets a new transdisciplinary standard regarding the recognition and support of the sexual rights of persons living with dementia. To achieve the complex balance between supporting access to sexual pleasure and achieving protection from harm will require further research and engagement with relevant stakeholders, including residents, families, and care providers. We hope that the imperative of this will be embraced by others who are equally committed to ensuring that persons living with dementia have equal opportunities to participate in all aspects of life to the fullest extent possible, including the pursuit of intimate and sexual relationships.

## References

Appel, J.M. (2010) Sex rights for the disabled? *Journal of Medical Ethics* 36(3), pp. 152–154.
Archibald, C. (1998) Sexuality, dementia and residential care: Managers' report and response. *Health & Social Care in the Community* 6(2), pp. 95–101.

Archibald, C. (2002) Sexuality and dementia in residential car: Whose responsibility? *Sexual and Relationship Therapy* 17(3), pp. 301–309.

Banerjee, A., Daly, T., Armstrong, P., Szebehely, M., Armstrong, H., & Lafrance, S. (2012) Structural violence in long-term, residential care for older people: Comparing Canada and Scandinavia. *Social Science & Medicine* 74(3), pp. 390–398.

Bartlett, R., & O'Connor, D. (2007) From personhood to citizenship: Broadening the lens for dementia practice and research. *Journal of Aging Studies* 21(2), pp. 107–118.

Bauer, M. (1999) Their only privacy is between their sheets: Privacy and the sexuality of elderly nursing home residents. *Journal of Gerontological Nursing* 25(8), pp. 37–41.

Bauer, M., Fetherstonhaugh, D., Tarzia, L., Nay, R., Wellman, D., & Beattie, E. (2013) "I always look under the bed for a man". Needs and barriers to the expression of sexuality in residential aged care: The views of residents with and without dementia. *Psychology & Sexuality* 4(3), pp. 296–309.

Bauer, M., Nay, R., Tarzia, L., Fetherstonhaugh, D., Wellman, D., & Beattie, E. (2014) "We need to know whats going on": Views of family members toward the sexual expression of people with dementia in residential aged care. *Dementia: The International Journal of Social Research and Practice* 13(5), pp. 571–585.

Behuniak, S.M. (2011) The living dead? The construction of people with Alzheimer's disease as zombies. *Ageing & Society* 31(1), pp. 70–92.

Benbow, S.M., & Beeston, D. (2012) Sexuality, aging, and dementia. *International Psychogeriatrics* 24(7), pp. 1026–1033.

Black, B., Muralee, S., & Tampi, R.R. (2005) Inappropriate sexual behaviors in dementia. *Journal of Geriatric Psychiatry and Neurology* 18(3), pp. 155–162.

Bourdieu, P. (1990) *The logic of practice.* Cambridge, UK: Polity Press.

Butler, J. (2000) *Undoing gender.* New York: Routledge.

Carlson, L. (2003) Rethinking normalcy, normalization, and cognitive disability. In R. Figueroa & S.E. Harding (Eds.), *Science and other cultures: Issues in philosophies of science and technology.* New York: Routledge (Taylor and Francis), pp. 154–171.

Cipriani, G., Ulivi, M., Danti, S., Lucetti, C., & Nuti, A. (2016) Sexual disinhibition and dementia. *Psychogeriatrics* 16(2), pp. 145–153.

Cornelison, L.J., & Doll, G.M. (2013) Management of sexual expression in long-term care: Ombudsmens perspectives. *The Gerontologist* 53, pp. 780–789.

Cossman, B. (2007) *Sexual citizens: The legal and cultural regulation of sex and belonging.* Stanford, CA: Stanford University Press.

De Giorgi, R., & Series, H. (2016) Treatment of inappropriate sexual behavior in dementia. *Current Treatment Options in Neurology* 18(9), p. 41.

Dixon-Mueller, R., Germain, A., Fredrick, B., & Bourne, K. (2009) Towards a sexual ethics of rights and responsibilities. *Reproductive Health Matters* 17(33), pp. 111–119.

Dupuis, S., Wiersma, E., & Loiselle, L. (2012) Pathologizing behavior: Meanings of behaviors in dementia care *Journal of Aging Studies* 26(2), pp. 162–173.

Duxbury, J., & Whittington, R. (2005) Causes and management of patient aggression and violence: Staff and patient perspectives. *Journal of Advanced Nursing* 50(5), pp. 469–478.

Earle, S. (2001) Disability, facilitated sex and the role of the nurse. *Journal of Advanced Nursing* 36(3), pp. 433–440.

Everett, B. (2007) Ethically managing sexual activity in long-term care. *Sexuality and Disability* 25(1), pp. 21–27.

Gardner, B. (2013) Care home call girls: Staff invite prostitutes for residents. *The Argus.*

Garland-Thompson, R. (2006) Integrating disability, transforming feminist theory. In L.J. Davis (Ed.), *The disability studies reader.* New York: Routledge, pp. 257–274.

Gatens, M. (1996) *Imaginary bodies: Ethics, power, and corporeality.* London, UK: Routledge.

Gibson, M.C., Bol, N., Woodbury, M.G., Beaton, C., & Janke, C. (1999) Comparison of caregivers, residents, and community-dwelling spouses opinions about expressing sexuality in an institutional setting. *Journal of Gerontological Nursing* 25(4), pp. 30–39.

Gill, M. (2015) *Already doing it: Intellectual disability and sexual agency.* Minneapolis: University of Minnesota Press.

Gott, M., & Hinchliff, S. (2003) How important is sex in later life? The views of older people. *Social Science & Medicine* 56(8), pp. 1617–1628.

Gott, M., Hinchliff, S., & Galena, E. (2004) General practitioner attitudes to discussing sexual health issues with older people. *Social Science & Medicine* 58(11), pp. 2093–2103.

Grigorovich, A., & Kontos, P. (2018) Advancing an ethic of embodied relational sexuality to guide decision-making in dementia care. *The Gerontologist* 58(2), pp. 219–225.

Grigorovich, A., Kontos, P., & Kontos, A.P. (2019) The "violent resident": A critical exploration of the ethics of resident-to-resident aggression. *Journal of Bioethical Inquiry* 16(2), pp. 173–183.

Guay, D.R.P. (2008) Inappropriate sexual behaviors in cognitively impaired older individuals. *The American Journal of Geriatric Pharmacotherapy* 6(5), pp. 269–288.

Gullette, M.M. (2004) *Aged by culture*. Chicago: University of Chicago Press.

Herskovitz, E. (1995) Struggling over subjectivity: Debates about the self and Alzheimer's disease. *Medical Anthropology Quarterly* 9(2), pp. 146–164.

Hill, E. (2014) Well always have shady pines: Surrogate decision-making tools for preserving sexual autonomy in elderly nursing home residents. *William & Mary Journal of Women & the Law* 20(2), pp. 469–490.

Holm, S. (2001) Autonomy, authenticity, or best interest: Everyday decision-making and persons with dementia. *Medicine, Health Care, and Philosophy* 4(2), pp. 153–159.

Innes, A., & Manthorpe, J. (2012) Developing theoretical understandings of dementia and their application to dementia care policy in the UK. *Dementia: The International Journal of Social Research and Practice* 12(6), pp. 682–696.

Kafer, A. (2003) Compulsory bodies: Reflections on heterosexuality and able-bodiedness. *Journal of Womens History* 15(3), pp. 77–89.

Kafer, A. (2013) *Feminist, queer, crip*. Indiana: Indiana University Press.

Kamel, H.K., & Hajjar, R.R. (2004) Sexuality in the nursing home, part 2: Managing abnormal behavior-legal and ethical issues. *Journal of the American Medical Directors Association* 5(2), pp. S49–S52.

Kaplan, M.S., & Krueger, R.B. (2010) Diagnosis, assessment, and treatment of hypersexuality. *Journal of Sex Research* 47(2–3), pp. 181–198.

Katz, S. (2012) Embodied memory: Ageing, neuroculture, and the genealogy of mind. *Occasion: Interdisciplinary Studies in the Humanities*. Available from http://occasion.stanford.edu/node/97

Katz, S. (2013) Dementia, personhood and embodiment: What can we learn from the medieval history of memory? *Dementia: The International Journal of Social Research and Practice* 12(3), pp. 303–314.

Kittay, E., & Carlson, L. (2010) *Cognitive disability and its challenge to moral philosophy*. UK: John Wiley & Sons.

Kontos, P. (2006) Embodied selfhood: An ethnographic exploration of Alzheimer's disease. In L. Cohen & A. Leibing (Eds.), *Thinking about dementia: Culture, loss, and the anthropology of senility*. New Brunswick, NJ: Rutgers University Press, pp. 195–217.

Kontos, P. (2012a) Alzheimer expressions or expressions despite Alzheimer's? Philosophical reflections on selfhood and embodiment. *Occasion: Interdisciplinary Studies in the Humanities*, pp. 1–12. Available from http://arcade.stanford.edu/sites/default/files/article_pdfs/OCCASION_v04_Kontos_05311 2_0.pdf

Kontos, P. (2012b) Rethinking sociability in long-term care: An embodied dimension of selfhood. *Dementia: The International Journal of Social Research and Practice* 11(3), pp. 329–346.

Kontos, P. (2014) Musical embodiment, selfhood, and dementia. In L.C. Hydén, J. Brockmeier, & H. Lindemann (Eds.), *Beyond loss*. New York: Oxford University Press, pp. 107–119.

Kontos, P., & Grigorovich, A. (2018a) Dancing with dementia: Citizenship, embodiment and everyday life in the context of long-term care. In S. Katz (Ed.), *Ageing in everyday life: Materialities and embodiments*. Bristol, UK: Bristol University Press, pp. 163–180.

Kontos, P., & Grigorovich, A. (2018b) Rethinking musicality in dementia as embodied and relational. *Journal of Aging Studies* (Special Issue: Aging, body and society. Key themes, critical perspectives, Eds., Martin, W., & Twigg, J.) 45, pp. 39–48.

Kontos, P., Grigorovich, A., Dupuis, S., Jonas-Simpson, C., Mitchell, G.J., & Gray, J. (2020) Raising the curtain on stigma associated with dementia: Fostering a new cultural imaginary for a more inclusive society. *Critical Public Health* 30(1), pp. 91–102. DOI: 10.1080/09581596.2018.1508822.

Kontos, P., Grigorovich, A., Kontos, A.P., & Miller, K.L. (2016) Citizenship, human rights, and dementia: Towards a new embodied relational ethic of sexuality. *Dementia: The International Journal of Social Research and Practice* 15(3), pp. 315–329.

Kontos, P., & Martin, W. (2013) Embodiment and dementia: Exploring critical narratives of selfhood, surveillance, and dementia care. *Dementia: The International Journal of Social Research and Practice* 12(3), pp. 288–302.

Kuczewski, M.G. (1999) Ethics in long-term care: Are the principles different? *Theoretical Medicine and Bioethics* 20(1), pp. 15–29.

Kulick, D., & Rydstrom, J. (2015) *Lonliness and its opposite: Sex, disability, and the ethics of engagement*. Durham, NC: Duke University Press.

Lachs, M.S., Teresi, J.A., Ramirez, M., Van Haitsma, K., Silver, S., Eimicke, J.P., Boratgis, G., Sukha, G., Kong, J., Besas, A.M., Luna, M.R., & Pillemer, K.A. (2016) The prevalence of resident-to-resident elder mistreatment in nursing homes. *Annals of Internal Medicine* 165(4), pp. 229–236.

Liddiard, K. (2017) *The intimate lives of disabled people*. London: Routledge.

Lindsay, J.R. (2010) The need for more specific legislation in sexual consent capacity assessments for nursing home residents: How grandpa got his groove back. *The Journal of Legal Medicine* 31(3), pp. 303–323.

Lottes, I.L. (2013) Sexual rights: Meanings, controversies, and sexual health promotion. *Journal of Sex Research* 50(3–4), pp. 367–391.

Mahieu, L., Van Elssen, K., & Gastmans, C. (2011) Nurses perceptions of sexuality in institutionalized elderly: A literature review. *International Journal of Nursing Studies* 48(9), pp. 1140–1154.

Mcclintock, A. (2013) *Imperial leather: Race, gender, and sexuality in the colonial contest*. New York: Routledge.

McRuer, R. (2006) *Crip theory: Cultural signs of queerness and disability*. New York: NYU Press.

McRuer, R., & Mollow, A. (2012) *Sex and disability*. Durham & London: Duke University Press.

Mendez, M.F., & Shapira, J.S. (2013) Hypersexual behavior in frontotemporal dementia: A comparison with early-onset Alzheimer's disease. *Archives of Sexual Behavior* 42(3), pp. 501–509.

Merleau-Ponty, M. (1962) *Phenomenology of perception*. London, UK: Routledge & K. Paul.

Miles, S.H., & Parker, K. (1999) Sexuality in the nursing home: Iatrogenic loneliness *Generations* 23(1), p. 36.

Mitchell, G., Dupuis, S., & Kontos, P. (2013) Dementia discourse: From imposed suffering to knowing other-wise. *Journal of Applied Hermeneutics*, pp. 1–19. Available from http://jah.journalhosting.ucalgary. ca/jah/index.php/jah/article/viewFile/41/pdf

Ni Lochlainn, M., & Kenny, R.A. (2013) Sexual activity and aging. *Journal of the American Medical Directors Association* 14(8), pp. 565–572.

Obermueller, N. (2011) *Assisted loving: Prostitutes and the elderly*. Available from http://www.exberliner.com /features/lifestyle/assisted-loving/

Ortega, F. (2009) The cerebral subject and the challenge of neurodiversity. *BioSocieties* 4, pp. 425–445.

Parker, S. (2007) What barriers to sexual expression are experienced by older people in 24-hour care facilities? *Reviews in Clinical Gerontology* 16(4), pp. 275–279.

Post, S. (1995) *The moral challenge of Alzheimer disease*. Baltimore, Maryland: Johns Hopkins University Press.

Richardson, D. (1998) Sexuality and citizenship. *Sociology* 32(1), pp. 83–100.

Richardson, D. (2000) Constructing sexual citizenship: Theorizing sexual rights. *Critical Social Policy* 20(1), pp. 105–135.

Richardson, D. (2015) Rethinking sexual citizenship. *Sociology* 51(2), pp. 208–224.

Richardson, J.P., & Lazur, A. (1995) Sexuality in the nursing home patient. *American Family Physician* 51(1), pp. 121–124.

Roach, S.M. (2004) Sexual behaviour of nursing home residents: Staff perceptions and responses. *Journal of Advanced Nursing* 48(4), pp. 371–379.

Sarikaya, S., & Sarikaya, B. (2018) Aripiprazole for the treatment of inappropriate sexual behavior: Case report of an Alzheimer's disease patient known as heterosexual with recently shifted sexual orientation to same gender. *Journal of Alzheimer's Disease Reports* 2(1), pp. 117–121.

Shakespeare, T. (2000) Disabled sexuality: Towards rights and recognition. *Sexuality and Disability* 18(3), pp. 159–166.

Shakespeare, T., Zeilig, H., & Mittler, P. (2019) Rights in mind: Thinking differently about dementia and disability. *Dementia: The International Journal of Social Research and Practice* 18(3), pp. 1075–1088.

Shildrick, M. (1999) This body which is not one: Dealing with differences. *Body & Society* 5(2–3), pp. 77–92.

Shildrick, M. (2005a) Beyond the body of bioethics: Challenging the conventions. In M. Shildrick & R. Mykitiuk (Eds.), *Ethics of the body: Postconventional challenges*. Cambridge, MA: MIT Press.

Shildrick, M. (2005b) Unreformed bodies: Normative anxiety and the denial of pleasure. *Womens Studies* 34(3–4), pp. 327–344.

Shildrick, M. (2007) Contested pleasures: The sociopolitical economy of disability and sexuality. *Sexuality Research & Social Policy* 4(1), pp. 53–66.

Shildrick, M. (2009) *Dangerous discourses of disability, subjectivity and sexuality*. New York: Springer.

Shuttleworth, R., Russell, C., Weerakoon, P., & Dune, T. (2010) Sexuality in residential aged care: A survey of perceptions and policies in Australian nursing homes. *Sexuality and Disability* 28(3), pp. 187–194.

Simpson, P., Horne, M., Brown, L., Wilson, C.B., Dickinson, T., & Torkington, K. (2015) Old (er) care home residents and sexual/intimate citizenship. *Ageing and Society* 37(2), pp. 1–23.

Stubbs, B. (2011) Displays of inappropriate sexual behaviour by patients with progressive cognitive impairment: The forgotten form of challenging behaviour? *Journal of Psychiatric and Mental Health Nursing* 18(7), pp. 602–607.

Syme, M.L. (2014) The evolving concept of older adult sexual behavior and its benefits. *Generations* 38(1), pp. 35–41.

Taylor, A., & Gosney, M.A. (2011) Sexuality in older age: Essential considerations for healthcare professionals. *Age and Ageing* 40(5), pp. 538–543.

Torrisi, M., Cacciola, A., Marra, A., De Luca, R., Bramanti, P., & Calabro, R.S. (2016) Inappropriate behaviors and hypersexuality in individuals with dementia: An overview of a neglected issue. *Geriatrics & Gerontology International* 17(6), pp. 865–874.

Tosh, J. (2011) The medicalisation of rape: A discursive analysis of "paraphilic coercive disorder" and the psychiatrisation of sexuality. *Psychology of Women Section Review* 13(2), pp. 2–12.

Tsatali, M., Tsolaki, M., Christodoulou, T., & Papaliagkas, V. (2011) The complex nature of inappropriate sexual behaviors in patients with dementia: Can we put it into a frame? *Sexuality and Disability* 29(2), pp. 143–156.

Tucker, I. (2010) Management of inappropriate sexual behaviors in dementia: A literature review. *International Psychogeriatrics* 22(5), pp. 683–692.

Tufford, F., Lowndes, R., Struthers, J., & Chivers, S. (2017) "Call security": Locks, risk, privacy and autonomy in long-term residential care. *Ageing International* 43(1), pp. 34–52.

Tyler, I. (2013) *Revolting subjects: Social abjection and resistance in neoliberal Britain*. London: Zed Books Ltd.

Vidal, F. (2009) Brainhood, anthropological figure of modernity. *History of the Human Sciences* 22, pp. 5–36.

Villar, F., Fabà, J., Serrat, R., & Celdrán, M. (2015) What happens in their bedrooms stays in their bedrooms: Staff and residents reactions toward male–female sexual intercourse in residential aged care facilities. *The Journal of Sex Research* 52(9), pp. 1054–1063.

Volicer, L., & Hurley, A.C. (2003) Management of behavioral symptoms in progressive degenerative dementias. *The Journals of Gerontology* 58(9), pp. 837–845.

Ward, R., Vass, A.A. Aggarwal, N., Garfield, C., & Cybyk, B. (2005) A kiss is still a kiss? *Dementia: The International Journal of Social Research and Practice* 4(1), pp. 49–72.

Wilkins, J.M. (2015) More than capacity: Alternatives for sexual decision making for individuals with dementia. *The Gerontologist* 55, pp. 716–723.

Yelland, E., & Hosier, A. (2017) Public attitudes toward sexual expression in long-term care: Does context matter? *Journal of Applied Gerontology* 36(8), pp. 1016–1031.

# 5

# COMBATING OLD IDEAS AND BUILDING IDENTITY

## Sexual identity development in people with disabilities

*Emily M. Lund, Anjali J. Forber-Pratt and Erin E. Andrews*

Sexuality, as a component of disability identity, has a complex but often ignored history. People with disabilities have had their sexuality pathologised or disregarded by family members, providers and the general public. People with disabilities, especially those with physical or intellectual disabilities,[1] are often desexualised by society. Keller and Galgay (2010) found that both desexualisation and a secondary pattern of hypersexualisation are a prominent microaggressive theme against disabled persons. Frequently, disabled people are erroneously assumed to be universally asexual (Azzopardi-Lane and Callus 2015; Milligan and Neufeldt 2001; Smart 2009)—that is, lacking in sexual desire, the physical ability to engage in sexual expression and/or intercourse or being so universally sexually undesirable that they would be unable to ever find a partner if they were indeed willing and able to engage in sexual expression or behaviour. People with intellectual disability, and some people with physical disability, may be cast as "eternal children" (Priestly 2000; Starke et al. 2016). Such ableist infantilisation results in these individuals being regarded as a people who are indeed never able to develop a sexual identity, much less a healthy or fulfilling one. These beliefs around asexuality and disability are problematic in that they influence not only the ways in which people with disabilities view themselves as sexual people but also those in which people with disabilities integrate disability and sexual identities.

Conversely, sexual expression by disabled individuals can be seen as inherently pathological or dangerous, and thus must be oppressed (Gomez 2012; Milligan and Neufeldt 2001). In this conceptualisation, any sexual behaviour by people with certain types of disabilities, especially intellectual disability and psychiatric disabilities, may be seen as both excessive and inherently deviant or threatening (Milligan and Neufeldt 2001). Such expressions of sexuality from disabled persons may elicit hostility, discomfort or even disgust (East and Orchard 2014). For example, direct support providers or parents operating under these assumptions may attempt to ban all expressions of sexual behaviour by someone with a disability. This is harmful when the message to people with disabilities becomes one where sexual behaviour for pleasure and sexual expression is something to be ashamed of. In fact, sexual expression and sexual behaviour for pleasure should be considered to be a human right and a part of basic human development—safe and mutually consensual sexual expression and behaviour should be supported, regardless of disability status. Although researchers have found that society is becoming more open to the idea that sexual behaviour by people with disabilities is not inherently deviant or wrong (Gomez 2012), these attitudes may be limited to certain types of sexual behaviour (e.g., approving of private masturbation but not partnered sexual activities; approving of heterosexual but not homosexual sexual behaviour).

This notion is further complicated when assumptions are made about individuals' sexual behaviour aimed at reproduction. The idea that sexual behaviour in people with disabilities should be suppressed has its roots largely in eugenicist policies and ideals, and disabled people have been historically subject to involuntary sterilisation (Fleischer and Zames 2001). Furthermore, this view is still present in many medical providers' default attitudes towards reproductive health among people with disabilities (Powell et al. 2016). Of course, these eugenicist approaches towards disability and sexuality have their roots in the desire to eliminate the existence of disabled people through eliminating the potential inheritance of any hereditary disabilities (Lombardo 2008). Such policies and attitudes in and of themselves reflect a deeply rooted belief that disability itself is inherently undesirable, wrong, dangerous and a detraction to one's worth and humanity.

On a less severe—but still undoubtedly harmful—level, the idea that disabled people could never be desirable as sexual or romantic partners also reflects the general social belief that disability is inherently bad and that attraction to a person with disability is exceptional, pathological or reflective of other qualities so attractive that they "compensate" for the inherent unattractiveness of the partner's disability (Aguilera 2000; Limoncin et al. 2014). There are clear and harmful implications to the ableist notion that any attraction to people with disabilities by those without disabilities represents not genuine attraction or affection but rather a type of sexual deviancy; the implication, of course, is that people with disabilities are inherently unsuitable and unattractive and that no "normal" (i.e., mentally and physically healthy/non-disabled) person would find a disabled person attractive or would pursue a relationship with a disabled partner.

## Disability identity development

Because disability has been historically pathologised and shunned as inherently undesirable and something to be hidden or eliminated, the concept of positive disability identity development has only been explored within the academic literature within the past few decades. However, one can assume that these concepts were, of course, explored and expressed by individuals with disabilities in great detail and nuance in other ways and media, that were often left unrecorded or ignored. In a seminal article on disability identity development, Carol Gill (1997) described a four-stage model of disability identity integration in which a disabled person comes to accept his, her or their disability as a positive part of their identity that they are willing and able to express to the broader world. Gill likened this process to the "coming out" process that many sexual minority individuals undergo where they come to accept and express their sexual orientation with pride as a key part of their identity, and described the four stages as: "(1) 'coming to feel we belong' (integrating into society); (2) 'coming home' (integrating with the disability community); (3) 'coming together' (internally integrating our sameness and differentness); and (4) 'coming out' (integrating how we feel with how we present ourselves)" (p. 39). Since then, other models of disability identity development have been developed, including models that focus on political disability identity development (Putnam 2005) and models that focus on the social and psychosocial aspects of disability identity development (Forber-Pratt and Zape 2017). As noted by Forber-Pratt et al. (2017) in their recent systematic review of disability identity development, disability identity development often involves the process of negotiating both one's own personal feelings about and relationship to disability and the broader social and cultural attitudes towards disability.

Of note, the literature on disability identity development—and indeed the concept of *disability identity* itself—is strongly anti-ableist in its approach to disability. In sharp contrast to the long-held and still pervasive beliefs that disability is inherently bad, pathological or shameful, disability identity theories frequently incorporate concepts of pride in one's disability and active membership of the disability community (Forber-Pratt et al. 2017). Indeed, researchers have consistently found that a positive disability identity is protective and is linked to higher self-esteem, lower psychological distress and greater sense of community and engagement (Bogart 2014; 2015; Bogart et al. 2018; Nario-Redmond et al. 2013; Nario-Redmond and Oleson 2016). Identifying as a person with disability (or a disabled person) represents the ability to witness society's negative attitudes towards and rejection of disability and nevertheless hold a sense of ownership—and often pride—in one's historically stigmatised identity. Bogart et al. (2018) found that disability pride partially mediated the relationship between stigma and self-esteem, again highlighting the protective nature of positive disability identity in the face of considerable social stigma and messages of worthlessness and inadequacy. Thus, although disability identity development is not necessarily or inherently a staged or linear process (Forber-Pratt et al. 2018), it does appear to play an important role in understanding and enhancing the well-being of people with disabilities.

## Sexual identity development

Sexual development is a process that spans the entire human lifespan and includes biological, psychological and social factors (DeLamater and Friedrich 2002; Worthington et al. 2002). Over the span of one's life, but perhaps most prominently in adolescence, people develop and become aware of patterns of sexual and romantic attraction, and engage in sexual and romantic behaviour and relationships (Rosario et al. 2004; 2011; Worthington et al. 2002). Although all individuals go through a process of sexual identity development in which they discover and define their personal sexual attractions, wants, likes, dislikes and needs (Worthington et al. 2002), the literature on sexual identity development has largely focused on identity development in sexual minority individuals—mostly those who identify as lesbian, gay, or bisexual (Rosario et al. 2011). As with disability, the process of recognising, accepting and integrating an aspect of one's being into one's identity is a more prominent and often fraught or complicated process when that identity is marginalised. In other words, individuals who are sexual minorities must navigate not only their sexual and romantic attractions but the social attitudes towards and the potential social consequences of acting on and embracing those attractions (Rosario et al. 2011). This can add complexity to the sexual identity development process that heterosexual, cisgender individuals would not experience. Additionally, the sexual identity development process can also intersect with other aspects of identity and identity development processes, such as gender (Savin-Williams and Diamond 2000) and racial and ethnicity identity development (Rosario et al. 2004), adding additional complexities and contingencies to the identity development process.

It is important to note that there is a distinction between sexual identity and sexual behaviour. Individuals may engage in behaviour that does not align with their current sexual identity and may adopt a sexual identity before engaging in congruent sexual behaviour (Savin-Williams and Diamond 2000). For example, an individual may engage in same-sex sexual behaviour while still identifying as heterosexual or straight or may identify as lesbian, gay or bisexual before or without engaging in same-sex sexual behaviour. Additionally, sexual identity development consists of both romantic and sexual feelings and behaviour, which are often—but not always—

congruent with one another (Lund et al. 2016). Thus, the development of sexual identity can be a complex and multi-layered process, that is shaped over many years and through a multitude of different experiences (Worthington et al. 2002).

As in the literature on disability identity development discussed previously, researchers have found that greater integration of sexual identity into one's overall identity is linked to improved social and psychological well-being, even when controlling for other stressors (Rosario et al. 2011). Likewise, sexual identity development has implications for intimate partner violence, gender role values and sexual health behaviour; facilitating a healthy and adaptive sexual identity development process may help address dangerous or problematic attitudes, beliefs and behaviour by creating an environment where information and support on these topics is readily accessible (Worthington et al. 2002).

## Sexual identity development in people with disabilities

Sexual identity development in people with disabilities is a relatively under-explored topic. Historical and still-prevalent attitudes towards disability and sexuality hold that the two are fundamentally incongruent with one another—that disability de facto entails the loss or absence of all sexual identity (Millingan and Neufeldt 2001). Equally, expressions of disabled sexuality, when encountered, are often feared and considered repulsive. Because of these biases, people with disabilities are often denied comprehensive and high-quality information and education on sexual health and development (Eisenberg et al. 2015; Medina-Rico et al. 2018). When information on sexuality is provided to people with disabilities, it is often provided to parents or caregivers rather than to the person directly, or is focused on preventing or suppressing sexual behaviour or characteristics (Powell et al. 2016). However, despite the instance of others, that sexual behaviour and identity is incompatible with disability, the vast majority of people with disabilities, as with people without disabilities, desire and engage in meaningful and satisfying sexual and romantic relationships (Eisenberg et al. 2015; Medina-Rico et al. 2018; Nosek et al. 2001).

The schism between the discomfort exhibited by care providers and society in general towards disabled sexuality and the inherent sexuality of people with disabilities can create considerable challenges for people with disabilities. For example, disabled people may lack access to information on safer sexual behaviour, healthy and abusive relationship behaviours and adapted strategies for sexual expression and behaviour (Eisenberg et al. 2015; Medina-Rico et al. 2018). This can increase the risk of undesired or dangerous outcomes, such as sexually transmitted infections, unintended pregnancies and abusive or violent relationships. Furthermore, people with disabilities experience higher rates of childhood (Jones et al. 2012), adult (Hughes et al. 2012) and lifetime (Hughes et al. 2011) sexual abuse, meaning that their early or first sexual experiences may be marked by abuse, trauma and physical or emotional harm, as opposed to a consensual experience with a caring partner.

This combination of lack of education and elevated rates of sexual abuse may lead to situations where disabled people's sexual development is devoid or severely lacking in positive sexual and romantic experiences, models and information. At the same time, however, an exclusive focus on abuse in the context of sexual education for people with disabilities may lead to the incorrect implication that people with disabilities are incapable of healthy or consensual sexual or romantic relationships. For this reason, some abuse psychoeducation curricula for people with disabilities also include units on healthy relationships and positive and adaptive sexuality (e.g., ASAP for Women [Hughes et al. 2010] and SAFE [Lund and Hammond 2014]). Additionally, more recent resources, that explicitly incorporate anti-ableist and sex-positive approaches to sexual education and support for people with disabilities, may be useful in promoting positive and healthy sexual development through improved support and sexual education (e.g., Owens 2014). Conversely, provider or parent discomfort when discussing sexuality with their disabled clients or children may inhibit healthy exploration and positive sexual identity development in people with congenital or earlier-acquired disability, again potentially reinforcing ableist ideas regarding disability and sexuality.

The process of sexual identity development may be different for people with acquired disabilities than it is for people with congenital disabilities, and may also differ for individuals depending on the age of onset or diagnosis of the disability or the change in functional limitations over time. Whereas people with congenital or early-onset disabilities may have the negative stereotypes of disabled (a)sexuality foisted upon them from the beginning of their sexual identity development process, people with later-acquired disability may suddenly face a set of stereotypes and assumptions that challenge their previously developed sexual identity. They may face negative, ableist reactions from current or potential partners and may feel as though they suddenly go from being a desired partner to someone who is inherently seen as unattractive, asexual or undesirable (Nosek et al. 2001). Similarly, they may also have to relearn how to navigate dating, relationships and sex in light of their disability and thus have to re-evaluate and reformulate their sexual identity in this regard (Nosek et al. 2001).

Furthermore, individuals with later-acquired disabilities may also contend with internalised ableism surrounding sexuality and disability, judging their newly disabled selves to be no longer capable of sex or attracting a partner (Nosek et al. 2001). Disabled individuals interact with many service providers throughout their lives. These individuals may include medical and mental health professionals (e.g., physicians, psychologists), rehabilitation service providers (e.g., physical, speech or occupational therapists), educators and paraprofessionals, independent living and community support professionals, and direct service providers, among others. Licensed providers, such as psychologists and other medical-related professionals, must consider all aspects of disabled clients' identities, including their sexuality. Other individuals who provide support to people with disabilities should also have access to professional development, training and support in order to address these issues as they arise in their work. Providers who are uninformed, unwilling or unable to talk about navigating sexuality with a person with a disability may only serve to unintentionally strengthen perceptions that individuals with disabilities can no longer have active and fulfilling romantic and sexual lives (Eisenberg et al. 2015). Thus, it may not be prudent to use a client's or patient's assumptions that he, she or they can or will no longer have sex or sexual or romantic fulfilment as a way to avoid the often-times uncomfortable topic. Rather, providers should be willing to explore these ideas and beliefs with clients and patients in a way that introduces or reinforces the idea that the acquisition of a disability need not be the end to sexual or romantic fulfilment.

Accordingly, it is incumbent upon non-disabled providers and disability community allies to promote both positive overall disability identity development (Forber-Pratt et al. 2018) and disability-positive sexual identity development through learning about and embracing disability culture and a disability-affirmative, anti-ableist approach to sexuality. Such an approach requires that providers are willing to critically examine their own attitudes, biases and knowledge gaps as they relate to both sexuality and disability, and to become comfortable in explicitly and openly discussing sexuality with clients, patients and individuals to whom they provide support (Mona et al. 2017; Worthington et al. 2002). This requires that providers have not only a medical or psychological knowledge of sex, sexuality and relationship issues, but also a willingness to have open and honest dialogues about the individual's needs, wants, concerns and ideas, regarding their sexual and romantic relationships and behaviours. Mona et al. (2017) outline specific strategies to improve the service provision of culturally competent sexual health care for disabled individuals, including disability-specific clinical training in areas of assessment and accessibility. As academics, when tasked with writing on topics related to sexual health, sexual violence, sexual identity development and other related areas, it is imperative that we demand inclusion of a disability perspective (e.g., Forber-Pratt and Espelage 2018). The more that disability perspectives are included in mainstream discussions, the more these conversations are normalised and stigmas are reduced. Ideally, such a relationship should promote positive sexual identity development and lead to the co-development of strategies that the person with a disability and their partners can employ in order to address these wants and needs (Worthington et al. 2002).

## Sexual identity development in sexual minority individuals with disabilities

People with disabilities, who are also members of sexual minority groups (i.e., are non-heterosexual), may face additional considerations and complications during their processes of sexual identity development. Scholars and individuals with these dually marginalised identities have written about the exclusion that they feel from both the disability and sexual minority communities while trying to integrate these two aspects of their identities. For example, Corbett O'Toole and colleagues (O'Toole 2000; O'Toole and Brown 2002) have written about the challenges of finding medical and mental health professionals who are affirmative of both disabled and lesbian sexuality and identity, and of facing ableism in the lesbian community and homophobia in the disability community. Similarly, Caldwell (2010) has written of the dual "in/visibility" of both bisexual and disabled identities. Social pressures and incentives to "pass" as both heterosexual, non-disabled, or both—along with social and cultural discomfort with different and non-binary identities—essentially push disabled and bisexual people (and disabled, bisexual people) into a state of forced invisibility. This may force them to either hide or suppress their identities or constantly restate and validate their identities in order to make themselves visible, only to face potential social rejection or invalidation from others.

Along similar lines, Lund and Johnson (2015) discuss the issues faced by people who identify as both disabled and asexual. The historic and persistent efforts within the disability community and the disability sexuality movements in particular to reject the myth of universal asexuality among people with disabilities also has the unintentional side effect of marginalising disabled people who do indeed identify as asexual in terms of their patterns of sexual attraction (i.e., not experiencing sexual attraction to individuals of any sex or gender). Such blanket insistence of compulsory sexuality within the disability community may make individuals with disabilities, who do identify as asexual, feel as though they are "going against" the disability community goals and principles when they identify as such, leading to potential isolation and rejection.

Additionally, disabled people who are also sexual minorities may face rejection from their sexual minority communities due, in part, to the desire of individuals in these communities to appear "normal" or "not weak" (Cuthbert 2017; Lund and Johnson 2015; O'Toole 2000). For example, the past association of sexual minority identity with mental illness—or, in the case of asexuality, presumed physical illness (Lund and Johnson 2015)—may create a stigma against people who are perceived as bolstering those old, invalidating assertations by indeed having a psychiatric, developmental or physical disability (Cuthbert 2017). This may create pressure on the disabled person to either leave sexuality minority spaces if their disability cannot be hidden, or to hide their disability, if possible, in order to be better accepted within their sexual minority community.

The dual forces of ableism and heterosexism create challenges for healthy sexual and disability identity development in people who are both sexual minorities and people with disabilities.

As with other people with multiple marginalised identities, these individuals may feel pressure to "pick one" aspect of their identity when receiving support (Lightfoot and Williams 2009). Such a strategy, while a potential and sometimes necessary "survival tactic," runs counter to positive identity development and integration, and ignores the unique experiences and needs created by the intersection of multiple marginalised aspects of identity. For example, a man with physical disability may want assistance in developing safe, comfortable and pleasurable ways to engage in sexual intercourse with a partner but will probably not receive helpful ideas and support if his providers all assume that his partner is female when his partner is actually male. Indeed, homophobia or discomfort with same-sex or same-gender sexual behaviour is unfortunately still common among providers of services to people with developmental disabilities, even those who hold more positive attitudes towards disabled sexuality in general (Gomez 2012), interfering with the ability of gay, lesbian or bisexual individuals to receive support in developing their sexual identity and learning safe, fulfilling means of sexual and romantic expression. Thus, these individuals may benefit more from social groups and providers who are actively disability-affirming as well as queer-positive so that they can explore both their disability identity and their sexual identity openly and without prejudice. Despite these barriers to positive identity development, integration of sexual minority identities with other aspects of identity is a vital aspect of sexual minority identity development (Gupta 2017) and should be supported by care providers and social networks.

## Conclusion

The very concept of sexual identity development in disabled people is itself a radical idea. People with disabilities have historically been denied opportunities to develop a sexual identity of any kind, much less one that is positive and integrated with a developed sense of disability identity. However, ensuring that people with disabilities receive the support and resources necessary to develop and integrate disability-affirmative sexual identities has the potential to enhance well-being, community and sense of self, as well as to promote healthy and fulfilling sexual and romantic expressions and relationships. More research should be conducted on the sexual identity development process of people with disabilities, and how providers, family members and allies can further support sexual identity development in this population. Finally, researchers should further investigate how sexual identity development intersects with other aspects of identity and identity development in people with disabilities.

## Note

1 The term "intellectual disability" is used in its singular form throughout the chapter, as this is the current accepted terminology. The term "intellectual disability" is used to refer to a disability characterised by significant limitations in both intellectual functioning and in adaptive behaviour, which covers many everyday social and daily living skills. This disability originates before the age of 18. In contrast, the term "developmental disabilities" is a broader characterisation of multiple types of disabilities and/or chronic conditions caused by an impairment in physical, learning, language or behavioural areas (see Schalock et al. 2010).

## References

Aguilera, R.J. (2000) Disability and delight: Staring back at the devotee community. *Sexuality and Disability* 18(4), pp. 255–261.

Azzopardi-Lane, C., & Callus, A.M. (2015) Constructing sexual identities: People with intellectual disability talking about sexuality. *British Journal of Learning Disabilities* 43(1), pp. 32–37.

Bogart, K.R. (2014) The role of disability self-concept in adaptation to congenital or acquired disability. *Rehabilitation Psychology* 59(1), pp. 107–115.

Bogart, K.R. (2015) Disability identity predicts lower anxiety and depression in multiple sclerosis. *Rehabilitation Psychology* 60(1), pp. 105–109.

Bogart, K.R., Lund, E.M., & Rottenstein, A. (2018) Disability pride protects self-esteem through the rejection-identification model. *Rehabilitation Psychology* 63(1), pp. 155–159.

Caldwell, K. (2010) We exist: Intersectional in/visibility in bisexuality & disability. *Disability Studies Quarterly* 30(3/4). Retrieved from http://dsq-sds.org/article/view/1273/1303

Cuthbert, K. (2017) You have to be normal to be abnormal: An empirically grounded exploration of the intersection of asexuality and disability. *Sociology* 51(2), pp. 241–257.

DeLamater, J., & Friedrich, W.N. (2002) Human sexual development. *Journal of Sex Research* 39(1), pp. 10–14.

East, L.J., & Orchard, T.R. (2014) Somebody else's job: Experiences of sex education among health professionals, parents and adolescents with physical disabilities in southwestern Ontario. *Sexuality and Disability* 32(3), pp. 335–350.

Eisenberg, N.W., Andreski, S.R., & Mona, L.R. (2015) Sexuality and physical disability: A disability-affirmative approach to assessment and intervention within health care. *Current Sexual Health Reports* 7(1), pp. 19–29.

Fleischer, D., & Zames, F. (2001) *The disability rights movement: From charity to confrontation.* Philadelphia, PA: Temple University Press.

Forber-Pratt, A.J., & Espelage, D.L. (2018) Sexual violence. In *K–12 settings. The Wiley handbook on violence in education: Forms, factors, and preventions*, pp. 375–391.

Forber-Pratt, A.J., Lyew, D.A., Mueller, C., & Samples, L.B. (2017) Disability identity development: A systematic review of the literature. *Rehabilitation Psychology* 62(2), pp. 198–207.

Forber-Pratt, A.J., & Zape, A.M. (2017) Disability identity development model: Voices from the ADA-generation. *Disability and Health Journal* 10(2), pp. 350–355.

Gill, C.J. (1997) Four types of integration in disability identity development. *Journal of Vocational Rehabilitation* 9(1), pp. 39–46.

Gomez, M.T. (2012) The S words: Sexuality, sensuality, sexual expression and people with intellectual disability. *Sexuality and Disability* 30(2), pp. 237–245.

Gupta, K. (2017) "And now I'm just different, but there's nothing actually wrong with me": Asexual marginalization and resistance. *Journal of Homosexuality* 64(8), pp. 991–1013.

Hughes, K., Bellis, M.A., Jones, L., Wood, S., Bates, G., Eckley, L., McCoy, E., Mikton, C., Shakespeare, T., & Officer, A. (2012) Prevalence and risk of violence against adults with disabilities: A systematic review and meta-analysis of observational studies. *The Lancet* 379(9826), pp. 1621–1629.

Hughes, R.B., Lund, E.M., Gabrielli, J., Powers, L.E., & Curry, M.A. (2011) Prevalence of interpersonal violence against community-living adults with disabilities: A literature review. *Rehabilitation Psychology* 56(4), pp. 302–319.

Hughes, R.B., Robinson-Whelen, S., Pepper, A.C., Gabrielli, J., Lund, E.M., Legerski, J., & Schwartz, M.S. (2010) Development of a safety awareness group intervention for women with diverse disabilities: A pilot study. *Rehabilitation Psychology* 55(3), pp. 263–271.

Jones, L., Bellis, M.A., Wood, S., Hughes, K., McCoy, E., Eckley, L., Bates, G., Mikton, C., Shakespeare, T., & Officer, A. (2012) Prevalence and risk of violence against children with disabilities: A systematic review and meta-analysis of observational studies. *The Lancet* 380(9845), pp. 899–907.

Keller, R.M., & Galgay, C.E. (2010) Microaggressive experiences of people with disabilities. In Sue DW (Ed.), *Microaggressions and marginality: Manifestation, dynamics, and impact.* U.S.: John Wiley & Sons, pp. 241–268.

Lightfoot, E., & Williams, O. (2009) The intersection of disability, diversity, and domestic violence: Results of national focus groups. *Journal of Aggression, Maltreatment & Trauma* 18(2), pp. 133–152.

Limoncin, E., Carta, R., Gravina, G.L., Carosa, E., Ciocca, G., Di Sante, S., Isidori, A.M., Lenzi, A., & Jannini, E.A. (2014) The sexual attraction toward disabilities: A preliminary internet-based study. *International Journal of Impotence Research* 26(2), pp. 51–54.

Lombardo, P. (2008) *Three generations, no imbeciles: Eugenics, the Supreme Court and Buck v. Bell.* Baltimore: John Hopkins University Press.

Lund, E.M., & Hammond, M. (2014) Single-session intervention for abuse awareness among people with developmental disabilities. *Sexuality and Disability* 32(1), pp. 99–105.

Lund, E.M., & Johnson, B.A. (2015) Asexuality and disability: Strange but compatible bedfellows. *Sexuality and Disability* 33(1), pp. 123–132.

Lund, E.M., Thomas, K.B., Sias, C.M., & Bradley, A.R. (2016) Examining concordant and discordant sexual and romantic attraction in American adults: Implications for counselors. *Journal of LGBT Issues in Counseling* 10(4), pp. 211–226.

Medina-Rico, M., López-Ramos, H., & Quiñonez, A. (2018) Sexuality in people with intellectual disability: Review of literature. *Sexuality and Disability* 36(3), pp. 231–2488.

Milligan, M.S., & Neufeldt, A.H. (2001) The myth of asexuality: A survey of social and empirical evidence. *Sexuality and Disability* 19(2), pp. 91–109.

Mona, L.R., Cameron, R.P., & Clemency Cordes, C. (2017) Disability culturally competent sexual healthcare. *American Psychologist* 72(9), 1000–1010.

Nario-Redmond, M.R., Noel, J.G., & Fern, E. (2013) Redefining disability, re-imagining the self: Disability identification predicts self-esteem and strategic responses to stigma. *Self and Identity* 12(5), pp. 468–488.

Nario-Redmond, M.R., & Oleson, K.C. (2016) Disability group identification and disability-rights advocacy: Contingencies among emerging and other adults. *Emerging Adulthood* 4(3), pp. 207–218.

Nosek, M.A., Howland, C.A., Rintala, D.H., Young, M.E., & Chanpong, G.F. (2001) National study of women with physical disabilities: Final report. *Sexuality and Disability* 19(1), pp. 5–39.

O'Toole, C.J. (2000) The view from below: Developing a knowledge base about an unknown population. *Sexuality and Disability* 18(2), pp. 207–224.

O'Toole, C.J., & Brown, A.A. (2002) No reflection in the mirror: Challenges for disabled lesbians accessing mental health services. *Journal of Lesbian Studies* 7(1), pp. 35–49.

Owens, T. (2014) *Supporting disabled people with their sexual lives: A clear guide for health and social care professionals.* London, UK: Jessica Kingsley Publishers.

Powell, R.M., Andrews, E.E., & Ayers, K. (2016) RE: Menstrual management for adolescents with disabilities. *Pediatrics* 138(6), p. e20163112A.

Priestley, M. (2000) Adults only: Disability, social policy and the life course. *Journal of Social Policy* 29(3), pp. 421–439.

Putnam, M. (2005) Conceptualizing disability: Developing a framework for political disability identity. *Journal of Disability Policy Studies* 16(3), pp. 188–198.

Rosario, M., Schrimshaw, E.W., & Hunter, J. (2004) Ethnic/racial differences in the coming-out process of lesbian, gay, and bisexual youths: A comparison of sexual identity development over time. *Cultural Diversity and Ethnic Minority Psychology* 10(3), pp. 215–228.

Rosario, M., Schrimshaw, E.W., & Hunter, J. (2011) Different patterns of sexual identity development over time: Implications for the psychological adjustment of lesbian, gay, and bisexual youths. *Journal of Sex Research* 48(1), pp. 3–15.

Savin-Williams, R.C., & Diamond, L.M. (2000) Sexual identity trajectories among sexual-minority youths: Gender comparisons. *Archives of Sexual Behaviour* 29(6), pp. 607–627.

Schalock, R.L., Borthwick-Duffy, S.A., Bradley, V.J., Buntinx, W.H., Coulter, D.L., Craig, E.M., Gomez, S.C., Lachapelle, Y., Luckasson, R., Reeve, A., & Shogren, K.A. (2010) *Intellectual disability: Definition, classification, and systems of supports.* Silver Spring, MD: American Association on Intellectual and Developmental Disabilities.

Smart, J.F. (2009) *Disability, society, and the individual* (2nd ed.). Austin: Pro Ed.

Starke, M., Rosqvist, H.B., & Kuosmanen, J. (2016) Eternal children? Professionals' constructions of women with an intellectual disability who are victims of sexual crime. *Sexuality and Disability* 34(3), pp. 315–328.

Worthington, R.L., Savoy, H.B., Dillon, F.R., & Vernaglia, E.R. (2002) Heterosexual identity development: A multidimensional model of individual and social identity. *The Counseling Psychologist* 30(4), pp. 496–531.

# 6

# SEXUALITY AND DISABILITY IN BRAZIL

## Contributions to the promotion of agency and social justice

*Marivete Gesser*

## Introduction

Sexual and reproductive rights for disabled people are human rights and social justice issues. Viewing disability as a category of analysis, and based on studies and research conducted in Brazil, this chapter aims to characterise the contributions of critical disability studies, especially from a feminist approach, towards the construction of practices geared to the realisation of sexual and reproductive rights for disabled people. Therefore, this chapter will be divided into three sections. In the first section, the results of research performed in Brazil on the theme will be briefly presented in order to demonstrate: a) the main barriers that disabled people face every day in exercising their sexuality and reproductive rights; and b) the relationship between violations of sexual and reproductive rights and the social markers of difference, particularly the issues of gender and social class. In the second section, there is an investigation into the relationship these barriers have to the notion of disability, coming from the biomedical model, which, through the knowledge that establishes normative and deviating sexualities, produces ableist practices and delegitimises the sexual and political agency of disabled people. Finally, the third section presents some fundamental theoretical–methodological assumptions to be incorporated into professional practices committed to ensuring the sexual and reproductive rights of disabled people, in compliance with the UN Convention on the Rights of Persons with Disabilities (CRPD) (2006) and the perspective of social justice. These are: a) break with the normalisation processes; b) incorporate the intersectional perspective; c) consider the relations of dependence and interdependence; d) consider sexuality as a process of ethical, aesthetic and political creation; and e) promote the participation of disabled people in issues concerning sexuality and reproduction.

After approval of the UN CRPD, and its enactment into the Brazilian legal system as a constitutional amendment through federal decree (Federal Decree n. 6.949/2009), there were many advances from the legal perspective related to the guarantee of the sexual and reproductive rights of disabled people. This legal instrument, resulting from the international mobilisation of disabled people, aims to ensure all disabled people are free of oppression in their experience of sexuality. Different points of the UN Convention serve to guarantee the sexual and reproductive rights of disabled people through international human rights instruments. The first of them relates to the right to equality and non-discrimination based upon disability. Other rights assured to disabled people are to marry and form a family, if they so wish, in addition to having access to information on reproductive family planning and the means necessary to exercise these rights. The UN CRPD also ensures that disabled people are protected against all forms of exploitation, violence and abuse, including aspects concerning gender and sexuality, as well as the recognition that girls and disabled women are more vulnerable to sexual violence, and the stipulation that the States should construct forms of prevention.

Although the UN CRPD has been in effect in Brazil for more than a decade, disabled people still face numerous barriers that hinder the guarantee of sexual and reproductive rights. Many Brazilian studies on disabled people have shown that they have suffered countless forms of oppression and exclusion. Maia (2016), Nicolau et al. (2013) and Simões (2015) highlight that disabled people do not have access to information about sexuality and reproduction. Bastos and Deslandes (2012) and Simões (2015) point to the existence of conceptions of disability that oppress and delegitimise disabled bodies in their experience of sexuality. Studies by Bastos and Deslandes (2012), Gesser et al. (2014), Lopes (2018), Nicolau et al. (2013) and Regis (2013) indicate that there is an intersectional dimension that hinders the guarantee of sexual and reproductive rights to disabled people, based on gender. Thus, disabled women tend to face barriers to this guarantee of their right to exercise their sexuality and reproductive rights as a result of their bodies being discordant with the standards of beauty and also of not being considered capable of performing the duties of a housewife, worker and mother in accordance with that prescribed by the dominant social narratives. These studies corroborate the information contained in the World Report on Disability (WHO 2011).

According to Rosemarie Garland-Thomson (2002; 2005), disability is a category of analysis that, at the intersection with other categories such as gender, sexuality, ethnicity, race and class, exerts a tremendous social pressure in formatting, regulating and standardising all bodies. Based on this premise, the author highlights that "disability is the most human of experiences, touching every family and—if we live long enough—touching us all" (2002, p. 5). Predicated on the reflections of Linton (1998a), Garland-Thomson stresses that the study of disability is a prism through which a broader understanding of society and human experience may be obtained. Besides this, "understanding how disability operates as an identity category and cultural concept will enhance how we understand what it is to be human, our relationships with one another, and the experience of embodiment" (2002, p. 5).

Although the contributions presented here provide more support to the construction of professional practices in tandem with persons with physical and intellectual disabilities, this work starts from the assumption that narratives on disability concern bodies with different functional and aesthetic conditions. Thus, although we don't deny the specificities of different disability types, this work will not focus on a specific disability, but consider disability as a category of analysis, marked by the oppression of those characterised as deviant with reference to all manner of norms and barriers. The framing of disability as a category of analysis facilitates the comprehension of how the narratives on deviating bodies, in intersection with gender, race, ethnicity, age and social class, produce bodies more or less legitimised to exercise sexual and reproductive rights with participation and agency, as argued by Wilkerson (2002). They also indicate the necessity for deconstructing discourses and practices that, based on normative and aesthetic ideals, produce barriers to this social group's experience of sexuality. This deconstruction must be carried out in practices involving disabled people and in the training of professionals who work in the different sectors geared towards education, justice, social assistance and health.

Several authors linked to the field of disability studies, based on the intersectional under-standing, have shown the multiple relationships of disability with gender, sexuality, social class, race, ethnicity and generation (Banks 2018; Ben-Moshe and Magaña 2014; Gesser et al.2019; Gill and Erevelles 2017; Linton 1998b), as well as the effects of concepts of sexuality and dis-ability on the debasement of sexual and reproductive rights (Gesser et al. 2014; Lopes 2018). The following section will introduce some recent studies on sexuality performed in Brazil.

## Sexuality and disability in Brazil: A brief overview

In this section, several Brazilian studies will be presented, which have shown that disabled people tend to be denied their sexual and reproductive rights, and the main factors that con-tribute to this process will be identified. There are also indications that the architectural and, principally, attitudinal barriers (e.g., myths, stereotypes, taboos and prejudices) are elements that substantiate the constitution of this phenomenon (Gesser et al. 2014; Maia and Ribeiro 2010; Nicolau et al. 2013; World Health Organization (WHO) 2011). The mapping of these barriers is fundamental to implementing practices aimed at ensuring the sexual and repro-ductive rights provided for in the UN CRPD, since, as claimed by Block et al. (2012), these practices must be planned, bearing in mind the sociocultural and political context in which the subjects are situated.

Regarding the attitudinal barriers, many Brazilian authors have highlighted the myths related to the sexuality of disabled people as producers of rights violations. Research by Bastos and Deslandes (2012), Gesser and Nuernberg (2014) and Maia (2012) point out that one of these myths is that disabled people are asexual and don't have sexual thoughts, feelings and needs. This myth is intrinsically linked to the process of infantilising this social group, based on a standard of sexuality legitimised mainly by religious and biomedical discourses, which reduce it to penis/vagina coitus and reproduction. The sexuality of children, the elderly and disabled people is delegitimised, rendering them as infantile and therefore in no need of exercising their sexuality.

Maia and Ribeiro (2010) also draws attention to the myth that disabled people are hyper-sexual, displaying uncontrollable and rampant desires, mainly in relation to those with intel-lectual disabilities. These data were also identified in a study performed by Simões (2015) with teachers from a special education school, who viewed students with intellectual disability as out-of-control and dangerous. The data gathered by the author show that the teachers classify all the spheres of life of the intellectually disabled person as "abnormal," "overwrought" and having a "dangerous power," with students classified as "less intellectually developed" are perceived as more uncontrolled and compromised.

In Brazilian research (Maia and Ribeiro 2010), conceptions of disability that represent disa-bled people as unattractive, undesirable and incapable of maintaining a loving and sexual rela-tionship, are sterile, have offspring with disabilities, and/or do not have the means to take care of children also predominate. As we shall see in this section, the conceptions of disability are reflective of the professional practices involving disabled people, in so far as they legitimise the production of violence. Finally, the authors also highlight the idea, already identified in studies by Tepper (2000) in the United States, that disabled people can't enjoy normal sex and have sexual dysfunctions related to desire, arousal and orgasm.

The above-mentioned ideas about disabled people—which are strongly related to the per-spective of subject and sexuality reiterated by the biomedical model—have hindered the reali-sation of sexuality as per current Brazilian legislation advocates in different social contexts. They have contributed to people with different types of disability suffering from experiences of oppression and vulnerability.

Studies carried out with disabled people have shown that the professionals working in the areas of health and education have not been prepared to provide information related to sexuality and reproduction to disabled people. This was identified in studies by Denari (2010), Lopes (2018), Maia (2016), Nicolau et al. (2013) and Simões (2015). Maia (2016) performed a qualitative-descriptive survey with the aim of investigating the experience of sexuality and sexual health based on reports by 12 people, men and women, with intellectual disability. Results indicated that the participants, having received merely superficial information and an ineffective sex education, had still begun an active sexual and affective life and had been through situations of vulnerability, since the sexual relations reported occurred in places with little privacy and almost always without the use of a condom. A study by Soares et al. (2008) discussed the quality of life of young people with spina bifida in Brazil and the United States, while focusing on sexuality, and also found a major lack of preparedness in the health services regarding access to information related to sexuality and reproduction. The information gathered indicates a lack of discussions on sexuality and physical disability, both within the families and their respective health services. This was reported by survey participants as a barrier that hindered their experience of sexuality.

Many studies have shown the effect of the barriers that disabled people face regarding sexuality and their intersection with gender issues, which produces an intensification of vulnerability (Bastos and Deslandes 2012; Gomes et al. 2019; Nicolau et al. 2013). One of these studies, performed by Gesser et al. (2014) in the south of Brazil, aimed to characterise the experience of sexuality in women with disability, based upon a survey of eight women with disability, aged between 24 and 68. The interviewees had undergone a process of infantilisation and asexualisation, which was mainly reflected in them being unable to establish affective relationships. Associating physical disability with an inability to discern the intentions of possible suitors was also in evidence among the families of certain participants. Another element that appeared in the participants' testimonies, and identified as responsible for suffering, was the delegitimisation of the status of wife and mother, as a result of the women in question not being able to reproduce the gender specifications that referred to housekeeping, and looking after kids and husbands. The discourse that disability is a medical tragedy showed up in the subtexts of some interviewees. The de-eroticising of the body with disability was a constitutive element of the women's relation with the dimension of sexuality, in so far as the lesions and impediments were perceived as elements that make the bodies deviate from the socially enshrined standard of erotic, that is widely expressed by the media. Other elements that also surfaced in the accounts of participants as having an influence in their relation with sexuality were: issues of gender, which assign women the role of fragility and defencelessness; the myth that a person with disability doesn't have the same needs as a person without a disability (particularly those needs related to sexuality); the idea that disabled people can be defined by their lesion and are lacking in attractiveness, and who a person without a disability would only approach in order to take advantage.

Bastos and Deslandes (2012) investigated 14 parents of teenagers with intellectual disabilities and enquired about the perceived sexual manifestations exhibited by their children. Results showed that there were differences between the parents of sons and daughters. With regard to auto-erotic practices, the parents'/guardians' accounts pointed to their non-approval to the degree that, when this behaviour is demonstrated by females, it is seen as wholly inappropriate. The desire to have an active sex life tends to be questioned in women with intellectual disability but accepted in men, based on a biological explanation that men naturally have sexual needs. The result of this research demonstrates the importance of incorporating an intersectional viewpoint in professional practices, since disabled women face more barriers than disabled men in achieving their sexual and reproductive rights.

Nicolau et al. (2013) performed a study on 15 women with different severities and types of disability, with the aim of identifying individual, social and programmatic dimensions of vulnerability, in three basic healthcare services in the city of São Paulo. Results obtained indicated experiences of rejection or overprotection by families, difficulties in acquiring equipment for their autonomy, little investment in study or professional qualification, less social participation, barriers to their realisation of sexuality and maternity, lack of physical accessibility and less-than-welcoming attitudes in health services, characterising total vulnerability.

A study by Lopes (2018), focusing on the experiences of maternity among women with physical disabilities (e.g., osteogenesis imperfecta and paraplegia) and visual disabilities, showed that many obstacles were faced. During their pregnancies, the participants reported difficulties in accessing information, whereby the women with private health plans had better experiences with pre-natal care, providing evidence of how disability relates to social class. Besides this, blind women were also questioned on their capacity to take care of their children, which led to one of them not feeling secure enough to report a situation of sexual abuse she was suffering out of fear she might lose her child. The eugenics discourse appears in the health services too, where women with osteogenesis imperfecta were advised by their families, friends and health professionals not to have children. Attitudinal barriers and the lack of accessibility negatively affected the maternity experience for all those interviewed. Difficulties were faced bringing their kids to school, on trips and medical appointments. All participants cited the relations of interdependence as fundamental, both during pregnancy and in the experience of maternity.

Another interesting study, related to the field of reproductive rights for disabled people in the south of Brazil, was carried out by Regis (2013), studying parental discourse on the involuntary sterilisation of 23 women with intellectual disabilities. Reasons that were given to justify these procedures were stopping menstruation, preventing pregnancy and protection from sexual violence. Furthermore, the eugenic character of the sterilisation may be identified in the accounts of participants, since the possibility of having a grandchild with a disability was seen, without exception, as a problem for these families. Sterilisation surgeries were mostly performed by the National Health Service (SUS). In only three cases were due legal processes followed, via court authorisation. The participation of the women was practically insignificant, with some of them making it known that they wished to be mothers.

Past research on the sexual and reproductive rights for disabled people point to the existence of many social barriers that hinder full access to these rights. These barriers are reproduced in different social spaces, including public education policies and in the health services. There is a strong correlation between these barriers and questions of gender, as shown in studies by Bastos and Deslandes (2012), Gesser et al. (2014), Lopes (2018), Nicolau et al. (2013) and Regis (2013), and social class (Lopes 2018), although the dimensions of race and age still haven't been sufficiently covered in Brazilian research. Although it was not the focus of this topic, there is a strong relation between the oppression suffered by disabled people and the biomedical discourse, since it transforms disabled people into bodies that don't function, maintaining the connection that associates disability with tragedy. The form in which the normative ideas of body and sexuality produce deviant sexualities and contribute to the debasement of rights for disabled people will now be addressed.

## Biomedical model: Normalisation, pathologisation and medicalisation

The biomedical model, which is strongly based on the normal/deviant binomial, has contributed greatly to the persistence of the barriers faced on a daily basis by people with disability, hindering the realisation of their sexuality. Through this model, the person with disability is seen as deviating from an ideal subject, and his/her sexuality is transformed into a medical tragedy, and his/her reproduction, from the perspective of eugenics, is viewed as a danger to a society founded on ideas of normality and on the interpretation of deviation as something that should be forcibly avoided. Furthermore, by creating a normative ideal of a healthy, functional and performing body, the biomedical model fosters practices aimed at the pathologisation and medicalisation of deviating sexualities, lessening the sexual and political agency of disabled people.

Davis (2006) highlights that, although the ideal of human being has been identified since ancient times, the norms and process of disabling associated with it "arrived with industrialisation and with the set of practices and discourses that are linked to late eighteenth—and nineteenth—century notions of nationality, race, gender, criminality, sexual orientation and so on." The author stresses that this concept of norms is strongly related to the field of statistics, which began to be applied to distinguish human characteristics and, based on the belief that the measurement of these characteristics could produce an improved human race, statistics began to be applied for this purpose, since the vast majority of the statisticians of that time were eugenicists.[1] Finally, Rafter (1988) underscores that the State took advantage of modern medicine, based on the notion of normality, to prevent the dissemination of genetic disease and degenerate family types in the wider population. Eugenic studies were the key to institutionalisation and sterilisation in the United States, whereas, in Germany, they were used to justify the extermination of many disabled people.

The biomedical discourse, founded on the notion of normality, had the effect of transforming disabled people into deviating bodies, characterised as sick, dysfunctional and in need of treatment to return to normality. Therefore, the understanding emerged of disability in the body, with lesions and impediments, and actions were geared towards disabled people in the field of body medicalisation. As such, this diverges from the proposal of Block et al. (2012, p. 164) of considering the experiences of disabled people in "the sociocultural and policy contexts that affect disabled people's everyday occupations, including that of their sexual lives."

Sexuality was also the object of biomedical discourse. Foucault (1978) points out that sexuality was used as a population control device, being reduced to sex between a heterosexual couple and aimed at reproduction. This process of normalising sexuality, with the advance of sexology studies, became increasingly standardised and being considered as the normal cycle of sexual response, disqualifying diverse forms of experiencing pleasure. One of the groups impacted by this reductionist understanding of sexuality was that of physically disabled people who can't manage to complete the cycle of sexual response in accordance with the accepted medical discourse or who are compromised in their reproductive function (Shakespeare 2007). The reduction of sex to family and reproduction, associated with the eugenic discourse, led to the sterilisation of many people in Brazil (Regis 2013) and in countries such as the United States (Carey and Ben-Moshe 2014).

An effect of the biomedical model is ableism. Campbell (2008) is an important author, who has conceptualised ableism, as well as showing the relation of this process as a normative body ideal, that tends to be compulsorily considered. The author highlights that:

> Ableism refers to ... a network of beliefs, processes and practices that produces a particular kind of self and body (the corporeal standard) that is projected as the perfect, species-typical and therefore essential and fully human. Disability, then, is cast as a diminished state of being human.
>
> *(Campbell 2001, p. 44)*

Similarly, Mello (2016) emphasised that this situation is reflected through prejudiced attitudes and discrimination, that rank people according to the conformity of their bodies to standards of beauty and functional capacity. The studies that have been completed in the field of sexuality for disabled people indicated ableism as an element that hindered the sexual and reproductive rights of disabled people, in so far as it delegitimises the non-normative experiences of sexuality. The ableist perspective, which is based on the biomedical model, tends to reiterate a normative view of sexuality, reducing it to the cycle of sexual response described in the Diagnostic and Statistical Manual of Mental Disorders (DSM-5), the standard classification of mental disorders, and to reproduction. It also reinforces the maintenance of social practices aimed at curtailing the reproductive rights of disabled people, by using eugenic principles. Moreover, based on the normality/deviation binomial, the biomedical model considers disability to be a universal and incapacitating experience, producing negative effects or limitations in terms of sex life and erotic pleasure (Gesser and Nuernberg 2014).

Wilkerson (2002) has also highlighted how the biomedical model has lessened the sexual and political agency of disabled people. This comes about due to the expression of erotophobia, a type of phobia towards people who express sexuality in a manner that differs from the socially established norm. The author emphasises that erotophobia marginalises cultural minorities in that it is not merely a taboo against open discussions about sexuality and exhibitions of sexual behaviour, "but a very effective means of creating and maintaining social hierarchies, not only those of sexuality, but those of gender, race, class, age and physical and mental ability" (p. 41). Through it, oppressed groups face restrictions, penalties and coercion, while being denied access to important information on their sexuality, with people who challenge the taboo being stigmatised.

Thus, the medical discourse—considering its scientific, systematic and technological character—represents a source of information that dehumanises disabled people, promoting moral stigmas that pathologise bodies and reduce their capacity for agency by viewing them as deviating from the norm (dangerous from the reproductive perspective) and sick (Wilkerson 2002). The shame attached to exercising deviant sexuality and the infantilisation process that disabled people go through are points to be considered in comprehending the relation between sexuality and agency. Therefore, the biomedical discourse, which perpetuates practices founded on the elimination of deviating bodies, whether by eugenic, corrective and/or even preventive methods, operates by lessening the agency of the subject. The author stresses the relation between sexual agency and political agency. This is because the discourses that imagine those bodies that deviate from the ideal of normality to be asexual, infantilised and/or associated, through the exercise of sexuality, with danger, reduce the political legitimacy of this social group.

Furthermore, Wilkerson (2002), based on the thinking of Rubin (1993), underlines the relevance of taking an intersectional perspective when discussing sexual agency, since the social aspects of sexual agency constitute a hierarchy in which those who are socially privileged in the various strands of social difference (including sexual orientation, along with race, class, age and expression of gender, among others) will probably be more respectable and dignified. Rubin (1993) also emphasises that there are "relationships between stigmatised erotic populations and the social forces which regulate them" (p. 150). In this way, a poor, disabled woman will have less legitimacy to live sexually than a rich, white, young man. Thus, the intersectional perspective offers elements with which to frame the experiences related to sexuality and reproduction within the different social and political contexts, which may corroborate the qualification of services geared towards guaranteeing social and reproductive rights.

The reflections above demonstrate how the biomedical, norm-centred model contributes to the production of processes that oppress disabled people, and have significantly lessened their capacity for agency, whether by provoking shame through the pathologisation of sexual expression or by seeking the medicalisation of deviations. Below, based on an anti-ableist view of sexuality and disability, some principles will be presented that may help to deconstruct these oppressive forms of interpretation of the sexuality of disabled people.

## Theoretical and methodological contributions: Promoting agency and social justice in sexuality

Grounded on the premise that the guarantee of sexual and reproductive rights for disabled people is a question of social justice, this section will put forward some theoretical and methodological considerations concerning this issue. These could contribute to the construction of practices in the different fields of professional activity involving disabled people, that are guided by the assurance of these rights. They are interrelated and complementary, having, as the central point, the rupture with modern ideals of the subject that marginalise those who deviate from it, according to the existence of a normative ideal.

The first assumption, which is related to all the rest, refers to the need to *break away from the normalisation processes in the sexuality of disabled people*, which transform human variations into pathologies. Ableist practices, which emerge from these processes, operate as oppressors of the subjects through the delegitimisation of their sexual expression and also by the compulsory medicalisation of their bodies, striving to mould them and normalise them according to what is considered to be typical. Through the reflections of Foucault and Wilkerson (2002), the previous section already demonstrated the effects of the biomedical discourse, which has been widely centred on the processes of evaluation and pathologisation of deviating sexualities. Studies related to the sexuality of disabled people have shown that the curtailment and medicalisation of sexuality based on normative standards have seriously limited the guarantee of sexual and reproductive rights for disabled people. As such, without denying the functional conditions of disabled people, it is vitally important that programmes geared towards the sexuality and reproduction of disabled people also consider the contextually localised experiences, instead of the established standard.

Linked to considerations arising out of feminist disability studies, the second assumption is to *incorporate the intersectional perspective in research and professional practices related to sexuality and reproduction*. Corroborating the thinking of Wilkerson (2002), it is highlighted that this perspective makes it possible to understand how the intersection between oppressive systems such as racism, sexism, ableism, ageism and capitalism have effects to stigmatize disabled people, delegitimizing sexual and reproductive rights. In the words of Garland-Thomson: "Together, the gender, race, ethnicity, sexuality, class and ability systems exert tremendous social pressures to shape, regulate and normalise subjugated bodies. Such disciplining is enacted primarily through the two interrelated cultural discourses of medicine and appearance" (2002, p. 10). Moreover, the intersectional perspective contributes to the deconstruction of the biomedical model which, based on the norm/deviation binomial, sees disability as an experience that necessarily leads to, for example, disadvantages or limitations in sex life and enjoyment of erotic pleasure.

Along these lines, Wilkerson (2002), based on Rubin (1993), points out that the sexual agency of disabled people cannot be understood outside this system of oppressions, seeing as the individuals, depending on their position on the social ladder, will have more or less legitimacy for agency. The intersectional dimension of the sexuality experience for disabled people complicates professional work on this issue, besides pointing to the need for political strategies that will lead to the deconstruction of social discourses that identify certain subjects as inferior to others. In other words, working on sex education with young, rich, white men with disabilities is different to doing so with elderly, poor, lesbian black women with disabilities, since the latter have less legitimacy to express their sexuality.

The third assumption refers to the need to *consider the relations of dependence and interdependence as fundamental in ensuring the sexual and reproductive rights* of disabled people. Many people with a physical disability, for example, cannot experience their sexuality within the limits that are established as the standard of functionality. For persons with intellectual disability, there need to be adjustments to their comprehension of information on sexuality and prevention including, in some cases, as Shakespeare highlights (2014), protection against sexual abuse. Furthermore, there are situations in which no matter how many adaptations in spaces and forms of providing information are offered, certain people will need support and relations of interdependence to exercise their sexuality. Therefore, a major challenge in guaranteeing sexual and reproductive rights is in deconstructing the modern ideal of independence that constitutes all subjects and considers relations of dependence and interdependence as an ethical principle and a human right, as proposed in the field of studies, the Ethics of Care (Kittay 2011). This principle is very valid, principally for people who depend on personal assistants to organise their space and help them to prepare for sex or even provide a sexual assistance service so that they can discover and/or amplify their erotic potential. When we think about reproductive rights specifically, the ethics of care viewpoint, which Kittay (2011) has developed, is useful since one of the challenges it proposes is that of not making disabled people more dependent than they really are. Many disabled people, with the appropriate support to help them exercise their sexual and reproductive rights, may, if they wish, not only have sexual relations but also get married and start a family. However, there needs to be an effectively implemented perspective of care as a public and a human right.

Another important assumption aimed at ensuring sexual and reproductive rights is to *consider sexuality as a process of ethical, aesthetic and political creation*. To this end, it is essential to deconstruct conceptions of sexuality that restrict it to penis/vagina coitus and reproduction or limited to the cycle of sexual response framed by desire, arousal and orgasm. This principle is in agreement with that proposed by Gesser and Nuernberg (2014), when they identify the need to guide disabled people to the infinite erotic possibilities that each body possesses, which may be discovered with the support of many others, among which are sexual assistants, when necessary. As such, the previously presented principle that we should consider relations of dependence and interdependence is vitally important, since many people could discover new erotic potential through relations established with other people, whether boyfriends/girlfriends, husbands/wives or even sexual assistants. Thus, based on Ethics of Care studies and the philosophy of Judith Butler (2009), it is proposed that practices related to the field of sexuality concerning disabled people should aim to break away from ableism and incorporate the relations of dependence and interdependence previously presented.

To expand the creative potential related to sexuality, it is necessary to transcend the idea that restricts relationships to monogamous and heterosexual ones, already criticised by Foucault (1978). In this sense, Sakellariou (2011), Shakespeare (2007) and Tepper (2000) and highlight that disabled people, either in a stable relationship or not, can feel pleasure through sex, whereby it need not be restricted to penis/vagina coitus. In order for us to promote practices based on the principle that sexuality is a process of creation, we need to rethink the definition of sexuality and eroticism, seeing as these concepts carry a normocentric bias in which corporeal and pleasure possibilities are guided by the hegemony of the body without impediments of a physical, sensory or intellectual nature. Along these lines, Shakespeare (2007) critiques the quite medicalised view of the literature and practice related to sexuality, which is centred on the biological deficiencies and functional incapacities of disabled people and the pursuit of medical treatment. The author proposes that the focus be relational and centred on the context, corroborating the proposal by Block et al. (2012). Moreover, Shakespeare suggests:

Just as the threat of HIV forced gay men and others to think more imaginatively about sex, so the limitations of impairment mean that some disabled people have to reorient their sexual lives, perhaps away from penetrative sex and toward more adventurous and varied ways of achieving and giving pleasure. Others may rely on personal assistants to get them ready for sex and sometimes even need to use sexual surrogates or prostitutes to achieve sexual expression.

*(Shakespeare 2007, p. 56)*

As Gesser and Nuernberg (2014) have already highlighted, the experience of disability offers the potential to redefine sexuality and eroticism. Certain people with physical disabilities, for example, those with urinary incontinence, which can often lead to desexualisation, may bring their partners with or without a disability to another plane of affectivity in which there is an openness and comprehension of such difficulties, and this leads to greater sexual intimacy. Blind people can develop a relationship with the voice of their partners and the oral expressions at the moment of affective and sexual contact that involve more eroticism than most of those who see. To a degree, these examples focus on certain human potential that is revealed in the sexual experience of disability, contrary to the medical outlook, that seeks what is lacking or which diverges negatively. Instead of seeking a priori gaps and impossibilities for disabled people in the sexual sphere, there should be a willingness to discover the new potential that corporal and functional variation allows in affective, erotic and sexual life.

This erotic potential may be discovered through groups of disabled people, who yearn for spaces to discuss sexuality and reproduction. An example of this practice refers to work by Gesser and Nuernberg (2012) involving a group of women with disabilities who are in an association that fights for the rights of persons with physical disabilities. One of the objectives of this work was to contribute to the recognition, by women with physical disabilities, of sexual and reproductive rights. Results showed that the activities involving women with disabilities helped to highlight the strangeness of the oppressive discourses on disability and reinvented some aspects regarding the experience of sexuality, especially those relating to knowledge of the body and the construction of new possibilities for feeling pleasure. Participants also began to identify episodes of violence they had gone through in their daily lives, which had been naturalised previously.

Therefore, a professional approach to working with groups of disabled people, based on an ethical/political/anti-ableist view of disability, has a fundamental role both in the supply of information on sexuality and in the construction of collective spaces in which disabled people can swap experiences, related to the process of knowing their bodies and feeling pleasure.

Finally, the last assumption addresses the need to *consider the person with a disability as a subject capable of participating in the whole process of constructing strategies related to his/her sexuality,* whether in the field of education, health services or community practices. This premise is aligned with the motto of disabled people "Nothing about us without us," the UN CRPD and with the emancipatory perspective of disability, proposed by Fontes et al. (2014), whose overriding idea is that disabled people must actively participate in the decision-making processes in relation to their sexuality, even in situations of complex dependence. On this point, we need to take care that the strategies used in the development of sex education programmes consider the diverse forms of participation and strive to break away from the barriers present in the different social contexts, whether they be architectural, attitudinal, communicational, pedagogical or methodological. In this manner, breaking with the normative perspectives of participating and investing in an inclusive concept, based not on the autonomous subject, but on the possibilities of interdependence and support of the community (Erevelles 2011) is a strategy of great value for the promotion of sexual agency and the guarantee of the sexual and reproductive rights of disabled people.

Accordingly, in order for disabled people to feel the social legitimacy to participate, professional practices need to have a political commitment to deconstructing the still- predominant framing of disability which, based on the biomedical model's discourse, views it as lack, deficiency and tragedy. It is believed that the premise that disability is a human variation and that sexuality is a process of ethical, aesthetic and political creation offers clues to the construction of practices to promote the agency of disabled people.

## Conclusion

This paper started from the assumption that guaranteeing the sexual and reproductive rights of disabled people is a human rights and social justice issue, and that the biomedical model of disability hinders the guarantee of this right through the processes of normalisation, pathologisation and medicalisation of sexuality. Analysis of recent developments in the field of sexual and reproductive rights in Brazil points to the importance of the intersectional perspective in understanding disability, since it offers elements with which to frame the experiences related to sexuality and reproduction within the different social and political contexts. Analysis of research also indicates that sexual and reproductive rights cannot be thought of as separate from the other rights advocated by the UN CRPD, seeing as determiners, such as gender and social class, for example, have an important impact on guaranteeing these rights.

Considering the numerous barriers in the areas of education and the health services that disabled people face in order to gain access to information about sexuality, it is especially important that the theme related to the sexual and reproductive rights of disabled people is incorporated into the initial and continued training of professionals working in these areas.

## Note

1 The eugenic movement started in England in the 1860s when Francis Galton, a cousin of Charles Darwin, formulated the movement's principles and, simultaneously, began to develop the family-study method (Rafter 1988).

## References

Banks, J. (2018) Invisible man: Examining the intersectionality of disability, race, and gender in an urban community. *Disability & Society* 33 (6), pp. 894–908. Available at: https://doi.org/10.1080/09687599.2018.1456912 (Downloaded: 14 September 2018).

Bastos, O. M., and Deslandes, S. F. (2012) Sexualidade e deficiência intelectual: narrativas de pais de adolescentes' [Sexuality and intellectual handicap: Parents' narratives]. *Physis Revista de Saúde Coletiva* 22 (3), pp. 1031–1046.

Ben-Moshe, L., and Magaña, S. (2014) An introduction to race, gender, and disability: Intersectionality, disability studies, and families of color. *Women, Gender, and Families of Color* 2 (2), pp. 105–114. Available at: https://www.jstor.org/stable/10.5406/womgenfamcol.2.2.0105 (Downloaded: 18 March 2018).

Block, P., Shuttleworth, R., Pratt, J., Block, H., and Rammler, L. (2012) Disability, sexuality and intimacy. In N. Pollard and D. Sakellariou (eds.). *Politics of occupation-centred practice: Reflections on occupational engagement across cultures.* Hoboken: John Wiley & Sons, pp. 162–179.

Butler, J. (2009) *Frames of war: When is life grievable?* New York, NY: Verso.

Campbell, F. K. (2001) Inciting legal fiction: Disability's date with ontology and the ableist body of the law. *Griffith Law Review* 10, pp. 42–62.

Campbell, F. K. (2008) 'Able' refusing able(ness): A preliminary conversation about ableism. *M/C Journal* 11 (3), pp. 1–9. Available at: http://journal.media-culture.org.au/index.php/mcjournal/article/view/4 6 (Downloaded: 24 September 2019).

Carey, A. C., and Ben-Moshe, L. (2014) Preface: An overview of disability incarcerated. In L. Ben-Moshe, C. Chapman, and A. C. Carey (eds.). *Disability incarcerated: Imprisonment and disability in the United States and Canada*, pp. ix–xiv. New York, NY: Palgrave Macmillan.

Davis, L. J. (2006) Constructing normalcy: The bell curve, the novel, and the invention of the disabled body in the nineteenth century. In L. J. Davis (ed.). *The disability studies reader*. New York, London: Routledge, pp. 3–16.

Decreto n° 6.949/2009, de 25 de agosto de 2009 (2009, 25 de agosto) Promulga a Convenção Internacional sobre os Direitos das Pessoas com Deficiência e seu Protocolo Facultativo, assinados em Nova York, em 30 de março de 2007 [Publishes the text of the Convention on the Rights of Disabled People and its Optional Protocol, signed in New York on March 30, 2007]. Brasília: Diário Oficial da União.

Denari, F. E. (2010) Adolescência, afetividade, sexualidade e deficiência intelectual; o direito ao ser/estar [Adolescence, affection, sexuality and intellectual disability; the right to be]. *Revista Ibero-Americana de Estudos em Educação* 5, pp. 1–9.

Erevelles, N. (2011) *Disability and difference in global contexts: Enabling a transformative body politic*. New York, NY: Palgrave Macmillan.

Fontes, F., Martins, B. S., and Hespanha, P. (2014) The emancipation of disability studies in Portugal. *Disability & Society* 29 (6), pp. 849–862. Available at: https://doi.org/10.1080/09687599.2014.880332 (Downloaded: 14 September 2018).

Foucault, M. (1978) *The history of sexuality volume 1: An introduction*. London: Allen Lane.

Garland-Thomson, R. (2002) Integrating disability, transforming feminist theory. *NWSA Journal* 14 (3), pp. 1–32. Available at: https://muse.jhu.edu/article/37970 (Downloaded: 9 April 2012).

Garland-Thomson, R. (2005) Feminist disability studies. *Signs* 30 (2), pp. 1557–1587.

Gesser, M., Block, P., and Nuernberg, A. H. (2019) Participation, agency and disability in Brazil: Transforming psychological practices into public policy from a human rights perspective. *Disability and the Global South* 6 (2), pp. 1772–1791.

Gesser, M., and Nuernberg, A. H. (2012) Gênero, direitos humanos e cidadania: A psicologia contribuindo para a ressignificação da experiência da deficiência em mulheres de camadas populares [Gender, human rights and citizenship: The psychology contributing to the redefinition of the experience of disability of low-income women]. In *Prêmio Profissional: "Democracia e Cidadania Plena das Mulheres"*. Brasília: Conselho Federal de Psicologia, pp. 15–41.

Gesser, M., and Nuernberg, A. H. (2014) Psicologia, sexualidade e deficiência: Novas perspectivas em direitos humanos [Psychology, sexuality and disability: New perspectives on human rights]. *Psicologia: Ciência e Profissão* 34 (4), pp. 850–863. Available at: http://dx.doi.org/10.1590/1982–370000552013 (Downloaded: 14 December 2014).

Gesser, M., Nuernberg, A. H., and Toneli, M. J. F. (2014) Gender, sexuality, and experience of disability in women in Southern Brazil. *Annual Review of Critical Psychology* 11, pp. 417–432. Available at: https://th ediscourseunit.files.wordpress.com/2016/05/23-gender.pdf (Downloaded: 14 January 2014).

Gill, M., and Erevelles, N. (2017) The absent presence of Elsie Lacks: Hauntings at the intersection of race, class, gender, and disability. *African American Review* 50 (2), pp. 123–137. Available at: https://doi. org/10.1353/afa.2017.0017 (Downloaded: 14 December 2018).

Gomes, R. B., Lopes, P. H., Gesser, M., and Toneli, M. J. F. (2019) New dialogues in feminist disability studies. *Estudos Feministas* 27 (1), pp. 1–13.

Kittay, E. F. (2011) The ethics of care, dependence, and disability. *Ratio Juris* 24 (1), pp. 49–58. Available at: 10.1111/j.1467-9337.2010.00473.x (Downloaded: 14 December 2018).

Linton, S. (1998a) *Claiming disability: Knowledge and identity*. Available at: http://www.amazon.co.uk/kindlestore (Downloaded: 15 June 2018).

Linton, S. (1998b) *Claiming disability: Knowledge and identity* (Kindle Version). Available at: http://www.amazon.com/

Lopes, P. H. (2018) "Eu posso ser mãe, sim": Processos de significação acerca da gestação e da maternidade de mulheres com deficiência' ["Yes, I can be a mother": Processes of signification about the gestation and maternity of disabled women. Dissertation (Master's in Psychology). Florianópolis: UFSC.

Maia, A. C. B. (2012) A sexualidade depois da lesão medular: uma análise qualitativa-descritiva de uma narrativa biográfica [Sexuality after spinal cord injury: An analysis of biographical narrative]. *Interação em*

*Psicologia* 16 (2), pp. 227–237. Available at: http://dx.doi.org/10.5380/psi.v16i2.21212 (Downloaded: 14 December 2018).

Maia, A. C. B. (2016) Vivência da sexualidade a partir do relato de pessoas com deficiência intelectual [Sexuality experience as from report of people with intellectual disability]. *Psicologia em Estudo* 21 (1), pp. 77–88.

Maia, A. C. B., and Ribeiro, P. R. M. (2010) Desfazendo mitos para minimizar o preconceito sobre a sexualidade de pessoas com deficiências [Dispelling myths to minimize prejudice about the sexuality of people with disability]. *Rev. Bras. Educ. Espec.* 16 (2), pp. 159–176.

Mello, A. G. (2016) Deficiência, incapacidade e vulnerabilidade: Do capacitismo ou a preeminência capacitista e biomédica do Comitê de Ética em Pesquisa da UFSC [Disability, inability and vulnerability: On ableism or the pre-eminence of ableist and biomedical approaches of the Human Subjects Ethics Committee of UFSC]. *Ciência & Saúde Coletiva* 21 (10), pp. 3265–3276.

Nicolau, S. M., Schraiber, L. B., and Ayres, J. R. C. M. (2013) Mulheres com deficiência e sua dupla vulnerabilidade: Contribuições para a construção da integralidade em saúde [Women with disabilities and their double vulnerability: Contributions for setting up comprehensive health care practices]. *Ciência & Saúde Coletiva* 18 (3), pp. 863–872. Available at: https://dx.doi.org/10.1590/S1413-812320130003 00032 (Downloaded: 14 December 2018).

Rafter, N. H. (1988) White trash: Eugenics as social ideology. *Transaction/Society* 26 (1), pp. 43–49.

Regis, H. (2013) Mulheres com deficiência intelectual e a esterilização involuntária: De quem é esse corpo? [Women with intellectual disability and involuntary sterilization: Whom this body belongs to?]. Dissertation (Master's in Psychology). Florianópolis: UFSC.

Rubin, G. (1993) Thinking sex: Notes for a radical theory of the politics of sexuality. In Henry Abelove, Michele Aina Barale, and David M. Halperin (eds.). *The lesbian and gay studies reader.* New York: Routledge, pp. 3–44.

Sakellariou, D. (2011) Sexuality and disability: A discussion on care of the self. *Sexuality and Disability* 30 (2), pp. 187–197.

Shakespeare, T. (2007) Disability, normality, and difference. In *Psychological challenges in obstetrics and gynecology.* London: Springer, pp. 51–59.

Shakespeare, T. (2014) *Disability rights and wrongs revisited.* 2nd Edition. New York: Routledge.

Simões, J. (2015) Deficiência intelectual, gênero e sexualidade: Algumas notas etnográficas em uma APAE do interior do estado de São Paulo-Brasil [Intellectual Disability, Gender and Sexuality: Some ethnographic notes on a APAE in the state of São Paulo-Brazil]. *Rev. Fac. Med.* 63 (Suppl. 1), pp. 143–148.

Soares, A. H. R., Moreira, M. C. N., and Monteiro, L. M. C. (2008) Jovens portadores de deficiência: sexualidade e estigma. *Ciência & Saúde Coletiva* 13 (1), pp. 185–194. https://dx.doi.org/10.1590/S1413-81 232008000100023

Tepper, M. S. (2000) Sexuality and disability: The missing discourse of pleasure. *Sexuality and Disability* 18 (4), pp. 283–290.

United Nations (2006) Convention the rights of persons with disabilities. Adopted on 13 December 2006 at the United Nations Headquarters in New York. Available at: https://www.un.org/development/desa /disabilities/convention-on-the-rights-of-persons-with-disabilities.html

WHO (2011) *World report on disability.* Geneva: World Health Organization. Available at: https://www.who .int/disabilities/world_report/2011/report.pdf (Downloaded: 14 September 2012).

Wilkerson, A. L. (2002) Disability, sex radicalism, and political agency. *NWSA Journal* 14 (3), pp. 33–57. Available at: muse.jhu.edu/article/37983 (Downloaded: 14 September 2018).

# PART II

# Subjugated histories and negotiating traditional discourses

# 7

# SEXUALITY, DISABILITY AND MADNESS IN CALIFORNIA'S EUGENICS ERA

*Jess Whatcott*

In the early twentieth century, reformist administrators urged the government in California in the western United States of America (USA) to take action to contain the "defective classes" (*Board of Charities and Corrections* 1915, p. 35). In an annual report to the state legislature, the state-appointed Board of Charities and Corrections (1919, p. 62) declared:

> We must make it our business to awaken the people to a realisation of the fact that it is as foolish to permit human defectives to reproduce themselves as to permit defective domestic animals to beget offspring. The whole stream of human life is constantly polluted by the admixture of the tainted blood of the extremely defective. If this source of contamination could be cut off, the beneficial effects would begin to show in a single generation, and, in a very few generations, the average level of human society would be very materially lifted.

According to this theory of eugenics adopted by the board, the defective classes passed on weaknesses to their children, and were consequently responsible for causing a number of social problems, including juvenile delinquency, criminality, insanity, poverty and general immorality. Consequently, the California legislature enacted a number of eugenic policies to contain the spread of defectiveness within the state's population. Similar eugenicist language can be found in state laws that authorised the construction and expansion of institutions of confinement, including state hospitals, state homes for the feeble-minded and juvenile and adult reformatories. Additionally, eugenic language provided the rationale for the state's *Asexualization Law 1909*, which permitted the non-consensual reproductive sterilisation of inmates in state institutions.

In 2003, the California state government apologised for its role in coercively sterilising approximately 20,000 institutionalised people under the *Asexualization Law 1909* and the *Asexualization Law 1913 (1917)*. The California Senate Concurrent Resolution (2003) reads in part:

> Under California's eugenics law, and subsequent amendments to it, inmates of state hospitals and institutions for the mentally retarded, and certain prison inmates who were considered sexual or moral 'degenerates' or were serving life sentences, and epi-

leptics could be involuntarily sterilised so that they would not produce similarly disabled offspring…

In practice, the eugenics laws were used to target virtually any human shortcoming or illness, including alcoholism, drug addiction, pauperism, syphilis and criminal behaviour.

According to this apology, the sterilisation laws were designed to target those people who would today be categorised as mad ("inmates of state hospitals"), intellectually disabled (inmates of "institutions for the mentally retarded"), sexually deviant ("degenerates") and epileptic. Yet the state's apology diffuses the impact on these specific groups by describing how a wide variety of people were targeted for "virtually any human shortcoming or illness." A generous read of the apology is that recognising the impact of eugenics policy on diverse people will increase the number of people who view themselves as having a stake in remembering this history of injustice. However, this rhetorical strategy simultaneously diminishes an understanding of the ways that the category of "defective" specifically impacted those that today might identify as disabled and mad, as well as those—disabled, mad and otherwise—who identify as a part of sexual minority groups. Additionally, this strategy neglects the gendered discourse surrounding eugenics laws, and, consequently, the unique impact of eugenics programmes on assigned women and girls.[1] Furthermore, rather than arbitrary application of eugenics, the rhetoric of the apology obscures the role that eugenics discourse played in constructing the very categories and parameters of intellectual disability, physical disability, madness and sexual deviance. Whereas these identity categories are often taken for granted as biological properties of an individual, this analysis concurs with earlier evaluations of eugenics policies as a practice of socially constructing gender, sexual deviance, disability and madness (Kline 2001; Rembis 2011; Stern 2005).

While other analyses of eugenics have tended to focus on a single category of difference, this chapter insists that California's eugenics history illustrates precisely how these categories of difference were intimately and irrevocably co-produced. I review archival evidence that shows how disability, madness and sexuality for assigned women and girls were co-constituted under the discourse of eugenics. Firstly, the sexually deviant behaviour of women and girls was articulated by eugenicists as a disability of the gendered body-mind. In the present case study, those people assigned to the category of girls and women who engaged in sex outside of marriage, or expressed any other kind of "deviant sexual behaviour," were labelled "defective" and were targeted by eugenics policies of institutionalisation and sterilisation. Secondly, assigned women's and girls' physical disabilities, intellectual disabilities, and madness were conceived of as being inherently sexual deviant. That is, the possibility of women and girls with disabled and/or "mad" body-minds having sex and expressing sexuality, was construed as a threat to the health of the normative body politic. This took place, as I will discuss below, even when disabled and mad assigned women and girls had sex within the confines of heterosexual marriage.

Whereas I argue that we cannot understand the constitution of disability, madness or sexuality as separate categories under eugenics, this is also a politically fraught orientation. This is because it forces a re-examination of the mid-twentieth century effort in the US to re-brand homosexual identities as normal and healthy. The effort to normalise lesbian and gay identities (and later bisexual, transgender and queer identities, or LGBTQ) challenged earlier understandings of sexual minorities as abnormal, pathological, and requiring medical treatment (Kunzel 2017). Re-linking disability, madness and sexual deviance through the kind of analysis I call for, problematises both the strategy and outcome of this work to de-pathologise and de-medicalise queer sexualities. However, revisiting disability, madness and sexuality as co-constituted categories also creates new possibilities that are politically urgent. The shared history of eugenics invites scholars of queer studies, disability studies, and madness studies to learn from each other and creates opportunities for political coalitions and alliances to address urgent questions of health and well-being affecting sexual minorities.

## California eugenics

In the first three decades of the twentieth century, administrators of California state institutions actively promoted eugenics theory (Kline 2001). Scholars trace the term "eugenics" to a British statistician named Sir Francis Galton in 1883, who applied his cousin Charles Darwin's theory of evolution to humans (Kline 2001; Lombardo 2008; Stern 2005). Galton theorised that through statistical analysis of inherited traits, and the implementation of selective breeding programmes, the positive qualities of the human race could be enhanced, while the negative qualities could be diminished. Adopting Galton's concept, but adapting it for a variety of purposes, eugenicist reformers believed they could save the human race from degradation and protect the social order from ruin through government-sponsored programmes driven by eugenics (Kline 2001; Ordover 2003).

Eugenics ideas became widespread in the United States during the first three decades of the twentieth century and were especially taken up by a social movement of middle-class whites in the United States called Progressives (Kline 2001). According to Wendy Kline (2001), Progressives imagined a secular, scientific, public sphere that could grapple with the social and economic changes wrought by the industrial revolution, including rapid urbanisation and migration. In California, Progressivism offered a counter to the political power of monopoly capitalists, who held control of natural resources, infrastructure and labour. Progressives took over the state government in the 1910's to break the iron grip of a gigantic railroad monopoly that muckraker Frank Norris called "the octopus" (Norris 1901). Overlapping with the moral purity movement, Progressives were equally alarmed at social changes that rapid urbanisation had caused, including vagrancy, the spread of venereal disease, commercial sex, slum conditions, gambling, child labour and child abandonment (D'Emilio and Freedman 1997; Kline 2001; Rembis 2011). Countering possible radical critiques that the poor quality of life in urban slums was caused by monopoly capitalism, Progressives blamed poor people for their own misery. Harkening back to the early nineteenth century theories of Thomas Malthus (Ordover 2003), Progressives theorised that the conditions facing poor and working-class people in urban areas were caused by their unchecked breeding, and the passing on of weakness and undesirable traits to masses of children.

Like other reformers across the United States, Californians implemented eugenics programmes to protect and improve the human race. Such eugenics programmes are described in a report by the Board of Charities and Corrections (1915, p. 35) to the California legislature demanding:

(a) More definitive and stringent laws for the commitment of feeble-minded persons,
(b) establishment of farm colonies for the feeble-minded, (c) segregation of the sexes,
(d) sterilisation when necessary, (e) laws preventing the marriage of the feeble-minded,
(f) immigration laws to exclude the defective classes, and (g) special schools for the backward child.

By the time of this report, the state had already implemented some of these proposals, including aiding the federal enforcement of restrictive immigration laws (Hernandez 2017).

At the state level, the California legislature passed additional eugenics laws. Regarding proposal (a) in the list above, the state intensified efforts to commit the defective classes to institutions, starting with the *Insanity Law 1897*. California's *Insanity Law 1897* was designed to "provide a uniform government and management of the state hospitals for the insane, and to provide for the care, custody and apprehension of persons believed to be insane, and the commitment of insane persons." The capaciousness of the *Insanity Law 1897* caused the number of inmates in the state hospitals to double over the next 20 years. According to annual reports by the state's Commission in Lunacy, the daily population in state hospitals measured at mid-year almost doubled from 5,276 in 1900 to 10,119 in 1920. The swelling population caused serious overcrowding in the state hospitals, and administrators and oversight boards begged the state to expand existing facilities and to build a new hospital in southern California.

Originally governed by an application system for open beds rather than a legal commitment process, the state's home for the feeble-minded in Eldridge, Sonoma County also experienced rapid population growth and extreme overcrowding over the same period. According to annual reports to the legislature written by the Board of Charities and Corrections, the average daily population of the home tripled from 552 inmates in 1900 to 1537 in 1920. Originally a single building dedicated to housing so-called feeble-minded children whose families could not care for them, the Sonoma State Home had by 1914 grown into a farm colony that was over fifteen hundred acres in size and housed people of all ages from 0 to 80. Narrative descriptions and photographs of the home show large brick administration and hospital buildings mixed in with small cottage dormitories, along with provisional outdoor tent sleeping areas. The home included a dairy, vegetable gardens, fruit orchards, a laundry, a bakery, and a kitchen, each staffed partially, if not predominantly, by the inmates themselves. The overcrowded conditions at Sonoma, and distance between that home and the growing southern part of the state, compelled the government to authorise funds to build a second home for the feeble-minded, named the Pacific Colony, which opened permanently in 1927 near the city of Pomona, Los Angeles County.

Administrative reports and other documents in the state archives insist that life at the state hospitals and state homes for the feeble-minded was intended to be therapeutic, or, at the very least, to provide custodial care. However, reading between the lines in the archive, and supplementing this research with newspaper accounts, indicates that inmates were denied autonomy at best, and subjected to outright abuse at worst. Analysis of the *Insanity Law 1897* indicates that eugenics policies authorised a practice of forcible commitment that mimicked conviction and detention for crime. Newspaper accounts following the passage of the law indicate repeated accusations of abuse by staff and neglect of inmates, as well as the use of invasive and experimental medical therapies that doubled as punishments. Additionally, an assessment of the use of "occupational therapy" at multiple institutions shows that inmates who were capable of physical labour were put to work sustaining life on the farm colonies, including working in both on-site and off-site laundries, bakeries, kitchens and in agriculture and care work for other inmates. This work was billed as a kind of beneficial treatment for inmates, rather than as labour that should be paid. An assessment of the recreational opportunities at the facilities suggests that for those perceived to be unable to work, life on the farm colonies was probably filled with minimal intellectual or physical stimulation, and long bouts of unrequited boredom.

Despite this evidence, administrators claimed in annual reports that inmates were "apparently happy and decidedly grateful to the state for what is being done for them," *(Department of Institutions* 1929, p. 12). Unfortunately, there are limited materials in the state archives composed by early twentieth century inmates themselves that can confirm or deny this claim. However, one note written by a woman inmate of Stockton State Hospital dated 26 April 1937 reads:

Any more of this 'prison' existence will stifle me—I thought I could take it—but now I want to get out…I hold the highest regard for you and doctor Paine and most of your staff—I think you have a wonderful hospital (but many things needed) but I beg of you to please release me.

While the author of this brief note expresses some regard for the staff of the hospital, her plea for release from the prison-like setting of the hospital at the very least complicates the state's assertion that inmates were grateful for their detention.

In addition to these concerns about eugenics institutionalisation, the state passed the first version of the *Asexualization Law 1909*, taking away another level of autonomy from inmates, as well as creating more opportunities for abuse. The revised *Asexualization Law 1917* stipulated:

Before any person who has been lawfully committed to any state hospital for the insane, or who has been an inmate of the Sonoma State home, and who is afflicted with mental disease which may have been inherited and is likely to be transmitted to descendants, the various grades of feeble-mindedness, those suffering from perversion or marked departures from normal mentality or from disease of a syphilitic nature, shall be released or discharged therefrom, the state commission in lunacy may, at its discretion…cause such person to be asexualised.

The law went on to detail that medical superintendents at the state homes and the state hospitals were explicitly authorised to reproductively sterilise almost any inmate at any time, whether or not the patient or their next of kin consented.[2] Because the *Asexualization Law 1917* gave wide discretion to institutional administrators, some hospitals and homes invested serious resources and time into sterilisation, while others sterilised very few inmates (Wellerstein 2011). In the case of the Sonoma State Home, for example, one administrator named F.O. Butler was particularly active in sterilisation, going as far as committing girls and women to the institution apparently solely for the purpose of sterilising them before release (Kline 2001; Stern 2005).

A common perception is that eugenics beliefs and practices disappeared, or at least went underground, following the revelations about the Holocaust in Europe under the Nazi regime (Ordover 2003). However, feminist scholars have argued that eugenics beliefs and practices persisted in a variety of forms in California into the latter part of the twentieth century (Roberts 1997; Stern 2005). Even as the state of California was at the forefront of the movement to deinstitutionalise psychiatric care in the 1960s, the *Asexualization Law* was not officially repealed until 1978 (Stern 2011). Although sterilisations at the state level had declined over the 1950s, the repeal followed another case of coercive sterilisations, targeting Mexican and Central American women in Los Angeles County Hospital (Stern 2011). Eugenics rationale also emerged in a case of sterilisations of people in one of California's large prisons for women taking place as recently as 2010 (Whatcott 2018). As ideas and practices oriented toward the perfection of the human race continue to emerge in the twenty-first century, the practice of state-sponsored eugenics in California arguably cannot be understood to be restricted to the early twentieth century. Rather, California eugenics has continued relevance to the present day.

## "Moral imbeciles": Girls and women targeted by eugenics

Variation in bodies and minds was targeted by eugenics-based institutionalisation and sterilisation policies in California. A report compiled by the state- appointed Commission in Lunacy (1913, p. 22) described the inmates of the state hospital system as:

A vast class of nervous and mental conditions, embracing insanity, epilepsy and fee-blemindedness on the one hand; alcoholic tendencies, vice, eccentricities, absence of the moral sense, undue excitability and various anomalies of conduct and disposition on the other.

This compilation is somewhat shocking in the present day, linking together what are now understood to be very distinct categories of madness, physical disability, neurodivergence, drug addiction, queerness, and failure to conform to social norms. Yet rather than reject these early twentieth century categorisations as nonsensical, I argue for re-examining them for what they might reveal about the entanglements of present-day categories of disability, madness and queerness. From the records of early twentieth century California institutions emerges a consistent pattern at the intersection of gender, sexuality, disability and madness, regarding the targets of state-sponsored eugenics-based intervention.

One case that helps to tease apart the gendered and sexualised patterns of eugenics in California is the case of Elise, committed to Sonoma State Home for the Feeble-Minded in 1914.[3] Because she did not know, or would not share her age, staff at the home guessed that Elise was approximately 20 years old. One staff member wrote in the casebook entry, "Little is known about [Elise]. She came from Los Angeles County Hospital. They received her from Whittier Reform School…Was then tried twice for insanity and both times dismissed." The last line is a reference to a failed application of the *Insanity Law 1897* described above, and signals that a judge twice ruled that Elise was considered not insane. Despite the failure to either commit Elise to a state hospital, or to keep her at the state's reform school, a spot was eventually secured for Elise at Sonoma State Home. Once committed to a state institution, Elise's stay was indefinite, as, according to policy, release was premised on the assessment of a medical superintendent that she had "recovered." As of 1921, from which point the casebook was no longer updated, Elise was approximately 27 years old and still institutionalised at the home. Once institutionalised, Elise also became a candidate for reproductive sterilisation under the *Asexualization Law*, amended in 1917. Soon after this version of law was passed, Elise, then approximately age 23, underwent an irreversible, and medically unnecessary "double partial salpingectomy for sterilisation." The most advanced sterilisation procedure at the time, salpingectomy involved the complete removal of both fallopian tubes, and would have required several weeks of recovery.

Elise lacked both a legal diagnosis of insanity under the terms of the *Insanity Law 1897*, and also a juvenile delinquency or juvenile dependency ruling under the *Juvenile Court Law 1909* that could keep her at a reform school, allowable for girls up to age 23. However, something drove administrators to persist on placing Elise in state care. The staff member completing the casebook entry at Sonoma State Home inadvertently reveals the urgency by diagnosing Elise as a "moral imbecile." Analysis of the use of this diagnosis in the first two decades of the twentieth century indicates that Elise was either believed to be already engaged in immoral behaviour, or was presumed incapable of protecting herself from being seduced into immorality. As I discuss further below, both of these possibilities—that Elise had chosen a life of disrepute or was at risk of being forced into one—were conflated under the discourse of eugenics in California.

Elise's case illustrates how people assigned to the category of girls and women were targeted for state-sponsored eugenics intervention through a conflation of feminine sexual deviance, physical condition and mental defect. Terms like "moral imbecile" marked a particularly gendered component of the broader "defective class." Similarly, the term "feeble-minded" was applied to those "who, by reason of incomplete development of brain or mind, are unable to perform their duties as members of society in the position of life to which they were born and live a normal life" (*Commission in Lunacy* 1915, p. 15). For those assigned to the category of men and boys, feeble-mindedness referred to those who were "unable to compete with their fellows in the struggle for existence…[and] unable to manage their own affairs with ordinary prudence." However, for those assigned to the category of girls and women, feeble-mindedness signalled something less about inability to compete in the capitalist order, and more about the feminine tendency toward immorality. Namely, feeble-minded women and girls were those who were "unable to exercise that self-control which is necessary to live a moral life" (*Commission in Lunacy* 1915, p. 15).

As this discourse attests, in the case of people assigned by the state to the category of girls and women, "defective" was synonymous with sexual activity. Nowhere is this more explicit than in the discussion about the girls detained at the California School for Girls.[4] A list from the Superintendent of the California School for Girls (1919, p. 19) described the girls and women that required state intervention in a report to the state legislature:

> Ninety-five per cent of the girls committed to this school are sexually immoral.
>
> …71 per cent have been treated for gonorrhea.
>
> Fourteen per cent are or have been married.
>
> Twelve and a half per cent of those married have bourn [sic] children (most of them were forced marriages and the girls continued their delinquency).
>
> Seven per cent of the 283 have given birth to illegitimate children. None of them have married…
>
> Six per cent have been victims of the lusts of their own fathers or brothers, or both…
>
> Another feeble-minded girl, with the knowledge of her parents, had one child by a white man and a second one by a Chinaman.
>
> Semisexualism is not uncommon.

As other scholars of the period have observed, understandings of problematic behaviour for girls and young women included sex before marriage, pregnancy outside of marriage, abortion, interracial sex and same-sex sex, here named as "semi-sexualism" (Kunzel 1993). However, two more observations are worth analysing further.

First, under the discourse and practice of eugenics, there was no distinction made between consensual sexual behaviour and sexual violence victimisation, including incest and child sexual abuse. In the case of Elise, for example, it is unclear whether the diagnosis of "moral imbecile" was a reference to her choice to engage in sex, or whether she was merely viewed as being at risk of sexual violence. Included in the list are indications that assigned girls may have been detained in whole or in part because they were "known to be victims of deplorable sex practices." Sex outside narrow confines for girls and women was construed as immoral, whether it was consensual or not. This is because the danger of feeble-minded women under the discourse of eugenics was through human reproduction, which could occur through consensual sex just as easily as non-consensual sex. Arguments were made for the institutionalisation of women and girls through their portrayals as potential "victim[s] of passionate, brutal men" (*Board of Charities and Corrections* 1907, p. 69). Rather than contain male perpetrators of sexual violence, the risk of feeble-minded reproduction became yet another argument for containing girls and women in institutions like the Sonoma State Home.

The second observation is that the list gestures towards a pattern where even sex and pregnancy within the confines of heterosexual marriage could at times be included under the rubric of girls' and women's sexual immorality. Consider the institutionalisation of assigned girls who were married, and those who had borne children within the confines of marriage. According to the Superintendent, these girls were detained because "most of them were forced marriages and the girls continued their delinquency." This indicates that forced marriage was designed to be a form of containment for the female moral imbecile, but if the girl continued to engage in gender non-conforming or otherwise deviant embodiment, then her sexual activity continued to be threatening under the terms of eugenics. Consequently, the possibility of defective and feeble-minded body-minds having sex and sexuality posed a queer threat to the health of the normative body politic, creating the need for state intervention.

Although not included in the list compiled by the girls' school superintendent, further analysis of state documents indicates that girls' and women's participation in commercial sex was a primary marker of immorality. Whereas commercial sex was treated as a criminal matter in the early twentieth century, the discourse of eugenics also pathologised girls' and women's participation in commercial sex work. First, commercial sex workers were assumed to be mentally defective. For example, the *Commission in Lunacy* (1915, p. 16) insisted in a 1915 report:

> There is no doubt but that the higher-grade defectives, the subnormally dull, and those with specialized defects, are responsible for much drunkenness, much prostitution, much crime, and much that tends to disturb public welfare... There is no doubt but that a very large proportion of the prostitutes are mentally deficient.

Sexual deviance in the form of commercial sex was here conceived of as an inevitable symptom of feeble-minded and mentally defective women and girls.

Once sex work and defect were linked, assigned girls and women engaged in commercial sex work could be legitimately institutionalised. This practice was heightened during the two World Wars, when Sonoma State Home was proposed as a repository for women and girls who were perceived to threaten soldiers and sailors by spreading both moral impurity and venereal disease. The Commission in Lunacy called for "the construction of a cottage for the housing of delinquent feeble-minded females from around the army and navy camps as a protection to enlisted men. This change in the use of the appropriation was made as an urgent war measure" (1919, p. 75). In the same year, the Board of Charities and Corrections argued that an appropriation of funds during war-time resource scarcity was required because "This need has been forcibly shown in the last two years of the war, when prostitutes have been temporarily detained in the counties, released and floated on to the next community with no attempt of any kind for their social, mental and physical rehabilitation' (1919, p. 67). Engaging in commercial sex itself, rather than just the venereal disease associated with it, was construed as a kind of chronic illness spreadable to men that were needed as able-bodied and able-minded in order to support the war effort.

In sum, tracking the commitment and diagnosis of a "moral imbecile" like Elise illustrates how the discourse and practice of eugenics in early twentieth century California took place at the gendered intersection of sexuality, disability and madness. Eugenics institutionalisation and sterilisation are not stories about disabled people who had no sexuality, nor are they about the sexual immorality of assigned women and girls who were otherwise able-bodied and neurotypical. Rather, eugenics practices illustrate how sexual deviance was conceptualised in the early twentieth century as a disability of the body-mind, and how sex by disabled and mad women and girls was construed as inherently deviant. The process of constructing human variation as forms of intellectual disability and mental illness was intimately entangled in eugenicist concerns about the sexual deviance of people assigned to the category of women and girls. Simultaneously, programmes to target the sexual deviance of women and girls operated by constructing these populations as part of a defective class that was biologically inferior and pathologically dangerous.

## A co-history of disability, madness and sexuality

Today, identity-based social movements work separately toward rights for lesbian, gay, bisexual, transgender and queer (LGBTQ) individuals, disabled people, psychiatric survivors and self-identified mad people. Disability rights and disability justice activists have created cultural space for discussing the sexual health and sexual desires of disabled people, including claiming queer and other sexual minority identities (Kafer 2013). Yet the mainstream LGBTQ movement in the United States, at least, continues to overlook the issues of disabled and mad queers, prioritising campaigns such as legalising same-sex marriage and access to military careers, rather than considering how these campaigns might impact disabled couples and families, much less paying attention to more pressing issues for disabled and mad queers, such as poverty or police violence.

Queer historian Regina Kunzel argues that this disavowal of disability and madness among LGBTQ people in the US is at least partially rooted in the effort to depathologise and demedicalise homosexuality (Kunzel 2017). Specifically, this effort took the form of a campaign to remove the diagnosis of homosexuality from the third edition of the Diagnostic and Statistical Manual (DSM), published in 1973. In the USA, the DSM is the governing text for both health statistics and for clinical psychological evaluation. Early editions of the DSM included "homosexuality" as a category of mental disorder, legitimating dominant views of same-sex sex as a pathology that required so-called homosexuals to be subjected to corrective treatments, such as electroshock therapy. Kunzel (2017) recalls that the campaign to remove homosexuality from the DSM relied heavily on social scientific research that sought to prove that gay people were "healthy," "happy" and otherwise normal. The unintended consequence of this strategy to gain political legitimacy, argues Kunzel, is that it distanced queerness from disability in ways that reinforced both heteronormativity and able-normativity. In short, the project to depathologise homosexuality as an abnormal sexuality left in place dominant assumptions about what constitutes normal bodies and normal minds.

Revisiting eugenics institutionalisation revives the shared, entangled history of sexual deviance, gender normativity, disability and madness. Failure to acknowledge this shared history is probably grounded in the internalised assumption that illness, madness, and disability are undesirable qualities that need to be overcome and disavowed. As Douglas Baynton (2001) has argued, disability continues to act as a signifier for legitimate discrimination and unequal treatment. This leads to hesitation to build (historical) solidarity with disabled and mad people who continue to experience marginalisation, because their exclusion is for seemingly justifiable reasons.

The case for reviving the shared, entangled history of sexual deviance, disability and madness is based on both theoretical possibility and political strategy. Theoretically, scholars are developing new insights about concepts that are explored separately by queer and crip theory, including normativity and cure, by bringing these theoretical fields together (Clare 2017; Kafer 2013). Extending this work, through an examination of shared queer and disability history would expand understanding of how in the present day physical and mental disability are conceived of as forms of sexual deviance, as well as the ways that sexually deviant behaviour continues to be conceptualised as a kind of disability of the body-mind. This historicisation would push queer studies away from reproducing able-body-minded normativity, and would push disability studies further toward its radical engagement with sexual health, queerness and gender non-conformity.

Politically, while disabled and self-identified mad activists have made clear their stakes in LGBTQ organising, LGBTQ movements have not yet recognised their deep stakes in disability activism. As this chapter has shown, sexual minorities are also disabled and mad people. Although gendered sexual deviance is not pathologised in precisely the same ways as it was in the past, queers have a stake in recognising themselves as disabled and mad, especially as they continue to experience the effects of ageing, chronic health conditions (which are associated with increasing air and water pollution) disproportionate levels of mental unwellness, including suicidal ideation and high rates of drug addiction and drug overdose. Disabled and mad people are on the frontlines of fighting for access to affordable and dignified health care, access that all but the most privileged queers desperately need. If queer people are brave enough to recognise themselves as disabled and mad people—a project further developed by recognition of the shared history of disability and gendered sexual deviance—the potential political coalitions could be transformative.

## Notes

1  Throughout the chapter, the phrase "assigned women and girls" draws attention to the way that the state conferred these identity categories onto people who may have identified differently.
2  Section 2 of California's *Asexualization Law 1917* allowed for the sterilisation of state prisoners under slightly different criteria, focused explicitly on sexual depravity.
3  Name has been changed to protect the confidentiality of patients in compliance with California state law governing archived patient records more than 75 years old.
4  The California School for Girls was a juvenile reform facility established in 1913 near Ventura, California. Assigned girls housed at Whittier Reform School were transferred to the new facility after Elise exited the programme.

## References

### *Archives*

Sacramento, California
  California State Archives
    Appendices to the Journals of the Senate and Assembly of the Legislature of the State of California
      Reports of the Board of Charities and Corrections
      Reports of the Board of Prison Directors
      Reports of the California School for Girls
      Reports of the Commission in Lunacy
      Reports of the Department of Institutions
    Mendocino State Hospital Patient Case Files
    Sonoma State Home Case Books
    Stockton State Hospital Commitment Register
    Stockton State Hospital Correspondence Files
    Stockton State Hospital Records of Operations
    Ventura School for Girls Inmate History Register

### *Government Documents*

California State Senate (2003) Senate Concurrent Resolution No. 47: Relative to Eugenics. Resolution Chapter 130, Statutes of 2003. https://leginfo.legislature.ca.gov/faces/billTextClient.xhtml?bill_id=200320040SCR47

### *Newspapers*

*Healdsburg Tribune*
*Los Angeles Herald*
*The Marin Journal*
*The Sacramento Union*
*San Francisco Call*

## *Secondary References*

Baynton, D. (2001) Disability and the justification of inequality in American history. In D. Baynton & L. Umansky (Eds.), *The new disability history*. New York: New York University Press.

Clare, E. (2017) *Brilliant imperfection: Grappling with cure*. Durham: Duke University Press.

D'Emilio, J., & Freedman, E.B. (1997) *Intimate matters: A history of sexuality in America*. 2nd edn. Chicago: University of Chicago Press.

Hernandez, K.L. (2017) *City of inmates: Conquest, rebellion, and the rise of human caging in Los Angeles, 1771–1965*. Durham: The University of North Carolina Press.

Kafer, A. (2013) *Feminist, queer, crip*. Bloomington: Indiana University Press.

Kline, W. (2001) *Building a better race: Gender, sexuality, and eugenics from the turn of the century to the baby boom*. Berkeley: University of California Press.

Kunzel, R.G. (1993) *Fallen women, problem girls: Unmarried mothers and the professionalization of benevolence, 1890–1945*. New Haven: Yale University Press.

Kunzel, R.G. (2017) Queer history, mad history, and the politics of health. *American Quarterly* 69, pp. 315–319.

Lombardo, P.A. (2008) *Three generations, no imbeciles: Eugenics, the Supreme Court, and Buck v. Bell*. Baltimore: Johns Hopkins University Press.

Norris, F. (1901) *The octopus: A story of California*. New York: Doubleday, Page & Co.

Ordover, N. (2003) *American eugenics: Race, queer anatomy, and the science of nationalism*. Minneapolis: University of Minnesota Press.

Rembis, M.A. (2011) *Defining deviance: Sex, science, and delinquent girls, 1890–1960*. Urbana: University of Illinois Press

Roberts, D.E. (1997) *Killing the black body: Race, reproduction, and the meaning of liberty*. New York: Pantheon Books.

Stern, A. (2005) *Eugenic nation: Faults and frontiers of better breeding in modern America*. Berkeley: University of California Press.

Stern, A. (2011) From legislation to lived experience: Eugenic sterilization in California and Indiana, 1907–1979. In P.A. Lombardo (Ed.), *A century of eugenics in America: From the Indiana experiment to the human genome era, bioethics and the humanities*. Bloomington: Indiana University Press.

Wellerstein, A. (2011) States of eugenics: Institutions and practices of compulsory sterilization in California. In S. Jasanoff (Ed.), *Reframing rights: Bioconstitutionalism in the genetic age*. Cambridge: MIT Press.

Whatcott, J. (2018) No selves to consent: Women's prisons, sterilization, and the biopolitics of informed consent. *Signs: Journal of Women in Culture and Society* 44, pp. 131–153. https://doi.org/10.1086/698280

# 8

# DISABILITY RIGHTS THROUGH REPRODUCTIVE JUSTICE

## Eugenic legacies in the abortion wars

*Michelle Jarman*

## Introduction

As many feminist disability studies scholars have documented, tensions have long existed between dominant pro-choice movements and disability rights perspectives on reproduction and abortion. In mainstream abortion rights discourse, disability has been primarily featured as a crucial rationale for maintaining legal access to abortion, including termination decisions based upon prenatal screening. From the 1980s to the early 2000s, disability studies scholars developed strong and effective social and bioethical critiques by challenging selective abortion based upon disability as *neo-eugenic*. Widely cited scholars, such as Ruth Hubbard (2010), Marsha Saxton (2010) and the bioethicist Adrienne Asch (1999), have been especially influential in interrogating ableist belief systems underlying pre-natal screening and selective abortion, as well as in suggesting social, attitudinal and practice-based interventions to address biases—strategies that highlight the importance of input and leadership from disability rights groups. Hubbard's (2010) critique connects selective abortion directly to eugenic doctrine that rose to prominence in many modern industrialising nations in the early 20th century. Tracing eugenic categorisations of the "fit" and the "unfit," constituted by varying designations of disability, she demonstrates the ways eugenicists from the United States, Britain, and, most profoundly, Germany were attempting to proscribe "who should and who should not inhabit the world" (Arendt 1977, cited in Hubbard 2010, p. 114). While not equating eugenics with prenatal screening, Hubbard (2010, p. 115) suggests researchers creating 21st century screening technologies are similarly "engaged in developing the means to decide what lives are worth living," and, by extension, what lives are not. Similarly, Marsha Saxton (2010) has challenged pro-choice movements to pay more attention to disability by connecting the legacy of eugenics to current medical pressure to test and terminate. She argues that healthcare professionals often have too narrow a view of disability, seeing it as the ultimate source of suffering, with little understanding of the "*social factors* that contribute to suffering" (Saxton 2010, p. 125, italics in original). This narrow view, she points out, has been widely expressed in pro-choice rhetoric, but Saxton (2010, p. 127) argues that "choice doesn't exist as a neutral option when 'choice' is so constrained by oppressive values and attitudes" about disability. Indeed, choice arguments have long depended upon disability as a key justification for abortion rights. Adrienne Asch (1999), who contributed immensely to integrating disability

132

rights into reproduction debates in her lifetime, argued that prenatal screening situated disability as a stigmatised master status, used to negatively predict a future life. She interrogated medical pressure and parental decision making that viewed disability as "the only relevant characteristic," and "such a problematic characteristic that people eagerly awaiting a new baby should terminate the pregnancy and 'try again' for a healthy child" (Asch 1999, p. 1652). Whereas these scholars actively distance themselves from anti-abortion arguments, such work has been effective in pushing clinical practitioners to gain more insight into disability in order to offer unbiased information in prenatal counselling, and in demanding that critical attention be paid to ableism in mainstream abortion rights advocacy. At this juncture, however, these anti-eugenic arguments have proven so powerful that they are being appropriated by pro-life activists to pursue state-level abortion bans in cases based upon genetic anomaly—widely touted as protection of foetuses identified with Down syndrome. In the last few years, several U.S. states have pursued such bans: North Dakota is the only state that has enacted legislation to prohibit abortions based on foetal anomaly; Louisiana and Ohio have had their legislation enjoined by court order. In 2018, other states, including Missouri, Pennsylvania, Utah and Kentucky (Guttmacher Institute 2019) considered or introduced similar legislation, and the collective efforts are intended to contribute ultimately to challenging *Roe v. Wade* in the Supreme Court.

This chapter engages with these current debates, especially the use of disability rights and non-discrimination rhetoric by anti-abortion activists, to argue for further bridging of feminist disability studies with reproductive justice frameworks, specifically by situating eugenic histories of racial and disability injustice together. Reproductive justice, a framework developed by women of colour feminists, is capacious enough to not only include disability, but to engage with the complex concerns disability communities bring to these conversations. I build on recent scholarship by Alison Kafer (2013), Alison Piepmeier (2013), Dorothy Roberts (1997) and Sujatha Jesudason (2011), as well as my previous work (Jarman 2015) to highlight the importance of engaging with and continuing to chart the shared territory between disability studies and reproductive justice. To bridge some of this terrain, I focus on three areas: first, by looking primarily at eugenic history, intricate connections between racial and disability injustice are traced; second, bringing that history forward, the chapter explores parallel and overlapping political strategies at the nexus of racial and disability "discrimination"; and, finally, returning to the purported valuing of Down syndrome as a site of analysis, I argue that, in order to advocate for the value of disabled lives, we must move beyond critiques of individual reproductive decision making, and insist on addressing neo-eugenics with demands for political and social structures of support for people with disabilities—beyond the womb.

## Historical interdependencies: Racial, disability and reproductive oppression

Reproductive justice was formally defined as a movement in 1994 by women of colour feminists working to articulate their goals in a holistic, historically grounded context. At the time, women of colour had grown frustrated with white women leaders of pro-choice organisations for not integrating their social and policy concerns; further, they saw the mainstream focus on abortion rights as too limiting an approach to address the myriad impacts of racial reproductive oppression. In a recent collection edited by Loretta Ross, Erika Derkas, Whitney Peoples, Lynn Roberts, Pamela Bridgewater and Dorothy Roberts (2017), all leaders and scholars in this movement, they define the framework in straightforward terms:

> Reproductive justice is not difficult to understand. It is both a theoretical paradigm shift and a model for activist organising centring three interconnected human rights

values: the right *not to have children* using safe birth control, abortion, or abstinence; the right *to have children* under the conditions we choose; and the right to *parent the children we have* in safe and healthy environments.

*(2017, loc 208, italics in original)*

These three tenets are grounded in a vision of human rights, and the authors stress that this is an "intersectional" and "dialectical process in which individual, group, corporate and government actions are interdependent to achieve reproductive freedom and control" (Ross et al. 2017, loc 217). This framework supports continued access to abortion but situates this in relation to access to health care and to safe, supportive environments. Reproductive justice also provides a larger context to examine historical oppression and progress; to advocate for health education and care; and to connect specific group struggles through common efforts of national and global reproductive justice efforts. The three tenets of reproductive justice—*the right not to have children, the right to have children and the right to parent the children we have*—are capacious enough to include the unique historical context and contemporary reproductive issues impacting multiple diverse communities, including people with disabilities and their relational networks. Created and led by women of colour feminists, this framework stresses intersectionality as both process and material experience. My premise is that greater attention to intersectionality—as we look at historical oppression and contemporary debates—will strengthen efforts toward disability reproductive justice.

Reproductive justice finds one of its great strengths and parallels with disability rights in a deep rootedness in history. African American women have a long history in the United States of fighting for autonomy, dignity and control of their own bodies. After all, the management and economic success of slavery depended upon controlling the bodies of those in bondage, and reproductive control was essential to slaveholders' power. Reproductive control of women took many violent forms: sexual abuse, rape, forced pregnancies, breaking of kinship bonds through removal of enslaved children, and disregard for chosen marriages or partnerships. Enslaved women were treated as sexual property and exploited as vessels of successive generations of enslaved labour. As Dorothy Roberts (1997) points out in her seminal text, *Killing the Black Body*, reproductive control was not purely driven by economics: "Domination of reproduction was the most effective means of subjugating enslaved women, of denying them the power to govern their own bodies and to determine the course of their own destiny" (1997, p. 55). Even in slavery, however, women resisted control and worked within their slave communities to protect children and forge kinship bonds beyond nuclear family systems, which were often impossible to sustain. This legacy of communal bonds has endured in some forms and continues to shape the meaning of family within some African American communities. As Roberts explains, "This flexible family structure has proven to be an adaptive strategy for surviving racial injustice" (1997, p. 54). Situated in the context of reproductive justice, legacies of bondage and racial oppression have shaped rights demands, especially the right to *parent the children we have*. Reproductive justice insists upon historical recognition and strategies that emerge out of resistance to enduring legacies of injustice based upon white supremacy.

Within the history of state-sanctioned reproductive coercion, especially at the intersection of disability and race, the eugenic era represents a dramatic period that ushered in national surveillance of marked body-minds, as well as systematised reproductive and population control. Eugenic ideology, in its focus on racial betterment, was inherently a white supremacist movement; however, in pursuing scientific credibility, eugenicists developed tools for ranking human biology—and human value—based widely upon disability. Eugenic social interventions mobilised in two directions: positive eugenics, promoting hereditarily beneficial marriages and

134

voluntary reproductive management; and negative eugenics, which used ever-expanding tax-onomies of "deviancy" to enact reproductive controls.

Harry H. Laughlin (1919), a zealous eugenicist, described positive eugenics as the "construc-tive, aristogenic, or eugenics-proper phase—which aims to secure the high-fertility and fittest matings among the more talented families" (1919, p. 1). Positive eugenics was reinforced across the country with "fitter family" contests that rewarded white families for aesthetic superiority and hereditary fitness; in other words, they demonstrated an absence of illness and disability and embraced eugenic self-regulation. Positive eugenics also pursued racial purity. Eugenicists were singularly preoccupied with anti-miscegenation. Dorothy Roberts (1997) provides a revealing example from the Second International Congress of Eugenics in 1921: more than half the pres-entations focused on "the biological and social consequences of marriages between people from different ethnic groups" (Roberts 1997, p. 71). These concerns reinforced racial divisions, and were codified in legislation; by 1940, thirty states had enacted laws barring interracial marriage (Roberts 1997, p. 71). Buttressed by the ideological tenets of positive eugenics, negative eugen-ics focused upon cutting off "dysgenic" bloodlines; in other words, hereditary lines thought to carry any number of eugenically identified traits increasingly used to mark individuals as socially inadequate. Ultimately, negative eugenics focused upon restricting reproduction, and sought "to cut off the descent-lines of those individuals who are so meagerly or defectively endowed by nature that their offspring are bound to...entail a drag upon the more effective members of society" (Laughlin 1919, p. 1).

By positioning the existence of cognitive and physical impairment as a growing national threat, eugenicists developed methodologies of defining, diagnosing and regulating "deficient" bodies and minds. Disability studies scholars have rightly brought attention to the ways eugenicists pro-duced and leveraged a negative potency of disability to justify discrimination and extreme social control of not only disabled people, but also of immigrants, people of colour, women, the poor and anyone caught up in criminal justice or detention systems. In *Cultural Locations of Disability*, Sharon Snyder and David Mitchell (2006, p. 71) argue that eugenics oversaw a crucial "historical transition from a 'curative' promise of rehabilitation to an increasingly 'custodial' proposition," in which people labelled pathological were brought into ever more restrictive environments. They refer to this trajectory of eugenic regulation as the emergence of "diagnostic regimes" (Snyder and Mitchell 2006, p. 70), a phrase that aptly describes the authority eugenicists attempted to gain over people's lives. Indeed, eugenicists solidified their authority with diagnostic categories and precise terminology. They actively sought to dismantle overly broad designations in favour of greater specificity. For example, they replaced wide-ranging categories of "degeneracy" and "dependency" with particular, putatively medical designations of the "socially unfit": "idiot," "imbecile," "feebleminded" and "defective" became accepted eugenic diagnostic terms. Eugenic categories, however, like their predecessors, proved to have very blurry boundaries. Social issues such as poverty, homelessness, alcoholism and many forms of criminal activity were suddenly attributed to eugenic causes, and individuals were easily categorised under the broad umbrella of "feebleminded" (Davenport 1912, pp. 20–29).

With this background in eugenic ideology, I want to focus on two domains of reproductive control to further examine intersections of disability and race in the early to mid-20th century: the expansion of birth control under the leadership of Margaret Sanger, and the growth of state-sponsored eugenic sterilisation programs. Margaret Sanger occupies a complicated histori-cal position in reproductive rights. She devoted her professional life to promoting birth control and securing access to contraception for women throughout the world, but, as many scholars have noted, she did so by forging questionable partnerships with eugenicists, and certainly by dehumanising people with disabilities. Indeed, Sanger mobilised the threat of disability and disease—which she often equated with poverty—to promote greater restrictive oversight from middle-class progressives over the burgeoning ranks of the poor and those marked by mental and physical disability. Even her rhetoric of liberating women from involuntary motherhood depended upon what she considered the greater promise of eradicating disability. In public addresses, Sanger offered eugenic birth control as a panacea to myriad social problems, which she increasingly described in economic and biological terms:

> [T]he example of the inferior classes, the fertility of the feeble-minded, the mentally defective, the poverty-stricken classes, should not be held up for emulation to the mentally and physically fit, though less fertile, parents of the educated and well-to-do classes. On the contrary, the most urgent problem today is how to limit and discourage the over-fertility of the mentally and physically defective.
>
> *(Sanger 1921, p. 5)*

Like eugenicists, Sanger often conflated disability with poverty, criminality and disease, implying that birth control could function as a cure-all for the social problems of her day.

Her orientation toward eugenics, disavowal of disability, and support of population control contributed to a complicated record on race. In notable ways, Sanger worked to empower African American communities, and she resisted using designations of disability, as a disqualifier, to promote racism. In fact, in the 1930s and 40s, she worked closely with leaders of the African American community to support the establishment of locally run family planning clinics in Harlem and the Bronx (Roberts 1997, pp. 79–82). At the same time, Sanger actively promoted two eugenic ideas that negatively affected communities of colour and people with disabilities. First, she relentlessly linked social problems to excessive reproduction and, although her suggested solutions involved empowering women, these arguments were easily appropriated by social reformers zealous to control specific populations. Second, Sanger unapologetically figured disability as the ultimate marker of the foreign, which, in her usage, implied unwanted and threatening. These philosophical orientations allowed her to justify population control measures and unsafe birth control trials, including sterilisation programmes outside of mainland United States, including Puerto Rico, as discussed below. Ultimately, Sanger provides a context for thinking about justice; her belief that certain types of people should not reproduce or be produced rendered it impossible for her to pursue justice in reproduction. In these eugenic social calculations, empowerment of some depended upon injustice for countless *others*.

One of the most destructive and enduring legacies of eugenics has been involuntary sterilisation. From the early years of the movement, eugenicists promoted surgical reproductive control over the "feebleminded." For example, physician S.D. Risley (1905, p. 97) warned, with the hyperbolic flair common to eugenicists, that allowing feebleminded people to procreate would unleash a "Pandora's box" that would "permit the escape and free riot of monstrous, indescribable things." People housed in hospitals and schools for the feebleminded, people caught within the "diagnostic regimes" of eugenics, were already vulnerable to extra-legal sterilisations, but eugenicists wanted to extend their reach. Concerned about the population of "borderline" individuals, who could pass as "normal" and threatened society with their disproportionate fecundity, eugenicists believed additional methods, beyond institutional segregation, had to be pursued. By 1911, Henry Goddard, the Director of the Eugenics Records Office, urged political leaders to put a sterilisation law "upon the book of every State." In his estimation, this would be the only means by which "normal" society could "get control of the situation" (p. 514).

Eugenicists enjoyed impressive early success in the establishment of sterilisation laws. The state of Indiana pioneered the first law in 1907, and, within the next decade, another eleven states had followed suit. By the end of World War I, fifteen states had legalised sterilisation, and, by 1932, that number had doubled. The most significant event in U.S. sterilisation history took place in 1927 when the Virginia law authorising the sterilisation of inmates "inflicted with hereditary forms of insanity that are recurrent, idiocy, imbecility, feeblemindedness or epilepsy" (Landman 1932, p. 84) was upheld by the United States Supreme Court in the (in)famous case of *Buck v. Bell*. Nancy Ordover (2003) suggests that *Buck v. Bell* was part of a specific strategy to target women, rather than men, as the primary inmates to be sterilised. Into the 1920s, as eugenicists became more concerned with the reproductive threat of feebleminded, especially "borderline" women, sterilisation seemed to offer the most flexible and effective means of social control. The trend toward using the new legislation for female control was markedly evident within the first five years of the court's decision. In 1927, before *Buck v. Bell*, 53 percent of all the sterilisations in the United States had been performed on men; by 1932, this proportion dropped to 33 percent (Ordover, 2003, p. 135), a dramatic shift that continued in some states for many decades. In addition to granting states tremendous power over women's reproduction, *Buck v. Bell* provided institutional superintendents with a surgical solution to overpopulation and parole. Sterilisation allowed them to release inmates who would have otherwise remained in the institution for years. In this way, doctors could quickly and legally make room for new inmates and save state resources in the process. Further, as women with disabilities continued to be institutionalised into the 1960s and 1970s, they were completely vulnerable to sterilisation.

A more familiar history of sterilisation involves the systemic targeting of women of colour. As sterilisation programmes expanded in the 1930s and 1940s, the focus shifted from poor and disabled white men and women to women of colour. As Dorothy Roberts (1997) points out, the slow demise of Jim Crow in the South ironically opened the doors of institutions to African Americans. North Carolina provides a case in point, earning the dubious honour of developing one of the most extensive sterilisation programmes in the country. Between 1930 and 1940, of the 8000 eugenic sterilisations performed in the state, 5000 of them were on Black women (Roberts 1997, p. 90). These legal procedures were precursors to massive sterilisation abuses enacted upon communities of colour. I highlight a few of the most egregious examples to provide a glimpse into the ease with which eugenic ideas were again mobilised, well after the atrocities of World War II. Poor, African American women who were on welfare or receiving Medicaid were targeted by doctors across the United States to be sterilised without their knowledge; these "Mississippi appendectomies" became so common, they were often performed with student audiences in teaching hospitals around the country. In the 1950s and 60s, in Puerto Rico, a federally funded educational campaign, in partnership with Planned Parenthood, encouraged women to elect to be sterilised—by not offering other options and promising low-cost procedures. This campaign was so "successful" that over thirty percent of the women of childbearing age were sterilised. During the same period, the US government targeted indigenous families for reproductive and cultural control. The legacy of removing Native children to state boarding schools has caused generational trauma that continues to impact Native families and communities. Adding to the cultural violence of removal, in the 1970s, Native women were targeted for large-scale, federally funded sterilisation initiatives. These programmes, driven by settler colonial legacies, resulted in tens of thousands of indigenous women being sterilised, and in locations where tribal populations were already small, they were, quite literally, genocidal campaigns (Roberts 1997, pp. 90–99; Ross et al. 2017, loc 1133). As we think about these histories in relation to current attacks on reproductive rights, it is crucial to pay attention to state-sponsored, systemic practices that enacted eugenic violence in the past, and that threaten to enable neo-eugenic control over reproduction today.

## Eugenic legacies in disability reproductive justice

As this brief history demonstrates, the reproductive control of the eugenic period targeted many different groups, using similar rationales of weeding out deficient traits: promoting reproduction of white, middle-class women while forcibly curtailing the reproduction of women with disabilities, women of colour, poor women, indigenous women, LGBTQ and non-cisgender individuals, immigrant communities and others. The legacies of eugenics continue to impact all of these groups in overlapping, intersecting, yet distinct ways. While it is not within the scope of this chapter to examine the eugenic impact upon all of these groups, the intersections of disability and race/ethnicity are inextricably linked. Because eugenics was driven by a rhetoric of disability and grounded in white supremacy, when disability studies scholars investigate eugenic history to provide insight into contemporary beliefs, we should be vigilant not to understate the racial dimensions of eugenic violence. Reproductive justice, grounded in histories of racial (in) justice, provides a theoretical foundation for intersectional disability scholarship. This approach allows a stronger understanding of individuals with multiple, intersecting identities, and a more robust appreciation of the shared experiences and concerns of diverse communities that have been targeted for reproductive control. For example, after egregious sterilisation abuses in communities of colour, even in recent years, there have been government-sponsored and private incentive programmes for poor (predominantly black and brown) women with substance addictions, as well as for indigenous women, to agree to surgical sterilisation or long-acting contraceptives such as Norplant or Depo Provera (Price 2010, pp. 58–59). Similarly, women with disabilities are often encouraged to undergo sterilisation or to use such contraceptives, and, if these women have guardians, they may or may not be involved in these decisions.

Reproductive justice advocates encourage distinct communities to trace and understand interconnected histories in order to highlight overlaps while amplifying their unique concerns. Turning to specific reproductive issues related to disability, in this case, disabled women and parents (including prospective parents) of children with disabilities, the utility of the framework's three reproductive justice tenets becomes evident. The first reproductive claim is "the right *not to have children* using safe birth control, abortion or abstinence" (Ross et al. 2017, loc 208). This assertion invokes a complicated history in disability communities because it brings into focus a deep and enduring social assumption that many individuals with disabilities do not belong in reproductive conversations; they are rarely seen as sexual beings or as potential parents. In *Fading Scars*, Corbett O'Toole cites earlier research with colleague Tanis Doe which found that, with rare exception, "people with disabilities do not get asked if they want to have children. They don't get asked if they want to be sexual" (2012, National Council on Disabilities, cited in O'Toole 2015, p. 254). Although this is changing somewhat, as disabled people assert their sexual and reproductive agency, such cultural and professional ableism remains pervasive. Often the silence around sexuality, romance and parenting includes people closest to individuals with disabilities (O'Toole 2015, p. 254). If family members, teachers, counsellors, doctors and other professionals in a disabled person's life assume asexuality or disinterest in parenthood, they will not effectively provide or seek out sexual and reproductive education. This is especially true for women with intellectual or developmental disabilities, whose reproductive decisions are often made by parents and guardians. These women also suffer much higher rates of sexual abuse than other women, seven times higher according to Bureau of Justice statistics (Shapiro 2018), which can be another factor leading to unwanted pregnancy. Not having children may be appropriate for some disabled women, but having accurate medical information and exercising individual agency in sexual and reproductive decision making—to the greatest extent possible—is also paramount.

On another side of this issue, as more disabled women have asserted their embodied agency, and as disability rights proponents have demanded reproductive choices to have children with disabilities, including hereditary conditions, it is important to resist framing individual reproductive decisions *not* to have children as a rejection or repudiation of disability justice more broadly. If reproductive justice includes the right to not have children, disabled and nondisabled women should be supported to make decisions for themselves. These decisions are uniquely complicated for individuals with disabilities and their families. The deaf community and people with dwarfism, for example, often embrace passing on these traits, and welcoming children into distinct *cultural* communities; in fact, some members of these groups have advocated in specific cases to genetically select *for* these traits (Kafer 2013; OToole 2015). By contrast, other people with disabilities make decisions not to have children partly because they live with disability. Katie O'Connell, a proponent of reproductive justice, explains that her experience with disability, especially having severe migraines, distinctly shapes her reproductive decisions:

> Controlling my own reproductive future is absolutely vital to me as a disabled woman. It ensures I can stay on my medication guilt-free. It means I don't have to worry about passing a genetic disability onto future children. It means I can continue to afford my medications and not worry about how that money impacts my family.
>
> *(2017, loc 5916)*

O'Connell points to important, interacting elements that impact her thinking about reproduction: her own physical well-being, a desire not to pass on her disability coupled with the socio-economic realities of managing her condition. As a disabled woman, O'Connell is also acutely aware of how disability is (mis)used to disqualify women from being parents, so she underscores that her decision should not be appropriated to control other individuals: "Choosing not to have children due to my disability does not mean that I think other people with disabilities should not have children" (O'Connell 2017, loc 5916). Reproductive decisions are complex and contextual, and many factors shape them. Supporting every woman and pregnant person's right to determine their reproductive future also means engaging in complicated discussions, and understanding that individual, materially informed decisions, even those based in part or wholly on not having a child with a disability, may not be ableist or easily categorised as neo-eugenic.

One of the pernicious ableist legacies of the eugenic era is the assumption that disabled people *should not* parent, so the second and third values, "the right *to have children* under the conditions we choose; and the right to *parent the children we have* in safe and healthy environments" (Ross et al. 2017, loc 208, italics in original), are crucial to racial *and* disability justice. Samantha Walsh (2011) provides a telling example of the quotidian presumption that she, as a disabled woman, would *never become* a parent. In a casual conversation with an aesthetician in a nail salon, Walsh began talking about the potential challenges of becoming a parent, but the nail stylist dismissed the topic: "Listen, you don't want kids anyway, so what does it matter?" Walsh transforms this offhand question into a meditation about how she, as a wheelchair user, is *not seen* as a potential parent. The conversation, in her mind, exposed the troubling presumption "that disability is not something that should intersect with the experience of motherhood" (2011, p. 82).

As disability rights activist and scholar Corbett OToole (2015) argues, such attitudes are pernicious and widespread. Her research reveals the scope of the problem in stark material terms:

> Disabled people are told repeatedly that they should not be parents, that they are not safe with children and that they should not pass on their disability to the next generation. Disabled people are sterilised to stop them from ever getting pregnant. If they do

have children, many disabled parents lose them because someone complains to Child Protective Services, and the people evaluating them as parents do not believe that disabled people should have children.

*(OToole 2015, p. 246)*

Furthermore, OToole cites research done by the organisation Through the Looking Glass, which critiques the body of scholarly literature on disabled parents on the grounds that most researchers investigate the *problems* in these families; in other words, the research itself is driven by a bias of "presumed incompetence." Their report goes on to state: "As this stereotype becomes enacted through custody and policy practices, disabled parents experience extraordinarily high rates of family disruption through actual or threatened loss of custody" (2012, National Council on Disabilities, cited in OToole 2015, p. 251). OToole recounts one devastating but all too common example of a physically disabled mother who had devised a safe method of changing her infant's diaper on her bed, where the child would not be at risk of rolling off. When a social worker visited, she insisted the mother change the baby on a table, but this made the task unsafe—so the social worker used this "fact" to remove the child. Even with an excellent lawyer who provided video evidence of the mother safely changing the baby, this disabled mother lost custody of her child—as well as that of the baby she was expecting.

These few examples are meant to illustrate that parents with disabilities have valuable insights into the coercive, oppressive systems impacting their ability to parent. People of colour with disabilities and people occupying multiple intersecting identities have much to contribute to these intersecting justice movements. As well, disabled people have crucial social knowledge and strategies for creatively navigating gatekeeping systems, building communities of support and establishing new kinship models. Disability reproductive justice provides a framework to integrate the unique insights of disability communities into policy discussions, as well as to demand and integrate stronger networks of support for disabled parents, children and families.

## Conclusion: Interdependent reproductive justice

The intersectional history of eugenics and broader reproductive concerns within disability communities provide a crucial context for challenging contemporary efforts to ban abortions based on genetic anomaly—most often based upon a positive screening for Down syndrome. Appropriating the rhetoric of disability rights, supporters of these bans claim to be addressing disability discrimination by resisting eugenic beliefs and practices. Karianne Lisonbee, sponsor of a bill in the U.S. state of Utah, boldly proclaimed, "Utah's message to the world is that we will not tolerate discrimination" (quoted in Thiessen 2018). Individuals with Down syndrome have also advocated for abortion bans as a way of asserting their inherent value. Frank Stephens, in his testimony before Congress, compared abortion based upon prenatal screening to Hitler's eugenic programme: "I completely understand that the people pushing this particular 'final solution' are saying that people like me should not exist," then he encouraged representatives to "pursue inclusion, not termination" (quoted in Thiessen 2018). Exposing discrimination and valuing people with Down syndrome are foundational to disability rights, and concerns from individuals such as Stephens must be taken seriously. In fact, exposing the neo-eugenic dangers in biased or ableist promotion of pre-natal screening has been at the heart of disability studies critiques, as demonstrated in the discussion of Hubbard, Asch and Saxton. A crucial distinction, however, is that anti-abortion arguments situate the origin and the responsibility for discrimination with pregnant women; individual reproductive decisions, not larger economic, social and structural support of people with disabilities, become the battleground. This strategy decontextualises reproductive decisions, while condoning state-sponsored control and criminalisation of reproduction—a more troubling legacy of eugenics.

Indeed, as anti-abortion groups claim to be opposing eugenics through legislative bans, a distinction must be made between belief systems that inform individual decisions and state-sponsored reproductive control initiatives. After all, one of the primary objectives of the eugenics movement in the United States was to control women's reproduction, and in the case of targeted groups such as women of colour and women with disabilities, to violate their bodily integrity and self-determination. In proposing abortion bans, anti-abortion supporters are enacting new forms of reproductive oppression and attempting to criminalise pregnant women and the doctors who provide care. If these groups really want to address disability discrimination and social inclusion, they could look at their state funding levels for Medicaid, at programmes supporting people with intellectual and developmental disabilities, at competitive employment programmes and other structural support of disabled people. Rachel Adams, a disability scholar and mother of a son with Down syndrome, made a similar point in an editorial in 2016. At the time, Missouri and Ohio were moving to pass abortion bans, slated as "non-discrimination" laws through their state legislatures. Notably, these same governing bodies slashed funding for programmes serving people with disabilities during that session (Adams 2016). Such state policies could be described thus: non-discrimination in the womb; indifference from birth forwards. Rather than policing pregnancy, governments could play a much more empowering role in reproduction. As Sujatha Jesudason (2011, p. 524) suggests, "a reproductive justice approach advocates for an affirmative role for the government to play in assuring that all women have the social, political and economic power and resources" to bring to bear in making the best and most appropriate decisions for themselves and their loved ones. Government does have a role, but that role should not be coercion, control or criminalisation.

Those concerned with disability and reproductive justice might also learn from women of colour who organised to oppose a similar "anti-discrimination" campaign in their communities. Loretta Ross describes an anti-abortion campaign aimed at women of colour that parallels more recent abortion bans based on Down syndrome. Beginning in 2008, groups organised to introduce abortion bans on the basis of race and sex. Drawing upon statistics documenting that African American women were having higher rates of abortions than other racial groups, anti-abortion organisations began a public campaign claiming that black mothers seeking abortions were enacting eugenic tactics and were selecting to abort based on race. As reproductive justice scholars explain: "Claiming to campaign against 'reproductive racism', conservatives use the bizarre 'abortion is racist' narrative that disregards the rights, wishes, and needs of women of colour" (Ross et al. 2017, loc 1322). Wealthy, white donors partnered with conservative, pro-life African American leaders to attack African American women seeking abortions. Echoing previous malicious attacks, such as calling black women "welfare queens" or "hyperfertile," this campaign "placed black women as the destroyers of the black family" (Ross et al. 2017, loc 1300). Conservative legislators, who had always voted against civil rights legislation, were suddenly introducing bills named after civil rights leaders to address the supposed racism of abortion—as if African American women were aborting because their foetuses were black. The strategies in this case are very similar to those in the Down syndrome campaign. Mothers are blamed for enacting eugenics while anti-abortionists cast themselves as champions of justice. In these cases, we need to contextualise the real racial and disability histories of eugenics, and call attention to the eugenic nature of reproductive control and criminalisation of pregnant bodies. Furthermore, like women of colour who confronted abortion bans based upon race, reproductive and disability justice advocates must pursue larger conversations about creating sustainable, supportive environments for children and parents, as well as resources, within communities that will allow women to make actual "choices" about their bodies, minds and potential families.

In the case of Down syndrome, many women choose to terminate because they see this condition as non-optimal; however, this perception may be driven by ableism, neo-eugenic beliefs or by the material realities and lack of social support afforded people with disabilities. The solution to changing perceptions and challenging ableism in such cases is to educate people about disability rights, insight and experience, not to criminalise reproductive decisions. Alison Piepmeier's (2013) research revealed that assumptions of ableism, discrimination or neo-eugenic ideology failed to capture the complications of reproductive choices. Piepmeier did a series of interviews with women who had received prenatal diagnoses of Down syndrome during their pregnancies and found that their decision making was very complex. After the screening results, most of the women became concerned about their networks of support, about the social opportunities, prejudices and resources they might, or might not, have available. Piepmeier (2013) determined that reproductive justice frameworks acknowledge this complexity more than pro-choice framings: "Reproductive justice makes room for messier questions and concerns. It emphasises social justice, which removes this decision from an individualised space and makes it part of a broader set of community priorities" (2013, p. 176). Loretta Ross and her co-editors echo this idea: "Reproductive justice is collective and interdependent by definition" (Ross et al. 2017, loc 3251).

As this exploration into the intersectional legacies of eugenics in current reproductive debates demonstrates, the interests of disability and reproductive justice are intertwined, and both movements are enhanced through this interdependence. As recent anti-abortion efforts demonstrate, appropriation of disability rights discourse will continue to be weaponised to enact reproductive control, but disability scholars must resist efforts to pit disability rights against women's reproductive freedoms. Supporting this process, the three tenets of reproductive justice broaden the focus and insist upon intersectional and community approaches. Ultimately, individual decisions about reproduction depend upon the social and material contexts of disability, gender, race and access to resources and supports—and these contexts will improve only if we focus on supporting lives, not mandating births.

# References

Adams, R. (2016) My son is not a mascot for abortion restrictions. *Washington Post*, 19 February.

Arendt, H. (1977) *Eichmann in Jerusalem: A report of on the banality of evil*. New York: Penguin.

Asch, A. (1999) Prenatal diagnosis and selective abortion: A challenge to practice and policy. *American Journal of Public Health* 89 (11), pp. 1649–1657.

Davenport, C. (1912) The trait book. In *Eugenics record office bulletin no. 6*. Cold Spring Harbor, NY: Eugenics Record Office Archive.

Goddard, H. (1911) The elimination of feeble-mindedness. *Annals of the American Academy* (March), pp. 505–16.

Guttmacher Institute (2019) Abortion bans in cases of sex or race selection or genetic anomaly. Available at: https://www.guttmacher.org/state-policy/explore/abortion-bans-cases-sex-or-race-selection-or-ge netic-anomaly (Accessed 1 January 2019).

Hubbard, R. (2010) Abortion and disability: Who should and should not inhabit the world? In Davis, L.J. (ed.), *The disability studies reader*. 3rd edn. New York: Routledge, pp. 107–119.

Jarman, M. (2015) Relations of abortion: Crip approaches to reproductive justice. *Feminist Formations* 27 (1), pp. 46–66.

Jesudason, S. (2011) The paradox of disability in abortion debates: Bringing the pro-choice and disability rights groups together. *Contraception* 84, pp. 541–543.

Kafer, A. (2013) *Feminist, queer, crip*. Bloomington, IN: Indiana University Press.

Landman, J.H. (1932) *Human sterilization: The history of the sexual sterilization movement*. New York: Macmillan.

Laughlin, H.H. (1919) The relation of eugenics to other sciences. *Eugenics Review*, pp. 1–12. Charles Davenport Papers. Philadelphia: American Philosophical Society Library.

National Council on Disabilities (2012) Rocking the cradle: Ensuring the rights of parents with disabilities and their children. Available at: https://www.ncd.gov/publications/2012/Sep272012 (Accessed 27 August 2019).

O'Connell, K. (2017) We need to talk about disability as a reproductive justice issue. In Ross, L., Derkas, E., Peoples, W., Roberts, L., Bridgewater, P., & Roberts, D. (eds.), *Radical reproductive justice: Foundation, theory, practice, critique*. New York: Feminist Press at City University of New York, pp. 5878–5937.

Ordover, N. (2003) *American eugenics: Race, queer anatomy, and the science of nationalism*. Minneapolis: University of Minnesota Press.

OToole, C.J. (2015) *Fading scars: My queer disability history*. Fort Worth, TX: Autonomous Press.

Piepmeier, A. (2013) The inadequacy of choice: Disability and what's wrong with feminist framings of reproduction. *Feminist Studies* 39 (1), pp. 159–186.

Price, K. (2010) What is reproductive justice? How women of colour are redefining the pro-choice paradigm. *Meridians* 10 (2), pp. 42–65.

Risley, S.D. (1905) Is asexualization ever justifiable in the case of imbecile children? *Journal of Psycho-Asthenics* 9 (4), pp. 92–97.

Roberts, D. (1997) *Killing the black body: Race, reproduction, and the meaning of liberty*. New York: Vintage.

Ross, L., Derkas, E., Peoples, W., Roberts, L., Bridgewater, P., & Roberts, D. (eds.) (2017) *Radical reproductive justice: Foundation, theory, practice, critique*. Available at: https://www.amazon.com/Kindle-eBooks (Accessed 23 May 2018).

Sanger, M. (1921) The eugenic value of birth control propaganda. *Birth Control Review* 5 (10), pp. 4–7. American Eugenic Society Papers. Philadelphia: American Philosophical Society Library.

Saxton, M. (2010) Disability rights and selective abortion. In Davis, L.J. (ed.), *The disability studies reader*. 3rd edn. New York: Routledge, pp. 120–132.

Shapiro, J. (2018) The sexual assault epidemic no one talks about. *All Things Considered*, National Public Radio, 1 January.

Snyder, S.L., & Mitchell, D. (2006) *Cultural locations of disability*. Chicago: University of Chicago Press.

Thiessen, M. (2018) When will we stop killing humans with Down syndrome? Washington Post, 8 March.

Walsh, S. (2011) "What does it matter?" a meditation on the social positioning of disability and motherhood. In Lewiecki-Wilson, C., & Cellio, J. (eds.), *Disability and mothering: Liminal spaces of embodied knowledge*. Syracuse, NY: Syracuse University Press, pp. 78–84.

# 9

# SEXUALITY AND THE DISREGARD OF LIVED REALITY

## The sexual abuse of children and young people with disabilities

*Gwynnyth Llewellyn*

## Introduction

In mid-2016, I was surprised in a meeting at the Royal Commission for Institutional Responses to Child Sexual Abuse (hereafter, the Royal Commission). I was sharing a figure produced by Jones et al. (2012) in a meta-analysis commissioned by the World Health Organisation and published in *The Lancet*. These authors found that children and young people with disabilities were nearly three times more likely to be sexually abused than were their non-disabled peers. On hearing this figure, one lawyer exclaimed: "I cannot believe this happens, that we don't protect our most vulnerable." Her startled response was a timely reminder of a broader ignorance about children and young people with disabilities. This chapter questions why children and young people with disabilities are virtually absent from discourse on preventing sexual abuse, given their significantly increased risk of sexual abuse.

The chapter brings an historical-socio-cultural perspective to the lives of children and young people with disabilities in Australia, to examine understandings about child sexual abuse. It begins by examining two time periods in Australia. The first period is the early decades from the first colony in 1788, when colonial governments were forced to confront the ever-growing group of children in need of care. The second time period is the present, a time of national disability reform in Australia. The final section of the chapter presents contemporary responses to keeping all young Australians safe from sexual abuse, examining these through an analytic lens informed by the materials available to the Royal Commission. Following definitions employed by the Royal Commission, the term "children" is taken to mean children and young people up to the age of 18 years and institutions are understood to be public or private entities or groups of entities now or in the past where adults have contact with children. Historical terms now considered derogatory are used, where relevant, in single quotation marks and with no intention to cause offence.

## The early decades of the colonies

The history of European Australia is one of transportation of convicts from the British Isles to the Great South Land, creating what was to become Australia (Blainey 1988). Although near starvation in the early days of the 1788 colony resulted in high adult and child mortality within a few years, observers were commenting on the healthy, strong and muscular nature of colonial children and youth (Gandevia 1978). The children referred to, however, were the progeny of the governing and settler classes. Indigenous children, if noticed at all, did not appear in the European colonisers' narratives. The same was true for children with disabilities.

The primary concern in the historical record at the time was the 'moral blight' of a rapidly increasing number of destitute children (Garton 1990). The destitute children included orphans, children abandoned by parents and children of parents who were incarcerated or mentally ill. At first, the ruling class of government and military officials and their wives 'rescued' these children, who were then 'boarded out' with extra rations given to 'foster' families willing to take in a child. As the number of abandoned children grew, voluntary committees of free settlers were funded by 'subscriber' charities and subsidised by the government to remove abandoned children and young girls at risk of sexual exploitation and put them in institutions (Garton 1990). Sexual exploitation of females who were not 'ladies' of the governing class was rife in the colony (Summers 1975). The first asylum in New South Wales (NSW) housed deserted women and children with homeless older men and the mentally ill. It was not long before orphanages appeared. These catered for 'true' orphans and also for the children of widows, deserted wives and unmarried mothers (Liddell 1993).

This was a time of moral treatment for society's outcasts (Garton 1990). Parentless children in institutional care were subjected to strict, regimented, daily routines to turn them into industrious citizens. Industrial schools took in the neglected children of living parents and aimed to reduce their risk of becoming future criminals. To prevent a new generation of paupers, the children of the poor were removed into training homes—domestic training for girls and farm training for boys—which also ensured a supply of domestic servants and farm labourers for free settlers (Swain 2014a). Children with disabilities were not mentioned in the accounts of these early institutions. It is reasonable to assume, however, they would be among the children left parentless, orphaned, neglected or living in poverty.

The first mention of children with disabilities comes with a second wave of government intervention. The "new" solution followed the overwhelming increase in neglected children living on the streets and vulnerable to sexual and financial exploitation after the 1850s gold rush, particularly in NSW and Victoria (Garton 1990). The first response was to expand existing institutions and, when this failed to meet demand, to build more. However, growing public concern about institutional conditions resulted in two Royal Commissions, one in Victoria (1872) and the other in NSW (1873–1874) (Gandevia 1978, p. 105). Outrage about the conditions exposed led to the boarding out system regaining favour and becoming legislated government policy in Victoria (1874) and in NSW (1881). To ensure the success of boarding out, host families received a care subsidy that was, at least in NSW, according to Gandevia (1978, p. 108), "higher for delicate children, and higher still for sick or disabled ones."

It is unclear what 'disabled ones' means in this context. Several events in the closing years of the nineteenth century suggest some possibilities. For example, Lancaster (as cited in Gandevia 1978, p. 92) noted the rubella epidemic of 1898 resulted in a five-fold increase in expected admissions to the NSW Institute for the Deaf and Dumb. Other forces were also at work. Dawson (no date, as cited in Gandevia 1978, p. 138) noted that the economic depression of the 1890s "obliged families to throw their idiot children and older imbeciles on the state," an early reference to differentiating those less socially able from those deemed 'lunatic.' There was also increasing interest in moral training to reintegrate 'ineducable' children into society. Kew Idiot Asylum, established in 1887, later renamed Kew Cottages, was the first hospital institution for this purpose in Australia.

By the turn of the twentieth century, most parents of children with disabilities had only two choices. These choices were to care for their children at home or, if this was not possible, to commit them to a state mental hospital such as Kew in Victoria or Watt Street Mental Hospital in NSW (Cummins 2003; Swain 2014a). However, for parents with greater financial resources, there was a third choice: To place their child in a charitable or privately-owned institution such as the NSW Institute for the Deaf and Dumb. Allowah Babies Hospital continues today as an example of many such institutions across the country. Initially established in 1954 to cater for "severely handicapped" children under two years of age and now the Allowah Presbyterian Children's Hospital, providing care for children and young people with disability up to 18 years (Allowah Presbyterian Children's Hospital n.d.).

In the early days, little thought was given to the consequences of children and adults cohabiting in government or charitable institutions. The Victorian 1894–1896 Royal Commission on Asylums for the Insane and Inebriate marked a discursive turn in thinking on this matter. The minutes of this Commission were "unanimous in condemning the 'indiscriminate mixing' of the children with adult patients in asylums" (cited in Monk and Manning 2012, p. 76). Although knowledge of sexual abuse of 'inmates' may have driven these concerns, Monk and Manning (2012) note that contemporaries only ever hinted at the likely abusive consequences of indiscriminate mixing. When immorality began to be mentioned in government inquiries, as Swain (2014b, p. 8) has observed, it was "taken as referring to sexual behaviour among children and the risk of moral contagion when innocent children were placed amongst the already 'depraved.'" In this way, responsibility for immorality was deemed to reside in the children whose nature was thought to be duplicitous or gullible in matters of sexuality.

With the advent of testimonial-based inquiries in the 1990s, survivor accounts catapulted the lived reality of sexual abuse in institutional settings into the public domain. Their accounts, including those of survivors with disabilities, spoke of adults unleashing their power over child inmates. Most frequently, this was by force, less commonly by the practice now known as grooming. As Swain (2014b, pp. 9–11) reports, inquiries from this time began to include whole sections dedicated to sexual abuse, although not yet included as a term of reference. For example, *Forgotten Australians: A report on Australians who experienced institutional or out-of-home care as children* (Commonwealth of Australia 2004, pp. 103–105) and later the specific inquiry into institutional conditions for people with disabilities in the report, *Violence, abuse and neglect against people with disability in institutional and residential settings, including the gender- and age-related dimensions, and the particular situation of Aboriginal and Torres Strait Islander people with disability, and culturally and linguistically diverse people with disability* (Commonwealth of Australia 2015). That said, it took another decade of advocacy and political lobbying before the Australian Government called for a Royal Commission into Institutional Responses to Child Sexual Abuse in late 2012 (Commonwealth of Australia 2017a). The chapter now turns to examining how children with disabilities have been positioned in Australia's disability reform agenda of the twenty-first century.

## The present and the disability reform agenda

Australia embarked on an ambitious national disability reform agenda pursuant to signing and ratifying the UN Convention on the Rights of Persons with Disabilities (CRPD) (United Nations 2006) in March 2007. The overall aim of this agenda was to ensure maximum social and economic participation for all Australians with disabilities. In short order, the reforms resulted in the National Disability Strategy 2010–2020 (Commonwealth of Australia 2011a), an updated National Disability Agreement (COAG 2009), a National Carer Strategy (Commonwealth of Australia 2011c), a National Disability Research and Development Agenda (Department of

Social Services 2011), and the landmark Productivity Commission Inquiry into a National Disability Long-term Care and Support Scheme (Commonwealth of Australia 2011b), resulting in the National Disability Insurance Scheme (NDIS) (Commonwealth of Australia 2013).

The centrepiece of the national disability reform agenda was the *National Disability Strategy 2010–2020*, the fundamental purpose of which was to include Australians with disabilities as full and equal citizens. The reform agenda dictated that people with disabilities participate in the mainstream of society—in health, education, transport, housing, citizenship activities and so on—and access specialised supports and services only when required. The *National Disability Insurance Scheme* sets out to bring this vision to fruition. As Llewellyn et al. (2016, p. 67) noted: "To achieve the promise of mainstreaming requires inclusive and accessible communities... to achieve inclusive and accessible communities requires a paradigm shift in community attitudes. Communities and institutional contexts can only become inclusive and accessible when they too take on board that people with disability are citizens first and foremost, and entitled to a respected place in society, due process and protection from harm." Two areas of concern have emerged in relation to children with disabilities in the process of aligning Australia's regulatory frameworks with the CRPD. First is the application of CRPD terminology *persons with disabilities*, and second is competing understandings of disability in the Australian context.

## Persons with disabilities

The dominant terminology of CRPD is *persons with disabilities*. This term aims to be inclusive of all, irrespective of age, gender or ethnicity. Persons is generally assumed to mean adults and children. At face value, using this term seems reasonable for the purpose of aligning national regulatory frameworks with the international convention. However, Llewellyn et al. (2016, pp. 24–26) found that, when the term is used, the needs and concerns of adults with disabilities are privileged. Readers of the reform documents are left with the strong impression that it is as if individuals with disabilities arrive at adulthood without having lived through childhood and adolescence. This makes little sense. The lack of attention paid to children with disabilities ignores societal conditioning, identity formation and childhood and adolescent experiences which shape these children, and which also shape broader community awareness about their lives.

Australia has followed the international trend in using the term "persons," which, as Sabatello (2013) has also noted, silences the particularities of children with disabilities. Furthermore, when adults as family take pride of place as the institutional unit for supporting children, children with disabilities become passive dependants beholden to the family. For example, the Preamble to CRPD (United Nations 2006, para x, p. 4) states:

> *Convinced* (italics in original) that the family is the natural and fundamental group unit of society and is entitled to protection by society and the State, and that persons with disabilities and their family members should receive the necessary protection and assistance to enable families to contribute towards the full and equal enjoyment of the rights of persons with disabilities.

Australian disability reform documents follow the Preamble. These documents refer to children with disabilities most frequently in relation to their families, and their families' need for support, not as agents in their own right (Llewellyn et al. 2016, pp. 24–26). Putting the emphasis on families presumes that families will act in their children's best interests. This is not always the case as the evidence on maltreatment of children with disabilities presented later in this chapter demonstrates. Discussion of competing understandings of disability in the Australian context will now be discussed.

## Understandings of disability

The description of disability which appears in Article 1 of CRPD was drawn from the International Classification of Functioning, Disability and Health (ICF) (WHO 2001), already well established in Australian statistical collections, though less so in regulatory frameworks. In the international standard language of the ICF, functioning and disability are understood as umbrella terms denoting the positive and negative aspects of functioning from a biological, individual and societal perspective. Article 1 of the CRPD (United Nations 2006, p. 4) describes disability as "Persons with disabilities include those who have long-term physical, mental, intellectual or sensory impairments *which, in interaction with various barriers* (my emphasis), may hinder their full and effective participation in society on an equal basis with others." However, analysis over the longer time frame suggests that quite different understandings of childhood disability have continued to flourish.

Childhood disability still remains understood as a condition 'lodged' within the individual and, as such, the defining element in their lives. As Fraser (1998) and Young (1998) explain, there is a traditional private/public divide in which families are thought to be responsible for what happens 'inside' the family (the private arena) and the state is responsible for that which occurs 'outside.' Jarratt et al. (2014) found that, when society 'bestows' disability on the individual child, they and their family come to 'own' disability. Disability becomes embodied in the child and their family. The Westminster system reinforces the private/public divide, which, in turn, reinforces family ownership of disability. Children with disabilities are relegated to being non-actors, dependent on their families or on the state as a last resort. The children have become 'special.' This perceived (negative) specialness is further emphasised by segregating them from the community either 'out of sight' within family homes, or in institutional care. This process of 'othering' according to Robinson (2012, p. 5) leads to children with disabilities being, on the one hand, more vulnerable to maltreatment including sexual abuse, and, on the other, to disinterest in terms of abuse of these children, with less being done to prevent abuse occurring or in appropriate response after the event.

Furthermore, when childhood disability is 'known' by conditions associated with negative difference, such as cerebral palsy, epilepsy, fragile X and Down syndrome, with the 'bearers' of these conditions being seen to deviate from the norm. Condition status comes from medical diagnoses; accordingly, medical knowledge about disability is privileged. Solutions become medicalised. Medical conditions that require technical, medical solutions are deemed outside the responsibility of society more generally. An early example is Kew Idiot Asylum. Medical diagnosis secured a place in that institution. Despite being established to train the 'ineducable' for reintegration into society, it was governed by doctors and set up in the grounds of a 'lunatic' asylum. NSW also established hospitals for children with disabilities. The Asylum for Imbeciles and an Institution for Idiots established in Newcastle in 1882 opened its doors for the 'idiot' and 'imbecile' children of families unable to care for them, retaining this purpose until 1967 as the (renamed) Watt Street Mental Hospital (Cummins 2003, pp. 40–41). This condition-based perspective also remains embedded in national legislation. Llewellyn et al. (2016, p. 11) pointed to the example of the Commonwealth Disability Discrimination Act 1992 (still current), in which disability is defined by ten separate points on loss, disorder and malfunction, the last stating that disability is imputed to a person.

The legacy of the past practices of hospital-based institutional care, disability-specific charitable organisations and disabilities embedded in individuals continues to influence practices today.

The deeply rooted socio-cultural and community beliefs about disability stand in contradiction to the understanding of disability arising from *in interaction with various barriers* contained within the CRPD and Australia's national disability reform documents. The relatively recent focus on children with disabilities as actors in their own right, and their right to take their place in the community alongside their non-disabled peers is apparent only in reform documents, not in everyday reality (Llewellyn et al. 2016, pp. 65–66; Robinson 2012). In this contested context, there are likely to be significant conservative pressures brought to bear on families to treat their children as part of a 'special group.' Recent evidence of a trend towards more specialised educational settings for students with disabilities in Australia suggest that this may be the case. In 2015, 15% of students with disabilities overall, and 26% of students with severe or profound limitation attended a special school. At the same time, fewer students attended special classes within a mainstream school environment—decreasing since 2003 by 22% for students with disabilities overall, and by 8% for students with severe or profound limitation (AIHW 2017a).

## Understandings of sexuality for children and young people with disabilities

How societies consider sexuality for children and young people with disabilities derives from the pervading social and cultural constructions of childhood, sexuality and disability at the time and the socio-economic positioning and social status or class. For example, in the late eighteenth and early nineteenth centuries in Australia, girls and women were particularly vulnerable to sexual exploitation but not if they were 'ladies.' Sexual exploitation of girls and young women who were 'feeble-minded' was also likely under those conditions; however, the historical record is silent on this point. Their absence from the record presupposes their inherent sexuality was not of concern; rather, they were seen in a utilitarian manner, with bodies readily available as objects for the sexual pleasure of others (Borsay and Dale 2012; Swain 2014b).

The eugenics period of the late 1800s through to mid-20th century brought forth a different denial response to the sexuality of people with disabilities, dedicated at this time to ensuring they did not procreate and 'sap the strength of society.' This is clearly evident in the case of Carrie Buck, a 17-year-old Virginian girl, who became pregnant after being raped by the visiting nephew of her foster parents with whom she lived as a 'domestic.' Powerful men in the governing class conspired (successfully) to have her rendered 'feeble-minded' as her mother had been before her, then sterilised and institutionalised. This legal (at that time) act was in defiance of the testimony of her school records demonstrating that Carrie, although from a poor and at times homeless family, was an attentive and well-liked student with good grades (Cohen 2016). In the words of the court decision penned by Justice Wendell Holmes in *Buck v. Bell,* 1927 cited in Cohen 2016, p. 6, "It is better for all the world, if instead of waiting to execute degenerate *offspring* for crime, or to let them starve for their imbecility, society can prevent those who are manifestly unfit from continuing their kind."

The denial of personhood and agency to children and young people with disabilities, which allowed the sexual exploitation of Carrie Buck and permits ongoing abuse and neglect, is, as noted by Robinson (2012, p. 5) "...a longstanding and pervasive social problem. It is underpinned by social and systemic practices and attitudes which set low expectations for children and young people with disability and which frequently leave them on the margins in both practice and policy. Too often, we allow practices for children with disability which would cause community outcry if used for children without disability. We fail to take action as bystanders to address concerns about neglect, possible abuse, or professional poor practice which can result in abuse. We prioritise other concerns over the rights and needs of children and young people with disability to be safe, which can result in them remaining in situations of risk or of actual abuse."

The historical record in Australia demonstrates that 'care' of children and young people with disabilities in community organisations and institutions, as well as in some families, has been and, for many, continues to be exploitative and abusive. The earlier constructions of the lives of children and young people with disabilities as expendable continues to 'permit' perpetrators of sexual abuse to act with expectations of impunity (Llewellyn et al. 2016).

## Abuse in the lives of children with disabilities in Australia

In Australia, there are no national figures on risk or prevalence of sexual abuse of children with disabilities and its related impact on the child and, more broadly, their family. This means that much less attention has been paid than needed to the very serious societal problem of maltreatment of children with disabilities. Abuse and neglect of children with disabilities is not consistently recorded across the Australian states and territories. It is not possible to monitor the success or otherwise of any prevention or intervention strategy put in place or to evaluate policy initiatives. For robust evidence, Australian commentators have no alternative but to rely on the international figure presented at the beginning of this chapter (Llewellyn et al. 2016 pp. 33–51). To reiterate, Jones et al. (2012) found that children with disabilities are nearly three times more likely than their non-disabled peers to experience sexual abuse. This means that, for every 100 children with disabilities, 9 to 14 of these children are likely to experience sexual abuse. In the absence of national data, the following section pieces together research findings from maltreatment substantiations, care and protection proceedings, reportable conduct matters and out-of-home care to shed light on the breadth and depth of maltreatment of children with disabilities in Australia.

Maclean et al. (2017) reported findings from a linked administrative dataset of all children born in Western Australia between 1990 and 2010. These authors showed that children with disabilities comprised just over one-quarter (25.9%) of maltreatment allegations and over one-quarter (29%) of substantiated allegations. The highest over-representation in substantiated allegations was for children with mental/behavioural disorder (15.6%), intellectual disability (6.7%) or conduct disorder (4.5%). These findings on elevated risk for children with intellectual disability and mental/behavioural problems echo the results of the meta-analysis by Jones et al. (2012). Jones et al. (2012, p. 904) were only able to report risk ratios for children with mental/intellectual disability as there were insufficient numbers to do so for other groups of children with disabilities. The risk ratios demonstrated that children with mental/intellectual disability were 4.3 times more likely to experience any type of violence, 4.6 times more likely to experience sexual violence, 4.3 times more likely to experience emotional violence, and 3.1 times more likely to experience physical violence than their non-disabled peers.

Data from care and protection proceedings and reportable conduct investigations also demonstrate that families and/or the state do not always fulfil their obligations to serve the best interests of children with disabilities. In a study conducted over a nine-month period in the NSW Children's Court, McConnell et al. (2000) found that 24% of reportable conduct investigations involved children with an identified disability or medical condition. Ombudsman NSW (2016) has reported the over-representation of children with disabilities in reportable conduct matters. Over one-quarter (29%, 588) of all notifications of reportable conduct in a two-year period (2013–2015) involved children with disabilities or additional support needs. In the out-of-home care sector, 36% of reportable conduct cases involved children with disabilities, a three-fold over-representation of these children. In the education sector, 21% of the notifications involved children with disabilities, representing a nearly two-fold over-representation of these children.

It was not until 2017 that the Australian Institute of Health and Welfare (AIHW) was able to report the proportion of children with disabilities in out-of-home care from the Australian Child Protection Minimum Dataset (AIHW 2017b, p. 46). The data were available for only six Australian jurisdictions, representing 71% of children in out-of-home care at 30 June 2017 (AIHW 2017b, p. 46). Overall, almost 15% (n=5,020) of children in out-of-home care were reported as having a disability. (The remaining 85.3% did not report any disability; Table S41, AIHW 2017b). This is just over twice the proportion of children with disabilities between 0 and14 years in the Australian population, as reported by the Australian Bureau of Statistics (2012), at 7.2%.

The data from these four sources separately and together demonstrate significant over-representation of children with disabilities, compared to their non-disabled peers. It could be expected that national childhood protection frameworks would highlight this over-representation and recommend strategies to address the significant societal failure to protect children with disabilities. However, in *Protecting Children is Everyone's Business: National Framework for Protecting Australia's Children 2009–2020* (Commonwealth of Australia 2009, p. 21), childhood disability gets only one specific mention. The only solutions offered (inappropriately) are service supports under a subsection of Strategy 3.3, namely "Increase services and support for people with mental illness." Each of the five recommended actions speak to managing risk by way of increasing supports to families rather than directly addressing the contexts in which abuse is known to occur. This continues the tradition of locating the 'problem' in the child, for whom the family bears the primary responsibility. As Llewellyn et al. (2016, p. 31) wrote in their research report for the Royal Commission, "If children with disability were regarded as children first, they would have a strong presence in policy frameworks designed to keep children safe. One would expect to see explicit detailing of the ways in which societal attitudes play out in institutional contexts frequented by children with disability and the ways in which these children may be at potential risk of maltreatment, including sexual abuse. This is not the case."

## Royal Commission into Institutional Responses to Child Sexual Abuse

The primary task of the Royal Commission was to inquire into how institutions with a responsibility for children have managed and responded to allegations and instances of child sexual abuse. Issues relating to children with disabilities were examined in four public hearings of some two weeks each (Case Study 9: St. Anne's Special School; Case Study 41: Disability Service Providers; Case Study 38: Criminal Justice Issues; and Case Study 57: Nature, Causes and Impact of Child Sexual Abuse). The publicly available report runs to 17 volumes, informed by an extensive document bank of research studies, policy documents and case study reports (Royal Commission into Institutional Responses to Child Sexual Abuse Commonwealth of Australia 2017a). Of the nearly 8000 survivors of child sexual abuse who told their stories in private sessions, 4.2% reported having a disability at the time of the abuse. The survivors with disabilities were more likely to be male (61.4%); 16% were Aboriginal or Torres Strait Islander; and more than one-quarter (25.9%) reported being first sexually abused after 1990. Many survivors with disabilities talked about experiencing other forms of abuse, most commonly mentioning emotional and physical abuse. Of the survivors who talked about the type of sexual abuse, where it happened, and who carried out the abuse, an overwhelming majority (86.1%) talked about experiencing multiple episodes of sexual abuse; nearly three-quarters (62.8%) reported that the abuse lasted up to a year; about one-third (32.4%) said they were abused in school; and just over one-third (34.1%) talked about being abused in historical out-of-home care (pre-1990) (Royal Commission into Institutional Responses to Child Sexual Abuse Commonwealth of Australia 2017b).

The chapter turns now to distilling the reasons why these children are at greater risk of sexual abuse. This section draws on my analysis of survivor testimony, case studies, organisational disclosures, research reports and expert advice provided to the Royal Commission (Llewellyn personal papers 2019). The findings from the Royal Commission are considered within the context of the everyday lives of children with disabilities in Australia.

## Everyday lives of children with disabilities

The everyday circumstances of children with disabilities are quite different in many ways to those of their non-disabled peers. Children with disabilities spend a great deal of time with adults in institutional contexts and much more so than their non-disabled peers. Some contexts, where children with disabilities receive support, are not yet 'normalised.' They remain specialist settings as these are only for children with disabilities. Children with disabilities receive specialist disability services, attending hospital clinics for children with disabilities or related medical conditions; they go to respite care in group homes for children with disabilities or have respite carers in their own home; they are enrolled in special schools or attend special classes in mainstream schools; and, although some may participate in mainstream sport and recreation activities, many attend disability-specific sports organisations. Whether children participate in inclusive or segregated (specialist) contexts, depends on their impairment and the capacity of mainstream services to provide support. No matter the institutional context, however, children with disabilities are with adults who they are not related to or do not know for a large proportion of their day.

Specialist services remain separated physically and ideologically from services for other children, with a corresponding separation of staff. Segregated services are less open to informal scrutiny by the general community as they are out of the public eye. For example, special classes in schools are often in separate buildings with their own teachers, teacher's aides and dedicated playgrounds. Children with disabilities are also in closer contact with supervisory adults than are non-disabled children, due to the well-established practice of higher staff-to-child ratios.

The daily routines for children with high physical care needs involve close, frequent and repetitive contact with adults. This occurs in intimate personal activities, such as toileting, showering, bathing and being dressed. These activities typically happen in private, reducing the likelihood of detection of sexual abuse. Many children receive frequent intimate and personal care from one or more adults. Ideally, intimate personal care routines would always be carried out by a known and trusted adult. This is not always the case. Staff call in sick, replacements are organised, and the child may not know the replacement carer. This means that children with high physical care needs may have multiple adults carrying out their personal care over the course of a week, month, or year. Staffing services for children with disabilities presents a management challenge. This means that staff behaviours, which would normally be considered to be unacceptable, may be overlooked by managers struggling to maintain fully staffed rosters. Staff may be poorly trained to understand what constitutes sexual abuse of children with disabilities, and therefore minimise or dismiss their colleague's behaviour, rather than recognising this as grooming or other sexual predatory behaviours.

Children with disabilities also live in closed environments where there is no external visibility and little supervision (Kaufman and Erooga 2016). In these settings, there is a more complete separation of children from the community and, in turn, an enhanced power differential between staff and children. The focus in closed environments is on containment under legislation, rather than responding to children's support needs. The staff in these environments rarely have training to work with children with disabilities. This can lead to mistreatment of all kinds, including punitive measures for what staff think of as 'misdeeds.' Abusive actions can become normalised as an 'appropriate' way to manage 'misdeeds' or behaviours that staff find challenging. Children with non-visible disabilities are most at risk; they also represent the greater proportion of children with disabilities in detention centres and juvenile justice institutions. *Protecting Australia's vulnerable children report: A national challenge* (Parliament of Australia 2005, pp. 168–175) reported that more than one-quarter (28%) of the young people in juvenile justice institutions had borderline or below-average intellectual functioning. Baldry (2013) revealed evidence of a direct relationship between children with disabilities spending time in out-of-home care then moving into juvenile justice, and, later, as young people, being more likely to enter adult incarceration.

Over the two time periods addressed in this chapter, it is hard not to conclude that children with disabilities have been "othered" into absence from the community by their 'specialness.' They have been physically and socially excluded and lost to community consciousness. Their absence from community and ignorance about the realities of their lives plays out in ways which can both facilitate and conceal sexual abuse. The 'specialness' of children with disabilities in needing (particularly) medical assistance works to absolve the community from direct responsibility. Uncertainty on the part of community members, about what to do or the fear of doing the wrong thing, leaves children with disabilities vulnerable and virtually alone (even if not physically) in public spaces with strangers. Specialness and perceived negative differences lead to thinking that children with disabilities, and especially those with cognitive impairment, do not feel trauma and are not harmed by sexual abuse. Because children with disabilities can face additional barriers to disclosure, they are vulnerable to continuing abuse and therefore cumulative harm. As found by the Royal Commission (2017b, p. 5), "of the survivors with disability who told us about barriers to disclosure, almost one-third did not disclose due to shame and embarrassment and a similar proportion also feared retribution from the institution. Just over one in five survivors had no one to tell. One in five survivors were frightened that they would not be believed."

## Contemporary responses to keeping children safe

The contemporary response is to create child-safe organisations. As the Royal Commission (2017a, pp. 24–26) found, the community and the organisations within it need to be held responsible for keeping all children safe. To that end, the Council of Australian Governments recently endorsed a set of *National Principles for Child Safe Organisations* (Australian Human Rights Commission 2019). The principles provide a top-down situational prevention approach to addressing children's safety in organisational contexts peopled by adults (Kaufman et al. 2012). Each of the ten *National Principles for Child Safe Organisations* is designed to apply to everyone, requiring that all adults are accountable to all young Australians, including children with disabilities. National Principle four specifically focuses on diversity, stating that "Equity is upheld, and diverse needs respected in policy and practice." This is a welcome step forward. However, when societies relegate children with disabilities to exclusionary settings (even though these may be in community locations), they are less likely to be seen, heard, known and respected by the community. The community is where children with disabilities need to be.

To be effective in meeting the needs of children with disabilities a whole-of-community approach requires that adults, especially policy makers, organisational leaders and service providers, become informed about why and under what conditions sexual abuse occurs. It means exercising due diligence in understanding and responding to the particularities of the everyday lives of children with disabilities. When organisations are ignorant about children with disabilities and their vulnerability to sexual abuse and do not adapt their practices to accommodate ways of communicating or have codes of conduct with expectations of staff behaviour, children with disabilities are not seen as children first, but through the lens of their impairment or health condition. When organisations regard children with disabilities through one lens only—the disabled lens—they are typically seen as less able, less in need of care and protection, and their rights to education about protective behaviours and being protected in child-safe organisations are neglected.

The *National Principles* focus on much needed change at an institutional level. A first step for organisations is to have processes in place to challenge the culture of closed communication that surrounds notification of abuse of children with disabilities, vividly brought to the Royal Commission through the testimony of adult survivors and parents speaking on behalf of their children. Specialist disability services appear remarkably isolated from the children's sector, where keeping children safe has a longer history and a deep knowledge base. There are organisational cultures in the disability sector that dismiss the likelihood—and reality—of sexual abuse of children with disabilities. Some do so through staff ignorance and lack of training, poor leadership or inactive or inadequate internal processes. Others actively deny organisational culpability or the culpability of their staff. The end-result is the same: Devastating personal impact on children with disabilities, with their families being maligned, and their complaints demolished to protect organisational reputations. Specialist disability organisations would benefit from looking outwards and adopting the community of practice strategy now well established in the care and protection sector.

The starting point for organisations is recognising children with disabilities as active participants in being safe. Children with disabilities, like all children, have the right to receive evidence-based education about what is and what is not safe touching and learning about and practicing protective behaviours, including training on body privacy. This means that protective behaviour learning materials need to be available in accessible, culturally appropriate and diverse formats, including Easy English, community languages, Auslan, Braille, audio, video, graphic, large print and screen reader-accessible. As Wayland and Hindmarsh (2017, p. 7) wrote "regrettably children with disabilities are usually 'left out' of prevention programmes, on the assumption that these programmes are not relevant, that they will not understand the content or learn what to do, or on the naïve assumption that it is unlikely children with disabilities will be abused."

Children with disabilities have the right to have their views respected, including on environments and practices that make them feel safe. Robinson and McGovern (2014) have called for a prevention framework that listens to and respects the voices of children with disabilities on staying safe. This approach respects the agency of children with disabilities in matters affecting their lives, as required by the CRPD. Families also have the right to information about the vulnerability of children with disabilities to sexual abuse, particularly, but not only, in institutional contexts. Families need to be involved with education programmes for their children to ensure that they are familiar with the content and their child's understanding about protective behaviours. Families or carers are best placed to be the first to notice any change in a child's behaviours. They have the right to be taught and supported to recognise changes in child behaviour that are associated with child sexual abuse.

Attention needs to be focused on disability-specific settings, including respite care, school transport, personal care services and community participation activities. There is also a larger job to do in closed environments, such as juvenile justice and detention centres, where children with disabilities are doubly disadvantaged: By the wielding of adult power over child inmates in a custodial context, and by ignorance or naïvety about the particularity of children with disabilities.

# Conclusion

This chapter aimed to contribute to overturning the current practice of keeping the sexuality of children and young people with disabilities out of sight and out of mind. The chapter presented an historical-socio-cultural perspective on sexual abuse of children and young people with disabilities in Australia, arguing that practices from the early days continue to play out in daily life for these children in the twenty-first century. Children with disabilities are missing from community consciousness, despite reform efforts to achieve equity and participation of persons with disabilities in Australian society. Time will tell whether the inclusive approach taken in the *National Principles for Child Safe Organisations* will be sufficient to address the reasons why children with disabilities are sexually abused. Ongoing, nationally consistent recording and reporting of sexual abuse against children and young people with disabilities is fundamental to preventing sexual abuse continuing to occur. Implementation of the current policy initiative to create child-safe organisations will need to be examined rigorously and in timely fashion. Then, if required and based on sound evidence, policy settings can be adjusted to make good the promise of keeping all young Australians safe.

# References

Allowah Presbyterian Childrens Hospital (n.d.) viewed 3rd March 2009, https://www.allowah.org.au

Australian Bureau of Statistics, (2012) *Children with a disability*, Catalogue No.4102.0, ABS Canberra, ACT, viewed 3rd March 2019, http://www.abs.gov.au/AUSSTATS/abs@.nsf/allprimarymainfeatures/E0094F F470CFB2A0CA257A840015FA3D?opendocument

Australian Human Rights Commission (2019) *National principles for child safe organisations.* Australian Human Rights Commission and National Framework for Protecting Australias Children 2009–2020, viewed 3rd March 2019, https://childsafe.humanrights.gov.au/national-principles/download-national-principles

Australian Institute of Health and Welfare (2017a) *Disability in Australia; changes over time in inclusion and participation in education.* Australian Institute of Health and Welfare Fact Sheet, Canberra, Act, viewed 3rd March 2019, https://www.aihw.gov.au/getmedia/34f09557-0acf-4adf-837d-eada7b74d466/Education-20905.pdf.aspx

Australian Institute of Health and Welfare (2017b) *Child protection Australia 2016–17*, Child Welfare Series No. 68. Cat. No. CWS 63. Australian Institute of Health and Welfare, Canberra. Table S41 Data supplement, viewed 3rd March 2019, https://www.aihw.gov.au/reports/child-protection/child-protection-australia-2016-17/data

Baldry E. (2013) *Compounding disability and disadvantage: Pathways into the criminal justice system*, viewed 3rd March 2019, https://lha.uow.edu.au/content/groups/public/@web/@law/@lirc/documents/doc/uow166210.pdf

Blainey G. (1988) *The tyranny of distance: How distance shaped Australias history.* South Melbourne, Victoria: Macmillan Company of Australia.

Borsay A. and Dale P. (Eds.) (2012) *Disabled children: Contested caring, 1850–1979.* London: Routledge.

Cohen A. (2016) *Imbeciles: The Supreme Court, American eugenics, and the sterilization of Carrie Buck.* New York, New York State: Penguin Random House LLC.

Commonwealth of Australia (2004) *Forgotten Australians: A report on Australians who experienced institutional or out-of-home care as children.* Senate Community Affairs Reference Committee, Commonwealth of Australia, Canberra, viewed 3rd March 2019, http://www.aph.gov.au/Parliamentary_Business/Committees/Senate/Commun ity_Affairs/Completed_inquiries/2004-07/inst_care/report/index

Commonwealth of Australia (2005) *Protecting Australias vulnerable children: A national challenge.* The Senate Community Affairs Reference Committee, Second report on the inquiry into children in institutional or out-of-home care Commonwealth of Australia, Canberra, viewed 3rd March 2019, http://www.aph.gov.au/Parliamentary_Business/Committees/Senate/Commun ity_Affairs/Completed_inquiries/2004-07/inst_care/report2/index

Commonwealth of Australia (2009) *Protecting children is everyones business: National framework for protecting Australias children 2009–2020*. An initiative of the Council of Australian Governments, viewed 3rd March 2019, https://www.dss.gov.au/sites/default/files/documents/child_protection_framew ork.pdf

Commonwealth of Australia (2011a) *2010–2020 National Disability Strategy*. An initiative of the Council of Australian Governments, viewed 3rd March 2019, https://www.coag.gov.au/sites/default/files/nation al_disability_strategy_2010-2020.pdf

Commonwealth of Australia (2011b) *Disability care and support productivity commission inquiry report no. 54, 31 July 2011*. Canberra, viewed 3rd March 2019, http://www.pc.gov.au/inquiries/completed/disabilit y-support/report

Commonwealth of Australia (2011c) *National carer strategy*. Australian Government, Canberra, viewed 3rd March 2019, http://www.seslhd.health.nsw.gov.au/Carer_Support_Program/Documents/20 11% 20national_carer_strategy.pdf

Commonwealth of Australia (2013) *National Disability Insurance Scheme Act 2013*, in Force as Act No. 126, 2015 Canberra, viewed 3rd March 2019, https://www.legislation.gov.au/Details/C2016C00213

Commonwealth of Australia (2015) *Violence, abuse and neglect against people with disability in institutional and residential settings, including the gender and age related dimensions, and the particular situation of Aboriginal and Torres Strait Islander people with disability, and culturally and linguistically diverse people with disability*. The Senate Community Affairs Reference Committee, Canberra, ACT, viewed 3rd March 2019, https:// www.aph.gov.au/Parliamentary_Business/Committees/Senate/Community_Affairs/Violence_abuse_ neglect/Report

Council of Australian Governments (2009) *National disability agreement: Intergovernmental agreement on federal financial relations*. Canberra, viewed 3rd March 2019, http://www.federalfinancialrelations.gov.au/ content/npa/disability/national- agreement.pdf

Cummins C.J. (2003) *A history of medical administration in NSW 1788–1973*, 2nd ed. Sydney, NSW: NSW Department of Health.

Department of Social Services Disability Policy and Research Working Group (2011) *National disability research and development agenda November 2011*. Department of Social Services, Canberra, viewed 3rd March 2019, https://www.dss.gov.au/our-responsibilities/disability-and-carers/program-services/ government-international/national-disability-agreement/national- disability-research-and-develo pment-agenda

Fraser N. (1998) From redistribution to recognition? Dilemmas of justice in a post-socialist age, in A. Phillips (Ed.), *Feminism and politics*. Oxford, UK: Oxford University Press.

Gandevia B. (1978) *Tears often shed: Child health and welfare in Australia from 1788*. Rushcutters Bay, NSW: Pergamon Press.

Garton S. (1990) *Out of luck: Poor Australians and social welfare 1788–1988*. Sydney, NSW: Allen & Unwin.

Jarrett C.F., Mayes R. and Llewellyn G. (2014) The impact of disablism on the psycho-emotional well-being of families with a child with impairment. *Scandinavian Journal of Disability Research* 16(3), pp. 195–210. DOI:10.1080/15017419.2013.865671.

Jones L., Bellis M.A., Wood S., Hughes S., McCoy E., Bates G., Mikton C., Shakespeare T. and Officer P. (2012) Prevalence and risk of violence against children with disabilities: A systematic review and meta-analysis of observational studies *Lancet* 380, pp. 899–907. DOI:10.1016/S0140-6736(12)60692-8.

Kaufman K. and Erooga M. (2016) *Risk profiles for institutional child sexual abuse: A literature review*. Royal Commission into Institutional Responses to Child Sexual Abuse, Commonwealth of Australia, viewed 3rd March 2019, https://www.childabuseroyalcommission.gov.au/sites/default/files/file-list/Research%20Report%20-%20Risk%20profiles%20for%20institutional%20child%20sexual%2 0abuse%20-%20Causes.pdf

Kaufman K.L., Tews H., Schuett J.M. and Kaufman B.R. (2012) Prevention is better than cure: The value of situational prevention in organisations, in M. Erooga (Ed.), *Creating safer organisations: Practical steps to prevent the abuse of children by those working with them*. Chichester, UK: John Wiley & Sons.

Liddell M. (1993) Child welfare and care in Australia: Understanding the past to influence the future, in C. Goddard and R. Carew (Eds.), *Responding to children: Child welfare practice*. Melbourne, Victoria: Longman Cheshire.

Llewellyn G., Wayland S. and Hindmarsh G. (2016) *Disability and child sexual abuse in institutional contexts*. Royal Commission into Institutional Responses to Child Sexual Abuse, Commonwealth of Australia, viewed 3rd March 2019, www.childabuseroyalcommission.gov.au/sites/default/files/file-list/Research Report - Disability and child sexual abuse in institutional contexts - Causes.pdf

Maclean M.J., Sims S., Bower C., Leonard H., Stanley F.J. and ODonnell M. (2017) Maltreatment risk among children with disabilities. *Pediatrics* 139(4), pp. e1–e10. DOI:10.1542/peds.2016-2017.

McConnell D., Llewellyn G. and Ferronato L. (2000) *Parents with a disability and the NSW Childrens Court.* Sydney: The Family Support and Services Project University of Sydney, http://sydney.edu.au/health-sciences/afdsrc/docs/mcconnell-parents.pdf

Monk L. and Manning C. (2012) Exploring patient experience in an Australian institution for children with learning disabilities, 1887–1933. In A. Borsay and P. Dale (Eds.), *Disabled children: Contested caring, 1850–1979.* London: Routledge.

Ombudsman NSW (2016) Disability reportable incidents.16 years of Reportable Conduct Forum, viewed 3rd March 2019, https://www.ombo.nsw.gov.au/what-we-do/our-work/community-and- disability -services/part-3c-reportable-incidents

Robinson S. (2012) *Enabling and protecting: Proactive approaches to addressing the abuse and neglect of children and young people with disability.* Children With Disability Australia, Clayton, Victoria, viewed 3rd March 2019, http://epubs.scu.edu.au/ccyp_pubs/90/

Robinson S. and McGovern D. (2014) *Safe at school? Exploring safety and harm of students with cognitive disability in and around school.* Centre for Children and Young People Southern Cross University, Lismore, NSW, viewed 3rd March 2019, http://www.lawfoundation.net.au/ljf/site/templates/Grants_Reports/ $file/SCU_ Safe_At_School_Final_2014.pdf

Royal Commission into Institutional Responses to Child Sexual Abuse Commonwealth of Australia (2017a) *Final report.* Royal Commission into Institutional Responses to Child Sexual Abuse, viewed 3rd March 2019, https://www.childabuseroyalcommission.gov.au/final-report

Royal Commission into Institutional Responses to Child Sexual Abuse Commonwealth of Australia (2017b) *A brief guide to the final report disability*, viewed 3rd March 2019, https://www.childabuseroy alcommission.gov.au/sites/default/files/a_brief_guide_to_the_final_report_-_disability.pdf

Royal Commission on Asylums for the Insane and Inebriate. *Victorian Parliamentary Papers*, 2, 1886, Minutes of Evidence Q.8677–80, p. 349.

Sabatello M. (2013) Children with disabilities: A critical appraisal. *The International Journal of Childrens Rights* 21(3), pp. 464–487.

Summers A. (1975). *Damned whores and Gods police: The colonization of women in Australia.* Ringwood, Victoria: Penguin Books.

Swain S. (2014a) *History of institutions providing out-of-home care residential care for children.* Royal Commission into Institutional Responses to Child Sexual Abuse, Commonwealth of Australia, viewed 3rd March 2019.

Swain S. (2014b) *History of inquiries reviewing institutions providing care for children.* Royal Commission into Institutional Responses to Child Sexual Abuse, Commonwealth of Australia, viewed 3rd March 2019, http://www.childabuseroyalcommission.gov.au/documents/published- research/historical-perspectiv es-report-3-history-of-inquir.pdf

United Nations (2006) *Convention on the rights of persons with disabilities*, viewed 3rd March 2019, http:// www.un.org/disabilities/convention/conventionfull.shtml

Wayland S. and Hindmarsh G. (2017) Understanding safeguarding practices for children with disability when engaging with organisations. CFCA Practitioner Resource, viewed 3rd March 2019, https://aifs.gov.au/ cfca/publications/understanding-safeguarding-practices-children-disability-when-engaging

World Health Organisation (2001) *International classification of functioning, disability and health.* World Health Organisation, Geneva, viewed 3rd March 2019, https://apps.who.int/iris/bitstream/handle/10665/ 42407/9241545429.pdf;jsessionid=FDC3EFE9475B99C63109DE16A7A0E3B5?sequence=1

Young I. (1998) Polity and group difference: A critique of the ideal of universal citizenship, in G. Shafir (Ed.), *The citizenship debates: A reader.* Minneapolis: University of Minnesota Press.

# 10

# SEXUALITY AND PHYSICAL DISABILITY

## Perspectives and practice within Orthodox Judaism

*Ethan Eisen*

## Introduction

This chapter comprises a discussion of the three-pronged intersection of Judaism, sexuality and physical disability. Despite the apparent relevance of this combination of topics to the generations of Jewish communities over the past 2,000+ years, there is, to my knowledge, no previous rabbinic or academic literature providing a detailed overview of the salient issues. This reality presents unique opportunities and challenges. This chapter will provide an important foundation from which further research and community changes can emerge. Relevant aspects of this intersection could fill many chapters with previously unaddressed examinations, and the topics that are discussed below may act to provoke deeper critical examination of the ways in which Judaism can further explore the inclusiveness of sexuality and intimacy among people with physical disabilities (PPDs).

## Jewish historical and legal context

A few notes to the reader are important to clarify some issues before discussing the content of the material. Orthodox Jewish life is governed by what is known as *halakha,* which is the system of Jewish law that generally guides daily living and ritual. The foundations of this system are derived from the Talmud, which was compiled in several stages roughly 1,500–2,000 years ago, and the basic principles and practical applications have been codified over the centuries by scholars. As a result, this chapter will be different stylistically from standard academic articles or chapters in this volume in a number of ways. First, some of the writing may seem unusually legalistic for an article related to culture, disability and sexuality; the reason for this is that the Orthodox Jewish culture, to a large extent, relates to questions about disability and sexuality through *halakha,* its legal system. In contemporary western society, issues of sexuality are often centred on desire, emotions, rights and fulfilment, whereas, in *halakha*, the language used is more often about obligations and prohibitions.

Secondly, many of the citations that I use do not come from contemporary scholarly sources. Instead, I use primary sources from which the principles at work in contemporary Orthodox

Jewish society are derived. I chose to do this because it allows the reader to gain a more authentic view of how these issues may be seen in Orthodox Jewish society, and because the dearth of contemporary academic literature dealing with these issues does not allow for substantial citations. Although this style may make it more challenging for readers to look up sources and reach their own conclusions, in order to provide the most comprehensive and authentic view of these issues, it is necessary to use the sources that would be included among discussions in the Orthodox Jewish world. Appendix I provides a list of important works and scholars included in this chapter.

Finally, it is crucial to note that not all people who identify as Orthodox Jews adhere to or apply *halakha* in the same way, and these differences apply to both clergy and laypeople. Just as different judges apply legal principles differently, so, too, scholars of *halakha* have many disagreements about how principles should be applied, particularly as they relate to new questions that arise in modern times. Laypeople may also have differing levels of commitment to following *halakha*. Some will ask their rabbi for guidance related to the smallest details of *halakha*, while others may only rarely seek guidance from their religious leaders. Many people may identify nominally with the Orthodox Jewish community, but they choose not to accept the rigours of *halakha*, at least in certain domains of life, such as certain strict limitations on intimacy, as will be discussed below. A larger discussion of the diversity of thought and practice within the Orthodox Jewish community is well beyond the scope of this chapter; nevertheless, it is important for the reader to be aware that the Orthodox Jewish population is far from monolithic, and ideas discussed here may not apply to all members or communities who identify as Orthodox Jewish.

## Sexuality and physical disability in Jewish thought and practice

The intersection of sexuality and physical disability in the Orthodox Jewish community has received little attention by scholars, and, as such, this chapter aims to provide an overview of the major relevant issues. First, values and principles related to sexuality will be presented, followed by the ways in which those principles interface with issues of physical disability. Notably, I herein refer to my understanding of the Orthodox Jewish perspective without the qualifier "Orthodox," except when I think the sentence may not be clear without this qualifier, or that other denominations would strongly disagree with the perspective I am taking; it is possible that other denominations of Judaism would agree or disagree with various points or nuances, and exploring those similarities or differences is beyond the scope of this chapter.

### *Sexuality*

Before providing an overview of how sexuality is seen in traditional Jewish texts, certain aspects of Orthodox Jewish thought and practice should be clarified, particularly in how they may differ from thought and practice in contemporary western culture. Homosexual intercourse is prohibited, and homosexual marriage is not recognized by *halakha*. In traditional Jewish sources, the terms "sex" or "intercourse" typically refer to penile-vaginal intercourse, which is what they mean in this chapter, unless otherwise indicated. There exists an emerging body of literature in the Orthodox Jewish world offering perspectives on homosexuality and its role in Orthodox Jewish life (e.g., Ariel 2007; Greenberg 2004; Rappaport 2004; Slomowitz and Feit 2015), and that discussion is beyond the scope of this chapter.

Contemporary Jewish writers on sexuality tend to adopt a sex-positive stance toward sexuality and intimacy (Boteach 1999; Grushcow 2014; Rosenfeld and Ribner 2011; Ruttenberg

2009). Despite this trend, sexuality, as represented in traditional sources, is perhaps the most polarising construct in Jewish thought (see Lichtenstein 2005 for an excellent review of the opposing viewpoints). When expressed in a proscribed context, the physical urges and lust related to sexual desire are regarded as animalistic and appealing to humanity's most basic desires, which should be avoided, perhaps even eradicated. The male circumcision, perhaps the most widespread ritual that dates back nearly 4,000 years to Abraham, the founder of the Jewish faith, is understood by some to be an act that curbs sexual desire (Maimonides Moreh Nevuchim 3:49). Indeed, the Rabbinic tradition (Sifrei Bamidbar 115) understands that, in addition to the prohibitions related to illicit sexual relationships, the lustful pursuit of sexuality is also a Biblical prohibition (Maimonides Sefer Hamitzvot 47), based on the verse (Numbers 15:39), "and do not explore after your heart and after your eyes after which you stray."

Despite the strong condemnation of the pursuit of physical pleasure, when expressed in a permitted context, sexuality is seen by many authorities as the most holy of expressions, paralleling the potential for intimacy between God and human beings. Song of Songs, the Biblical text that, taken at face value, appears to be an erotic poem between two lovers, is understood to be an allegory for the love between God and human beings, and is regarded by the Talmudic Sages as the "holy of holies" (Mishna Yadayim 3:5). In practical terms, in addition to the act of heterosexual intercourse, a married couple's sharing of the pleasure of sexual intimacy is regarded positively, as least by most contemporary religious leaders, beyond the practical necessity of sexual intercourse to fulfil the Biblical obligation to sire offspring (Talmud Bavli ("TB") Yevamot 61b).

The Rabbis view the provision of physical intimacy, including sexual intercourse, as a Biblical obligation upon a husband (Maimonides Sefer Hamitzvot 262), based on the verse (Exodus 21:10) "…he shall not diminish her food, her clothing, or her marital relationship." So great is this obligation that, at least theoretically, a husband must ask permission from his wife to travel for work or change professions, if these changes will lead to a decline in their intimate time together (Maimonides Hilchot Ishut 14:2). Additionally, neither spouse is permitted to withhold intimacy out of spite, and such a practice may be grounds for divorce (Maimonides Hilchot Ishut 14:7–8).

Notably, although there is very limited discussion in Rabbinic literature regarding the specific applications of this obligation of intimacy as it relates to PPDs, some principles do emerge from the *halakha*. Most fundamentally, physical disability does not exempt a person from the obligations of intimacy, nor does it disqualify a person's right to expect intimacy from one's spouse. This point may seem like a minor feature, but its ramifications are quite significant. Research has demonstrated that PPDs tend to experience lower levels of sexual esteem and sexual satisfaction and engage in mutual sexual activity less frequently than those without physical disability (e.g., McCabe and Taleporos 2003). Studies have also consistently found that PPDs often feel a social stigma that considers them to be asexual or not sexually desirable (e.g., Milligan and Neufeldt 2001). In contrast to those sentiments, *halakha*'s stance, that PPDs have the right and obligation to be a fully participating partner in a sexual relationship with a spouse, may have profound positive effects for PPDs. As such, *halakha* does not distinguish between the sexuality of persons with or without physical disabilities, which may serve to reduce the stigma of not being seen as sexual; similarly, the basic obligation to provide one's spouse with sexual intimacy may serve to boost each other's sexual esteem and satisfaction.

This fundamental principle, encouraging the participation in a sexual relationship, does have certain limitations if disability affects one's ability to engage in sexual intimacy. For example, one ruling in *Shulchan Arukch* (Even HaEzer 76:11) seems to be particularly relevant: "…if he is ill or weak and cannot have intercourse, he should wait six months until he regains his health, as this is the maximum interval for intimacy, and, after that, he needs to receive permission from her, or he should grant her a divorce"; this ruling is similar to the text in Maimonides's ruling. Later authorities debate whether this law is meant to apply to people whose condition will remit within the six months, where the expected duration of the condition is unknown, or it is assumed that the condition will continue much longer (e.g., Avnei Nezer Choshen Mishpat 83). This law seemingly has some bearing on the issue of sexuality among PPDs. It is clear that, whereas the law recognises that chronic illness or physical disability may change the frequency of sexual intercourse, these conditions do not absolve PPDs entirely from the obligation. It is also clear that the couple may come up with an alternative to intercourse as a suitable substitute, indicated by the notion that he can receive permission from her to not have intercourse for a longer period of time.

Another area of *halakha* also demonstrates the positive stance of Jewish tradition, particularly as it contrasts to recent western history. Western countries, including the United States, have long histories of discouraging PPDs from having children. Healthcare professionals typically guided PPDs against having children, and involuntary sterilisation was a common approach to reach this end (e.g., Waxman 1994). This approach is in stark contrast to *halakha*, which strictly prohibits any type of forced sterilisation (e.g., Maimonides Hilchot Issurei Bi'ah 16:11). Whereas it is true that some authorities permitted voluntary chemical self-sterilisation, this leniency was in the context of women who experienced great pain during childbirth and wanted to avoid experiencing such physical distress; nowhere, to my knowledge, did any authority recommend this approach for PPDs. Instead, as noted above, *halakha* seems to assume that PPDs are meant to be full participants in intimacy and bearing children, assuming there is not too great a risk to their own health.

Despite these implications from *halakha*, perhaps the most significant message is that there is a tremendous gap in the Rabbinic literature, with virtually no *halakhic* discourse directly addressing the law as it relates to PPDs and sexuality. Indeed, even recent literature cited above (Boteach 1999; Grushcow 2014; Rosenfeld and Ribner 2011; Ruttenberg 2009), that has addressed issues of Jewish practice and sexuality, has largely omitted any discussion of PPDs. As such, an attempt will be made in this chapter to address some aspects of Jewish thought and practice that have the greatest impact on issues of sexuality for PPDs.

## Intersectional challenges

### Family purity

There exist many restrictions related to sex in Jewish thought and practice, with prohibitions which are both Biblical and Rabbinic in origin. In addition to the commandment outlawing adultery (Exodus 20:13; Deuteronomy 5:17), there are whole sections of the Torah (Pentateuch) listing the forbidden sexual relationships (Leviticus 18:6–23; 20:8–21). Two specific restrictions are particularly relevant to this issue of sexuality and physical disability, and an appreciation of some of their details and nuances is important for a more complete understanding.

The first restriction, based on Leviticus 18:19, is known as the laws of "*niddah*," or colloquially as "family purity," and refers to the prohibition against sexual contact with a woman who is on her menstrual period. Biblical law requires a husband and wife to desist from sexual contact for one week starting with when she sees menstrual blood, during which she is considered "*tamei*," translated as "ritually impure." In order to restore ritual purity, she must immerse herself in a ritual bath ("*mikvah*") after those seven days. Due to complex factors, the Rabbinic tradition requires a woman to have seven "clean" days, during which she does not see any menstrual blood. As a result, a woman is typically ritually impure for 12 or 13 days per month—unless she is pregnant or using a method of birth control that delays her period—accounting for five or six days of menstrual bleeding, and seven clean days; once she achieves the seven clean days, she immerses in the *mikvah* and is considered ritually pure, allowing the couple to resume sexual contact.

In addition to a prohibition against intercourse during the time of *niddah*, Jewish law also requires that a husband and wife not engage in activity that is intimate, or may lead to intimacy during this period, as a safeguard against leading to sexual contact. Some relevant examples include the laws that spouses do not have any physical contact, and that they sleep in separate beds (couples typically furnish their master bedroom with two mattresses so that they can separate the beds during this period).

For PPDs, two primary challenges arise from the laws of family purity. The first, which Jewish legal authorities largely solved hundreds of years ago, is that PPDs may need help from a personal assistant, family member and/or spouse to perform activities of daily living such as dressing, showering and/or using the bathroom independently. Indeed, all of these activities require close physical contact. In most cases, Rabbinic guidance would allow for one spouse to help another, based on the rulings of *Shulchan Aruch* (Yoreh De'ah 195:16–17).

The second challenge relates to the immersion in a *mikvah*. The laws regarding construction of a *mikvah* are complex and far beyond the scope of this chapter, but some relevant practical information is necessary. A *mikvah* requires a substantial amount of rainwater collected directly into a cavity in the ground that does not leak or allow water to seep (Yoreh De'ah 201). To fulfil these requirements, a *mikvah* can cost tens or hundreds of thousands of dollars to construct; in addition to the immersion pool or pools, the building typically has rooms available for women to prepare for the ritual immersion in the prescribed way. A properly constructed *mikvah* is an essential feature to any Orthodox Jewish community, and they are typically funded by the donations of community members.

For PPDs, the process of immersion, which requires complete immersion at one time with water being able to reach all parts of the body, presents a range of potential difficulties related to accessibility, as noted by previous authors (e.g., Ratzon et al. 2006). In addition to the general concerns of accessibility to the building and ritual bath, special equipment is necessary that can secure the woman safely and immerse her completely in the water. The area around the ritual bath must be designed in a way that can either permanently hold this equipment, or at least allow for the attachment of this equipment when necessary. The community also needs to train the women who serve as *mikvah* attendants to operate this equipment to ensure the safety of the woman who is immersing herself.

As regular immersion in a *mikvah* is an unavoidable aspect of married life, PPDs within the Orthodox Jewish community may find themselves with severe geographic restrictions due to the need for proximity to an accessible *mikvah*. This geographic limitation may be compounded by the reality that, on average, it is wealthier communities that build *mikvahs* in an accessible way; at the same time, research has shown that labour participation is far lower for PPDs, and the poverty rate is higher (e.g., Stapleton et al, 2006). When combining the potential for higher living costs associated with disability (e.g., She and Livermore 2007), the cost of living in wealthier neighbourhoods may put an increasing burden on PPDs. Of course, one need not live in a community to make use of its *mikvah*; nevertheless, travelling to neighbouring towns or cities to use the *mikvah* can be a substantial challenge for some PPDs. On a practical level, for example, if the PPDs required a driver, friends or family may be reluctant to drive long distances on a regular basis, and higher rates for rides in accessible vehicles may be a substantial financial burden.

In recent years, an increasing number of Jewish communities have built *mikvahs* accessible to PPDs that include ramps, adequately designed bathroom facilities and immersion pools containing the necessary equipment for safe immersion. Despite this progress, it is clear that there remains an insufficient number that are consistently accessible, and no provision/law requires accessibility. One website (Mikvah.org 2019) has a searchable map that features a filter to show only disability-accessible *mikvahs*. Notably, there are only roughly 15 locations listed with accessible *mikvahs* in North America, most of which are in major Jewish centres. Additionally, it is clear that the website is not current, as this author is aware of other locations that are accessible to PPDs and are not listed on the website. Another website (Orthodox Union 2019) lists many Jewish communities across the USA, and their listing includes the number of *mikvahs* in the respective communities; however, most of the listings do not make clear whether the *mikvah* is accessible to PPDs, and someone seeking that information would have to make phone calls to each site. Additionally, the person answering the phone calls would need to be sufficiently trained to answer questions relevant to PPDs.

A recent news article highlighted the issue of cost as it relates to making *mikvahs* fully accessible to PPDs (Ginsberg 2017). The author surveyed a number of communities that had recently built or renovated a *mikvah;* some communities did decide to include hydraulic lifts or other mechanisms to make the *mikvah* fully accessible, but others, particularly in smaller communities, found that the additional cost was not justified by the need in their respective communities. In a separate interview with the director of a recently-built fully-accessible mikvah, the estimated cost of adding a lift was several thousand dollars (Ruderman Foundation 2016), in addition to whatever the costs are to make the rest of the building fully accessible. It is a positive development in the Jewish community that issues surrounding accessibility to the *mikvah* are receiving increased attention. Nevertheless, the community as a whole is still far away from removing this issue as a barrier for PPDs.

## *Sexual expression*

A second potential challenge for PPDs relates to restrictions around sexual expression, particularly for men to ejaculate not in the context of penetrative vaginal intercourse. The basic principle is that sex, and more specifically the emission of semen, is meant to be facilitated in a manner that could lead to the creation of a child. Despite this generally accepted principle, the source and level of this restriction is not entirely obvious, although many scholars identify a Biblical (Genesis 38) story as a source. Judah found a wife, Tamar, for his son, Ehr, who, as the verse records, "was evil in the eyes of God, and God caused him to die." As was the practice, Ehr's brother, Onan, was meant to marry Tamar in lieu of Ehr, and any son born of this second union would be considered Ehr's, not Onan's. The verse states as follows: "And Onan knew that the seed would not be his; and it came to pass when he went in unto his brother's wife, that he spilled it on the ground, lest he should give seed to his brother." God's displeasure with and punishment of Onan for his action are commonly cited as the source for the prohibition against masturbation and the severity with which this act is viewed. Notably, it is from this story that the English word "onanism" came about.

Although an understanding of the nuances of Jewish legal reasoning is not necessary for this discussion, it is important to add that some prohibitions are considered Biblical, while others are considered Rabbinic. Generally speaking, Rabbinically based prohibitions tend to have more leniencies. Regarding masturbation, there is disagreement among leading mediaeval scholars about the basis of the prohibition, with some arguing that it is included in the Biblical prohibition against adultery, while others argue that, whereas masturbation is viewed very severely, the level of the prohibition is Rabbinic. Some authorities seem to see two distinct problems with masturbation (e.g., Igrot Moshe Even Haezer 1): the first is with "wasting seed," whereas the second relates to the act of causing oneself to ejaculate with one's own hand (TB Ketubot 13b). Other passages raise additional problems with engaging in sexual acts or lustful thoughts outside the context of marital relations, such as the verse "you shall guard from anything evil" (Deuteronomy 23:10, based on the interpretation of TB Ketubot 46a), and "and do not explore after your heart and after your eyes after which you stray" (Numbers 15:39). The prohibition against masturbation is regarded as more severe for men, particularly as the first two problems listed above are generally thought to be specific to men; however, later authorities have pointed out that the latter two probably apply to women as well.

This discussion has substantial ramifications for PPDs, some of whom may find it very difficult or even impossible to assume the physical positions necessary for intercourse. Two aspects of this rule seem relevant to PPDs. The first is that such restrictions may limit the ability of PPDs to experience pleasure from sexuality, which has been recognised as a problem not limited to religious communities (e.g., Tepper 2000). The second is that, for an observant Jewish PPD, these restrictions may appear to disqualify her or him from fulfilling the requirement of sharing intimacy with one's spouse. In the secular world, mental health professionals working with PPDs will often recommend alternatives to intercourse for couples to engage in sexual intimacy. Some methods, such as nude cuddling or deep kissing, would certainly be permitted and encouraged by Rabbinic authorities, just as these activities would be encouraged between spouses who do not have physical disability (Igrot Moshe Even Haezer 4:66). Other methods, such as oral sex or masturbation, may run contrary to the *halakha*. In a personal communication with experts who offer an online service related to issues of sexuality in *halakha*, and who also tend to be lenient in a number of related matters, a variety of leniencies was offered to allow for shared sexual stimulation, but they reiterated that ejaculation without intercourse is not permitted (Zimmerman 2018). For example, the expert advised that, assuming ejaculation outside the vagina does not take place, stimulation by hand is encouraged, and oral stimulation could be permitted in many circumstances. They also added that, when physical limitations restrict penile-vaginal intercourse, anal sex, including ejaculation, may be permissible. Other leniencies that the expert recommended require explanations that are beyond the scope of this chapter.

There does not seem to be a consensus among Jewish legal authorities regarding guidance for couples who live with disability, while not all authorities would agree with all of the leniencies presented above. Although there is no research literature on the practice of seeking religious guidance under these circumstances, it seems likely that, in many instances, couples will not seek or follow rabbinic instruction when there is a strong chance the guidance will create a serious barrier to their intimate life. Scholars should give increased attention to this issue to see whether conditions may permit *halakhic* leniencies, which would allow couples feel that this part of their lives, as well, is part of their spiritual life. There does exist one Talmudic example allowing for ejaculation outside the context of intercourse (TB Yevamot 76a), and some later authorities have expanded this leniency in very limited ways (Beit Shemuel 25:2). Theoretically, this ruling could provide a basis for Rabbinic leniency regarding non-vaginal ejaculation among PPDs; however, this ruling has not yet been discussed by authorities in this context. Additionally, some people do not ejaculate due to physical disability; it is likely that many of the prohibitions described above would not apply to them.

## Marriage and community life

A third challenge faced by PPDs, which may be considered a prerequisite for sexual expression among members of the Orthodox Jewish community, relates to the practice of dating and marriage. Although this issue is not directly related to sexuality per se, in many, if not most,

Orthodox Jewish communities, communal life revolves around family-centred activities and milestones. A full discussion of the courtship traditions in various segments of the Orthodox community is well beyond the scope of this chapter; nevertheless, some background information is necessary. In most segments of the Orthodox Jewish community, dating is understood to be specifically in the context of finding a marriage partner. In many communities, matchmakers, either formal or informal (such as mutual friends), play an important role in introducing single people to one another. It is not uncommon for young people to have "dating resumés" to aid these matchmakers in finding suitable pairs. The duration of dating until engagement also varies, ranging from several weeks, particularly in ultra-Orthodox circles, to months or years for those in more modern Orthodox circles.

This process can present significant barriers for PPDs to enter relationships. Before people date, it is common that a great deal is known about the other person, from their educational history, nuanced understanding of their religious affiliations, family history, professional aspirations and other personal information, that is used in an attempt to find compatible matches. Although this system may streamline the dating process in some ways, in this framework the presence of physical disability may overshadow other qualities, and PPDs may not be set up on dates. Whereas this reality may exist across the general population, this concern is especially relevant in some segments of the Orthodox Jewish community. Perhaps due to the expectedly short courtship in the ultra-Orthodox community, factors that may be deemed less than ideal on paper can lead to a potential match not being offered in the first place. In addition, due to the emphasis on having and raising children, factors that may be considered to potentially affect one's ability to bear or raise children may disqualify PPDs in this system.

An internet search will reveal professional matchmakers in the Orthodox Jewish world who specifically focus on finding dates for PPDs (Craig 2016). On the one hand, it is positive that the community has taken steps to assist PPDs to find suitable partners. On the other hand, this highlights the reality that PPDs are viewed differently when it comes to dating and marriage and are not yet fully included in the dating process.

## Conclusion and future directions

Jewish traditions and laws have a great deal to offer in terms of recognising the full rights and responsibilities of PPDs participating in marital sexual relationships. Practically speaking, the community itself has also made substantial strides in recent decades to fulfil these values. For example, in a recent theatrical production entitled "Mikvah the Musical" focusing on issues surrounding the *mikvah*, one of the main characters is a woman with physical disabilities (Mayim Hayyim 2019); Mayim Hayyim's website includes videos and a suggested curriculum for address issues of inclusivity in the Jewish community. Despite these positive advancements, much progress still needs to be made to allow PPDs to be full participating members.

From a systems perspective, the more open and inclusive the Jewish community is generally, the more PPDs will become fully participating members in many domains of life, including sexuality. For example, if communities place a priority on having inclusive and accessible classrooms, camps, synagogues and social gatherings, PPDs will have more opportunities to be equal members of the social groups and experiences that create the tight-knit communities often seen in the Jewish world. As such, it is likely that any stigma associated with physical disability will decline, and PPDs have the potential to be seen as viable partners for marriage and sexual intimacy. Similarly, as PPDs are more integrated into social circles, communities will also be more invested in ensuring that *mikvahs* and other communal buildings are inclusive and accessible.

In addition to bottom-up grassroots advocacy, it is crucial for rabbinic leadership to receive training in the needs of PPDs, including their needs relating to sexuality. Rabbinical students, including myself, may spend a year or more studying the laws of family purity; yet, to my knowledge, rabbinical schools do not include any direct training on issues of sexuality and physical disability. There could be many reasons why this omission is present: there is insufficient traditional rabbinic literature to provide adequate training, and the teachers themselves are not familiar with the issues; few of the rabbis-in-training expect to receive questions about these issues, so it may seem like an inefficient use of time; discussions of sexuality and intimacy are somewhat taboo, and this topic requires the student to leave the world of the abstract and make use of very specific language and concepts; alternatively, some other reasons may limit the interest or ability to present these topics in a training setting.

Despite these reasons for this topic's widespread exclusion from the curriculum, it is crucial for rabbis and scholars to advance the theory and practice in a top-down approach as well. Particularly in Orthodox Jewish communities, the rabbi or other leaders may have considerable influence over the priorities of their membership, and their advocacy can have a profound impact on the inclusion of PPDs. Congregants often follow the lead of rabbinic leadership, regarding how to donate their charity funds—affecting the available funds for building accessible communal buildings—and what type of programming to run in their synagogues. The fundamental values of the Jewish community are in line with the basic human right of PPDs to be full participants in fulfilling physical intimate and sexual relationships. With increased awareness and attention, the community has the ability to continue to make changes to their institutions to ensure that they are more in line with their guiding principles.

## Appendix I

Below are primary Jewish texts and scholars mentioned in the chapter. They are listed below in chronological order.

Talmud—the Talmud forms the foundation of Jewish law and practice. It comprises two main works: the Mishna and the Gemara. The Mishna contains relatively short rulings that are the basis for the law, and it was compiled by Rabbi Yehuda HaNasi, a leading scholar, roughly 2,000 years ago. There were two versions of the Gemara completed roughly 1,500 years ago, one composed in ancient Palestine ("Talmud Yerushalmi"), the other in ancient Babylon ("Talmud Bavli"); the Babylonian text is considered to be the more authoritative. The Gemara is an extensive text, divided into tractates, that examines the sources and applications of the laws in the Mishna and records many disagreements among the Rabbis regarding the final ruling of many laws.

Sifrei—the Sifrei is a *halakchic* text composed by Rabbis at the time of Mishna, whose rulings were not compiled in the Mishna.

Rabbi Moshe ben Maimon ("Maimonides" or "Rambam")—Rambam was a leading 12th century scholar who codified the laws of the Talmud topically, unlike previous works codifying the law. His primary text is divided into 14 sections, as well as a volume listing the 613 Biblical commandments.

Sefer HaChinuch—this 13th century text, providing descriptions of the 613 Biblical commandments, was written anonymously. Despite not being certain of its authorship, later scholars recognize it as a primary source for interpreting various aspects of the commandments.

Shulchan Arukh—this 16th century text was an additional authoritative codification of Jewish law, based on the views of Rambam and other leading mediaeval scholars. It was composed primarily by Rabbi Yosef Karo, with emendations by a contemporary scholar, Rabbi Moshe Isserles. Since that time, Shulchan Arukh has served as the primary codification of Jewish law for Jewish communities around the world. The work contains four main divisions, each focusing on specific domains within Jewish law: Orach Chaim, Even Ha'ezer, Yoreh De'ah and Choshen Mishpat. A primary commentary on the Even Ha'ezer section, as mentioned in the text of this chapter, is called B*eit Shmuel.*

Avnei Nezer—this 19th century text was written by a leading scholar, Rabbi Avraham Bornsztain.

Igrot Moshe—this 20th century text was written by Rabbi Moshe Feinstein, a leading scholar in *halakha*.

# References

Ariel, Y. (2007) Gay, orthodox, and trembling: The rise of Jewish orthodox gay consciousness, 1970s–2000s. *Journal of Homosexuality* 52(3–4), pp. 91–109.

'Avnei Nezer Choshen Mishpat 8'3. Available from https://www.hebrewbooks.org/1347, p. 113.

Beit Shmuel Even Ha'ezer 25:2 (ed. Rosh Pinah, p. 32).

Boteach, S. (1999) *Kosher sex: A recipe for passion and intimacy.* New York: Doubleday.

Craig, Kohn (2016) *Jewish special-need shadchanim.* Available from http://craigkohntheaspergerscoach. blogspot.com/2016/06/jewish-special-need-shadchanim.html (Accessed 15th November 2018).

Ginsberg, J.R. (2017) Submerging safely: Accessing the mikvah with a disability. *New Jersey Jewish Times.* Available from https://njjewishnews.timesofisrael.com/submerging-safely-accessing-the-mikvah-with-a-disability/ (Accessed 1st November 2018).

Greenberg, S. (2004) *Wrestling with god and men: Homosexuality in the Jewish tradition.* Madison: University of Wisconsin Press.

Grushcow, L.L. (Ed.) (2014) *Sacred encounter: Jewish perspectives on sexuality.* New York: CCAR Press.

Igrot Moshe Even Ha'ezer Vol 1:64 (p. 163).

Igrot Moshe Even Ha'ezer Vol 4:66 (p. 135).

Lichtenstein, A. (2005) Of marriage: Relationship and relations tradition. *A Journal of Orthodox Jewish Thought* 39(2), pp. 7–35.

Maimonides Hilchot Ishut 14:2. Available from https://www.sefaria.org.il/Mishneh_Torah%2C_Marriage. 14?lang=bi, pp. 7–8.

Maimonides Hilchot Issurei Bi'ah 16:11. Available from https://www.sefaria.org.il/Mishneh_Torah%2C_ Forbidden_Intercourse.16?lang=bi.

Maimonides Moreh Nevuchim 3:49 (ed. Friedlander). Available from http://files.libertyfund.org/files/ 1256/0739_Bk.pdf, p. 378.

Maimonides Sefer HaMitzvot, Negative Commandment 47, 262. Available from https://www.sefaria.org. il/Mishneh_Torah,_Negative_Mitzvot?lang=bi.

Mayim Hayyim (2019) *Accessibility and inclusion discussion guide.* Available from https://www.mayyimhayyim. org/resources/accessibility-inclusion-discussion-guide/ (Accessed 18th June 2019).

McCabe, M.P., & Taleporos, G. (2003) Sexual esteem, sexual satisfaction, and sexual behavior among people with physical disability. *Archives of Sexual Behavior* 32(4), pp. 359–369.

Mikvah.org (2019) *Mikvah directory.* Available from www.mikvah.org/directory (Accessed 18th June 2019).

Milligan, M.S., & Neufeldt, A.H. (2001) The myth of asexuality: A survey of social and empirical evidence. *Sexuality and Disability* 19(2), pp. 91–109.

Mishna Yadayim 3:5. Available from https://www.sefaria.org.il/Mishnah_Yadayim.3?lang=bi

Orthodox Union (2019) *Community finder.* Available from https://www.ou.org/communities/ (Accessed 17th June 2019).

Rapoport, C. (2004) *Judaism and homosexuality: An authentic Orthodox view.* London: Vallentine Mitchell.

Ratzon, N., Bar, M., & Halevy, Z. (2006) Accessibility surveys as a teaching tool and as a way to promote community accessibility/קרי נגישות ככלי הוראה וכמכשיר לקידום נושא הנגישות בקהילה. IJOT. *The Israeli Journal of Occupational Therapy* 15(3), pp. H83–H91. Available from http://www.jstor.org/stable/23468969

Rosenfeld, J., & Ribner, D. (2011) *The newlywed guide to physical intimacy.* Jerusalem: Gefen Publishing House.

Ruderman Foundation (2016) *Innovative and inclusive mikveh: Mayyim Hayyim.* Available from http://rudermanfoundation.org/innovative-and-inclusive-mikveh-mayyim-hayyim/ (Accessed 18th November 2018).

Ruttenberg, D. (Ed.) (2009) *The passionate Torah: Sex and Judaism*. New York: NYU Press.

She, P., & Livermore, G. A. (2007) Material hardship, poverty, and disability among working-age adults. *Social Science Quarterly* 88(4), pp. 970–989.

Shulchan Arukch Even Ha'ezer 76:11 (ed. Rosh Pinah, p. 82).

Shulchan Arukch Yoreh De'ah 195:16–17 (ed. Friedman, pp. 160–162).

Shulchan Arukch Yoreh De'ah 201 (ed. Friedman, pp. 225–276).

Sifrei Bamidbar 115. Available from https://www.sefaria.org.il/Sifrei_Bamidbar.115?lang=bi

Slomowitz, A., & Feit, A. (2015) Does God make referrals? Orthodox Judaism and homosexuality. *Journal of Gay & Lesbian Mental Health* 19(1), pp. 100–111.

Stapleton, D.C., O'Day, B.L., Livermore, G.A., & Imparato, A.J. (2006) Dismantling the poverty trap: Disability policy for the twenty-first century. *The Milbank Quarterly* 84(4), pp. 701–732.

Talmud Bavli Ketubot 46a. Available from https://www.sefaria.org.il/Ketubot.46a?lang=bi

Talmud Bavli Niddah 13b. Available from https://www.sefaria.org.il/Niddah.13b?lang=bi

Talmud Bavli Yevamot 61b, 76a. Available from https://www.sefaria.org.il/Yevamot.61b?lang=bi

Tepper, M.S. (2000) Sexuality and disability: The missing discourse of pleasure. *Sexuality and Disability* 18(4), pp. 283–290.

Waxman, B.F. (1994) Up against eugenics: Disabled women's challenge to receive reproductive health services. *Sexuality and Disability* 12(2), pp. 155–171.

Zimmerman, D. Site supervisor of Nishmat.net. (Personal communication, 2nd September 2018).

# PART III

# Politics, policies and legal frames across the world

# 11

# SEXUAL CITIZENSHIP, DISABILITY POLICY AND FACILITATED SEX IN SWEDEN[1]

*Julia Bahner*

## Introduction

This chapter explores sexual facilitation within the context of personal assistance services (PAS) in Sweden by detailing the experiences of people with mobility impairments, personal assistants, managers and policy stakeholders. The focus is on how sexual facilitation is conceptualised, responded to and organised—and what consequences different approaches have on assistance users' sexual lives. By studying the phenomenon of 'sexual facilitation' from these four perspectives, a broad and complex picture is presented. The research illuminates the different types of experiences that the concerned parties have around sexual facilitation.

Sexual facilitation is the support needed by some disabled people to be able to express and explore their sexualities. It can include preparations before sex (intimate grooming, undressing), assistance during sex (help with using sex toys, positioning, communication between sex partners), and personal hygiene after sex (Earle 1999; 2001; Mona 2003). Furthermore, assistance can comprise provision of accessible information, purchase of sex aids, toys, contraception and pornography, arranging meetings with partners or sex workers (if legal), or whatever activities are relevant to an individual's sexual life and expression.

The Swedish disability services law does not mention anything related to sexuality, making sexual facilitation an unregulated issue. It is therefore up for negotiation among service users and personal assistants, or for managers to provide instructions. Needless to say, this results in varying approaches in different local contexts (see below).

The research comprised four empirical studies: 1. Interviews with ten assistance users and observations of discussions on an online forum for disability issues, created by and primarily for disabled people (Bahner 2012). Examples will also be given from a handbook produced by The Swedish Youth Federation of the Mobility Impaired, on the topic of sexual facilitation (Svensk 2011); 2. Interviews with 15 personal assistants and observations of discussions on an online forum for personal assistance issues, created by and primarily for personal assistants (Bahner 2013); 3. Three focus group discussions with ten personal assistance managers (Bahner 2015a); and 4. An analysis of a statement about how sexual facilitation in PAS can be understood and handled, published by the Social Committee of the Swedish National Board of Health and Welfare's Ethics Board (Bahner 2015b).

## *The policy framework*

In Sweden, the main disability policy is the "Act concerning support and service for persons with certain functional impairments" [*Lag om stöd och service till vissa funktionshindrade*] (SFS 1993:387), abbreviated to LSS. This law governs PAS with the aim to make it possible for service users "to live as normal a life as possible," "under good living conditions" and on the same terms as non-disabled citizens. The law marks an ideological shift from 'care' to 'service,' based on Independent Living (IL) ideology championed by the disability movement. However, there are many problems in realising the rights to "live like others," as this is largely left to the service providers' discretion.

The relationship with personal assistants (PAs) is also influential. The work of PAs is characterised by low levels of standardisation and high levels of flexibility and individualisation. There are few prescribed instructions for how to execute tasks, meaning that PAs may experience new situations on a daily basis, which they need to decide on their own how to handle (Egard 2011). Since the work is so individualised and variable in character, it is hard to regulate, standardise and systematically describe. Agencies may have guidelines, but general skills for assistants are often described as being flexible and learning to adapt to individual service users' needs.

A complicating factor in personal assistance relationships is to what extent the assistance user can demand certain services before breaching the Work Environmental Law (1977:1169). The law states that working conditions must allow workers to remain physically and psychologically well. However, the LSS law gives service users the right to dismiss employees on the grounds of, for example, "failing personal chemistry." This allows for lower employment security, and some claim that the strong service user rights have overshadowed employee's needs (Guldvik et al. 2014).

Lastly, the Penal Code (SFS 1962:700), which makes it illegal to purchase sexual services, further complicates issues around sexual facilitation. As personal assistance services are funded by the state or the municipality, some fear that sexual facilitation could be interpreted as an indirect purchase of sexual services (Kulick and Rydström 2015). Even though such fear is unfounded, the Swedish anti-prostitution/sex work discourse has a strong influence in interpreting and handling sexual situations.

In other words, the policy framework for sexual facilitation within PAS is lacking, to say the least. As the empirical sections will show, disabled people and staff have different ways of managing it, and some do not acknowledge the need to provide such services at all. Disabled people's sexual citizenship is therefore at danger of not being met on an equal and systematic basis.

## Sexual citizenship: From theory to practice

> Sexuality is the result of the whole person's life situation. Without the prerequisites for self-esteem and personal independence there can be no sound sexuality... [F]or many of us who have extensive disabilities our sexual liberation does not so much depend on sex counselling or mechanical sex aids but on the availability of tax-funded personal assistance services which empower us to take control over our own lives.
>
> *(Ratzka 1998)*

The Independent Living movement began as a political struggle for disabled people's equal rights to choose their own ways of living, with personalised support that was based on their own needs rather than what had been deemed relevant by a medical professional. Autonomy and self-determination were—and still are—central tenets. However, the issue of sexuality has been left largely undiscussed, leaving many disabled people, as well as assistants, insecure about how sexual supports from an independent living perspective would best be provided. Adolf Ratzka, one of the founders of the Swedish IL movement, wrote a rare article on the topic, as seen in the extract above. To him, sexual expression and opportunity was closely tied to the availability of welfare provision, and PAS specifically. What he was talking about can be interpreted as sexual citizenship in practice.

The concept of sexual citizenship was developed in relation to the LGBT movement's activism and struggle for equal rights to sexual expression (Weeks 1998). Central aims included the right to marriage, the abolishment of laws that made certain homosexual acts illegal and the right to express one's identity openly without fear of discrimination, harassment or violence. According to Diane Richardson (2000), sexual citizenship can be understood as a framework for rights claims, divided into three categories: conduct-based, identity-based and relationship-based rights. Ken Plummer (1995, p. 151), who prefers the term intimate citizenship, has explained it in a similar way:

> the *control (or not) over* one's body, feelings and relationships; *access (or not) to* representations, relationships, public spaces, etc.; and *socially grounded choices (or not)* about identities, gender experiences, erotic experiences [original emphasis].

In summary, sexual citizenship is a concept that challenges heteronormativity and the ways in which it structures society and relationships. I use it as a social justice framework to illustrate how individual opportunities are impacted by structural conditions, often limiting for those outside of the norm. The concept thereby also challenges the public/private binary, which will be familiar to disability studies. However, sexual citizenship studies have rarely included disabled subjectivities (for a recent exception, see Kirsty Liddiard's (2018) work on intimate citizenship). This chapter aims to develop an understanding of how the concept needs to be elaborated in order to better encompass disabled people's sexual realities.

## Disabled peoples' perspectives

> I would like it to be as natural to discuss the body, sexuality and feelings, as it is to discuss how to sit right in the chair!
>
> *(forum discussant)*

People with mobility impairments often experience a lack of interest in, or knowledge about, their sexual needs from professionals. The comment from the forum discussant above reflects on their regular use of habilitation services growing up—not once having discussed any potential support needs relating to sexual activity, even though these could also be seen as "activities of daily living" (ADL). This is mentioned by several other interview participants as well, both in general comments and in more personal stories of struggling with body-image and self-esteem. The professionals that surround disabled people, especially in adolescence, have an important part to play in socio-sexual development and sexual health knowledge. This should not be different from the other areas of life and bodily issues that they help with in their occupations.

With regards to participants' experiences of PAS, one interviewee talked about the negative aspect of her identity and life becoming more focused on disability now that assistance was a big part of it, in that she had to take her assistants' needs into account in planning her life, as well as figuring out the logistics of the assistance, which was involved in everything she did—from the way she got dressed to travelling. For those who grew up with assistance, there were examples of assistants becoming close friends. Either way, an assistant's presence meant an additional consideration in their daily life—including in the sexual sphere.

For young disabled people wanting to explore their bodies and sexualities, being reliant on PAs can pose an obstacle, as explained by The Swedish Youth Federation of the Mobility Impaired in the introduction to their handbook *Secrets Known by Many*:

> For most young people who have impairments, the thoughts, feelings and dreams are like others'. We wondered how one handles having one or several personal assistants around when these processes are going on? For example, how do you discover your own body when you can't undress on your own? How do you go about flirting or having sex when assistants are around?
>
> *(Svensk 2011, p. 7)*

The organisation set out to explore how their members dealt with these issues. It turned out that many had never discussed or even thought about it before, partly owing to it being taboo even within the organisation. And before embarking on discussions about sexual facilitation practices, members' low self-esteem, negative body image and lack of knowledge about sex, sexuality and how the body works needed to be dealt with. Later, conversations were held around how to develop good communication with assistants and how to set boundaries in relationships where assistants are also involved. In other words, young disabled people may need substantial support before having the confidence and knowledge to manage their personal assistance in relation to their sexual expression. Evident, too, are structural issues around equal access to sex education, as well as the impact of body normativity.

Among adults, the following examples from the disabled people's forum show how experiences of not being recognised as sexual could be exacerbated by assistants' attitudes:

> I've really got the feeling that people don't think we disabled people have sex. Doesn't get easier living with assistance. Something happened that I'd like to share here to get some advice, but I'd like to remain anonymous. My boyfriend and I had sex one morning and when my assistant came into my bedroom to start helping with my morning routine, she says 'did you guys have sex or what? Do you really have to do that when we work?' Well, the thing is I have assistance around the clock.

Several other forum discussants replied, saying that the assistant's behaviour was unacceptable, as they work for the assistance user, in their home, and should not dictate how they live their life. The demeaning attitude was felt to be especially problematic as the assistant was not even being asked to facilitate the couple's sexual activity but was required to do nothing more than her usual duties.

Another discussant asked:

> What should I do not to feel embarrassed about asking my assistant to put me naked in bed? We disabled people also have the need to pleasure ourselves, don't we? I find this so embarrassing I refrain from this because I can't undress and put myself to bed. Anyone have any tips on what to do?

In this case, the assistance would essentially be the same as a nightly routine, but since the aim is to masturbate rather than sleep, it is implied that this activity is almost shameful. The absence of discussion around the sexual lives of assistance users can therefore be as much of an obstacle in sexual expression as direct confrontation. Other similar discussions showed that the uncertainty whether sexual facilitation was allowed was common. One discussant asked if they could ask their assistant "to have a dildo put into place," and if so, how to handle it if the assistant declines. As in the replies above, fellow assistance users thought that such sexual facilitation would be more than reasonable.

An interview participant told of strategies to achieve privacy with their partner around their sexual activities. Since they were both assistance users, it could be rather crowded in their small apartment at times. For example, they would ask the assistants to watch a movie with headphones on, or ask the assistant on duty to leave earlier and the next assistant to come in a bit later, leaving them with a few hours of assistant-free time. These strategies were shared by forum discussants too. Another interviewee said that she and her husband had started to sex chat on their phones when they were not able to be physically alone.

In other words, it is not so much the sexual activity that troubles participants, which is the topic that has often been the focus in previous research on sexuality and disability (Shuttleworth 2010). Instead, it is the practical logistics of sex, which include attitudinal barriers, such as ignorance or judgemental comments from staff, which can impact upon assistance users' opportunities for sexual expression. Such misrecognition can lead to feelings of insecurity about voicing needs for sexual facilitation to begin with. Other influential factors include architectural and economic constraints, such as living in a small apartment with limited opportunities for privacy. Several participants developed strategies to be and feel sexual, whereas others were yet to overcome their personal insecurities or to change negative relationships with their assistants in order to accomplish this.

## Personal assistants' perspectives

Several of the personal assistants who I interviewed had never thought of the assistance users having a sexual life that they needed to consider in their work. For one participant, this interview was the first occasion that they had discussed issues around assistance users' sexuality. The view of disabled people as not being sexual or of being unable to have sex prevails, as can be seen even more explicitly in another interview participant's statement:

> Sex life hasn't been an issue. They've all been very disabled.

Fortunately, this was not the case with all assistants. A woman who had studied sexology had wanted to discuss with the service provider manager her idea of developing a company policy and value statement on working with sexuality issues. But, considering the silence around such issues from her manager and colleagues, she had refrained in the end because "it felt so taboo."

On the forum, there were several examples of negative attitudes towards needing to consider disabled people's sex lives:

> Yes, most people have sexual needs, but that doesn't mean they have the right to sex or someone else helping them with it. [. . .] Sex is a need but not as strong as food, because if we don't eat, we die.

This statement highlights how an assistant's conceptualisation of sexuality, and more specifically of sexual needs, impacts upon their understanding of sexual facilitation. These kinds of statements were relatively common on the forum. They were often based on arguments revolving around personal assistants' insecure working conditions, low status as workers and the risk of being seen as *prostitutes* if associated with work tasks relating to sexuality.[2] Several of the discussions concerning sexual facilitation occurred in the part of the forum dealing with work environment issues; there was no separate section for discussing sexuality issues like in the disability forum.

Another difference compared with the disabled people's forum, where discussants strongly argued for the rights for service users to lead sexual lives, in the personal assistants' forum, discussants emphasised workers' rights and vulnerability when encountering assistance users' sexual expression. The disabled people's viewpoints were based on the right to live like others in society (as stated in the LSS law), but the assistants' perspectives were based on their right to a good work environment, negative attitudes about prostitution/sex work and their personal discomfort or lack of knowledge about disabled people's sexual rights. In other words, they had a different conception of their professional obligations in relation to sexual facilitation than did the assistance users.

A similarity between the forums as well as from interview participants, though, was the commonly felt insecurity about whether facilitating assistance users' sexual lives was even allowed. If the manager has not given any instructions, assistants are left to figure out how to deal with sexual facilitation themselves. Many of the assistants who took part in the discussions expressed insecurity, and some asked further questions relating to specific situations, such as assistance with masturbation, in BDSM practices, or accompanying the assistance user to a nightclub where they wanted to find someone to have sex with. Evident in these discussions was also the need to define what could count as sexual, and, thereby, where to draw the line between acceptable and unacceptable assistance; several discussants deemed sexual support both immoral and unacceptable.

But there were also several good examples offered by participants of successfully organised sexual facilitation:

> I've worked as an assistant for a service user who had very natural needs together with her husband, I saw absolutely nothing strange or repulsive about that. (…) She and I had a code (to avoid unnecessary dialogue) and when she showed it to me I just "disappeared into a wall" and stayed away. It worked well and the service user was very happy.
>
> *(forum discussant)*

This example suggests that having an open discussion about the type of assistance needed beforehand is a successful strategy that leaves both parties at ease when handling such situations as they occur. However, the quote also shows that this type of assistance is not uncontroversial as she feels the need to argue about "natural" needs that are not "strange" or "repulsive." The fact that she did not have to be more intimately involved in the assistance user's sex life, and perhaps also because it concerned a married heterosexual couple, might have made it easier or more acceptable.

To sum up, proper introduction to what the work entails, including acknowledgement of assistance users' sexuality, and opportunities to discuss with colleagues, are clearly needed. Maintaining the current status quo is not surprising within the context of PAS, however, where sexual expression is seen as a less important need than eating—almost a luxury. And, if personal assistance is viewed as a service catering to assistance users' basic needs, then sexual facilitation does not have a valid place in service provision.

It is also interesting that, although PAs are often deeply involved in the private and intimate lives of assistance users, for example concerning toileting, showering, catheterising, dressing/

undressing, menstrual hygiene and so on—the sexual domain remains "too intimate," "too private" to broach—mostly for assistants but also for some assistance users. This poses interesting questions about sexuality as a phenomenon. Assistants are not completely clear on what it is with sex and intimate lives that evokes this unease, discomfort and even repulsion that other intimate tasks do not, other than fear of being labelled a sex worker.

## Managers' perspectives

Few of the ten managers who participated in the three focus group discussions in this study had thought about sexual facilitation. Some of them said that these focus groups were the first time that they spent actually thinking about and discussing such issues. The focus groups also provided them with the opportunity to meet with other managers, which they rarely took the time to do, especially across service provider types (municipality, private company or user cooperative). When asked to discuss their initial thoughts around sexual facilitation, the following exchange took place:

*Manager 1:* I believe that *how* you deal with sexuality within this area is dependent upon the individual assistant but also the management—how we as managers approach it; is it okay to do these things, is it okay to talk about it as an open subject, to have a dialogue? (…)
*Manager 2:* Yes, because the issue often arises. (…)
*Manager 1:* And there is nothing in the law supporting it.
*Manager 2:* Exactly.
*Manager 1:* That's why it gets complicated. A lot of our work is controlled by laws and regulations, and we know where to set our boundaries, but this becomes a very ambiguous area, many ethical and moral, like, positions, making it much more difficult.

This discussion illuminates the managers' insecurity about where to set the boundaries around sexual facilitation, since it is not stated in any law or regulation. They also mentioned the complex nature of sexuality with regards to individual ethical and moral values as a reason for their insecurity. Sexuality was described as different from other aspects of their work in that it, first, was not expected of them to deal with, according to laws or regulations, and therefore they did not feel sufficiently competent to take any initiatives. Second, the sexual domain was perceived as being too private and morally ambiguous to handle, especially without policy support.

Despite the issue arising often, only one manager had developed guidelines to support assistants in handling assistance users' sexuality. However, these guidelines only concerned one particular service user, prompted by staff questions about how to handle the user's particular needs. The manager specified that the assistants were not allowed to do more than help with preparations before sexual activity, which the service user would then have to manage on their own. She had also invited an external psychologist to provide professional supervision of the personnel (most assistants are generally not provided with any supervision).

Another manager had recently included sexuality in the induction for new employees, as one of several themes that they were expected to know about in their work. By doing so, she had raised awareness of assistance users' sexual lives, and let workers know that they could raise any concerns they had over this. Although this did not prepare them sufficiently for dealing with sexual facilitation in practice, they were at least better prepared for encountering assistance users' sexual expression than if sexuality had not been mentioned at all.

In one focus group, when questioned about what managers did in the area of sexuality, the managers started to laugh. The discussion turned into an argument about how their heavy workloads and responsibilities for budget, administration and employees required more immediate attention than issues relating to service users' sexual lives. Another argument was about "assistants hav[ing] great respect concerning sexual integrity" and that they, as managers, do not really discuss any issues that they "do not have anything to do with," as one participant expressed it. In other words, managers' respect for assistance users' privacy and integrity was used to avoid doing anything at all, which is not uncommon in disability services (cf. Kulick and Rydström 2015). Finally, someone suggested that it was the responsibility of the Swedish Social Insurance Agency [*Försäkringskassan*] to include sexual needs when assessing eligibility for personal assistance services. In that way, service providers would have directions on what type of assistance to include in the individual care plans:

> I mean, as long as it's not taken into consideration on a higher level, it gets harder for us to implement and even harder to consider as natural.
>
> *(focus group participant)*

But even this discussion prompted laughter and joking from the managers. How would it be possible to assess sexual needs, in terms of time or in advance, in the same way as needs for other types of assistance? And what would count as a "normal" amount of sexual activity, since all other needs are assessed in relation to a "normal lifestyle"? These are, of course, relevant questions, but the discussion did more to ridicule them than to seriously discuss practical possibilities.

Managers also discussed instances of abusive, manipulative and sexually frustrated service users, who sexually harassed/exploited inexperienced and vulnerable (young) assistants. Their strategy to not employ young women for such men was seen as necessary risk management, based on conceptualisation of sexual activity as a work environmental risk. Thereby, they never considered attending to the root cause of the problem, namely how to support service users to express their sexuality in a way that benefits their health and well-being (cf. Kulick and Rydström 2015). Such support would likely decrease or even prevent sexual harassment towards staff, resulting in better working conditions as well. Furthermore, the "othering" of disabled (male) sexuality would cease to be an issue.

In another focus group, a manager explained how she had handled a request for sexual facilitation by a female assistance user. The woman had asked her assistants for facilitated masturbation and for help to "find a man." The assistants then asked the manager for advice. The manager decided to arrange a meeting to discuss it with all assistants concerned, at which they agreed to provide assistance by purchasing sexual aids, and with preparations for her to be able to masturbate.

One of the assistance user's preconditions for facilitated masturbation was that she be allowed to choose which particular assistants to involve, as she did not want substitutes, for example. The manager was very content with how the situation was eventually handled:

*Manager 1:* Since the assistants didn't have any objections to it and since it was her [the service user's] wish, we only saw it as part of her needs. And because this help was nothing more than we do when helping someone with a catheter, I didn't think it was anything to worry about.
*Manager 2:* But it seems like the group had great maturity in that, to be able to, somehow…
*Manager 1:* Yes, they were women aged between 35 and 40 and they had known her for a few years, and liked her, and overall it was a fun and nice work environment so they didn't have any problem with it at all.

As the exchange shows, the manager considered that some of the reasons behind the successful handling of the situation were the well-functioning and trustful relationship between the assistance user and her assistants, the likeability of the service user and the assistants' age and/or maturity. The gendered and aged nature of care-service work becomes visible in how these female assistants, by virtue of being of a certain age, are considered to be competent to handle complex issues. But also, the manager's own conceptualisation of sexual facilitation as something equal to managing an assistance user's catheterisation meant that it was considered to be a 'normal' part of assistance service provision.

Managers' discretionary judgements were often based on their personal values and normative conceptualisations of "appropriate" sexual activity, which informed their decisions as to what could count as acceptable sexual facilitation. Judgements and decisions were sometimes heteronormative, illuminating the intersectional nature of these interactions. Their decision-making power exhibits the power dynamics between manager and service user, as well as between a non-disabled person and a disabled person. What can be seen is how the managerial conduct in many cases makes invisible service users' sexual support needs. In most cases, this happens by misrecognition of disabled people as asexual.[3] In other cases, especially those concerning male service users, they are "managed" as sexual predators, stripping them of (acceptable forms of) sexual agency, which serves to dehumanise them.

Independent of the reasons behind and the results of certain managerial strategies, disabled people are "drowned" in stereotypes that make "normal" sexuality unimaginable. Instead of being considered to be individuals with impairments, who have varying support needs, they become subject to management according to a standardised service user category. By not paying sexual support needs any attention and by handling upcoming situations on a case-by-case basis, instead of as general issues, managers deny assistance users choice, unless they are capable of voicing it themselves. This is counter to IL ideology and the aims of the LSS law, which puts personalisation and self-determination at the centre.

## The Social Committee of the Swedish National Board of Health and Welfare's Ethics Board

As the chapter introduction showed, there are no policies, regulations or other types of advisory materials concerning sexual facilitation issued by national authorities. For service providers seeking advice around these issues, there was a period of time where it was possible to submit questions to the Social Committee of the Swedish National Board of Health and Welfare's Ethics Board. The Committee was comprised of medical doctors, jurists, social welfare officials and professors but has since been incorporated into the Board and is not answering questions anymore. They met every three months to discuss ethical dilemmas from personnel in all areas of the social services. Their discussions were then published online as "discussions to be regarded as examples of how one can think about the specific issue and hence *not to be seen as advisory*" (emphasis added). In other words, despite their formal position within the National Board of Health and Welfare, a powerful governmental authority that, among other things, issued advice and regulations, these particular statements were not issued as formal advice.

They published two discussion statements on sexual facilitation, following two questions that staff had submitted: "Is the personal assistant allowed to facilitate purchase of sexual favours?" and "Can the personal assistant help a disabled person to masturbate?" The discussions were published on the Swedish National Board of Health and Welfare's website in 2011 and 2012, respectively, but has since been deleted (see Bahner 2016, appendixes 10–11 for saved copies). I focus here on the second discussion dealing with facilitation with masturbation. The case dealt with an assistance user and their assistant who had come to an agreement about facilitated masturbation (more detail was not provided). The first part of the question concerned legal boundaries: on the one hand, whether this could be considered facilitating the purchase of a sexual service, and hence be illegal, and, on the other hand, whether the disability services law allows for assistance with sexual activity. The Committee began by stating that facilitated masturbation could not be considered an illegal sex purchase, according to the law's definitions of a "temporary sexual relation." When it came to the disability services law, they expressed a more insecure stance:

> [T]here is no mention [in LSS] of sexual needs, which generally ought to be considered basic human needs. It is therefore unclear what the lawmaker intended in terms of the boundaries around what kind of services should be included.

In other words, they do not provide a clear answer to the questions that so many assistance users, personal assistants and managers have asked about facilitating masturbation. However, they do mention that the aim of the disability services law is to make it possible for the assistance user to "live a life like others" and be fully included in society.

The second part of the question dealt with the employee's dependency on the assistance user and whether an assistant in such a dependency relationship could be considered able to consent to sexual facilitation. The Committee confirmed that there were no legal obstacles to facilitating masturbation in a personal assistance relationship, as the assistant could not be considered to be in such a dependency state as to prohibit consent. On the contrary, they wrote that the agreement "seems to be reached between equal parties."

However, in the discussion part of their statement, they raise contradictory concerns.

> Can sexual needs be equated with other basic needs? Can help to masturbate be reduced to a technical issue, detached from feelings between the one giving the help and the one receiving it? Can the assistants give such help in a professional manner? [...] Even if the assistant only helps the service user to masturbate on their own, this help may be experienced as sensitive and private, which is why the assistant's voluntariness to help must be emphasised, discussed and assessed from case to case.

Contrary to their previous statement that sexual needs could be considered part of basic human needs in general, they now question whether it is or not. They do not discuss it further, so we are left to wonder how to interpret this almost rhetorical question.

They pose a question that is actually an interesting, theoretical one: whether sexual facilitation is necessarily imbued with feelings, or if it can be offered with emotional detachment. The answer is surely dependent on the individual assistant—although the other questions lean more towards an understanding of sexual facilitation as something very sensitive and hard to handle in a professional manner. And yet again, the Committee contradicts their previous statement about the equality of the relationship in which the agreement had been reached:

> [T]here is a dependency relationship inasmuch as the helping assistant is employed to give health and social care [sic]. That their voluntariness is questionable can therefore not be ruled out.

Without mentioning the assistance user's needs, wishes, rights or vulnerabilities and how these should be addressed, they emphasise the assistant's vulnerable position.

Lastly, the Committee discussed the boundaries around sexual facilitation and whether the assistant could "in different ways, and with great imagination, facilitate the service user to masturbate on their own" or if the assistant is also expected to "actively execute this act." The Committee concludes that there is a crucial difference between

> facilitating someone to masturbate on their own, which can be considered okay and something that can be included in professional health and social care [sic]. But to also actively execute the sexual act means that the staff and the service user begin a sexual relationship, which cannot be seen as compatible with how professional health and social care [sic] is to be conducted.

However, one member of the Committee meant that they also thought the execution of the sexual act to be part of the staff's task, as long as the agreement to do so had been fully voluntary. With that last comment, the readers are once again left in ambiguity. On the one hand, we are reassured that there are no legal obstacles, but, on the other hand, there are moral boundaries and staff's personal values to consider.

What is also remarkable in the Committee's discussion is the description of personal assistance as "health and social care." This is contrary to the independent living ideology, which frames assistance as service, precisely in opposition to care. So, not only do the discussions lack consideration of the assistance users' sexual needs and rights, but also frame their situation as something very different than what they are supposed to be.

## Discussion

This chapter explored the perspectives of Swedish disabled people, personal assistants, managers and policy stakeholders regarding sexual facilitation within the context of PAS. Among all concerned parties, there is insecurity in how sexual needs and sexual facilitation should be conceptualised and handled in practice. This insecurity is often related to the fact that the disability services law (LSS) does not mention sexuality, notwithstanding the observation that the LSS aims to provide the assistance users with the possibility to "live like others," support self-determination and enable equal participation in society. In other words, although disabled service users enjoyed social citizenship by having access to PAS, this did not include recognition of their sexual needs. Their sexual citizenship was not recognised in policy, making it difficult to implement sexual support in practice. Instead, there were examples of how assistance users' sexual needs were routinely denied or went unrecognised.

One example is the power of assistants to practically veto the sexual expression of their clients via informal complaints and comments which reflect their discomfort. Such comments are not only degrading but can result in assistance users not being able to express and explore their sexualities in the desired ways. Furthermore, it can lead to fear of losing essential services, should assistants complain or wish to end their employment on grounds of not wanting to work with sexual facilitation. As previous research has shown, such circumstances can have severe effects on (sexual) self-esteem, identity formation and confidence to socialise with potential partners (cf. Liddiard 2018; Shuttleworth 2000; Wiseman 2014).

One unintended effect of the Independent Living philosophy's anti-professionalism agenda is that there are no professional standards around the roles of assistants in the service user's right to sexual expression. The central tenet, that assistance users are the experts, and should be able to direct how they want their services delivered, implies that this is also the case when it comes to sexuality. However, when being faced with silence around one's sexuality, or with outright demeaning attitudes, this is surely more difficult—especially if one already has low self-esteem. This difficulty is exacerbated when there are in fact no policies on whether or not assistance in sexual expression is even allowed.

Personal assistants furthermore have great occupational discretion in deciding who to help (or not) and in what ways, based on their personal values and conceptualisations of sexuality and sexual facilitation. This is made possible by three factors: the lack of formal regulations around sexual facilitation, the power advantage that some assistants have in relation to assistance users and the lack of professionalisation pertaining to job requirements and qualifications. Most of the assistants seem unaware of how their burdens of judgement impact upon their reasoning and decisions—instead they emphasise their own vulnerable positions. They are struggling to handle complex situations in the best way that they can, given the prerequisite.

Managers also wield significant power in determining if (and how) to address the sexual needs of service users. In one focus group, managers ridiculed sexual issues, almost showing an unwillingness to engage in a serious discussion about how it could be handled professionally. Managers also used a strategy of reformulating requests for sexual facilitation to fit into organisational and administrative procedures more easily. One assistance user's wish for assistance to "find a man" and to masturbate resulted in organisation of assistance to purchase sexual aids and to facilitate their use. Although such sexual facilitation is indeed helpful and appropriate, it did not exactly meet the assistance user's initial wish.

An additional problem is that, for many managers and policy makers, the potential to support assistance users' in their sexual activities is reformulated as work environment hazards, that is, risks for the assistants. This perspective is fuelled by an insecurity over how to balance assistance users' rights, according to the disability services law, with workers' rights to safe working conditions, according to the Work Environment Law. However, rather than discussing such dilemmas and trying to resolve them in ways that took both parties' needs into account, their "solutions" most often catered to assistants' needs rather than those of the assistance users. By managing the "risk" of assistance users' sexual expression as primarily a work environment issue, the right to "live like others" by expressing sexuality was effectively denied.

Even those managers or policy makers, who recognised and discussed sexual needs, focused on specific situations, rather than on educating all assistants, or developing a policy that encompasses all assistance users, informing them about their sexual rights. Similarly, the "non-policy making" by official Government entities demonstrated that displacement of responsibility was commonplace. The Youth Federation was the only (disability) organisation that had carried out any type of comprehensive work around not only sexuality issues in general but focusing specifically on sexual facilitation. Their handbook, *Secrets Known by Many*, is unique in that respect and has not been followed by other work since its publication in 2011.[4] Unfortunately, the silence from the disability rights movement makes it easier for other parties to avoid prioritising sexuality, assistants, managers and policy makers.

The research findings discussed in this chapter also demonstrate the need to develop the concept of sexual citizenship to better incorporate disability perspectives. Whereas the LGBT movement wanted to be seen as an acceptable sexual minority, not deviant and dangerous, people with mobility impairments are seldom recognised as sexual to begin with (Shuttleworth 2007). Their struggles involve the fight for sexual recognition on a deeper level. This struggle forms the basis for other rights claims—for sexual conduct and institutional provisions. In order for some disabled people to be able to have the sex life they desire, sexual facilitation is needed. This kind of sexual support involves a third party, in this case a personal assistant, who represents a service provider and a welfare organisation. In other words, assistance users' sexual expression is dependent upon politically determined structures and institutional provisions. The feminist slogan "the personal is political" is a fitting description. But the other way around works as well: when policies do not recognise the sexual lives of disabled people, their potential needs for sexual support are not recognised as being relevant to service provision.

Whereas Adolf Ratzka (1998) made an important point about the need for PAS in order for disabled people to be able to lead independent lives and become sexual agents, I believe that more concrete measures are needed for this to be realised. We need policies that recognise all citizens as sexual, and organisational structures that allow these policies to be implemented. For disabled service users, new and creative practices may be needed to develop sexual supports that are sensitive to both service users' and workers' rights. Disabled people need to be included and involved in the development and implementation of these policies and practices, in accordance with the important slogan: "Nothing about us without us" (Charlton 1997). Independent living principles must guide sexual support provision so that it does not become overly standardised but takes account of every service user's unique circumstances and requirements. On one hand, sexual citizenship is a very general concept, dealing with the right to sexual recognition and expression, but, on the other hand, it is deeply personal and private. Unless we can ensure that the needs of those people who need sexual support from a third party are met, which challenges heteronormative sexual discourse, we cannot proclaim that sexual citizenship has been equally achieved for all.

## Notes

1  This chapter is based on Chapter 3 in Bahner, J 2019, *Sexual Citizenship and Disability: Understanding Sexual Support in Policy, Practice and Theory*, London: Routledge.
2  'Prostitute' and 'prostitution' are the most commonly used words in Sweden; however, I will hereafter use 'sex worker' and 'sex work' in line with the international sex workers' rights communities' recommendations.
3  That is not to deny the possibilities of a sexual identity involving asexuality, but assistance users are not given the choice to identify either way.
4  A recent project, 'Sex in movement' (2016–2019), dealt with various issues around sexuality for young people with mobility impairments, although not the intimate forms of sexual facilitation specifically. It was initiated by the Stockholm branch of the Swedish Association for Sexuality Education (RFSU) and conducted in collaboration with the Stockholm branch of the Swedish Youth Federation of Mobility Impaired. See website https://sexirorelse.se

## References

Bahner, J. (2012) Legal rights or simply wishes? The struggle for sexual recognition of people with physical disabilities using personal assistance in Sweden. *Sexuality and Disability* 30(3), pp. 337–356.

Bahner, J. (2013) The power of discretion and the discretion of power: Personal assistants and sexual facilitation in disability services. *Vulnerable Groups & Inclusion* 4:1. Available at https://www.tandfonline.com/doi/pdf/10.3402/vgi.v4i0.20673?needAccess=true

Bahner, J. (2015a) Risky business? Organizing sexual facilitation in Swedish personal assistance services. *Scandinavian Journal of Disability Research* 18(2), pp. 164–175.

Bahner, J. (2015b) Sexual professionalism: For whom? The case of sexual facilitation in Swedish personal assistance services. *Disability & Society* 30(5), pp. 788–801.

Bahner, J. (2016) *Så nära får ingen gå? En studie om sexualitet, funktionshinder och personlig assistans [Too Close for Comfort? A Study of Sexuality, Disability and Personal Assistance]*. Göteborg: University of Gothenburg.

Charlton, J. (1997) *Nothing about us without us: Disability oppression and empowerment*. Berkeley: University of California Press.

Earle, S. (1999) Facilitated sex and the concept of sexual need: Disabled students and their personal assistants. *Disability & Society* 14(3), pp. 309–323.

Earle, S. (2001) Disability, facilitated sex and the role of the nurse. *Journal of Advanced Nursing* 36(3), pp. 433–440.

Egard, H. (2011) *Personlig assistans i praktiken. Beredskap, initiativ och vänskaplighet [Personal assistance in practice: Preparedness, initiative and friendliness]*. Lund: Lund University.

Guldvik, I., Christensen, K., & Larsson, M. (2014) Towards solidarity: Working relations in personal assistance. *Scandinavian Journal of Disability Research* 16(suppl 1), pp. 48–61.

Kulick, D., & Rydström, J. (2015) *Loneliness and its opposite: Sex, disability, and the ethics of engagement*. Durham: Duke University Press.

Liddiard, K. (2018) *The intimate lives of disabled people*. London; New York, NY: Routledge.

Mona, L. (2003) Sexual options for people with disabilities. *Women & Therapy* 26(3–4), pp. 211–221.

Plummer, K. (1995) *Telling sexual stories: Power, change, and social worlds*. London: Routledge.

Ratzka, A. (1998) *Sexuality and people with disabilities: What experts often are not aware of*. Independent Living Institute, accessed 2 September 2020, https://www.independentliving.org/docs5/sexuality.html

Richardson, D. (2000) Constructing sexual citizenship: Theorizing sexual rights. *Critical Social Policy* 20(1), pp. 105–135.

SFS (1962):700, Brottsbalk [Penal Code].

SFS (1977):1160, Arbetsmiljölag [Work Environment Law].

SFS (1993):387, Lag om stöd och service till vissa funktionshindrade, LSS [Law on support and service for certain disabled people].

Shuttleworth, R. (2000) The search for sexual intimacy for men with cerebral palsy. *Sexuality and Disability* 18(4), pp. 263–282.

Shuttleworth, R. (2007) Disability and sexuality: Toward a constructionist focus on access and the inclusion of disabled people in the sexual rights movement. In N. Teunis & G. Herdt (eds.), *Sexual inequalities and social justice*. Berkeley and Los Angeles, CA: University of California Press, pp. 174–207.

Shuttleworth, R. (2010) Towards an inclusive sexuality and disability research agenda. In R. Shuttleworth & T. Sanders (eds.), *Sex and disability: Politics, identity and access*. Leeds: The Disability Press, pp. 1–20.

Svensk, V. (2011) *Hemligheter kända av många–En metod och handbok för dig som har personlig assistans [Secrets known by many–A method and handbook for personal assistance users]*. Farsta: Förbundet Unga Rörelsehindrade.

Weeks, J. (1998) The sexual citizen. *Theory, Culture & Society* 15(3), pp. 35–52.

Wiseman, P. (2014) *Reconciling the private and public: Disabled young peoples experiences of everyday embodied citizenship*. Glasgow: University of Glasgow.

# 12

# SEXUAL HEALTH AND DISABILITY IN ZIMBABWE

*Tafadzwa Rugoho and France Maphosa*

## Introduction

Fifteen percent of the world population is estimated to have some sort of disability, according to the World Health Organisation (WHO) 2011), with the WHO World Report further noting that the majority of people with disabilities are found in low-income countries. Women constitute the majority of the people with disabilities in low-income countries (Phillips 2012). Whereas people with disabilities constitute a significant proportion of these populations, their sexual and reproductive rights remain a contested terrain, especially in low-income countries (DeBeaudrap et al. 2019) People with disabilities, especially women, face various challenges in accessing their sexual and reproductive rights. In most African countries, disability is viewed from the medical and charitable perspectives. The medical model assumes that people with disabilities are sick people who need the intervention of medical facilities for them to function and fit well into the dictates of societies (Retief and Letšosa 2018). This "sick role" that is given to people with disabilities has resulted in their marginalisation and exclusion from discourse on sexual and reproductive health issues. Put simply, people with disabilities are considered to be sick people who do not have the mental and physical stamina to indulge in sexual activities. As observed by Rugoho (2017), disability models that are promoted by societies and governments have a bearing on the promotion of the rights of people with disabilities. He noted that countries which views disabilities from a medical perspective have retrogressive laws on disability, such as treating them as charity cases, not as rights holders. Those countries which used a rights-based approach have progressive laws on disability. Countries using the medical and charitable models are lagging behind those with a rights-based approach in the promotion of disability rights.

There are multiple reasons that might explain why societies do not want people with disabilities to engage in sexual and reproductive issues. Some communities believe that people with disabilities, especially those with intellectual disabilities, are promiscuous and hyper-sexed (Rohleder and Swartz 2009). The fact that they are seen as hyper-sexual means they should not be exposed to sexual knowledge and information. This school of thought considers that people with disabilities will have sex recklessly. From this perspective, it is clear that people with disabilities are regarded as people without the mental capacity to make decisions concerning their sexual lives. Communities have a tendency of generalising that all people with disabilities have no decision-making capacities when it comes to sexual issues. Some communities simply view people with disabilities as being asexual (Kim 2011; Shah 2017). Hence, it may be thought that there is no need to provide people with disabilities with sexual opportunities because they will not utilise them. In the context of Zimbabwe, Peta (2017) found that communities do not approve of people with disabilities being sexually active. Young women with disabilities find it difficult to express their sexuality because of how they are viewed by the community. The negative stereotypes and discrimination have forced many into celibacy. Peta (2017) further noted that men with disabilities were in a better position, compared with their female counterparts, when it comes to being potential marital partners. Men with disabilities are more likely to get married to non-disabled women. Zimbabwean communities do not have the knowledge and understanding of sexual rights for disabled people (Rugoho and Maphosa 2017). In most cases, they deny people with disabilities access to sexual education. Some still do not approve of people with disabilities getting married.

In most countries, there are a lot of cultural and religious beliefs which hamper the promotion of disability rights. There are many superstitions that surround disability issues (Bunning et al. 2017; Rohleder et al. 2018). For example, disability is closely associated with witchcraft (Rugoho and Maphosa 2017). People with disabilities are seen as victims of witchcraft. This view has resulted in societies not treating disability as part of normal diversity. Instead of promoting the rights of people with disabilities, more effort is put towards a curative approach. Marrying or having sexual relationships with a person with a disability can stimulate many negative reactions in the community. Some believe that marrying women with disabilities can bring bad omens to the family. Some communities also prohibit people with disabilities from indulging in sexual and reproductive activity, for they fear that the disability can be transferred via sexual intercourse. They believe that parents with disabilities can transfer "the disability gene" to their children, with women with disabilities being more likely to give birth to children with disabilities. In this chapter, specific issues relevant to the sexual and reproductive rights of women with disabilities will be discussed.

## Models of disability in Zimbabwe

As has already been highlighted, models used by a country in managing disability issues impacts on the promotion of the rights of people with disabilities. Rugoho (2017) observed the importance of using the right-based approach rather than the charitable or medical models in disability management. Disability models which are used in a country provide lenses on the commitment of governments in protecting and guaranteeing the rights of persons with disabilities. Countries which promote the medical or charitable models are generally lagging behind with respect to disability rights, compared with those using the rights-based models. Zimbabwe and Uganda are two typical countries which can be used to illustrate this assertion. Zimbabwe uses the medical and charitable models in disability management. As such, people with disabilities have little opportunities in all spheres of life in Zimbabwe, including sexual and reproductive health. Uganda, on the other side, uses the rights-based model, with people with disabilities having more opportunities. In Uganda, people with disabilities are more involved in political and social participation. They are represented in critical decision-making boards such as the parliament and government ministries.

Utilising the charitable model of disability is toxic to the general promotion of rights for people with disability in Zimbabwe, especially in the area of sexual and reproductive health

(Rugoho 2019). The Zimbabwean government views disability issues as the preserve of the church and non-governmental bodies. The premise is that that non-governmental organisations and the church are better equipped to address the challenges faced by people with disabilities. The government in Zimbabwe abrogates its responsibilities as a primary duty bearer in guaranteeing the rights of people with disabilities. Non-governmental organisations can do little to improve the situation of people with disabilities. The government has been accused of perpetuating the marginalisation of people with disabilities, because of their welfare approach to disability (Rugoho and Maphosa 2017).

The medical model approach has also affected the inclusion of all people with disabilities with respect to development and policy issues. The Disability Act (1992) takes a narrow medical definition of disability. The Disability Act had been accused of excluding other disabilities. The Zimbabwe constitution (2013) section 22 (1) states that "the State and all institutions and agencies of government at every level must recognise the rights of persons with physical or mental disabilities, in particular their right to be treated with respect and dignity." This, however, is dismissed by section 83 of the constitution, which clearly states that people with disabilities can only access their rights based on availability of resources. Other diverse groups of people, such as older adults, war veterans and/or women and children, were never mentioned in the constitution . This clearly shows that the Government of Zimbabwe views disability as expensive. The charitable model of disability is also popular in Zimbabwe, where, for example, people with disabilities are treated with pity. It is the duty of the society to provide for people with disabilities. Others argue that those people with disabilities, who are viewed from a pity perspective, cannot be given sexual and reproductive rights. Once you are pitied, you are surviving at other people's mercy, and it is very hard to demand your sexual and reproductive rights.

## Political will in promotion of rights, including sexual rights, of people with disabilities

Political will is important for the success of the promotion and implementation of any policy. Governments need to commit to the rights of people with disabilities to avoid their rights remaining at the bottom of the public policy heap. There is need for a radical approach to disability rights. Many brilliant ideas and initiatives have been unsuccessful because of a lack of the political will to implement them. Zimbabwe is one of the countries in the world which has been commended for coming up with brilliant policy documents. However, it has been criticised for failing to implement disability-related initiatives. Rugoho and Maphosa (2017) observed that the government of Zimbabwe lacks the political will with respect to the promotion of policies towards people with disabilities in general. Zimbabwe was one of the first countries in the world to adopt and create a disability-related law, With the Disability Act being adopted in 1992. This Act resulted in a number of countries learning from Zimbabwe on how to create disability-related laws. The Disability Act of 1992, although considered to be a progressive law in the promotion of the people with disabilities in Zimbabwe, suffered a still birth because it was never fully implemented (Mtetwa 2011). The Act prohibited discrimination on the grounds of disability. It further stated that no people with disabilities should be denied services due to restrictive environments, such as physical, cultural or social barriers. Reading the law, one would assume that health and sexuality were among the services which were being referred to in the act.

Another example which clearly shows that Government is not committed to the promotion of people with disability rights is the ratification and implementation of the Convention on the Rights of Persons with Disabilities (CRPD) (UN General Assembly 2007). The CRPD is a legally binding international human rights treaty which addresses the rights of people with disabilities. There had been promises by Zimbabwe governments year after year on the implementation of the CRPD but nothing has been done on the ground. During international peer review of the CRPD, Zimbabwe continued to present fabricated results on the status of the disability situation in the country. The first author of this chapter attended the deliberations focused on discussing the degree to which Zimbabwe had improved conditions for people with disabilities. He observed that the government ministers lied on what was actually occurring within their respective ministries. From reading interministerial reports on disability, one would get the impression that people with disabilities in Zimbabwe were being included in policy developments.

Moyo Delight (2010) observed that, after Zimbabwe had adopted the Disability Act 1992, it shaped disability laws in many African countries. Countries like Botswana, Malawi, Uganda and many others learned from the Zimbabwean experience in promoting the rights of people with disabilities. Whereas Zimbabwe seems to have adopted an impressive piece of legislation, the law was not implemented or amended. The ruling party, the Zimbabwe African National Union–Patriotic Front, which has been in government since 1980, had been promising people with disabilities improvement in their rights during their political rallies.

A number of Disability Expos and related events had been held in Zimbabwe, led by the Office of the President. These Disability Expo events appeared to be a platform created by the government with the goal of showcasing opportunities available for people with disabilities. People with disabilities were invited from all over the country to these events, receiving the message that the government would include them in decisions. Rugoho (2017) questioned whether or not there was any intent to genuinely improve the rights of people with disabilities. During these events, different ministries were invited to listen to the concerns of people with disabilities. However, government ministries were represented by low-level government officials. These lower-level officials do not participate in the deliberations of the meetings of their ministries where important policy decisions are made.

## Policy on sexual and reproductive health

The constitution of Zimbabwe is the supreme document in the promotion of citizens' rights. Howie (2018) further observed that the constitution spells out the civil and political rights of its citizens. The constitution should guarantee the promotion and protection of fundamental rights. Access to sexual and reproductive health is a fundamental right to all human beings (Jones 2015; Uberoi 2013). However, people with disabilities were not included in the conceptualisation of these rights until the adoption of the CRPD by the UN on 13 December 2006, after which people with disabilities began to be represented under human rights initiatives (Hoffman 2016; Scholten 2018). The CRPD was also one of the first international laws on disabilities to explicitly recognise the need for sexual and reproductive health rights for people with disabilities (Degener 2016). Several articles of the CRPD recognise specific areas which prevent people with disabilities from realising their full potential. It also recognises that women with disabilities face several challenges in accessing sexual and reproductive health information. It called for States to put necessary measures in place to ensure disabled people's full and equal opportunity of rights, as well as their full development, advancement and empowerment. Furthermore, governments must systematically mainstream the rights of girls with disabilities in all their national plans and policies, as well as in their sectoral plans.

The constitution of Zimbabwe does not advocate the promotion of disability rights. It does not promote the rights of people with disabilities because it continues to view them from a charitable perspective. The constitution does not clearly indicate the sexual and reproductive rights of people with disabilities. Section 83 (d) notes that people with disabilities will have access to medical, psychological and functional treatment. Relevant department and ministries were supposed to formulate policies and practices which facilitate access to sexual and reproductive health for people with disabilities. More than six years after the adoption of the new constitution in 2013, nothing has been done by the Ministry of Health to develop inclusive policies for people with disabilities. In Zimbabwe, sexual and reproductive health issues came into the headlines in 2006, when the government formulated the National Reproductive Health Policy (2006). The policy offers services such as maternal health, family planning, treatment for sexually transmitted infections, including HIV and AIDS, and adolescent reproductive health. Surprisingly, the policy proffered few interventions for women with disabilities.

## Sex education and people with disabilities

Sexual education is widely taught in schools the world over. Decula (2014) observed that the majority of children with disabilities, however, are not in school. This was also previously observed by UNICEF (2007). Mpofu et al. (2017) and Mukhopadhyay (2012) observed that people with disabilities in Zimbabwe generally face challenges in accessing formal education. There had been strong arguments that education facilitates access to sexual and reproductive services (Godia 2013; Morris 2015), with people who are educated being better able to access information on sexual issues. Furthermore, they are well able to make informed decisions. Education has been used to empower people in all facets of life. The fact that a lot of people with disabilities in Zimbabwe do not have access to primary education is a signal that they face a lot of challenges in accessing sexual and reproductive health information.

Whereas a number of non-governmental organisations in Zimbabwe, such as Sign of Hope and Leonard Cheshire, have tried to provide information on sexual and reproductive issues, the number of people with disabilities they have reached is still not sufficient to make any impact on the ground (Rugoho in press). Most people with disabilities are reluctant to be seen attending discussions run by these groups because of the discrimination and stigma attached to them by society. The authors argue that people with disabilities may be shy to attend group discussions because of their low self-esteem. Low levels of education amongst people with disabilities in Zimbabwe may be another reason for such low self-esteem. Rugoho and Maphosa (2015), on studies on gender-based violence, noted that it is females with disabilities who are more affected by low self-esteem, in comparison with their male counterparts.

## Attitudinal barriers by services providers in Zimbabwe

Negative attitudes about and towards people with disabilities exists in almost all communities and these have the potential to create major barriers in the lives of these community members in Zimbabwe, as reported by Rugoho and Maphosa (2017). Most of these negative attitudes are found in the areas to do with the sexual and reproductive health of people with disabilities. Negative attitudinal barriers by health service providers have been noted across a number of countries (Badu 2018; DeBeaudrap et al. 2019; Rugoho 2017). Stereotypes and lack of knowledge concerning disability are some of the major causes by which people form negative attitudinal barriers. In an empirical study conducted by Rugoho and Maphosa (2017), on access to sexual and reproductive health information for women with disabilities, one of the major challenges noted by participants was negative attitudes by healthcare personnel. Healthcare providers in Zimbabwe reported negative attitudes towards sexual and reproductive health issues affecting people with disabilities. This had a detrimental effect on the subsequent desire of these people to seek out sexual and reproductive services. One of the participants in that study said:

> I visited the clinic when I was pregnant with my fourth child. The nurses said very hurtful things to me. They said I was giving birth like a dog. They said they pitied the men who introduced me to sex because I was no longer able to control my sexual feelings. I will never go back to that clinic again.

The government of Zimbabwe has not made attempts to improve the attitudes of its healthcare providers regarding sexuality issues for people with disabilities. For example, there does not appear to be any specific information or training on sexuality with respect to people with disabilities.

## Physical barriers

Most of the health centres in Zimbabwe were built without people with disabilities in mind. In particular, people with physical and visual disabilities face a lot of accessibility challenges in these facilities. Owusu and Owusu-Ansah (2011) acknowledged that physical barriers may prohibit accessibility for people with disabilities from seeking healthcare. For example, most of the health centres in Zimbabwe are not wheelchair accessible. Another participant in the study by Rugoho and Maphosa (2017) said:

> The clinics do not have ramps to help those in wheelchairs, like me. One day, I decided to go to the clinic to ask for information on sexual and reproductive health. I had problems negotiating my way around the buildings. When I asked for help, the nurses told me that they could not help and that I should have come with my relatives to aid me. I was so humiliated and frustrated that I developed a headache, for which I ended up getting treatment and forgot about the information on sexual and reproductive health.
>
> *(Rugoho and Maphosa 2017)*

It is not only wheelchair users who face physical challenges at these health centres. Documents related to these health centres are not offered in Braille. This means that people who are blind or visually impaired have to rely on the help of other people to access the information in these documents.

## Inaccessible information

Access to sexual and reproductive health information is critical to all, and the UNCRPD (2007) called on governments to make the information accessible to people with disabilities. Specifically, sexual and reproductive health information should be accessible in a variety of formats for people across the disability community. In Zimbabwe, there is a challenge in accessing information by deaf, blind and visually impaired individuals. Most of the information that is available is in print or audio format. Peta (2017) acknowledged that the absence of accessible information is a danger to other groups of people with disabilities. For example, Banks et al. (2017) examined knowledge of HIV and AIDS among the deaf. The results indicated that the deaf participants had difficulties understanding specific information about being HIV positive. In fact, most of the study subjects believed that being HIV positive was a good healthy test result. Others have argued that the unavailability of accessible information has detrimental global efforts in the fight against HIV and AIDS. Rugoho et al. (in press) observed that some young groups of women with disabilities did not know how to use condoms properly, because they were not able to access the instructions for their use.

# Privacy

One of the golden rules in health provision is the guarantee of privacy. Beltran-Aroca et al. (2016) noted that confidentiality is important in safeguarding the well-being of patients and ensuring the confidence of society in healthcare providers. Other scholars (Noroozi et al. 2018) view confidentiality as the cornerstone of ethics in healthcare provision. When patients share their stories with healthcare providers, they need to be confident that the information that they are sharing remains confidential. Cohen (2013) noted that confidentiality is a patient's right which ought to be respected and guaranteed. As shown by the vast literature, patients with disabilities have long had their right to confidentiality violated (Dyke et al. 2016; Geist 2016; Joffe 2010). The confidentiality of patients with disabilities is frequently violated by healthcare providers in Zimbabwe, as noted by Rugoho and Maphosa (2017).

In Zimbabwe, a study conducted by Rugoho and Maphosa (2017) on women with disabilities highlighted that their right to confidentiality was violated when they visited health centres. Violation of the confidentiality of people with disabilities in health centres is rife. For example, when deaf women visiting hospital are unable to find personnel who are able to communicate in sign language, the deaf patient's nurses usually seek assistance from other hospital staff. Deaf patients have argued that this results in their right to privacy being violated. Disabled patients are more visible when they visit the hospital. In most cases, people ask patients the purpose of their visit to health centres, especially opportunistic infections clinics. The research of Rugoho (in press) on young women with disabilities, who visited the opportunistic infections department, observed that when people with disabilities visit the departments, staff have negative attitudes towards these patients. Patients with disabilities have reported that they receive negative attitudinal treatment from healthcare providers, which might result in no formal sexual health assessment from healthcare providers. This lack of treatment has forced some people with disabilities to seek advice from street drug merchants or traditional healers. One young woman with a disability said:

> When they received my results which confirmed that I was infected by sexually transmitted diseases. The nurse laughed at me and called other nurses and staff. Everyone was made aware that I had sexually transmitted diseases. They scolded me. They called me a loose woman. While their intention was to reprimand me, I did not like the treatment. They did not respect my rights to privacy. Why did the nurse call others? I thought my results were private and confidential. But I guess as person with a disability, I do not have a right [sic]. I have also noted this with my friends. Our results as people with disabilities are never confidential. Nurses share our results with one another. They think that we should not be sexually active.

The above experience, from a deaf participant in the research, demonstrates that the right to confidentiality is not always protected when accessing sexual and reproductive health. Rugoho and Maphosa (2017) pointed out that, when healthcare providers are being trained in Zimbabwe, no element of sign language is incorporated into their training. The authors recommended that healthcare personnel be trained in sign language. This would remove the need to seek assistance from other people, which violates the privacy of patients who are Deaf. Currently, the

Government of Zimbabwe seems to be doing little to address the plight of Deaf patients. The recently adopted Sustainable Development Goals (United Nations Development Programme (UNDP) 2018) have called for no one to be left behind. This includes making sure that all people with disabilities are included in development agendas in all spheres, particularly health.

Consultation rooms are not accessible to wheelchairs users in most cases. Even the arrangement of furniture inside the consultation rooms is not always user friendly to people with disabilities, who are using a wheelchair for mobility purposes. Inaccessible consultation rooms have resulted in patients in wheelchairs being consulted in the corridors. In some cases, the consultations take place at the room entrance, with the door open. This makes it difficult for wheelchair users to have confidentiality. Their health issues will end up in the public domain because of patient-unfriendly consultation rooms. The Disability Act (1992) had stated that such health facilities needed to be modified to accommodate people with disabilities. More than twenty years down the line, nothing tangible has been achieved in Zimbabwe.

Another challenge in Zimbabwe is that people with disabilities continue to be treated as children. When they are addressed in communications, they are usually referred to via a third person. Most professionals do not speak directly to people with disabilities. Instead, the information is communicated through a relative or their helper. This practice by healthcare providers illustrates their belief that people with disabilities do not have the capacity to talk for themselves. Those that visit health centres are often asked questions such as "Where is your helper?", "Why do you come alone?" or "Don't you have someone to help you?" People with disabilities are further directed to attend health care appointments with an assistant.

## Disabled people organisations

Internationally, disabled people's organisations (DPOs) have made an impact on the welfare and rights of people with disabilities. Young et al. (2016) observed that disabled people's organisations can challenge the old approach to disability, replacing it with the rights-based approach. They can further lobby governments to treat people with disability with dignity and respect. DPOs can also help in the identification of the unique needs of people with disabilities. They should be able to provide a voice for people with disabilities to express their views and priorities. The World Health Organisation (2011) noted that DPOs can also evaluate services available and offer recommendations. In the areas of sexual and reproductive health, DPOs in Zimbabwe have been lagging behind in terms of lobbying and advocating for polices which address the sexual and reproductive needs of people with disabilities. As such, there is a need for developmental partners to build capacity, to be able to impart sexual and reproductive education to their members.

When it comes to identifying important services for people with disabilities, issues that affect people with disabilities have been generalised by DPOs in Zimbabwe. The authors noted that, when DPOs give presentations in parliament or within government departments, sexual and reproductive health issues tend to be minimised. DPOs in Zimbabwe lack specialisation in specific areas, generally being involved with all areas that affect people with disabilities, from education to livelihood and many others (Rugoho and Maphosa 2017). There is no DPO in Zimbabwe known to the authors which specialises only in sexual and reproductive health issues for people with disabilities. This lack of specialisation by DPOs compromises their effectiveness and the efficiency of DPOs in lobbying and advocacy strategies for a policy on sexual and reproductive health issues.

Since independence in 1980, the government of Zimbabwe has been forced to implement some of the recommendation of DPOs, because of their lobbying and advocacy skills. Some ill-treatment of people with disabilities, with regard to their sexual rights, has been outlawed because of this strong advocacy. The DPOs were credited with successful lobbying for the rights of people with disabilities, especially soon after independence. It is noteworthy that the majority of the people who were at the forefront of the disability movement at independence were people who had been injured during the liberation war against Britain.

## *Poverty*

The World Disability Report (WHO 2011) stated that the majority of people with disabilities are poor. In Zimbabwe, people with disabilities also represent one of the low-income populations (McCluskey 2017; Pirrie 2012), and Sundari Ravindran (2014) noted that poverty affects the attainment of rights. Poor populations face challenges in accessing most of their fundamental rights. Grut et al. (2012) observed that the area in which the poor suffers most is in accessing medical services. In Zimbabwe, the majority of people with disabilities live in rural areas and they experience unique problems in accessing health services. Their situation is influenced by multiple factors, which develop and interplay throughout the course of the person's life. Most of the barriers they experience are rooted in a life of poverty. The poverty is also further sustained by social, economic and political factors, in which people with disabilities are regarded as perpetual children or people without the ability to direct their own lives.

In Zimbabwe, schools and health centres are distant from much of the population. People in the rural areas have to travel kilometres to access basic health services. Access to sexual and reproductive health information is difficult for most people with disabilities. For those who rely on helpers for mobility assistance, they are dependent upon the availability of these helpers to visit health centres, and, as a consequence, health issues are prioritised, with sexual and reproductive health issues being seen as a luxury. In towns, the majority of health centres demand fees from people to receive health services. The payments demanded are, in most cases, beyond the reach of many people with disabilities, who are, in most cases, unemployed.

## Research by government and academia

One of the reasons behind the slow improvement of the rights and access to sexual and reproductive health information for people with disabilities is that Zimbabwe has had an absence of research and relevant information focusing on this topic. Whereas the significance of research in sexual and reproductive health issues has been discussed widely (Hunt et al. 2017; Kassa et al. 2016), very little research has targeted the sexual and reproductive rights for people with disabilities in Zimbabwe (Rugoho and Maphosa 2017; Rugoho 2017; Peta 2017). Most of the research on disability conducted in Zimbabwe is biased towards inclusive education (Majoko and Dunn 2018; Mpofu et al. 2017). Attempts have also been made to understand the experiences of students with disabilities in higher education in Zimbabwe (Chataika 2007; Rugoho 2017), whereas Rugoho and Chindimba (2017) have also attempted to research women entrepreneurs with disabilities. There has been a lack of interest on sexual and reproductive health issues for people with disabilities from the academic sector. This may be attributed to few or no opportunities to study disability disciplines (e.g., Disability Studies, Medical Anthropology) in universities in Zimbabwe.

Whereas the Government of Zimbabwe commissioned a research project entitled Living Conditions among Persons with Disability Survey Report, the report did not address sexual and reproductive health. Choruma (2007) noted that there is very little research content which is linked to the sexual and reproductive health issues of people with disabilities. Other reports by other organisations have followed a similar pattern. This is disheartening because there is little evidence to show to relevant stakeholders the barriers that are currently being faced by people with disabilities with regard to their sexual and reproductive rights. The Government has not commissioned any research on the area because it does not have the political will to improve the rights of people with disabilities, as argued by Rugoho and Maphosa (2017).

Other African countries such as Uganda, South Africa and others have made great strides in the improvement of access to sexual and reproductive health rights, and are always commissioning relevant research. Such research helps to identify gaps in policy and practice. Disability research also helps the governments to adopt best practices, which have been implemented in other countries (World Health Organization 2011). Universities are supposed to be inventors of knowledge. They are supposed to offer research which challenges stereotypes, myths and outdated cultural practices and beliefs (Rugoho 2017). This has been lacking with Zimbabwean universities. They are also doing little to empower people with disabilities to be scholars themselves, so that they could research the issues affecting themselves and people in similar situations.

## Conclusion and recommendations

This chapter has illustrated that people with disabilities in Zimbabwe face numerous challenges in obtaining sexual and reproductive rights. The government, which has the primary duty of guaranteeing the rights of people with disabilities, lacks the political will to improve the rights of people with disabilities. Although the country ratified the UNCRPD in 2013, this has remained only an idea, since the convention has not yet been implemented in Zimbabwe. Accessing information is still a challenge for the Deaf, the visually impaired and users of augmentative and alternative communication devices. The government has further marginalised people with disabilities by not prioritising policies that promote their sexual and reproductive health. We propose that the government of Zimbabwe makes greater efforts to ensure that the sexual and reproductive rights of people with disabilities are realised by implementing the UNCRPD, creating provision for sexual education to people with disabilities, and addressing the environmental and logistical access issues for people with disabilities at health centres across the country.

## References

Banks, L.M., Zuurmond, M., Ferrand, R., Kuper, H. (2017) Knowledge of HIV-related disabilities and challenges in accessing care: Qualitative research from Zimbabwe. *PLoS ONE* 12(8) pp. 1–9.

Beltran-Aroca, C., Girela-Lopez, E., Collazo-Chao, E., Montero-Pérez-Barquero, M., & Muñoz-Villanueva, M.C. (2016) Confidentiality breaches in clinical practice: What happens in hospitals? *BMC Medical Ethics* 17(1), pp. 1–12.

Bhavisha Virendrakumar, Emma Jolley, Eric Badu, & Elena Schmidt (2018) Disability inclusive elections in Africa: a systematic review of published and unpublished literature. *Disability & Society*, pp. 1–11.

Bunning, K., Gona, J.K., Newton, C.R., & Hartley, S. (2017) The perception of disability by community groups: Stories of local understanding, beliefs and challenges in a rural part of Kenya. *PLOS One* 12(8), pp. 23–33.

Chataika, T. (2007) *Inclusion of disabled students in higher education in Zimbabwe: From idealism to reality—A social ecosystem perspective*, PhD thesis. Sheffield: University of Sheffield.

Choruma, T. (2007) *The forgotten tribe: People with disabilities in Zimbabwe*. London: Progressio.

Cohen, J., & Ezer, T. (2013). Human rights in patient care: A theoretical and practical framework. *Health and Human Rights* 15(2), pp. 1–7.

DeBeaudrap, P., Mouté, C., Pasquier, E., Mac-Seing, M., Mukangwije, P.U., & Beninguisse, G. (2019) Disability and access to sexual and reproductive health services in Cameroon: A mediation analysis of the role of socioeconomic factors. *International Journal of Environmental Research and Public Health* 16(3), pp. 417–432.

Dyke, S.O.M, Saulnier, K.M., Pastinen, T., Bourque, G., & Joly, Y. (2016) Evolving data access policy: the Canadian context. *Facets* 1(1), 138–147.

Geist, M. (2016) The policy battle over information and digital policy regulation: A Canadian perspective. *Theoretical Inquiries in Law* 17, pp. 415–449.

Godia, P.M., Olenja, J.M., Lavussa, J.A. et al. (2013) Sexual reproductive health service provision to young people in Kenya; health service providers' experiences. *BMC Health Services Research* 13, 476, pp. 1–11.

Government of Zimbabwe (1992) *Disability act*. Harare: Government of Zimbabwe.

Government of Zimbabwe (2006) *National reproductive health policy*. Harare: Government of Zimbabwe.

Government of Zimbabwe (2013) *Constitution of Zimbabwe (Amendment 20)*. Harare: Government of Zimbabwe.

Government of Zimbabwe (2014) *Living Conditions among Persons with Disability Survey Report*. Harare: Government of Zimbabwe.

Grut, L., Mji, G., Braathen, S.H., & Ingstad, B. (2012) Accessing community health services: Challenges faced by poor people with disabilities in a rural community in South Africa. *African Journal of Disability* 1(19), pp. 1–13.

Hoffmann, Stefan-Ludwig (2016) Human rights and history. *Past & Present* 232(1), pp. 279–310.

Hunt, X., Carew, M.T., Braathen, S.H. Swartz, L., Chiwaula, M., & Rohleder, P. (2017) The sexual and reproductive rights and benefit derived from sexual and reproductive health services of people with physical disabilities in South Africa: Beliefs of non-disabled people. *Reproductive Health Matters* 25(50), pp. 66–79.

Joffe, K. (2010) Enforcing the rights of people with disabilities in Ontario's developmental services system. In *The Law as it affects persons with disabilities*. Ottawa: Law Commission of Ontario.

Jones, N., Abu Hamad, B., Odeh, K., Pereznieto, P., Abu Al Ghaib, O., Plank, G., Presler-Marshall, E., & Shaheen, M. (2016) Every child counts: Understanding the needs and perspectives of children with disabilities in the State of Palestine. London and Jerusalem: ODI and UNICEF.

Kassa, T.A., Luck, T., Bekele, A. et al. (2016) Sexual and reproductive health of young people with disability in Ethiopia: A study on knowledge, attitude and practice: A cross-sectional study. *Global Health* 12(5), pp. 87–99.

Kim, E. (2011) Asexuality in disability narratives. *Sexualities* 14(4), pp. 479–493.

Majoko, T., & Dunn, M.W. (2018) Participation in higher education: Voices of students with disabilities. *Cogent Education* 5(1), pp. 1–17.

McCluskey, G. (2017) Closing the attainment gap in Scottish schools: Three challenges in an unequal society. *Education, Citizenship and Social Justice* 12(1), pp. 24–35.

Morris, Z. (2015) Disability benefit reform in Great Britain from the perspective of the United States. *International Social Security Review* 68(1), pp. 47–67.

Mpofu, J., Sefotho, M.M., & Maree, J.G. (2017) Psychological well-being of adolescents with physical disabilities in Zimbabwean inclusive community settings: An exploratory study. *African Journal of Disability* 8(1), pp. 1–9.

Mtetwa, E. (2011) Cross-cutting issues: Disability and the constitution making process for Zimbabwe. *The Indian Journal of Social Work* 72(2), pp. 257–275.

Mukhopadhyay, S., & Musengi, M. (2012). Contrasting Visions of Inclusive Education: Comparisons from Rural and Urban Settings in Botswana and Zimbabwe. *Electronic Journal for Inclusive Education* 2(10), pp. 1–30.

Mutswanga, P., & Sithole, C. (2014) Perceptions of people who are deaf on sign language teaching and communication by hearing people: Harare Urban, Zimbabwe. *Greener Journal of Education and Training Studies* 2(2), pp. 025–037.

Noroozi, M., Zahedi, L., Bathaei, F.S., & Salari, P. (2018) Challenges of confidentiality in clinical settings: Compilation of an ethical guideline. *Iranian Journal of Public Health* 47(6), pp. 875–883.

Owusu-Ansah, F.E. (2015) Sharing in the life of the person with disability: A Ghanaian perspective. *African Journal of Disability* 4(1), Art. #185, 1–8 pages.

Peta, C. (2017) Disability is not asexuality: The childbearing experiences and aspirations of women with disability in Zimbabwe. *Reproductive Health Matters* 25(50), pp. 10–19.

Phillips, S. (2012) *Women with disabilities in the Europe & Eurasia region*. Chicago: United States Agency for International Development.

Pirrie, A., & Hocking, E. (2012) Poverty, educational attainment and achievement in Scotland: A critical review of the literature, Edinburgh: Scotland's Commissioner for Children and Young People.

Ravindran, S. Brentnall, J., & Gilroy, J. (2014) Conceptualising disability: A critical comparison between Indigenous people in Australia and New South Wales disability service agencies. *Australian Journal of Social Sciences* 52(4), pp. 367–387.

Retief, M., & Letšosa, R. (2018) Models of disability: A brief overview. *HTS Teologiese Studies/Theological Studies* 74(1), pp. 1–8.

Rohleder, P., Braathen, S.H., Hunt, X., Carew, M.T., & Swartz, L. (2018) Sexuality erased, questioned, and explored: The experiences of South Africans with physical disabilities. *Psychology & Sexuality* 9(4), pp. 369–379.

Rohleder, P., & Swartz, L. (2009) Providing sex education to persons with learning disabilities in the era of HIV/AIDS: Tensions between discourses of human rights and restriction. *Journal of Health Psychology* 14(4), pp. 601–610.

Rugoho, T. (2017) Fishing in deep waters: Sex workers with disabilities. *International Journal of Gender Studies in Developing Societies* 2(3), pp. 227–240.

Rugoho, T., & Chindimba, A. (2017) Experience of female entrepreneurs with disabilities in Zimbabwe. In *Examining the role of women entrepreneurs in emerging economies*. Pennsylvania: IGI Global, pp. 145–163.

Rugoho, T., & Maphosa, F. (2015) Gender-based violence amongst women with disabilities: A case study of Mwenezi district, Zimbabwe. *Gender Questions* 7(2), pp. 12–27.

Rugoho, T., & Maphosa, F. (2017) Challenges faced by women with disabilities in accessing sexual and reproductive health in Zimbabwe: The case of Chitungwiza town. *African Journal of Disability* 60(1), pp. 1–8.

Rugoho, T., Wright, P., Stein, M.A., & Broerse, J.E.W. (2020) Dealing with sexual and reproductive health barriers for youth with disabilities in Zimbabwe. *African Disability Rights Yearbook*.

Scholten, M., & Gather, J. (2018) Adverse consequences of article 12 of the UN Convention on the Rights of persons with disabilities for persons with mental disabilities and an alternative way forward. *Journal of Medical Ethics* 44, pp. 226–233.

Shah, S. (2017) Disabled people are sexual citizens too: Supporting sexual identity, well-being, and safety for disabled young people. *Frontiers in Education* 2(46), pp. 1–9.

Uberoi, V., & Modood, T. (2013) Inclusive britishness: A multiculturalist advance. *Political Studies* 61(1), pp. 23–41.

UNDP (2018) What does it mean to leave no one behind? A UNDP discussion paper and framework for implementation, Geneva. Available at: https://www.undp.org/content/dam/undp/library/Sustainab le%20Development/2030%20Agenda/Discussion_Paper_LNOB_EN_lres.pdf

UNICEF (2008) *Violence in Zimbabwe affecting children and relief effort.* Nairobi, Namibia.

World Health Organisation (2011) *World report on disability.* Malta: World Health Organisation.

Young, R., Reeve, M., & Grills, N. (2016) The functions of Disabled People's Organisations (DPOs) in low and middle-income countries: A literature review. *Disability, CBR and Inclusive Development* 27(3), pp. 45–71.

# 13

# "TICK THE STRAIGHT BOX"

## Lesbian, gay, bisexual and transgender (LGBT+) people with intellectual disabilities in the UK

*David Abbott*

### Introduction

James belongs to a support group for lesbian, gay, bisexual and trans (LGBT+) people with learning disabilities in a large town in England. The group is going to a local Pride event and James is able to attend the event with his paid support worker. The festival lasts several days and he is able to stay the whole period with support. The support worker offers practical help (e.g., arranging travel and accommodation and buying tickets) but there has been discussion that James also sometimes needs a bit of emotional support as he is worried about feeling overwhelmed by the event. At Pride he meets someone who later becomes his boyfriend. The following year they attend Pride together, supporting each other.

Creating a chapter about LGBT+ men and women with learning disabilities (the term used in the UK which usually compares with the term "intellectual disabilities" in other parts of the world) is complex for many reasons. A conventional approach would be to rehearse and repeat the different reasons why the intersection of sex, sexuality and learning disability has been so problematic and oppressive; to highlight the limited, but gloomy, research; and to point to the lack of or ineffectiveness of policy. These are important issues, but I and others have started too many pieces of writing of this nature in that way. Pleasure, risk, fun, sex, adventure—these become peripheral and 'aspirational asides.' So, I start with James who, by all accounts, had the time of his life at Pride. I start here too because I am a culpable part of a tradition which writes something like, "...not much is known about the experiences of being an LGBT+ person with learning disabilities." But who is it that does not know? LGBT+ people with learning disabilities have known about their own experiences—and sometimes those of others—for all of time. Perhaps their knowledge is not privileged or considered valid until it has been told or 'colonised' by non-disabled people? Conscious of that, my aim as a non-disabled, gay researcher and academic is to be an ally to the community of LGBT+ people with learning disabilities and to tell and amplify their stories in the hope of effecting social change. This chapter is part of that endeavour.

It is nearly twenty years since I participated in one of the very first studies to hear directly from LGBT+ people with learning disabilities in the UK. "Secret Loves, Hidden Lives?" (Abbott and Howarth 2005) came about because we were told about a gay man with learning disabilities who was being bullied and called names in his day service. We went in search of some life stories of other LGBT+ people with learning disabilities to inform staff training and found none. The results of the research that ensued were overwhelmingly heart-breaking (Abbott and Burns 2007; Abbott and Howarth 2007). The lives we heard about were often secret and hidden, and characterised by loneliness, discrimination and depression. Interspersed within these stories were moments of joy, of emotional and sexual connection and of relationship and good support.

Our research (Abbott and Howarth 2005) coincided with a Labour government in the UK and increases in public spending alongside stronger legal and policy frameworks in relation to all disabled people. Same-sex relationships were referred to in the main policy driving force for people with learning disabilities, "Valuing People" (Department of Health 2001). There were a number of local social/support groups for LGBT+ people with learning disabilities. These were set up to combat loneliness, social isolation and discrimination, and to give LGBT+ people with learning disabilities safe places in which to explore their identity. Most of these were run by individual champions on a basis that did not seem massively sustainable. As the political climate changed and in a rather speculative update to this research published 15 years ago, I suggested that austerity had eroded much of the work that had been done and that sex and sexuality would return to the back burner, labelled, "nice but not necessary" (Abbott 2015). The UK is more than ten years into the political project called austerity and of course LGBT+ people with learning disabilities continue to exist and to have the whole range of aspirations, hopes and challenges as non-disabled LGBT+ people. Their voices surface in quite ad hoc ways on social media, in tweets and news stories, in local initiatives and re-established groups. In a spirit of trying to understand the current climate of issues related to sexuality and LGBT+ people with learning disabilities, I visited six of these social/support groups for LGBT+ people with learning disabilities in England. These were visited because organisers and some members confirmed their willingness to participate and meet with me. (I was aware of one other group that I was not able to visit.) These visits were focused on getting a sense of whether things were changing positively for LGBT+ people with learning disabilities and to learn more about how the groups were operating, and are the focus of this chapter. There follows a brief commentary on the 'state of play' with regard to research, policy and practice, and then a more detailed look at current approaches in a number of social/support groups for LGBT+ people with learning disabilities in England.

## Research

There have been important research studies and texts with a primary focus on sex, sexuality and the broad category of disability (for example, Brown et al. 2000; Liddiard 2017; Shakespeare and Richardson 2018; Shuttleworth and Mona 2002; Toft et al. 2019). This is not a section informed by a systematic review but, to my knowledge, there remain very few other empirical studies which foreground the voices of LGBT+ people with learning disabilities (for one example, see Stoffelen et al. 2018). Is this inherently bad? Perhaps not. On one hand, why would we need more research as opposed to change and action? But on the other, our research belonged to a particular point in time and failed to capture many aspects of LGBT+ identity: parenthood, marriage, sexual risk taking, intersections of sexuality with ethnicity, perspectives of people with higher levels of support needs, detailed discussion of Trans issues etc. I also have a suspicion that voices which are deemed easier to access (staff, family members) are foregrounded and privileged, whereas the most important voices continue to go missing. Research in this area continues to focus on staff/family attitudes or reviews of published research (e.g., Brown and McCann 2018; Saxe and Flanagan 2014). Wilson et al. (2018) highlighted the lack of intervention-based and evaluative research which could inform a social or educational intervention for LGBT+ people with learning disabilities, with some measurable outcomes. Their comprehensive review also highlights the lack of empirical work with LGBT+ people with learning disabilities themselves.

## Policy and practice

Policy relating specifically to learning disability in England has stalled significantly, since the heyday of a fairly comprehensive policy called "Valuing People" (Department of Health 2001) and the significant preoccupation with Brexit. "Valuing People" was billed as "a new vision for learning disability services based on the principles of rights, independence, choice and inclusion." Since the demise of the programme, there has been no driving policy force relating to the lives of people with learning disabilities. Government-funded programmes of work aimed at improving care and support in community settings for people with learning disabilities (as opposed to hospital or institutional settings) continue at the time of writing (NHS England 2015). This is seen as a response—effective or otherwise—to the ongoing premature deaths of people with learning disabilities and marked differences in life expectancy from preventable causes between people with learning disabilities and non-disabled people (Heslop and Glover 2015). The overarching question of how to provide and pay for sustainable and high-quality adult social care has evaded successive governments and a long-awaited government policy document on the future of adult social care has faced delay after delay.

Of course, general legal protections remain in place such as the Equality Act (2010) and the UK-ratified UN Convention on the Rights of Persons with Disabilities, UNCRPD (2006). These provide general duties and legal protections in relation to domains such as family life, inclusion, participation and privacy. It is worth noting, however, that there were some considerable difficulties in negotiating the place of sexuality as the Convention was written (Schaaf 2011).) These high-level frameworks are, however, hard for people to draw upon or enforce in their own lives (Quinn 2009). There is a strong policy imperative in England to deliver social care which maximises the choice and control of service users and, for example, the Care Act (2014) is clear about the kinds of personalised and high-quality support it wants to realise for disabled people, including those living with long-term conditions. The Care Act (2014), for example, focuses on well-being, which is defined as a broad concept which includes domestic, family and personal domains. To meet obligations under the Act, professionals assessing a person's social care needs are obliged to consider equality and diversity issues (including sexual identity/orientation). However, in recent research with LGBT+ disabled people with physical, sensory or intellectual support needs (Abbott et al. 2018), we asked respondents if they had mentioned their sexual orientation or gender identity during the last assessment they had had for their social care support needs. Forty-six per cent said that they had mentioned it, 41% said they did not mention it and 13% said they could not remember. Beyond a disclosure of sexual orientation or gender identity, we wanted to know if any support needs related to being LGBT+ were considered or acted upon. Sixty-seven per cent of respondents said that they were not considered.

In England, there is also practice guidance for how to support relationships and sexuality in adult health and social care (Blackburn et al. 2017; CQC 2019). This is welcomed but it remains uncertain how far guidance translates into practice. In the research cited above, one interview proved especially memorable and distressing. A gay man with learning disabilities lived alone in a flat and had support workers, whom he employed, coming in to give support a couple of hours a day. The man had left out a copy of a magazine, *Gay Times* but was told by one support worker to put it away and had said they didn't want to see, "that kind of thing." So, the man had made a "secret cupboard where I put all my gay stuff in." He said that he would love to have a rainbow flag on his wall, "but I can't, I just can't."

## Social/support groups for LGBT+ people with learning disabilities in England

Between October 2018 and January 2019, six visits were made to investigate support/social groups for LGBT+ people with learning disabilities (Table 13.1). At most of the visits, I met a combination of people who had set up the groups—mostly non-disabled, and some of the group members themselves. We met in informal, community settings and talked for a couple of hours. I asked questions about the nature and purpose of the group; how it had come about; the activities of the group; the membership of the group; funding for the group; challenges; aspirations for the future; reflections on opportunities and barriers for LGBT+ people with learning disabilities. I took notes and changed the names of anyone mentioned to protect confidentiality and anonymity. I was upfront that my visits were designed to inform the writing of this chapter. Conscious that this was an output that might be of limited use to them, I also undertook to write a blog outlining some of the issues that came out of the meetings. This seemed to be an attractive idea as (see below) several of the groups wanted to know more about other groups and any similarities/differences.

The set-up and organisation of the six groups, all of whom I identified on social media, can be described as follows.

### *Group beginnings*

With the exception of one, the groups developed in informal and relatively ad hoc ways, based on LGBT+ individuals identifying that, "something needed to be done." This was sometimes based on the fact that there were LGBT+ people with learning disabilities in local services and sometimes a response to particular events—positive or negative. One group leader had been to a workshop about sexuality and learning disability and said that there had been, "an electric atmosphere. It really inspired me." Other groups had been set up following reports of LGBT+ people with learning disabilities not having had a good reception in the commercial gay scene or trying to attend generic support/social groups for LGBT+ people. In another area, the group leader had felt prompted to act having felt that there was pervasive heterosexism in services:

> There were two men in a shared house and the support workers kept saying to them, 'We've got to find you a girlfriend' and I asked, 'Why?' It was the same in another service when I knew the woman was a lesbian and the staff kept trying to fix her up with a man.

*Table 13.1* Group member and participation data

| Group | Group process | Meeting frequency | Number of members |
|---|---|---|---|
| 1 | By a volunteer | Monthly | 9 |
| 2 | As part of an advocacy organisation | Monthly | 8 |
| 3 | By a volunteer | Monthly | 9 |
| 4 | As part of an advocacy organisation | Monthly | 4 |
| 5 | By a volunteer | Planned Monthly | First meeting pending |
| 6 | As part of an advocacy organisation | Fortnightly | 3 |

All of the groups, with one exception, were initiated by non-disabled people in professional roles. Groups were set up without having much certainty about who would come or how groups would, or wouldn't, work. "We were making it up as we went along," said one group coordinator. Social media was key in advertising the groups (especially Twitter and Facebook), as well as sending information to services. Groups used a variety of settings to meet, including accessible community venues. One group was able to use the offices of a gay men's sexual health organisation. Although there was uncertainty about how many people would come, group co-ordinators fixed meetings ahead of time with a degree of regularity. They felt this would make it easier than people having to try and find out if a meeting was happening that month or not.

## Group process

There was a lot of similarity in terms of how groups functioned. For most, there was a chance to be together and talk, share experiences, play games. The emphasis was on a relaxed atmosphere, "we have pizza and hot dogs and chill." Then a part of the meeting might be a speaker, or a topic and examples of speaker topics had included hate crimes, identity, trans issues and accessing services. Sometimes a member, or their LGBT+ support worker would lead a talk based on their own experiences of being LGBT+. Then, a chance to go out together socially might be introduced, often but not always to a LGBT+ venue. Group organisers admitted being unsure about how to structure sessions. They all wanted to work towards a situation where the groups were led by attendees.

It's hard to know how to start and run a group. There's no manual!

All of the groups said very similar things about what people wanted from the group—social support, a safe place, a place to meet other people, to have fun.

People come wanting social stuff, a sense of belonging, to meet a boyfriend or a girlfriend. They want a place where they can talk openly about themselves.

In those groups that had a post-meeting social, some individuals researched meeting logistics in advance, including the accessibility of the event. Some venues chosen were deliberately alcohol free and some where the acoustics would work for people who struggled with excessive noise. Some groups responded to members who were more openly seeking romance, love and/or sex by, for example, helping them use on-line dating apps. Interestingly but perhaps unsurprisingly, in groups where love and/or sex was more proactively raised as a topic, more members said it was something that they wanted.

We had one session where people could write any question about sex and put it in a hat and then they were pulled out and discussed. There were some dark horses and surprising questions like, 'Is it okay to be naked in bed?' It felt like some of the questions had been on their minds for a long time and suppressed. It's good to have frank conversations in the group. We talk about sex in a normal and relaxed way, not sat in front of a psychologist.

There was a mix of people coming into the groups, with or without paid support staff. Many of those staff members who did attend were, according to group coordinators, LGBT+ -identified themselves. It was more likely than not that they were out to the people that they were supporting, although some were not, "there is still a lot of reticence about coming out to service users."

## Barriers

One of the main challenges faced by groups related to their size—a perception that they were very small and that there must be more people who would benefit from the service. There was a recognition that promoting the groups on social media could exclude quite a lot of people with learning disabilities. Getting referrals from services was described at times as being particularly problematic. Linked to this, it could appear that staff in services were acting as fairly active impediments.

> There can be problems with house managers. One guy tried to come three times but, on each occasion, they were always "short staffed" at the last moment. We had one guy come to Pride but he had to leave early as the staff member said their shift was over.
>
> You can organise an event about which people are super enthusiastic and then, as it gets nearer, the calls start to come in, '…there's no driver, no support, it's too late' and so on. One support worker rang and said, 'don't like driving on the motorway, she won't be coming.'

Two group co-ordinators referred to what they called an "obsession" with determining capacity whenever the issue was related to sex/sexuality. "If there's any talk about sexuality, it's always about concerns and capacity."

Some groups collected equality and diversity information when meeting new members. Often, this was said to be misunderstood with a lot of people not knowing either what "sexuality" or "LGBT" meant. In some instances, support workers intervened to direct the person with learning disabilities to fill in a form one way or another:

> People can look at our easy read info and still say they don't know what 'LGBT' stands for. When we fill in the equality and diversity form they don't always understand the sexuality question. Often, carers will answer for them and nine times out of ten they'll say, 'Tick the straight box.'

Groups that were more formally funded were worried about size as they did not know if they would be able to make a case for continued funding if they could not prove that they were meeting a demand: "If we don't up the numbers, we probably won't get refunded." Other groups which had no such constraint said that they would continue to run even if numbers remained small:

> My viewpoint is that even if one person came then we'd continue as where else would that person go?

There were very mixed fortunes in making connections with the wider, largely non-disabled focused LGBT+ community, either in the community or the commercial sector. Some group members had tried to attend disability non-specific groups but found them difficult, inaccessible or unwelcoming. One group had approached local LGBT+ commercial organisations for some (small amount of) sponsorship but with no luck. There had been approaches to national organisations like Stonewall and the Terrence Higgins Trust: "…thought they would have been more interested, but no…" In some instances, groups were not getting any support on social media to promote or retweet things from local or national LGBT+ organisations. One group coordinator had started out thinking that finding ways of incorporating LGBT+ people with learning disabilities into existing LGBT+ groups would be best, to reduce segregation:

When I started to think about the group, I thought about the fact that there was an existing LGBT group and also a young people's LGBT group and I asked them both about opening them up to people with learning disabilities. They both found polite ways to say why it wouldn't work.

There was some scepticism about being "wooed" by large learning disability organisations. There were concerns that the work would lose its informality, fun and risk and become "corporatized":

I'm worried that the big learning disability groups might try and steal it. It's all 'big lights' with them but I don't think it'd end up being ground-breaking.

I wouldn't want it to be hijacked by [large national learning disability organisation] as they'd use it for their publicity and claim to be LGBT inclusive.

Doing it as a volunteer does means freedom from service-related things that make your toes curl. But, if it was more formal, they might have more money to help with things.

Open opposition to the groups had surfaced only on social media, with a small number of people protesting that groups "like this" should not exist.

## Are things getting better?

There were reasons for some optimism after hearing many disconcerting themes when observing the groups. One group leader pointed out that the local college for young people with learning disabilities was leading inclusive and positive conversations about sexuality. She hoped that this signalled a better chance for a younger generation of LGBT+ young people with learning disabilities. There were plenty of examples of staff and parents being supportive and encouraging of their sons, daughters, folks they supported. One man's parents drove him a significant distance so that he could attend a group as there were no local ones. The general sense was that there was more openness to thinking about and discussing the intersection of learning disability and sex and sexuality. This had not, it was suggested, translated easily into widespread ease and confidence about talking about or supporting LGBT+ people with learning disabilities.

The fact that these groups existed at all is some reason to be cheerful. In some groups, numbers attending are small but it is important not to under-estimate what work might have to go on for a person with learning disabilities to be able to attend. That said, there are large parts of the country without such support/social groups. However, visibility is important and this was a point I wanted to stress in my discussions with groups. There is a slow, "chipping-away" at the presumption of heterosexuality and, in most interviews, I recalled an interview from our "Secret Lives, Hidden Loves?" (2005, p. 29) study. This is an edited version of the excerpt from an interview with a manager in an advocacy organisation for people with learning disabilities. She is talking about one of the men with learning disabilities who comes to the group; his name has been changed:

Jim was just a quiet bloke, very nice and he was often in the office helping us do stuff on the computers. We had some images on the wall to show that being gay was okay. One of them was a postcard that said, 'Pride not Prejudice' on it and had a picture of two men. And Jim kept looking at this, he kept taking it off the board and looking at it. And he said, 'That's what I fancy…men,' he said, 'This is me.' He told us a lot about his harrowing experiences. He'd had a close relationship with this guy in a day service who died. He just didn't turn up one day. So, Jim asked where he was. The relation-ship wasn't open so no-one told Jim the man had died. So, he had to ask again. And of course, then he couldn't be open about his grief, you know, none of the closeness of the relationship was known about. And Jim had to carry all of that; he couldn't really [tell] anyone about it. He didn't tell anyone till he told us. And it was years ago. And when he told us, we were crying, he was crying and that suffering really was a conse-quence of it not being okay for him to be gay. You can't have rules about people not having sexuality because they have it anyway.

The intention of highlighting this story was to highlight that a postcard on a wall of an office had changed a man's life. People involved in running the groups were sometimes worried about the small numbers and whether or not they were making a difference. It is clear that any efforts towards the inclusion of LGBT+ individuals with learning disabilities have the potential to change lives. Questions about sustainability of these social/support groups remain pressing. The commitment of individual champions and volunteers is impressive but groups may not be viable long-term without more people becoming involved. However, that is not necessarily a sugges-tion that making groups more formally part of or embedded in services is the best way forward. Spaces outside of services can be more creative and risk averse. Most of the groups had very similar aspirations for the future: to grow, to become more user led, to enable more romantic/sexual relationships, to continue to create a safe and sociable space.

## Discussion and conclusions

In "austerity Britain," romance, love and sex continue to exist for LGBT+ people with learning disabilities. The absence of policy and the existence of equality legislation make little or no dif-ference. People's needs, wants and desires make their way to the fore. Staff with their radar out— often because of being LGBT+ too— try and make spaces open up. There were many positive facts relayed by the groups, even from those who did not always feel confident about how much difference they were making. That said, we can also see that the concerns of LGBT+ people with learning disabilities might easily be overlooked in the face of the squeeze on eligibility for state-funded support and the pressure to reduce spending on public services. Malli et al. (2018, p. 1430) point to the impact of austerity on the ability of people with learning disabilities to be involved in society, be with friends, be part of social networks:

in the current climate of economic austerity, the funding that is made available to sup-port people with learning disabilities is increasingly poorly aligned to the care needs of this population.... The narrative testimonies of participants in the research studies provided extensive accounts of multiple dimensions of loss due to austerity. Such losses were evident not only in economic terms, but also in reduced social networks and participation, and consequently in increased isolation.

This is a chapter, which does not foreground the voices of LGBT+ people with learning dis-abilities, and, in that sense, only adds to the growing literature with the most important actors missing. However, as mentioned in the Introduction, perhaps social action and change is needed more than research. That said, there remain gaps in "our" knowledge about the intersection of learning disability and being LGBT+. One gap relates to work with people with learning dis-abilities with much more profound levels of support needs. This is just one group who are not going to be able to access social media to find out about groups which would be able to easily attend. Vehmas (2019, p. 17) argues for action and the sharing of knowledge with a view to enhancing sexual pleasure for this group and he wonders why there have been so few descriptive studies, as opposed to theoretical imaginings:

> The focus of these studies is the formation of cultural norms and sociocultural imagery around sexuality and disability. Too often, the result is the kind of account where the carnal elements of sexuality are evaporated into [theory].

His concern mirrors mine in so much as there are only so many times we need research that suggests that staff members have a range of different views about LGBT+ people with learning disabilities (or LGBT+ sexuality in general); or that staff/families are unsure about how to respond to the "problem" of masturbation, sexual expression, sexual fetish, "risky" sex and so on. There have been enough systematic reviews of a limited number of studies. There have been enough critiques of the very few studies that foreground the voices of LGBT+ men and women with learning disabilities as "descriptive" or "limited." Meantime, away from academic and often non-disabled gaze, people get on with leading/trying to lead happy and fulfilling sexual lives against many of the odds. Is a gap in "our" knowledge filled only when something is validated by research? Alexander and Gomez (2017, p. 117) make this point and argues that the whole discussion is often skewed:

> In addition, there is a persistent lack of role models and absence of representation in the media of people with intellectual disabilities being sexually desirable, unless it is being portrayed as vulnerability and abuse. Whilst [we] acknowledge the vulnerability to sexual exploitation, people with intellectual disability are rarely afforded the agency to make their own adult decisions, and we fail to acknowledge that people with disability are already having sex.

Individual stories continue to change perceptions. They do the work that Plummer (2004) believes is transformative, i.e., their power makes for unassailable common ground between the story tellers and the listeners. They expand our understandings of what constitutes intimate citizenship (Ignagni et al. 2016). The research interactions that sometimes magnify these stories also challenge us/me in how we organise positionality and ethics in so-called "sensitive" areas of research. There is an imperative to collaboratively uncover stories from an ethical and reflective value base and this is something Liddiard (2017, p. 179) does beautifully in her book, *The Intimate Lives of Disabled People* in which she sets a bar for research endeavour in areas like this:

> When people let you into their lives in such ways, into their intimate spaces, thoughts, feelings, exposing themselves to give voice to that which, for most, is deeply intimate, often painful, shameful and baring, you understand how important it is for stories to be treated faithfully and in line with an ethics of care that acknowledges both the preciousness and power of such stories and the truly collaborative nature of storytelling.

In an early review of our work, *Secret Loves, Hidden Lives?* (Abbott and Howarth 2005), there was criticism from one reviewer that, as authors, we had not discussed our own sexual orientation and any impact this might have had on the work. This irked me at the time and I reflected that my heterosexual colleagues were generally not asked to "out" themselves in their writing on sex and sexuality. Later, I reflected that, to my shame, at the point that I embarked upon the research, some fifteen years ago, I did not think of LGBT+ men and women with learning disabilities as being "my people" or "my tribe." Listening to their stories about coming out—or not—and about their struggles made me realise that they were telling my story too (Abbott 2019). But the continued experience of being "othered" even by other marginalised communities is one of the gloomier aspects of my recent meetings with social and support groups. The overall lack of interest or engagement from the wider LGBT+ community is a source of sadness. The bonds of solidarity (Formby 2017) between the so-called "mainstream" LGBT+ community and other parts of the LGBT+ community, notably disabled and Black/minority ethnic groups, are woefully lacking (Martino 2017).

It was not my intention to end the chapter on a gloomy note and I am aware that the writing is based largely upon a snap-shot of interactions. The continued existence of groups such as these is a reason for hope. We must tell the story of James as per the Introduction to this chapter, alongside those whose stories are less positive. And we must keep asking questions about whose stories remain untold, or secret or hidden.

## Acknowledgments

I am grateful to all the people and groups that met with me and gave me their valuable time and insights to help with preparing this chapter. I remain indebted to my co-author of *Secret Loves, Hidden Lives?*, Joyce Howarth, and to all the research participants in that study which was funded by the Big Lottery Fund. Resources to make the visits outlined in this chapter were provided by National Institute for Health Research (NIHR) School for Social Care Research. The views expressed in this chapter are those of the author and not necessarily those of the NIHR School for Social Care Research or the Department of Health and Social Care/NIHR.

## References

Abbott, D. (2015) Love in a cold climate: Changes in the fortunes of LGBT men and women with learning disabilities? *British Journal of Learning Disabilities* 43(2), pp. 100–105.

Abbott, D. (2019) *Gay's the word: Bookshop saves young life* [online]. Available from: https://blogsfrom thetyla.home.blog/2019/01/19/gays-the-word-bookshop-saves-young-life/ [Accessed 20 August 2019].

Abbott, D., & Burns, J. (2007) "What's love got to do with it?": Experiences of lesbian, gay, and bisexual people with intellectual disabilities in the United Kingdom and views of the staff who support them. *Sexuality Research & Social Policy* 4(1), p. 27.

Abbott, D., & Howarth, J. (2005) *Secret loves, hidden lives? Exploring issues for people with learning difficulties who are gay, lesbian or bisexual*. Bristol: The Policy Press.

Abbott, D., & Howarth, J. (2007) Still off-limits? Staff views on supporting gay, lesbian and bisexual people with intellectual disabilities to develop sexual and intimate relationships? *Journal of Applied Research in Intellectual Disabilities* 20(2), pp. 116–126.

Abbott, D., Ottaway, H., Gosling, J., Fleischmann, P., & Morrison, T. (2018) *Self-directed social care support & LGBT+ disabled people* [online]. Available from: https://www.sscr.nihr.ac.uk/PDF/Findings/RF77.pdf [Accessed 20 August 2019].

Alexander, N., & Taylor Gomez, M. (2017) Pleasure, sex, prohibition, intellectual disability, and dangerous ideas. *Reproductive Health Matters* 25(50), pp. 114–120.

Blackburn, M., Chambers, L., & Earle, S. (2017) Talking about sex, relationships and intimacy: New guidance and standards for nurses and other health and social care professionals working with young people with life-limiting and life-threatening conditions. *Journal of Advanced Nursing* 73(10), pp. 2265–2267.

Brown, H., Croft-White, C., Wilson, C., & Stein, J. (2000) *Taking the initiative: Supporting the sexual rights of disabled people*. Brighton: Pavilion.

Brown, M., & McCann, E. (2018) Sexuality issues and the voices of adults with intellectual disabilities: A systematic review of the literature. *Research in Developmental Disabilities* 74, pp. 124–138.

Care Act 2014.

Care Quality Commission (2019) *Relationships and sexuality in adult social care services* [online]. Available from: https://www.cqc.org.uk/sites/default/files/20190221-Relationships-and-sexuality-in-social-care-PUBLICATION.pdf [Accessed 20 August 2019].

Department of Health (2001) *Valuing people: A new strategy for learning disability for the 21st century*. London: Department of Health.

Equality Act 2010.

Formby, E. (2017) *Exploring LGBT spaces and communities: Contrasting identities, belongings and wellbeing*. London: Routledge.

Heslop, P., & Glover, G. (2015) Mortality of people with intellectual disabilities in England: A comparison of data from existing sources. *Journal of Applied Research in Intellectual Disabilities* 28(5), pp. 414–422.

Ignagni, E., Fudge Schormans, A., Liddiard, K., & Runswick-Cole, K. (2016) "Some people are not allowed to love": Intimate citizenship in the lives of people labelled with intellectual disabilities. *Disability & Society* 31(1), pp. 131–135.

Liddiard, K. (2017) *The intimate lives of disabled people*. London: Routledge.

Malli, M., Sams, L., Forrester-Jones, R., Murphy, G., & Henwood, M. (2018) Austerity and the lives of people with learning disabilities: A thematic synthesis of current literature. *Disability & Society* 33(9), pp. 1412–1435.

Martino, A. (2017) Cripping sexualities: An analytic review of theoretical and empirical writing on the intersection of disabilities and sexualities. *Sociology Compass* 11(5), p. e12471.

NHS England (2015) *Building the right support* [online]. Available from: https://www.england.nhs.uk/learning-disabilities/natplan/ [Accessed 20 August 2019].

Plummer, K. (2004) *Intimate citizenship: Private decisions and public dialogues*. Montreal: McGill-Queen's University Press.

Quinn, G. (2009) Resisting the 'temptation of elegance': Can the convention on the rights of persons with disabilities socialise states to right behaviour? In: O. Arnardottir & G. Quinn (eds.) *The UN convention on the rights of persons with disabilities: European and Scandinavian perspectives*. London/Boston: Martinus Nijhoff Publishers, pp. 215–256.

Saxe, A., & Flanagan, T. (2014) Factors that impact support workers' perceptions of the sexuality of adults with developmental disabilities: A quantitative analysis. *Sexuality and Disability* 32(1), pp. 45–63.

Schaaf, M. (2011) Negotiating sexuality in the convention on the rights of persons with disabilities. *International Journal on Human Rights* 14, p. 113.

Shakespeare, T., & Richardson, S. (2018) The sexual politics of disability, twenty years on. *Scandinavian Journal of Disability Research* 20(1), pp. 82–91.

Shuttleworth, R., & Mona, L. (2002) Disability and sexuality: Toward a focus on sexual access. *Disability Studies Quarterly* 22(4), pp. 2–9.

Stoffelen, J., Schaafsma, D., Kok, G., & Curfs, L. (2018) Women who love: An explorative study on experiences of lesbian and bisexual women with a mild intellectual disability in the Netherlands. *Sexuality and Disability* 36(3), pp. 249–264.

Toft, A., Franklin, A., & Langley, E. (2019) Young disabled and LGBT+: Negotiating identity. *Journal of LGBT Youth* 16(2), pp. 157–172.

UN Convention on the Rights of Persons with Disabilities 2006.

Vehmas, S. (2019) Persons with profound intellectual disability and their right to sex. *Disability & Society* 34(4), pp. 519–539.

Wilson, N., Macdonald, J., Hayman, B., Bright, A., Frawley, P., & Gallego, G. (2018) A narrative review of the literature about people with intellectual disability who identify as lesbian, gay, bisexual, transgender, intersex or questioning. *Journal of Intellectual Disabilities* 22(2), pp. 171–196.

# 14

# SEXUALITY AND SEXUAL RIGHTS OF YOUNG ADULTS WITH INTELLECTUAL DISABILITY IN CENTRAL JAVA, INDONESIA

*Dian Ramawati and Pamela Block*

## Introduction

Sexuality is one of the essential parts of human life and begins in the early stages of life, from childhood through adulthood. Sexuality is not only a matter of romanticism and sexual behaviour between two persons, but also about the perception and emotional states of one person directed towards another. The World Health Organization (WHO) defines sexuality "as a central aspect of being human." Furthermore, they explain that sexuality embraces "sex, gender identities and roles, sexual orientation, eroticism, pleasure, intimacy, reproduction." Sexual health relates to a person's state of physical, emotional, mental and social well-being, with regards to sexuality. All individuals are entitled to a positive and self-assertive opportunity to maintain sexual health (WHO 2006). Sexuality is thus a natural process in human development. Yet sexuality is still an under-discussed topic for disabled people. Disability, like sexuality, is a multidimensional and complex aspect in the lives of millions of people. Disability is considered to be rooted in cultural beliefs. Therefore, the term disability has taken various meanings, based on the perspectives of the individuals, cultures or social communities who are defining it (Black 2005; Henry et al. 2010).

According to the Americans with Disabilities Act of 1990 (ADA 1990), individuals with disability may have physical or mental limitations that hinder performing daily activities, such as self-care, walking, hearing, thinking, communicating and working. Intellectual disability (ID) is the term currently preferred for those with the latter kind of impairment, replacing earlier designations such as mental retardation. For many years, people with intellectual disabilities were assumed to be asexual and not needing to be loved or involved in fulfilling relationships with others. Their rights to sexuality, which are undisputed for the rest of society, have traditionally been neglected or denied (Milligan and Neufeldt 2001). Although young people with intellectual disabilities may learn at a slower pace, they are able to learn abstract concepts, such as sexuality, and they also experience sexual development at the same pace as their counterparts without intellectual disabilities (Murphy and Young 2005; Tepper 2005).

The first issue in addressing the sexuality of people with intellectual disabilities, a topic which is still taboo in many cultures, is the difficulty in educating them due to their cognitive limitations.

Secondly, even if parents and caregivers or educators are willing to discuss romantic relationships and sexuality topics, they often do not feel confident enough because they lack knowledge about sexual development in people with intellectual disability. They also do not know how to deliver the topics to people with intellectual disability in ways that are easy to comprehend. A final obstacle is that the laws and policies regarding sexual education and sexual rights for people with ID are not well detailed, written or implemented. This chapter will provide research findings from an ethnographic research project still underway about the sexuality of young adults with intellectual disability in Central Java, Indonesia. The research presented in the chapter is based on policy data, court cases that have entered public discourse through news media and interview research with teachers and parents of young adults with intellectual disability (ID). Other components of this study include interviews with and direct observations of young adults, with the goal of representing their perspectives and to advocate for the services and supports that they desire, but those data will not be presented here. Three main questions on how young adults with intellectual disability attain sexual knowledge and learn about sexuality in Indonesia drive the larger study. Specific overarching research questions for the project include:

1) How do young adults with ID learn about sexuality and in what ways do young adults with ID talk and express their sexual needs?
2) What influences learning about sexuality for young adults with ID?
3) What are the roles of parents, educators and policy makers in providing sexual knowledge for young adults with ID?

In the study as a whole, our goal is to "to listen to the voices" of young people with ID through their life experiences with sexuality in Central Java, Indonesia, and to understand how their beliefs and ideas are influenced by culture, religion, parental roles and the education system. We are also exploring how parents and educators talk about sexuality with teenagers and young adults with ID, detailing the language and definitions they use to discuss sexuality with these young adults, including the colloquial words that they use to refer to sexuality among peers, parents and educators. We are also focusing on the roles of parents, educators and policy makers that provide and support sexual education for young adults with ID in special education schools. At the same time, we wish to also understand how state policy impacts on the lives of young people with ID and their families. The primary goal of this study is to seek to improve the quality of education and health services for people with ID by providing data from the experiences of individuals with ID, their parents and educators.

In this chapter, we focus specifically on the intersection of sexuality, disability and sexual health for young people with intellectual disability in terms of relevant policy, court cases and the perspectives of parents and teachers. These perspectives include their preferred strategies to deliver sexual knowledge to young adults with intellectual disabilities at home and school. In this area of research, we are influenced by the notion of "conscientização" or critical consciousness, as articulated by Paulo Freire (1970), and disability and sexuality theories developed by Michael Gill (2015). Freire discusses the relationship between the teachers and students in a critique of what he calls a "banking system" of knowledge. The "banking system" of knowledge means that education becomes the act of "depositing," in which the students are the depositories and the teacher is the depositor. In this system, the students receive information, memorise and repeat it. Freire argues that education is not merely a "depositing" process but more about a "transforming" structure so that the students can expand their knowledge and make it their own. Freire's approach is consistent with the observed teacher–student interactions in Indonesian special education schools. The teachers need to negotiate students' individual conditions, that are sometimes quite different from one to another. Moreover, students with intellectual disabilities have different abilities to understand the topics within a school's curriculum, including comprehension of sexual education. Therefore, Gill's concept of recognising the experience and sexual expression of people with intellectual disabilities becomes crucial in developing and delivering sexual education in the classrooms. Furthermore, Gill proposes that the sexual education materials should be not only about "harm reduction" or "stranger danger," but also about enhancing the student's capabilities to acknowledge sexual rights and make decisions.

Indonesia is unique in the variety of ethnic, cultural and linguistic variations across the region. Therefore, to explore and improve the quality of education and health services for people with intellectual disabilities in Indonesia, it is important to engage with the local cultures and beliefs, especially for a sensitive topic such as sexuality and sexual education. Using interview data from educators and parents, we explore how young adults with ID, their parents and teachers negotiate with cultural beliefs to learn about sexuality through sexual education at home and school. As such, we further investigate how educational curricula and policies may influence the sexual possibilities and practices among young adults with ID in Indonesia.

A related study has been conducted by Kulick and Rydstrom (2015), comparing data on social and political discourse between Denmark and Sweden. The study examines how policies support or repress the sexual rights of people with disabilities in both countries. Denmark and Sweden are recognised as countries which have made significant advances in social welfare and disability services; however, the researchers found that the "role of language" in the political domain, either in the past or in contemporary practice, can result in promoting or prohibition of the sexual rights of people with disabilities. Denmark has facilitated the sexual expression of disabled people by producing and implementing policies to support sexual education for people with disabilities. The policies have been created in consultation with organisations and human rights movements. In contrast, Sweden has identified the sexuality of people with disabilities as potentially "dangerous," so that the system creates barriers to romantic relationship and offers limited opportunities for people with disabilities to express their sexuality. Indonesia has an approach similar to that of Sweden, with respect to sexuality for people with disabilities, especially people with intellectual disabilities. Indonesia has both national and local policies to accommodate and support the lives of disabled people, namely *Peraturan Pemerintah (PP) Republik Indonesia* [Indonesian Governmental Policy] No. 52, 2019 on Social Welfare for People with Disabilities, *Undang-undang* [Regulations/Law] No. 8, 2016, article 96 on People with Disabilities, *Undang-undang* [Regulations/Law] No. 16, 2016 on Inclusive Programs, Governor of Central Java Regulations No. 11, 2014 on People with Disabilities' Rights, and Banyumas Mayor's Regulations No. 19, 2014 on Protection and Facilitation for People with Disabilities. These regulations focus mostly on education and social welfare for disabled people in Indonesia, but do not do specifically mention the sexual rights and health for people with disabilities; even these educational rights and social supports, though existing on paper, are not often put into systematic practice. Thus, the sexual lives of people with disabilities remain controlled by family members, caregivers or educators.

This chapter includes consideration of the intersection of sexuality, sexual education and sexual health policies for people with intellectual disabilities in the Indonesian context, as well as providing information on the implementation and impacts of those policies on the sexuality of people with disabilities in Indonesia. The chapter culminates in a discussion of research findings, and we then suggest some directions for further research. Note that, in our work, we consider individuals with ID as not only a vulnerable group at higher risk of sexual abuse and crime, but also as active members of society, who have voices that need to be heard and who need to be reckoned with, despite their differences.

## Regional context

Indonesia's population of 260 million, including 1,128 ethnicities and 746 local dialects, makes Indonesia the country with the fourth-highest populations and the largest archipelagic state in the world. With multiple ethnicities, traditions, habits and perspectives from many regions, Indonesia has great diversities in culture, as well as challenging complex social distinctions. The population density on every island in Indonesia varies greatly, with the greatest population concentrated on large islands, such as Java, the most populated island, as well as Sumatra, Kalimantan, Sulawesi and Papua. This uneven population distribution has led to different paces of development and unequal availability of public services, including transportation and health facilities for persons with disabilities (Indonesia Disability Convention Team 2017).

This study is based in the southern part of Central Java Province, Indonesia, which consists of seven districts or counties of suburban cities and took place in special education schools for students with intellectual disabilities (formerly known as "mental retardation" in Indonesia). The seven counties are Cilacap, Banyumas, Purbalingga, Banjarnegara, Kebumen, Purworejo and Wonosobo, with a total population of approximately 7.7 million people (*Badan Pusat Statistik Provinsi Jawa Tengah* [Statistics of Central Java Province] 2014). Central Java Province has 47 special education schools, with only a few of those being for students with intellectual disabilities (*tuna grahita*/slow learner). The number of students with ID in Central Java is recorded as 290,837 students (13.68% of the student population in the same province) and students with a combination of ID and physical disabilities is 149,458 (7.05%) (*Pusat Data Informasi/Pusdatin* [Centre of Data Information] *dan Direktorat Rehabilitasi Sosial Orang dengan Kecacatan* [Directorate Social Rehabilitation for People with Disabilities, Ministry of Social Welfare] 2012). Furthermore, people in Central Java mostly live in rural areas (29 regencies/districts and 8,558 villages) rather than urban areas (six cities), with the greatest proportion of its citizens being in the age range 15–64 years (*Profil Kesehatan Provinsi Jawa Tengah* [Central Java Province Health Profile] 2016).

According to the 2015 Disabilities in Intercensal Population Survey (SUPAS), almost half of the people with disabilities (48.51%) in Indonesia have multiple disabilities. Intellectual disability is the second-largest group of people with disabilities in Indonesia, after physical disability, with percentages of 15.41% and 21.86%, respectively (Amannullah 2016). Intellectual disability (ID) is defined in Indonesia as an intellectual impairment due to low intellectual ability, caused by various factors expressed before the age of 18. The impairment can include slow learning, learning disabilities and Down syndrome (Undang-undang No. 8 Tahun 2016 [Indonesian Persons with Disability Law of 2016]). The proportion of people with disabilities in Indonesia has increased from 0.92% to 2.45% (6,515,500) of the population. In 2012, the estimated number of people with ID was 213,033, with the proportion of people with disabilities who live in the countryside being larger than the proportion who live in the city: 2.71% vs. 2.20%, respectively.

Winarni et al. (2018) studied the attitudes of parents and caregivers from a rehabilitation centre for people with disabilities, religious organisation leaders and members of the community. The study explored participants' concerns about sexuality and people with intellectual disabilities. This research has revealed that Indonesian understandings of sexuality for people with ID is very conservative, due to strong cultural and religious beliefs in Indonesia. Winarni et al. recruited 30 people in Temanggung, Central Java, who filled out the Attitude to Sexuality Questionnaires towards individuals with ID (ASQ-ID), based on the response to sexual expression of individuals with ID. Respondents showed negative attitudes towards sexuality of males and females with ID, and only a slightly positive response to sexual rights, compared with other subscales (parenting, self-control and non-reproductive sexual behaviours). These findings suggest that the acknowledgement of sexuality and sexual rights for people with ID is still very low among non-disabled people in Indonesia. The study did not explore the factors contributing to these negative attitudes or the reasons underlying these perspectives. A study to explore parents' or caregivers' viewpoints on sexuality of people with ID is needed to determine their attitude towards sexual rights for people with ID.

## Disability policy and disability experiences in Indonesia

Discrimination and unequal availability of public facilities in Indonesia has led to the lack of implementation of rights for persons with disabilities as part of society and the recognition that they have equal value and dignity as do other citizens. People with disabilities have to deal with unsupportive public facilities and health services to meet their daily needs. The problems experienced by people with disabilities in Indonesia can range from a scarcity of job opportunities, basic needs such as food, clothing and housing, to education opportunities and an increased risk of being abused (physically, mentally or sexually) by their counterparts in everyday life.

Indonesia signed the UN Convention on the Rights of Persons with Disabilities (UN CRPD) on 30 March 2007 (Undang-undang Republik Indonesia No. 19 Tahun 2011 [Indonesian Law No. 19, 2011]). Since then, after continuous facilitation efforts by the *Organisasi Penyandang Disabilitas* (Disabled People's Organizations, DPO), the Government of Indonesia finally ratify the Convention on the Rights of Persons with Disabilities into national Law No. 19 of 2011. As a consequence, Indonesia is now bound and obliged to implement the content of the UN CRPD in all sectors of life.

After the ratification in 2011, the Government of Indonesia then passed a new law on Persons with Disabilities, namely Law No. 8 of 2016. The new law replaced Law No. 4 of 1997 on People with Impairment that had been in place since long before the UN CRPD was ratified. The ratification of the UN CRPD, and the enactment of Law No. 8 of 2016 can be considered to be a new chapter in the implementation of the rights of persons with disabilities in Indonesia.

In this regard, public officials working for disability-related central government agencies observed, "Even after the convention, superstition and bias against disabilities are still strong in some rural areas, and because of its psychological effect, interviewees tend to hide the existence of their family members with disabilities from public investigation." Public officials reason that the actual number of persons with disabilities is much higher than the official statistical figures (Data Collection Survey on Disability and Development in Indonesia 2015).

Indonesia has experienced an evolution of disability policies in recent decades. The previous law for Persons with Disabilities was The Law of 1997 (Undang-undang No. 4 tentang Penyandang Cacat Tahun 1997), which mentioned the definition and the rights of people with disabilities. In the 1997 law, the word used to refer to people with disabilities was "*penyandang cacat*" or "handicapped persons." The 1997 law clearly had not included or stated its purposes, as well as not mentioning women's and children's rights, spiritual, sexuality, health and equality in justice or law in its text.

In 2016, the Indonesian government agreed to revise the 1997 law, according to the UN CRPD 2008 ratification. The 2016 law changed the language from *penyandang cacat* to *penyandang disabilitas*, which can be translated as "person with disabilities." The 2016 law also broadened the definition of disability, covering not just physical and mental impairments, but also social and environment barriers. Furthermore, the law of 2016 clearly stated its purpose and broadened what was previously a single category of disability into four categories, including intellectual and sensory disabilities. The purpose of this new law is to: 1) Provide advances, protection and implementation of human rights and fundamental freedoms in full and equal measure for people with disabilities; 2) to ensure respect for their inherent dignity; 3) to achieve a better quality of life for people with disabilities; 4) to protect them from neglect and exploitation, from discrimination of their human rights and from violence; and 5) to ensure that people with disabilities have safe, free, and respectful opportunities for their self-development in all areas of life (Indonesian's Persons with Disability Laws 2016).

Intellectual disability terminology is not as widely known in Indonesia as the term mental retardation or *tuna grahita*. In the medical and health field, most clinicians still use "mental retardation" as part of their diagnosis, whereas, in education, professionals use *tuna grahita* or slow learner to refer to people with ID. There are many Indonesians who still do not understand the meaning of these terminologies or recognise an intellectual disability as a medical or educational category. As a consequence, many parents do not know if their children have an intellectual disability and the impact of this lack of knowledge has resulted in children with ID being treated negatively, which alters their growth, development and life experiences. Children or individuals with ID have often been treated as useless persons, who cannot work properly and are unproductive. In contrast, there are families who accept and treat their child with ID positively and are proactive, such as in trying to get more information about ID from professionals, treating their child with ID the same as they would a non-ID child and not separating the child from social activities. These differences may result from the readiness of the family to understand what it means to have a child with ID prior to the birth of the child. This situation can also be influenced positively by the information given by healthcare professionals during prenatal and postnatal care (Hendriani et al. 2006).

Indonesia and Central Java Province have made efforts to provide equal opportunities and accommodation for people with disabilities by introducing regulations and laws. As mentioned earlier, the Central Java government has produced policies and laws to support the lives of disabled people, one of those being *Peraturan Gubernur* [Governor's Regulation/Policy] No. 19/2014 on the rights of people with disabilities, whereas Banyumas district local government has also issued the Mayor's regulation No. 19/2014 on the protection and facilitation for people with disabilities. However, a survey conducted by Indonesian Corruption Watch (ICW, in Harsono 2019) in four inclusive cities, which means disabled people–friendly cities, Surakarta, Bandung, Makasar and Kupang), with 800 disabled people as participants, found that the support for people with disabilities is still lacking, with many disabled people, especially in rural areas, being unaware of regulations and benefits available to them. ICW states that 88.1% of participants claimed they had never been visited or recorded by the census officers. As a consequence, many people with disabilities do not have *Kartu Penyandang Disabilitas (KPD)* [Disabled People Identification Card]. This card is the first step to obtaining the health and social welfare benefits for people with disabilities, because this card is used as evidence that a disabled individual has been documented and verified by the authorities; without this card, people with disabilities cannot get any benefits from the government (Harsono 2019 on https://liputan6.com).

A case study by Ratnawati, Martha, & Fasya (2019), entitled "The Policy of Reproductive Health Education for Adolescents with Intellectual Disabilities in Indonesia" notes that not all stakeholders in this study have formulated and developed policies on reproductive health guidelines for adolescents with intellectual disabilities. This is a qualitative study with interviewees from governmental agencies, teachers or educators from Special Education Schools, and national organisations (NGOs) of women and children with disabilities. This study reveals that the policy makers face many obstacles in formulating the reproductive health education policies/education curriculum due to lack of knowledge and information on sexuality and sexual development of adolescents with ID. The Ministry of Education and Culture is the only stakeholder that has formulated any policies on reproductive health, with the assistance of Rutgers World Population

Foundation (WPF) Indonesia. The Rutgers WPF is focusing on the improvement of sexual rights and equality, including sexual and reproductive health, across Europe, Africa and Asia.

This means that people with disabilities, especially intellectually disabled individuals, in Indonesia are presently experiencing unequal access and discrimination in education, health services and social welfare. This happens because many people, including policy makers, lack the knowledge and the will to promote the health and rights of people with ID, let alone to facilitate sexual justice and expression for young people with ID. Indonesia needs to improve its national policies systems in order to provide equal rights and treatment for all citizens, including disabled citizens.

## Sexuality and disability in Indonesia

Based on the limited information and statistics available in Indonesia, people with disabilities, including people with intellectual disability (ID), have greater vulnerability to physical and sexual abuse or violence, due to their physical, mental and intellectual status. People with moderate ID are at greater risk than their non-disabled peers, with the incidence of abuse being up to three times as high (Commissie Samson 2012; Reiter et al. 2007; van Berlo et al. 2011). The perpetrators are usually people known to the victims: Those who live or work near them, peers, staff members or family members, with most perpetrators identified as male (van Berlo et al. 2011). Interviews with parents and educators that Ramawati conducted in Indonesia in 2018 show similar vulnerabilities.

International research has shown that individuals with ID also have limited opportunity to have positive sexual experiences, as compared to their non-disabled peers, including experiences of friendship, kissing, cuddling and holding hands, due to the fact that such sexual expression is often prohibited by parents or discouraged by caregivers and teachers (Leutar and Mihokovic 2007; Löfgren-Mårtenson 2004; Siebelink et al. 2006). Ramawati's research in 2018 also revealed that parents tend to "keep" the young people with ID at home in order to "protect" them from harm or to prevent them feeling discouraged by society.

In Indonesia, there is a long list of documented sexual crimes against people with disabilities. Most of the cases remain unsolved due to lack of evidence or due to the fact that victims were too scared to testify. In 2016, *Sasana dan Gerakan Advokasi* (SIGAB) or the Inclusion and Advocacy Movement documented 25 sexual assaults cases against people with disabilities but only 2–3 cases reached trial, with a verdict against the perpetrators. Then, in 2017, SIGAB received 35 cases but only two cases went to trial and achieved a verdict (*Majalah Komisi Yudisial* [Judicial Commission Bulletin] 2018). One of the cases happened in Sukoharjo, Central Java in 2013, where the victim was a student in a Special Education School (SLB) who was molested by their teacher. The perpetrator was punished with 8.5 years in prison. Significantly, the court accepted testimony only from non-disabled witnesses, even though there were also similar statements available from other, disabled students. The most recent case of sexual assault was allegedly experienced by students in a special educational school in Tanjung Jabung Timur, Jambi Province, Sumatra. The perpetrator was a staff member who worked in the students' dormitory and ten female students claimed they were sexually assaulted. The case was reported by one of the victims who was frequently targeted. The staff member had been working in the dormitory for five years (Rosyid 2019 on https://www.gatra.com/rubrik/nasional/pemerintahan-daer ah/382604-Siswa-SLB-Diduga-jadi-Korban-Pelecehan-Seksual). However, many parents and teachers remain wilfully unaware or reject consideration of the risks disabled young adults face in this area. Parents and teachers have been known to hide cases because they feel embarrassed about their children's disability or refuse to believe their story (Tempo 2014).

There is not much existing data or research about sexuality among teenagers or young adults with intellectual disabilities in Indonesia. However, a case study by Sudiar (2010) of a 16-year-old boy with autism in Yogyakarta, Indonesia, based on an interview with his parents and caregivers, reported that the teen had sexual attraction for women, experienced sexual desires and could understand simple explanations about sexual organs given by his parents. This study also revealed that, most of the time, when the teen had sexual urges, he would be redirected, asked to masturbate in his bedroom or focus on other activities. Another study by Rokhmah and Warsiti (2015) on female teenagers with intellectual disabilities identified the needs for knowledge about reproductive health and sex education as part of special education school curricula. These results challenge parents, teachers, caregivers and researchers to develop culturally sensitive interventions, to improve the awareness and knowledge of the sexual needs of people with intellectual disability (ID). Another study in Yogyakarta by Ariadni et al. (2017) explores parents' perceptions about providing sex education to their children with ID. In this study, the parents revealed that they have limited knowledge of the sexual development of their children and this made it difficult to provide sex education or to answer the children's questions about sex. Parents also stated that they were confused about how to start and what kind of language to use in order to be understood by their children with ID. The parents also wanted to do it properly and at the right time, but they were unsure of how to do this or when the right time might be. This study also found that the parents, especially mothers, who have the most important role in providing sex education, are aware that sex education is not a taboo topic anymore; however, they still use religion as a foundational life context for delivering sex education.

Religion and the education system are believed to be the two main interrelated influences towards sexual attitudes in Indonesia. In Utomo's (1999) work on the history of sexuality in Indonesia, he mentioned that the change of attitudes towards sexuality in modern Indonesia was due to influences from the Dutch educational system during the colonisation era, and the coming of Islam to Indonesia (Supomo 1996, cited in Utomo 1999). He states that conservative attitudes towards sexuality have a strong influence among the upper and middle class of Indonesian people who were educated by the Dutch, but less so in rural communities which did not have the same opportunity to receive education at the time. Islam also contributes to sexual conservatism in line with its spread through Indonesia, being especially dominant in the area where the movement of Islamic reformation is strong (e.g., Sumatra) and less dominant in the eastern areas of Indonesia, which have weak exposure to Islam. Before the Dutch colonization and the coming of Islam, Indonesia was ruled by many *kratons* or small kingdoms on almost every island. In the island of Java, Jogjakarta and Surakarta Kratons were the largest two. Utomo (1999) quotes a story from *Centini*, a book about Javanese culture written in 1820 by Pangeran Adipati Anom (Pakubuana V) from Surakarta Kratons, Central Java, which tells a story concerning the attitudes of Javanese women at that time towards sexuality. It was stated that the women in the book were "very open" and had the power to initiate sexual activities with their partners. *Centini* describes the process by which Javanese women from Surakarta Kratons learned about sexual activities. It mentioned that, most of the time, women learned from their parents and later, they experimented on their own with their partners. At that time, sexuality was perceived as something "natural," with women not being afraid of men and being able to make decisions regarding sexual activities. In contrast, more recent Javanese literature is written with men in the "active" role and women playing a passive part in sexual activities. Javanese society represents ideal women as those who are beautiful, do not talk too much, show patience and obedience to their husbands, as well as being sexually passive (Utomo 1999). The roles of men and women in Javanese traditions are also mentioned by Hayati et al. (2014) in a study conducted in Central Java on the views of masculinity. The study revealed that the participants held religious values as the basic reference for their views on marital roles and considered the division of the roles in the family as crucial.

Furthermore, Sari (2011) states in her work that there are four modern myths in Indonesia related to sexuality: (1) Sexuality is a natural process, so does not need to be taught to the "youngsters," (2) sexuality is a dangerous matter and "dirty," (3) marriage is a safe place for sexuality, and (4) sexuality is a taboo and private matter (Sari 2011) (*Empat mitos seks di Indonesia* [Four sexual myths in Indonesia], on http://www.gaya.tempo.co, October 31, 2011). As a result, the sexuality of women in the Indonesian context is now defined via discussions of virginity or married women's needs, which is related to myths more than knowledge, and it is *taboo* to talk about sex in public spaces. Teenagers and young adults with disabilities lack knowledge about sexuality, such as the function of reproductive organs, the sanctioned range of sexual activities and the extent of assault or abuse (Adolescent reproductive health in Indonesia: Culture, religion, tradition and myths, on http://www.k4health.com, funded by USAID). In this context, it can be perceived that education and religion may play a significant role in sexual attitudes among people in modern Indonesia. Education can contribute by shaping how people think about sexuality and religion and may influence how people react to sexuality issues. Research and cultural-based studies focusing on the intersection of education, religion, and sexuality in Indonesia are needed to provide sexual knowledge and promote strategies for young adults to understand sexuality.

According to the Indonesia Disability Convention Team (2017), the absence of laws concerning sexual violence against persons with disabilities makes punishing the perpetrators difficult, if not impossible. Furthermore, the Indonesian Criminal Code only regulates rape and sexual abuse, but it does not legislate against sexual harassment. Disabled women continuously experience discrimination in the fields of education, work, leadership participation and decision-making positions. They often suffer, both physically and mentally, due to sexual abuse. Women and girls with disabilities must endure discriminations and stigma which makes them vulnerable to being exploited by members of their family and society. The Fact Sheet of the 2014 Annual Notes of the National Commission on Violence against Women, recorded 40 cases of violence involving women with disabilities, of which 37 cases were sexual abuse. The sexual crimes and injustices most often experienced by women with disabilities in Indonesia are sexual abuse, forced abortion, forced sterilisation (including at special education schools), limited access to reproductive health and sexual education, as well as injustice or victimisation in the forms of sexual slavery, domestic violence and sexual trafficking. One example is when a 17-year-old girl with an intellectual disability living in Sukoharjo District was repeatedly raped by her 70-year-old neighbour in May 2016, who got her pregnant. The girl's family were forbidden by the neighbours and village officers to report the incident to the police and were threatened with shunning. Her family was forced to accept compensation of 25 million rupiahs, which was paid in instalments. Another example was a deaf girl with an intellectual disability, a student at a Special Education School (SLB), who was sexually abused six times by a teacher; however, the school tried to suppress the case. After some facilitation by the paralegal, the case was reported to the police, but the police didn't provide any sign language interpreter, making it difficult to file the lawsuit (Indonesia Disability Convention Team 2017).

These conditions show that the Indonesian government has long been negligent, not giving equal treatment in justice and healthcare for young people with disabilities compared to their non-disabled counterparts; this includes the applications of physical and non-physical traditional therapies to young people with disabilities, which are undertaken without proper monitoring, and thus can be detrimental and may even add additional disability experiences to the younger generation. Moreover, young people with disabilities experience discrimination and are being harmed as a result of government policies that are poorly implemented. In short, society does not listen to or understand young people with disabilities and their needs.

## Discussion of research findings

Ramawati's research has so far explored the perception of parents and educators regarding disability, intellectual disability and sexuality in a special education school in Banyumas District, Central Java, Indonesia. This phase of the study included a focus group discussion with twelve parents who each have a child aged 15–25 with ID, and a focus group discussion with eight educators who teach students with ID in the same age range. The parents reveal that they all have limited knowledge about sexuality and intellectual disability. Most of these parent participants question the existence of any sexual development in their children and infantilise them due to their intellectual disability. In contrast, the educators recognise students' need for sexual expression; however, they feel reluctant to provide sexual knowledge to the students and argue that it is the parents' responsibility to educate their children about this topic. Participant (T) claims, "I am a teacher for high schoolers, and I am worried to teach sex education to them because students with intellectual disabilities might have different abilities to comprehend the topics." Another teacher adds that sexuality is still a taboo topic in Indonesia and it is better not to talk about it in the classrooms. R and Y also reiterate, "sexuality is still taboo in Indonesia [...] [We] do not need to talk about it, maybe parents will talk about it right before marriage." Moreover, only one teacher claims that sexual education is embedded in the school's curriculum, with all the others stating the opposite.

The teachers' arguments against giving sexual education to their students with ID are in line with opposition to the "banking education system." Freire (1970) wrote that education is often based on the paternalistic pattern in social life and it becomes political. Teachers or parents decide what they want to teach the students, ignoring the actual needs of the students. The teachers reasoned that sexuality was better taught by the parents because they feared that their teachings might be misunderstood by the students. This attitude also has the same pattern as the "banking education system." In the "banking education" system, teachers play the role as those who think, talk, choose, and know everything, whilst the students act the passive or receiving role. Teachers and parents become the "authorities" in deciding what to teach and when to teach their students and children, without any confirmation or assertion from the students.

Interviews with parents reveal their hopes and concerns about their children's future life. Two parents mention that they want their children to get married in order to be accepted as "normal" persons in society, and also for the parents to have someone to take care of them in the future. As one informant (B) states, "Sexual education is important for our children with intellectual disabilities so they can have their own children in the future, just like the normal children." However, the parents mention that they do not have the knowledge to teach their children about sexuality. But other parents argue that they, as parents, are more entitled to teach sexual education to their children. Some parents, however, hope that sexual education can be taught both at home and at school, since teachers are more trusted by their children (the students). Interviewee (A) mentions this with "Parents are more entitled to teach sex education but it's better if the teachers also teach the students about sexuality at schools, since the children trust the teachers more. So, parents and teachers can complement each other."

The research is also showing a major difference in educational background between the parents and the educators. The parents have less education, most of them are high school graduates or lower, whereas all the educators have higher education backgrounds. Thus, the parents have limited ability to access information about sexuality, especially when the information is provided in an on-line format. The parents also have limited resources and often cannot operate computers or access websites to obtain information related to sexuality or sexual education for young adults with ID, but most of them do have smartphones and can use them to access the internet. This limited ability to access and obtain information on sexuality for people with intellectual disabilities is perceived as the main obstacle for the parents to negotiate when trying to understand the sexual needs of young people with ID. Educators, whom the parents hope will teach sexuality to the youth, are still reluctant to do so, with the lack of detailed sexual knowledge and resources in the school curriculum being the main reason.

These findings suggest that parents' and educators' perspectives on sexual education and sexuality for young adults with ID, as well as schools' policies on sexual education, need to be further explored and discussed openly in local contexts in Central Java Province, and in Indonesia more generally. Both parents and educators have limited knowledge and resources to deliver sexual education to young adults with ID. However, parents were more eager to learn about sexuality to provide sexual education to their children. On the other hand, teachers are reluctant to deliver sexual education to their students, although they have more access and ability to obtain information about sexuality for young people with ID. For all these reasons, the sexual education curriculum for students with ID has been neglected for many years, although it also has a significant influence on the decisions students will make in their future lives. Policy makers, scientists and researchers should work together to create laws and develop knowledge that this population needs. As Kulick and Rydstrom (2015) mention in their work, political assertiveness by or on behalf of people with disabilities is needed to open more opportunities for equal treatment in all sectors of life. This includes equal treatment in justice and the implementation of laws.

As discussed earlier in the chapter, Ramawati is continuing ethnographic research in Central Java, Indonesia with parents and educators, to obtain a better understanding of their points of view. She is also interviewing young adults with ID to record their beliefs and experiences. The views of these young people with ID on sexual education and sexuality will be presented in future publications.

## Conclusion

This chapter provides a review of some important issues relating to sexuality and intellectual disability in Indonesia. It provides a summary of policies relevant to disabled people, as well as the cultural context in which people with intellectual disability live, with particular focus on people living in the region of Central Java. We have reviewed the sparse available research on the topic of sexuality and disability in Indonesia and have reviewed court cases that have been publicised in the Indonesian news media. Finally, using interview data with parents and educators, we have explored how young adults with ID, their parents and teachers negotiate local cultural beliefs in learning/teaching about sexuality through sexual education in their homes and schools. We also discussed how sexual education policy can influence teachers' motivation to deliver sexual education to their students with ID. It appears that the teachers avoid teaching about sexual education to young people with ID because they do not have sufficient resources, just like the parents. Despite the conservative religious and cultural restrictions, parents appear eager to educate themselves and their children about sexuality and disability. This is motivated primarily by the hope that their children will eventually be able to marry and enjoy family life, but also because it may guarantee the care of the parents in their old age. Significant efforts are therefore urgently needed to develop comprehensive guidelines for teachers, parents and other stakeholders in the Indonesian context in order to facilitate the sexual rights and health for individuals with ID. In the long run, education for sexual rights should become not just a limited and prohibited resource of knowledge, but available to everyone through the efforts of all stakeholders and the involvement of people with ID. Through this one aspect of life, we can create an inclusive society for all in every sector: Education, health, laws and sexuality.

# References

*Adolescent reproductive health in Indonesia: Culture, religion, tradition and myths.* Funded by USAID. Available at: http://www.k4health.com

Amannullah, Gantjang (2016) Measuring disability in Indonesia, *Social meeting in Bangkok*, July 27, 2016. (Author is the Director of People Welfare Statistics, BPS-Statistics Indonesia), on https://unstast.un.org/demographic-social/meeting/2016/bangkok-disability-measurement-and-statistics/Session-6/Indonesia.pdf

Ariadni, D.K., Prabandari, Y.S., & Sumarni, D.W. (2017) Parents perception having children with intellectual disability providing sex education: A qualitative study in Yogyakarta, Galore. *International Journal of Health Sciences and Research* 2(5), pp. 164–169.

Badan Pusat Statistik Provinsi Jawa Tengah (The Centre of Bureau of Statistics of Central Java Province) (2014) *Statistik dasar kependudukan [Baseline of population].* Available at: https://jateng.bps.go.id/subject/12/kependudukan.html#subjekViewTab3

Black, K. (2005) Disability and sexuality: Holistic care for adolescents. *Pediatric Nursing* 17(5), pp. 34–37.

Buletin Jendela Data dan Informasi Kesehatan [Bulletin of Health Data and Information] (2014) *Situasi penyandang disabilitas* [People with disabilities situation], Kementerian Kesehatan RI [Ministry of Health Republic of Indonesia], semester II, ISSN 2088-270X.

Cambridge, P. & Mellan, B. (2000) Reconstructing the sexuality of men with learning disabilities: Empirical evidence and thereotical interpretation of needs. *Journal Disability and Society*, vol. 15, Issue 2.

Commissie Samson (2012) *Omringd door zorg, toch niet veilig: Seksueel misbruik van door de overheid uit huis geplaatste kinderen, 1945 tot heden* [Sexual abuse among children who have been placed out of their homes by the government, 1945 until now]. Available at: http://www.rijksoverheid.nl/documenten-en-publicaties/rapporten/2014/02/06/rapport-commissie-samson-omringd-door-zorg-toch-niet-veilig.html

Dinas Kesehatan Jawa Tengah (2016) *Profil kesehatan Provinsi Jawa Tengah Tahun 2016* [Central Java Province Health Profile of 2016]. Available at: http://www.depkes.go.id/resources/download/profil/PROFIL_KES_PROVINSI_2016/13_Jateng2016.pdf

Freire, P. (1970) *Pedagogy of the oppressed.* Myra Bergman Ramos (trans.). New York: Bloomsburg.

Gill, M. (2015) *Already doing it: Intellectual disability and sexual agency.* Minneapolis: University of Minnesota Press.

Harsono, F.H. (2019) 88.1 persen penyandang disabilitas tidak punya kartu identitas khusus [88.1 percent of disabled people do not have people with disabilities identification card]. Available at: https://www.liputan6.com/health/read/4042140/881-persen-penyandang-disabilitas-tidak-punya-kartu-identitas-khusus, 20 Agustus 2019.

Hayati, E.N., Emmelin, M., & Eriksson, M. (2014) "We no longer live in the old days": A qualitative study on the role of masculinity and religion for men's view on violence within marriage in rural Java, Indonesia. *BMC Womens Health* 14, 58. Available at: https://bmcwomenshealth.biomedcentral.com/articles/10.1186/1472-6874-14-58

Hendriani, W., Handariyati, R., & Sakti, T.M. (2006) Penerimaan keluarga terhadap individu yang mengalami keterbelakangan mental [Familys acceptance of individuals with mental retardation]. *INSAN* 8(2). Fakultas Psikologi Universitas Airlangga.

Henry, W.J., Fuerth, K., & Figliozzi, J. (2010) Gay with a disability: A college students multiple cultural journey. *College Student Journal* 44(2), pp. 377–388.

Indonesia Disability Convention Team (2017) *Indonesia disability convention team: Implementation of the United Nations convention on the rights of persons with disabilities.* The United Nations Committee on the Rights of Persons with Disabilities.

Kulick, D., & Rydstrom, J. (2015) *Loneliness and its opposite: Sex, disability, and the ethics of engagement.* Durham and London: Duke University Press.

Leutar, Z., & Mihokovic, M. (2007) Level of knowledge about sexuality of people with mental disabilities. *Sexuality and Disability* 25, pp. 93–109. doi: 10.1007/s11195-007-9046-8.

Löfgren-Mårtenson, L. (2004) "May I?" About sexuality and love in the new generation with intellectual disabilities. *Sexuality and Disability* 22, pp. 197–207. doi: 10.1023/B:SEDI.0000039062.73691.cb.

Majalah Komisi Yudisial [Judicial Commission Magazine] (2018) *Akses hukum dan keadilan bagi difabel* [Law and justice accessibility for people with different ability (diffability)]. Komisi Yudisial Republik Indonesia, ISSN 1978–1458. Available at: http://www.komisiyudisial.go.id/assets/uploads/files/Majalah-KY-April-Juni-2018.pdf, Juni 2018.

Martino, A.S., & Schormans, A.F. (2018) When good intentions backfire: University research ethics review and the intimate lives of people labeled with intellectual disability. *Qualitative Research Forum* 19(3), pp. 375–392.

Milligan, M.S., & Neufeldt, A.H. (2001) The myth of asexuality: A survey of social and empirical evidence. *Sexuality and Disability* 19(2), pp. 91–109.

Moras, R. (2013) Feminism, rape culture, and intellectual disability: Incorporating sexual self-advocacy and sexual consent capacity. In M. Wapper & K Arndt (Eds.), *Emerging perspectives on disability studies*. New York, NY: Palgrave Macmillan, pp. 189–208.

Murphy, N., & Young, P.C. (2005) Sexuality in children and adolescents with disabilities. *Developmental Medicine & Child Neurology* 47, pp. 640–644.

Nathanson, C.A. (1991) Dangerous passage the social control of sexuality in women's adolescence. Available at: https://philpapers.org/rec/NATDPT

Ratnawati, D., Martha, E., & Fasya, S. (2019) The policy of reproductive health education for the adolescence with intellectual disability in Indonesia: A qualitative study. *Indian Journal of Public Health Research & Development* 10(3), pp. 78–88.

Reiter, S., Bryen, D.N., & Schachar, I. (2007) Adolescents with intellectual disabilities as victims abuse. *Journal of Intellectual Disability* 11(4), pp. 371–387.

Rembis, M.A. (2010) Beyond the binary: Rethinking the social model of disabled sexuality. *Sex and Disability* 28, pp. 51–60. doi: 10.1007/s 11195-009-9133-0.

Reynolds, T., Zupanick, C.E., & Dombeck, M. (2013) History of stigmatizing names for intellectual disabilities. *Intellectual Disability*. Available at: http://www.mentalhelp.net/articles/

Rokhmah, I., & Warsiti, W. (2015) *Identification for adolescent reproductive health needs of women with mental disabilities in Public Special Education School Yogyakarta*. STIKES Aisyiyah Yogyakarta.

Rosyid, I. (2019) *Siswa SLB diduga jadi korban pelecehan seksual* [Special educational schools students allegedly experienced sexual assaults]. Available at: https://www.gatra.com/rubrik/nasional/pemerintahan-daerah/382604-Siswa-SLB-Diduga-jadi-Korban-Pelecehan-Seksual, 21 January 2019.

Sari, D. (2011) *Empat mitos seks di Indonesia* [Four sexual myths in Indonesia]. Available at: http://www.gaya.tempo.co, 31 October 2011.

Siebelink, E.M., de Jong, M.D.T., Taal, E., & Roelvink, L. (2006) Sexuality and people with intellectual disabilities: Assessment of knowledge, attitudes, experiences, and needs. *Mental Retardation* 44, pp. 283–294. doi: 10.1352/0047-6765.

Sudiar, R.S. (2010) Sexual behaviour management on autistic teenager in Special Education School in Yogyakarta: A case study of a 16-year-old male teenager. Universitas Islam Negeri Sunan Kalijaga Yogyakarta.

Tempo, R. (2014) *Difabel kerap jadi korban diskriminasi seksualitas* [People with disabilities as the victims of sexual discrimination]. Available at: https://nasional.tempo.co/read/555622/difabel-kerap-jadi-korban-diskriminasi-seksualitas

Tepper, M. (2005) Becoming sexually able: Education for adolescents and young adults with disabilities. *Contemporary Sexuality* 39(9), pp. 1–7.

The Americans with Disabilities Act (1990) *Information and technical assistance on the Americans with Disabilities Act (ADA): Introduction to the ADA*. Available at: https://www.ada.gov/ada_intro.htm

Trent, J.W. (1994) *Inventing the feeble mind: A history of mental retardation in the United States*. Berkeley: The Regents of the University of California.

Undang-undang Republik Indonesia, No. 4 Tahun 1997 tentang Penyandang Cacat [Indonesian law for handicap people], on Media Elektronik Sekretariat Negara Tahun 1997.

Undang-undang Republik Indonesia, No. 8 Tahun 2016 tentang Penyandang Disabilitas [Indonesian law for persons with disabilities]. Available at: www.hukumonline.com

Undang-undang Republik Indonesia (RI) No. 19 Tahun 2011 tentang Pengesahan Konvensi mengenai Hak-hak Penyandang Disabilitas [Laws on The ratification of UN Covention on The Rights of Persons with Disabilities], on https://www.bphn.go.id/data/documents/11uu019.pdf

United Nations (2008) *Convention on the rights of persons with disabilities and optional protocols*. Available at: http://www.un.org

Utomo, I.D. (1999) Sexuality and relationship between the sexes in Indonesia: A historical perspective. *The European Population Conference*, The Hague, Netherlands, 30 August to 3 September 1999.

van Berlo, W., de Haas, S., van Oosten, N., van Dijk, L., Brants, L., Tonnon, S., & Storms, O. (2011) *Beperkt weerbaar: Een onderzoek naar seksueel geweld bij mensen met een lichamelijk, zintuiglijke of een verstandelijke beperking [Sexual violence among individuals with a physical, sensory, or intellectual disability]*. Utrecht, The Netherlands: van Berlo.

Winarni, T.I., Hardian, H., Suharta, S., & Ediati, A. (2018) Attitudes towards sexuality in males and females with intellectual disabilities: Indonesian setting. *Journal of Intellectual Disability: Diagnosis and Treatment* 6, pp. 43–48.

World Health Organization (2006) Sexual and reproductive health: Defining sexual health. Available at: http://www.who.int/reproductivehealth/topics/sexual_health/sh_definitions/en/

# 15

# ADVANCE CONSENT AND NETWORK CONSENT[1]

*Alexander A. Boni-Saenz*

## Capacity and its legal regulation

*Mental incapacity* is the condition of lacking the requisite psychological abilities to engage in autonomous decision making. There are four primary types of mental incapacity, which are primarily differentiated by their temporal scopes (Boni-Saenz 2015). *Temporary extended incapacity*, such as minority (i.e., of being a minor), has a long duration but a definite end point. The mental incapacity is due to the fact that the individual's psychological faculties are not yet mature, but almost everyone will eventually grow out of this condition. In contrast, *temporary transient incapacity* is relatively short-lived—it comes and goes. Examples include intoxication, episodic mental illness or bouts of delirium. *Persistent lifelong incapacity* does not go away and exists from a very early age. There is no "growing out of it," and there is no "wearing off." The paradigmatic example is intellectual disability, an umbrella term for a variety of cognitive impairments, including genetic conditions such as Down Syndrome. Finally, *persistent acquired incapacity* exists when a person suffers an impairment that does not go away, but which arises after a period of relatively unimpaired functioning. The paradigmatic example is dementia. Capacity and incapacity are functional concepts. This means that incapacity is assessed with respect to the ability to make a particular decision, rather than as a general status (Frolik and Radford 2006). For instance, one might lack the capacity to engage in complex financial transactions, but still have the capacity to decide whether to eat broccoli or asparagus for lunch.

*Legal incapacity* is the condition of lacking the requisite legal authority to engage in autonomous decision making. A *legal incapacity doctrine* refers to any civil or criminal legal doctrine that deems an individual to lack decisional capacity in a particular domain. *Sexual incapacity doctrines* refer to those doctrines that do this with respect to sexual decision making. In other words, even if one gives unambiguous, verbal, affirmative consent to sex, this "Yes" may be transformed into a legal "No" by the sexual incapacity doctrine, rendering the sexual conduct criminal or tortious. While mental and legal incapacity converge in many cases, they may also diverge. In other words, one may be deemed legally capable when one does not have the mental capacity for certain decisions, or, more frequently, one may be deemed legally incapable when one still possesses the requisite mental capacity.

Whereas the law generally deals with the temporary extended incapacity of minors using a variety of bright-line age-of-consent rules, adult incapacity doctrines are less determinate. The legal test for capacity to consent to sex in most jurisdictions requires that one has the mental capacity to understand the "nature and consequences" of the sexual decision. Courts vary, however, on which consequences are relevant to the legal inquiry. For example, in *People v. Breck*, the Appeals Court in Michigan held that "the statutory language in question is meant to encompass not only an understanding of the physical act but also an appreciation of the nonphysical factors, including the moral quality of the act." In contrast, the Supreme Court of North Dakota in *State v. Mosbrucker* held that an individual must "understand the nature of the sexual act as well as its consequences, such as pregnancy and sexually transmitted diseases, but not the moral nature of their participation in the act of intercourse." Thus, in order to have the capacity to consent to sex, one might have to possess the capacity to understand that a particular sexual act could entail such things as pleasure, pregnancy, and sexually transmitted diseases. Alternatively, one might need to understand the social or moral consequences of the sex act as well. Courts will examine a wide range of evidence in performing the capacity inquiry, including diagnostic medical evidence, lay and expert testimony about functional capacity, and assessments of sexual knowledge or understanding (Harris 2018).

## Liability and social norms

The nature and consequences test has its merits, especially in its application to situations of temporary transient incapacity. In those situations, it serves to protect a baseline non-impaired self (who will re-emerge) against sexual exploitation by others while one is in a temporarily altered mental state (Harmon 1990). Because of the relatively short duration of the incapacity in that context, the restriction on sexual opportunities is relatively minor. In contrast, people with persistent incapacity, either lifelong or acquired, may face lifetime or otherwise extensive restrictions on sexual activity if their cognitive abilities are deemed inadequate.

These restrictions are the result of the legal and social effects of sexual incapacity doctrines, which exert regulatory control by imposing legal liability. The primary target for liability is the sexual partner of the person lacking legal capacity. Secondary targets are individuals or institutions that, in some way, have responsibility for individuals deemed to lack capacity. Criminal liability arises from prohibitions on rape or sexual assault, which are applied to the sexual partner of the person lacking legal capacity (Torcia 1995). Civil liability arises from tortious battery, which involves a tortfeasor engaging in harmful or offensive contact (Moore 2012). In addition to that direct civil liability, there may be vicarious civil liability for institutions (Kapp 1991). For example, the family of a nursing home resident might press a claim for negligent supervision if that resident was sexually battered while in the institution's care. Institutions also face a body of regulatory law that punishes them for inappropriately caring for people who lack capacity and who are in their care. Repercussions may include the loss of state funding, which can have disastrous consequences for institutions that are reliant on such money. It could further include loss of accreditation by relevant quasi-governmental regulatory bodies. Finally, individuals that work at such institutions might face sanctions in the form of revocation of professional licensure.

In addition to these explicit legal and regulatory effects, the law has expressive effects when it invalidates the apparent consent choices of those who are deemed to lack legal capacity (Sunstein 1996). This is clearest in the case of criminal prohibitions, which carry the weight of societal condemnation for the acts that are deemed criminal. But both civil and criminal doctrines act to construct the sexualities of citizens by demarcating the boundaries of acceptable and unacceptable sex and reinforcing existing understandings of the sexuality of certain groups (Frug 1992). To the degree that a class of individuals is deemed to lack sexual consent capacity, this can devalue that class and construct its members either as asexual and undesirable or as hypersexual and in need of control (Emens 2009). Thus, the sanction and expressive functions of the law work in mutually reinforcing ways to regulate sexuality both in the behaviour and attitudes of the people they govern.

An example might help illustrate these harms (Gruley 2014). Henry and Donna Rayhons met in their late 60s, and neither of them expected to find love again after being widowed. They first flirted in church while singing for the choir. Two years later, they were getting married in front of over 350 guests. Now in their 70s, they enjoyed several activities together, such as bee-keeping, farming, and long leisurely drives. They also had sex. In 2010, Donna was diagnosed with Alzheimer's disease. As her condition worsened, two of her daughters from a previous marriage moved her to a residential care facility. Henry would regularly visit her, and on one visit in May 2014, Donna's roommate thought she heard sexual noises coming from across the privacy curtain in their shared room. This led Donna's daughters to seek guardianship over Donna and to limit Henry's interactions with her. On August 8, 2014, Donna passed away. A week later, Henry was arrested and charged with felony sexual abuse on the basis that Donna Rayhons suffered a "mental defect" that made her unable to consent. After two days of deliberations, the jury acquitted Henry of wrongdoing, but not after a week-long trial that garnered national media attention and exposed details of Donna and Henry's relationship, Donna's medical condition, and their alleged sexual encounter.

The current legal regime empowered the state to pursue Henry Rayhons for his alleged sexual contact, even if his court case ended up in acquittal. It will further have the effect of discouraging institutions from permitting sexual contact for people with Alzheimer's disease or other cognitive impairments, regardless of whether those interactions are desired by or beneficial to the person with cognitive impairments. In other words, the legal regulation of sexual capacity serves as a barrier to a flourishing sexual life for people with cognitive impairments. The key issue is that the individual with cognitive impairments may lack the legal capacity to consent to sex at the time of the sexual act. The next sections explore two possibilities for legal reform—advance consent and network consent.

## Advance consent

Sexual consent is normally considered to be more or less contemporaneous with the sexual act it authorises. In other words, individuals cannot consent in advance to sex. This is legally unusual, as the law affords individuals the ability to make decisions in advance or to delegate decision-making authority in advance of incapacity in a host of other contexts through the use of advance directives. In other words, advance directives facilitate decision making across time. At Time 1, an individual declares a decision or appoints a surrogate decision maker. At Time 2—typically when the individual lacks legal capacity to make decisions—the advance decision is executed or the decision of the surrogate decision maker is implemented.

There are two general types of advance directives. The first is an *instructional directive*, which sets out particular decisions in advance. The second type of advance directive is the *proxy directive*, which sets out a particular *surrogate decision maker* in advance. *Hybrid directives* designate a proxy decision maker but also provide written guidance about the individual's beliefs in varying levels of mandatory language. Advance directives are in widespread use throughout the legal regime, from wills that allocate property after death to health care powers of attorney that designate a health care decision maker to trusts, which give a trustee control over assets, with guidance from the trust's provisions.

Sexual advance directives would apply the traditional legal device of advance directives to the novel decision-making domain of sexuality in situations of persistent incapacity (Boni-

Saenz 2016). They would be limited to individuals with persistent acquired incapacity living in long-term care institutions, such as assisted-living facilities, nursing homes, and continuing-care retirement communities, in order to provide some third-party oversight of the implementation of the legal device. In these contexts, sexual advance directives would permit individuals to consent at Time 1 to specific sexual acts at Time 2, when they lacked legal capacity to consent to sex. Alternatively, they could empower agents at Time 1 to make certain sexual consent decisions at Time 2, with different types of guidance or parameters around that decision-making authority.

Thus, sexual advance directives have the effect of fragmenting sexual consent over time and across people. There is the consent to a particular sexual act or decision maker at Time 1, when a person still possesses the legal capacity to consent to sex. This prospective consent at Time 1 is useful because the individual still has mental and legal capacity, but it is also given without full knowledge of the sexual facts in the future moment in which they will arise. Second, there is the token of consent from the individual with cognitive impairments at Time 2. This consent is good for assessing sexual desire in the moment, but it is not a consent that necessarily comes from a deliberation of the nature and consequences of the sexual act. Third, there is potentially the consent given at Time 2 by an agent, if one was so empowered at Time 1. A sexual agent empowered at Time 1 and assessing the sexual situation at Time 2 is qualified to deliberate about it on some level, but this agent does not have first-hand knowledge of the sexual feelings of the person with cognitive impairments. In addition, there is always the possibility that the agent may not be acting dutifully in her proxy decision making.

There is significant philosophical disagreement about which consent decision should dominate in the context of health care advance directives (Dresser 2003; Dworkin 1994). There is, however, a way to diffuse the tension between these views with respect to sexual advance directives. The debates over advance directives paint a picture of the Time 1 and Time 2 selves in conflict, but it is quite possible that the two temporal selves' interests will actually align in many situations. When this occurs, there is a *consensus of consents* (Boni-Saenz 2016). Thus, only when all of the fragments of consent are aligned should sexual advance directives have legal force and should an associated sexual act to be legally authorised.

For example, consider an individual who fills out a sexual advance directive at Time 1, consenting to certain sexual acts with a spouse and empowering a relative to act as a sexual agent. Then, at Time 2, when that individual lacks legal capacity, she actively seeks out or otherwise expresses desire for sexual contact with her spouse, and her sexual agent consents to such sexual contact. All of the partial consents in this situation align, creating a consensus of consents, and that sexual act should be legally permissible. In contrast, consider the same facts, but the individual at Time 2 does not seek out or otherwise express desire for sexual contact with her spouse. Even if all of the other partial consents align, the sexual act should not be legally authorised. In other words, under no circumstances should one of the other partial consents override the consent decision of the person with cognitive impairments in the present. Legally valuing these tokens of consent provided by the person with cognitive impairments alongside the other partial consents preserves the agency of that individual and centres the analysis on her desires.

Because sexual advance directives are premised on the consensus of consents, there must be a high degree of certainty that the relevant consents at Time 1 and Time 2 are valid. First, we want to be reasonably sure that the sexual advance directive is actually a reflection of the views of the person who created it. In addition, because the document will be examined in the context of a civil or criminal trial with serious allegations such as rape or sexual battery, it is important that the sexual advance directive provide good evidence of the person's intent with respect to sexual consent or the delegation of the sexual consent decision. Sexual advance directives should thus be executed with the heightened level of formalities that are traditionally required of wills—

typically a writing, signature, and attestation by two witnesses—to ensure that the document is authentic, voluntarily entered into, and easily processed by a court (Langbein 1975). Many states include a statutory form that is suggested for execution of health care advance directives, and the same could be done for sexual advance directives. Such statutory forms should encourage specificity, noting the sexual acts authorised and the permissible sexual partners.

Second, we want to be reasonably certain that the consent given at the time of the sexual act reflects accurately the person's sexual desires. As a threshold matter, one must have the ability to express volition freely. While many people with Alzheimer's disease or dementia will possess this capacity, there are many who will not. Beyond this, there must be affirmative consent on the part of the person with cognitive impairments. This could be a verbal "Yes," but it could also come in nonverbal form as the initiation or active pursuit of sexual expression (Tuerkheimer 2016). It may also be specific to the individual, as every person has unique forms of communication or ways of initiating sexual encounters. While the affirmative consent standard has not been without controversy, the case for its application to this context is particularly strong. This more demanding consent standard is needed to incentivise sexual partners to ensure that there is a true mental state of acquiescence to the sexual act, as embodied in words or action. Because of the vulnerability of this population, it is reasonable for the law to structure the sexual transaction so that sexual partners without impairments must engage with a sense of care and caution. In other words, any risk-shifting that the standard implies is justified by the disadvantaged social position of one of the sexual partners (Baker 2005).

The Time 2 contemporaneous consent will take place in the context of an institution in which the person with persistent cognitive impairments resides. This highlights a potential privacy trade-off in implementing sexual advance directives. Many institutions will understandably want both to protect residents from sexual abuse and to protect themselves from legal liability. To accomplish these objectives, institutions may decide to engage in monitoring of resident sexual activity to ensure that there is contemporaneous consent. The privacy cost that this type of monitoring entails is one that some individuals or their partners will not be willing to stomach. Said individuals are free to make the valid choice of privacy over sexual expression. For those who are willing to sacrifice some privacy for the opportunity to engage in sexual expression and the physical and social benefits it provides, however, the trade-off may be worth it.

Third, we want to be reasonably convinced that the proxy decision maker at Time 2, if there is one, is acting well. There are some natural checks on agents empowered by advance directives of all kinds, as they must interact with third parties in order to actualise their authority under the directive. For example, health care agents must cooperate with the medical profession to implement health care decisions, and financial agents acting under a power of attorney must often deal with financial institutions in order to manage the financial resources of the principal. Even if there is not a designated sexual agent, instructional advance directives are not self-executing, creating a need for someone or some entity to interpret and implement them. The law leans heavily on trusted institutions or professions to perform this implementation function in other areas. For example, doctors are supposed to honour the treatment preferences expressed in a living will, and they are bound by a code of ethics and are subject to state licensing regimes. Similarly, wills are addressed to the probate court, and judges are supposed to honour the wishes of the testator as embodied in the distributional directives in the will.

Long-term care facilities, such as assisted-living facilities, nursing homes, and continuing-care retirement communities, fill this role for sexual advance directives. They house a large proportion of adults with cognitive impairments, thus granting them familiarity with this population and its needs. In addition, these institutions already actively manage both the social environment and the intimate care of residents. In serving residents' various needs, these institutions should treat sexual expression similarly to other issues that are routinely managed, such as nutrition, pressure ulcers or medical treatment. Practically, this means protecting the individual with cognitive impairments from the objective welfare threats entailed by sexual contact. The content of this protection depends on the situation, but examples might include providing a physical space that minimises risk of falls or physical injury during sexual activity and guarding against threats to the health of the person with cognitive impairments, such as sexually transmitted diseases.

This is not to say that long-term care institutions are perfectly suited to perform these tasks; in fact, many institutions are substandard. Sexual advance directives will work best in a quality system of long-term care, and many institutions may not be prepared to implement them yet. Facilitating the sexual lives of residents is possible only if long-term care institutions are cognizant of the needs for sexual expression among residents and adopt policies that facilitate sexual environments with sexual choice and freedom from sexual abuse. Underlying the ability of long-term care institutions to facilitate sexual expression is the basic economic issue of providing adequate financing for long-term care services (Kaplan 2007). Such funding would assist in addressing one of the largest chronic problems among nursing homes, which is understaffing. If long-term care institutions are expected to manage the sexual lives of residents in line with sexual advance directives, they must have sufficient staff to conduct capacity assessments and to manage residents' social and physical environments.

Sexual agents empowered by sexual advance directives must also interact with these institutions to convey and implement sexual decisions. In exercising their power under sexual advance directives, these sexual agents bear the burden of fiduciary duty. A fiduciary is an individual who is in a position of power and trust with respect to another person, putting that other person at risk if the fiduciary does not act in her interests (Frankel 2011). Because of their position of power and trust, fiduciaries have certain duties. The first is the duty of care, which requires the agent to perform their services with prudence, attention and proficiency. The second is the duty of loyalty, which requires that an agent act in the interests of the person for whom the agent is a fiduciary. A sexual agent's duty of care mimics the requirement that institutions adhere to a standard of care with respect to their residents, requiring that sexual agents work to eliminate objective harms to the principal generated by sexual expression. Thus, there is a convergence of interests between the sexual agent and the institution with which they deal, all geared towards protecting the principal from harm.

The duty of loyalty requires that an agent must act within the scope of the consent granted in the sexual advance directive and the consent granted at the time of the sexual transaction. Thus, in this context, it duplicates the requirement to honour the consensus of consents. There is an additional wrinkle, however, in the duty of loyalty for many sexual agents. In numerous cases, the sexual agent will also be a primary sexual partner of the person with cognitive impairments, such as a spouse. This technically constitutes a conflict of interest, and the traditional understanding of the duty of loyalty prohibits such conflicts. The consequence of adopting the traditional view would be to prohibit sexual contact and deprive many people of a continued long-term sexual relationship that was likely the goal in executing a sexual advance directive in the first place. This problem is not unique to the sexual sphere, as powers of attorney often empower agents, whose economic or other interests intersect with those of the principal as well (Langbein 2005). Thus, the modern view, embodied in the Uniform Power of Attorney Act, permits such conflicts if they are in the best interests of the principal. This recognises that proxy directives are meant to be a more informal way of handling the delegation of decision making, and that such conflicts may be an inevitable consequence of delegation to family members or loved ones.

## Network consent[2]

Sexual advance directives might be helpful for those who possess legal capacity at some point in their adult lives and plan in advance. However, they do not address the situation of those with lifelong persistent incapacity or those who do not plan in advance. Enter network consent, a concept that is premised on two conceptual pillars (Boni-Saenz 2015). The first is the capabilities approach, which views living as a combination of functionings, which are a series of "doings" and "beings." These "doings" and "beings" are the various activities that one could engage in, or the various states of "being" that one could achieve (Sen 1985). Certain capabilities are fundamental in the sense that they should be guaranteed in some fashion to all members of society (Nussbaum 2000). Sexual capability, in turn, represents the opportunity to achieve functionings associated with sex and sexuality (Boni-Saenz 2015). This could include having sexual pleasure, forming a sexual identity, or feeling sexy. This is connected to the fundamental capability of bodily integrity but could also be a part of other fundamental capabilities, such as senses, imagination and thought, emotions, practical reason, affiliation and play (Kulick and Rydström 2015; Nussbaum 2000).

The second conceptual pillar is relational autonomy, a feminist reconfiguration of the traditional autonomy concept. Its key insight is that our sense of self and autonomy is experienced and interpreted relationally (Nedelsky 2011). This creates an inevitable dependency on others in order to exercise our self-determination (Fineman 2000). As children, we construct a sense of self in relation to our parents or other loved ones who might surround us. They help us to develop our capacities for individuality and choice, and this process continues as we age. Thus, our social relationships can enhance the exercise of our autonomy or hinder it. This is also and especially true of people with cognitive impairments, who can and often do exercise decision-making potential through supportive decision-making networks (Kohn et al. 2013). These networks can consist of a single caregiver, a set of family members, or an institution's staff. The type of support provided will vary in accordance with the impairments the network is helping to address. This could be support in formulating one's purposes, in engaging in the decision-making process with other parties or in acting on the decisions one has made (Bach and Kerzner 2010). Supported decision making broadens our understanding of the decision-making apparatus and of personhood from the individual and her body to the individual nested in a series of relationships, that facilitate meaningful decision making and flourishing.

Network consent represents the legal implementation of the sexual capability value for people with persistent cognitive impairments. It derives its name from a recognition that some individuals achieve sexual decision-making capacity through the assistance of a decision-making support network. The test proceeds in three general steps. The first step is to gauge whether the individual has the threshold capacity to express volition with respect to a sexual decision. This volition is traditionally expressed as verbally saying "Yes" to sex. People with cognitive impairments, however, may have difficulty with standard communication. In this case, volition might be inferred in a variety of ways, which will often be specific to the person. It might come in the form of initiation and active pursuit of sexual expression. Alternatively, it might require an interpretation of cues by someone familiar with the person's communication methods, which could include nonverbal signals or facial expressions. If one is not capable of even this basic level of communication of volition, one cannot proceed to be a sexual agent.

If the first step is satisfied, the second step is to assess whether the individual has the necessary mental capacities to understand and reason about the nature and consequences of a given sexual decision. At a minimum, this requires an understanding that the person has the power to make a choice, to consent or not. Beyond this, the capacity to understand and judge consequences is necessary to perform a subjective welfare calculus for oneself. Thus, at an abstract level, the set of consequences of sex, that one should have the capacity to understand, should start large, encompassing both its physical and non-physical effects. This starting point is justified by the fact that sex has many meanings and many effects, any of which might be relevant to a given decision maker. One might want to achieve pleasure with sex, forge a specific identity, solidify social relationships with others, or all of the above. This set of consequences, however, should not include the "moral" consequences of sex. First, it is not clear that there is a consensus on the moral quality of different sexual acts in society, making a determination of that consensus an impossible cognitive task. Second, since sexual activities are often more private than other activities, the moral views of society as a whole would not be a relevant consequence for most individuals engaged in sexual relations.

This relatively large set of consequences must then be calibrated to the particular sexual situation. In other words, because capacity is determined on a functional basis, one must consider each sexual decision at a particular point in time under a particular set of circumstances. Physical and non-physical consequences vary significantly with sexual behaviour; thus, the type and quantity of consequences that one must be capable of understanding should vary as well. For example, holding hands and kissing do not involve significant risks of negative physical consequences, while penile-vaginal intercourse poses more significant risks of pregnancy or sexually transmitted diseases. Since there are more consequences at issue, the latter will likely require a higher level of cognitive capacities than the former. In sum, the context of the sexual encounter must be examined to determine which consequences are actually present for a given sexual decision maker.

If one meets this requirement, then one has sexual consent capacity without the need for assistance. If a person with persistent cognitive impairments does not have the requisite mental capacities on their own to reason about a specific sexual decision and its consequences, then the court must proceed to the third step and broaden the inquiry to determine whether an adequate decision-making support system is in place and participated in the making of the relevant sexual decision. This support system can take many forms, including friends, family or institutional staff. The network will often include people who have been legally appointed to make decisions for a person with cognitive impairments, such as a guardian or attorney-in-fact. That legal authorization to act as a surrogate decision maker, however, is not sufficient to establish that a valid decision-making support system is in place. In other words, a decision-making support system does not exist to make the sexual decision as a surrogate for the person with cognitive impairments, but instead to facilitate her wishes and desires. It is possible that there will be many individuals who are potential members of the decision-making support system, and they might disagree on how best to actualise the sexual desires of a person with cognitive impairments. Ideally, such disagreements will be worked out before the sexual activity takes place. However, it is not the role of the court to determine the one true decision-making support system. As the court would be deciding the case ex post in a civil or criminal proceeding, its only task is to determine whether the supportive network or portion of the network that facilitated the sexual relations in question was adequate.

After verifying the existence and involvement of a decision-making support system, the court should assess its quality to ensure that it is adequate. This is essentially an inquiry into the health of the decision-making apparatus as a whole, similar to the inquiry into the individual's mental capacities. The principles governing fiduciary relationships provide useful guideposts for conducting this fact-intensive and contextual inquiry, just as they do for sexual advance directives (Frankel 2011). With respect to the duty of loyalty, courts should be sceptical of members of a supportive network who are also engaging in sexual expressions with the person who has cognitive impairments. This indicates a conflict of interest that could potentially reflect that the sexual decisions being made do not reflect the preferences of the person with cognitive impairments. However, just as in the case of sexual agents in sexual advance directives, to rule out all sexual encounters with members of supportive networks may restrict the desirable sexual options of people with cognitive impairments completely.

At the same time, conflicted networks should not get a pass; they should be subjected to a rebuttable presumption of network inadequacy, which can be overcome if sufficient evidence of loyalty and care is supplied to the court. The goal, in sifting through the evidence, is to ascertain whether members of the supportive network have tried to avoid conflicts and whether they have adopted an orientation of selflessness towards the person with cognitive impairments at the centre of the network. The evidence would be particularly important to overcome the presumption of network inadequacy in situations of conflicts of interest. One valuable type of evidence would be whether the person with cognitive impairments put trust in the loyalty of individuals who might be in the supportive network. This could include her acceptance of an individual as a marital partner or her appointment of someone as an agent for decision making through a sexual advance directive or a health care or financial durable power of attorney (Boni-Saenz 2013). This is not to say that spouses, health care proxy agents, or attorneys-in-fact will always act loyally, but that the person who selected them has already put her trust in them, which is significant. Other evidence might be testimonial or documentary in nature, indicating whether or not the person was acting in a trustworthy and loyal way towards the person with cognitive impairments.

The supportive network's duty of care entails knowing the subject they are assisting and providing a safe space for the actualization of sexual desire. In this context, evidence of care could include the degree of familiarity the network has with the person's history, preferences and forms of communication, assuming that this knowledge was put to good use. Sufficient care ensures that the supportive network can actualise the subjective elements of the sexual experience for the person with cognitive impairments. For an institution, an inquiry into the level of care should seek to determine whether it performed a thorough analysis of the resident's capacity, whether it gathered information about the resident's history, preferences, and forms of communication, and whether it kept adequate records of these efforts to be reviewed by the court, if necessary.

In addition to acquiring and operationalising knowledge about the individual, care also involves providing a safe space for sexual expression to take place and taking reasonable steps to protect the individual with cognitive impairments from physical harm. This guarantees that the supportive network has recognised and dealt with the consequences of sex that entail more objective welfare threats, such as sexually transmitted diseases and unwanted pregnancy. Evidence of care with respect to these objective welfare effects could include efforts to enact various consequence-diminishing interventions, such as birth control to protect against pregnancy or Truvada to protect against HIV infection. Supportive networks can take precautions with respect to the physical environment as well, to prevent the risk of fall or physical injury during sexual activity. In sum, the evidentiary inquiry on the care axis would focus on whether members of the supportive network have acted as a reasonable or prudent person or institution would have.

Having reviewed the types of evidence that would be relevant to the capacity analysis, it is important to note the types of evidence that would not be relevant for a network consent analysis. Certain facts that are integral to the sexual encounter would not be per se relevant to the capacity analysis. Examples include the sex of a subject's sexual partner or the particular sex act engaged in with said partner. Similarly, other contextual facts that could trigger moral evaluation of the sexual situation, such as whether the person with cognitive impairment is engaging in adultery, or whether the motives of the sexual partner are benign, would likewise be irrelevant to the capacity analysis unless some specific link to a relevant step of the network consent test could be established.

There are at least two plausible routes to relevance. First, it is possible for such facts to enter the inquiry in a limited way in step two. This step requires assessing the relevant consequences of a particular sexual decision, as those consequences will dictate the level of mental capacities needed to process them. For example, the sex of the partner coupled with particular sexual acts may create the risk of a pregnancy consequence. Or the fact that the sex is adulterous may create the risk of fracturing an important relationship with a spouse–caregiver. Thus, these types of facts may enter the analysis because they affect the relevant consequences that one must have the capacity to consider.

Second, certain facts could be relevant in step three of the network consent test, which requires a deeper assessment of the loyalty of members of the supportive network in situations of conflict of interest. Thus, the motives of a conflicted member of supportive network could be relevant to the loyalty inquiry. In short, these types of facts only enter the analysis in a limited way for purposes of analysing consequences or network adequacy, rather than for purposes of moral judgment. This excludes normative judgments about sex from the formal analysis of capacity in the legal test, and thus permits people with persistent cognitive impairments to pursue the wide range of sexual relationships and practices that those without impairments are entitled to pursue. For this reason, the network consent test is at least nominally sex-positive.

## Conclusion

The law exerts powerful regulatory control over the sexual lives of people with cognitive impairments, but it is not the only determinant of a flourishing sexual life. The legal solutions provided in this Chapter may serve to remove legal barriers to sex, but the law is only one of many disabling features of the socio-sexual environment (Shakespeare et al. 1996). Thus, changing the law is a necessary, but not sufficient, condition for realising the sexual capability of people with cognitive impairments. It needs to be part of a multi-pronged approach that includes guarding against sexual abuse, and exploitation, allocating sufficient societal resources to individuals with disabilities and pursuing cultural change.

## Notes

1 This chapter reprints material originally found in Boni-Saenz (2015; 2016).
2 In past work, I referred to "network consent" as the "cognition-plus test" (Boni-Saenz 2015). I now believe that the term "network consent" better captures the substance of the legal test.

## References

Bach, M., & Kerzner, L. (2010) *A new paradigm for protecting autonomy and the right to legal capacity*. Toronto: Law Commission of Ontario.

Baker, K.K. (2005) Gender and emotion in criminal law. *Harvard Journal of Law and Gender* 28(2), pp. 447–466.

Boni-Saenz, A.A. (2013) Personal delegations. *Brooklyn Law Review* 78(4), pp. 1231–1278.

Boni-Saenz, A.A. (2015) Sexuality and incapacity, *Ohio State Law Journal* 76(6), pp. 1201–1253.

Boni-Saenz, A.A. (2016) Sexual advance directives. *Alabama Law Review* 68(1), pp. 1–47.

Dresser, R. (2003). Precommitment: A misguided strategy for securing death with dignity. *Texas Law Review* 81(7), pp. 1823–1847.

Dworkin, R. (1994) *Life's dominion: An argument about abortion, euthanasia, and individual freedom.* New York: Vintage Books.

Emens, E.F. (2009) Intimate discrimination: The state's role in the accidents of sex and love. *Harvard Law Review* 122(5), pp. 1307–1402.

Fineman, M.A. (2000) Cracking the foundational myths: Independence, autonomy, and self-sufficiency. *American University Journal of Gender, Social Policy & the Law* 8(1), pp. 13–29.

Frankel, T.T. (2011) *Fiduciary law.* Oxford: Oxford University Press.

Frolik, L.A., & Radford, M.F. (2006) "Sufficient" capacity: The contrasting capacity requirements for different documents. *NAELA Journal* 2, pp. 303–323.

Frug, M.J. (1992) *Postmodern legal feminism.* New York: Routledge.

Gruley, B. (2014) Rape case asks if wife with dementia can say yes to her husband, *Bloomberg* [online]. Available at: https://perma.cc/S7SZ-LPJ4 [Accessed 30 Jan. 2019].

Harmon, L. (1990) Falling off the vine: Legal fictions and the doctrine of substituted judgment. *Yale Law Journal* 100(1), pp. 1–71.

Harris, J.E. (2018) Sexual consent and disability. *New York University Law Review* 93(3), pp. 480–557.

Kaplan, R.L. (2007) Retirement plannings greatest gap: Funding long-term care. *Lewis & Clark Law Review* 11(2), pp. 407–450.

Kapp, M.B. (1991) Malpractice liability in long-term care: A changing environment. *Creighton Law Review* 24(4), pp. 1235–1260.

Kohn, N.A., Blumenthal, J.A., & Campbell, A.T. (2013) Supported decision-making: A viable alternative to guardianship? *Penn State Law Review* 117(4), pp. 1111–1157.

Kulick, D., & Rydström, J. (2015) *Loneliness and its opposite: Sex, disability, and the ethics of engagement.* Durham: Duke University Press.

Langbein, J.H. (1975) Substantial compliance with the Wills Act. *Harvard Law Review* 88(3), pp. 489–531.

Langbein, J.H. (2005) Questioning the trust law duty of loyalty: Sole interest or best interest? *Yale Law Journal* 114(5), pp. 929–990.

Moore, N.J. (2012) Intent and consent in the tort of battery: Confusion and controversy. *American University Law Review* 61(6), pp. 1585–1655.

Nedelsky, J. (2011) *Laws relations: A relational theory of self, autonomy, and law.* Oxford: Oxford University Press.

Nussbaum, M.C. (2000) *Women and human development: The capabilities approach.* Cambridge: Cambridge University Press.

*People v. Breck* (1998) 584 N.W.2d (Mich. Ct. App.); 602.

Sen, A. (1985) *Commodities and capabilities.* Amsterdam: North-Holland.

Shakespeare, T., Gillespie-Sells, K., & Davies, D. (1996) *The sexual politics of disability: Untold desires.* London: Cassell.

*State v. Mosbrucker* (2008) 758 N.W.2d (S. Ct. N.D.); 663.

Sunstein, C.R. (1996) On the expressive function of law. *University of Pennsylvania Law Review* 144(5), pp. 2021–2053.

Torcia, C.E. (1995) *Wharton's criminal law.* 15th ed. Eagan: Thomson Reuters.

Tuerkheimer, D. (2016) Affirmative consent. *Ohio State Journal of Criminal Law* 13(2), pp. 441–468.

# PART IV

# Representation, performance and media

# 16

# MISSING IN ACTION

## Desire, dwarfism and getting it on/off/up—a critique and extension of disability aesthetics

*Debra Keenahan*

"We demand the right to be hot and sexy" (2000, p. 164), sociologist Tom Shakespeare states in his landmark essay on "disabled sexuality." The question can be asked—has this demand been met nearly two decades later? One group all-too-frequently excluded from consideration as desirable sexual partners are people with disability (Liberman 2018). Liberman came to this conclusion after reviewing political and medical ethical literature in the UK, appealing for sex rights for the disabled. The purpose of the review was to provide support for the legal/political argument that disabled people have a right to sex, but, because of social exclusion and thus limited opportunity to fulfil this right, disabled individuals should receive special exemption to the general prohibition against prostitution. Liberman (2018) describes the findings from extensive public surveys in the UK, showing significant resistance (70% of survey respondents) to considering that the aged, the obese and the disabled can in Stanford's (2005, p. 24) words, "get their kit off and get it on." Liberman's (2018) findings that the aged, obese and disabled are frequently excluded from consideration as desirable sexual partners is because these corporealities are not being readily equated with beauty. Stanford's (2005) study of commercial images in advertising and cinema yields similar findings of exclusion due to what he calls the "body fascism" of contemporary culture, which reinforces negative attitudes towards sex as being only for the able-bodied—or rather—"the young and the beautiful."

In the visual arts, the "aesthetic equation" between disability and the value judgement of beauty has been critically examined by a number of authors (e.g., Siebers 2005; 2010; Garland-Thomson 2012; Millett-Gallant 2010). This emerging area of study, referred to as disability aesthetics, critiqued the paucity of positive visual representations of disability but also the extent to which cultural critique is oblivious to the existence of any such positive visual representations. A key focus of disability aesthetics is to challenge these exclusionary trends in the visual arts and a critical aim of this chapter is to extend the concept of disability aesthetics to an examination of contemporary art works addressing the all too long-ignored subject of disability and sexuality.

An exhibition that overtly challenges this exclusionary trend of body fascism is *Intimate Encounters* by the photographer Belinda Mason. The purpose of this exhibition is to explore the intricate nuances between disability and sexuality. In this chapter, I will be discussing the representation of and responses to disability and sex in the exhibition *Intimate Encounters*. However, examinations of the impact of these images have virtually, without exception, focused upon the responses of the viewers to the exhibition (e.g., Murray and Jacobs 2010). In this chapter, I want to focus instead upon the process of the artist in relation to the experience of the subject of the art works. As such, I will be analysing the subjectivity of disability and sex in the visual arts by comparing the representation of self and sex in portraiture by another with the representation of self and sex in self-representations. It is essential at this point that I declare myself as an "insider" to this subject area in that I am a woman with achondroplasia dwarfism and I am also an artist. As such, I will be simultaneously the researcher and subject of the research. That is, through critical reflexivity I will provide insight into and analysis of Belinda Mason's process as she captures my portrait as the 41st image for inclusion in *Intimate Encounters*. I will then compare Mason's process and image of me with my self-representation of my sexuality. Through this comparative self-reflective analysis, I aim to address three interlinked questions:

- What is the representation of and public response to disability and sex in the exhibition *Intimate Encounters*?
- How does the representation of self by another differ from self-representation as a sexual being?
- How does the application of disability aesthetics, as currently conceptualised, represent those with dwarfism as sexual beings?

In addressing these questions, I will propose future directions for disability aesthetics in the visual arts to meet Tom Shakespeare's initial demand that people with disability "have the right to be hot and sexy." But prior to this artistic documentation and reflection, it is essential to provide conceptual definitions and the theoretical framework which form the basis for this analysis and critique of the representation of disability and sex in the visual arts.

## Defining and theoretically framing disability in the visual arts

Disability is a fundamental part of the human condition, which is complex, dynamic and multidimensional, but also contested (WHO 2011). As Garland-Thomson (2002; 2012) shows, impairment does not discriminate—it is inclusive of all groups and an inevitability with old age. As such, impairment is inherent in the human condition and the oppressive act of exclusion of people with disability from such a basic pleasure as sex has the potential to negatively impact every individual. But there is an important nuanced distinction between impairment and disability, which is critical to this chapter. Impairment refers to the physical/psychological/emotional condition/state, which has a negative effect upon the individual's behaviour/performance/life, whereas disability refers to the social imputations overlaid upon those conditions/states, which encumber the individual beyond the limitations of the impairment (Garland-Thomson 2012). In making this distinction between the physical condition of impairment and the social status of disability, this chapter adheres to the social model of disability, a key element of which is that disabled people are viewed as an oppressed group, with non-disabled people and organisations contributing to that oppression (Shakespeare 2013). As Shakespeare (1996; 1999; 2000) shows in his critical approach to the politics of disability, identity and sexuality, this social oppression can be vehemently expressed publicly but acutely felt and impactful in the most personal and private interpersonal relations and sense of self. In this chapter, my focus is upon the site of the most intimate of interactions being a critical example of the theoretical distinction between the physical condition of impairment and the social status of disability as illustrated through the condition of dwarfism.

Dwarfism is a readily recognisable physical condition. A person with dwarfism may be able to perform the full range of bodily movements and activities relative to the average population (such as sex) but their restricted height may limit the speed and agility with which they perform some movements and activities within a physical environment, relative to the average-sized individual (as such, the performance of sex isn't necessarily impaired because of dwarfism). However, if confronted with situations/institutions/relationships in which the dwarfed individual is excluded from equal participation and denied enjoyments and benefits that could be provided and that exclusion is on' the basis of their dwarfism, then the individual has become disabled. In other words, impairment refers to the corporeal difference and disability refers to the social status/identity of the corporeal difference. In this chapter, as I have dwarfism and I am both the researcher and subject of the research, I will be using this corporeality to illustrate this social dynamic of disability. Specifically, I am a disability theorist and an artist focused on researching the dynamic of being "seen" and, with regards to this chapter, that is specifically in the expression and representation of sexuality.

The desire for sex is no different for most people with disability as for most people without disability. As the opening lines by Tom Shakespeare (2013) indicate, but as he goes on to describe, what is of greatest import is intimacy, warmth, validation and connection. In other words, it is the qualitative dynamic of "being in relationship with" and "relating with" another, in which patterns of recognition occur with each other, that forms the basis of desire. Given the exclusionary practice of "body fascism" in cultural representations and language, such exclusion can have unintended and adverse consequences for those with disability about their identity as being sexual (Andrews et al. 2019). Therefore, a particular and alternative focus of this chapter is upon the subjectivity of the visual representation of disability (dwarfism) and sexuality.

## The art of being sexual in *Intimate Encounters*

It was the demand for the aged and those with disability to have their right to be sexual that was the inspiration for Belinda Mason's international exhibition *Intimate Encounters*. Since 1998, this exhibition has been shown across every metropolitan and key regional city in Australia (32 venues in 2001–2007). It has been exhibited internationally in nine cities across five countries (Canada, New Zealand, Spain, the UK, the US) and held in six institutions, including The Museum of Sex in New York and ShapeArts in London. This body of work currently contains 41 photographic portraits of people with a variety of disabilities in portrayals of their sexuality and intimate relationships. The portraits are diverse across disabilities and varied imagery, with viewers presented with scenes of fantasy figures (e.g., the Little Mermaid), open nakedness, suggestive sensuality and tender expressions of emotion. But despite the variety of images, there is a specific intention with each portrait that is best expressed in Mason's (2019) own words:

> The exhibition became a form of social activism as all participants chose to expose not only their bodies, but also their souls in order to demystify the perception that having a disability means you have no need or desire for sex, love, family, friends or intimacy. The commonality was the wish to be acknowledged as having a sexual identity and sexual expression.
>
> *(personal communication)*

That is, the subjects of these portraits were "demanding the right to be hot and sexy." Three works in *Intimate Encounters* are of people with dwarfism. Tom Shakespeare is one of those portraits, along with Margherita Coppolino. This exhibition originally consisted of 40 portraits. Another has been added in 2019—21 years after commencement of *Intimate Encounters*. The third portrait of a person with dwarfism is of me.

I will first present Mason's representations of others (Tom and Margherita), then describe her process in capturing my image for *Intimate Encounters*, followed by the account of my process and self-representations (Figure 16.1).

In this image, the subject is viewed from the perspective of an average height person. Tom's openly naked stance communicates a sense of comfort with his body and sexuality without declaration or suggestion of sexual inclination or proclivity. The starkness of the white background broken only by his shadow ensures the viewers' gaze focuses upon and is met by the post-production special effect of Tom's other portrait within his portrait. In this image, Tom is shown to be figuratively and literally "in his head," which is clear reference to his long and distinguished academic career, but also representative of his sense of self. As with all the portraits in the exhibition, "Tom" is captioned with a few of his own thoughts:

> I have worked as an academic, an activist and a performer, exploring ideas about disability and about genetics. Three generations of my family are affected by restricted growth. The condition doesn't usually prevent you doing things, but everyone stops to look at you. That's not always easy to deal with. Short people usually use humour to turn the tables, and we often appear very confident and extrovert. I feel very positive about myself and my sexuality. I am happy to be photographed naked. But despite all that I've achieved, there's still a vulnerable person lurking inside, with a tremendous need for love.

There is an additional brief descriptor of the subject's profession/affiliation or role, and concludes with a single sentence describing their disability (e.g., "Tom has achondroplasia").

*Figure 16.1* "Tom." Source: Image reproduced with permission from Belinda Mason and Tom Shakespeare.

*Figure 16.2* "Rennovating." Source: Image reproduced with permission from Belinda Mason and Margherita Coppolino.

Another portrait in *Intimate Encounters* of a person with dwarfism is of Margherita Coppolino. The work is titled "Rennovating" (Figure 16.2).

In this work, the subject stands face-on to the viewer, gazing out of the work with an expression that is simultaneously quizzical and challenging. Margherita is neither coy nor apologetic of her nakedness, nor of her dwarfism. She stands in the lower left of the frame and, though she is not central, her figure is larger in scale to the architectural renovation site that forms the background. This composition, as with "Tom," is a visual metaphor of Margherita's personal history and perspective on interpersonal relationships. It is the themes of relationship and intimacy, as indicated in Margherita's words, which recur in the majority of narratives attached to the works:

> I grew up in a children's home, I was used to people going in and out of my life on a regular basis. It wasn't until a person said to me, in my mid-twenties, "I am not going anywhere and you better get used to having me around" that they encouraged me to start letting people in and taught me how to start trusting and loving people. I was not good at doing this in the first instance as I was taken advantage of in many different ways when I was younger. By not letting people get close to me, I could not be hurt, abused and abandoned. Since that time, I have been in positions where I have allowed friends, lovers have relationships on a deeper level, into my life. I sometimes still get scared and hide, however, this happens less and less. The more okay I become with myself, the more I invite people to do the same with me.

Margherita has dwarfism

Public responses to the exhibition were varied in both strength and their positivity or negativity. Moreover, the focus of the negative responses differed distinctly between members of the viewing public with disability and those without disability. Interestingly, it was this alternative practice of providing a narrative by the subject, a brief biography and description of the subject's disability, which met the strongest criticism from members of the public with disability. In their account of curating *Intimate Encounters* for the Museum of Sex in New York, Elizabeth Mariko Murray and Sarah Helaine Jacobs (2010) describe a particularly public criticism distributed through social media which interpreted the narratives as a "strategy" targeting a non-disabled audience, which prioritised the medical model. Another criticism suggested the inclusion of these narratives was too directive and contextual, when the images should have stood on their own for the viewer to interpret. However, negative responses from the viewing public, were at the most extreme, acrimonious calls for the closure of the exhibition, to paternalistic expressions about the subjects' "exploitation."

In the 20-year retrospective documentary of *Intimate Encounters* by Dieter Knierim (2018), Margherita Coppolina describes an encounter at the opening of the exhibition. A "well-meaning" member of the public complained—"You can't show people with disabilities like that!" Such negative responses reflect Stanford's (2005) earlier account of public resistance to the disabled, aged and obese being able to "get their kit off and get it on." Similarly, Winterflood (2004), reporting public responses to the exhibition's opening in the Northern Territory at the Charles Darwin University (CDU) library, describes a member of the general public proclaiming "pornography is not art." Interestingly, this latter response to the sometimes graphic, unapologetic images of nudity predominated the reasons for criticism (Knierim 2018). However, without expressly referring to the subjects as having disability, such rhetorical criticism raises the equally rhetorical question as to whether such negative sentiments would be expressed if the images were of normative bodies (Knierim 2018).

In comparison to these criticisms, the majority of responses to the images themselves were unsurprisingly supportive. Patrons with a disability expressed delight and revelled in members of the disability community being represented in sexually positive imagery (Murray and Jacobs 2010). Moreover, the power of the lived experience of the subjects was a consistent comment from patrons—with every work giving voice to a unique perspective. On the other side of the globe, the positive comments were similar. As one member of the public wrote in the CDU library visitor's book, "I applaud anything that challenges our understanding and impressions of 'normality'…It's good to be taken out of your comfort zone and see life from a different perspective other than *Dolly* magazine" (Winterflood 2004, p. 2). Being "taken out of your comfort zone" is a statement that could equally describe the experience for some of the subjects of *Intimate Encounters*, as it certainly did for me.

## The process and product of my "intimate encounter"

Belinda Mason commenced work on the images for *Intimate Encounters* in 1998 and her process has remained consistent. Mason focuses on the subjectivity of the sitter and developing an interpersonal relationship of respect, which entails extensive interviews with the prospective subject. The goal for her images is to represent the story that the subject wants conveyed. It is because the aim of each portrait is to represent the lived experience of the subject that Mason describes each portrait as a collaboration because as she simply stated to me, "I'm interested in capturing people's perspective of themselves."

During our first face-to-face meeting, I asked myself—"What was it about the images in *Intimate Encounters* that made me feel 'safe' to do this?" I am not an exhibitionist, nor am I shy and retiring. But I have a fundamental belief that it is necessary to move out of one's comfort zone to challenge unfairness, inequality and oppression for social change to occur. I also have an unwavering belief in the power of the visual arts to be an instrument for such change. So, when I saw the essentially individualised humanity of Mason's work, I considered "my demand for the right to be hot and sexy" had a chance to be visually expressed in a manner respectful of me without fear of perverse emphasis upon real or imagined peccadillos. Hence, I found myself sat in a cafe with the photographer, sharing stories from my past, which were formative and sometimes critical incidences, that shaped my sense of self and sexuality.

In the beginning, there was much talking and no photographing, which is essential to Mason's process. But this was more than an information-gathering exercise because there was a development of mutuality. I asked, "Why do you do this work?" Her response was inarguably straightforward, "this (sex and sexuality) should be a part of everyone's life." It was this essential belief that set Mason on this decades-long course, when she realised, through her attendance as a photo journalist at a public forum about aged homes, that this basic human right was denied to too many, particularly the aged and disabled. Therefore, with *Intimate Encounters*, for Mason and those who chose to participate, "it was very clear from the outset that we were wanting to smash existing stereotypes." This pushback against stereotypes, with a focus on collaboration, is evident when viewing the exhibition. Because there is both a diversity of subjects and a diversity in their representation, a frequent comment from the viewing public is their initial thoughts that the images in the exhibition were captured by different artists. This diversity in representation speaks to the collaborative process in which the voices of the individual subjects are figuratively and literally being heard and seen in the variation of works. The images are not uniform because Mason is conscious of the need to avoid homogenising disability.

Once our mutual comfort level was achieved, developing, realising, capturing and producing the image was an incredibly quick process—achieved in a single sitting. However, prior to the actual shoot I was asked two direct and strategic questions by the photographer: Would I like to confront the audience? My immediate response—Yes! Did I want my portrait to have confronting nudity? My slightly hesitant response—no. The reason for my hesitation was based in concern that the alternative could breach employer protocols and, though these answers seem mutually exclusive, for me, the "confrontation" I envisaged at the time of answering would manifest through gaze. Moreover, I experience suggestion, seduction and subtle sensuality to be my desirable expression of sexuality, which I conveyed, in an early interview.

By the time of the "shoot," my comfort level was such that I viewed this experience as an opportunity to freely express myself. Mason's intentions are to be a conduit—in her own words:

> I enjoyed translating the emotions they (the subjects) expressed into a visual image…I was a blank canvas and they guided what was created. Having the participants be pro-active…was just awesome…I created a vehicle for them to use to send their message.
>
> *(Mason 2019, personal communication)*

Below is the image, which is mine and Belinda's collaboration (Figure 16.3).

The caption accompanying my portrait reads:

> For me, sexuality is an instinctive, expressive energy that is paradoxical because it is intensely personal and private but also overtly public. My personal desires I choose to share and fulfil selectively. But social mores, practices and pressures sometimes place strong restrictions on my options. I look different from the average woman and that has sometimes meant negative attitudes to my difference have led to attempts to

*Figure 16.3* "Basic Instinct." Source: Image reproduced with permission from Belinda Mason and Debra Keenahan.

> restrict me from fulfilling my desires, or unfettered perverse interest in my capacities. My personal, but also very public desire is to be respected as a human being (which inherently means I am desirous and desirable sexually). Because, as someone very near and dear once said, […] "we are all the same height lying down," to which I would add […] or sitting in a chair, wearing stilettos, slathering on oil, body painting, wrestling in mud, tickling with feathers…
>
> *(Debra Keenahan 2019)*

My portrait is also accompanied by a brief biography and the statement, "Debra has achondroplasia."

The title for my portrait was suggested by Mason, based on my opening statement and choice of outfit. At all times, the photographer prioritised my comfort. At no time was there any discussion about the representation of my dwarfism. The focus was always about the expression and representation of desire, desirability, sexuality and sensuality. After our familiarisation and the sharing of my stories, Mason described her impression of me, which she felt would be best communicated through an image that made direct eye contact with the viewer, with a subtle cheeky smile, and a relaxed demeanour. The chair was one of a number of options, which I chose due to what I consider to be its sensual shape. The stark background was chosen because, for exhibition, the intention is to present the image life-size, on a reflective background such that, as viewers approach the work, they will gradually see a reflection of themselves in my image. The reason for such presentation—it represents how my own work challenges people to reflect upon their own and society's treatment of others. There was minimal post-production and the image was completed and printed in less than an hour of the sitting, unlike the time taken in the process and product of my self-representation.

## The process and product of "being sexually me"

In setting the task of representing myself sexually, my immediate thought was to incorporate private symbolism through jewellery (wedding ring), colour (red) and costume (lingerie). Initial attempts with my husband behind the camera focused upon technicalities of lighting, whereas my focus was always upon the communication of sensuality, suggestion and mood. As a result of this confliction, these efforts resulted in images I detested and rejected because they "weren't me," despite the technical elements being in place. My daughter then asked the simple question, "what makes you feel sexy?" I immediately thought of my husband and realised that my focus upon the symbolism in the initial images was an effort to incorporate our relationship into the portrait, albeit metaphorically. With this realisation, I made a significant shift in composition, with my husband brought in front of the camera (Figure 16.4, Figure 16.5).

The power of these images, for me, is in communicating the warmth, safety, fun and sense of freedom I feel with my husband. The viewer is a voyeur as we are cocooned in our private, intimate world—we know each other, our secrets and desires—and it is that intimate knowledge that matters. His being clothed, compared with my being exposed, mirrors our public experience when we are together—my body being the focus for those who see us, whilst looking past him. It is in relation with others that I experience the ecstasy of acceptance and the devasta-

*Figure 16.4* "Sealed."

*Figure 16.5* "Enwrapped."

tion of rejection. This unceasing dynamic tension between others' acceptance/rejection impacts self-worth. My greatest desire is for reciprocated respect, hence my self-representation captures a relationship because the resilience to "demand my right to be hot and sexy" despite societal pressures, has developed from relationships of respect.

Immediate impressions of the image of me by another and the self-representations show clear similarities (subject and costume) and significant differences (composition, gaze of the subject, lighting, post-production effects). But as the subject of these works, I am comfortable and confident these images appropriately and accurately communicate me as "hot and sexy." In other words, I am responding to the aesthetic of these images with the impression they were intended to evoke. Moreover, the exhibition *Intimate Encounters* has received international recognition and praise, which is counter to expectations, given the predominant social and sexual exclusion of the disabled, aged and obese. To try to understand this contradiction between expectations of responses and actual outcomes as regards the aesthetic communicated, I will discuss these works as manifestations of disability aesthetics within the theoretical framework of Critical Disability Studies.

## Disability aesthetics during *Intimate Encounters*, "being sexually me"

Just as sex is rhetorically "only for the young and the beautiful," so too has aesthetics—the foundation of both the appreciation and criticism of art—been rhetorically synonymous with the visual arts and the valuation of beauty (Weitz 1956). Moreover, the body and its representation is inarguably the most enduring theme in the visual arts. In discussing the concept of embodiment and its relationship to aesthetics, Noland (2014, p. 469) states, "It could be claimed that all thinking about art begins with the Body," and the capacity of perception and feelings to be affected by art. Similarly, the representation of physical difference is evidenced continuously throughout art history, with the visual arts frequently used as an anthropological indicator of social attitudes towards and the social standing of those with disability (Kozma 2006). But just as sex is performed and enjoyed by many more than just the young and the beautiful, so too has aesthetics a much broader reference than just the valuation of beauty. Rather, contemporaneously, aesthetics has been reinvigorated as a critical concept applicable to a broad spectrum of disciplines and social phenomena (Bennett 2012). It is in the discipline of Disability Studies that this term receives a re-examination and remains highly contested.

The concept of disability aesthetics was introduced by Tobin Siebers in 2005 and expounded upon in his watershed work of 2010. His main argument was that disability presents an aesthetic, which requires recognition, appreciation and extrapolation, and challenges conventional aesthetics. According to Siebers, disability aesthetics consider physical and mental difference as inherently valuable and seeks to emphasise the presence of disability in the tradition of aesthetic representation which "embraces beauty that seems to be broken, and yet it is not less beautiful, but more so, as a result" (2010, p. 3). As such, disability aesthetics do not adhere to the notions of harmony, bodily integrity and health as standards of beauty. Rather, they embrace an appreciation of disability, such that in Siebers' (2010, p. 19) words, "the appearance of the disabled or wounded body signals the presence of the aesthetic itself." Therefore, from this alternative perspective, disability becomes more than just a subject or theme of art; nor is it merely a personal or autobiographical response embedded within an artwork, or simply a political act. Instead, disability aesthetics affirm disability as operating both as a conceptual framework, from which to critique aesthetic presuppositions in the history of art, and as a value in its own right. In being considered an aesthetic value in itself, Siebers (2010, p. 20) claims disability then actively participates in "a system of knowledge that provides materials for and increases critical consciousness about the way that some bodies make other bodies feel."

Siebers' call for a shift in the aesthetic valuation of disability is an erudite sociopolitical proclamation. However, his account employs metaphor as the method of describing disability aesthetics—that is, the perceptual experience of "looking into a broken mirror"— and, as such, does not provide a definition enabling clear meaning or understanding of the concept. Moreover, in framing disability aesthetics as representation which "embraces beauty that seems to be broken" (2010), he is adhering to the very same beautiful/ugly dialectic which he claims disability aesthetics disavows. Such disavowal is explicit in his claim that "Disability aesthetics refuses to recognise the representation of the healthy body—and its definition of harmony, integrity and beauty—as the sole determination of the aesthetic" (2010, p. 64). As such, within the framework of the social model of disability, disability aesthetics aims to critique how the visual representation of corporeal difference effectively contributes to the social discourse on disability. Furthermore, disability aesthetics adhere to the perspective that this contribution to the social discourse has been powerfully positive, especially through the development of modern art (Siebers 2010), but this assumption can be seriously questioned when considering the visual arts' representation of disabled people as sexual beings.

Anthropologists consider eroticism to be the significant feature that distinguishes the sexual activity of humans, and it has become a cultural form and a medium of human creativity (Fellmann and Walsh 2016). To consider eroticism a medium of creativity in intimate relations is to focus upon the emotional communication in which feelings are cultivated and expressed through bodily sensation (Giddens 1992). Such embodiment includes affect and feelings— reflective of the previous definition of aesthetics as "the way one body makes another body feel," with an ethical dynamic. Therefore, what is considered erotic, or sexy, is highly subjective, mapped within the cultural context, and aimed at achieving a mutual understanding of feelings and intentions of intimate relations (Fellman and Walsh 2016). The expression of sexuality and eroticism is described by Finger (1992, p. 9) as "often the source of our deepest oppression." Such oppression is evidenced in the exclusion of people with disability by many in the consideration of sexual partners. Therefore, if disability aesthetics is to adequately repudiate the all-too-ready conflation of disability and ugliness, the concept of disability aesthetics needs to disengage with the categorical imperatives of beauty and ugliness altogether because it is a debate which, in Weiss' words, is "logically doomed to failure" due to the subjective nature of the judgements of beauty and ugliness, and it could be argued, even the category of disability itself.

Aesthetics are neither neutral nor objective but rather they are a medium for the construction and maintenance of socio-political privilege and oppression. But more importantly, as argued by Rancière (2013), despite aesthetics being a system of social organisation and implicit bias toward white supremacy, patriarchy and heteronormativity, they are also the site for resistance to such oppression. For James (2013, p. 112), "any critical political approach to aesthetics must de-centre both the political *and the apparently apolitical philosophical norms* that have traditionally governed aesthetics." James goes on to suggest that one of the most straightforward approaches to such resistance is to "skew" the aesthetic through the artwork or artist's positioning in raced, hetero-gendered (and disability) systems of social organisation.

Such "skewing" is the approach exemplified in the respective processes of Belinda Mason and myself in our representations of disability and sex. That is, central to Mason's and my own processes was the subject "taking control of the gaze." As Tom Shakespeare mentions in the caption with his image "The condition doesn't usually prevent you doing things, but everyone stops to look at you." In the images for *Intimate Encounters* Mason captures what Hooks (2003) first referred to as the "oppositional gaze"— where agency is reclaimed by a traditionally passive marginalised group member returning the gaze from members of the hegemonic group.

In my self-representations, I do not adopt the oppositional gaze. However, by privileging myself (with dwarfism) as the subject, as in Mason's collaborative approach, I am "taking control of the gaze." In this way, these images re-frame stigmatised representations of dwarf subjects and thus aim to "encourage the viewer's compassionate involvement, as opposed to attracting a historically prevalent, morbid and reductive curiosity" (Cachia 2015, p. 1). Moreover, such positive self-image-making resists what Hevey (1992, p. 2) refers to as "internalised oppression" and imperialist, over-protectionist "hand-wringing" ("You can't show people with disabilities like that!"). In this political act of "taking control of the gaze," the visual arts can become a powerful weapon in the push for the empowerment of people with disability to "demand their right to be hot and sexy."

From a practice-based perspective, an appropriate representation of the transactional phenomenon of the sexuality of people with disability (i.e., intimate relations) is the capture of what Saito (2014) refers to as "social aesthetics" of interactions. That is, the character, composition and compilation of interpersonal relationships and social environments and the diverse ways in which people with disability manage these aspects of everyday life, of which sex and sexuality are integral parts. But the particular import of social aesthetics for the representation of disability and sex is recognition of the role and agency of the practitioner (artist/subject), because:

> We judge others' actions and also perform our own actions accordingly. Social aesthetics thus highlights the double roles we play in everyday life: as a spectator and a practitioner.
>
> *(Saito 2014, p. 526)*

In prioritising the voice of the subject in her collaborative process, both figuratively through imagery, and literally in the accompanying captions, Belinda Mason consciously resists traditional tropes of disability and sex in the visual arts. As she shared with me:

> This thought (that I was in control of the participants and what they did or said) was really quite insulting and paternalistic towards all the participants…I was told that the viewer needed to interpret the image for themselves…It (removing accompanying quotes by the subjects) is a way of silencing the participant and reinforcing old ideas… This personal text made the viewer accountable for their thoughts and feelings when they looked at the image.
>
> *(Mason 2019, personal communication)*

My self-representations of sexuality were realised when they captured the dynamic of being-in-relationship-with. Though this is a different dynamic to that realised in my portrait, "Basic Instinct," particularly in the employment of an alternative gaze, both processes are the same in prioritising the subject as "taking control of the gaze."

For too long in the visual arts, the sexuality, eroticism and desires of people with disability have been "missing in action." The images by Belinda Mason and my self-representations in this chapter, I argue, meet Tom Shakespeare's initial "demand for the right to be hot and sexy." Because, though these works were produced by separate processes, they were fundamentally similar in being subject (as human)-led rather than stereotypically (as other)-driven. I think it is appropriate that the rationale for this practice is presented in Mason's (2019) words:

> We cannot argue when someone says, "I feel"—it is not our right. It is part of our own journey to learn empathy, rather than compassion. Our own reaction exposes us to ourselves, and our ability to listen when someone lays their naked soul in our path.
>
> *(personal communication)*

In other words, we (people with disability/people with dwarfism) are hot and sexy. Visual arts practices that are disability-subject-led have a great capacity to represent us as such.

The judgement of the effectiveness of a practice that is focused on the subjectivity of the sitter, I want to leave to my husband. I could not be with him when he first viewed my image "Basic Instinct" for inclusion in the exhibition *Intimate Encounters*—he texted me his totally biased, subjective response…☺

# References

Andrews, E. Forber-Pratt, A., Mona, L., Lund, E., Pilarski, C., & Balter, R. (2019) #SaytheWord: A disability culture commentary on the erasure of "disability." *Rehabilitation Psychology 64*(2), pp. 111–118.

Bennett, J. (2012) *Practical aesthetics: Events, affect and art after 9/11.* London: IB Tauris.

Cachia, A. (2015) Composing dwarfism: Reframing short stature in contemporary photography. *Review of Disability Studies: An International Journal 10*(3–4), pp. 6–19.

Fellmann, F., & Walsh, R. (2016) From sexuality to eroticism: The making of the human mind. *Advances in Anthropology 6*(1), pp. 11–24.

Finger, A. (1992) Forbidden fruit. *New Internationalist 233*(9), pp. 8–10.

Garland-Thomson, R. (2002) Integrating disability, transforming feminist theory. *NWSA Journal 14*(3), pp. 1–32.

Garland-Thomson, R. (2012) The case for conserving disability. *Journal of Bioethical Inquiry 9*(3), pp. 339–355.

Giddens, A. (1992) *The transformation of intimacy: Sexuality, love and intimacy in modern societies.* Cambridge: Polity.

Hevey, D. (1992) *The Creatures Time Forgot*. London: Routledge.

Hooks, B. (2003) The oppositional gaze: Black female spectators. In A. Jones (ed.) *The feminism and visual cultural reader*. London: Routledge, pp. 94–105.

James, R. (2013) Oppression, privilege, & aesthetics: The use of the aesthetic in theories of race, gender, and sexuality, and the role of race, gender, and sexuality in philosophical aesthetics. *Philosophy Compass 8*(2), pp. 101–116.

Kozma, C. (2006) Dwarfs in ancient Egypt. *American Journal of Medical Genetics Part A 140*(4), pp. 303–311.

Liberman, A. (2018) Disability, sex rights and the scope of sexual exclusion. *Journal of Medical Ethics 44*(4), pp. 253–256.

Millett-Gallant, A. (2010) *The disabled body in contemporary art*. Basingstoke: Palgrave Macmillan.

Murray, E.M., & Jacobs, S.H. (2010) Revealing moments. In R. Sandell, J. Dodd, & R. Garland-Thomson (eds.) *Re-presenting disability: Activism and agency in the museum*. London: Routledge.

Noland, C. (2014) Body. In M. Kelly (ed.) *Encyclopaedia of aesthetics*. Oxford: Oxford University Press.

Rancière, J. (2013) *The politics of aesthetics*. Cornwall: Continuum.

Saito, Y. (2014) Everyday aesthetics. In M. Kelly (ed.) *Encyclopaedia of aesthetics*. Oxford: Oxford University Press.

Shakespeare, T. (1996) Power and prejudice: Issues of gender, sexuality and disability. In L. Barton (ed.) *Disability and society: Emerging issues and insights*. Essex: Addison, Wesley Longman, pp. 191–214.

Shakespeare, T. (1999) The sexual politics of disabled masculinity. *Sexuality and Disability 17*(1), pp. 53–64.

Shakespeare, T. (2000) Disabled sexuality: Toward rights and recognition. *Sexuality and Disability 18*(3), pp. 159–166.

Shakespeare, T. (2013) The social model of disability. In L.J. Davies (ed.) *The disability studies reader*. 4th edn. New York: Routledge, pp. 214–221.

Siebers, T. (2005) Disability aesthetics. *PMLA 120*(2), pp. 542–546.

Siebers, T. (2010) *Disability aesthetics*. Ann Arbor, Michigan: University of Michigan Press.

Stanford, P. (2005) Attitudes to sex and disability shame us. *Third Sector 380*, p. 24.

Weitz, M. (1956) The role of theory in aesthetics. *The Journal of Aesthetics and Art Criticism 15*(1), pp. 27–35.

Winterflood, J. (2004) School sex pix set off uproar. *Alice Spring News*, March 24, http://www.alicespringsnews.com.au/1108.html

World Health Organization (2011) *World report on disability 2011*.

## Digital and Media Resources

Belinda Mason Photography http://www.belindamason.com/

Knierim, D. (2018) *'Intimate Encounters' 20 Years On*. Sydney Film Festival.

# 17

# SEX, LOVE AND DISABILITY ON SCREEN

*Courtney Andree*

As the 2015 Oscar nominations were announced, critics were abuzz over Eddie Redmayne's portrayal of Stephen Hawking in the biopic *The Theory of Everything* (2014). Outlets ranging from *The New Yorker* to *The Times* (UK) and *Rolling Stone* latched onto a single minute-long scene from the film—a scene in which Hawking reveals to a Cambridge friend that his sexual organs have been unaffected by motor neurone disease. While the film shows Stephen lovingly cradling his children and exchanging banter and affection with his wife Jane (Felicity Jones), this brief sequence acts as one of the most explicit reminders that Hawking maintains sexual desire *and* sexual function after the onset of amyotrophic lateral sclerosis (ALS). Responses to *The Theory of Everything* reveal that the sexual lives of people with disabilities continue to remain under wraps in most mainstream films; at best, sexuality is tacitly and cursorily acknowledged, even as it goes unrepresented, as is the case here. Critics have invariably read the scene as light-hearted; *The Times*'s film critic Kate Muir considers it "a funny scene [where] Hawking explains his prolific reproduction to university friends" (Muir 2015, p. 7), whereas *Slate*'s Dana Stevens remarks that in the scene "Stephen impishly confesses to a male friend, not every part of his body is disabled" (Stevens 2014, n. pag.). While many considered the film to mark a watershed moment in disability representation, physician and *The Telegraph* columnist Max Pemberton was perplexed to find that, "[A]t the end of the film no one was discussing whether the laws governing quantum mechanics and relativity can ever be reconciled. Instead, it was how his disease affected his bedroom performance" (Pemberton 2015, n. pag.). The very fact that people with disabilities have a sexuality (the fact that they desire and act upon their desire) continues to jar, disarm and fascinate the public, however circumscribed these representations may be.

The scene in question begins on the steps of St. John's College, Cambridge, on the evening following Stephen's successful presentation to the leading scientists in his field. Hawking is accompanied by a group of intoxicated school cronies who navigate his wheelchair through a wayward and inaccessible course. The camera continues to track fluidly alongside them until they reach a tall and imposing stone staircase. As Stephen and his friends approach this obstacle in the built environment, the camera remains behind (stationary, for the moment), framing the men in an extreme long shot with heavy stonework and a massive bronze statue looming overhead. Brian (Harry Lloyd) remains behind with Stephen as the rest of the men run upstairs, and, after bending to pick his friend up, he and Stephen make their slow way to the top together, with Stephen's arm draped over his neck and his body cradled carefully in his friend's arms.

The pair are intimately framed in medium-close-up, with shallow focus, and, as the handheld camera follows their progress to the landing, Brian asks, "Stephen, your 'motor mouth' disease; does it affect…um…everything?" He pauses for a moment, breathing heavily from the exertion and looks down at Stephen, whose face is just inches from his own. With a wry smile, Hawking replies "What? No…different system. Automatic." After this oh-so-brief revelation, the men are able to continue their progress upstairs, and walk off screen as Brian remarks that it "certainly explains a lot about men." Stephen remains out of view as his friend retrieves his wheelchair, but the final shot of the sequence reveals that he has been placed in the outstretched arms of an enormous and grotesque Queen Victoria statue fabricated for the film—a visual gag that reinserts this scene (and Hawking's very serious revelation) into a comedic framework. This brief, yet intimate, sequence serves to assuage the audience's anxieties and curiosities about Hawking's sexual "abilities," even while making light of the pervasive stereotypes that continue to curtail the sexual lives of people with disabilities. Hawking's closest friend is given the license to ask the questions that are apparently on the tip of every tongue: Can Hawking get it up? Is he still having sex? Can people with disabilities have "healthy" sexualities?

While the brief conversation between Hawking and his friend inspired critics and viewers alike to reflect upon the complexities of the physicist's sexual life, this minute-long scene fits jaggedly into the fabric of a film that chronicles the romance of Stephen and Jane Hawking. Without providing more than a single passing reference to their sexual relationship, director James Marsh draws a portrait of Hawking as a kind and caring family man, and, although the arrival of their three children points to the fact that his and Jane's sexual relationship must have continued after the onset of his symptoms, the film offers virtually no evidence of this intimacy.

There is an admitted trend toward more open and frank discussions of sex and disability on screen in recent years, yet intimate acts are rarely presented in instances where one or both partners have disabilities. It is notable that most fiction films that place love, sex and disability at their centre continue to cast able-bodied actors in leading roles; unsurprisingly, this has tended to dictate the types of bodies and disabilities represented on screen. Historically, mainstream film has favoured the representation of sensory disabilities (e.g., blindness and deafness), cognitive disabilities, mental illness and paralysis, those disabilities that preserve the supposed integrity and "wholeness" of the body for potential cure.

This chapter will trace the complex history of disability and sexuality on screen—from more recent blockbusters to eugenics and exploitation films of the 1920s and 30s, which posed that the desires and sexual activities of people with disabilities imperilled the nation, and women's pictures of the mid-century, which fetishised the sexual violation of women with disabilities and connected love to cure. In order to better understand how bodies and differences have been visualised over the past century, I draw upon recent scholarship on disability and film and ground this discussion in considerations of film industry practices and representational strategies, including de facto and official forms of censorship; shifting legal and medical outlooks on disability, particularly in conjunction with eugenic ideology; and overlapping social histories of sexuality, gender and race.

## Historicising sex and disability on- and off-screen

In the past couple of decades, discussions of the sexual rights of people with physical, cognitive and emotional disabilities have finally begun to reach mainstream media in Britain and the US, inspired in no small part by independent films and major studio releases like *The Theory of Everything, Pumpkin* (2002), and *The Other Sister* (1999)—and for an earlier generation of filmgoers, innovative films like *Coming Home* (1978) and *Children of a Lesser God* (1986). Nonetheless, as Margrit Shildrick acknowledges, "Despite increasing portrayals of disability in the media in the interest of diversity, relatively few of these provide positive representations of disabled people in a sexual context…[S]uch depictions more often invoke an implicit pity stemming from the notion that disability signals that sexuality has either been lost or never experienced" (Shildrick 2012, p. 66). Even films that seek to chronicle true stories of overcoming social boundaries and biases, like *My Left Foot* (1988), starring Daniel Day-Lewis, or *Born on the Fourth of July* (1989), featuring Tom Cruise, record social alienation, continual romantic rejection, and restricted sexual privacy and autonomy. By prioritising scenes of sexual "failure" and humiliation over representations of sexual and romantic fulfilment, these films circulate negative disability stereotypes—dangerous stereotypes that continue to shape public discourse around disability rights and sexual access, perpetuating the myth that people with disabilities are undesiring, undeserving of love or sexually dysfunctional.

Disability rights activists have begun to advocate more aggressively for sexual rights, and scholarship on disability and sexuality has grown apace. As Tom Shakespeare, Kath Gillespie-Sells and Dominic Davies suggest in their ground-breaking study, *The Sexual Politics of Disability: Untold Desires* (1996, p. 3), a "medical tragedy model" dominates many discussions of disability and sexuality. A unique double-bind exists, as disability stereotypes result in the over- or under-sexualisation of people with disabilities—stereotypes that, by and large, extend throughout the history of mainstream cinema. These stereotypes influence how disability is perceived, problematised and consumed in the public sphere, resulting in what Shildrick has called a "strange paradox," which "alternates between denying that sexual pleasure has any place in the lives of disabled people, and fetishising it" (2012, p. 60). If sex is taboo, reproduction is even more contentious for people with disabilities. Tobin Siebers argues that, "It is reproduction, then, that marks sexuality as a privileged index of human ability…They will not reproduce, but if they do so, the expectation is that the results will be tainted" (Siebers 2012, p. 41). Unsurprisingly, parenthood is rarely presented as a viable option for disabled characters in mainstream film, particularly if they are congenitally or cognitively disabled.

Whereas more and more interdisciplinary scholarship on film, media and disability has begun to emerge in recent years in important works like Julie Passanante Elman's *Chronic Youth: Disability, Sexuality, and U.S. Media Cultures of Rehabilitation* (2014), Angela M. Smith's *Hideous Progeny: Disability, Eugenics, and Classic Horror Cinema* (2012), Sally Chivers's *The Silvering Screen: Old Age and Disability in Cinema* (2011), and expansive collections like *The Problem Body: Projecting Disability on Film* (2015) and *Different Bodies: Essays on Disability in Film and Television* (2013), sexuality has remained a tangential concern in most studies of disability and mainstream film. Martin F. Norden's brief discussion of the sexual stereotypes that govern representations of disability on screen in *The Cinema of Isolation: A History of Physical Disabilities in the Movies*, still represents one of the most cogent attempts to come to terms with Hollywood's representations of disabled bodies and sexualities to date. As Norden argues, although the film industry "seldom if ever constructed physically disabled people as sexual beings in its productions' surface stories, it exhibited a tremendous concern for sexual differences among such characters" (Norden 1994, p. 315). Historically, mainstream cinema has tended to represent women with disabilities as "asexual," "childlike objects requiring domination," who "by virtue of their femaleness and disabled status appear doubly disempowered" (1994, p. 316). Disabled women have been alternately depicted either as the asexual "Sweet Innocent" that Norden describes, as potential victims of male abuse and aggression, or as patients in need of a loving physician's intervention and cure.

Historic representations of male sexuality tend to fall into two general categories. From the First World War onward, a subset of American and British films depicted the return of the "broken" soldier from the battlefield, and his subsequent struggles to find love and acceptance. With the bodily basis of his masculinity in question and his romantic and sexual confidence shaken, the disabled male war veteran returns home to find the landscapes of home, family, work and love dramatically altered. As Paul Longmore recognises, "Even when disability does not limit sexual functioning" for the disabled veteran, "it may impair the person emotionally. Disabled characters may be quite capable of physical love making, but spurn opportunities for romance because of a lack of self-acceptance…" (Longmore 1987, p. 73). Fact and fiction converge in *The Best Years of Our Lives* (1946) as recently-disabled US Army veteran Harold Russell takes on the role of Homer Parrish, a Navy recruit who loses both of his hands after his ship catches fire. Parrish is outfitted with a pair of split hooks, trained by the Navy hospital on how to use them, and eventually dispatched home to rebuild his life. It seems significant that his romantic anxieties and triumphs bookend the film. As Homer embarks on his journey home, he attempts to remain upbeat and carefree about the loss of his hands, but when conversation turns to the topic of his high school sweetheart, Wilma (Cathy O'Donnell), his anxiety and self-pity become palpable. As Parrish remarks, she's a "kid," "She's never seen anything like these hooks."

A brief sequence midway through the film attracted much attention from critics and viewers alike. When Homer learns that Wilma's family intends to send her away in the hopes that she will forget about him and move on, he decides to make her fully aware of what marriage with him would entail. After telling her that he doesn't want her to be "tied down, just because [she has] a kind heart," he asks her to come upstairs with him to "see for yourself what happens" at night. Wilma slowly and cautiously follows him upstairs—not quite knowing what to expect as he begins to disrobe. Instead of seeking intimacy, Homer takes off his hooks and proceeds to tell her how helpless he is: "I'm as dependent as a baby who doesn't know how to get anything except to cry for it." Unlike actor Harold Russell, who could put his hooks on and take them off without assistance, his character is shown to be fully dependent upon his family's care. In spite of Homer's attempts to scare Wilma away, she passionately kisses him and vows to honour her commitment. Rather than taking their intimacy any further, the scene ends with her tucking him in like a child as we fade to black. The fact remained that if *all* sex was taboo onscreen in mid-century Hollywood, the potential sexual union of disabled characters was even more objectionable. The end of the bedroom scene in *The Best Years of Our Lives* serves to contain Homer's desires and constructs Wilma as perpetual caregiver—as the person who will tuck him in at night from now on, as a woman who must make sacrifices to repair home and nation. As Sonya Michel argues, post-war films like *The Best Years of Our Lives,* "showed that veterans needed both sexual and maternal attention…ministrations by wives and sweethearts who knew how to balance loyalty, deference and support with a discreet sensuality" (Michel 1992, p. 114).

Images of returning veterans' supposed sexual, romantic and emotional incapacity differ sharply from other representations of excessive, queer and "deviant" disabled male sexuality in film noir, the horror genre and drama films like *Of Mice and Men* (1939). In his discussion of disability and film noir, Michael Davidson argues that, "In numerous noir films, a physical or cognitive disability marks a sexual inscrutability, otherwise unspeakable" (Davidson 2011, p. 61), representing a potential "crossing of medical and sexual closets during the early cold war" (77). In films noir like *The Lady from Shanghai* (1948) and *Double Indemnity* (1944), the continued presence of the femme fatale's physically disabled husband complicates her unfolding relationship with the noir hero—creating an imbalanced love triangle.

The horror film is another important site of consideration. As Angela Smith contends, "Classic horror films kept eugenic dramas in the public eye and encouraged viewers to look at both disabled bodies and the ugliness of eugenic rhetoric and practice" (Smith 2012, p. 22). In innumerable films from the 1930s onward, "the monster's incursion into the bedrooms of young, marriageable women intimates the sexual nature of his threat, while the use of bodily signifiers …associates this disruption of normative procreation with physical degeneracy" (2012, p. 38). Like the eugenics movement itself, classic horror films like *Dracula* (1931), *Frankenstein* (1931) and *The Creature from the Black Lagoon* (1954) were constructed around the premise that sexually and physically aberrant bodies must be vanquished or neutralised—figuring the monster's threat in reproductive terms.

The de facto prohibition against the screening of sex and disability seems surprising given the fact that, after the breakdown of the Hollywood Production Code in the 1960s, there was an explosion of interest in displaying sex on screen—what Linda Williams has characterised as "a much larger proliferation of all manners of screening sex, from chaste kisses to the most graphic and frenetic of penetrations" (Williams 2008, p. 7).[1] As Williams notes, it was not just society's outcasts, "the lower classes, the unmarried, the criminal, the homosexual, or the colored"—and by extension, the disabled—"whose sexual contacts were made taboo by the Code, but also those of married reproductive heterosexuals whose pregnancies, births and sexual relations also became unrepresentable" (2008, p. 34). The genres that most frequently presented disabled bodies were particularly subject to censorship's demands—including eugenics or venereal disease films, films recording medical scenes, the horror genre, and exploitation films, which hoped to skirt censorship altogether. The ways in which the larger histories of eugenics, sexual hygiene, censorship and war influenced disability's representation in mainstream cinema will now be explored, through case studies of the exploitation and sex hygiene film *Tomorrow's Children* (1934) and the women's melodrama *Johnny Belinda* (1948).

## Exploiting sex and disability in *Tomorrow's Children* (1934)

In an internal memo circulated in May 1934, Production Code Administration consultant J. B. Lewis reported back to office head Joe Breen on a film that he'd been asked to investigate. In the opening of his screening summary for Bryan Foy's *Tomorrow's Children*, Lewis suggests that the film goes most astray when it "depicts gruesome scenes of feeble-mindedness, if not actual insanity, two operating table sequences and deals with a subject that is highly controversial"—namely, enforced sterilisation (Inter-office Memo Re: Tomorrow's Children 1934, n. pag.). Whereas a 1930 addendum to the Code would specify that "sex hygiene and venereal diseases" were both unfit subjects for motion pictures, as were "surgical operations," it is striking that Lewis's strongest objections to the film concern the visual spectacle of disability—the "gruesome scenes of disability," "deformity" and "insanity" that he proceeds to describe in detail in his report ("Code Addenda," reprinted in Doherty 1999, p. 363). The representation of disability was not officially codified as grounds for deletion or banning in the US, but it is notable that other films with a disability focus would similarly fall prey to state censors in the pre-Code era—perhaps none more so than Tod Browning's *Freaks* (1932), which helped to bolster campaigns for federal censorship.

At the time that *Tomorrow's Children* was released, a short-lived exploitation boom was just winding down; over thirty new exploitation features had been produced and distributed in the US over the period from 1931 to 1934, offering up sex education, drug use and nudity to eager audiences (Schaefer 2001, p. 150). Thomas Doherty has characterised this period as a "fascinating and anomalous passage in American motion picture history," a period where "Code commandments were violated with impunity and inventiveness," laying bare what would be "concealed, subterranean and repressed…under the code" (Doherty 1999, p. 2). Although *Tomorrow's Children* is concerned with reproduction, heredity and negative eugenics, it also attempts to capitalise on the furore over domestic and international sterilisation programmes. The film's opening preface features an inflammatory statement from the Human Betterment

Foundation in Pasadena, in which they recommend mass sterilisation as a "means of preventing reproduction by the insane and criminal classes," and put forth a proposal for the sterilisation of eighteen million "unfit" Americans. This proposal is countered by another statement, signed by Father O'Dwyer of the Catholic Diocese of Los Angeles, who attacks sterilisation as a "false scientific theory." Although these opening titles make note of the fact that sterilisation (whether mandatory or voluntary) was the law of the land in 27 states and spends screen time airing the views of both sides on the issue, the film's runner also draws an uneasy comparison between the situation in the US and the practices of the Nazis as it closely frames a newspaper headline reading "Newsflash! Hitler Decrees All Unfit to be Sterilised."

Even as *Tomorrow's Children* was critical of sterilisation practices in the US and abroad, it relied upon the spectacle of "real" disabled bodies for narrative coherence, depicting disability as visual evidence of sexuality and reproduction gone awry, of poor partnerships being made. As Annette Kuhn recognises, VD, eugenics and sex hygiene pictures had a long history of presenting disabled bodies and actual medical patients (Kuhn 1988, p. 65). *Tomorrow's Children* built on the examples of infamous eugenics film *Are You Fit to Marry?* (1927) and VD films like *Damaged Goods* (1933) and *Open Your Eyes* (1919), which featured conspicuous and clinical displays of a range of disabilities and diseases, a representational practice that fuelled calls for censorship and inspired negative critical responses. In *Are You Fit to Marry?* (the 1927 re-release of *The Black Stork* [1917]), eugenicist and doctor Horace Dickey (Dr Harry Haiselden) hopes to derail the dysgenic union of Claude Leffingwell (Hamilton Revelle) and Anne Schultz (Elsie Esmond). In order to better convince them (and, in turn, the audience) that they shouldn't enter into marriage without first taking proper stock of their medical and family histories, Dr Dickey showcases a gallery of patients who have resulted from other eugenically "imprudent" marriages. Stark and closely framed views of a crying baby with microcephaly are presented, before the child is passed off screen without much ado. Next, a young boy with Down syndrome is pulled in front of the visitors, staring directly into the camera in medium shot. Then, Dr Dickey's next subject is quickly brought into the frame—a pleasant-looking girl in a simple white dress, who smiles and laughs directly at the viewer in single-shot. Martin Pernick argues that it is possible to read these images against the grain—rather than producing only horror and disgust, as its eugenicist filmmaker intended, when real patients are introduced, "their brave hopes and shy smiles at the camera convey an immediately appealing humanity" (Pernick 1999, p. 145).

*Tomorrow's Children* makes the abstract problems of disease and dysgenic matches more personal as we follow the story of Alice Mason (Diane Sinclair), who is brought to the brink of sterilisation after her family is reported to the local welfare board. Initial views of the family reveal that each family member, with the sole exception of Alice, is a perfect "degenerate" type: we have our alcoholic father; overly fecund mother, who brings miserable life after miserable life into the world; criminal older brother; physically disabled younger brother; an emotionally disabled boy; and a cognitively disabled infant. After Alice's mother delivers a stillborn baby, the well-intentioned but misguided Dr Brooks (Crane Wilbur) attempts to intervene, wanting to improve Alice's situation by relieving some of the burdens that have been placed on her as their sole means of support. The local welfare board pays the family a visit and tells them they will only offer them relief if Mr. and Mrs. Mason and Alice undergo sterilisation, and if they place all of the remaining children in institutions. A complicating factor is introduced when it is revealed that the apparently productive and "normal" Alice is not actually the biological child of the Masons. Her efforts to evade sterilisation dominate the remainder of the film, taking her before the corrupt courts of the land, and into the heart of the hospital ward—a locked ward filled with the criminally insane, the cognitively disabled and the microcephalic. *Tomorrow's Children* finally ends with Alice's last-minute reprieve on the operating table, after she has already been prepped for surgery and the surgeon is on the verge of cutting into her bare abdomen with his scalpel, a reprieve that the other inmates of the hospital will not be lucky enough to receive. It is remarkable that even a nominally anti-sterilisation picture like *Tomorrow's Children* would uphold eugenic distinctions between "fit" and "unfit." The younger cognitively, physically and emotionally disabled Mason children are *not* rescued from surgeon's blade and the institution's walls, while their conspicuously healthy sister is saved for romance, marriage and parenthood.

## Fetishizing female disability and victimhood in *Johnny Belinda* (1948)

While eugenics and sex hygiene films deployed images of disability in their attempts to educate the public about the potential outcomes of prudent and imprudent sexual unions and frequently presented disabled bodies in the context of the hospital ward or institution, doctors were most often the vehicle for medical and romantic "cures" in the women's picture and melodrama. In *Johnny Belinda,* Jane Wyman takes on the role of Belinda MacDonald, a girl who is deaf and mute from birth, cut off from education and sign language, and insulated from virtually any interaction with the Cape Breton community outside of her family's farm. Dr Robert Richardson (Lew Ayres), who comes to the island from the outside world, discovers Belinda and begins to teach her sign language. The intelligent Belinda thrives under his tutelage and the seeds of affection are planted. Just as she is beginning to feel comfortable communicating with her family and starts to attend village events, she attracts the notice of village ne'er-do-well Locky McCormick (Stephen McNally), who follows her home after the town dance and brutally rapes her. The truth of the attack is only revealed after Dr Richardson discovers that she is with child—a child that the village eventually seeks to remove from Belinda's care, after they voice their eugenically-inflected concerns about her "fitness" as a parent. The village decides to give her child to Locky and his new wife Stella (not realizing that he is the biological father) and the pair goes to Belinda's farm to claim the baby, at which point a struggle ensues and Belinda shoots Locky to prevent him from taking her son. With Dr Richardson's love and intervention and Stella's last-minute confession, Belinda is reprieved and empowered to marry. Whereas the doctor promises Belinda a life, a home and his love, it is notable that we don't see this life—we only see Belinda, baby Johnny, her aunt, and the doctor riding away from the trial together.

At the time *Johnny Belinda* was produced, Hollywood's Production Code Administration (PCA) had very clear strictures in place to prevent the representation of rape on screen. The Code specified that rape should "never be introduced as subject matter *unless* absolutely essential to the plot," and, when essential, it "must not be more than *suggested*" as "even the struggles preceding rape should not be shown" ("Motion Picture Code of 1930," reprinted in Doherty 1999, p. 354). When Joseph Breen first reviewed the original play by Elmer Harris in 1940, he was incensed by its representations of disability, rape, pregnancy and illegitimacy. Calling it both "morbid" and "sordid" in a letter to RKO studio head J. R. McDonough, Breen suggests that "Normal people will undoubtedly experience a feeling of revulsion as to any suggestion of the raping of a deaf-mute girl," a feeling compounded by her subsequent pregnancy, the illegitimate child produced, and the murder that transpires (Letter from Joe Breen to J. R. McDonough 1940, n.pag.). By the time that the PCA office advised Warner Bros. on the script for the film in 1947, Breen had changed his tune, convinced that the story of *Johnny Belinda* could be told within the bounds of mainstream film. He reminded them that they needed to tread carefully and recommended that the prelude to Locky's attack be cut down (Letter from Joe Breen to J.L. Warner 1947, n.pag.). Although Warner Bros. would follow Breen's general recommendations, the rape scene that resulted—and which was granted approval—was one of the most frank and suggestive up to that point under the Code.

It seems curious that *Johnny Belinda,* a film that deals with the sexual violation of a young woman with disabilities, was the test case that loosened requirements for future films treating sexual abuse. Although women with disabilities are frequently desexualised in the public sphere, the fact remains that they are disproportionately subject to sexual abuse, and sexualised images of disabled women's violation have become common in film and media. As John Schuchman traces in *Hollywood Speaks* (1988), the staging of rape, abuse and murder of deaf and blind characters became almost de rigueur over the second half of the twentieth century. To list just a few examples: *The Story of Esther Costello* (1957) depicts the rape of a young deaf-blind girl; in *The Curse of the Werewolf* (1960), a deaf servant girl is raped by a beggar; in *Eyes of a Stranger* (1981), a deaf, blind and mute woman falls prey to a serial rapist and murderer; and in *Wait Until Dark* (1967), Audrey Hepburn plays a victimised blind woman. It is evident that stereotyped images of disabled female victimhood, passivity, dependency and vulnerability sell. As Margrit Shildrick argues, there is a "highly evident strand of voyeurism—and in some cases identification—that spills over into a fetishistic focus on disabled bodies precisely as sexual" (2012, p. 96).

The film's scenes of sexual violence operated through suggestion and the build-up of tension and fear, as Locky grabs Belinda, kisses her roughly and continues to chase her through the mill as she vainly attempts to escape. After he finally overpowers her, she is obscured by shadows, and the scene ends with a fade out. While the film was widely billed as a women's picture and actively promoted in women's magazines, Warner Brothers developed an advertising campaign calculated to draw in male viewers. Focused on the crucial moment in which Locky overpowers Belinda, the ads almost invariably depicted a prone Belinda, her body lithe and partially obscured in shadows, as Locky looms over her (see Fig. 17.1). The tagline reads, "There was temptation in her helpless silence…and then torment." As *Showmen's Trade Review* reported at the time of the film's release, this campaign was "calculated to reach 40,689,397 men," striving to convey "all possible romantic impact" ("Leading Exploitation" 1948, p. 106). Seeking to capture the post-war male viewer with images of the lovely and vulnerable female victim and suggestions of her "helpless silence," the studio dealt in rape fantasies, myths of masculine power and stereotypes of disabled women's weakness.

## Narrating disability, sex and love

Moving beyond a discussion of the problematic sites and particular genres where disability and sexuality have most consistently appeared across the past century, it is equally important to consider how disabled body minds have been narratively and visually constructed. Sharon L. Snyder and David T. Mitchell suggest that we must remember that film was "born during the classical eugenics era," and, as such, "that era's film grammar assumes that an audience will be automatically repulsed and riveted by the display of *any* disability on-screen" (Snyder and Mitchell 2010, p. 194). Film not only "produces interest in its objects through the promise of providing bodily differences as an exotic spectacle," it "promotes its status as a desirable cultural product largely through its willingness to recirculate bodies typically concealed from view" (2010, p. 181). In many disability-focused films, we oscillate from moments of pure spectacle (moments that don't necessarily further the plot), in which the disabled body is displayed for the viewer's titillation and amazement, to images of loss, rejection and humiliation. Tod Browning's *Freaks* (1932) can be seen as a useful case in point. Unlike most films of that era or the present day, which frequently feature able-bodied actors in disabled roles, *Freaks* made use of real performers with disabilities. We are presented with moments of spectacle that verge on the carnivalesque—including closely and starkly framed shots of the "freaks," as their oddities and talents are narrated by the circus barker, and ordinary domestic and romantic scenes that are designed to amaze. Backstage, the viewer is offered access to the engagement and marriage of

*Figure 17.1  Johnny Belinda* advertisement from *Modern Screen*'s November 1948 issue, p. 9. Source: Courtesy of the Media History Digital Library.

the conjoined Hilton sisters; views of the bearded lady's (Olga Roderick) new-born daughter; and tales of the romantic exploits of the Bird Girl (Elizabeth Green). The story of the romantic entanglement of the dwarf Hans (Harry Earles) with the beautiful and diabolical "big woman" Cleopatra (Olga Baclanova) overlays the body-based spectacle that dominates the remainder of the film, and offers a record of the able-bodied woman's manipulation, betrayal and attempted murder of Hans, as well as his unrequited love and the infamous steps he takes to obtain revenge.

While a classic love plot progresses through the stages of flirtation, dating, marriage and consummation, protagonists with disabilities often have very different narrative arcs. In film's first decades, characters with disabilities were most frequently "cured" and reabsorbed into the social fabric, or killed off, leaving the scene with desires unfulfilled and the threat of potential reproduction shut down. For instance, the British charity appeal film *If Thou Wert Blind* (1917) stages the restoration of blind sculptor Haydn Strong to both sight and love, while *Siege* (1925) tells the story of its deaf protagonist's thwarted romantic desires, humiliation, depression and ultimate decision to take his own life. Martin Norden has observed that, of the approximately 430 features to focus on disability between 1912 and 1930, about "35 percent had their disabled characters eventually gain or regain the use of their sight, hearing, legs, etc." through miracle cures, heavenly favour, or medical intervention (1994, p. 58). In the years following the First World War, this trend was even more pronounced, with almost half of the features produced staging the cure and social and romantic reintegration of disabled characters (1994, p. 59). Films of this era almost invariably communicated that love, sex, marriage and reproduction were off limits for people with disabilities. To be entitled to the full rewards of heterosexual courtship, they must be "cured"—an impossible demand for people with permanent impairments.

Curiously, even more recent disability features have preferred to leave sex, marriage and long-term commitment unrepresented. *My Left Foot*, which is based on the life story of Christy Brown, who lived with cerebral palsy, openly represents his struggles to achieve romantic attention and social parity and dwells on his repeated rejections by Sheila (Alison Whelan) in his youth and Dr Eileen Cole (Fiona Shaw) in his adult life. The film omits Brown's long-term love affair with American Beth Moore and condenses his tumultuous relationship with his future wife Mary Carr (Ruth McCabe) into a brief, idyllic scene on a hillside overlooking Dublin, and a final title that announces that "Christy Brown married Mary Carr on the 5th October, 1972." *The Sessions* (2012) similarly opts to devote only a few seconds of screen time to the romance that would dominate the last eleven years of protagonist Mark O'Brien's life. Showing only his initial meeting with Susan Fernbach (Robin Weigert), the film abruptly cuts to O'Brien's funeral and reveals that he had finally found the committed, loving and sexually fulfilling relationship that he so desired. It seems curious that, even when a disabled individual's biography records romantic success and sustained love, biopics devote most of their screen time to recording the protagonist's sexual struggles and misadventures—giving only a cursory mention of long-term relationships, sexual fulfilment and happy (or unhappy) marriages.

If the love plot has been narrated in formulaic ways for decades on screen, so too has the sex act. In *Screening Sex*, film theorist Linda Williams recognises that there's a given order of events for the modern sex scene: the "usual temporality of a sex act that would progress through…a modernist 'jabbing, thrusting eroticism,' and predictably end 'with male orgasm presumed to signal the end of the female's pleasure as well'" (2008, p. 173). The conventional Hollywood bedroom scene is meant to appear spontaneous and unpremeditated, it is staged in a private setting, and, after variable amounts of foreplay, the main course is almost always the heterosexual, "jabbing, thrusting," penetrative sex act. It is important to acknowledge that sites of eroticism can shift when disability is a factor and the temporality of the sex act can also be very different—on and off the screen. As Tobin Siebers suggests,

> Because disabled people sometimes require advanced planning to have sex, as their sexual activity tends to be embedded in thinking about the day, not partitioned as a separate event. Among disabled people, the so-called sex act does not always qualify as an action or performance possessing distinct phases such as beginning, middle and end.
>
> *(2012, p. 49)*

Moreover, when individuals live in institutional settings or make use of the services of personal attendants, it can be difficult to find private spaces for sexual expression (with or without a partner), and challenging to self-direct how, when and where sex can play out.

For people with intellectual disabilities, it can be even more difficult to obtain reliable information about sex—particularly since some schools do not offer equal access to sexual education classes for students with disabilities, and parents do not always consider their children to be "adult" enough for sex. Recent productions, like *The Other Sister* (1999), the short film *Yolk* (2007) and the documentary *Monica and David* (2009), represent characters' efforts to educate themselves about the sex act, contraception and the possibilities of parenthood. Protagonists seek answers from manuals like *The Joy of Sex*, nearby sexual health clinics and their friends even when their families have attempted to disavow their desires. When sex finally happens in *The Other Sister*, it is on Carla (Juliette Lewis) and Daniel's (Giovanni Ribisi) terms, after months of careful planning and growing intimacy. Michael Gill argues that, "Even with the passion and spontaneity missing, it is remarkable that the film respects the sexual lives and choices of the characters" (2015, 161). Nonetheless, it is notable that, when Carla and Daniel finally take the plunge, the camera doesn't follow them into the bedroom; it remains in Carla's living room, tightly focused on her fishbowl as marching band music suggestively swells.

Recent films like *The Other Sister*, *The Theory of Everything* and *The Sessions* reveal that there have been real efforts to represent the sexual exclusion, stereotyping and discrimination that people with disabilities face and to initiate more frank conversations about disability and sexuality. These films offer the able-bodied public the much-needed reminder that people with disabilities desire and act upon their desires. Nevertheless, a majority of mainstream directors have preferred to keep sex and the naked, disabled body off screen—making use of the power of suggestion, discussions of pleasures unseen and final title sequences to describe relationships that last decades. In many ways, these representational strategies seem more in line with practices developed to cope with and to contain sex during cinema's first decades, operating through elision, misdirection and omission. This reticence surrounding sex and disability is curious, since the industry as a whole has continued to become increasingly permissive and open from the sexual revolution onward—a period that coincides with the breakdown of censorship and the Hollywood Production Code and the ascendancy of the adult film.

Unsurprisingly, the mainstream film industry continues to under- and misrepresent disability as a whole—with fewer than 2 percent of characters in blockbusters having a sensory, physical or cognitive impairment. Within this small subset, 73 percent of characters with disabilities are male, and LGBTQI characters with disabilities are virtually unrepresented in fiction film (Smith 2019, p. 2.). Hollywood has also continued the practice of almost always casting able-bodied actors in these roles and favouring plots focused on disabilities that preserve the "beauty" and integrity of the body. Even more troubling is the fact that many of the gendered disability stereotypes that were forged in cinema's golden age persist. Although there are some notable exceptions, like *The Shape of Water* (2017) and *Rust and Bone* (2012), which show women with disabilities taking control of their lives and desires, mainstream film still frequently depicts women with disabilities as vulnerable, sexually inexperienced and physically weak, diminishing women's agency and feeding fantasies of disabled women's victimhood. This is a damaging trend, given the fact that 83% of women with disabilities experience sexual abuse during their lifetimes (Dolmage 2017, p. 158). Men with disabilities continue to be represented as either "undersexed" or asexual, as in *The Theory of Everything,* as sexually and emotionally dysfunctional, as in *The Sessions* and *My Left Foot* or as figures in need of technological restoration before they can find love and sexual fulfilment, as in films like *Avatar* (2009) or *Strange Days* (1995).

More complex representations of disability have come out of new online series directed by and for the disability community, in alternative and world cinema and even adult film—representations that offer young people and adults with disabilities the important reminder that they should feel empowered to love and desire as they choose. But the fact remains that the mainstream film industry very seldom depicts happy and stable relationships for characters with disabilities, rarely records disabled characters' sexual fulfilment on screen and almost never shows us characters pursuing non-heterosexual desires. While television series like *Special*, *Friday Night Lights*, *Born this Way* and even *Game of Thrones* have broken new ground in the past decade, challenging the status quo and proving that viewers are ready for more nuanced and accurate representations of sex, love and disability, the film industry has been slower to recognise this fact and will need to keep pace with other media in order to remain relevant and to better serve *all* viewers.

# Note

1  The Production Code Administration (PCA) was in full operation in the US from July 1934 through to the breakdown of the Code in 1968. Unlike the British Board of Film Censors, which was under the control of the national government from 1913 onward, the PCA was a voluntary organisation established by the Motion Picture Producers and Distributors Association (MPPDA).

# References

Davidson, M. (2011) *Concerto for the left hand: Disability and the defamiliar body*. Ann Arbor: University of Michigan Press.

Doherty, T. (1999) *Pre-code Hollywood: Sex, immorality, and insurrection in American cinema, 1930–1934*. New York: Columbia UP.

Dolmage, Jay T. (2017) *Academic ableism: Disability and higher education*. Ann Arbor: University of Michigan Press.

Gill, M. (2015) *Already Doing It: Intellectual Disability and Sexual Agency*. Minneapolis: University of Minnesota Press.

Inter-office memo regarding *Tomorrow's Children* from J.B. Lewis to Mr. Breen dated 10 May 1934, *Tomorrow's Children* file. Motion Picture Association of America, Production Code Administration records. Margaret Herrick Library, Beverly Hills, California.

Kuhn, A. (1988) *Cinema, censorship and sexuality, 1909–1925*. London: Routledge.

'Leading exploitation' (1948) *Showmen's trade review*, 25 December, p. 106.

Letter from Joe Breen to J.L. Warner at Warner Bros. Pictures, regarding script for *Johnny Belinda* dated 29 August 1947, *Johnny Belinda* file, Motion Picture Association of America, Production Code Administration records, Margaret Herrick Library, Beverly Hills, California.

Letter from Joe Breen to J.R. McDonough at RKO Pictures, regarding stage play of *Johnny Belinda* dated 13 December 1940, *Johnny Belinda* file. Motion Picture Association of America, Production Code Administration records. Margaret Herrick Library, Beverly Hills, California.

Longmore, P. (1987) Screening stereotypes: Images of disabled people in television and motion pictures. In Gartner, A., and Joe, T. (eds.) *Images of the disabled, disabling images*. London: Praeger, pp. 65–78.

Michel, S. (1992) Danger on the home front: Motherhood, sexuality, and disabled veterans in American postwar films. *Journal of the History of Sexuality* 3(1), pp. 109–128.

Muir, K. (2015) The theory of everything. *The Times of London*, 2 Jan. [online]. Available at: http://www.thetimes.co.uk/tto/arts/film/reviews/article4310618.ece

Norden, M. (1994) *The cinema of isolation: A history of physical disability in the movies*. New Brunswick: Rutgers UP.

Pemberton, M. (2015) Sex and the disabled: Why are the able-bodied so coy? The Telegraph, 19 Jan. [online]. Available at: http://www.telegraph.co.uk/news/health/11350578/Sex-and-the-disabled-why-are-the-able-bodied-so-coy.html

Pernick, M. (1999) *The black stork: Eugenics and the death of 'defective' babies in American medicine and motion pictures since 1915*. Oxford: Oxford University Press.

Schaefer, E. (2001) *Bold! Daring! Shocking! True!: A history of exploitation films, 1919–1959*. Durham: Duke UP.

Schuchman, J. (1988) *Hollywood speaks: Deafness and the film entertainment industry*. Urbana: University of Illinois Press.

Shakespeare, T., Gillespie-Sells, K., and Davies, D. (1996) *The sexual politics of disability: Untold desires.* London: Cassell.

Shildrick, M. (2012) *Dangerous discourses of disability, subjectivity and sexuality.* Basingstoke: Palgrave Macmillan.

Siebers, T. (2012) A sexual culture for disabled people. In McRuer, R., and Mollow, A. (eds.) *Sex and disability.* Durham: Duke University Press, pp. 37–53.

Smith, A. (2012) *Hideous progeny: Disability, eugenics, and classic horror cinema.* New York: Columbia University Press.

Smith, S.L. (2019) Inequality in 1,200 popular films: Examining portrayals of gender, race/ethnicity, LGBTQ & disability from 2007 to 2018. *Annenberg Inclusion Initiative.* Los Angeles: University of Southern California.

Snyder, S., and Mitchell, D.T. (2010) Body genres: An anatomy of disability in film. In Chivers, S., and Markotić, N. (eds.) *The problem body: Projecting disability on film.* Columbus: Ohio State UP.

Stevens, D. (2014) Cosmic relief: In the theory of everything, Stephen Hawking's marriage is as complex as the universe. *Slate,* 6 Nov. [online]. Available at: http://www.slate.com/articles/arts/movies/2014/11/the_theory_of_everything_a_film _about_stephen_hawking_and_his_marriage_reviewed.html

Williams, L. (2008) *Screening sex.* Durham: Duke UP.

# 18

# DYNAMICS OF DISABILITY AND SEXUALITY

## Some African literary representations

### Omolola A. Ladele

### Disability and sexual rights in Africa: A background survey

Disability is a complexly heterogeneous discourse and its trajectories transcend national, histori-cal and cultural boundaries. As a socially constructed phenomenon, disability, like gender and sexuality, may be defined by dynamic political, economic, social and cultural imperatives which may alter over time and across space. According to 2011 World Health Organization (WHO) reports, more than one billion people live with disability around the world. With the enormity of its prevalence and with the possibilities of almost everyone having at least one form of dis-ability with age, it should be a growing concern. A human rights approach to disability would therefore indicate that the rights of people with disability, including their rights to sexual inti-macy and well-being, should be statutorily protected by the appropriate state authorities and institutions.

In an important document, "Human Rights and Persons with Disabilities in Developing Nations of Africa," David Anderson (2004) makes the point that whereas recent decades in the Global North witnessed improved awareness of the various issues with which persons with dis-abilities have to contend, this contrasts with the situation in Africa and other locations in the Global South, where persons with disabilities are "devalued, dehumanised and rejected." Similarly, Raymond Lang and Lucy Upah in their 2008 study of the Nigerian context reinforce Andersons's observation. Lang and Upah aver that contemporary Nigerian society fails to understand that dis-ability rights is "inexorably linked to and rooted in human rights" (Lang & Upah, p. 6). Also, Helen Meekosha (2008), as well as Tom Shakespeare (2000), variously remark on the dominance of the Global North in the propagation of disability discourses. Thus, while the Global North continues to be at the forefront in addressing the needs of persons with disability, disability is still considered a taboo in many cultural contexts of the developing world. It then becomes clear that the concept of and attitudes toward disability is socio-culturally differentiated and these have direct implications for the lived realities of persons with disabilities. Therefore, various forms or degrees of abilities/disabilities affect the statuses of people differently.

In many parts of Africa, several factors in various combinations complexly intertwine to affect the lives of persons with disabilities. For instance, there is the prevalence of negative super-stitious stereotypes and loosely held cultural beliefs about disability. Also, there is the lack of supportive infrastructural and affordable medical/healthcare facilities for persons with disabilities. These factors often combine with widespread poverty and lack of political and economic support for the subsisting legal documents and protocols on the rights of persons with disabilities. The result is that people living with disabilities are further excluded and pushed to the lowest rungs of the socio-economic ladder in developing nations. Furthermore, the WHO document *World Report on Disability* of 2011 states that: "negative attitudes, beliefs and prejudices constitute barriers to educational, employment, healthcare and social participation" of persons with disabilities (p. 3). These disabling barriers and the prejudicial perceptions that underlie them disregard the humanity of persons with disabilities, ignoring and delegitimising their corporeal existence. Hebron L Ndlovu (2016) complexifies this argument, holding that:

> Indigenous African beliefs concerning people with disabilities are ambivalent: on one hand, some African beliefs promote the stigmatisation and marginalisation of persons with disabilities through exclusion and depiction of them as objects of pity and ridicule, and as victims of evil forces; alternatively, other African beliefs inculcate ethical teaching aimed at protecting and empowering those living with disabilities by depicting them as full human beings who have the same rights, obligations and responsibilities as "normal" persons.
>
> *(p. 29)*

One of the most important defining aspects of a person is their sexuality. Sexuality has different components, which may include the act of sex, fantasies, attitudes, eroticism, desires, sexual practices, preferences, gender and identity. The WHO (2006a) report on sexuality suggests that various interactions and dimensions of biology, culture, economics, ethics, histories, psychology and other factors may influence sexual practices, roles, intimacies, relationships and beliefs. Therefore, beyond a person's physicality (regardless of an individual's ability or disabilities), several other factors may influence or nuance an individual's sexuality.

In spite of the centrality of sexuality as an integral aspect of being human, and despite the fact that disability is part of our humanity, the reality is that the sexuality of persons with disabilities, their sexual fantasies, desires, drives and feelings, is often repressed, ignored or discountenanced by preconditioned social and cultural prejudices and attitudes. Indeed, dominant discourses on disability in many societies have tended to historically exclude sexuality. For instance, observing the British context over two decades ago, Tom Shakespeare (2000) notes that "issues of sexuality and relationships had a low profile in the British disability movement and the developing field of disability studies" (p. 4), while other scholars, including Jenny Morris (1991), Abby Wilkerson (2002), and Neville Parkes (2006), variously emphasise this point in their work from that same time period. However, in recent time, issues of sexuality are now being slowly foregrounded, albeit mostly in the Global North. The recent studies by Matthew Hall (2018), Fran Vicary (2014) and Russell Shuttleworth (2014) all centralise the place of sex and sexualities in the lives of disabled persons.

In contemporary Africa, culture plays significant conditioning roles in defining and prescribing the boundaries of the sexualities of individuals or groups, and dominant discourses and socio-cultural constructions of disability often preclude sexuality issues. As previously stated, the realities of persons with disabilities in many African societies are shrouded in superstitions, taboos and cultural exclusions, leading to their marginalisation and dehumanisation. Sexuality, like disability in Africa, has also, historically, not been the subject of mainstream discourses; it has for long been shrouded by layers of taboos and demeaning cultural beliefs. With their sexualities so obscured, it becomes easy to misconstrue and misrepresent their sexualities.[1] And, as for persons with disabilities, their sexualities are often misconceptualised, as they are considered to be asexual, desexualised and incapable of handling, expressing or feeling profoundly affective emotions. Although sexuality is one of the most significantly humanising aspects of a person, the sexual lives of Africans with disabilities remain largely ignored and treated as inconsequential. Indeed, there is a dearth of rigorous and scientific studies of the sexualities of persons with disabilities in Africa, making conversations about the sexuality of such persons problematic.

## Mapping the theoretical frame

Disability has been variously imagined and recreated in the fictional works of African writers, and this chapter is an attempt to contextualise disability and sexuality within the human rights matrix of typical African cultural contexts. For instance, in considering the twin subjects of disability and sexuality within the conceptual structure of the Empire/colony interface in the novels of Tsitsi Dangarembga, Omolola Ladele (2016) calls attention to the "abject disfigurement of the characters in those novels and illustrates how the profoundly enervating political processes [of colonisation and decolonisation] alter the identities and sexualities of once-colonised people" (p. 29). Ladele illustrates how the "titles of both novels are highly suggestive of the overriding tropes of illness, disease, amputation or disfigurement in physical as well as in psychological terms through which the author creates the tableau for reading disability in the novels" (p. 28). Thus, with misshapen bodies, several characters in those novels become disabled and sexually undesirable. Ladele's position corroborates Makuchi Nfah-Abbenyi (1997) summation that: "there can be no clear-cut separation between sexuality, history, economics and politics in texts that are written about women's lives in a postcolonial context" (p. 30). Thus, to follow both Ladele and Nfah-Abbenyi, African women's sexualities cannot be de-politicised precisely because the politics of their history, race, gender, class and even religion are important in defining them. And, as Signe Arnfred (2004) suggests, "discussions of sexuality are so noticeably absent in analytical works of African feminists" (p. 59). Some of the effort of this chapter is, therefore, to centralise sexuality—in particular female sexuality, making it more visible because, as Arnfred puts it: "the conceptual terrain [of sexuality in Africa] is carved and cut in all sorts of often invisible ways" (p. 60).

It is important to note that as a cultural artefact, literature encapsulates and precipitates the production and dissemination of cultural idioms, ethos and mores. By directly observing the experiential dynamics of some disabled characters in the selected texts, the reader may begin to understand some of the material conditions that produce and underlie the experiences of persons living with disabilities and observe how their sexualities play out. Thus, the capacity for interpretive pluralities in literature helps the reader viscerally confront and re-think the various socio-cultural factors that underpin the experiences of other persons, including those with disabilities. This process creates avenues though which, to follow Clare Barker (2011), we are able to uncover the "materiality of embodied difference" (p. 2).

In this chapter, I examine two novels: Aminata Sow-Fall's second novel, *The Beggars' Strike*, in which I reconstruct the sexual realities of the disabled, poor people (mostly men)—the *borrom bàttu*—in that novel; and Petina Gappah's short story, "An Elegy for Easterly," in the collection of stories by that title. Gappah's short story provides, in a more visceral manner, another dimension to this study of sexuality. Since sexuality touches the totality of a person, the representations in these two novels present a somewhat holistic picture as we observe both male and female differently disabled persons at close range. This study ultimately illustrates the ramifications of gender and the implications of these for the sexual rights of persons with disabilities.

Feminist and Disability Studies theorist, Rosemarie Garland-Thomson (2002), suggests among other things, that "disability—like gender—is a concept that pervades all aspects of culture: its structuring institutions, social identities, cultural practices, political positions, historical communities and the shared human experience of embodiment" (p. 5). As already established, too, sexuality, like disability and gender, also deeply connects with and permeates other planes of human interactions. This therefore informs a materialist reading through a feminist lens of the selected texts.

Feminism, as a movement, originally emerged as a response to various forms of oppressions experienced by women. In terms of its various ideological underpinnings, feminism is also concerned with oppressive socio-cultural, economic and political institutions, the assignation and formation of identity roles along gender lines and the subordination of one gender to the other, as well as the conscious or unconscious prejudices that, inevitably, resonate within these formations. Anita Ghai's (2002) accounts of her personal experiences as a disabled, Indian woman are relevant as she holds that feminist notions are helpful to disability discourse. Since feminist discourse may be culturally nuanced, the interweaving of feminist notions with disability discourses inevitably reflect the specific, oppressive cultural nuances that significantly mark the lives of people with disabilities, with a view to dismantling them.

Indeed, as feminist scholarship also intersects and resonates with several other subject and identity formations including diaspora and disability, as well as migrancy and other positions that pre-condition isolation/vulnerability and "otherisation," it illuminates the positions of these lives lived on the margins. The application of feminist perspectives to disability studies also illustrates the cultural, historical and economic overlaps between gender and disability, which helps us rethink how women with disabilities negotiate their socio-cultural and political positioning as they seek to express their sexualities. For instance, in a typically patriarchal African society, where the dynamics of power versus privilege are skewed to the disadvantage of women, female bodies are often the sites of exploitation, oppression, marginality and exclusion, with this being particularly more so for disabled women. Thus, feminist interrogations of patriarchal structures and institutions uncover the underlying, skewed power relations of such structures and institutions.

The influential studies of Susan Wendell (1989), Rosemarie Garland-Thomson (2002) and Kim Hall (2011) show how feminist studies are pivotal in shaping studies on both sexuality and disability since their epistemologies are founded in socio-political formulations. Wendell's early study, for instance, pushes for more radical engagement of both disability and feminist studies, suggesting that the work of feminist theorists reveal not only the gender disparities of persons with disabilities but also the:

> Cultural attitudes to the body, suggesting that some of the same attitudes about the body which contribute to women's oppression generally also contribute to the social and psychological disablement of people who have physical disabilities.
>
> *(Wendell 105)*

Wendell's argument becomes even more acute when situated within the hemispheric context of the Global North/South dialectic.[2]

## Intersections of disability and sexuality in Gappah's "An Elegy for Easterly"

Zimbabwean female novelist, Petina Gappah, in her debut collection of short stories, writes compellingly of stories of women with disabilities revealing the ambivalence and ostracisation that circumscribe their lives. One such story is that of Martha Mupengo, a deranged, lonely woman in "An Elegy for Easterly." Martha is the heroine of the story; the fictional image of Martha and her mental illness are compelling, and, although "her name and memory, past and dreams, were lost in the foggy corners of her mind" ("Elegy," p. 30), Martha is the central figure around whom the entire story revolves and she evokes an important metaphor in the study of disability and sexuality within the racially defined national context of Zimbabwe.

In Gappah's story, Martha, is so named, because she looked like another Martha—some uncanny reincarnation of Mai James' niece-in-law "in the last days when her illness had spread to the brain." Thus, the later Martha's identity is instantly characterised and framed by a back-reference to another woman with a similar psychological or mental illness. Thus, the present Martha both embodies and conjures the image of lunacy and her identity appears fully realised and legitimised and, as the novelist points out:

> It was the children who called her Mupengo, Mudunyaz and other variations on lunacy...The name Martha Mupengo stuck more than the others, becoming as much a part of her as the dresses of flamboyantly coloured material, bright with exotic flowers...
>
> *(p. 31)*

Significantly, one of the representational strategies the author deploys is the act of naming Martha. Naming in the African socio-cultural context is important, and it refers to the way people may be perceived since it allows us to place and identify them in the social world. Thus, names may serve political, social or psychological purposes. Typically, parents or adults name their children but, in other senses, children may name adults. One of such instances is when children nickname adults, but such cases are not quite normalised as acceptable behaviour. Gappah captures this effectively:

> The children understood that Martha's memory was frozen in the time before they could remember, the time of once upon a time, of good times that their parents had known, of days when it was normal to have more than leftovers for breakfast. ...Like Martha's madness, the Christmas records and bonuses were added to the games of Easterly Farm.
>
> *(p. 34)*

Martha is one of the last local migrants to the Easterly neighbourhood—a sprawling ghetto that springs up when the government dispossesses the people of their land and homes to make way for new developments in the township. This dislocation, in particular, is for the Queen of England's three-day visit to Harare. Living on the fringes of the community, and with only children—"the Under-Eight's" led by Tobias—as her occasional companions, Gappa powerfully delineates her protagonist's separation and isolation from the rest of the Easterly Farm community. Martha's isolation is accentuated by the distance, apathy and caution with which the adults of Easterly Farm treat her, thus making her life more perplexing.

In spite of her isolation, Martha's life story innocuously and routinely intersects with that of other members of the community in both private and public dimensions, foregrounding an important representational strategy the novelist deploys to comment on the socio-cultural realities of all the people. Martha's life is conditioned by her socio-cultural and political positioning as a woman in a traditionally patriarchal setting. Then also, in quite a bizarre, yet profound manner, the novelist deploys Martha's lunacy as a metaphor representing the universally bad times that had befallen the country, forcing large numbers of Zimbabweans to live in such dehumanising, sordid squalor and abject poverty found on Easterly Farm and the other slums of Porta Farm, Hatcliffe, and Dzivaresekwa Extension. More particularly, Martha is both female and disabled, and she is thus doubly jeopardised by these circumstances.

Death and its various other symbolisms—loss, destruction—in physical, material and deeply psychological terms is perhaps the most pervasive literary trope Gappah deploys to critique life on Easterly Farm. As one of the late-comers to Easterly, Martha simply arrives and sets up residence in the vacant house of the late Titus Zunguza, where a triple homicide had previously occurred. Completely oblivious to the horror, stench and stains of blood in the deserted house, Martha starts a new life that significantly challenges the normative patriarchy of many African cultures. Since Titus Zunguza is dead ,and unlike the lives of the other women in the story, whose husbands still dominate their lives and psyche, Martha is totally free from any form of patriarchal domination or authority, both physically and psychologically. This, therefore, spells for Martha a certain degree of sexual freedom that the other women in the story do not seem to experience.

Typically represented in this story as marginal, Martha's sexual life appears contradictorily dominant, yet simultaneously subsumed in the vague references to it. As noted earlier, sexuality in its most rudimentary form refers to a person's sexual identity, which is, of course, framed by the background of a particular history or culture. In particular, sexuality is embodied, and it is realisable or performed through relationships or interactions with other persons. In Africa, the sexuality of persons with disabilities are often conceptualised as non-existent, so that Martha's sexuality is not conveyed in bold, definitive terms by the author. Rather, it is described in largely subtle and merely suggestive undertones. Indeed, Martha's sexuality is often obscured and depicted in infantile terms in the narrative. This is because she is mostly in the company of children who are not conscious of their sexual realities at this stage. Constantly, as she gently lifts her colourful dresses when she is teased by her regular companions—the children—Martha may be expressing her sexual fantasies or yearnings, but this is, for the children, no more than playful sport. As far as the surrounding adult world is concerned, Martha's gestures are merely childlike. At any rate, she is imprisoned by her disability, so they cannot possibly conceive of any connection between her reported gestures and any sexual yearnings because, in their traditional orientation/imagination, her disability rules out such possibilities.

In typical African cosmology, life and death cycles are never rigidly separated. Thus, in death, life is often subsumed. In Gappah's narrative, life and death seem to be inexorably intertwined through Martha and this is dramatically played out at least twice in the story. Although Martha may be seen to embody death as she lives in Zunguza's house, she also carries within her sexuality a potentially life-giving force. As previously noted, Martha's sexual well-being is not a primary concern to the adults in Easterly [she is hardly seen as human]. Yet, her sexuality at a critical moment profoundly connects with that of Josephat, the miner. Indeed, in a somewhat paradoxical manner, we encounter Martha, not as a 'monstrous freak', but as one whose body provides Josephat with warmth and succour. Regardless of the momentariness of the occasion, the experience validates Martha's sexuality as well as her personhood. And although Martha becomes pregnant, no one in the neighbourhood shows any concern for her consequent reproductive health issues.

In a moment of overwhelming intense desire for sex, Josephat's unassuageable sexual needs are, momentarily, of utmost importance—seeming out of control and yet, paradoxically, in control— Josephat defies strong cultural taboo, the stench and horror that haunt the late Zunguza's hut, and even the fear of a lunatic woman, to have deep intimacy with Martha. This singular encounter powerfully suggests that libidinal forces may compel sexual desires without a priori reasoning. And this incident is neither indiscretion nor rape; rather, it is the author's powerful symbolisation of a barrier-breaking reconstruction of traditional narratives about sexuality and persons with disabilities. In this powerful symbol, the author deconstructs the concept of normalcy to demonstrate the humanity of disabled persons. Indeed, in that overpowering moment, Josephat imbues Martha with humanity and she becomes for him, a sexual being—challenging and completely deconstructing the image of a perfect, desirable, normal and beautiful mind and body that is usually acceptable or attractive for intimate relations.

The central narrative at this point is that Martha's body is acted upon; she is a passive receptacle for Josephat's forceful libido. It is not Martha who seeks intimate, sexual relations with Josephat. On the contrary, Josephat seeks her out, although her sexuality appears repressed all along. In spite of the surrounding social attitudes that ignore or discountenance Martha's femininity, Josephat seems momentarily preoccupied with a mere body, anybody: even one devoid of any capacity for intelligent sexual expressions. This is underscored by the commonly held notion that disabled people are incapable of expressing deep emotions and intimacy and, to this extent, she may be seen as the victim of a violation.

The author also prefigures the intersection of Martha's life with that of Josephat's wife. In a very brief but ominous encounter, the two women brush shoulders, instantly mirroring their connections. For the first time, Josephat's wife sees "the full contours of pregnancy" in Martha, unknown to her that it was her husband who had uncannily connected the two women. As the novelist describes it, "the winter of Martha's baby was the winter of Josephat's leave from the mine. It was Easterly's last winter" (p. 45). In this way, Martha's disability, directly or indirectly, affects the lives of normative others in the society. Auspiciously, Josephat's wife midwife's Martha; delivering her of the baby just as Martha faints, again dramatising the life/death cycle of African cosmology.

Josephat's wife performs an incredibly humanising act of kindness. Able to, at least, save the baby which she takes in to be nursed by her and completely oblivious that she is carrying her husband's child in her arms, Josephat's wife's actions proves not only humanising—they become a life-giving idiom that deconstruct the barriers that alienate people with disabilities. By fostering the new-born child, Josephat's wife provides the possibility of a better future for the child. Thus, in the ironic twist of their connecting fates, Martha provides the emotional link and resolution for Josephat and his wife—humanely resolving the anxieties and stigmatisation of lunacy.

## *The Beggars' Strike*: **The dirt trope**

Aminata Sow-Fall is one of the earliest female writers from francophone West Africa. Her novel, *The Beggars' Strike*—first published in English by Longman in 1981 and possibly set in Dakar, Senegal in West Africa—provides a complementary pastiche to Gappah's singular story considered in this chapter. Traditionally, Sow-Fall's slight novel has not been seen from an overt disability perspective. I therefore centralise the experience and treatment of disabled people to comment on their sexualities. Inextricably woven into the novel's subject-matter is the notion of environmental degradation which appears to frame a complex web of existential and ecological concerns as I consider disability and sexuality in this novel.

An important thematic focus of the novelist pertains to the rights to human dignity of people with disabilities as the author explores how normative socio-cultural, religious and political institutions often deny and neglect such persons. The novelist explores these contradictions, uniquely, through the dramatization of the cumulative tensions and turmoil of the private, personal ambition of Mour Ndiaye, Director of Public Health and Hygiene, as he attempts to rid the Capital of the "filth" and "menace" of the *borrom bàttu,* hoping also to clinch the position of Vice-President of his nation, should he accomplish this task.

From the opening of the novel, the author, with great attention to detail, describes the *talibés,* that is, "beggars," in the most despicable, undignified terms—she describes them as—"lepers," and "cripples," "derelicts," "parodies of human beings," "the dregs of society" (p. 1). The layers of description of the beggars in these terms are literally and metaphorically indicative of their amputated/fractionalised humanity. Represented as mere fragments and depicted in these unwholesome terms, their sexuality is occluded and completely non-existent. Nonetheless, their overwhelming physical presence as well as the cumulative visual images they present appear immediately indicative of the prominence the novelist wishes to give to her concerns with disability in the novel.

Also, the overriding dirt trope in the novel frames the social interactions of everyone in the novel—able-bodied and disabled persons alike. Sow-Fall's novel appears to recycle the dirt, decadence and filth textualised in many African literary texts; perhaps most evidently typified in Ayi Kwei Armah's early novel, *The Beautyful Ones Are Not Yet Born* (1969), and Meja Mwangi's *Kill Me Quick* (1973). The cycle of filth and dirt in these novels are demonstrative of the long genealogy of the notion of filth and dirt in African political and literary discourse. Filth in these three novels is a metaphor for the decadent postcolonial, post-independence failures and corruption in the socio-political systems of both Ghana and Kenya, as well as in Sow-Fall's fictional Capital.

Given the African colonial experience, particularly in countries where there were white settler communities, space is often a contentious site especially as colonialism disrupts and remaps people in material, psychological and economic terms. Polarised spaces are thus often interlocked into the imaginaries of writers who experience these profound disruptions. Unmistakably, discriminatory divides were ostensibly to prevent or avoid contamination from the "otherised" "filthy blacks." One of these polarised spaces that clearly illustrates this point is in Tsitsi Dangarembga's novel, *Nervous Conditions*, in which Jeremiah's dilapidated and squalid hut is contrasted with Babamukuru's clean but sterile house.

And, as Patrick Oloko (2018) recently suggests, dirt, since the new millennium, seems to acquire new meaning and metaphorical applicability in which dirt seems to:

> mesh the literal and the metaphorical when applied to social behaviours and existential conditions in certain urban environments. In this regard, the principal issue is the question of what results can be obtained when frameworks relating to extracting and disposing material dirt (in order to secure a pleasing material atmosphere, for instance) is applied in new contexts, such as the body politic.
>
> *(p. 2)*

In Sow-Fall's novel, the beggars are just like the often-overwhelming natural presence of dirt in developing urban spaces of contemporary Africa. Dirt in *The Beggars' Strike* signals a symbiotic correlation between the environment and the people. The beggars are the "running sores," the flotsam, and the dirt of the Capital—that Mour Ndiaye, Director of the Health Ministry, needs to get rid of. Keba, Ndiaye's hard-working subordinate, is the one on whose shoulders rests this onerous and daunting task of ridding the city of this "conglomeration of humanity." But this "conglomeration of humanity" are not treated with humanness. And this creates the central conflictual tensions in the novel. Seen as the dregs, the "running sores" of society, it is difficult not to treat them as more than inconsequential. Yet, for all their incapacities and disabilities, the beggars display a strong sense of their self-worth and a right to dignifying existence under the leadership and direction of Salla Niang, who also acts as their banker.

The reader's first encounter with the beggars is in Salla's capacious courtyard where they gather for the daily draw of their "subscription scheme" and, as the author notes, it was a mixed crowd:

> [t]he crowd comprises men and women of all ages and sizes, some crippled, some hale and hearty, all depending on their outstretched hands for their daily pittance...Among them is blind Nguriane Sarr, always correctly dressed with his tie, soiled starched collar, dark, gold rimmed spectacles, invariable navy-blue suit and white stick. Nguriane Sarr has a somewhat distinguished air...among the hoary heads and ulcerated limbs, covered with the pustules of scabies or eaten away with leprosy among the rags which leave half-naked bodies which have long been innocent of any contact with water, among the beggars' crutches, sticks and *bàttu*, there are some adorable little tots...
>
> *(p. 8–9)*

This is truly a mixed multitude, a motley collection of persons with or without prosthetics. The "ulcerated limbs" and "half-naked bodies" are completely desexualised and distinctly oppositional to the wholesome and desirable bodies of Salla Niang and her husband on one hand, as well as Keba Dabo and his playful mistress, Sagar, on the other.

Within the proximate space of Salla's compound, the author subtly calls attention to the contrasts of the differentiated bodies, depicting two antithetical sexual images. In the concatenation of amputated bodies in Salla's yard, the author shows Salla's husband's lustful yearning for his wife's body, which only he knows is adorned with traditional waist beads. Salla and her husband are both able-bodied and their virility is intact. For the beggars, on the other hand, their dehumanisation is so overwhelming, it is deeply etched into their psyche, such that they have no thoughts for their sexuality. For instance, we learn of Madiabel's emotive story—of how he had to migrate from the village to the City because his profession as a tinker could no longer sustain his family of two wives and eight children. He therefore comes to the City to become a "bàttu-bearer." Madiabel is not only a beggar, he is also lame and in the process of fleeing one of the raids or round-ups, he fatally runs into a moving vehicle.

From the point of view of the beggars, the officials [who] "lay about like madmen; when they get worked up like that, [they] seem to forget that we are human beings" (p. 11). In this manner, the beggars are constantly and routinely hounded, brutalised and mistreated. They are so demeaned that they must carry a "certificate of indigence" to identify them as paupers. Madiabel does not have such an identification paper when he is brutalised, and so he is left to die of his injuries in the government hospital.

Yet, the beggars only want to be treated with some humanity and dignity as we note in Nguriane Sarr's dignified posture or in Papa Gorgui Diop—

> the old man, who has a knack of winking an extra mite out of the donors, thanks to his extraordinary talent: he is a perfect scream, the way he acts an old man in love with a young girl...Gorgui Diop is well known all over the City and people come from a great distance to see him do his act in his accustomed pitch, in front of 'his' bank.
>
> *(p. 9)*

The novelist appears to suggestively allude to the possibilities of the beggars attaining some sexual delight once they are, at least, able to gain some dignifying respite from the onslaught of government officials. Embedded in her playfulness, Sagar seems to convey the traditional attitudes toward the beggars as she attempts to dissuade Keba from embarking on his mission against the beggars, suggesting that it is fait accompli for:

> The beggars have been here since the time of our great-great-grandparents. They were there when you came into this world and they'll be there when you leave it. You can't do anything about them...religion teaches us we must help the poor.
>
> *(p. 15)*

In a profoundly illuminating moment, the author deftly uncovers Keba's disdain for Sagar, which appear to correlate with the same kind of reaction he feels toward to the beggars— it's the "familiar suffocating" feeling and he has to restrain himself from going "wild" with her. Thus, paradoxically crippling this relationship is the underlying obnoxious image of the beggars: a simultaneously sweet and sour metaphor!

Finally, pushed to their tether's end by the disregard and disrespect for their humanity, Nguriane calls for the organisation of the beggars. A revolutionary reaction to their mistreatment by government thus begins in Salla's courtyard. Following the death and burial of Gorgui Diop, Nguriane Sarr's perceptive persuasion stirs his colleagues; Salla, also deeply touched by the passing of Diop, joins her voice and previous experience with Sarr, saying:

> Now my friends, the hour has come to make our choice: to live like dogs, pursued, hunted, tracked down, rough handled, or to live like men. Gorgui Diop's reason for living was always to bring a little cheer to men's hearts. But these madmen have forgotten the meaning of cheer. Since Gorgui Diop has not been spared—Gorgui Diop who made people laugh—no one will be spared...
>
> *(p. 38)*

Strategies are mapped and the process of the revolution is set in motion. Now relocated to the "new Slum-Clearance Resettlement Area"—the outskirts of the City, where people have to come and give them donations, the beggars are the ones who decide what they will have. This revolutionary posture is antithetical to Mour Ndiaye's ambition and it becomes a crisis point. Ndiaye's rising profile and the possibility of him becoming the Vice-President requires him to make "sacrifices" to the beggars. His dilemma and desperation is played out when he sends his young son, then Salla, his maid, to look for a beggar with specific prescriptions to give some cola nuts. Soon the beggars realise that they have effected a change in the status quo:

> They [beggars] are now living like princes; they are even beginning to get bored at having nothing to do from morning to evening. Donations rain from heaven and lately they have noticed greater crowds and more generous gifts...
>
> *(p. 60)*

Eventually, it gets to the point that Mour Ndiaye has to make a donation, personally, and he goes in search of the beggars to invite them back to the City, to take up their usual positions, for this to happen according to the specifications of his marabouts. In that encounter he is not only shocked by Nguriane's impertinence, and Salla's "ostentatious air of difference," he is also amazed by the:

271

image of so many physical defects, so much physical decrepitude and human disin-tegration from which, it is true, some patches of light stand out, like this Salla Niang, whose face gleams like a bronze bust, fashioned by a master sculptor.

*(p. 80)*

That encounter ends in a disaster that puts an end to Mour Ndiaye's ambition and the beggars wrest their dignity from the clutches of the inhuman, able-bodied society.

## Conclusion

As seen in this discussion, literature, as cultural artefact, powerfully captures and resonates with the cultural symbolisms that affect the dual subjects of disability and sexuality in Africa. Although they are fictional stories, the two novelists imaginatively reconstruct some of the social, religious/metaphysical and cultural attitudes and prejudices that constitute barriers of denial, exclusion, stigmatisation, mistreatment and anxieties which circumscribe the lives of persons with disabilities. In these narrations, we observe, first-hand, how the several layers of oppressive exclusions, stigmas and dehumanisation place such people living with disabilities in disadvantaged or marginal positions.

From Gappah's Zimbabwe in southern Africa to Sow-Fall's Capital in West Africa, the stories by both writers strategically deploy disabled characters as narrative figures—offering from their experiences, insights to their lived realities. For instance, the pervasive death and dirt tropes uncovered in both novels are aesthetic and narrative metaphors of the "sub-human" status ascribed to disabled persons in many contemporary African contexts. Thus, with bodies captured in such despicable and disgusting terms, they are hardly regarded as human, such that their impairments become sites for the discrimination, oppression and mistreatment for persons living with disabilities. Take for instance, the *borrom bàttu* in *The Beggars' Strike*; the beggars are mostly poor. But beyond their acute poverty and their disability, which is also mostly physical, they are regarded as flotsam, dirt, filth—mere fragments of humanity and, in this way, they are treated in the most despicable, undignified terms.

Although sexuality constitutes an important defining feature of humanity, contemporary African societies do not, normally, subject issues of sex and sexuality to open discourse in public spaces. They therefore remain as contested sites in public discourse. It is also important to note that we cannot point to what may be described as a universal "African sexuality." This would, of course, throw up some racial implications of, for instance, an essentialised image of hypersexual or profligate "other." However, we may differentiate sexualities according to other imperatives, including gender.

For instance, Aminata Sow-Fall's beggars are mostly hapless males and their masculinity is mostly eroded. In African cultures, masculinity is normatively associated with vigour, strength and virility. In the novel, however, the beggars are represented as amputated sub-humans who are depicted as sexless, their sexualities completely removed from public purview. Several factors, their penury (economic status), their mistreatment by society (negative socio-cultural beliefs) and condescending religious beliefs, seem to castrate the beggars. They are more immediately concerned with matters that emasculate them—they insist on being treated with dignity, as humans. Notice, for instance, Nguriane Sarr's insistence that they should be treated "as men." The point is made in a significantly non-obtrusive manner; the import, it appears, is that the restoration of their dignity has direct implications for the restoration of their virility and mas-culinity. Thus, when treated as men, they can variously express their latent sexual feelings and desires. This illustrates the point that, in considerations of sex and sexuality, the psychological is connected with the physiological.

In the two novels, both authors seem to discount the sexuality of persons with disabilities. Through Martha's life story, it appears that female sexuality may be associated with violations, as seen in her treatment by Josephat. Also, in Martha's world, all of the adults in that novel pay no attention to either her femininity or her personhood. Female sexuality and, in particular, the sexuality of a disabled woman is completely ignored in the story. Perhaps the burdens of the dire circumstances of even the able-bodied women in Gappah's story prevents a clear delineation of their sexualities. With no meaningful role, place or function in her society, (she does not arrive as either "wife of" or "mother to") Martha's place is with the children. No one thinks of her in sexual terms and her sexuality is thus amputated in profoundly psychological and material terms from the psyche of her society. Precisely because of Martha's mental disability, she is ostracised and de-sexualised by her society.

Therefore, irreducible from the two narratives, is that, whether physical or mental, any disability sets up such people for a life on the margins—an incomplete life, devoid of and incapable of fulfilling sexual activities. And, as demonstrated in this study, the prevailing socio-cultural contexts, more than the various disabilities/impairments, form the barriers which typically prevent persons with disabilities from actualising the possibilities of their full potentials and sexualities.

## Notes

1 For instance, one such misconstruction is the typical commodification or objectification of African women's bodies within the Euro-American imaginary, which often represents them as having wild, profligate, hyper-sexualities. In particular, I note as relevant what Anne McClintock (1995) refers to as the 'porno tropics' of European imagination. Also, the Senghorean depiction of the idealized image of the archetypal African woman as Mother of Africa, inadvertently, denigrates the African woman.

2 The contextual Global North versus Global South critical dialectic in this study further underscores the paucity of critical attention on disability studies so far. Beyond this, disability studies in Africa is also inexorably nuanced by several overarching issues including racial (given the postcolonial context of almost all of Africa and Asia), gender, religious and cultural factors and it would be instructive for western knowledge produced on disability studies to accommodate these.

## References

Anderson, D. (2004) *Human rights and persons with disability in developing nations of Africa.* Paper presented at the Fourth Annual Lilly Fellows Program. National Research Conference: "Christianity and Human Rights", Samford University, Birmingham.

Armah, A.K. (1969) *The beautyful ones are not yet born.* London: Heinemann.

Arnfred, S. (2004) African sexuality'/sexuality in Africa: Tales and silences. In *Re-thinking sexualities in Africa.* Uppsala, Sweden, pp. 59–78.

Barker, C. (2011) *Postcolonial fiction and disability: Exceptional children, metaphor and materiality.* New York: Palgrave Macmillan.

Garland-Thompson, R. (2002) Integrating disability, transforming feminist theory. *NWSA Journal* 14(3), pp. 1–32.

Ghai, A. (2002) Disabled women: An excluded agenda of Indian feminism (discrimination in India). *Hypatia* 17(3), pp. 49–66.

Hall, K.Q. (ed.) (2011) *Feminist disability studies.* Bloomington, Indiana: UP.

Hall, M. (2018) Disability discourse and desire: Analyzing online talk by people with disabilities. *Sexualities.* http://doi.org/10.1177/1363460716688675

Ladele, O.A. (2016) Disabling sexualities: Embodiments of a colonial past. *Graduate Journal of Social Science* 12(1), pp. 23–39. http://gjss.org

Meekosha, H. (2008) *Contextualising disability: Developing southern/global theory*. Keynote at Biennial Disabilities Conference Lancaster University, pp. 1–20. www.wda.org.au/Meekosha

Morris, J. (1991) *Pride against prejudice: Transforming attitudes to disability*. London: The Women's Press Ltd.

Mwangi, M. (1973) *Kill me quick*. London: Heinemann.

Ndlovu, L.H. (2016) African beliefs concerning people with disabilities: Implications for theological education. *Journal of Disability and Religion* 20(1–2), pp. 29–39.

Nfah-Abbenyi, J.M. (1997) *Gender in African women's writing: identity, sexuality, and difference*. Bloomington: Indiana University Press.

Oloko, P. (2018) Human waste/wasting humans: Dirt, disposable bodies and power relations in Nigerian newspaper reports. *Social Dynamics: A Journal of African Studies*, pp. 1–14. http://doi.org/10.1080/02533952.2018.1441111

Parekh, P.N. (2007) Gender, disability and the postcolonial nexus. *Wagadu* 4, pp. 142–161.

Parkes, N. (2006) Sexual issues and people with a learning disability. *Learning Disability Practice* 9(3), pp. 32–37. http://doi.org/10.7748/idp2006.04.9.3.32

Shakespeare, T. (2000) Disabled sexuality: Towards rights and recognition. *Sexuality and Disability* 18(3), pp. 159–166.

Shuttleworth, R. (2014) Conceptualising disabled sexual subjectivity. In M. Pallotta-Chiarolli & B. Pease (eds.) *The politics of recognition and social justice: Transforming subjectivities and new forms of resistance*. London: Routledge, pp. 153–188.

Sow-Fall, A. (1985) *The beggars' strike*. Trans. by Dorothy Blair. Essex: Longman.

Vicary, F. (2014) Sex and disability: Yes, the two can and should go together. *The Guardian*, 20 February. http://www.theguardian.com/commentsfree2014feb21/sex-and-disability-yes-the-two-can-and-should-go-together

Wendell, S. (1989) Towards a feminist theory of disability. *Hypatia* 4(2), pp. 104–124.

Wilkerson, A. (2002) Disability, sex radicalism and political agency. *NWS Journal* 14(3), pp. 33–57.

World Health Organization (WHO) (2011) *World report on disability*. Geneva: WHO.

# 19

# FLAUNTING TOWARDS OTHERWISE

## Queercrip porn, access intimacy and leaving evidence

*Loree Erickson*

Tiny, fluffy purple feathers cover the sofa and the floor. The purple dollar store boa, a delicate teacup perfectly matching the boa, sci-fi comic books, and graphic novels all sit carefully on a small table nearby. A radiant blue cane rests on the couch, joined by a dismantled boom pole. Laughter, ideas, and warmth fill the room. And of course, my cat is sitting right in the middle of everything. My living room has been transformed. I have been transformed. My collaborators and the other people involved have been transformed. And when I say transformed, I do not mean in a way that is finished, measurable, or finite; I mean something subtle yet significant, deeply felt and always moving…

The above statement describes a moment following the shooting of one of the queercrip porn scenes that were part of my dissertation research project, *Unbreaking Our Hearts: Cultures of Un/Desirability and the Transformative Potential of Queercrip Porn*. This research works to illuminate and interrogate the operation of cultures of undesirability and the ways that the collaborative production of queercrip presents opportunities for disruption of the dominant narratives of un/desirability and participates in cultivating, recording and sharing fluffy purple feathers of possibilities, knowledges and imaginaries vital to queercrip flourishing. This chapter highlights just a small taste of the significant outcomes of interviews and the collaborative production of six porn scenes with an incredible group of nine queercrip collaborators: Kylie, Nomy, Lisa, Isobel, Juba, Romham, Corrinne, Mia and Allen. My focus here is to highlight how, by enacting radical access (Withers, 2012), sharing in moments of access intimacy (Mingus 2011; 2017) and building interdependent community through practices of leaving evidence (Mingus 2010), queercrip porn makes necessary epistemological and political interventions that open opportunities to build, live and flaunt otherwise, pushing against the harm, erasure and exclusion of cultures of undesirability.

## Introducing cultures of undesirability

"Cultures of undesirability" attempts to talk about the multitude of lived experiences and structural practices that undermine and determine the collective worth and well-being of marginalised communities. The concept emerged out of my work on the conjunction of disability and sexuality and attempts to complicate the normative limitations of interrogating ideas of sexiness. How we determine who is valued, understood as desirable and granted the status of personhood is always a complex interlocking socio-political matrix. Cultures of undesirability as an interdependent concept seeks to bear witness to the historic and current systemic and interpersonal impacts of the eugenic construction of "undesirable others" across and through multiple simultaneous regulatory systems. As marginalised people are subject to an ever-quickening cycle of poverty, violence, isolation, criminalisation and medicalisation (Withers 2012), I hope that cultures of undesirability provides an intersectional frame that enables us to address these violent cycles by recognising the complex and messy multiplicity and interplay of identities, bodies and systems of oppression.

My research aims to articulate a collective conceptualisation of cultures of undesirability that centres queercrip community as knowledge and cultural producers, adding necessary depth to current understandings of the manifestations and impacts of this concept. As highlighted in the quotations below, my research collaborators not only regularly navigate isolation, separation and exclusion, they also interrupt these and other key lived impacts of cultures of undesirability. They question the unquestionable.

This chapter expands on four interrelated themes that emerged from the collaborative research on cultures of undesirability: in/visibility, shame, exclusion and control. According to Romham:

> Cultures of undesirability is about the whole social, structural, institutional, and…generational process of rendering folks (including variously disabled folks) undesirable based on bullshit ideas about worth and hotness and so on. But it's not just ideas—it's actions.
>
> *(interview 2013)*

Cultures of undesirability do not manifest as a one-off experience. Romham states, "…they are literally everything." Collaborators shared multiple stories testifying to the omnipresence of cultures of undesirability: on the street, public transit, at work, in doctor's offices, prisons, classrooms and community spaces or with friends, lovers and allies, not to mention in our homes and hearts.

## Managing in/visibility

Avery Gordon, using the work of Laura Kipnis, describes visibility as a "complex system of permission and prohibition, of presence and absence" (Gordon 1996). This complex system is never complete: various collaborators spoke to moments of (mis)recognition in which queercrips are made to appear even in their disappearance. We appear and move through the world with a constant sense that there is something wrong with us. Bodies and ways of living are made noticeable by the disappearance of differences amongst normative bodies and the normalisation of ways of communicating, moving and feeling that adhere to the dominant terms of desirability. Kylie speaks to the ways larger systems of oppression create standards of desirability and practices of reading bodies and identities. These practices are critical in terms of recognition and ways of being and appearing in the world:

> Society has standards of belief of people according to a white perspective, and that standard and that belief of what is perceived as the norm, affects trans women, and trans women of colour. It forces all of us to try to conform to that expectation, so people who don't fit that are oppressed…People perceive me as a guy. They assume my gender.
>
> *(Interview 2014)*

Nomy discusses how desirability is erased or rendered invisible for those who do not conform to normative ideals:

> I was never crushed on, I was never asked out. I grew up in a completely white, middle-class area. I had very beautiful lifeguard, soccer, athlete friends, who just had dates across the board forever, and I was always the one that was kind of chubby, so I was just really put as an "other" from way back. I think for me, I just un-thought of myself that way. I just turned off thinking that I was crushable…And it was this long stretch of numbness, or something, around that.
>
> *(Interview 2013)*

In this way, non-normative bodies, when they appear, are hypervisible. Gordon describes hypervisibility as a "kind of obscenity that abolishes the distinctions between permission and prohibition, presence and absence" (Gordon 1996, p.). I understand my collaborators' storying of their experiences as a technology of the self within economies of desirability, particularly how the terms of legibility, subjectivity and possibility are forever entangled (Butler 2005; Martin et al. 1988). Corrinne's response on what being a queer, black, chronically ill femme means to her touched on many of the tangled threads between managing visibility and normative exclusions:

> I think it means that there's a lot of intersections of things…like, when I'm experiencing something and I don't necessarily know if I'm coming up against racism or homophobia or ableism or what's happening. And I think fundamentally it's kind of like a mix of all of those things, because I can't really separate my identities. What else does it mean? It means that I'm resilient as fuck [laughter], because the world doesn't really account for me to be in it sometimes, like Audre Lorde says, "existing in a world that wasn't meant for you to survive."
>
> *(Interview 2014)*

Queercrips continuously navigate the mixed messages of visibility. Corrinne talked extensively about the ways that the amount of sick time she takes off work or what she does on her days off are policed by co-workers and friends. Corrinne's story is a very typical experience for people living with chronic illness and certain experiences of queer femmeness:[1] Both of these experiences are often shaped by invisibility. Some of us, or some parts of us, are devalued by cultures of undesirability and thus erased; experiences such as chronic illness, chronic pain and certain expressions of madness and neurodiversity were understood by my collaborators as being erased by dominant culture. At the same time, some of the ways we fall outside terms of desirability are hypervisible. All of the collaborators of colour in this research project were Black, and they each spoke to the ways that their blackness marked them as hypervisible. Corrinne followed her example of invisibility with this telling story:

> I was having a good body day, not a lot of pain. I had just had a really nice meeting with some other Black colleagues and I was walking down the street and, several white men started yelling at me. And I was like, "What is happening?" But this one man was like, "Oh, you think you own this street," And yeah, I was wearing this big fancy coat and just feeling really good, and just saw how intensely threatening that was to, like, be myself and be, um, y'know, confident and proud of who I was.
>
> *(Interview 2014)*

This story clearly illustrates what happens when noticeable bodies exceed or are seen to exceed the space of marginality. In this particular example, in dominator culture, Corrinne's blackness and femininity are closely associated as angry and dangerous. Part of what is threatening is when marginalised people are feeling good, confident in all our ways of being, in our noticeable bodies; in owning visibility—not shrinking but flaunting—we are threatening. In order to feel safe, we are encouraged to fall into alignment (Ahmed 2006); in other words, disappear through conformity.

Often, in order to become legible as human and to access resources, we must participate in reinforcing the terms of normativity, possibility and desirability. Marginalised communities are allowed to show up as the model minority, as the supermom, as the supercrip. The supercrip makes dominant culture, and, in particular, non-disabled people, feel safe by compelling us to desire "ability" and promoting the belief that embodied difference and the associated devaluing and marginalisation are avoidable, if you can overcome (Peers 2012). Visibility that grants us legibility is often predicated on a politic of respectability that instils a norm-based achievement occurring in spite of the very things that make us different, not because of.

Collaborators also named the policing of fatness, gender nonconformity as additional sites of hypervisibility. Here, I will focus on the stories surrounding adaptive devices. Pens, sunglasses, earplugs, cell phones, shoes, clothing—all of these items could be considered adaptive devices because they shift bodily capacities with their use; of course, what distinguishes these adaptive devices, from canes, wheelchairs, hearing aids, and so on, is the combined normalisation of the former and the medicalisation of the latter.

The powerful mix of medicalisation and hypervisibility means that our adaptive devices, as the symbol of disability and thus accommodation, are viewed with pity and pathology. As conveyed by Isobel's story:

> I had my cane, and it matched my outfit. I was super fancied-up... there was this younger black guy that I think was working there, and I was sitting down, not visibly with my cane. And he was giving me the eyes, and I was not interested, but he was doing his thing, and I was like cool. I walked somewhere else with the cane and he came up to me and had this whole thing about "you're so beautiful, *except*," and "I thought it was only for these kinds of people," and, "oh, it's an illness—it's okay, you're going to get better." [There was this] massive queer love crip party happening behind me, and I was having this tiny interaction that was just shattering all of the confidence I'd had the whole night.
>
> *(Interview 2014)*

In this moment, Isobel's desirability is completely dependent upon the erasure of her cane as the marker that makes her undesirable.

Being marked as "one of those kinds of people" works in tandem with the simultaneous process of exulting certain body–mind experiences and communities (Thobani 2007). Isobel shares the following insights about the operation of white supremacy within mainstream disability discourse and communities:

> [I'm thinking about] about how the media and criminal injustice system and prison industrial complex work together with mainstream disability discourse [to] make room for and make compassion for humanising white folks who are crazy or mad identified or neuro-divergent in other ways, that commit crimes, whereas many black folks who are disabled in similar ways who may or may not have committed "criminal acts" aren't given the opportunity to be human and [are] often killed institutionally or

otherwise. And so the erasure of black, indigenous, and [people of colour]…in main-stream disability discourse, which is mostly white, means there is less room for us to take up space as neuro-divergent or crazy folks because the risk is that we are going to get shot or arrested and then unduly tried.

*(Interview 2014)*

Present in Isobel's quote and through so many of these stories shared by collaborators is the way cultures of undesirability produce an incredibly dangerous and harmful world for marginalised people along lines of in/visibility. Systemic exclusion from the terms of worth and desirability, and, by extension, humanity, creates the conditions for immense violence and oppression.

## Shame

The combination of shame, naturalisation, and internalisation works to perpetuate and justify the terms of legibility and ways of knowing and living created by cultures of undesirability. Cultures of undesirability rely on the working of shame as a panoptical device (Foucault 1995, pp. 202–203). Within cultures of undesirability, we become ashamed about parts of ourselves that we perceive to diverge from normative standards of worth and desirability and, in turn, we monitor those shameful parts of ourselves, hiding them away or attempting to rid ourselves of them; as we monitor ourselves, we also monitor the potentially shameful parts of each other, and we sense others monitoring the potentially shameful parts of ourselves. This experience of shame functions behind the erasure, exclusion and general tendency toward disconnection that characterise cultures of undesirability. Romham discussed the struggle involved in resisting the constant pressure to disconnect and disappear:

I fight (myself and others) to allow parts of me to show up. I get embarrassed so easily, mortified by myself, by my body, by the stuttering that happens sometimes, by my leg cramps, and balding head, the way my fat has changed since using the mobility scooter, my tits, etc. I used to think to myself "well, at least I've got strong arms from using the forearm crutches, and in a queer sexual context people (whoever *they* are!) seem to like strong arms," that kinda shit; now I don't have such strong arms, and sometimes I wonder what I bring to the table.

*(Interview 2013)*

Romham's experiences speak to how deeply cultures of undesirability are internalised and individualised, how they make us question what we 'bring to the table'.

## Exclusion

It is also important to note that the standards of judging the worth of what we bring is set so that our contributions never even appear at the table. The system of determining desirability—and thereby worth—requires exclusion. All of my collaborators spoke to the pressure to be productive under capitalism as a foundational structural logic of economies of desirability. Under capitalism, bodies become legible as human and as valuable as long as they are productive labouring bodies or consumable bodies. If one is not productive in a capitalist system, one is disposable; this is justified by the idea that bodies unable or unwilling to labour are a burden to the capitalist system. Economies of desirability situate disability and embodied differences as individual deficits, rather than beneficial and important. The internalisation of the personal inferiority model and normative desirability teaches us that the problem is located within ourselves—our ways of being. Therefore, we must either change (or rehabilitate) those things about us that are different, or, as Isobel and Corrinne put it, shut down (or exterminate) pieces of ourselves.

Capitalism not only creates an inequitable distribution of value and worth, it also produces the idea that resources are finite and, as a result, it forces us into a system of competition. This system sets individual well-being against collective well-being, creating a sense of scarcity and justifying exclusion, as named by Juba:

> [T]hat there's not enough for all of us to have enough love, sex, food, money—whatever…[F]or everybody to get and have…somebody has to not just be excluded, but to be not wanted.
>
> *(Interview 2014)*

These processes combine to produce a powerful normativity, where we are pressured to disconnect from all that is determined to be not worthy, not good enough, or not deserving; as my collaborators revealed, there are severe consequences for diverging from this normativity. While this process happens in part through individual, tangible messaging, it also occurs at the level of epistemological frameworks upon which social organisation rests. Our relational fabric, under the terms of cultures of undesirability, is necessarily constrained. We face exclusion from legibility, from resources, from pieces of our complex personhood (Gordon 1996), and from community.

Disability often gets constructed as a singular experience: we are only ever disabled, not disabled and black or disabled and queer or black, queer and disabled. As a result, our complex personhood has a hard time belonging anywhere in its entirety. Collaborators shared painful moments of exclusion from various communities, such as navigating disablism in black communities and racism in disabled communities. Multiple collaborators shared feelings of being excluded from queer community due to narrow ideas of who is a desirable queer subject.

## Control

Once constructed as disposable by cultures of undesirability, marginalised people become objects for control and consumption, to the benefit of dominant power structures. The invasive ownership of marginalised bodies shows up on a structural level in several ways; examples include, but are by no means limited to: Supposed "therapeutic" sterilisation, custodial care approaches to 'managing' our sexuality, corrective surgeries at birth, freak shows and the negation of consent when being forcibly committed to psychiatric facilities. We see this structural impact through the individual examples shared by collaborators. Romham's experiences highlight the subtlety and constancy of boundary violations that disabled people often face:

> People touch my body, move my equipment, etcetera, without asking; or they ask and then do it before I have a chance to respond; or they ask, hear me say 'no' and they do it anyways, because I guess they know better. This happens basically every day, and it sends a clear message that my body is not my own, that my words don't have meaning, and that I don't know what I need and want and desire.
>
> *(Interview 2014)*

The internalisation of the message that we do not know our own bodies, needs and desires impacts our ability to consent. Exclusion and the politics of scarcity combine to compel marginalised people to accept abuse and neglect out of fear of not ever being chosen—of feeling undesirable. When we say no or assert ourselves, we are aware that we will likely either be ignored or disposed of.

## Cultivating otherwise: Access intimacy and queercrip porn

In the following story by Nomy, we see an act of imagining otherwise and the importance of envisioning new worlds of possibility. She says:

> As early as I could remember, I didn't ever see myself ever getting married. So, I made up this story to keep myself alive about how at some point I would have a community of people that were artists that were trying to change the world through like music, and a key aspect of it was that you wouldn't be able to tell who was a boy and who was a girl. [laughs] Umm, so that was kind of a vision that I invented that maybe I was practicing for down the road because I could never picture there being like a man and I knew that was like the normal thing.
>
> *(Interview 2013)*

When ways of being and imagining worlds don't fit in our reality or are causing us significant harm, it's time to tell new and different stories: the very stories that cultures of undesirability make largely unknowable. Just as cultures of undesirability create a world that is violent and difficult, if not impossible, for marginalised people to navigate, radical access and fostering moments of access intimacy are key to creating queercrip porn worlds that make space for the things written out and written over by cultures of undesirability; worlds where it is okay for us to show up and flaunt our complexity.

In "Access Intimacy: The Missing Link," Mia Mingus first articulates access intimacy as "the way your body relaxes and opens up with someone when all your access needs are being met" (Mingus 2011b, n.p.). Mingus later elaborates how "access intimacy at once recognises and understands the relational and human quality of access, while simultaneously deepening the relationships involved. It moves the work of access out of the realm of only logistics and into the realm of relationships" (Mingus 2017, n.p). Many of my collaborators came to this project with a strong sense of how to cultivate access intimacy. For example, Mia Gimp reflected on filming *Krutch* (Matthews 2014):

> So often in a non-disabled community…you find yourself explaining things and using your energy to explain your needs without actually getting what you need accomplished. [Director] Mat Fraser and I had this moment where I kept having to…take my shoes off and put my shoes back on… We did three or four takes of that. And Mat says to me, "Do you want me to help you?" And I go, "No," and he goes, "Yeah, I know, because it'll take too long for you to explain to me exactly what it is you need." I didn't have to tell him [why I didn't want help]; he already knew because he lived it and experienced it.
>
> *(Interview 2014)*

In dominant culture, ease of movement characterises what privilege *feels* like: Ease is taken for granted and assumed, and the world is built on this assumption. This assumption results in huge amounts of extra work that must be done on a daily basis for people moving through a world that was not built with us in mind. Radical access compels us to start from a different point as we begin to imagine new worlds and new ways of being and living, where ease is built into the structure rather than attached to particular ways of being. When I asked Isobel about how this prioritisation of access intimacy and radical access shaped her experience filming, she replied:

> Magic. Just, who knew? The fact that I knew mostly everybody in the room and that I had performed in one way or another for folks…there was kind of an intro that was already gone, we had a bit of comfort. And the people [who] I didn't know were so good, were so easy to be around. And the fact that it was different bodies, different brains; that there was trauma talk in the room; that there was food sensitivities and needs.
>
> *(Interview 2014)*

Access intimacy is larger than feeling supported or getting access needs met—it's a connective moment of "interdependence in action," where all involved are recognised and held in all of their complexity (Mingus 2017, n.p). Access intimacy calls us to cultivate a space "of steel vulnerability" where we can hold all of the barriers, pain, logistics and needs for and with each other (Mingus 2011b, n.p.).

## Introducing queercrip porn

*want* (Erickson 2006a) and *sexxxy* (Erickson 2006b) had a formative influence in the naming and beginning of queercrip porn as a category of counter-public porn. Plummer's (1995) analysis of the interdependence of emergence, distribution and reception of sexual storytelling applies here: it's not that queercrip porn didn't exist before, as queercrips have been making porn in a variety of contexts, but my films were among the first to be widely viewed, distributed and understood *as queercrip porn*. According to Romham:

> Total honesty? I don't really know [laughter]. But it's something I want to see, want to *be part of* in some way. When I've seen the things you've made, there is always something (many things!) in it that really *connects for me*, there's always such hotness and creativity and humour.
>
> *(Interview 2013)*

While there may seem to be a touch of "I know it when I see it" occurring here, reminiscent of Justice Stewart's words used to describe and identify "hard-core pornography" in *Jacobellis v. Ohio* (Williams 1999, p. 5), I think there is also an "I know it when I *feel* it" element, which signals an important shift. Rather than the distanced and otherising voices of Stewart and other so-called experts on porn, Romham's words emphasize connection: Romham wants to be a part of the making of queercrip porn.

## Disruption

In showing up and becoming visible as queercrip, we intentionally shift the frame of desirability. This intentionality goes beyond a liberal concept of disability rights that demands access and inclusion into an oppressive system. Rather than demand to be recognized within dominant ideologies of citizenship, we disrupt these ideologies as conditions for gaining recognition as desirable subjects. Moments of disruption to dominant ideologies were important to each collaborator. This sense of disruption is conveyed by Juba's understanding of queercrip porn as "a sort of cliff dive for people" (Interview 2014). Queercrip porn interrupts the naturalised and taken-for-granted order of things by intentionally being out of alignment with normative culture, consciously doing and being something noticeably different from normalised ways of doing and being. Indeed, over the course of producing porn with my collaborators, my own understanding of queercrip porn as a counter-public strategy, seeking to disrupt and interrogate normative ideologies, grew. Claiming one's own experience and being disruptively 'out' are crucial elements of queercrip porn. Corrinne adds, "redefining things for ourselves, like we are doing here, is a really important part of transformation" (Interview 2014). At the same time, collaborators complicated a liberal concept of coming out: queercrip porn is about more than just being out; it is about flaunting those very things about us that disrupt the normativity so critical to economies of desirability. In other places, I have articulated this flaunting elsewhere using the conceptual frame "sites of shame as sites of resistance" (Erickson 2010; 2007).

## Flaunting our whole messy selves, "tires and all"

In reflecting on how queercrip porn can provide a space to practice bringing in our messy, vibrant, vulnerable, powerful whole selves, I am reminded of a story Romham shared about fucking in their scooter during the pre-shoot. That experience was some of the best sex they'd ever had. It was one of the first times that they felt like they could bring their whole self forward, "tires and all," and were really seen and held by their partner and themselves. The queercrip porn stories of adaptive devices featured in these scenes signal an opportunity for contemplation on the interrelation of crip embodiment, visibility and desirability. In Isobel and Juba's scene, shot by Allen, Isobel stretches her cane across the doorframe for support while being spanked. Below, Mia explains the idea for *Krutch* (Matthews 2014), a film she made outside of this project but which she reflected on heavily in her interviews:

> I watched *want* and...*sexxxy*, and it seemed to me like there was commentary on the ways in which our adaptive devices can be sexy too. And then there was this bondage workshop that I was taking where...there was a piece of bamboo that was tied between the model's legs and they tied a vibrator to the bamboo. And I was like, "Holy crap! I could tie my crutch to my body." I practice bondage play as a form of accessibility. My favourite quote is, "Tie my legs down to the bed so I don't kick my partner in the head."
>
> *(Interview 2014)*

Lisa added:

> I noticed, since I moved here and Nomy is my partner, that I see people with canes, and I am totally looking. Like, fat femme presenting people. It's broadening the scope of what I think is hot, right? And totally identifying with a cane is hot. My friends that are like, "I should be using this cane, but I am not going to because I don't want to." It would make things more accessible for them to use. And it's like, "the canes are sexy!"
>
> *(Interview 2014)*

These moments together reframe associations with adaptive devices, both as parts of ourselves that we use to facilitate access and as sources of pleasure.

Our access devices are deeply significant to our sense of possibility in terms of survival, what our day-to-day looks like, and how we come to understand ourselves, our relationships and the worlds we live in. In addition to this reframing of adaptive devices, the reframing of sites of shame as queercrip skills and embodied gifts shifts the conversation away from disability as deficit. An important component of both Nomy and Corrinne's scenes involved flaunting skills and embodied differences which depart from normative desirability. For Corrinne, sitting in pain during meditation posed very little challenge for her because she is very familiar with meditating in pain; however, people who do not experience chronic pain would need to acquire these skilful practices. In *Wall of Fire*, Nomy uses her little leg to penetrate her partner. She uses the term "sexual abilities" to describe "things that I can do with my body that people who do not have a little leg cannot do" (Interview 2014).

Queercrip porn seeks more than a simple reversal of the dominant narrative, where the triumphant overcomer—"the supercrip" (Clare 2015)—replaces tragic victim, reaffirming and rehabilitating the individual at the expense of collectivity. Mia emphasised how sharing the hard stuff is necessary when presenting our complex whole selves:

> Because what I can't do also informs the way that I live in this world and the ways in which my body informs what I do. What I can't do actually sometimes helps me get things done because I've learned other ways to do things that are actually more adaptable and save time; or sometimes it doesn't save time, and I've learned to budget my time. I know how to do that because of my disability.
>
> *(Interview 2014)*

Creating this whole picture isn't easy work to do. Due to the internalisation of cultures of undesirability, the impulse to hide away any aspect of ourselves seen as disposable, shameful or unworthy—is also always present, mixing in with this taste of possibility we are creating. And it kind of has to be. Bringing queercrip worlds into existence further creates a space to understand and claim sites of shame as potential sites of resistance, opening up shared opportunities to practice living otherwise. Corrinne shares, "I practice ignoring the voices that say no. I practice not feeling bad or giving into shame. I was worried about not looking good. [Creating my scene] gave me a moment to reflect. If I see something I don't like I get to challenge myself around that. It's a process." Similar to the effects of access intimacy, seeing something in porn that resonates with our experiences and embodiment means we don't have to work as hard to explain our needs and justify our existence, freeing up time and capacity for rest, fun, pleasure, which are all crucial ingredients for sustainability. We are still living and negotiating cultures of undesirability every day. Queercrip porn does not solve systemic oppression, but it does provide openings for disruption, recognition and transformation. It gives us something to hold onto: New stories to disrupt and counteract the din of undesirability.

This transformation has to occur on multiple levels; we have to not only think differently, we need to feel differently. I began making porn precisely because it makes us feel things— good, bad and everything in between. As queercrips, we need to feel new possibilities by embodying different ways of living. In this way, shifting a site of shame to a site of resistance happens at the level of ontology, where shame transforms from something to move away from, cover up or disconnect with into something to move toward, connect with and feel out. In spending time within sites of shame and understanding them differently, we open up the potential to produce new knowledges, ways of living, being and feeling. We need the haunting, Avery Gordon's concept attending to the ethereal, murky, absent presences that inhabit and shape our worlds, in order to move towards flaunting (Gordon 1996).

## Leaving evidence

Creating a record of moments when complex personhood appears and is flaunted allows us to shape how others see us as well as how we see ourselves. Juba shared a moment of powerful resonance while watching *TransEntities* (Diamond 2007):

> At the end of this real hot scene, Will pulls out his Albuterol and like takes a couple of puffs on the inhaler, and I was just like "DAMN!" I'm certain that's not the first time where someone did that—but that got edited out. And that [the director] left it in there, it was powerful for me, as this moment where this Black, masculine spectrum person with respiratory issues and it's like, "I'm layin' this hot fuck, but I still need my [inhaler] and I'm not hiding that; my masculinity and the way that my masculinity is structured, I don't need it to be this immutable…container that doesn't have any cracks in it…I'm going to be all of these things in this moment and that's okay."
>
> *(Interview 2014)*

Several collaborators conceptualised queercrip porn as facilitating the presence of multiple intersecting positionalities, movements, cultures and histories as illuminated in Juba's example. Many discussed the importance of creating a record and an archive of otherwise to interrupt the erasure and hypervisibility so critical to the functioning of cultures of undesirability. Queercrip porn is essentially "leaving evidence," to evoke the powerful title of Mia Mingus' seminal blog. This concept was a crucial reference to all of us involved in this project. Mingus writes:

> We must leave evidence. Evidence that we were here, that we existed, that we survived and loved and ached. Evidence of the wholeness we never felt and the immense sense of fullness we gave to each other. Evidence of who we were, who we thought we were, who we never should have been. Evidence for each other that there are other ways to live—past survival; past isolation.
>
> *(2008, n.p.)*

Queercrip porn, by leaving evidence, works as an art of life (Champagne 1991). It seeks to connect with subjugated knowledge, generating new relationship systems and new ways of understanding bodies, sex, sexuality, desirability and identity. I think it significant that the evidence we are producing as we work to expose the oppressive hegemonic worldview that subjugates our experiences and knowledge grants permission to subjugated subjects to understand our ways of living and being as valuable. As Isobel stated:

> You know, it's not a normal thing that these kinds of bodies get to do these kinds of things and are celebrated… [In] having this multi-faceted experience of talking about it, planning it, filming it, talking about it again and then whatever's going to happen in terms of viewing it, we've made this strength-web kind of thing and whatever happens with it, whoever sees it, I feel like it's going to fucking change their lives, because it changes; every time I see something that's like, that's kind of like me… It changes my day, it changes everything.
>
> *(Interview 2014)*

As Isobel's words highlight, the interdependency and connection between the makers and the audiences of queercrip porn is both disruptive and transformative. Collaborators named multiple community exclusions they experienced and the ever-creeping pressure towards norma-tivity running throughout even communities of resistance. Despite the limits of community, collaborators spoke to the way that connecting with queercrip community provided support and recognition, which were essential in creating counter-publics and new ways of living.

We are all interdependent beings: We need each other to survive. This project itself would not be possible without direct community support, not only from my collaborators, but also from people who let us stay at their houses, and from my care collective, who lift me out of bed every morning. Because social exclusion is such a huge part of how cultures of undesirability operate to tear us apart from each other and ourselves, community must be central to any effort at transfor-mation. Flaunting interdependent connection blows apart dominant ideologies that work to cover over the complexity, wisdom, pleasure and potentiality derived through interdependence. A criti-cal difference between the queercrip porn produced for this project and mainstream porn is that, even while part of the pleasure derived from making and watching it happens on an individual or interpersonal level, queercrip porn centres a commitment to collective liberation—fostering not just self-worth but a communal sense of worth. According to Romham, it is in the moment of collective disruption and community building that queercrip becomes possible, giving rise to new relational systems and ontologies that are actively formed and nurtured through interdependence:

> When I read [Mia Mingus'], "Wherever You Are Is Where I Want to Be: Crip Solidarity," I cried because that's it, right? Whether we're talking porn or a protest, or both—or play parties, food sharing, advocating for one another—whatever it is, we need to be able to show up how we show up, we have to make each other priorities. None of it means shit if we're not doing this together. I think that's why I was like, "okay, y'know what? I need to show up for this [project], tyres and all, just fucking do this thing."
>
> *(Interview 2013)*

For me, these statements signalled an important pushback to the process of individualisation that underlies the functioning of cultures of undesirability.

## Flaunting back: Reclaiming agency or "yelling to the stars"

When asked what queercrip porn does to transform cultures of undesirability, Juba emphatically replied:

> It says, "*Yes*." It says, "fuck this shit! I *am* and I *want*, and I can *be*, and I can *do*." [It's] about the *capture* of it as a record.... This is something that just came to me, that so much of mainstream porn is about "this is what you're supposed to want and how it's supposed to look and how it's supposed to be…" Whereas queercrip porn, for me, just sort of yells to the stars: "This is where we was, right today…and we with it right now."
>
> *(Interview 2014)*

As Juba stated, so much of mainstream porn transports us to worlds of undesirability and, by dictating how we are "supposed" to experience sexual pleasure and what we are "supposed" to find desirable, moves so many of us away from ourselves. Queercrip porn, with the deliberate and campy playfulness of Muñoz's (1999) disidentification, says "yes" to newly imagined worlds. It does so by making use of porn's potential to tell and live subjugated sexual stories. The fact that these worlds are created, recorded and witnessed provides evidence that throws into ques-tion the supposed truths of dominant culture.

At the start of this essay, I discussed the impacts of being inundated by the effect these supposed truths has upon marginalised people, how cultures of undesirability lay the groundwork for abuse and violence. Multiple collaborators articulated the challenges of cultivating a sense of agency and bodily autonomy in the face of staggering structural inequality. Nomy shared several bad experiences with making porn, where her boundaries were not respected. In one such story her and her scene partner's naked bodies were literally up for grabs as, at an event, the director, without getting consent, had stacks of their porn for people to take home. After multiple negative experiences, she took a break to do deep healing work. She returned to porn for this project, because she felt that this would be a chance to make porn in a more consensual and supportive context, and because she wanted to support my work. Nomy explained that, through making and sharing queercrip porn, she was given something to hold—a record with witnesses—which enabled her to reflect upon intimacy, her body, agency and her relationship with Lisa:

> I think both of us, as artists and collaborators and people with trauma histories, being able to hold something and look at it and have some control and make some decisions about it is kind of a cool thing to be able to do around sex... Our desires are commodified and controlled by a system that is totally patriarchal and heteronormative, and the ways that we internalise those messages—and then ways that we transform them by making our own stuff and being in an economy or an ecology of artists who are supporting and exchanging this work with each other—it's a totally different model....For us to have agency around our own stories and desire...I feel that's been a big part of my work and my path and transforming cultures of undesirability was happening through my work to feel okay with myself, with love and desire, to be desired and connecting that to a political framework so that it's not just about me.
>
> *(Interview 2014)*

In other collaborators' accounts, making collaborative queercrip porn was disruptive precisely because they regained control in the face of domination: A control that reflected agency combined with collective responsibility and shared vulnerability. Isobel speaks in similar but broader terms about not only the extensive impacts that cultures of undesirability have on every aspect of our lives, but also the necessity of reconnecting with a sense of agency:

> There's so little power that my body has in the world most of the time. Like, transit, meds, health, all that stuff. I feel like that is an experience of a lot of marginalised folks too. So it feels strengthening and powerful to be able to decide how we're being seen and decide how we're being sexualised, and to participate in the ways that we're being perceived as sexual, because I feel that very much pushes against 89% of my experience of being catcalled and all the things that happen to me in the day just being out in the world...The way that we did it was really good for me—the *talk, talk, talk; stop; do it again; talk, talk, talk a little bit longer; stop; do it again; talk, talk, talk; stop; do it again*—I wasn't thinking about the camera a lot, which was great, even though it's a massive boom... Like, I was aware, and I could feel when Allen was moving around... But I think the fact that we talked about the kind of shots I feel comfortable with and what kind of parts of myself I did or didn't.
>
> *(Interview 2014)*

Claiming control over the process of making porn not only addresses the denial of agency queercrips so often experience but also powerfully speaks back to narratives of disposability and consumption. This project rested on the idea that we need to create spaces where the stories of dominant imaginary are brought forward, questioned and then disrupted in order to tell collective stories that produce embodied differences as poetic and as providing necessary ways of being and living. Mia explained:

> [U]sing the ways in which our bodies are poetic, and have a language, like the ways in which my crutch clicking has a soundscape that communicates something in it. And there's a … communication in the ways our wheels move on the ground, or the ways our bodies sit in our chairs. There's a language there.
>
> *(Interview 2014)*

There was something important—magical even—about the connections between access intimacy, radical disability politics, reimagining embodied difference and the process of recording these precious moments.

## Conclusion: Resilience, tiny purple feathers of queercrip possibility

I started this project because I felt like I was losing the epic battle with cultures of undesirability. I was desperate to find out how other queercrips navigated, survived systemic oppression, violence and abuse so present in many of our lives. I focused on porn in part because it is one of the ways I nurture resilience. Each of the collaborators, through their smiles, laughter, tears, brilliant insights and blazing hot scenes provided me with resilience through what were most certainly the worst two years of my life.

On a day when I was feeling particularly sick as well as defeated by a broken heart, I received a link from Nomy to her solo scene, *Time to Say Hello*. I watched the short, powerful video and cried. Nomy shared how this movie was pivotal in connecting to her own sexual agency and power:

> The whole point of that piece is: I'm feeling sad and it's impossible for me to get what I want. I don't see how it is possible for me to go to the person who I love most and ask for what I want, and so then I give it to myself. And that moment, I just made that little movie in one night… I focused on what I think is super-hot, which is people's faces. So it was just on my face. Since making that movie, it's kind of launched me on doing a different level of sexual healing where I feel like in the past I have been doing this sexual healing where I am like healing from trauma, [now] I'm energising the battery and creating that energy in me… [It] can engage with others but it doesn't have to because it's actually mine and it's for me. And that's like a really different worldview and not one that I grew up thinking I had a right to.
>
> *(Interview 2014)*

*Time to Say Hello* made me cry because it was a moment of glorious access intimacy. Here was this beautiful, sexy, moving, complicated queercrip porn that expressed so powerfully the ways that cultures of undesirability work to undermine our sense of being worthy of desire, love and celebration as whole complex persons. Similar to the emotions communicated by Nomy above, being able to feel worthy of asking for and receiving what I want and need from sexual partners will probably be something I'll always struggle with. Witnessing this moment of powerful vulnerability shared by Nomy made me so grateful to be a part of making this hard and necessary work. Queercrip porn fosters resilience by interweaving the four key qualities discussed above: Intentional and disruptive cultural production, creating space for complex personhood, interdependence, and leaving evidence of queercrip world-making. The interweaving of these qualities was active in each of the scenes we made, in the conversations we had, and in the connections we shared. Like my living room after shooting Juba and Isobel's scene, *Blackbeats &AfroCripLust [LightYears Away]*, queercrip porn disrupts and transforms. It leaves delightful purple feathers of evidence of queercrip flourishing and flaunting towards otherwise.

## Note

1  Some experiences associated with queered femininity or with queer femmes experience hypervisibility around their gender expression, whereas others experience invisibility. Hypervisibility and invisibility are not separate, rather they are interconnected, producing precarity and policing (both external and internal). Many trans women of colour writers, such as Kai Cheng Thom, and non-binary writers, such as Alok Vaid-Menon, have written about the complexity of managing visibility.

## References

Ahmed, S. (2006). *Queer phenomenology: Orientations, objects, others.* Durham, NC: Duke University Press.

Butler, J. (2005) *Giving an account of oneself.* New York: Fordham University Press.

Champagne, J. (1991) Interrupted pleasure: A Foucauldian reading of hard core: A hard core (mis)reading of Foucault. Boundary 2(18), pp. 181–206. https://doi.org/10.2307/303285

Clare, E. (2015) *Exile and pride: Disability, queerness, and liberation.* Durham, NC: Duke University Press. https://doi.org/10.1215/9780822374879

Diamond, M. (2007) *Trans entities: The nasty love of Papi and Wil.* Independently produced and distributed film.

Erickson, L. (2006a) *Want.* Independently produced and distributed film.

Erickson, L. (2006b) *Sexxxy.* Independently produced and distributed film.

Erickson, L. (2007) Revealing femmegimp: A sex-positive reflection on sites of shame as sites of resistance for people with Disabilities. *Atlantis: Critical Studies in Gender, Culture & Social Justice* 31(2), pp. 42–52. http://journals.msvu.ca/index.php/atlantis/article/view/679/669

Erickson, L. (2010) Out of line: The sexy femmegimp politics of flaunting it. In Stombler, M., Baunach, D., Simonds, W., Windsor, E., & Burgess, E. (eds.), *Sex matters: The sexuality and society reader.* Upper Saddle River, NJ: Prentice-Hall, pp. 157–162.

Foucault, M. (1995) *Discipline and punish: The birth of the prison.* REP edition. New York, NY: Vintage.

Gordon, A. (1996) *Ghostly matters: Haunting and the sociological imagination.* Minneapolis, MN: University of Minnesota Press.

Martin, L., Gutman, H., & Hutton, P. (1988) *Technologies of the self: A seminar with Michel Foucault.* Amherst, MA: University of Massachusetts Press.

Matthews, C. (2014) *Krutch.* Independently produced and distributed film.

Mingus, M. (2010) Wherever you are is where I want to be: Crip solidarity. *Leaving Evidence.* https://leavingevidence.wordpress.com/2010/05/03/where-ever-you-are-is-where-i-want-to-be-crip-solidarity/ (accessed 5.31.19).

Mingus, M. (2011) Access intimacy: The missing link. *Leaving Evidence.* https://leavingevidence.wordpress.com/2011/05/05/access-intimacy-the-missing-link/ (accessed 5.31.19).

Mingus, M. (2017) Access intimacy, interdependence and disability justice. *Leaving Evidence.* https://leavingevidence.wordpress.com/2017/04/12/access-intimacy-interdependence-and-disability-justice/ (accessed 1.2.20).

Muñoz, J.E. (1999) *Disidentifications: Queers of colour and the performance of politics.* Minneapolis, MN: University of Minnesota Press.

Peers, D., (2012) *Desiring disability: Roundtable discussion with Robert McRuer, Loree Erickson, and Daniel Peers.* Canada: University of Alberta, 18th October.

Plummer, K. (1995) *Telling sexual stories: Power, change and social change.* New York, NY: Routledge.

Thobani, Sunera (2007) *Exalted subjects: Studies in the making of race and nation in Canada.* Toronto, Canada: University of Toronto Press.

Williams, L. (1999) Hard core: Power, pleasure, and the "frenzy of the visible." Berkeley, CA: University of California Press.

Withers, A.J. (2012) *Disability politics and theory.* Winnipeg: Fernwood Publishing.

# 20

# DESEXUALISING DISABLED PEOPLE IN THE NEWS MEDIA

*Gerard Goggin and Helen Meekosha*

## Introduction

In this chapter, we look at the role of contemporary mainstream news media in shaping disability and sexuality. We do so through discussing the highly controversial case of Anna Stubblefield, former Professor of Philosophy at Rutgers University in the US. In late 2015, Stubblefield was convicted of two counts of aggravated sexual assault of Derrick Johnson, a 29-year-old man with cerebral palsy. A core issue was whether Johnson, who uses facilitated communication, was able to consent to a sexual relationship with Stubblefield, given his communicative mode. Another key issue was the nature and implications of Johnson's intellectual disability and cognitive impairment. Also core to the case and the representations of disability and sexuality it generated were matters of race, including the fact that Johnson is a person of African-American heritage, and Stubblefield had been married to an African man, and had a longstanding research interest in the philosophy of race (Baron 2018, p. 438; Stubblefield 2009).

The Stubblefield case was widely covered in mainstream news media, as well as across emergent online and social media platforms. It raised a bewildering array of issues for disabled people and their allies and has caused divisions within the disability communities in the English-speaking world. The Stubblefield case drew widespread international attention especially because of the concerns it drew and the debate it generated about disability and sexuality. For this reason, it offers us important lessons. Of particular concern is the way that sexuality was represented across a range of media over the period in which the case was publicly visible; hence the argument in this paper that the dominant representation amounted to a desexualising of disabled people, especially people whose bodies, communicative styles, intimacies and sexualities do not fit the "norm."

Furthermore, we link the concept of consent, extremely problematic when applied to many disabled people (Harris 2018; 2019), to media coverage of the case in question. The laws and norms on sexuality and disability (Perlin and Lynch 2016; Perlin et al. 2019) are front and centre in the Stubblefield case, as we shall discuss—especially as these are mediated in public and private spheres (Papacharissi 2014; Schweik 2009). The Stubblefield case offers a disturbing and striking picture of the state of play of sexual norms in relation to disabilities; and it also underscores the crucial role that media, even new digital platforms, can play in policing disabled sexualities—or, alternatively, offering new modes of expression and supporting the reimagining of sexuality.

To develop our argument, this chapter is structured as follows. Firstly, we offer an account of disability, sexuality and media representation. In particular, we discuss emerging conceptualisations of how media representation of disability has changed in recent years—due to the entangled dynamics of cultural and social transformations in disability, as well as new media technologies, and the social and political participation they support (or circumscribe). Secondly, we provide a brief chronicle of the Stubblefield case, and highlight its key and influential moments in terms of how disability and sexuality were represented in international public spheres. Thirdly, we discuss some key themes in the representation of disability and media in the case. Fourthly, we draw together our argument about how the Stubblefield case is shaped by, and in turn, contributes to the desexualisation and consequent isolation of disabled people.

## Disability, sexuality and contemporary media

Access for disabled people to sexual experiences and activation of their sexual rights and identities have been slowly improving—especially in urban settings. But there remains much to be accomplished (see, for example, Kulick and Rydström 2015). Disability rights remain neglected in rural and remote areas—with local shops and clubs inaccessible, young people incarcerated in nursing homes, and carers and parents demonstrating an aversion to the use of sex workers (Meekosha 2012).

Nevertheless, the strength of disability movements and shifting community attitudes has resulted in a shift away from medicalisation and institutionalisation of disabled people (WHO, 2011). A threshold issue has been the dominant framing of disability and sexuality, as Paul Chappell notes:

> The medicalised and apolitical focus on disabled sexualities has not only drawn attention away from the sexual agency of disabled people, but also from the socio-cultural meanings of disability and desirability, and the experiences of multiple barriers to sexual expression and relationships.
>
> *(Chappell 2019, p. 19)*

Among those actors interested and committed to transformations in the social relations of disability, sexuality has lagged behind as a priority or indeed acceptable focus, something long noted in relation to the Global North, and recently raised in relation to diverse sexualities in the Global South also (Chappell 2019, p. 19). Internationally there are now signs of change. Disabled people's organisations increasingly have policies and procedures that incorporate resources for development and training of their members in sexual rights. Mainstream organisations in Australia, such as Family Planning Australia, have long devoted time and energy into supporting training manuals that encourage disabled people to explore their sexuality. Dignity of risk and duty of care are principles that have overruled the more infantilising constraints of earlier decades. For instance, the Australian Department of Family and Community Services in New South Wales have developed supportive policies around sexual relationships and disabled people's use of sex workers (FACS 2016).

If change in disability and sexuality remains slow and is definitely unfinished business, one important resource in recent years has been transformations in the media—in which disability has played a vital, if uneven, and at times, fraught role (Ellcessor and Kirkpatrick 2017; 2019; Ellis et al. 2020). Increased visibility of disabled people has resulted in wider community familiarity, especially in TV series, movies and increasingly in terms of sporting prowess. Dylan Alcott, a paraplegic and an international wheelchair basketballer/tennis player, hailing from Australia, often appears on mainstream media to discuss contemporary social and political issues, not necessarily related to disability. There have been a number of disabled people featured in TV series, such as the iconic US drama series *Breaking Bad* and *Game of Thrones* (Ellis 2015). People with disabilities are also increasingly participating in news and current affairs programs, such as *Q and A* and *The Drum* in Australia (Goggin and Ellis 2015). Disability is also being represented in innovative, if still contradictory, ways in new formats such as reality television series that are influential in popular culture (Ellis 2015). These signs of change are encouraging; however, employing someone with a disability to portray a character with a disability remains a central struggle within the disability movement, and across the media and the communication industry globally (Ellis 2016).

Sexuality has been threaded through these shifts in media in important ways. Notably, we can see a growing diversity of voices, identities and representations of disability and sexuality challenging the disabled myths of attractiveness, narrow bodily norms, and the restricted range of sexual identities and practices critiqued by many disability scholars and activists (Gill 2015; Malinowska 2018; Shakespeare 2000; Shakespeare et al.1996; Shildrick 2009). It is commonplace to acknowledge that the growth of social media has allowed marginalised people to have their voice heard and claim agency and a presence—again with constraints, exclusions and downsides in this, as discussed by various scholars (Ellis and Kent 2017). Hopefully, growing number of disabled people are now more likely to be portrayed in ways that they have long desired. To give but one example, the twitter hashtag #DisabledPeopleAreHot allows disabled people to chat about their sexual issues, how to dress sexily and adapt their personal aids and appliances to reflect their sexuality (Burgess et al. 2016). The idea of "hot" suggests a totally different way of describing disabled people as sexual beings and is appealing to those of us who object to labels such as inspiring and motivational (Feldman 2019). Despite such affirmative possibilities of social media platforms (Miller 2017), there are many other challenges in disability and digital cultures. For instance, online dating remains inaccessible to most disabled people. It remains uncommon to see disabled people on dating sites. Research and anecdotal evidence suggest that when individuals on dating sites disclose their impairments to potential partners, they often experience anxiety or are cruelly dismissed, reinforcing their sense of isolation and marginalisation (Mazur 2017; Milbrodt 2019; Saltes 2013).

In summary, it is a mixed picture when it comes to the state of play of disability and sexuality in relation to the media. We can point to many positive developments, areas of progress, ground-breaking advances in media and representation, as well as the new possibilities of digital technologies and cultures. Set against these advances are the persistence of disabled images and metaphors, stereotypes, a restricted palette of media opportunities for people with disabilities as well as new kinds of digital exclusions. Media and communications are deeply entrenched in everyday life, even more so when it comes to how we experience ours and others' sexualities. With this background, we have an important framework for approaching and understanding one of the most controversial cases in disability and sexuality in recent times.

## Law and media

Johnson's brother, Wesley, first met Professor Anna Stubblefield when he attended her philosophy classes at the Rutgers University Newark campus in 2008. Apparently, Stubblefield showed her class the 2004 documentary *Autism World*, which includes a section portraying a non-verbal girl, who, due to the method of Facilitated Communication (FC), is able to attend college (Baron 2018; Engber 2015; 2016). Before we proceed further, a short note on FC is necessary.

The method of Facilitated Communication (FC), an approach in which a person with disability points to words and images on a communication device, screen or keyboard, and is assisted by a "facilitator" in interpreting and formulating communication (ASHA 2018). FC is key to the way disability and sexuality is represented in this case, so a brief explanation is in order. FC was brought to attention by Douglas Biklen in a 1990 qualitative study of educator Rosemary Crossley's work (Crossley and McDonald 1984) experimenting with communication training with people with severe communication impairments in Melbourne, Australia, in the 1970s and 1980s (Biklen 1990; Crossley 1992). Since this point, FC has been dogged by controversy (Biklen and Cardinal 1997; Hudson 1995). From early on, critics of FC pointed to its association with issues of inadequate communication, lack of consent by people with disabilities to sexual relationships, and the use of FC in potentially false allegations of sexual abuse (see, for instance, the Melbourne cases discussed in Hudson (1995) and Boynton (2012)). Some thirty years on, many influential researchers and clinicians, such as the American-Language-Hearing Association (ASHA) (ASHA 2018; see also Hemsley et al. 2018), regard FC as "pseudo-science," whereas others still believe it has an important role to play in developing and supporting independent communication (such as Cardinal and Falvey 2015).

In the Stubblefield case, what occurred is that Johnson's brother asked Stubblefield for more information on FC, hoping that it might be of assistance for Derrick. After she was introduced to Derrick Johnson by his brother, she worked with Derrick, using FC with apparently encouraging results. They then used FC to write two academic papers (Engber 2015; Wichert 2015b). Assisted by Stubblefield, Johnson presented a paper at the Society for Disability Studies conference, a leading US and international annual meeting in the field. Versions of the paper were published in the newsletter of the Autism National Committee and also the journal *Disability Studies Quarterly*, which subsequently retracted it due to major overlap with the first published version (Johnson 2011a; 2011b). According to the testimony from her trial, in spring 2010, Stubblefield began to have "romantic feelings" for Johnson. In May 2011, Stubblefield disclosed their sexual relationship to Johnson's mother.

Public discussion and concern surrounding the relationship developed with various legal suits and processes. Johnson's brother reported the relationship and his concerns about Stubblefield's conduct to Rutgers University. Rutgers referred the matter to Essex County prosecutors, a move which kicked off the investigation and proceedings for breach of criminal laws. In January 2013, Stubblefield was indicted on aggravated sexual assault charges and in April 2014, she appeared in court for the charges (Szteinbaum 2014; Zambito 2014). In October 2015, a jury found Stubblefield guilty of two counts of first-degree aggravated assault. In January 2016, Superior Court Judge Siobhan Teare sentenced Stubblefield to two consecutive detentions of 12 years in prison and lifetime parole supervision, describing her as "the perfect example of a predator preying on their prey" (quoted in Wichert 2016).

Judge Teare explained her view that Stubblefield "took advantage of her position of power over someone that she knew or should have known was mentally and physically disabled and had no means to resist" (quoted in Wichert 2016). (We would note here that, whereas women predominate as victims of sexual attack and assault, feminists are often labelled as predators, likely to falsely accuse men of rape; cf. Cohen 2014; Javaid, 2016). Stubblefield subsequently filed an appeal against the verdict. The American Civil Liberties Union (ACLU) and various disability rights organizations prepared an amicus brief, arguing that the way Johnson was presented to the court was a violation of his rights (noted in McMahan and Singer 2017). The aim of the ACLU was to have this heard in conjunction with Stubblefield's appeal. However, the appellate court refused to consider this brief. Regardless, in June 2017, the appellate court overturned her convictions, deciding she did not receive a fair trial. The court ordered Stubblefield receive a fresh trial with a new judge (*State v. Stubblefield* 2017).

In March 2018, in the lead-up to the retrial, Stubblefield struck a deal with prosecutors to plead guilty to a third-degree charge of aggravated sexual conduct. Under its terms, she did not have to return to prison. Rather, Stubblefield would be obliged to spend the rest of her life under parole supervision, register as a sex offender under Megan's Law conditions and be forbidden to have contact with the victim or his family. The media have been noticeably quiet on what effect these conditions would have on the rest of her life.

As well as the criminal proceedings, Stubblefield also faced civil suits. Relatively early on, Johnson's brother and mother were clearly not satisfied with Rutgers' response to their complaints, and, in February 2013, filed suit against both Rutgers and Stubblefield. The case was moved to the Federal Court, which dismissed the complaint against Rutgers. The civil case against Stubblefield continued in the superior State Court, with Stubblefield representing herself. On 15 Oct 2016, the Court determined a default judgement, and awarded Johnson's brother and mother $4 million in damages—$2 million in compensatory damages and $2 million in punitive damages.

As this complex case unfolded publicly, we witness a close, if tense, relationship between the operations of law and media. Legal practitioners, institutions and actors shape, constrain and enable, indirectly or directly, the lives of people with disabilities—making the justice system an important project for disability transformations and rights advancement. As we are both academics from Australia, it is worth noting that we live in a country where sexual relations with disabled people are a crime in the law of least one state, Queensland. Section 216 of the Queensland *Criminal Code Act 1899* (the Criminal Code) effectively (or, at least, additionally) criminalises sexual activity with people with disability: "Any person who has or attempts to have unlawful carnal knowledge of a person with an impairment of the mind is... guilty of a crime" (s. 216, subsection 1). It is also a crime to engage in other sexual behaviours with a person with an "impairment of the mind" (with a very broad definition applying). The Office of the Public Advocate notes that, whereas such a provision may aim to provide specific protection for people with disabilities, it raises questions about the current law being "unduly restrictive of a person's freedom," and, more broadly, criminalises support and education programmes and indeed sexual relationships of people with disabilities—if "even giving consent may not mitigate the unlawfulness of the act" (Office of Public Advocate 2016, p. 11). Such laws exist in other countries, and have been the subject of ongoing reform, as well as discussion and debate concerning their implementation. As recently as 2017, for instance, the *Irish Criminal Law (Sexual Offences) Act*, which made a radical move decriminalising homosexuality, also replaced a provision in the previous 1993 Act that included the stipulation that any act of sexual intercourse involving a "mentally impaired person" is an offence (O'Malley 2017). Such legal reforms are engaged in a difficult balancing act (O'Malley 2017), and these changes also raise challenging questions of consent (Onstot 2019), information and communication.

What this discussion hopefully indicates is how ideas of disability, and how these are embedded in and shape culture and society, also exert a strong, reciprocal and still largely unexamined influence on the law (Couser 2007; Muller 2011; Sullivan 2017). The larger stakes in the interplay between law and disability are revealed in the way these dynamics shape general ideas of normality (*Goggin* et al. 2018; Kumari Campbell 2005; 2009; Muller 2011; Rioux et al. 2011;). For its part, the media plays a complex and vital role in communicating and shaping law on disability and normality (Ellcessor 2016; Goggin & Ellis 2015; Haller 2000; Scully 2011).

In the Stubblefield case, we can trace its widespread media attention internationally, structured around the twists and turns in the highly publicised legal proceedings. The case was reported consistently by local media in New Jersey (NJ), such as the NJ.com. This was important for publicity, because the legal proceedings, transcriptions, official verdicts and so on were not readily available. There were limited reporters in the courtroom, doing the necessary basic reporting as well as the investigative journalism and featuring writing that then followed (cf. Giles and Snyder 2009; Johnston and Breit 2010). The discourse on disability and sexuality was engaged with and, in turn, shaped the case—especially in relation to how particular details, remarks, scenes and quotations from the proceedings were reported, circulated and shared, discussed and received. The international interest in the case was spurred by a thoughtful and lengthy feature article in the *New York Times Magazine*, by journalist Daniel Engber (Engber 2015)—perhaps the most widely read, appreciated and discussed piece on the matter. Globally, high profile op-eds fuelled the discussion, including a piece in *The New York Times* by philosophers Jeff McMahan and Peter Singer (McMahan and Singer 2017).

Much of the media action surrounding the case—and key contours and dynamics of the issues—took shape via new media outlets, platforms and channels. Engber himself is best known as a journalist for the online magazine *Slate* and wrote two follow-up pieces at different stages of the saga for that publication (Engber 2017a; 2017b). The academic community debated the case vociferously on various online channels, ranging from email listservs (a technology still actively used) of the disability studies community through blog posts crossing various scholarly, activist and wider audiences to Facebook, Twitter, YouTube and social media channels.

As a result of this furious blend of traditional and new media representation, the issues raised by the case—especially the issues of communication and consent–were widely debated across different settings. In particular, the Stubblefield case served as a beacon for debates on sexual consent. The op-ed by McMahan and Singer was republished along pieces on other controversial cases, especially those such as the Harvey Weinstein case and others associated with the #Metoo movement, in a 2019 *New York Times* reader on *Defining sexual consent* (NYT 2019).

Woven into the fabric of the debate on consent were the striking issues of communication. At the heart of the discussion on communication was the question of Johnson's voice. In particular, the arguments revolved around whether Johnson was able to express himself and especially signify his consent to intimacy with Stubblefield via Facilitated Communication (FC). The complicating factor was the role that Stubblefield had played, and continued to play, in their relationship in supporting and assisting his communication—as his facilitator, when their romantic and sexual relationship was developing. In order to provide some clarity on these issues of communication, the defence called Rosemary Crossley herself as an expert witness (Engber 2015). In turn, the prosecution called a key witness who questioned the validity of FC. Before the trial commenced, as Engber notes, the judge disallowed Crossley's evidence, ruling:

> [T]hat facilitated communication failed New Jersey's test for scientific evidence …. The judge ruled that Anna, and only Anna, could testify about the typing and why she thought it worked.
>
> *(Engber 2015)*

Outside the courtroom, there was furious contention on FC and its role in the case (Sherry 2016). Disability scholars also discussed the wider complex and challenging issues of communication and listening, raised by the interactions between Johnson and Stubblefield (Mintz 2017; Savarese 2015). This included the issue of race, which Shelley Tremain argued was overlooked in discussion—hence her suggestion of various questions that should be posed concerning the "racialised reception" of the case: "How has race configured the reception of and responses to the verdict within the feminist philosophical community and within the disability studies community?" (Tremain 2015). Specifically concerning FC, she asked:

> Why have "facilitated communication" and its white facilitator, Anna Stubblefield, been effectively rendered the victims of the real victim's black family? Has the stubborn preoccupation with the alleged merits of a discredited technique actually concealed the white skin privilege and class privilege that continue to shape the fields of disability studies and feminist philosophy?
>
> *(Tremain 2015)*

There was also heated debate about whether Johnson would be permitted a voice in the courtroom—specifically to appear as a witness to give testimony (Grace 2015). In relation to this matter, disability scholar Devva Kasnitz gives this telling characterisation of the situation:

> Attending the trial taught me the power of the judge. The judge determined who could appear, what they could be asked, and what they could say in their answer. As an anthropologist, and a speech-impaired student of communication impairment in context, the trial transcripts were fascinating in their de-contextualization. Hearing and then reading them never answered my questions. Mr. Johnson's appearance as an exhibit, marched in the room like a silent toddler by his mother, who introduced him to the Judge and then marched him silently out, was a classic example of objectification. Although everyone's bias was clear, no one's truth was in that room.
>
> *(Devva Kasnitz, personal communication 2019)*

These matters encapsulated by Kasnitz thread through the issue of how representation of disability and sexuality plays out.

## Representing disability and sexuality

Over the period of the Stubblefield case, there were a number of aspects of disability and sexuality that loomed large in how the media reported and represented it. Via a complex range of strategies, Derrick Johnson was represented as lacking agency, capacity and attractiveness and being incapable of giving consent. In short, Johnson was portrayed as someone who was not able to be imagined, embodied or capable of being or becoming a sexual subject.

For her part, Stubblefield was often portrayed as strange, bizarre or weird for finding Johnson romantically and sexually attractive, and importantly she was a feminist. Feminists are often described as militant, radical and stereotypically as man hating, unfeminine and being lesbians. She was depicted as not caring, a sexual predator or at best being negligent in finding out whether Johnson was attracted to her and consented to their sexual relationship. Stubblefield's known feminist viewpoints were used negatively in the media and elsewhere. She was seen as using Johnson for her own political purposes and career advancement. Gender politics played a large role in the conviction of Stubblefield, but interestingly other feminist scholars with disabilities were present at her trial. These included Society for Disability Studies (SDS) activist and scholar Devva Kasnitz (present to testify about speech disability, but in the event not called as a witness) as well as psychology professor Michelle Fine (present as a friend), who argued in public debates that Johnson be allowed to testify.

Stubblefield was especially criticised for how she communicated with Johnson—notably that she relied on FC where she often was the facilitator, and so effectively passed off her voice as his voice, taking over his expression. All in all, various phrases, images and situations were combined and represented, often in cruel and sensational ways, to accentuate how sexually immature, incapable or unattractive Johnson was, and how far beyond the boundaries of intelligible, plausible and acceptable Stubblefield's behaviour had been.

Consider, for instance, that at the height of interest in the trial in September 2015, the UK *Daily Mail Online* encapsulated the various elements in this headline: "Rutgers professor accused of raping disabled, nonverbal, diaper-wearing man with cerebral palsy says he wanted sex and they were in love" Boyle 2015b). When Stubblefield was found guilty, *The Daily Mail* turned the volume up a notch with the headline: "Jury finds female professor GUILTY of raping mute, cerebral palsy-stricken, diaper-wearing man after she claimed the two were 'in love'" (Boyle 2015a).

Various aspects of Johnson were foregrounded by reporters in drawing a picture of the severity of his impairments: his relative stature (Stubblefield is a foot taller than him; he is 5 feet tall), his dependency upon others, his perceived powerlessness and lack of capacity to consent and communicate his consent. The headline of a piece in the UK *Daily Mirror* stated: "Professor sexually abused severely disabled man with cerebral palsy who is unable to walk or talk) (Dupuis 2015). Relatively early in the trial, coverage made prominent reference to his cerebral palsy and linked it to perceptions of his lack of capacity. A wire report, for instance, in April 2014 ran with the headline "Rutgers-Newark philosophy department chairwoman to appear in court for alleged sexual abuse of mentally handicapped man" (Szteinbaum 2014). The opening paragraph read:

> The chairwoman of the Department of Philosophy at Rutgers-Newark is expected to appear in court on Thursday for charges of aggravated sexual assault after she allegedly molested a man with cerebral palsy in 2011. The alleged victim, DJ, is said to have the mental capacity of an 18 month old.
>
> *(Szteinbaum 2014)*

Media reported the testimony of Dr Paul Fulford at the trial, who judged that Johnson was not able to consent to sexual activity (Wichert 2015a). Fulford first examined Johnson some 14 years earlier (Wichert 2015a). Fulford was then asked by prosecutors in 2011 to assess DJ's ability to consent and conduct an examination. Fulford recalled that:

> the man [DJ] provided "almost no significant responses" during the examination. "There were sounds, but not interpretable by me," said Fulford, adding that "there was no communicating or attempt to communicate."
>
> *(Wichert 2015a)*

Stubblefield's attorney, James Patton, queried Fulford on whether he conducted tests in this 2011 examination, who replied that this was not possible (Wichert 2015a).

A major thread in the coverage, then, was that Johnson was portrayed as lacking mental capacity to 1) understand sexual activity; and 2) be able to consent. In addition, Johnson was also portrayed as being unable to effectively communicate. In addition, much media coverage focussed on his lacking independence, as conveyed by various signs often associated with people with disabilities' lack of full agency, participation and capacity.

The single most striking, often-repeated image was the phrase characterising Johnson as wearing diapers. Media reported the prosecution's narrative on the first sexual encounter between Johnson and Stubblefield, in which the diaper figured prominently: "The cross-examination portion of Thursday's hearing included graphic descriptions of how Stubblefield allegedly pulled down DJ's diaper to perform oral sex on him" (Boyle 2015b). The rather more careful and decorous version of this was phrased in this way in an article by *The Independent*'s Caroline Mortimer, drawing on *The Daily Mail*'s reporting on the verdict. The headline read "Female university professor convicted of raping severely disabled man," with the opening line: "Dr Anna Stubblefield was found guilty of two counts of aggravated sexual assault on a 34-year-old man with cerebral palsy" (Mortimer 2015). The article explains: "The man, known as DJ, suffers from cerebral palsy and is unable to speak or feed himself and is forced to wear diapers" (Mortimer 2015). The phrase is "forced to wear diapers" is drawn, with acknowledgement, by Mortimer from Boyle's article, where she characterises Johnson in these terms: "DJ, who is unable to speak, needs help eating and walking and is forced to wear diapers, is intellectually disabled, according to his mother and brother who act as his legal guardians" (Boyle 2015a). The opening line of *The Daily Mirror* piece mentioned above also used a less strident, more formal tone: "Dr Anna Stubblefield, 45, faces up to 40 years behind bars for sexually assaulting the victim, who has to wear incontinence pads and has learning difficulties" (Dupuis 2015). The signifier of the diaper as an emblem of the lack of capacity and powerlessness of a person recurs across other media and discussion of the case. It is a small but telling detail in the coverage of Daniel Engber, including his reflective piece on the case where he argues for the importance of its "racial context": "If the roles had been reversed—if the victim had been a small, white woman in a diaper who could not speak or dress herself, and if the defendant had been a black man in a position of authority—would things have ended this way?" (Engber 2018). The negative, miserable connotations of "diaper" were criticised by many commentators. The rhetorical stakes were fittingly called out by writer Emily Brooks, who noted that:

> it doesn't matter one bit that he "wears diapers"—independent toileting has absolutely no correlation to intelligence level…We should not consider the fact that he wears diapers to mean he cannot have or want a sexual or romantic life.
>
> *(Brooks 2015)*

Another controversial detail of Johnson and Stubblefield's encounter was the yoga mat, and whether this was conducive to or unpleasant as a locale for romance.

The various details, themes and images in the predominant media coverage of the legal proceedings reinforced the portrayal of Johnson as variously, or in combination, lacking capacity, independence, ability to communicate or ability to understand sexual activity, being unattractive, and without power and agency. The alternative representation of Johnson, Stubblefield and the dynamics of their relationship and what had occurred was much less prominent in media representation. The nub of what was reported is summed up by Stubblefield's key comments at trial, regarding Derrick's personhood and agency: "He's very much his own person. He makes his own decisions," Stubblefield said (Boyle 2015b). Stubblefield argued that the pair were in a consensual relationship and "in love" (Boyle 2015b). Pressed by the prosecutor on why she had not communicated with Johnson's mother and brother (his guardians), she responded: "When people fall in love with each other, the last thing you're thinking about is, 'Hey, let's go ask our moms if it's okay'" (Boyle 2015b).

## Conclusion: Reimagining communication and consent in disability and sexuality

In this chapter, we have approached the entangled case of Anna Stubblefield and Derrick Johnson largely via a small selection of the available media representations. There is a great deal more to be said and dissected (especially in relation to the heated and wide-ranging public discussions of the case) in order to put together a full account of what transpired and its signifi-cance. However, we have focused on what the case and the media coverage have to tell us about the social relations and cultures of disability and sexuality, especially as these intimate matters are shaped by implacable structural, political and deeply affective forces. In this conclusion, we will offer some closing remarks on three areas we have touched upon: media; future directions; and challenges, conceptual, political and practical.

Firstly, we have sketched the contours of how disability and sexuality is imagined and re-imagined via mainstream media. Although we see some shifts, improvement in understanding and accuracy, and a little openness to contemporary realities and accounts of sexualities when it comes to disability, the Stubblefield case verged upon a moral panic—suggesting that there is a long way to go. Anna Malinowska has argued that "disabled sex still needs and deserves more thorough, less fragmented, less mediated, less compromising representations in popular culture" (Malinowska 2018, p. 376). Otherwise, disability will still, as Malinowska puts it, be "lost in representation" (Malinowska 2018). The point here is that disability and sexuality need to enjoy a flowering across cultural levels, with many more representations and voices to be available, received, listened to and engaged with—to shift the dominant paradigm. Here, journalism and reporting have a foundation role to play, in how we acknowledge, imagine and embrace the great variety of human life, communication, relationships and intimacy which an expanded recognition of disability, impairment and sexuality bring.

There are considerable conceptual and political challenges, especially in relation to two of the key intertwined ideas at the heart of the Stubblefield case: communication and consent. Consent is often referred to in the wide-ranging changes in sexual harassment and assault poli-cies, reporting and frameworks in a wide range of organisations. Sometimes, however, "consent" is seemingly used without context and in a very generalised way that does nothing to reduce sexual and disability inequality but much to make the very act of sex fearful, criminal and con-strained. As the screw has tightened, rightly, on non-consensual sex in the wider community, what are the implications and opportunities for disabled people who are non-verbal?

So, whereas consent is clearly a major conceptual challenge for Critical Disability Studies, in the wider context consent is a growing issue. This wider issue has largely emerged out of the sexual politics of feminism and the #MeToo movement, but the movement has failed to address the issues facing disabled people in this debate. In another case of alleged sexual assault, against a physically and cognitively disabled woman in Connecticut, it has been argued by Joseph Fischel and Hilary O'Connell that sexual autonomy rather than consent should have been the ethical barometer (Fischel and O'Connell 2019). They reject the equation of autonomy with consent and prefer to use a reconstructed concept of access (Mingus 2011; Shuttleworth 2007; Shuttleworth and Mona 2002). In the Connecticut case, the conviction relied on whether the disabled woman was helpless and thus, according to the law, unable to consent. As she was not able to consent, she was deemed more like a child or nonhuman animal. This poses the ques-tion: if disabled people are unable to consent, how are they able to be autonomous and access sexual cultures? Here feminist notions of autonomy and crip reconstruction of access overlap and could provide an answer to the conflicting issue of consent (Fischel and O'Connell 2019).

Communication sits alongside consent as another conceptual challenge, but also at the bed-rock is sexuality as part of personhood, as Kasnitz powerfully suggests:

> Communication becomes reduced to speech. We know anecdotally that disabled American men have used sex workers, a crime in the USA. We also know that disabled people are raped more often than non-disabled. In a sense, Johnson was absolutely a victim of at least *statutory* rape, regardless of whether he consented or not, in speech or

other communication. Like a 14-year-old, he probably didn't have the right to consent by NJ [New Jersey] law—or the NJ guardianship *statute* is so archaic that it denies people who fall under its "protection" of the right to pursue interpersonal sexual pleasure—or both. NJ is an American backwater in its approach to guardianship. That is an untold part of the story. I think there is an underlying trope that it doesn't matter if he consented, people like him should not have sex. I've had people tell me I'm not physically strong enough for sex. Johnson's mother declared that he had good food and his own room, "*What more could he want?*"

*(Devva Kasnitz, personal communication 2019)*

In conclusion, when it comes to disability and sexuality, and how it is represented, imagined and lived in private and public life, the political challenges raised by the Stubblefield case are enormous. This is especially the case as we move into an ultra-conservative climate with communities reflecting more traditional roles and reflecting less support for a diverse population. For the most part, the news media responded with stigmatising and unsupportive commentary, despite a lively commentary and debate across email lists, the blogosphere and social media. It is disappointing that, despite innovations in media, allied with social and political change in relation to disability, little appears to have changed in this regard.

## Acknowledgements

Our thanks to Dr Devva Kasnitz for her careful reading and critique of earlier versions of this paper, and for allowing us to quote from her analysis of the case and its implications. Thanks also to Rachel Wotton for her discussion of the general issues covered in this paper.

## References

American Speech-Language-Hearing Association (ASHA) (2018) *Facilitated communication* [Position Statement]. Available: https://www.asha.org/policy/PS2018-00352

Baron, M. (2018) Sexual consent, reasonable mistakes, and the case of Anna Stubblefield. *Ohio State Journal of Criminal Law* 15(2), pp. 429–450.

Biklen, D. (1990) Communication unbound: Autism and praxis. *Harvard Education Review* 60(3), pp. 291–314.

Biklen, D. & Cardinal, D. (1997) *Contested words contested science: Unraveling the facilitated communication controversy*. New York: Teachers College Press.

Boyle, L. (2015a) Jury finds female professor guilty … *Mail Online*, 2 October. Available: https://www.dailymail.co.uk/news/article-3257661/There-s-weird-relationship-Jury-deliberates-defense-argues-female-professor-did-NOT-rape-mute-cerebral-palsy-stricken-diaper-wearing-man-two-love.html

Boyle, L. (2015b) Rutgers professor accused … *Mail Online*, 25 September. Available: http://www.dailymail.co.uk/news/article-3249484/Professor-takes-stand-defense-against-claims-raped-disabled-nonverbal-diaper-wearing-man-cerebral-palsy-says-love.html

Boynton, J. (2012) Facilitated communication—What harm can it do: Confessions of a former facilitator. *Evidence-Based Communication Assessment and Intervention* 6(1), pp. 3–13.

Brooks, E. (2015) Disability rights on trial: What's really at stake in the Stubblefield case? *Emily Brooks Writer*, Blog Post, 21 September. Available: https://emilybrookswriter.wordpress.com/2015/09/21/disability-rights-on-trial-whats-really-at-stake-in-the-stubblefield-case/

Burgess, J., Cassidy, E., Duguay, S. & Light, B. (2016) Making digital cultures of gender and sexuality with social media. *Social Media + Society* 2(4), pp. 1–4. https://doi.org/10.1177/2056305116672487

Cardinal, D.N. & Falvey, M.A. (2015) The maturing of facilitated communication: A means towards independent communication. *Research and Practice for Persons with Severe Disabilities* 39(3), pp. 189–194.

Chappell, P. (2019) Situating disabled sexual voices in the Global South. In P. Chappell & M. de Beer (eds.), *Diverse voices of disabled sexualities in the Global South*. Cham: Springer, pp. 1–25.

Cohen, C. (2014) *Male rape is a feminist issue: Feminism, governmentality, and male rape.* Houndsmills, UK: Palgrave Macmillan.

Couser, G.T. (2007) Undoing hardship: Life writing and disability law. *Narrative* 15(1), pp. 71–84.

Crossley, R. (1992) Getting the words out: Case studies in facilitated communication training. *Topics in Language Disorders* 12(4), pp. 46–59.

Crossley, R. & McDonald, A. (1984) *Annie's coming out.* Melbourne: Penguin.

Dupuis, A. (2015) Professor sexually abused severely disabled man…. *Mirror*, 5 October. Available: https://www.mirror.co.uk/news/world-news/professor-sexually-abused-severely-disabled-6576429

Ellcessor, E. (2016) *Restricted access: Media, disability, and the politics of participation.* New York: New York University Press.

Ellcessor, E. & Kirkpatrick, B. (eds.) (2017) *Disability media studies.* New York: NYU Press.

Ellcessor, E. & Kirkpatrick, B. (2019) Studying disability for a better cinema and media studies. *JCMS: Journal of Cinema and Media Studies* 58(4), pp. 139–144.

Ellis, K. (2015) *Disability and popular culture: Focusing passion, creating community and expressing defiance.* Aldershot, UK: Ashgate.

Ellis, K. (2016) *Disability media work: Opportunities and obstacles.* New York: Palgrave Macmillan.

Ellis, K., Goggin, G., Haller, B. & Curtis, R. (eds.) (2020) *Routledge companion to disability and media.* New York: Routledge.

Elllis, K. & Kent, M. (eds.) (2017) *Disability and social media: Global perspectives.* London and New York: Routledge.

Engber, D. (2015) The strange case of Anna Stubblefield. *New York Times Magazine*, 20 October, viewed 22 August 2019. Available: https://www.nytimes.com/2015/10/25/magazine/the-strange-case-of-anna-stubblefield.html

Engber, D. (2016) What Anna Stubblefield believed she was doing. *New York Times Magazine*, 3 February, viewed 22 August 2019. Available: https://www.nytimes.com/2016/02/03/magazine/what-anna-stubblefield-believed-she-was-doing.html

Engber, D. (2017a) A second chance for Anna Stubblefield. *Slate*, 14 June. Available: https://slate.com/technology/2017/06/the-conviction-in-the-anna-stubblefield-facilitated-communication-case-has-been-overturned.html

Engber, D. (2017b) Talking without talking. *Slate*, 11 April. Available: https://slate.com/technology/2017/04/will-anna-stubblefield-get-a-new-trial-in-her-facilitated-communication-case.html

Engber, D. (2018) The strange case of Anna Stubblefield, revisited. *New York Times Magazine*, 5 April, viewed 23 August 2019. Available: https://www.nytimes.com/2018/04/05/magazine/the-strange-case-of-anna-stubblefield-revisited.html

Family & Community Services Department (FACS) (2016) Sexuality and relationship guidelines. Sydney, NSW, viewed 1 October 2019. Available: https://www.facs.nsw.gov.au/__data/assets/pdf_file/0008/590660/117-Sexuality_and_Relationship_Guidelines-accessible.pdf?fbclid=IwAR1mYB iEaJ-zdVGfyIkbMRZXk8-yFd5tuQgx7Pbqdwo8RFODi8YERu4bCd4

Feldman, J. (2019) This viral hashtag lets people with disabilities show off their sexy sides. *Huffington Post*, 12 March. Available: https://www.huffpost.com/entry/disabled-people-are-hot-hashtag_n_5c869242e4b08d5b78641b90

Fischel, J. & O'Connell, H. (2019) Cripping consent: Autonomy and access. In J. Fischel (ed.), *Screw consent.* Oakland, CA: University of California Press, pp. 135–171.

Giles, R.H. & Snyder, R.W. (eds.) (2009) *Covering the courts: Free press, fair trials & journalistic performance.* New Brunswick, NJ: Transaction.

Gill, M. (2015) *Already doing it: Intellectual disability and sexual agency.* Minneapolis, MN: University of Minnesota Press.

Goggin, G., Steele, L. & Cadwallader, J. (eds.) (2018) *Normality & disability: Intersections among norms, laws and culture.* New York: Routledge.

Grace, E.J. (Ibby) (2015) There is no case without D. Johnson's testimony. *Neuroqueer*, Blog Post, 16 September. Available: http://neuroqueer.blogspot.com/2015/09/there-is-no-case-without-d-johnsons.html

Haller, B. (2000) How the news frames disability: Print media coverage of the Americans with disabilities act. In B. Altman & S. Barnartt (eds.), *Expanding the scope of social science research on disability.* Bingley, UK: Emerald, pp. 55–83.

Harris, J.E. (2018) Sexual consent and disability. *New York University Law Review* 93(3), pp. 480–557.

Harris, J.E. (2019) The aesthetics of disability. *Columbia Law Review* 119(4), pp. 895–972.

Hemsley, B., Bryant, L., Schlosser, R.W., Shane, H.C., Lang, R., Paul, D., Banajee, M. & Ireland, M. (2018) Systematic review of facilitated communication 2014–2018 finds no new evidence that messages delivered using facilitated communication are authored by the person with disability. *Autism & Developmental Language Impairments.* https://doi.org/10.1177/2396941518821570

Hudson, A. (1995) Disability and facilitated communication. In T.H. Ollendick & R.J. Prinz (eds.), *Advances in clinical child psychology*, vol. 17. Boston, MA: Springer, pp. 197–232.

Javaid, A. (2016) Feminism, masculinity, and male rape: Bringing male rape 'out of the closet'. *Journal of Gender Studies* 25(3), pp. 283–293.

Johnson, DMan (2011a) The role of communication in thought. *Communicator* 20(1), pp. 4–5. Available: https://www.autcom.org/pdf/AutcomNLWinter2011.pdf

Johnson, DMan (2011b) The role of communication in thought, retracted article. *Disability Studies Quarterly* 31(4). Available: http://dsq-sds.org/article/view/1717/1765

Johnston, J. & Breit, R. (2010) Towards a narratology of court reporting. *Media International Australia* 137(1), pp. 47–57.

Kulick, D. & Rydström, J. (2015) *Loneliness and its opposite: Sex, disability, and the ethics of engagement.* Durham, NC: Duke University Press.

Kumari Campbell, F. (2005) Legislating disability: Negative ontologies and the government of legal identities. In S. Tremain (ed.), *Foucault and the government of disability.* Ann Arbor, MI: University of Michigan Press, pp. 108–130.

Kumari Campbell, F. (2009) *Contours of ableism.* Basingstoke, UK: Palgrave Macmillan.

Malinowska, A. (2018) Lost in representation: Disabled sex and the aesthetics of the "norm." *Sexualities* 21(3), pp. 364–378.

Mazur, E. (2017) Diverse disabilities and dating online. In M.F. Wright (ed.), *Identity, sexuality, and relationships among emerging adults in the digital age.* Hershey, PA: IGI Global, pp. 150–167.

McMahan, J. & Singer, P. (2017) Who is the victim in the Anna Stubblefield case? *New York Times*, 3 April. Available: https://www.nytimes.com/2017/04/03/opinion/who-is-the-victim-in-the-anna-stubblefield-case.html

Meekosha, H. (2012) Presentation to panel—*Rural women & girls with disabilities: Economic empowerment & political participation*, 58th session of the United Nations Commission on the Status of Women, New York. Available: http://wwda.org.au/wp-content/uploads/2013/12/CSW56MeekoshaPaper.pdf

Milbrodt, T. (2019) Dating websites and disability identity: Presentations of the disabled self in online dating. *Western Folklore* 78(1), pp. 68–100.

Miller, R. (2017) My voice is definitely strongest in online communities: Students using social media for queer and disability identity-making. *Journal of College Student Development* 58(4), pp. 509–525.

Mingus, M. (2011) Access intimacy: The missing link. *Leaving Evidence.* Available: https://leavingevidence.wordpress.com/2011/05/05/access-intimacy-the-missing-link/

Mintz, K. (2017) Ableism, ambiguity, and the Anna Stubblefield case. *Disability & Society* 32(10), pp. 1666–1670.

Muller, J.F. (2011) Disability, ambivalence, and the law. *American Journal of Law and Medicine* 37(4), pp. 469–521.

*New York Times* Editorial Staff (NYT) (eds.) (2019) *Defining sexual consent: Where the law falls short.* New York: New York Times Company.

Office of the Public Advocate (2016) Submission to the legal affairs and community safety committees. *Human Rights Inquiry.* Available: https://www.justice.qld.gov.au/__data/assets/pdf_file/0004/465160/opa-submission-human-rights-act-inquiry.pdf

OMalley, T. (2017) Criminal law (sexual offences) Act 2017—The offence provisions, paper presented to 18th annual national prosecutors conference, 25 November. Dublin Castle Conference Centre. Available: https://www.dppireland.ie/app/uploads/2017/11/PAPER_-_Tom_OMalley_BL.1.pdf

Onstot, A. (2019) Capacity to consent: Policies and practices that limit sexual consent for people with intellectual/developmental disabilities. *Sexuality and Disability.* Advance online publication. https://doi.org/10.1007/s11195-019-09580-9.

Papacharissi, Z. (2014) *Affective publics: Sentiment, technology, and politics.* New York: Oxford University Press.

Perlin, M.L. & Lynch, A.J. (2016) *Sexuality, disability, and the law.* Basingstoke, UK: Palgrave Macmillan.

Perlin, M.L., Lynch, A.J. & McClain, V.R. (2019) "Some things are too hot to touch": Competency, the right to sexual autonomy, and the roles of lawyers and expert witnesses. *Touro Law Review* 35(1), pp. 405–434.

Rioux, M., Basser, L.A. & Jones, M. (eds.) (2011) *Critical perspectives on human rights and disability law.* Leiden: Martinus Nijhoff.

Saltes, N. (2013) Disability, identity, and disclosure in the online dating environment *Disability & Society* 28(1), pp. 96–109.

Savarese, R. (2015) Anna Stubblefield and facilitated communication, 30 October. Available: https://stephenkuusisto.com/2015/10/30/anna-stubblefield-and-facilitated-communication/

Schweik, S. (2009) *The ugly laws: Disability in public.* New York: NYU Press.

Scully, J.L. (2011) "Choosing disability," symbolic law, and the media. *Medical Law International* 11, pp. 197–212.

Shakespeare, S., Gillespie-Sellis, K. & Davies, D. (1996) *The sexual politics of disability.* London: Cassell.

Shakespeare, T. (2000) Disabled sexuality: Towards rights and recognition. *Sexuality and Disability* 18(3), pp. 159–166.

Sherry, M. (2016) Facilitated communication, Anna Stubblefield and disability studies. *Disability & Society* 31(7), pp. 974–982.

Shildrick, M. (2009) *Dangerous discourses of disability, subjectivity and sexuality.* Basingstoke, UK: Palgrave Macmillan.

Shuttleworth, R. (2007) Disability and sexuality: Toward a constructionist focus on access and the inclusion of disabled people in the sexual rights movement. In N.F. Teunis & G. Herdt (eds.), *Sexuality inequalities and social justice.* Berkeley, CA: University of California Press, pp. 174–208.

Shuttleworth, R. & Mona, L. (2002) Toward a focus on sexual access in disability and sexuality advocacy and research. *Disability Studies Quarterly* 22(3), pp. 2–9.

*State v. Stubblefield* (2017) 162 A.3d 1074, 1075 (N.J. Super. Ct. App. Div. 2017).

Stubblefield, A. (2009) The entanglement of race and cognitive dis/ability. *Metaphilosophy* 40(3–4), pp. 531–551.

Sullivan, F. (2017) Not just language: An analysis of discursive constructions of disability in sentencing remarks. *Continuum* 31(3), pp. 411–421.

Szteinbaum, S. (2014) Rutgers-Newark philosophy department chairwoman to appear in court for alleged sexual abuse of mentally-handicapped man, 21 April. UWIRE.

Wichert, B. (2015a) Disabled man could not consent to sex in professors abuse case, expert says. NJ.com, 18 September. Available: https://www.nj.com/essex/2015/09/disabled_man_could_not_consent_to_sex_in_professor.html#incart_2box_nj-homepage-featured

Wichert, B. (2015b) Professor found guilty of sexually assaulting disabled man. NJ.com, 2 October. Available: https://www.nj.com/essex/2015/10/professor_found_guilty_of_sexually_assaulting_disa.html

Wichert, B. (2016) Professor sentenced to prison for sexual assault of disabled man. NJ.com, 16 January. Available: https://www.nj.com/essex/2016/01/professor_sentenced_to_prison_for_sexual_assault_o.html

World Health Organization (WHO) (2011) *World report on disability.* Geneva: WHO.

Zambito, T. (2014) Rutgers-Newark philosophy chairwomen fights sexual assault charges. 20 April. Available: https://www.nj.com/news/2014/04/chairwoman_of_rutgers-newark_philosophy_department_fighting_criminal_sexual_assault_charges_involvin.html

# PART V

# Sexual narratives and (inter)personal perspectives

# 21

# UNDERSTANDING THE LIVED EXPERIENCES OF TRANSGENDER YOUTH WITH DISABILITIES

*Angela K. Ingram*

## Introduction

Many adolescents struggle with transition into adulthood during high school, especially those with disabilities. Research has shown that adolescents with disabilities have lower rates of vocational and educational attainments when compared to their peers who don't have disabilities (Alsaman and Lee 2017). Adolescents with disabilities also face a higher risk of dropping out of school compared to their peers without disabilities (Stillwell and Sable 2013; Zablocki and Krezmien 2012). Furthermore, according to The National Longitudinal Transition Study 2012 (NLTS 2012), adolescents with disabilities face greater risks of being bullied in school, are suspended at higher rates and face discrimination (Lipscomb et al. 2017). Unquestionably, these factors and barriers can adversely affect these adolescents on different levels, including academic performance, access to employment and success at independent living (Dong et al. 2016; Newman et al., 2011).

Whereas adolescents with disabilities are more prone to experience negative post-school outcomes, research has identified numerous subgroups of individuals with disabilities (i.e. race, gender, sexual orientation) that are at even greater risk due to the intersection of marginalised identities (Kahn and Lindstrom 2015; Sorrells 2008; Trainor et al. 2008). Those who associate with multiple marginalised identities, such as lesbian, gay, bisexual, transgender and queer (LGBTQ) with disabilities, often face more difficulties (Duke 2011; Kahn and Lindstrom 2015). Compared to heterosexual youth with disabilities, LGBTQ adolescents are more likely to face discrimination and negative experiences (Balsam et al. 2005; Kahn and Lindstrom 2015). In the limited research that has focused on youth with disabilities who identify as LGBTQ (Duke 2011; Kahn and Lindstrom 2015; Robinson and Espelage 2012), findings have exposed compounding negative outcomes, including depression, suicide, harassment, discrimination, rejection and higher rates of school dropout (Morgan et al. 2011; Robinson and Espelage 2012).

While research is emerging on the experiences of LGBTQ youth with disabilities, findings cannot be generalised to all groups in the LGBTQ continuum, in particular, transgender youth with disabilities. Transgender is a term that refers to people whose gender identity differs from their assigned biological birth sex (Denny et al. 2007). Gender identity is a person's inner sense of being a man or a woman (or boy or girl). For transgender people, their biological sex assigned at birth and their own internal gender identity do not match. Hines and Sanger (2010) explain transgender as:

> A range of gender experiences, subjectivities and presentations that fall across, between or beyond stable categories of "man" and "woman"… [The term transgender includes] gender identities that have, more traditionally, been described as "transsexual," and a diversity of genders that call into question an assumed relationship between gender identity and presentation and the "sexed" body.
>
> *(p. 1)*

The limited research on transgender youth with disabilities has been entangled in LGBTQ literature; they are infrequently recognised as a unique group that should be studied outside the LGBTQ continuum (Hill, 2007; Kazyak 2011). The "T" in "LGBTQ" assumes that transgender individuals have the same experiences as lesbian, gay, and bisexual individuals. This is problematic because it assumes that gender identity and sexual identity are the same (Paxton et al. 2006). Although lesbian, gay and bisexual communities face similar discriminations and negative issues, gender identity and sexual orientation are different. Sexual orientation describes a person's physical attraction to another person (e.g., straight, gay, lesbian or bisexual), whereas gender identity describes a person's internal sense of being a man or a woman, or even outside that binary (Paxton et al. 2006). In other words, sexual orientation is about to whom you are attracted, whereas gender identity is about how people define themselves. Because sexual orientation and gender identity are different, there is a clear need and importance for researching and understanding the unique experiences of transgender individuals in isolation from the LGBTQ population. To date, no research has specifically examined the lived experiences of transgender youth with disabilities in high school settings outside of the LGBTQ continuum.

## Theoretical framework

Throughout this chapter, gender identity and sexual identity are viewed with the theoretical framework of transgender theory which incorporates (a) fluidity of gender, (b) importance of embodiment and (c) intersectionality. Transgender theory posits that the lived experiences of individuals and their negotiations of intersectional identities can be empowering to gender-diverse individuals (Monro 2000). The theory addresses the significance of the physical embodiment of intersecting identities along with an understanding of how lived experiences impact embodied aspects of identity (Roen 2001). Utilising transgender theory helps to challenge a categorical way of thinking by embracing a more fluid view of gender (Burdge 2007; Nagoshi and Brzuzy 2010) and an understanding of the embodied intersectional experiences of transgender individuals (Roen 2001).

### *Gender as fluid*

Transgender theory views gender as fluid on a spectrum. From this perspective, individuals fall based on their inner sense of gender, regardless of their biological sex assigned at birth (Roen 2001). Viewing gender as fluid allows for a broader understanding of gender (Lane 2009) and can help surpass how we view gender in our westernised society. This understanding of gender allows categories to be recognised as continuously shifting and never absolute. It also opens up an understanding of gender that is free of constricting categories. Broad (2002) proposes that:

Gender categories were destabilised not only through assertions of not fitting either gender, but also through claims to actually being a bit of both. It is the notion of transgender, meaning both man and woman, that drives many in the gender community to hold up intersexuality as perhaps the best way to describe transgender existence... The idea is that by being transgender, one really embodies an "intersexual" identity of being both man and woman.

*(p. 256–257)*

Similarly, Tauchert (2002) recognised a "fuzzy gender" or "shades of grey" approach to understanding gender. This concept of "fuzzy gender" emphasises the relationship between physical and mental characteristics of gender. With the concept of "fuzzy gender," variations of gender and sexuality are acknowledged as well as embodied experiences (Tauchert 2002). As Bornstein (1994) stated:

I'm constructing myself to be fluidly gendered now... I don't consider myself a man, and, quite frequently, I doubt that I'm a woman. And you—you still think gender is the issue! Gender is not the issue. Gender is the battlefield. Or the playground. The issue is us versus them. Any us versus any them. One day we may not need that.

*(p. 222)*

## Gender embodiment

Transgender theory leaves space open to consider the embodied aspects and experiences of being transgender. In our society, people have typically embodied characteristics that have been attached to the male/female binary. For example, women may express and embody their gender in more feminine ways, whereas men may express and embody their gender in more masculine ways. Transgender theory posits that gender can be experienced, felt and expressed in different ways in relation to one's felt internal sense of gender (Nagoshi and Brzuzy 2010). Transgender theory helps to embrace embodied aspects of gender, which can assist in exploring the relationships between identities related to transgender being and knowing. It opens up a space for a theoretical conversation about the ways in which corporeality matters in relation to embodiments of transgender identity. Moreover, understanding gender embodiment in the context of transgender lives and experiences is useful for explorations of transgender lived experiences.

## *Intersectionality*

Intersectionality is an important aspect of transgender theory, especially when it comes to understanding the lived experiences of individuals (Raftery and Valiulis 2008). Hines and Sanger (2010) expressed that there is a need for transgender scholarship to utilise an intersectional lens:

Much work on transgender [people] has lacked such an intersectional analysis with the effect that "trans people" are often represented as only that—as only trans. Hence trans people are disconnected from their intimate, material, geographical and spatial surroundings, and from other significant social signifiers. This problematic is not only (mis)representational, it also acts to homogenise and de-politicise. Thus privileging/ de-privileging forces, such as the economic resources to pay privately for surgery, geographical access to "trans-friendly" social spaces, levels of support from intimate networks [that] structure transgendered experiences are unaccounted for.

*(p. 12)*

Rooted in black feminism and critical race theory, intersectionality posits a framework that has been beneficial for explaining how an individual can be located within various social categories (i.e., race, class, gender, sexual orientation, disability) that define their experiences (Collins 1990; Crenshaw, 1989; King 1988; Sawyer et al. 2013). According to an intersectional perspective, inequities are not the result of a single factor, but the outcome of different social categories that are interrelated and merge to produce inequality (Crenshaw 1991; Schulz and Mullings 2006). The objective of an intersectional approach is to recognise how diversity that stems from other social categories affects individuals in different ways (Crenshaw 1991; McCall 2005). This communicates the idea that the intersections of categories are exposed to multiple forms of inequities and oppression (Gopaldas 2013).

## Method and participants

This chapter draws on data collected from institutional review board-approved semi-structured interviews with nine participants who identified as transgender and having a disability. The sample included individuals who (a) identified as being transgender, (b) identified as having a disability, (c) were between the ages of 16 and 22 and (d) had attended a high school in the Northwest Pacific part of the United States. Each participant completed three interviews over a two-month span. The participants responded to open-ended questions about their disability identity and gender identity. In addition to interviews, photography was utilised as an important data source. Participants were asked to take photos of things in nature that associated with their identities. At the end of the study, the photos were developed, and the participants engaged in a photo reflection of the photos they took. The photographs helped to have a visual insight into the experiences of transgender youth with disabilities. The visual and the linguistic complemented each other in capturing the lived experiences of transgender youth with disabilities. Moreover, the photographs provided the participants with another platform to express themselves in ways that the interviews could not capture. Data analysis consisted of the principal investigator transcribing each of the interviews verbatim and creating a list of relevant themes. Emergent themes were refined until no new themes could be identified. The author then selected statements illustrating each theme. In this chapter the following themes are discussed: (a) Gender as fluid, (b) conflation of identities, (c) society and identity and (d) difference as strength.

## Emerging themes

### *Gender as fluid*

All participants understood and embodied their gender identity to be fluid, as opposed to an essentialist, binary notion of gender. The participants unsettled and challenged oppressive notions of gender and expression, by their fluid statements of not fitting either of the binary genders, as well as their statements of being a bit of both genders. This notion of a more than male and female boxes were echoed across the interviews. Some of the participants expressed how they experience and express themselves as being both male and female, as being neither male nor female, or as falling completely outside the gender binary. They saw the division of "male" and "female" as socially constructed and at odds with their feelings about themselves. Consider the following statements from the participants:

> There's a disconnect between your body and your brain as far as your gender…It's just the way you were born. I think there is so much more than the male and female boxes."
>
> *(participant three)*

I identify as gender fluid, although I'm not sure if that's like an accurate description of my gender identity…I mean by definition of society and gender as a social construct, I am gender fluid. I go between male and female gender roles, sometimes a mix of everything and sometimes an absence of everything.

*(participant two)*

I'm male. I go by he/him/his. I use those pronouns and I don't find myself to be a macho man necessarily. But I also don't also find myself to be feminine. I do identify as being male, not non-binary or gender fluid. But I believe that no one is completely male or completely female, you know, we all lie somewhere on the spectrum.

*(participant four)*

Transgender theory has emphasised limitations with the gender binary and the importance of recognising fluidity and multiplicity within gender (Haritaworn 2008; Koyama 2020). Historically, westernised society has been built on the belief that gender is binary and biological, and has reinforced the gender binary through gender-related beliefs, rituals, rules and expectations (Thorne 1993). As transgender theory highlights, these gender-related beliefs in which institutions function is known as cisgenderism, which is recognised as a systematic ideology that devalues and pathologises genders that do not subscribe to gender-binary roles (Stryker and Whittle 2006). Lennon and Mistler (2014) explained, "This ideology endorses and perpetuates the belief that cisgender identities and expression are to be valued more than transgender identities and expression and creates an inherent system of associated power and privilege" (p. 63). In other words, this ideology perpetuates a system of power and privilege that suppresses transgender identities, expressions and experiences (Ansara and Hegarty 2012; Lennon and Mistler 2014). Throughout the interviews, it became clear that the participants were at odds with cisgenderism for not fitting into the gender binary. The photography the participants took further reflects this notion of gender as fluid:

The bush within the bush…The thing is that this pink bush was completely taking over the green bush. They were basically like in the same space, but one seemed to be more present than the other. I felt in that regard, I saw the bush that was inside the green bush, the pink bush. I saw the pink bush as like my gender in a way. It's internal, it's not like exactly outside of the green bush, but it's inside the green bush and it's a lot more prominent. It's a lot more out there. It's much more just there.

*(participant two)*

Transgender theorist Susan Stryker (2006) stressed that the fluid experiences of transgender embodiment fundamentally challenge cisgenderism and the gender binary. Furthermore, Bornstein (1994) asserts that gender binary is a class system, which can lead to negotiation of fluid gender identities. In line with Bornstein and Stryker, the transgender youth in this study unsettled and challenged these oppressive notions of gender, by their fluid statements of not fitting either of the binary genders, as well as their statements of being a bit of both genders. Thinking with transgender theory supports the participant's voices and helps us understand gender as being fluid and non-binary. This emphasis on gender as fluid, not only helps to move beyond a binary understanding of gender, but it helps to move away from the biological understanding of gender (Nagoshi and Brzuzy 2010). Recognising gender as fluid instead of fixed could help challenge cisgenderism in our society and help to support an environment of inclusion that is not tied to a binary gender (Figure 21.1).

*Figure 21.1*  Picture by participant two of a bush within a bush.

## Conflation of identities

Many participants spoke about how people blamed their gender identity on their disability. The participants were adamant that having a disability does not mean that that person is less capable of determining their gender identity and it does not mean that their disability caused their gender identity. Many of the participants spoke about their gender identity being discredited because of their disability. Consider the following statements made by the participants:

> I've had people also be like, "You're just saying you're trans, because you are retarded."
>
> *(participant two)*

> Just being discredited with regards to our gender identity, with regard to our disability. Because of the fact that we are disabled and we identify as transgender, people often times disregard that we could possibly know that we are transgender or that we can identify as such, because we are not "clear thinking or neurotypical people" and that's not the case a lot of times. We can think just fine."
>
> *(participant three)*

People who aren't understanding, they would blame my autism for being trans. They would say, "Oh, it's a phase; oh, your autism made you think that" or "You know your disability makes you think that way?" It's like, excuse me, it's like that has nothing to do with anything in general. So, they try to use my disability as a blame for who I am (participant nine).

The misconceptions towards disabled people meant that people struggled to understand how they could even identify as transgender. These fallacies acted to invalidate their gender identity. Societal misconceptions of disabled bodies as being non-normative has influenced how the gender identity of disabled people are seen and understood. Another social misconception relates to the ways in which transgender people and disabled people have been sexualised in our society. On one hand, people with disabilities have been de-sexualised and perceived as not capable of existing as a sexual being, which acts to maintain able-normativity and the pathology of disabled existence (Smith 2012). On the other hand, transgender people have been hyper-sexualised (Nagoshi and Brzuzy 2010). Many participants in this study made comments about how their transgender identity is not a fetish and should not be sexualised. For example, "One thing that I have been seeing around a lot, is people saying trans is a fetish…don't sexualise us" (participant two). In our society, transgender identities have been fetishised, which acts to invalidate and devalue their identity. These invalidating views of disabled bodies and transgender bodies are at odds with each other. It is not unreasonable to think that since disabled bodies have been constructed as nonsexual, that transgender identity of disabled people is not accepted and is problematic. As Shildrick (2013) asserts, "Both sex and disability threaten to breach certain bodily boundaries that are essential to categorical certainty and, as such, they provoke wide-spread anxiety" (p. 3).

This conflation of identities which participants experienced not only acts to invalidate their identity, but also further highlights intersectional barriers placed on them due to problematic understandings of disability and transgender identities. The statements expressed by the participants helps to expand our understanding of the unique forms of intersectional oppression that transgender people with disabilities experience. As Smith (2012) asserts "an awareness around the interconnectedness of the construction and pathology of queer and/or disabled people may be the best way to start to transform the harm that has been done and continues to be perpetuated" (p. 126). Thinking with intersectionality helps to acknowledge the ways gender identity and disability intersect, which allows for a more comprehensive and inclusive understanding of how identities can overlap and influence experiences. It can also help alleviate multiple and contradictory meanings that society has placed on gender identity and disability identity, because it shows how certain identities can be simultaneously positioned.

## Society and gender identity

Both gender and disability identity can impact the way individuals see themselves and their interactions in society. The experiences of transgender youth with disabilities in this study illuminated how societal meanings attached to their identities are rooted in hegemonic expectations. The experiences from the participants suggest that there are many intersectional parallels between the transgender and disabled experiences. A body with a disability is expected to strive toward able-bodied beliefs (García and Ortiz 2013), just as a transgender person is expected to strive toward cisgender expectations (Nagoshi and Brzuzy 2010). The participants had to navigate society's misconceptions attached to their gender identity and their disability identity. The participants all place themselves outside the conventional gender binary yet live in a society that has recognised only two genders.

First, participants spoke about how gender is a prescribed social construct attached to fixed expectations of ones' biological sex. Participants were all aware of the gender and sexual identity they were expected to embody according to society and strongly felt as though it was destructive. For example:

> I have friends, they are trans, they are trans women or trans men, they feel they have to be forced into the orientation of what a boy or girl is. I see many trans women, they have to feel as if they have to be the most feminine person to the point…the best way I can represent is like an action figure, like they embody the Barbie doll or they embody

the GI Joe of what a person would look like. I feel like they have destroyed themselves more then they're healing themselves.

*(participant seven)*

Before I ever understood anything about sexuality, gender, before I ever knew the words sexuality and gender, I was uncomfortable with parts of myself. Dysmorphia is part of myself, it's ingrained. But the dysphoria is socially, socially accumulated. It's bred by those interactions and those infrastructures that are binary. It's the…the dysphoria it comes from the interactions.

*(participant three)*

Some participants spoke about how they attempted to embody and express gender expectations attached to their biological sex, by growing their hair out or expressing themselves femininely, but eventually resisted and contested. The participants expressed frustration with how society places roles and expectations associated with one's identity and how someone's interests and expression should not be gendered. In line with transgender theory, participants embodied a more self-constructed and fluid identity and challenged gendered expectations and norms by embodying their true gender.

Participants in this study also expressed the structural consequences of not fitting into a society built for cisgender people. Power structures in our society benefit from maintaining gendered dichotomies, which render invisible transgender youth (Glenn 2002; Grossman and D'Augelli 2006;). As transgender scholar Susan Stryker (1994) stated:

Bodies are rendered meaningful only through some culturally and historically specific mode of grasping their physicality that transforms the flesh into a useful artefact. Gendering is the initial step in this transformation, inseparable from the process of forming an identity by means of which we're fitted to a system of exchange in a heterosexual economy.

*(p. 249–250)*

People who do not fit into gender categories have been perceived by society as threatening and have been subjected to consequences (Pascoe 2005). As participant nine expressed:

I don't identify as trans I am trans. I am a-gender. Nothing is built for trans people, everything is built for cis people. All things are designed for cis people; all interactions are scripted for cis people. Health care is scripted for cis people and just being trans makes everything harder. Finding a place to live is exceptionally, exponentially harder.

*(participant nine)*

Participants were subjected to structural discrimination in society for not fitting into dominant gender roles. Examples of structural discrimination include access to health care, housing and employment. As a result of structural discrimination towards transgender people, research has shown that transgender people are more likely to experience high rates of unemployment, homelessness and have limited access to healthcare (Nagoshi and Brzuzy 2010). In this study, four out of the nine participants were homeless and unemployed. The other five were dependent on family members for housing and financial support.

In addition, participants expressed how their transgender identity was viewed in pathological ways. Pathologising is as the name given to the labelling and treatment of people's identities as disordered (Ansara and Hegarty 2012). Systems of oppression develop in relation to our socio-historical context (Ehrenreich and English 1973). Western society has a history of the pathologisation and the construction of certain groups of people (i.e., women, people of colour), rendering them as unnatural or mentally ill (Ehrenreich and English 1973; Withers 2012). Likewise, transgender people have historically been pathologised and treated as if they were mentally ill, resulting in issues of stigma and misconceptions, and of discrimination (Bockting et al. 2013). As one participant expressed:

> I feel like a lot of people, like, have this predisposed ideology, that if there is something quote unquote wrong with someone that they have every right to just hurt them and they are completely justified because of the fact that that person had something wrong with them.
>
> *(participant six)*

The prevalent nature of cisgenderism has contributed to the normalisation and manifestation of the pathologisation of transgender identities, because it functions with the belief that those who are not cisgender are deviant and unnatural (Lennon and Mistler 2014). This pathological belief that there is something wrong and unnatural with transgender identities still lingers in our society, which further highlights the pervasive nature of cisgenderism, which allows and justifies discrimination against transgender people (Ansara and Hegarty 2012). Participants had a clear understanding of how their gender identity has been perceived as unnatural and wrong, which caused them to live under constant worry about how others might mistreat them and feel justified in doing so.

## Society and disability identity

Like transgender identity, disability identity has been socially constructed, pathologised and subjected to oppressive social mindsets (Withers 2012). The participants faced challenges due to preconceived and fixed ableist notions in society about their disabilities. Like their gender identity, the participants expressed statements about their disability being placed in boxes. In our society, disability has also been placed in a binary (i.e., abled/disabled), which can act to reinforce ableist beliefs. Similar to the concept of cisgenderism, ableism perpetuates discrimination and misconceptions against disabled people and deems disabled bodies to be less valuable (Longmore 1995). These beliefs have been tied to the medical model that seeks to "fix" or "cure" people with disabilities (Longmore 2003). Consider the following statements below:

> There are boxes that we are placed upon for people who have disabilities. For me, how they stereotype autistic people to be low on the spectrum, when, in reality, they can be anywhere on the spectrum.
>
> *(participant nine)*

> People don't know how to interact with people with autism... Just be patient and not like expect to be able to fix someone in my situation and not to expect to have a solution for people. Accept people for who they are and not be shit heads where "autism is bad and we need to cure it."
>
> *(participant eight)*

> My disability does not make me anything, I make me, me. I feel like people don't understand, I feel like society as a whole doesn't understand and society needs to

understand that. That my disability does not make me a bad person. It does not define me as a criminal or anything like that. I am not bad, I'm not. People automatically assume, that because I have a disability, there is something wrong with me and that they have every right to hurt me because of that.

*(participant six)*

People want to hear a capitalism version of: "What about your disability doesn't hold you back from being a productive member of society." My disability does hold me back from being a productive member of *your* version of society, because *your* version of society inherently devalues me. I wish people would understand and try to actually understand, that there is no better version of me…I am who I am and I have value because of who I am and what I contribute to my communities outside of capitalistic endeavours.

*(participant seven)*

This ableist notion of disability has the potential to create a misconception that an individual's disabilities place restrictions on what they can and cannot do. Participants in this study challenged this conception, explaining that their disability doesn't disable them, rather it's societal misconceptions of disability that are disabling. Consider the following photographs below:

I took the photo of the rock, because it looks as though the rock is actually snapped. "Oh, it's not pretty, it's not good." When in reality, I use the term "don't judge a book by its cover" and that's with my autism and my gender identity. With my autism, people will think about the stereotype, of course, about people who have disabilities, and they think that people with disabilities are "these things or those things," all these negative ways society has portrayed them, mostly. But in reality, we're more than that. We are filled with content, we're more than what the people who run our country perceive us as [Figure 21.2].

*(participant nine)*

It is almost like a silhouette kind of picture. I feel like a lot of that is just general contrast. I was very proud of this one because I thought about the idea of how… about the contrast it had, was…with the blue sky behind it with like the clouds stuff and the silhouette in front of it, I felt like a lot of that was actually related to me in a way. I am not certain how to explain that per se, but the contrast of, like, what is upfront and what is behind, you know, kinda thing. I feel like, it's kinda hard to explain in terms of my disability, but I do feel a connection to it, with regard to my disability. I feel like a lot of that is what people see my disability as, versus, like, what it actually is. Like basically just the contrast of, like, how I see the world and how people think I see the world [Figure 21.3].

*(participant three)*

## Intersecting identities

Participants were subjected to the pathologisation of their identities, as well as social constructs and structural barriers associated with ableism and cisgenderism. Intersectional theory has been useful to understand transgender experience because it allows for the examination of how mul-

*Figure 21.2*   Picture by participant nine of the rock.

*Figure 21.3*   Picture by participant three of a silhouette.

tiple categories overlap (Monro and Richardson 2010; Nagoshi and Brzuzy 2010). Many participants spoke about gender identity, disability and sexual orientation in an intersectional way. The similarities between how society sees gender, sexual orientation and disability are compounding and can be invalidating to one's experiences; however, participants resisted and negotiated these notions. Transgender theory posits that awareness of intersectional identities can allow individuals to respond to the social forces that impact one's identities (Somerville 2000). Consider the following statements and photographs:

> I do say there's intersectionality with disabilities with regard to social characteristics of our own self and gender, sexual orientation, race in general. I think that bringing more awareness, that people who have disabilities are also minorities too. We need to bring more awareness of them…I do think that not many voices of people with disabilities are being brought up. I had to teach myself to be way more independent. That's the way I had taught myself when it came to my autism and, when I came out, no one would accept me, I did most of the things by myself, honestly. I was a survivor, I was a lonely person who had no support or anything.
>
> *(participant eight)*

> This one I saw in and, the second I saw it on this, it was very, very bright, like in your face. I guess that kinda represented how everyone was to me about my gender identity and about my disability, was the fact that they were always in my face and "be like this or don't be at all" [Figure 21.4].
>
> *(participant three)*

> I like it because if you touch it the wrong way, it's not good. But it still has the prickly pear fruit. And if you learn how to work it right, it's an excellent cactus dish, if you learn how to make it appropriately, but, if you don't, if you don't prepare it exactly right, it's very disgusting and sharp. People not knowing how to interact with me as a disabled person and as a trans person [Figure 21.5].
>
> *(participant eight)*

*Figure 21.4*   Picture by participant two of in your face.

*Figure 21.5* Picture by participant eight of a cactus.

As a result of an intersectional awareness, other participants communicated their struggle to separate their own beliefs about their identities from what society was perpetuating. Explaining how eventually they realised there was nothing inherently bad about them as a person, rather there was something inherently wrong with how people interact with differences. The consequences of identity categories should not be understood independently from one another because these categories are multiplicative and mutually coexist. For example, disability, transgender identity and sexual orientation should not be conceptualised through a mono lens such as "disability+transgender+sexual orientation" rather they could be thought of in terms of "gendered ableism" or "cisgender ableism." Understanding with an intersectional lens shifts the focus from the pathological "unnatural" and "abnormal" conceptualisations of transgender and disability identity and provides a lens for considering institutionalised inequalities (Zinn and Dill, 2000). It also helps recognise that societal oppressive structures open up an understanding of how cisgenderism and ableism can affect transgender youth with disabilities. The voices of the participants call attention to the need to understand how transgender youth's identities are at constant odds with fitting into a society that was not made for them. Overall, this can help inform an intersectional approach to understanding transgender youth with disabilities lived experiences, because it makes visible aspects of inequality that have been linked to meanings of gender and disability in our society.

## Difference as strength

Findings from this study revealed that participants viewed their differences as a positive strength in a way that allowed them to empathise with marginalised people and social justice movements. For example, participants spoke about how they were active allies for anyone who was a sexual or gender minority. They also spoke about taking part in movements such as the Black Lives Matter movement. Transgender theory suggests that the lived experiences of individuals and their negotiations of intersectional identities may empower them. When asked about the

319

positives of identifying as transgender with a disability, participants spoke about how both their identities, of being transgender and disabled, empowered them to be empathetic and to understand people from multiple perspectives. Through their adversity and marginalised identities, participants in this study emphasised how they were able to have an intersectional understanding of perspectives.

> It gives me the experience to empathise with people who are in the same situation. Because we have gone through a lot. Both in terms of our gender identity and our disability, we can help people and empathise with people who are in those situations as well.
>
> *(participant three)*

> My empathy, I would say, just the idea that even though I have been through a lot, I have the ability to relate to people. I have the ability; I have a big heart and I want to share that heart with people. I want people to see that I am not just transgender, I'm not just disabled.
>
> *(participant nine)*

> I tend to educate myself with things that I don't identify with, but I know are a still a problem. I try to educate myself about racism and racial inequality, like Black Lives Matter and that kind of stuff.
>
> *(participant four)*

Their experiences existing at the intersection of marginalised identities allowed for a greater understanding and empathy for others with oppressed identities. Transgender theory suggests that the lived experiences of individuals and their negotiations of intersectional identities may empower them. When asked about the positives of identifying as transgender with a disability, participants spoke about how both their identities of being transgender and disabled empowered them to be empathetic and understand people from multiple perspectives. The photographs also revealed that, because of their experiences being transgender and disabled, they were able to see things from a different perspective.

> The first one I took was this one and the way I took this was I actually laid on the ground and took a picture from the ground up. I feel like in that regard, perspective matters. Because the way that this is, is first off, this tree is bright pink against the blue sky. And the way that I got this picture, the perspective that I got this picture from; people would not be able to see it, had I not taken the picture like that, had I not seen it this way. And I feel like that's important, that perspective matters. Not just because of…in fact because of my gender identity and because of my disability, I see the world from a different perspective and it's gorgeous [Figure 21.6].
>
> *(participant three)*

## Conclusion

The essence of this study was a shared experience of ongoing tenacity, tension and resistance to dominant discourses in society. The participants' identities in this study were constituted in heteronormative and able-bodied discourses across their high school and a wider societal landscape.

*Figure 21.6* Picture by participant three of the ground up.

The lived experiences of transgender youth with a disability extended beyond the walls of their high school and spilled into the realities of their everyday life. Their resistance existed, because of their intersecting marginalised identities, which meant a constant negotiation and awareness of how their identities of being disabled and transgender were situated in a society that was made for people to fall into a binary gender. Ultimately, their existence and experiences complicated the binary logics of male/female embedded in the fabrics of our institutional systems. The lived experiences expressed in this chapter help to challenge an ideological view of gender, disability and sexual orientation and open up an understanding of youth living with intersecting marginalised identities. Awareness is important to overcome any biased beliefs about transgender individuals. Educators should have an understanding of "gender identity" and "sexual orientation" since these terms are commonly conflated in the LGBTQ continuum (Brill and Pepper 2008; Haussmann et al. 2008). Moreover, it is important for educators to understand that there is a difference between biological sex and gender. Transforming these beliefs should include frameworks such as intersectionality and transgender theory, which can allow teachers to reflect on how they view and understand identities. Intersectionality, as a tenet of transgender theory, helps in understanding this multiplicity of gender identity, which shifts the focus from a fixed dichotomy of gender towards a more inclusive understanding of marginalised identities. Findings in this study suggest it is important to acknowledge that structures and social constructs are not just pre-existent, rather that we, as a society, produce them and we are responsible for how they are played out. Therefore, it is important that educators have an understanding of how cisgenderism and ableism are prevalent in our everyday life and schools. Knowledge of this can help influence school environments that are more inclusive of different types of gender identities and those with intersecting identities. In conclusion:

> If you are just looking at the big picture and seeing this big bush, that's all you are going to see, is the bush as a whole. But society is just not one bush. It's all these little

*Figure 21.7*   Picture by participant three of the big bush.

individual leaves that all need their different care. I feel like if you don't give each of those leaves their care, then they will eventually wither and they will die. People need to understand that [Figure 21.7].

*(participant three)*

# References

Ansara, Y.G., & Hegarty, P. (2012) Cisgenderism in psychology: Pathologizing and misgendering children from 1999 to 2008. *Psychology & Sexuality* 3, pp. 137–160.

Alsaman, M.A., & Lee, C. (2017) Employment outcomes of youth with disabilities in vocational rehabilitation. *Rehabilitation Counseling Bulletin* 60, pp. 107–198.

Balsam, K.F., Rothblum, E.D., & Beauchaine, T.P. (2005) Victimization over the life span: A comparison of lesbian, gay, bisexual, and heterosexual siblings. *Journal of Consulting and Clinical Psychology* 73(3), pp. 477–487. doi: org/10.1037/0022-006X.73.3.477.

Bockting, W.O., Miner, M.H., Swinburne Romine, R.E., Hamilton, A., & Coleman, E. (2013) Stigma, mental health, and resilience in an online sample of the US transgender population. *American Journal of Public Health* 103, pp. 943–951. doi: 10.2105/AJPH.2013.301241.

Bornstein, K. (1994) *Gender outlaw: On men, women, and the rest of us.* New York: Routledge.

Brill, S., & Pepper, R. (2008) *The transgender child: A handbook for families and professionals.* San Francisco, CA: Cleis Press.

Broad, K. (2002) GLB+T?: Gender/sexuality movements and transgender collective identity (de)constructions. *International Journal of Sexuality and Gender Studies* 7(4), pp. 241–264.

Collins, P.H. (1990) *Black feminist thought: Knowledge, consciousness, and the politics of empowerment.* New York: Routledge.

Crenshaw, Kimberle (1989) Demarginalizing the Intersection of Race and Sex: A Black Feminist Critique of Antidiscrimination Doctrine, Feminist Theory and Antiracist Politics. University of Chicago Legal Forum 1989, pp. 139–167.

Crenshaw, K.W. (1991) Mapping the margins: Intersectionality, identity politics, and violence against women of color. *Stanford Law Review* 43, pp. 1241–1299.

Denny, D., Green, J., & Cole, S. (2007) Gender variability: Transsexuals, crossdressers, and others. In A.F. Owens & M.S. Tepper (Eds.), *Sexual health volume 4: State- of-the-art treatments and research*, pp. 153–187. Westport, CT: Prager.

Duke, T.S. (2011). Lesbian, gay, bisexual, and transgender youth with disabilities: A meta-synthesis. *Journal of LGBT Youth* 8(1), pp. 1–52. doi: 10.1080/19361653.2011.519181.

Dong, S., Fabian, E., & Luecking, R.G. (2016) Impacts of school structural factors and student factors on employment outcomes for youth with disabilities in transition: A secondary data analysis. *Rehabilitation Counseling Bulletin* 59(4), pp. 224–234. doi: 10.1177/0034355215595515.

García, S.B., & Ortiz, A.A. (2013) Intersectionality as a framework for transformative research in special education. *Multiple Voices for Ethnically Diverse Exceptional Learners* 13(2), pp. 32–47.

Glenn, E.N. (2002) *Unequal freedom: How race and gender shaped American citizenship and labor*. Cambridge: Harvard University Press.

Gopaldas, A. (2013) Intersectionality 101. *Journal of Public Policy & Marketing* 32, pp. 90–94.

Grossman, A.H., & D'Augelli, A.R. (2006) Transgender youth: Invisible and vulnerable. *Journal of Homosexuality* 51(1), pp. 111–128. doi: 10.1300/J082v51n01_06.

Haussmann, C., Morrison, D., & Russian, E. (2008) Talking, gawking, or getting it done: Provider trainings to increase cultural and clinical competence for transgender and gender-nonconforming patients and clients. *Sexuality Research and Social Policy* 5(1), pp. 5–23.

Haritaworn, J. (2008) Shifting positionalities: Empirical reflections on a queer/trans of colour methodology. *Sociological Research Online* 13(1), pp. 1–12.

Hill, D.B. (2007) Trans/gender/sexuality: A research agenda. *Journal of Gay & Lesbian Social Services* 18(2), pp. 101–109.

Hines, S. (Ed.), Sanger, T. (Ed.). (2010) *Transgender identities*. New York: Routledge. https://doi.org/10.4324/9780203856147

Kahn, L., & Lindstrom, L. (2015) I just want to be myself: Adolescents with disabilities who identify as a sexual or gender minority. *The Educational Forum* 79(4), p. 362.

Kazyak, E. (2011) Disrupting cultural selves: Constructing gay and lesbian identities in rural locales. *Qualitative Sociology* 34(4), pp. 561–581. doi: 10.1007/s11133-011-9205-1.

King, D.K. (1988) Multiple Jeopardy. *Multiple Consciousness: The Context of a Black Feminist Ideology* 4(1), pp. 42–72.

Koyama, E. (2020) Whose feminism is it anyway? The unspoken racism of the trans inclusion debate. *The Sociological Review* 68(4), pp. 735–744. doi: 10.1177/0038026120934685.

Lane, R. (2009) Trans as bodily becoming: Rethinking the biological as diversity, not dichotomy. *Hypatia* 24(3), pp. 136–157.

Lennon, E., & Mistler, B. (2014) Cisgenderism. *Transgender Studies Quarterly* 1(1–2), pp. 63–64.

Lipscomb, S., Haimson, J., Liu, A.Y., Burghardt, J., Johnson, D.R., & Thurlow, M.L. (2017) *Preparing for life after high school: The characteristics and experiences of youth in special education. Findings from the National Longitudinal Transition Study 2012. Volume 1: Comparisons with other youth* (Full report: NCEE 2017-4016). Washington, DC: U.S. Department of Education, Institute of Education Sciences, National Center for Education Evaluation and Regional Assistance.

Longmore, P.K. (1995) Medical decision making and people with disabilities: A clash of cultures. *The Journal of Law, Medicine & Ethics* 23(1), pp. 82–87. doi: 10.1111/j.1748-720X.1995.tb01335.x.

Longmore, P. (2003) *Why I burned my book and other essays on disability* (American Subjects). Philadelphia: Temple University Press.

McCall, L. (2005) The complexity of intersectionality. *Signs: Journal of Women in Culture & Society* 30(3), pp. 1771–1800.

Monro, S. (2000) Theorizing transgender diversity: Towards a social model of health. *Sexual and Relationship Therapy* 15(1), p. 33.

Morgan, J.J., Mancl, D.B., Kaffar, B.J., & Ferreira, D. (2011) Creating safe environments for students with disabilities who identify as lesbian, gay, bisexual, or transgender. *Intervention in School and Clinic* 47(1), pp. 3–13. doi: 10.1177/1053451211406546.

Nagoshi, J., & Brzuzy, S. (2010) Transgender theory: Embodying research and practice. *Affilia* 25(4), pp. 431–443.

National Longitudinal Transition Study (2012) *Volume 1: Comparisons with other youth: Full report* (NCEE 2017-4016). Washington, DC: U.S. Department of Education, Institute of Education Sciences, National Center for Education Evaluation and Regional Assistance. Retrieved from: https://ies.ed.gov/ncee/projects/evaluation/disabilities_nlts2012.asp.

Newman, L., Wagner, M., Knokey, A., Marder, C., Nagle, K., Shaver, D., et al. (with Cameto, R., Contreras, E., Ferguson, K., Greene, S., & Schwarting, M. (2011) *The post-high school outcomes of young adults with disabilities up to 8 years after high school: A report from the National Longitudinal Transition Study-2 (NLTS2)* (NCSER 2011-3005). Menlo Park, CA: SRI International.

Pascoe, C.J. (2005) Dude, you're a fag: Adolescent masculinity and the fag discourse. *Sexualities* 8, pp. 329–346.

Paxton, K.C., Guentzel, H., & Trombacco, K. (2006) Lessons learned in developing a research partnership with the transgender community. *American Journal of Community Psychology* 37(3–4), pp. 349–356. doi: 10.1007/s10464-006-9049-0.

Raftery, D., & Valiulis, M. (2008) Gender balance/gender bias: Issues in education research. *Gender and Education* 20(4), pp. 303–307.

Robinson, J., & Espelage, D. (2012) Bullying explains only part of LGBTQ–heterosexual risk disparities. *Educational Researcher* 41(8), pp. 309–319.

Roen, K. (2001). Transgender theory and embodiment: The risk of racial marginalisation. *Journal of Gender Studies* 10(3), pp. 253–263.

Sawyer, K., Salter, N., & Thoroughgood, C. (2013) Studying individual identities is good, but examining intersectionality is better. *Industrial and Organizational Psychology* 6, pp. 80–84.

Smith, Tones, (2012) Pathology, bias and queer diagnosis: A crip queer consciousness. Masters Thesis, Smith College, Northampton, MA. https://scholarworks.smith.edu/theses/1074

Somerville, S.B. (2000) *Queering the color line: Race and the invention of homosexuality in American culture.* Durham, NC: Duke University Press.

Stillwell, R.A., & Sable, J. (2013) Public School Graduates and Dropouts from the Common Core of Data: School Year 2009-10. First Look (Provisional Data). NCES 2013-309. *National Center for Education Statistics.*

Stryker, S. (1994) My words to victor frankenstein above the village of chamounix: Performing transgender rage. *GLQ* 1(3), pp. 237–254. doi: 10.1215/10642684-1-3-237.

Stryker, S., & Whittle, S. (2006) *The transgender studies reader.* New York: Routledge.

Tauchert, A. (2002) Fuzzy gender: Between female-embodiment and intersex. *Journal of Gender Studies* 11, pp. 29–38.

Thorne, B. (1993) *Gender play: Girls and boys in school* (11th ed.). New Brunswick, NJ: Rutgers University Press.

Trainor, A., Lindstrom, L., Simon-Burroughs, M., Martin, J., & McCray Sorrells, A. (2008) From marginalized to maximized opportunities for diverse youths with disabilities: A position paper of the division on career development and transition. *Career Development for Exceptional Individuals* 31(1), pp. 56–64.

Withers, A.J. (2012) *Disability politics and theory.* Halifax, Canada: Fernwood Publishing.

Zablocki, M., & Krezmien, M.P. (2012) Dropout predictors among students with high incidence disabilities: A national longitudinal and transitional study 2 analysis. *Journal of Disability Policy Studies* 24(1), pp. 53–64.

# 22

# FLOWING DESIRES UNDERNEATH THE CHASTITY BELT

## Sexual re-exploration journeys of women with changed bodies

*Inge G. E. Blockmans, Elisabeth De Schauwer,*
*Geert Van Hove and Paul Enzlin*

## Introduction

I can go into a coffee shop and actually pick up the cup with my mouth and carry it to my table. But then that... that becomes almost more difficult because of the... just the normalizing standards of our movements, and the discomfort that that causes when I do things with body parts that aren't necessarily what we assume that they're there for. That seems to be even more hard for people to deal with... I would really try to make myself go out and just order a coffee by myself. And I would sit outside for hours beforehand in the park just trying to get up the nerve to do that.

*(Sunaura Taylor)*

On a walk in Astra Taylor's (2008) documentary *The Examined Life*, Judith Butler and Sunaura Taylor ponder "moving in social space, moving—all the movements you can do and which help you live, and which express you in various ways" (Butler). The quote above is Taylor's reply when Butler asks her, while she is moving in a power wheelchair with limited use of her arms and hands:[1] "Do you feel free to move in all the ways you want to move?" They discuss how normalised expectations of moving—and by extension, living and being—make it hard for people to feel free to be, live and move. They pose the question, not "what is the body?" but rather, for Deleuze, "the properly ethical question ... what can a body do?" (1990; in Guillaume and Hughes, 2011, p. 1). To answer this question, they approach people and their bodies as be(com)ing within relational "assemblages," i.e., ever-shifting constellations where materiality, society and semiotics flow together (Deleuze and Guattari, 1987, p. 22).

Along these lines, we research on the waves of the postmodern shift in Disability Studies, from approaching disability as an individual deficit to disability as a dynamic outcome of intra-actions between the materiality of one's body and of one's lifeworld intertwining with meaning-making by one's self, by intimate others (romantic partners, sex partners, friends, family, specific healthcare providers, etc.) and by more public, invisible others (discourses permeating and spread by media, education, healthcare policy, academia, etc.) (Shakespeare and Watson, 2001; Stroman, 2003; Watson, 2002).

We aim to write about the manoeuvring (i.e., moving with skill, care and caution) around picking up a cup of coffee with one's mouth (figuratively) in function of the search for sexual pleasure when it is impossible to comply with "the norms of normate sex" (Wilkerson, 2012, p. 187, cf. infra), here due to a body that functions and is seen differently from before, disrupting one's normality personally and within a society characterised by the normalisation of bodies as well as the normalisation of sex (Foucault, 1997). We have focused on the lived experiences of women with traumatic (i.e., non-congenital) spinal cord injury (SCI), who experienced a significant positional shift on the scale ranging from "normate" to non-normate or "extraordinary" (Garland-Thomson, 1997) and who once walked into the coffee shop unnoticed and picked up their coffee with their hands. This chapter will work with the material and discursive practices at play in the assemblages of being a woman with sexual desires living in and with a body affected by SCI.

On the one hand, their potential challenges in sexual pleasure and intimacy have been previously explored in research focusing on neurological functioning by drawing on obstacles following physical deficits, e.g., discussing spasticity, lubrication problems, lack of bladder or bowel control, absence of genital sensation, medication affecting arousal, risks of autonomic dysreflexia, difficulties in moving into certain positions, etc. (Sipski, 2007; Stoffel et al.2018; WHO, 2013). However, this research trend reinforces medicalised and performance-oriented views of sexuality and fosters healthcare practices that—once rehabilitation and medical interventions have reached their limits in fixing and modifying the body—ultimately run out of options in (re-exploring) sexual expression and pleasure (Tepper, 2000). On the other hand, research mainly fuelled by disability and social justice studies urges us to attend to the disabling impact of social/cultural/environmental factors on sexual identity formation and possibilities for sexual expression. In so doing, it risks leading to reverse essentialist approaches by neglecting the reality of living with physical limitations and discomfort and hence overlooking their potential impact on sexuality (Kool, 2010; Linton, 1998; Sakellariou, 2006; Snyder and Mitchell, 2001; Thomas, 2006).

Both research streams risk endorsing a deterministic view of living with a body that is often approached as undesirable in terms of "physical, cultural and social capital" (Houston, 2019; Hughes et al., 2005), a view in which there is no other role available than being either "victims of their malfunctioning bodies" or merely undergoing oppressive flows (Shakespeare, 2000, p. 162). Furthermore, SCI, or more broadly "disability," are often used as categories to group people assumed to share the same views, experiences and priorities, and to present category members as being different from temporarily able-bodied people. This subscribes to the "grand binary narrative" of people being either able-bodied or disabled/ill (rather than seeing people on a continuum of human embodiment) and their experiences of life as dominated by their physiological "condition" (Grosz, 1994), whereas medical or social labels do not necessarily serve as master "narrative arcs" to understand people's experiences of being in the world as a sexual being (Gallop, 2019, p. 9).

Research, that voices the lived sexual experiences and desires of women with SCI and positions them at the intersection of and in constant intra-action[2] with matter and normative practices, remains scarce (Kafer, 2003). Women with SCI still have the right to sexually intimate lives and they still desire it. The pressing question is: how? Where are the knots and openings in these women's search for and experience of sexual pleasure? What are the material, semiotic and social flows that constantly affect their desires and how can they (imagine to) experience their body freely as sources of pleasure and intimacy? This chapter is one of the outcomes of a wider doctoral study on "imaginative manoeuvrability" in the context of sexual well-being, i.e., "the *experience* of *feeling able* to move within and even create a particular context in a space with borders coming in the shape of matter and discourse, or rather, a space that is embedded in material-discursive practices about sexuality, (dis)ability, desirability, pleasure" (Blockmans, 2019, p. 138).

## Method

Throughout data gathering, analysis, and presentation of a study on sexual well-being of women with traumatic SCI, happening alongside autoethnographic trips of the first author, we have chosen to be guided by an ontology of becoming(s) rather than being, which "enables (even urges) us to see things differently—in terms of what they might become rather than as they currently are" (Martin and Kamberelis, 2013, p. 670). After ethical approval, we organised three data-gathering rounds between April 2015 and March 2017. The first round involved in-depth individual life story interviews with ten women with traumatic SCI (two or three times each, 45–90 minutes per interview) and explored the meaning of sexuality throughout their life and how these meanings came to be in relation to their environment, starting from one main question: "Could you please tell me your life story with a focus on your development of relationships and sexuality?" To foster the conversation, participants were also asked to bring an "object that marked" for them "an important stage or moment in their relational or sexual development." In all interviews, the spinal cord injury was the turning point between a past of satisfying sexual experiences and a present dominated by a changed, psychological and physical labour-consuming body, i.e., a body that demands attention and can no longer be taken for granted. The interviewees meandered around how they saw and lived with their bodies, but seldom talked about recent experiences of bodily pleasure. This led to a second data-gathering round with four participants, aiming to create a context for re-encountering their body. Encounters were arranged as "on-the-road" conversations during body-focused activities with a self-chosen friend ("a person with whom you feel you can discuss your body and desires"), including searching for clothes/accessories just outside their comfort zone with a stylist, make-up session, and photoshoot aiming at dynamic pictures; and follow-up duo-interviews that led organically towards discussing embodiment of difference, embodiment of femininity, struggles in maintaining ownership of their body, and how both SCI and body work had changed their relationship with their bodies and their perceived possibilities to achieve intimacy and sexual pleasure. The third round of encounters was a group discussion with the four women with SCI involved in the second round, about the preliminary findings of an inductive thematic analysis of the stories they had shared throughout the doctoral research, including sexuality as a journey, SCI as life changing, disownment of the body and the chastity belt as a metaphor for feeling (sexually) blocked. All conversations were audio-recorded and transcribed by the first author and a student for her master's dissertation in sexology.

For this chapter, we have chosen to focus on the four participants who participated in all three encounter rounds, although the stories of the other participants and the autoethnographic work of the first author were present as satellites, with stories overlapping and differing, contributing to our developing understandings. All participants were in a stable relationship and were mothers (one pregnant for the first time during data rounds two and three) aged 30, 38, 42 or 54 at the time of the life story interviews in 2015. Two were paraplegic and two tetraplegic. The four women's stories were re-analysed, drawing on Vagle's (2014) post-intentional phenomenology to make sense of how they experienced being a woman with sexual desires living in and with a body affected by SCI, with the experience of the body and the social/cultural world seen in a mutually constitutive relationship. The theoretical incentive in post-intentional phenomenological research does not concern whether something is or is not, but how the perception of something (here: the perception of the body as source of/vehicle for sexual pleasure) is produced (in particular moments) and how this production connects with other productions, assumptions and meanings associated with the object of study (Vagle, 2015, p. 607).

The analytical process was cyclical, involving reading the transcripts (finishing a first round of analysis of one participant's stories before continuing to the next participant's transcripts), commenting in the margins (on descriptive, paralinguistic, conceptual levels) on what was meaningful to the participants and what might be framing their meaning-making. During that process, we were also constantly questioning whether we were relapsing into thinking in binaries or staticity or decontextualisation in order to remain conscious of the ambivalence and shifts in the glimpses of lived experience that the participants shared with us. That is, conclusions about participants' lived experiences could not be as simple as "X has stopped searching for sexual pleasure," "X feels voiceless vs. Y is expressing her desires," etc. The analytical processs was inspired by Jackson and Mazzei's (2012) "plugging in" of concepts in data, interpreting participants' recounted experiences in the light of concepts considered relevant after first transcript readings. We gradually started to focus on "containment," with containment encompassing the "discursive, gendered and dis/ableist" process of controlling "both in terms of the microindividual level, such as learning to contain one's own bodily fluids, but also at the macro level through, for example, processes of incarceration and categorisation" (Liddiard and Slater, 2017, p. 320) as well as on sexual pleasure as "becoming" in context rather than as a continuous and static state of being (Deleuze and Guattari, 1987). Our focus was not so much on what it 'means' for participants to live with a changed, uncontrollable, not fully sensitive body that was less mobile than before, but rather on how their experience of and imaginable potential for sexual pleasure "becomes-in-the-world-with-others" (Price and Shildrick, 2002; Ellingson, 2017).

In the final stage of the analytical process—which overlapped with the write-up stage of this chapter—we took up Vagle's invitation to "explode beyond tradition" (2014, p. 132) and used our summaries per participant, selected rich citations, and our notes as building blocks in "crafting" this text in an attempt to engagingly animate the material-discursive processes of movement and captivity that are working on/through (living with) sexual desires with a changed and changing body. What follows has been written from what, for us, appears to be the spirit of the many encounters the first author (and the other authors through the readings of transcripts and exchanges of reflections on the shared stories) had with the research participants, after and whilst revis(it)ing our analyses and reflecting on the resonance of their "voices" within ourselves (Spry, 2009, iv).

The findings section below begins and ends with a fictionalised account of the first author getting home after re-visiting interviews, where she remembers fragments of encounters with research participants (presented in italics) while she is intra-acting with her partner and her self in the place she calls home. These fragments of encounters blend into a fictional narrator's analysis-based comments and questions (presented in bold or aligned right) responding to or announcing glimpses of participants' life stories (i.e., selected quotes from interview/field trip transcripts, presented following an indent). It is an attempt to share yet not to finalise our findings, and to share yet neither finalise our participants' voices nor create a template of "what it means to live with spinal cord injury." The text is an invitation to let the vibrations of meanings that make up their, our, your lives resonate and recognise parts of ourselves in the words and images about living in, with, through bodies.

## Insights

Let me tell you about the woman who once had over 40 pairs of heels and sexy lingerie, and occasionally sailed shamelessly naked around her bedroom. Presented as one, yet being many.

One question.
"How do you see yourself, when it comes to being sexual?"
She had whispered: *Yes... who am I, now?*
Returning home, our minds, our bodies, our souls were filled with echoes.
Echoes of her answers.

Echoes of her questions.
*What does sexuality mean to me?*
*It is a domain of life... something is not quite right if that domain is pushed to disappear out of someone's life completely, is it?*
*Tell me, can you tell me what is still possible?*

Echoes of questions she had never asked out loud.
*What do I desire?*
*What do I imagine bodily pleasure to be, for me, and for my partner?*
*How free do I feel to move, sexually?*

> Desire is about experimenting with "dare to become all that you cannot be."
> *(Massumi, 1992, p. 41)*

The moment you get home, the real rehab(il)itation starts.

> Rehabilitation.
> "The action of restoring something that has been damaged to its former condition." [3]

> Rehabitation.
> "The action or act of reinhabiting a country, area, house, etc."[4]... including one's body?

> Or: the process of actions in which old habits are revisited, current conditions are explored, and new ways of being are created?

*I am a living bust, positioned on a body that I am not conscious of,* she said.
I enter my flat, and nearly lose my balance when I feel a tender yet fiery kiss in my neck. Next, I find myself soaring through my flat and landing on the bed.

He starts unlacing my boots.
"Give me a minute," I whisper while covering his eyes.
"I know what that means," he grins excitedly.

Once in my bathroom, I try to rush out of my jeans skirt. As always, it gets stuck around my waist and I wiggle-waggle myself out of it. *It is that jeans again today?* I hear a participant's nurse telling me off, poking an accusing finger at me for making life more difficult. I jump on the toilet. I feel relieved. My impatient, sensitive bladder is not an obstacle... when I ask for a minute in the middle of a kissing session, he knows that I will be back for more of the same and beyond.

I notice the white support handles. The foldable shower chair melts and disappears in a hole in the wall… the hole becomes the hole in the middle of the seating of the wheelchairs I saw hovering over the toilet in hospitals and some of my participants' bathrooms. I am in my bathroom and yet I am in her bathroom. Sterile catheters sneakily pop out of my colourful storage boxes. I wash my hands, and the fragrant soap transforms into disinfectant. My hand lotion solidifies and covers my fingers with latex gloves.

**Imagine a woman**.

> You asked me to bring an object that was important in my sexual history.
> When I told my partner I needed to think of such an object, he replied I needed to take him with me. [laughs]

**She puts her bright red heels on the table.**

> My heels. The only pair I've kept. I've always worn high heels. Day in, day out.

**A woman with red heels she does not wear.**

> They make my feet fly now and they make my legs unstable. So I've promised my friends the remaining 39 pairs in my shoe closet, once I am ready in my heart to let go. I know it's ridiculous… they are just shoes… but they are/were once attributes that make/made me feel feminine. I can still wear them after struggling for 15 minutes, but I can't walk with them anymore… It's ambivalent… I love them, but I can't meet the criteria anymore. Somehow, though, I have grown into not needing them anymore to feel feminine.

> Shoes recalling memories of unre-reachable past moments. Between the feeling they used to give her and the loss they now symbolise. No longer supportive for be(com)ing a(nother) woman.

**A woman wearing red laced lingerie inside and a neutral set outside.**

> I need new nightwear, I decided. Everything I had was grey, grandmummalike. So many things in the windows that my partner would have loved, red, a bit of lace. If I had been mobile, I would have bought it for fun. Now, no… strangers see me all the way to my underwear. My closet cannot contain anything that should not be seen by the average man and woman. Basically, I should purchase things you can hang over the clothesline outside.

> I told myself:

> I'm not going to buy this, let's keep neutral colours. I left with something black and white.

**A woman with a past.**

> Either you crawl in a corner, or you continue living. But it's two periods, really. Before the spinal cord injury and after… It is a wholly different life, different like black and white.

Also relationshipwise, it's a search… simply because… you are not the same person… emotionally… sensory… physically…

## A woman with a body seemingly absent, yet very present…

Before the accident, I knew my body to perfection.

> Can you ever know every tiny particle?

I could bend and turn.
Lithe.

> How flexible can you be?

I felt good in and about my body.
I had a style of my own back then.

I was not meeting bodily ideals, but I didn't care. Now, I mirror myself to them and now there suddenly is something I cannot achieve anymore.

I don't feel as much of a woman anymore, not as I used to feel… I am slowly re-turning, but…

## and desires

I tell him, I want you to not only take care of me. [laughs]

Intimacy, sexuality. Being sweet towards each other. I don't feel it in hurried little kisses or routine, but in things you sincerely take time for… doing things together, and not to be settled quickly. Feeling attractive. The intimacy, feeling loved… the accident didn't make me asexual and without sensations… but it is a whole process with your partner… I need the physical touch, even if it's just holding each other. I need it, from my partner.

I have become convinced over time that sex is something for every human, something natural, part of life. Something that should be normal, and

## not absent, yet struggling to be present,

with me it is, of course, not so normal, given that… my body… is not what it used to be… Sex is also something to be done with two, preferably, with the same partner [laughs] in this stage of my life, with kids, but, in hindsight, I would have done well experimenting a bit more. If I want to experiment now, I need to do it with my husband.

De-routinising sex requires a constant search with changing bodies and changing wishes.

Because I have a partner, we [laughs] try to do something along those lines… but it remains difficult. Especially being spontaneous. I would love to wake up naked once in a while… But we don't do that. Imagine that the nurse sees you not wearing anything… [whispers] That's really not fun. Not for me, not for my partner.

> When is nakedness functional enough to pass as appropriate?

**trembling in a box.**

Sex remains sensitive stuff. I guess for most women it gets uncomfortable when you call it by its name, especially problems. Although sexuality has come more out in the open than before… sex as a disabled person is still in a dark little corner… it's still taboo.

There is shame. Before I was in a wheelchair, less. Now… I still struggle with my sexual experiences… because I cannot enjoy intimacy in the same way I used to…

It used to be difficult…

I guess I feel more vulnerable. When things don't go smoothly…you expose yourself.

But not anymore. No. Because for me it is not a problem anymore.

I'd rather talk about it with you than with friends, just because… bluntly… an orgasm is out of my range. They still have orgasms, I don't.

When they ask me whether I can still have sex, I reply I can, and then they quickly proceed with "can you still come?" People don't realise it is much more complex than "it is possible or not." There's the sensory stuff, there's the potential leaks…

and if you tell that… I really don't want the look of their "poor you" eyes. No thanks.

Orgasms feel different, but in one way or the other… I gradually have come to experience it is mainly mental now. It is not always about touching exactly the right spot or even registering touch… It is simply also about the visual imput and the feelings underlying intimacy.

**The traditional zone of sexuality has become a zone of shame, dirty leakiness, blemishes to hygiene, a zone that needs to be managed to have sex…**

It's weird, that that zone where you… are incontinent… is also the zone where sex needs to take place. It doesn't need to, there's other things related to sexuality, but it usually moves in that direction.

Of course, a pad or a diaper, it doesn't feel feminine, it doesn't feel attractive, it's not like "Come, honey, rip it off and let's start." In the beginning, I thought "No! Hands off!" Actually, I felt, in a particular way, dirty…

**The traditional tools do not suffice anymore.**

The most important stuff is supposedly to be found with men. Women have breasts, and that is supposedly important too…whereas I have the outer appearance of any other woman, but I don't feel my body… Men's sex lives should be splendid, but what a woman thinks or feels or wants, it doesn't seem to matter. It is mirrored everywhere around you: men always have sex and women don't.

A double sexual standard. Pressing more on women with "different" from "normal" sexual possibilities, who also have the right for pleasure?

**Female pleasure silently disappears out of sight.**

Whereas I disagree. But try and make that clear [sigh].

To a man, she says.

And what about making it clear to healthcare professionals?

**Pleasure and a body that urge for explanations.**

We got this booklet, with pictures of couples, where one of the two had a spinal cord injury to show what is still possible. But there was nothing about what a woman can feel. Not a word. Don't you believe that that is utterly childish?

Apparently a woman's sexual sensations are not important. What remains important is whether a man's penis can still get up after a spinal cord injury.

During rehab, it was like, find out for yourselves. Is it possible or not?

**A woman with questions.**

What exactly can you still feel?

If you cannot sense your skin.

Well, I want to know how it operates medically.

Can a woman with a spinal cord injury still have an orgasm?

Can she... hope?

**... or a woman who has stopped questioning.**
**Yet always with a past/memories of times gone by.**

It will never be like before. Too many nerves were damaged.

Irreversibility of change lined out in the body.

Acquaintances asked other friends, "how are they going to have sex?" It's like, you've ended up in a wheelchair, so now you are an asexual being... That is not true either, but it will never be like before. Your body simply responds completely differently. Imagine someone providing the same stimuli to my friend and to me. She will get horny and for me it will be "ah yes." Simply because the neurological wiring has become completely different. You keep being reminded of that. Because my partner... he has a healthy libido... And you know that you, on some levels... I'm not saying I'm failing him, but if I hadn't been in a wheelchair, it would have been completely different.

How can you use your body?

My role is more passive, so to speak. Whereas before I could do things more actively... before I could sit on top and now I can't. Some moments I just roll with it... but often there's a twist.

You talk about giving. What about receiving?

333

**A woman who feels blocked, sometimes, or often.**

You really want to know to what extent I am sexual or a woman? Well, I wear a chastity belt. A belt that makes it impossible to undertake anything with women. Men cannot touch women or it prevents women from having a relationship or being sexual altogether, and well, that counts for me too.

**A chastity belt materialising perceptions about body and sexual possibilities**

Dressing and undressing someone like me in a back-saving manner is a serious assignment. So, imagine the vibe feels good, and you happen to be both at home, as man and woman, or on a weekend trip or whatever, then *that* is still that major party pooper, which makes me think, even if my husband is up for it, like "yes, I will get those clothes off," then I sometimes give him this look, "I don't feel up for this." By the time my clothes are gone, my appetite is gone too, and definitely by the time I have everything on again. [laughs]

**A chastity belt coming in different shapes and sizes—materially, physically, psychologically, affecting manoeuvrability in space and time.**

We still sleep separately, separate beds, separate rooms, separate level in the house. If only we could sleep together, so you can hold each other…

**An unchosen belt encircling bodies, minds, desires**
And I feel ugly on top of all of that…especially when you need to get undressed.
Tell me about it. But are those men attractive when they are naked?

designed/tightened by internalised norms of desirability and the body beautiful

**and touching encounters.**

When my partner helps undressing or positioning my legs… it is part of foreplay… it is part of the game, also if you do not have any deficiencies. So being assisted is less a matter of giving up autonomy or dignity or self-worth…

Does him undressing you feel like assistance?

It's 50/50. I don't experience it as completely foreplay… because… it needs to happen a bit more carefully… and with turning my body from left to right… The caring seeps through. It is never completely gone…

**She experiences her body differently.**
**And so does her partner, she thinks.**

He needed to catheterise me when I wasn't able to… It must be so different to, as a man, to have to look at your wife that way, at her vagina. The need to be down there for a different reason, to do something different from normal. I'm not sure whether he really perceived it that way, but that's how I feel. I do wonder, how does he see me?

He was like "yes no no no I'll leave" whereas we used to go to the toilet in each other's presence. And now, it is surrounded by something medical, it seems.

I can imagine that it is terrifyingly hard for a partner to touch someone in a sexual or intimate manner when you, when that person does not even know herself how her body exactly functions.

## Bodies to be re-explored.

*What if every woman was treated the same, as in: unique and constantly evolving?*

## Desires to be negotiated or stripped bare

I am not paralysed completely am I?... I long for touching you again. It's not because I... that... I ask him, "am I not *allowed* to touch you anymore?" And he says yes but... it never happens. Is he having a hard time to relax and giving himself or does he find it hard for me... I guess he partly finds it hard for me... It really troubles him.

## Routines and definitions of sex life disrupted

In the past you were not capable of any less than someone else, and what you created with your partner, was done by yourselves, you know. [sighs] Whereas now, even if you want to create something together... I am basically blind with my body. Being paralysed is being blind with your body.

I can't
do
what I would desire to do.
You know, those spontaneous things
you would simply do
if you are yourself.

I would go and sit on his lap, spontaneously... that would make my man think "hmmmm." Now I'd need to tell him "come and sit on my lap." And then... I don't know how. What I also miss, I told you what my paralysis entailed and that my hands, my fingers are paralysed, and you said, "Can you still caress?" I thought, "Does she need to bring up exactly that topic?" I used to love massaging my partner. We focus on what I still can do, but there's obstacles to spontaneity.

Good sex... I miss it. It was not the most important aspect of our interactions, but it came automatically...

## Desires she needs to—

"Damn, they are there again." Actually I find they come way too early. Anyways, I need to accept it. Very well then. They come to put me in bed. The good thing is that my husband doesn't have to do it.

**—hide.**

You don't wear whichever night gown because you need to let yourself be put to bed by a man, a male nurse, so you do not, not once, put on a beautiful, fun, sexy thingie because you need to, you know, you do not have any privacy on that level. You need to be careful of what people find in your house. Because there's constantly people,

one, two, many,

in your house for cleaning and whatever. Also, you are not out of your clothes that quickly, you are not as quickly… in your clothes, you need to look decent for the nurses. Imagine you had been up to something during the night; you'd need to make sure you are clean and dressed again by the time the morning comes.

> The desire for sexual pleasure fuses worlds that are obstinately held separate.

Shame, I guess. I am too ashamed. Much more than necessary…

But I need to constantly lie here with my bum naked, do you know? Without me really wanting it? While really not liking it?

They do it all the time. To. You.

It is *their* domain. Really. When I lie there in bed and the nurse bathes me, it is *her* work territory, my *whole* body.

**—fight for?**

It was already too much for her to be asked, you have that tube here [shows stomach catheter] that needs to be attached to my bedside, so the urine flows in this bag. That tube was unimaginably short. Because my partner, if he'd simply want to pull me a bit closer, I'd be tied down to a cable. Alongside the fact that I can't turn towards my partner, I am tied with a cable, stuck to one side of my bed, with assist rails giving me the only chance to transfer myself a tiny bit, or to stretch my fingers when they are spastic. So if my partner desired to lie a bit closer, for once, because he is always tired [laughs], then he first needs to come and unfasten that cable. So I told the nurse—gosh, I really need to be articulate and assertive!—I would like you to attach a longer tube. "Why is that necessary? There is right enough space, isn't there?"

> Repeatedly and loudly denied.
> *(Kafer, 2003, p. 85)*

**Justifying desires.**
**Fighting for a voice**
**and keys to act on desires.**

Sometimes, though, I see something flickering ashore. My partner has already discovered things we hadn't discovered before, which made us both realise that it can still be fun, you know, as in: worth all the effort if we do come to… And you never know, there might be so much more possible than we have yet discovered… but where do you start searching?

I guess it feels like discovering sexuality for the first time…

Support handles turn into bamboo and entangling twigs of a willow tree.
The smell of rosemary and sea salt coffee scrub and the cold water enter my awareness.
With a not so elegant jump, I manage to get on the bed, automatically triggering me to start composing a smile to cover it up, but I roll over and I forget. A flash of passion and a promise for more, and then, like a soft stream his fingers trickle down to the upper band of my compression stockings. Gently, he unveils the previously contained terrain of lovemaking. The stockings have imprinted their presence, leaving dark red circles around my thighs and above my feet. We rub, tickle, and caress life back into them.

How do you touch?

"Let us agree that we will not focus on achieving something, shall we?"

What opens up when you let go of…?

Let us play and dive deeply into our closeness.


## (In)Conclusion

This chapter dug into the potential scope for manoeuvring that women with SCI recounted to experience in their journey of rehab(il)itation of their own body. The words above are all ripples of connection and uncertainty, dried sadness and flowing desire, feeling captive and feeling life ownership, experienced by women with changed bodies and changing expectations, but deep inside their ever-present needs for attraction and bodily intimacy. We hear the body scream to be seen as still a source of pleasure and not just as damaged vulnerable flesh covering breakable bones and freely flowing or constipated excreta to be regulated—a (medicalised) body that, for women with SCI, gains such a central focus in creating their living (together) yet at the cost of not only sexual pleasure, but also intimacy more broadly, i.e., intimacy with oneself and with others, including one's romantic partner.

Between the lines, we read how re-exploring changed bodies in the search for sexual pleasure and intimacy is challenged not only by the changed materiality of the body itself, but also by the women's own (and their partner's) internalised definitions of satisfying sexual pleasure. Their expectations about when, where, how, and to what goal to have sex as well as their meaning-making of intimate encounters built up throughout their life history reflect the performance-based and genital-focused approach to sex that is widespread in western society—*the chronic condition of being sexual in a world where sex is highly yet often invisibly normed and regulated*. These expectations include the constriction of sex to penetration with explosive sensations, preferably leading up to simultaneous orgasms, with a climax as necessary for satisfaction, the best sex as arising spontaneously, full of initiative of independently functioning and fit partners, etc. (Tepper, 2000). The confrontation with the inability to comply with these normative demands of sexual performance and sensations, that once were (perceived to be) within reach, feeds feelings of being overly abnormal or insufficient to be recognised as a sexual being and a satisfying sexual partner as well as feelings of being at loss, of not knowing how to move and manoeuvre in existing intimate spaces and how to create new, non-normative pathways.

Furthermore, *cure and care practices* bring in a focus on protection, preventing further bodily deterioration, controlling damage, looking after, etc. in the women's lives. While this perspective on the body is not inherently harmful in itself, it carries the risk of minimising one's felt ability to be intimate and to see the body as a potential source of pleasure and creation. These practices permeate bodily encounters and become so natural that the impact on one's relationship with one's body, one's sexual self, and one's romantic/sexual partner is rarely questioned or challenged. The medical(ised) body is present not only in the language of participants and their environment, but also in the materiality of their lives: clinically designed bathrooms and bedrooms, neatly timed body management by healthcare professionals with limited flexibility, clothes chosen for their practicality rather than aesthetics, etc.

Intersecting with these flows is the highly *gendered intimate labour* performed daily by the women. This intimate labour can be emotional (e.g., focusing on what they can give rather than receive in terms of pleasure, or as care receivers taking up a subordinate position both in romantic/sexual relationships and towards healthcare professionals), mental (e.g., negotiating priorities), and physical (e.g., perform actions to look "decent," cfr. erotic and thus 'non-functional' nakedness or clothing exposed to healthcare professionals). Much of the labour is shaped by or "rooted in their social and political positioning as disabled people and—as with the motivations of non-disabled heterosexual women—by normative notions of womanhood, femininity and (hetero)sexuality" (Liddiard, 2014, p. 125; Shildrick, 2009).

These (and other) practices affect women's relationships towards (living with) their body. The challenges to achieve intimate fulfilment and re-explore the body as a source of pleasure were captured by the metaphor of carrying a chastity belt: a locking object worn around the waist and genital area to prevent sexual intercourse. The feeling of wearing a chastity belt was mainly linked by participants to a body that required emotional and physical labour in their lives and thus was omnipresent, yet which did not feel completely theirs anymore, due to being different from what they had become used to (pre-SCI) and due to necessary daily assistance experienced as dependence on others to make things happen. Importantly, the chastity belt is locked and tightened both by others and the women themselves, in both discursive (cfr. negotiation about length of overnight catheter) and material practices (cfr. body that is moved and touched in particular ways) taking place on the flows of regulated sexuality, curing and caring, gendered intimate labour, etc. Hence, one's felt potential for movement does not stand on its own but is challenged and assembled by the material-discursive flows streaming underneath one's search towards intimacy and pleasure through one's body.

The entanglement of the flows described above, as well as how the women (and their close others) engage with them, fuel (de-)sexualisation of one's body/life and either give or draw away space and oxygen for desire to develop and grow, be it specifically sexual desire or more general desire to explore one's body for pleasure and as a source of pleasure. The risk resides in the fact that flows-potentially-becoming-bindings (i.e., flows that can tightly bind the chastity belt to/with the women['s bodies]) can be present every day, minute, second of people's lives yet remain barely noticeable as they have become so natural, unless they are questioned. The more affected by these flows, the more distant and abnormal that the (search for) expression of sexuality and the experience of the body as (a source of) pleasure becomes for the women and the people they are surrounded by (including romantic/sexual partners and people who move and work with the women outside the context of romance and sexuality), and the more challenging it becomes to manoeuvre in intimate spaces—intimate yet shaped by external-becoming-internal-flows that are not inviting to experiment, desire and imagine differently.

How, then, do we move on to the search for sexual pleasure? Will we continue to silently negotiate with the crowd of echoes in our head for hours before going for a coffee to try and predict what will happen when we enter the space where we can access something that will make us feel warm inside? We should not only ask how free people feel to express and enact what they desire (Foucault, 1997, p. 125–6). We should also ask how free they are to *imagine,* beyond what they are aware of. What makes people's imaginative manoeuvrability flow, close, open up?

## Acknowledgements

Our special thanks go to our research participants and to Maaike Boonstra, who defended her master's dissertation in sexology at KU Leuven based on our co-research ("The transformed body: The sexual rediscovery of women with a spinal cord injury"), for assisting with two data-gathering rounds and transcriptions, as well as preliminary analysis.

## Funding

The research on which this chapter is based was financially supported by a PhD Fellowship of the Research Foundation Flanders (FWO, grant number 11V8117N).

## Notes

1 These words, to describe features of Sunaura Taylor relevant to understanding the context of the quote, were taken from a statement written by Taylor herself for the Wynn Newhouse Awards: https://www.wnewhouseawards.com/sunaurataylor2.html (last accessed on 1 February 2019).
2 Intra-action, coined by Barad (2007), signals "the conceptual movement away from separate entities engaging with each other (interaction), toward the unfolding process of becoming, in relation to others, where each one is capable of affecting and being affected by the other (intra-action)" (De Schauwer et al., 2018, p. 609).
3 Definition taken from English Oxford Living Dictionaries (online).
4 Idem.

## References

Barad, K.M. (2007) *Meeting the universe halfway: Quantum physics and the entanglement of matter and meaning.* Durham, NC: Duke University Press.
Blockmans, I.G.E. (2019) *Manoeuvres in the dark: Re-creating (new) stories about sexuality and the body within/by women with a spinal cord injury* (Unpublished doctoral dissertation). Ghent University & KU Leuven, Ghent & Leuven, Belgium.
De Schauwer, E., Van de Putte, I., Blockmans, I., & Davies, B. (2018) The intra-active production of normativity and difference. *Gender & Education 30*(5), pp. 607–622.
Deleuze, G., & Guattari, F. (1987) *A thousand plateaus: Capitalism and schizophrenia* (B. Massumi, Trans.). Minneapolis: University of Minnesota Press.
Ellingson, L.L. (2017) *Embodiment in qualitative research.* New York & London: Routledge.
Foucault, M. (1997) Sexual choice, sexual act. In P. Rabinow (Ed.), *Ethics, subjectivity and truth: The essential works of Michel Foucault.* (R. Hurley, Trans.). New York: New Press.
Gallop, J. (2019) *Sexuality, disability, and aging: Queer temporalities of the phallus.* Durham & London: Duke University Press.
Garland-Thomson, R. (1997) *Extraordinary bodies: Figuring physical disability in American culture and literature.* New York: Columbia University Press.
Grosz, E. (1994) *Volatile bodies: Toward a corporeal feminism.* IN: Indiana University Press.
Guillaume, L., & Hughes, J. (Eds.) (2011) *Deleuze and the body.* Edinburgh, UK: Edinburgh University Press.
Houston, E. (2019) '"Risky" representation: The portrayal of women with mobility impairment in twenty-first-century advertising. *Disability & Society*, pp. 1–22. Published online: 8 March 2019.
Hughes, B., Russell, R., & Paterson, K. (2005) 'Nothing to be had 'off the peg': Consumption, identity and the immobilization of young disabled people. *Disability & Society 20*(1), pp. 3–17.
Jackson, Y.A., & Mazzei, L. (2012) *Thinking with theory in qualitative research: Viewing data across multiple perspectives.* New York & London: Routledge.
Kafer, A. (2003) Compulsory bodies: Reflections on heterosexuality and able-bodiedness. *Journal of Women's History 15*(3), pp. 77–89.
Kool, J. (2010) Ach arm, genegeerd lichaam…. *Medische Antropologie 22*(2), pp. 253–261.

Liddiard, K. (2014) The work of disabled identities in intimate relationships. *Disability & Society 29*(1), pp. 115–128.

Liddiard, K., & Slater, J. (2017) "Like, pissing yourself is not a particularly attractive quality, let's be honest": Learning to contain through youth, adulthood, disability and sexuality. *Sexualities 21*(3), pp. 319–333.

Linton, S. (1998) *Claiming disability: Knowledge and identity.* New York: New York University Press.

Martin, A.D., & Kamberelis, G. (2013) Mapping not tracing: Qualitative educational research with political teeth. *International Journal of Qualitative Studies in Education 26*, pp. 668–679.

Massumi, B. (1992) *A user's guide to capitalism and schizophrenia: Deviations from Deleuze and Guattari.* Cambridge, MA: MIT Press.

Price, J., & Shildrick, M. (2002) Bodies together: Touch, ethics and disability. In M. Corker, & T. Shakespeare (Eds.), *Disability/postmodernity: Embodying disability theory.* New York & London: Continuum, pp. 62–75.

Sakellariou, D. (2006) If not the disability, then what? Barriers to reclaiming sexuality following spinal cord injury. *Sexuality & Disability 24*, pp. 101–111.

Shakespeare, T. (2000) Disabled sexuality: Toward rights and recognition. *Sexuality & Disability 18*(3), pp. 159–166.

Shakespeare, T., & Watson, N. (2001) Making the difference: Disability, politics and recognition. In G. Albrecht, K. Seelman, & M. Bury (Eds.), *The handbook of disability studies.* Thousand Oaks, CA: SAGE, pp. 546–564.

Shildrick, M. (2009) *Dangerous discourses of disability, subjectivity and sexuality.* NY: Palgrave.

Sipski, M.L. (2007) Disabilities, psychophysiology, and sexual functioning. In E. Janssen (Ed.), *The psychophysiology of sex.* Bloomington & Indianapolis: Indiana University Press, pp. 410–424.

Snyder, S.L., & Mitchell, D. (2001) Re-engaging the body: Disability studies and the resistance to embodiment. *Public Culture 13*(3), pp. 367–389.

Spry, T. (2009) *Autoethnography and the other: Unsettling power through utopian performatives.* New York & London: Routledge.

Stoffel, J.T., Van der Aa, F., Wittmann, D., Yande, S., & Elliott, S. (2018) Fertility and sexuality in the spinal cord injury patient. *World Journal of Urology.* Published online: 14 June 2018.

Stroman, D. (2003) *The disability rights movement.* Lanham, MD: University Press of America.

Taylor, A. (2008) *The examined life.* Toronto: Sphynx Productions.

Tepper, M.S. (2000) Sexuality and disability: The missing discourse of pleasure. *Sexuality and Disability 18*(4), pp. 283–290.

Thomas, C. (2006) Disability and gender: Reflections on theory and research. *Scandinavian Journal of Disability Research 8*(2–3), pp. 177–185.

Vagle, M.D. (2014) *Crafting phenomenological research.* Thousand Oaks, CA: Left Coast Press.

Vagle, M.D. (2015) Curriculum as post-intentional phenomenological text: Working along the edges and margins of phenomenology using post-structuralist ideas. *Journal of Curriculum Studies 47*(5), pp. 594–612.

Watson, N. (2002) Well, I know this is going to sound very strange to you, but I don't see myself as a disabled person: Identity and disability. *Disability & Society 17*(5), pp. 509–527.

Wilkerson, A.L. (2012) Normate sex and its discontents. In R. McRuer & A. Mollow (Eds.), *Sex and disability.* Durham & London: Duke University Press, pp. 183–207.

World Health Organisation (WHO) (2013) *International perspectives on spinal cord injury.* Switzerland & Malta: WHO Press.

# 23

# (IL)LICIT SEX AMONG PWDS IN TRINIDAD AND TOBAGO

## Sexual negotiation or compromise?

*Sylette Henry-Buckmire*

### Introduction

As the sensual hedonistic dancing together of bodies to scintillating rhythmic soca music and revelry of the Trinidad and Tobago (T&T) Carnival season gives way to the calmer melodies, penance and solemnity of Lent, represented as a time of restriction and constraint, I reflect on the sexual (and reproductive health) rights of people with disabilities (PWD), specifically women, and their sexual (and reproductive) desires and hopes from an occupational justice perspective. By occupational justice, I mean the freedom these women must have to participate in activities that are meaningful to them, which contribute to their health and well-being. This chapter also considers what access to sex and intimacy looks like as disabled women reveal very personal experiences of love, desire and deep loss. It unearths how four women have manoeuvred within and without the labyrinth of relationships and their everyday lives to reveal how they define personal sexual freedom. As such, I develop the notion of sex/sexuality/intimacy as occupation and examine how intersections of gender, race/ethnicity, education, class, poverty and disability reveal sexual injustice.

I examine the lived experiences of four physically disabled women, all in sexual relationships, who use practical wisdom, stratagem and negotiations to ensure sexual expression. Anne Finger's (1992) statement rings as true today as it did over 25 years ago when she argued that, "Sexuality is often the source of our deepest oppression; it is also often the source of our deepest pain. It's easier for us to talk about—and formulate strategies for changing—discrimination in employment, education, and housing than to talk about our exclusion from sexuality and reproduction" (p. 9). My interlocutors reveal deep hurt over the controls placed over their sexuality, and the intriguing choreography they adapt with their sexual partners who are, in the case of two of the women, also their primary caregivers. Here, I introduce the concept of mobility-as-occupation (MAO). This notion of MAO capitalises on the permeability of the borders between the mobility of desire and the freedom of sexuality as occupation and encourages reflection on how my interlocutors' varied bodily abilities and physical capacities influence and serve to accomplish their occupational desires. I close this chapter using MAO considerations by describing what sexual justice can look like for these women.

## Carnival, colour and sex as occupation

The wild abandon, sexual expression, sexual freedom and social inclusion promised in a vivid array of Carnival colours, costumes and candidness contrast with the restrictive and differential access to disabled bodies on the nation's streetscapes and public spaces. These disabled bodies seem bound to perpetually experience the penance and fasting rules of Lent, evident in large measure by their absence on the nation's uneven streetscapes during the Carnival Monday and Tuesday national celebrations. Disability is a notion that has long permeated T&T culture, though which has often been left unarticulated. In this current era, their disabled bodyscapes have become canvasses on which the invisible systems of the colonial epoch are registered. These bodyscapes are illegitimate since they operate in crip time, the phenomenon which Alison Kafer (2013) argues, "[is] not only about a slower speed of movement but also about ableist barriers over which one has little control" (p. 26). These body-minds face temporalities of movement and are therefore unable to produce at the same rate, to the same quantity, or simply to produce, as other typical bodies. Hegemonic power undergirds these differential mobilities. What occurs when the body disobeys or operates outside the norm? When the body disobeys, when a disjuncture exists between form, spatial and cultural context, according to Garland-Thomson (2011), it becomes a misfit. And this scholar articulates, "To misfit into the public sphere is to be denied full citizenship" (p. 601). Yet, it is one thing to be a misfit in the public sphere, to become an un-citizen and to lack legitimacy in the public arena, but is this perpetuated in the privacy of the bedroom? What becomes of sexual citizenship?

To be a misfit is also to recognise that occupation is a moving target. Bodies have been deemed atypical because of how they occupy and/or are oriented in space in ways that contravene anthropometric and eugenicist principles. Thus, for these bodies that typically lack desirability (and desirable attributes), or bodies that are measured in standard deviations, occupation may be interpreted through the lens of occupying spaces of protest, of being occupied, or of occupying time and space (Block et al. 2016). Nevertheless, for the purpose of this chapter, occupation encompasses "all that people need, want or are obliged to do; what it means to them; and its ...potential as an agent of change" (Wilcock 2006). Here, occupation functions as a pre-requisite for health and well-being (Wilcock 1998), frames our personhood and identities (Laliberte-Rudman 2002; Magnus 2001) and is intricately bound up in defining who we are (Wilcock 1998). This makes sexuality an occupation of intrinsic value, evidenced also in Abraham Maslow's hierarchy of needs. Here, access to and participation in sex is a "first" tier or basic physiological need accompanied by the need for appreciation, with equal ranking to breathing, sleep and excretion. Persons with disabilities are often treated as the outliers within a normative system. The orientation of these body-minds in place and space often results in unevenness, that is, interaction with spaces, systems, objects, people, policies and attitudes that lack accessibility and do not promote inclusion and belonging for persons of different abilities. For example, what would prompt some department store personnel to 'unconsciously' conduct body audits by not allowing wheelchair users access to the stores, warding them off as they seek entry? On a more intimate level, some disabled women often exchange physical coitus for hopes for a loving, nurturing sexual and emotionally rich relationship but in reality are forced to endure relationships without being acknowledged as the significant other by their intimate partner among their friends and loved ones. It is this unevenness of experiences, where desired occupation is denied or somehow disrupted, across physical, relational, political, social, economic and other axes of differentiation that results in occupational injustice.

In this postcolonial era, the supposedly all-inclusive national cultural space of the Carnival festival celebrations fails to reveal the benevolence of inclusion. Rather, for many disabled people, the everyday lived experiences of occupational apartheid (OA) are palpable. Occupational therapists' and scholars Frank Kronenberg and Nick Pollard's (2005) definition of OA is described as "[the] more-or-less chronic established environmental (systemic) conditions that deny marginalised people rightful access to participation in occupations that they value as meaningful and useful to them" (p. 65). That Trini Carnival, as a representation of sexual agency, marks in a more pronounced manner how disabled bodies are systematically and routinely segregated from experiencing sexual expression and similar opportunities, in keeping with the notion of OA. Carnival is historically enjoyed as a "farewell to the flesh" prior to the Lenten season, but it is a rare sight to witness disability crossing the Carnival stage. Back in 2012, and for a few years prior, a national disability organisation engaged schools and homes for disabled children to participate in Carnival Saturday activities—Children's Parade of the Bands. Here, the Lady Hochoy Home for Physically Handicapped Children was represented, alongside similar organisations. With glee, children in their wheelchairs, crossed in front of the Memorial Park with their assistants moving to the sounds of soca music. Private disability organisation Carnival fetes continue today, to bring soca artistes and disabled people together in accessible locations. However, children grow up and the national infrastructural landscape often precludes engagement of disabled adults in typical Carnival celebrations. Disability justice demands collective access for all bodily abilities, such as providing even terrain, frequent rest stops, movement in crip time (acknowledging flexibility in temporality of movement), flexibility in time (preferably shorter times) outside on the streetscape in the hot sun, all at a manageable cost for the experience of being in a band and wearing a costume.

## Sexual apartheid and the disabled body

Sexual apartheid, my term, extending the work of occupational therapists and scholars Elizabeth Townsend and Ann A. Wilcock (2004), refers to the systematic disqualification that PWD experience from being denied access to sexuality, sexual freedom and expression, sexual decision making and any other aspects of meaningful participation in this aspect of their lives. In context, sexual apartheid upends the privilege of sexual expression and freedom, denying bodies marked as disabled their personal participation in sexual intimacy. Throughout this chapter, I will show how sexual apartheid becomes tangible lived experience for four women, exemplifying the four components of sexual apartheid: sexual alienation, sexual marginalisation, sexual imbalance and sexual deprivation.

Townsend and Wilcock (2004) identified four components that stymied meaningful occupation and well-being. These four occupational injustices include "occupational alienation, occupational deprivation, occupational marginalisation and occupational imbalance" (p. 80). Here, occupational alienation is understood to mean "prolonged experiences of disconnectedness, isolation, emptiness, lack of sense of identity, a limited or confined expression of spirit, or a sense of meaninglessness" (p. 80). Occupational deprivation, refers to "a state of prolonged preclusion from engagement in occupations of necessity and/or meaning, due to factors that stand outside the control of the individual" (p. 222). Occupational marginalisation operates invisibly and refers to "the need for humans to exert micro, everyday choices and decision-making power as we participate in occupations" (Townsend and Wilcock 2004, p. 81). Finally, occupational imbalance, according to Townsend and Wilcock, refers to the omission of population cohorts "that do not share in the labour and benefits of economic production" (p. 82). That said, how do we apply the concepts of occupational apartheid and the different forms of occupational injustices to access to sexuality?

Disabled bodies do different things and do things differently. In fact, occupational therapist and public health scholar Mansha Mirza and colleagues (2016) argue against the normalised expectation of participation. Defining participation as being "actively engaged in self-chosen activities and communities and having the freedom, opportunities and supports to do so …" (Mirza et al. 2016, p 161), Mirza and co-authors challenge the ableist nature of participation, critiquing the "obligation to participate" in expected modalities. As such, since they advocate for heterogeneity of participation, these body-minds are often deemed as non-conforming, perhaps even incapable of gyrating (action of dancing that begins with the sensual movement of the hips) to the syncopated beats of the soca music of Carnival. Feminist geographers Hansen and Philo (2007) argue, "If people's comportment seems out-of-the-ordinary, being too slow or taking too long, involving 'curious' jerks, postures or facial expressions, perhaps accompanied by 'odd' sounds or smells, then the risk is that they become treated with suspicion or hostility" (p. 496). Considering this, how is the sexuality of PWD living in T&T understood by themselves, by their family members and non-disabled others, and how might disabled participation in Carnival be envisioned?

Yet, the occupational therapy literature reveals that it is through our capabilities, our ability to function in the everyday, that our legitimacy and citizenship are concretised. Ann Wilcock (1993) describes occupation as "the mechanism by which individuals demonstrate the use of their capacities by achievements of value and worth to their society and to their world. It is only by their activities that people can demonstrate what they are, or what they hope to be" (p. 18). The physical, policy, attitudinal and institutional infrastructures of a nation's spaces articulate which bodies have legitimacy. More specifically, disabled body-minds are not welcome, and are denied participation in the sexual emancipation that historicises the Carnival celebrations emblematic of this cohort's existence on the periphery of society, lacking mobile and sexual citizenship.

## Methodology

Trinidad and Tobago represent a contradiction in terms of how we discuss sexual issues, who is involved, how and what type of sex is showcased, and represent a blurring of lines between private and public spaces. In a country that boasts 98% literacy levels, on the one hand, there is a level of cultural discomfort in the performative gendering of genitalia.

The methodology I adopted with these women was influenced by feminist theory which centred my interlocutors as experts of their own lived experiences, articulating for themselves the meaning of actions and events occurring in their lives. Disability and mobility frameworks (Brown & Durrheim 2009; Meth 2003; Price 2009) also influenced the methods adopted. As such, personal narratives gleaned from face-to-face in-depth interviews, completion of travel diaries and participant observations through the process of becoming mobile with my inter-locutors, were the three methods used to collect data. This was analysed using grounded theory (GT) to illuminate how my interlocutors contextualised everyday life events with meanings within the larger context of postcolonial T&T.

In the face-to-face interviews, the first level of engagement with my interlocutors, I listened for subjective experiences, a thought or feeling about a remembered activity or relationship, using a semi-structured interview guide. Based on the response and comfort level of my inter-locutors about intimacy, I adjusted my questions. This was critical as one loquacious interlocu-tor reminded me, "I think you need to revisit your questions. Some persons may see it as too much to digest. Sex IS a very sensitive issue." Here, I began the epistemological process of asking open-ended questions to obtain personal experiences. I then probed gently to inductively glean the meanings my interlocutors associated with these experiences. Feminist scholars Stanley and Wise (1990), have noted that women's experiences are "ontologically fractured and complex." Thus, one cannot assume that all women's experiences are similar or singular in scope, which resulted in my being alert to the multi-layered subjugated experiences that they shared.

The second phase involved inviting my research participants to complete travel diaries. Knowledge production is often a complex process. To alleviate any discomfort that may have arisen during the interview process, such as asking questions of a personal/intimate nature, which culturally is taboo, the travel diary, with a few writing prompts provided, placed in my interlocutors reach the opportunity for open sharing at their own pace. By asking my informants to keep travel diaries in this second phase of the fieldwork, they "mapped" the physical features, bureaucratic systems and interpersonal interactions that affected their capacity to move and to do. The completion of the diary was completely voluntary. Of the 35 interlocutors who completed the interview, only 19 persons returned travel diaries. The "body-maps" documented by two interlocutors within the travel diary, a drafting of how they were sexually abused by familial others, justified the relevance of this technique, as a follow-up to the face-to-face interview.

In personalising the third phase of data collection, participant observation, I introduced the empirical concept of becoming mobile, an interactive mobile ethnographic- influenced method that features the mobility of my interlocutors in public spaces, as I accompany them, taking on the role (loosely) of a care attendant. My movement with my interlocutors, always keeping a step behind, supporting them as they interact with the world in their everyday occupations allowed me opportunity to closely examine existing hierarchical power structures, from an insider/outsider perspective. To do this, I transported them, free of cost, to complete activities/errands of interest. This entire research project was paid out of pocket, and not grant funded.

Researching sexuality and disability was not a primary focus in my study and was not identified as a prevailing concern in the first few face-to-face interviews. However, the issue of sexual violence literally popped up in a travel diary submission which warranted a second interview with that interlocutor. The feedback from this second interview revealed that sexuality was very much a priority for discussion by some of my interlocutors, including the ones that I will profile below.

## Interlocutor profiles

With disability described as "… a mark on the body that can never be erased (Davis, 2006)," I introduce to you the four women that constitute the focus of this chapter. Occupational therapists and scientists (Pollard and Sakellariou 2007; Sakellariou and Algado 2006a; 2006b) have challenged the current notion that "sex belongs inescapably to the private domain" (Pollard and Sakellariou 2007, p. 363), and have written extensively about occupational therapy as a profession requiring a paradigm shift in supporting PWD to explore their potential as sexual beings in personal relationships. Embedded into the T&T national psyche is the cultural view ascribed on the female disabled body, as "leaking, flowing and out of control" (Plaskow 2008, p. 60). As such, these bodies bar them from some occupations. Relevant occupations in this context include sexual expression and sexual agency (that capacity to act independently and facilitate sexual decision making).

*Pseudonyms are used for each of my interlocutors.*

Cynthia is a feisty 47-year-old wheelchair user of mixed heritage, of African and Portuguese descent. She completed secondary school. Born able-bodied, she was involved in a vehicular accident at the age of 22 years from which she survived with her life, but in receipt of 3rd degree burns to most of her body. She self-identifies as differently-abled, arguing that, with her acquired disability, she has had to learn new ways of participating in meaningful occupation. Describing herself as "straight-leg and half-barbequed," Cynthia is involved in a visiting, sexual relationship with her boyfriend of 19 years. They have no children.

Rachel is a spirited 36-year-old grandmother of East Indian descent, who enjoyed a primary school education (up to age 12). Predominantly a wheelchair user, she longs for the day when her legs are strong enough to resume walking on a regular basis. Born able-bodied, after collapsing in her kitchen and remaining in a comatose state for one month in a public health facility (with no medical diagnosis offered), she regained consciousness and was discharged. She lives on her own in a galvanised structure on the same compound as her mother, step-father and other relatives. She has no access to a flush toilet and uses a latrine. She enjoys a visiting relationship with her husband.

Lisa, of African descent, is a vivacious 45-year-old wife and mother to two daughters (born prior to the accident). She completed tertiary level education. While transporting children back home from school, her preferred vocation, she was involved in a vehicular accident. Her spinal cord injury has resulted in loss of all feelings in her body from under her breast. She is currently able to use both of her arms. Her husband of 20 years is her primary caregiver.

Judy is an effervescent 54-year-old wheelchair user of mixed heritage, African and East Indian, who lives with cerebral palsy. A recipient of educational training from the local disability institution, she was sexually violated as a child. Judy never participated in formal employment but became a businesswoman by opening her own shop from her front yard selling snacks to put her sister through school after her mother passed away when her sister was 13 years old. She enjoys meeting people and occupies her time writing her autobiography. Her husband of fourteen years is her primary caregiver. They have no children.

## Sexual apartheid = sexual injustice

A significant commonality among my interlocutors is that their access to intimacy is/has been disrupted by several disabling forces—circumstance and ableism demonstrated at the familial and societal levels and relived daily. Personal narratives most meaningfully reflect experiences. Cynthia, a 47-year-old wheelchair user, shared:

> It started from my hospital bed, where I was told by nursing staff [that] I might as well give up and die, because no one would or could love someone like you, [one] who is so deeply scarred.

The eugenic principle continues to pervade both the medical and other rehabilitative disciplines. Eugenics era language and attitude reflected by medical staff revealed a callous attitude to disability. These public healthcare workers were actively denying my interlocutor the potential for enjoying a sexual relationship, as a key component of her life.

The notion that it was highly improbable that Cynthia could find love as a PWD is not surprising. These experiences are further compounded by her encounters with sexual repression, to be understood as a prolonged debarment from desired sexual engagement/expression. To this end, she shared personal narratives of her father actively seeking opportunities to personally embarrass her and to ward off potential partners. Cynthia's living quarters are located to the back of her parents' home, as an adjoining apartment. She lacks a key to the connecting door which is in her parents' access. As such, they often enter her apartment when she has male company under the guises of needing to empty her commode. These untimely and disconcerting intrusions by her parents are strategic as they seek to constantly keep her sexuality under surveillance.

Sexual and reproductive health scholars Addlakha et al (2017) reiterate that "While one may concede that sexuality is a basic human need, awareness and knowledge about sexuality are shaped through a range of contextually specific socio-cultural and religious ideas and practices. People with disabilities are systematically denied access to knowledge about sexuality, sexual behaviour and services leading to their sexual marginalisation" (p. 5). Not only are they denied access to sexuality but also access to reproductive health rights. On reproductive justice, Cynthia laments:

A doctor I went to when I became pregnant at 34 asked me what wicked man would do this! And he told my mother I would not be able to bear children. [He] never spoke to me about it. Whatever he did and gave me, he made sure I never got pregnant again.

The ambiguity surrounding this medical situation became clearer only a few years ago. What her mother revealed of that situation and her complicity with the medical doctor those many years earlier resulted in Cynthia's feelings of betrayal and hurt. In short, her mother knowingly gave permission to the physician to sterilise Cynthia. She revealed that she felt like "someone had cut me from the inside out." In her attempts to rationalise and cope with this duplicity, she considers that "people seem to think, especially in my family, that I don't need to know… It's as if I am incapable of understanding. The public health system treats me even worse." This experience is deeply painful but not uncommon for PWD in their sexual and reproductive lives. In the eugenics era, the law books denounced feeblemindedness. In 1927, for instance, the US Supreme Court famously ruled that Carrie Buck, who had been deemed "feebleminded" and institutionalised for "incorrigible" and "promiscuous" behaviour and who became pregnant after being raped, be compulsorily sterilised (Trent 1994). Cynthia exposed how her "friends" and even a close family member expressed surprise about, "How could I be in a wheelchair [and] have such a handsome man and they don't. My so-called girlfriends would reveal parts of their body (as in wearing short pants, skirts, etc.), anything that would reveal their beautiful skin."

With this pervasive attitude of sexual apartheid set against PWD, I invite you to examine with me the lives of my other interlocutors as they experience different types of sexual injustice. Judy had a complicated and violent introduction to sexuality, including bitter experiences of sexual violence and sexual alienation. By sexual violence, I refer to her being repeatedly raped for several of her teenage years until young adulthood by her father as well as her sister's sexual partner (the sister whom she "mothered"). Using the concept of sexual alienation, within the occupational injustice lens, Judy's agency and sense of identity was pulled from under her feet by those from whom it was least expected. This resulted in prolonged experiences of disconnectedness, and limited expression of spirit. She expressed to me her confusion when she sat in her chair bleeding, clutching her tummy, complaining of pain and her mother approached her, asking what was wrong. Judy yelled out, "Go ask _____" and her mother remained in place, doing nothing. Eventually, she called the police herself but received no justice. Trinidad and Tobago-based feminist scholar Patricia Mohammed (1991), in examining how calypso (oral song tradition in T&T) was used as a social instrument to foreground power especially at intersections of gender relations and sexual violence, showcases how cases of rape were responded to in the social domain in a song sung by calypsonian Lady Jane:

Send those rapermen to jail
Beat them with the birch 'til they wail
Then send in Calypso Jane
To throw some cat in dey tail.

Mohammed notes that "the genre of calypso is famous for its double entendre and sexual innuendo, which no doubt Lady Jane played on in the last line of her chorus. The male audience was not too pleased, however, with her suggestions for punishing convicted rapists" (p. 35). Although the introduction of the Sexual Offences Bill (1986) was fought for and applauded by women across the country, the cries for justice from the cohort of disabled women survivors of sexual violence still fall on deaf ears. While Judy demonstrated agency and sought redress to her abuse from family members, because the abuse was generated from within the home, there was no favourable response from the police. In the eyes of the police, this was "home business," an internal squabble to be rectified within the family. This unwillingness of the police to make a tangible intervention into this matter seems to perpetuate a level of ambivalence to sexual advantage, especially considering the reality of her disability and her inability to defend herself. Contrary to the police motto to "Protect and Serve," Judy existed in a setting that rendered disabled voices as lacking in both validity and rhetoricity (Price 2011; Yergeau 2016).

Due to her balance and coordination issues Judy needed full time assistance to complete her activities of daily living (ADLs). Margaret Nosek and colleagues (2001) conducted a study which revealed that "the need for personal assistance with daily living created additional vulnerability [to abuse]" (p. 177). For Judy, applicable vulnerabilities to abuse highlighted by Nosek et al. include, "perceptions of powerlessness, less education about appropriate and inappropriate sexuality, social isolation and risk of manipulation as well as dependence on perpetrator for daily survival activities, such as transferring from bed to wheelchair, eating…and taking medication" (p. 178). To this end, she commented on the transition from her family home and her concern for her husband:

> It's a struggle for him. I see it's a struggle because no family checks for us. You know. How you making out? How your husband making out? I was a burden taken away from them. Rather than congratulations, they only tell my husband, thank you, thank you.

Here, Judy reveals familial disaffection towards her and her husband as she describes how her family expressed their gratitude that she was no longer their responsibility. On her current experience with sexual intimacy, she proffered:

> Well, at a time I was excited about it. But it hasn't been much on my mind. But if my husband asked for it, I let him, just to satisfy him. It don't be all that enjoyable, maybe because of my age, 54. At this age ladies sometime don't have that urge. Then too, at times, I feel like my husband is just my caretaker, then, again, I think that he has no desire for intimacy also.

Judy attributed her sexual apathy to two realities, her age, perhaps pointing her towards a change of life and hormonal balance (menopause) and to the familiarity of her spouse in his role as caretaker. She also alludes to the possibility of her husband's loss of sexual desire in her, since she is his "patient." This serves as a segue into Lisa's lived experiences and sexuality as an occupation. Lisa's and Judy's experiences as married disabled women, fully dependent on their spouses, contrast with one another.

When I arrived at Lisa's two-storey home to conduct the interview, I met her husband, who indicated that he was just passing through for lunch. I later learned that, as a building contractor with his own business, Lisa's husband negotiated his time and his skills to make his wife's everyday life as comfortable as possible. As I considered her mobility access in her home, not only did her husband build the home with very wide doors and turning points, but he also built a lift to enhance her entry and exit from the home and also into and out of the pool that he built to meet her therapeutic needs. It was only the kitchen area that seemed built like on a dais, limiting access. Despite this familial support, Lisa experienced sexual marginalisation also, in terms of her inability to exert personal, decision-making power with regards to her sexual choices.

For example, since sexual intimacy is not only relational and processual but also political, my interlocutor felt that her agency was disrupted when she felt obliged to 'do it' for her husband since coitus was no longer mutually pleasurable? Lisa reflects on her husband's sexual advances:

*Hubby:* But yuh looking nice.
*Lisa:* Endless mamaguy [attempt to deceive, especially with flattery], but he does still do it.
*Hubby:* It's good therapy, good therapy.
*Lisa:* [sarcastic] Maybe on his part but not on my part. If I was getting any feelings, maybe.
    I just do it because of him. Make him feel happy. And as long as you are happy, I can
    get some money. But he say, you don't need any money...

Lisa, in her desire to express appreciation to her spouse and caregiver (relational), she reluctantly accepts the sexual encounter (processual) while negotiating potential benefits to herself for participation in sex (political). Lisa negotiated sex, which was not pleasurable for her, for a more palatable occupation—shopping. For Lisa, while she felt torn in her desire to demonstrate appreciation to her husband through sex, she proved her agency as she sought to actuate personal benefit. Notably, her tactics differed greatly from that of Judy's. Judy's sexual history helps serve as a foundation to her apathy but more so it amplifies the lack of negotiation and use of sexual agency. Lisa's early life experiences were empowering and provided her with a deep sense of personhood and self-identity that was not disrupted even after her mobility impairment.

Rachel's sexual experiences are framed at the intersection of her spirituality, familial relationships and vulnerabilities. As a former Hindu turned Christian, who believes in fatalism, she considers that her previous sexual acts with her brother-in-law may have resulted in her falling into a comatose state, with the resultant mobility impairment. Living in a communal environment, to the onlooker, Rachel's vulnerability to her brother-in-law, an adult, would have been considered rape at the age of 15 years old. However, she continues to live in abject poverty, revealing further helplessness. The level of poverty that Rachel continues to live in was not immediately abated since her husband was chased out of those living quarters, after a squabble with his in-laws. Her experiences of sexual imbalance are unearthed at the crossroads of deep poverty, especially considering expectations that, as a married woman. she would be adequately provided for. In this real-life account, where familial ties in a communal dwelling environment held stronger sway than even the marital relationship, the spasmodic presence of her husband defined the nature of financial support she received from him. Coupled with her inability to earn wages beyond the state-provided disability grant, Rachel's sexual encounters created an imbalance in her own life and negatively impacted the quality of relationship with significant others.

## Conclusion

This chapter discussed Trinidad Carnival and the level of inclusiveness of this national festival to persons of varying bodily abilities to sexual expression and freedom. The legitimacy of certain bodies over others to experience sexual abandon was also considered. In taking an intimate look at access to sex from different perspectives, I have reflected on how my interlocutors experienced disrupted sexual intimacy, as well as different forms of sexual and occupational apartheid.

What could sexual justice look like for my female disabled interlocutors? Introduced earlier, mobility-as-occupation (MAO) (Henry-Buckmire 2018) builds on the relational, processual and political nature of sexual desire and freedom. It first recognises the value and legitimacy that each of my interlocutors must first ascribe to themselves before they can relate with others external to themselves. MAO works as it develops the process of becoming, of identifying one's intrinsic value and actioning any and all changes necessary for others to acknowledge one's personal worth. The women highlighted in this chapter occupy different phases of this intimate growth experience.

The processes that facilitate the development of these relationships with themselves and others are varied. Nevertheless, all demonstrate how my interlocutors, in some cases, create and maintain positive (sexual) relationships with significant others. In other cases, they manoeuvre themselves to ameliorate their circumstances. The application of the political, as it relates to one's mobile or ever-changing desires, are critical. The crux of the justice component in the MAO comes in the harmonisation of these three factors, the relational, processual and political, with their personal capabilities and those of the supportive others within their networks.

For Lisa and Judy, their MAO moments were wholly dependent on the capabilities and beneficence of their partners. Using the lens of the MAO to better understand sexual intimacy, I argue that intimacy "is experienced through interdependence, acknowledging that you are a sexual being, seeking out a companion, reviewing one's embodied capacities, analysing the requirements for sex, and, by varied means of communication, articulating what assistance you may need to feel sexually satisfied" (Henry-Buckmire 2019). Within the frame of interdependence and collective justice, Rachel and Cynthia would have also experienced more meaningful sexual and reproductive lives. Of note, vulnerabilities to abuse must be addressed urgently for justice to be experienced by all.

Effecting sexual justice is not for the faint-hearted. It is not an overnight fix; it is a process. Sexual justice demands mainstreaming disability into the county, then achieving a national HIV/ AIDS response and making health promotion and sexual and reproductive health resources available for this often-marginalised cohort. Sexual justice challenges the protective services to create disability advocates in the penal system, to facilitate PWD making reports of a sexual nature and their continued presence through to the aftermath of the case, as necessary.

The discussions of access to intimacy reveal related but bifurcated occupations that demand interdependence and intertwining of individuals, professions and intersecting systems for improved sexual justice. This premise of interdependence underscores the need for relationality at all levels, of all those that occupy disability, that inspire that ray of hope and connectedness of which we, as humans, are in need.

# References

Addalakha, R., Price, J., & Heidari, S. (2017) Disability and sexuality: Claiming sexual and reproductive rights. *Reproductive Health Matters* 25(50), pp. 4–9.

Block, P., Kasnitz, D., Nishida, A., & Pollard, N. (2016) Occupying disability: An introduction. In P. Block, D. Kasnitz, A. Nishida, & N. Pollard (eds.) *Occupying disability: Critical approaches to community, justice, and decolonizing disability*. Dordrecht: Springer, pp. 3–14.

Brown, L., & Durrheim, K. (2009) Different kinds of knowing: Generating qualitative data through mobile interviewing. *Qualitative Inquiry* 15(5), pp. 911–930.

Davis, L. (2006) *Disability studies reader*. 2nd edn. London: Routledge.

Finger, A. (1992) Forbidden fruit. *New Internationalist* 233(9), pp. 8–10.

Garland-Thomson, R. (2011) Misfits: A feminist materialist disability concept. *Hypatia* 26(3), pp. 591–609.

Hansen, N., & Philo, C. (2007) The normality of doing things differently: Bodies, spaces and disability geography. *Journal of Economic and Social Geography* 98(4), pp. 493–506.

Henry-Buckmire, S. (2018) *Mobility-as-occupation: Justice manoeuvres and negotiations in T&T*. Doctoral dissertation, State University of New York at Stony Brook.

Henry-Buckmire, S. (2019) Intimacy and justice for all–Is it possible? Presentation at the University of the West Indies, St. Augustine, February 12.

Kafer, A. (2013) *Feminist, queer, crip*. Bloomington, IN: Indiana University Press.

Kronenberg, F., & Pollard, N. (2005) Overcoming occupational apartheid: A preliminary exploration of the political nature of occupational therapy. In F. Kronenberg, S.S. Algado, & N. Pollard (eds.) *Occupational therapy without borders-learning from the spirit of survivors*. Amsterdam: Elsevier.

Laliberte-Rudman, D. (2002) Linking occupation and identity: Lessons learned through qualitative exploration. *Journal of Occupational Science* 9(1), pp. 12–19.

Magnus, E. (2001) Everyday occupations and the process of redefinition: A study of how meaning in occupation influences redefinition of identity in women with a disability. *Scandinavian Journal of Occupational Therapy* 8(3), pp. 115–124.

Meth, P. (2003) Entries and omissions: Using solicited diaries in geographical research. *Area* 35, pp. 195–205.

Mirza, M., Magasi, S., & Hammel, J. (2016) Soul searching occupations: Critical reflections on occupational therapy's commitment to social justice, disability rights and participation. In P. Block (ed.) *Occupying disability: Critical approaches to community, justice, and decolonizing disability*. New York: Springer.

Mohammed, P. (1991) Reflections on the Women's Movement in Trinidad: Calypso's, changes and sexual violence. *Feminist Review* 38(1), pp. 33–47. doi: 10.1057/fr.1991.18.

Nosek, M.A., Howland, C., Rintala, D.H., Young, M.E., & Chanpong, G.F. (2001) National study of women with physical disabilities. *Sexuality and Disability* 19(1), pp. 5–40.

Plaskow, J. (2008) Embodiment, elimination, and the role of toilets in struggles for social justice. *Crosscurrents* 58(1), pp. 51–64.

Pollard, N., & Sakellariou, D. (2007) Sex and occupational therapy: Contradictions or contraindications? *British Journal of Occupational Therapy* 70(8), pp. 362–365.

Price, M. (2009) "Her Pronouns Wax and Wane": Psychosocial disability, autobiography, and counter–diagnosis. *Journal of Literary & Cultural Disability Studies* 3(1), pp. 11–33.

Price, M. (2011) *Mad at school: Rhetorics of mental disability and academic life*. University of Michigan Press.

Sakellariou, D., & Simo Algado, S. (2006a) Sexuality and occupational therapy: Exploring the link. *British Journal of Occupational Therapy* 69(8), pp. 350–356.

Sakellariou, D., & Simo Algado, S. (2006b) Sexuality and disability: A case of occupational injustice. *British Journal of Occupational Therapy* 69(2), pp. 69–76.

Stanley, L., & Wise, S. (1990) Method, methodology and epistemology in feminist research processes. In L. Stanley (ed.) *Feminist praxis*. London: Routledge, pp. 20–60.

Townsend, E., & Wilcock, A.A. (2004) Occupational justice and client-centred practice: A dialogue in progress. *Canadian Journal of Occupational Therapy* 71(2), pp. 75–87.

Trent, J.W. (1994) *Inventing the feeble mind: A history of mental retardation in the United States*. Berkeley: University of California Press.

Wilcock, A.A. (1993) A theory of the human need for occupation. *Journal of Occupational Science* 1(1), pp. 17–24. doi: 10.1080/14427591.1993.9686375.

Wilcock, A.A. (1998) Reflections on doing, being and becoming. *Canadian Journal of Occupational Therapy* 65(5), pp. 248–256.

Wilcock, A.A. (2006) *An occupational perspective of health*. Slack Incorporated.

Yergeau, M. (2016) Occupying autism: Rhetoric, involuntarity, and the meaning of autistic lives. In *Occupying disability: Critical approaches to community, justice, and decolonizing disability*. Dordrecht: Springer. pp. 83–95.

# 24

# REIMAGINING SEXUALITY IN THE DISABILITY DISCOURSE IN SOUTH ASIA

*Anita Ghai*

## Introduction

Sexuality is often the source of our deepest oppression; it is also often the source of our deepest pain. It's easier for us to talk about—and formulate strategies for changing—discrimination in employment, education, and housing than to talk about our exclusion from sexuality and reproduction.

*(Finger 1992 p. 9)*

This paper is written from my epistemic location as a woman with visible mobility impairment. It was in 1996 when I came across the quote above by Finger (1992), that marked the beginning of this paper. Finger's (1992) words resonated with my own apprehension about the pervasive absence of discussions around disability and sexuality in my own country. The salient issues regarding sexuality and disability in India will be addressed within the context of existing literature and my own personal perspective. Interviews conducted with disabled people about their own perceptions of their sexuality will also be highlighted in this chapter.

## Disability and sexuality in India

It is believed that disabled people are not complete "persons" with the same aspirations, desires and rights as the able-bodied in South Asia. Elsewhere I write about "children of a lesser god" (Ghai 2002, 52). It was indeed a coincidence that I came to know that a movie called *Children of a Lesser God* was made in 1986. It is interesting that in reality many disabled children would sit outside the temple. They were given alms even without the children asking for the offerings. Today, there is a growing appreciation of the rights of disabled citizens to lead lives of dignity. However, endeavours at our inclusion in society are primarily focused on education, vocational training and rehabilitation (Ghai 2018). The needs for intimacy, emotional connection and a fulfilling family life are seldom acknowledged or addressed. When sexuality is tagged with disability, the silence becomes even more "stinging" and our innate preconception about "difference" comes to the forefront. Disabled people do not have the privilege of having desire for intimacy and sexuality, to having the right to have children or to be perceived as sexual beings by others.

## *Beliefs about sexuality in India*

Taboos around sexuality in India are pervasive, as several feminist scholars have observed over the years. Nair and John (2000, p. 1) stated, "a focus on the conspiracy of silence regarding sexuality in India, whether within political and social movements or in scholarship, shields us from the multiple sites where 'sexuality' has long been embedded." Situated within this problem is the sexuality of disabled people whose desires are unsanctioned and who are perceived as non-sexual. It was decades before feminism in India took cognisance of the intersection of disability and sexuality or before the disability discourse recognised sexuality as a legitimate concern for disabled people. In the context of India, the advocacy focused on societal awareness in terms of "reasonable accommodation." Accessibility in both the built environment and augmentative communication were the core focus areas. The past twenty-five years became instrumental in bringing about change in the form of equal opportunities acts, but the discourse still continues to be framed within the frameworks of medicalisation and human rights.

As the issues about disability began to emerge, there was "amnesia" about sexuality. On a social media platform, Nisha[1] (2006) wrote about how she was reprimanded for her advocacy on sexuality and disability by a colleague who said, "you know better about the issues faced by the disabled people here than to waste your time on sex-obsessed western thinking." In the land of Kama sutra, such responses come from the incapacity to think of disabled people, and specifically disabled women, as sexual beings, in a society which gives respect to women only as long as they remain passive sexual objects. Thus, sexual desire in disabled women is very far from being recognised as a possibility, let alone a human and political issue. Furthermore,

> If the male gaze makes normal women feel like passive objects, the stare turns the disabled object into a grotesque sight… Disabled women contend not only with how men look at women but also with how an entire society stares at disabled people, stripping them of any semblance of resistance.
>
> *(Ghai 2015, p. 141)*

As Foucault (1979, p. 27) reminds us,

> Silence itself—the things one declines to say, or are forbidden to name, the discretion that is required between different speakers—is less the absolute limit of discourse, the other side from which it is separated by a strict boundary, than an element that functions alongside the things said, with them and in relation to them within overall strategies. There is not one but many silences, and they are an integral part of the strategies that underlie and permeate discourses.

As Addlakha (2007, p.4), a visually impaired scholar, observed, "Sexuality is an area of distress, exclusion and self-doubt for persons with disabilities." This observation about disability and sexuality has been observed by many scholars (e.g., Finger 1992; Ghai 2015; 2018; Shakespeare 1999). It was significant that Menon (2007), a well-known feminist scholar discussed the neglect of disable women's sexuality. She points out that "sexual desire in the work of scholars and activists working on the politics of disability in India makes evident that this is a promising albeit neglected field of research and feminist intervention" (2007 p. 35).

A focus on a deeper understanding of sexuality in India will now be explored. Hanisch's (2006) essay "Personal is political" transformed my understanding of the personal as political, where she emphasised that women's lives were impacted, not by the consequence of individual choices, but as a result of systematic patriarchal oppression. My personal journey regarding sexuality and disability within India will now be highlighted and connected with my political self in the following section.

## Person is political

During my childhood, a "cure" for my polio was indeed a mirage. A cure for pain and suffering had been a major preoccupation of Indian society. Across the 60 years of my life, I have negotiated with shamans, gurus, ojhas, tantric priests and faith healers, as well as miracle cures—all to ensure that I could become an able-bodied person finally! It took me years to banish associated anxieties and apprehensions. However, I was not left out in the familial interactions; the discourse within the family as well as my extended family and friends was that of the "unfortunate girl"—*itni soni hai par dekho na kismet ko* ("she is so pretty but look at her luck"). These words pulsated through my body and into my soul. My family did talk about marriage, but strangely my father was not very keen. In retrospect, I understand his concerns, as he did not want me to be tied to the difficult role of a married woman. I seemingly bought into this displacement of my womanhood by shifting my focus onto other pursuits including teaching, watching cricket and tennis, and nurturing a close association with TV soap operas and movies.

As a teacher who taught sexuality to students via psychoanalysis, my self-hood remained isolated from my sexual desires. I experienced my wishes for romance, sex and love life through Bollywood films and the stars of Bollywood. For me, it was a different ball game. Whereas other girls enjoyed the "wolf-whistle" and later detested it, I, on the other hand, yearned for the "wolf-whistle." Soon, I understood that, as a disabled person, I did not have the right to think about dating, intimacy or romantic relationships, as predominantly North Indian society thought of disabled people as non-sexual. Such a negative mental set was entrenched strongly in disabled men's and women's inner psyche. In previous writings, I note that, in several societies, including India, where any aberration from a normally accepted archetype is seen as a marked deviation, the impaired body becomes a symbol of imperfection.

> The ramifications of such historical renderings are to be found in the North Indian Punjabi culture where, although girls are allowed to interact with their male cousins, they are not allowed to sleep in the same room. Disabled girls, on the other hand, are under no such prohibitions.
>
> *(Ghai 2015 p. 141)*

This reflects what Hahn refers to as "asexual objectification" (as cited in Thomson 1997, p. 25), and is also evidenced by the disregard for the dangers of sexual violation to which disabled girls are exposed. What is significant is that when it comes to sexual relationships in India, the decisive conflict, apprehension and ambition is virtually always marriage. We need to know that any person's marriage is often seen as a family or public concern. Therefore, there is no scarcity of ideas when it came to the question of marriage. Sex within the institutions of marriage and procreation is acknowledged, arguably celebrated and facilitated. Family friends, as well as relatives, are ready to give suggestions for disabled persons. For example, when my family wanted new furniture for our living room, the designer counselled my father stating, and "Sir, you should get a poor boy and keep him in your own house and get him married to her. They both will be happy." The assumption was that I was "not good enough"[2] to marry a man of my own class and caste. This narrative is similar to that of many of my fellow disabled women and men, who were told to marry someone from a poor family. This is not to imply that someone from a poor family is somehow worth less, but it's telling of mind-sets that drive people to place the disabled alongside the economically marginalised and the class and caste prejudices against disabled people. It is expected that disabled people, and in particular disabled women, should either conceal their impairment or compensate for it with large dowries (Ghai 2003).

Disabled women exist in a strange and dangerous contradiction in India. Women are taught that their primary value is their sexual attractiveness; disabled women are constantly publicly desexualised. As a result, women with disabilities spend much of their life obsessively chasing something, which would always be, by definition, out of their reach. I am reminded of Cohen (1972) who referred to "moral panic" and argued that a moral panic occurs when "[a] condition, episode, person or group of persons emerges to become defined as a threat to societal values and interests" (Cohen 1972, p. 1). Moral panic is associated with any relationship that is not legitimate. The assertion of panic is sensationalised through social discourse (that is, media, society) to enhance hostility toward disabled people. Cohen (1972) places moral panic on idiosyncratic, didactic behaviours over structural societal factors, which result in reactionary, often punitive policies. This can lead to institutional, political and cultural shifts that drive societal injustice for disabled people. However, in a scenario where sexuality is intimately tied to marriage, opportunities for sexual exploration among disabled people in India, particularly the women, are very limited. If someone wants to be celibate and single, it would be fine. Marriage for able-bodied women, which is considered a safe haven for women, is not an easy option for disabled women.

## Searching for self and sexuality in therapy

It was at this juncture that I decided to go through psychoanalytical therapy. Free association recall became a tool for unearthing the pain and anguish and forgotten wishes and desires. Though it was a taxing process, I felt myself becoming both vibrant and alive, and the repressed memory reminded me of a male friend who belonged to a very traditional, conformist family. He was not allowed to bring girls to his home. However, he invited me for lunch with his family. Taken aback, I asked him whether it is appropriate. He replied in the affirmative, but the response unsettled me. Apparently, he shared my disability with his family and taking me home was like a good deed of the day. Being disabled, I evidently didn't count as a "woman." It wasn't an intentional and fraudulent act. He was trying to support me as a friend. It does seem like an intentional act, though. However, the damage was done, and I understood that the ableist gaze would never ever leave me. The idea that disability and sexuality cannot intersect illuminated my conscious. Insightful associations of my experience in the therapy made me realise that my fantasy of love from Bollywood cannot transform my reality. However, what was liberating was that I connected with my body now in a way in which I was not apologetic. As I played with the understanding of object relations[3] theory, I underscored Winnicott's (1991, p.104) notion of "potential space." It was this interpretation of space in which ways of being and relating enabled my venture into exploring my relationships with men. I should say that, while I moved, I really could not give up on the heterosexual mode; therapy invigorated my yearning for womanhood. Sexuality therefore embraced my mind and I understood that sexual desire is critical for the entirety of my being and subjectivity. The construction of resistance and insight enlightened my psyche. Nevertheless, there were many bridges that I had to cross. Theoretically, we accept that disabled women and men have a right to be sexual, but articulating my longing for pleasure and sexual gratification was a difficult task What matters is that the relationship between what we essentially do, or are compelled, permitted or forbidden to do as sexual beings is not a problem of fantasy. Rather, it's a problem of verbalisation. As Wilkerson (Wilkerson 2002, p. 46) argues, the repercussion for those with physical disabilities, like many others, may be silence and unintelligibility, their sexualities rendered incoherent, unrecognisable to others or perhaps even to themselves, a clear instance of cultural attitudes profoundly diminishing sexual agency and the sense of self and personal efficacy, which are part of it.

Notwithstanding my trepidation, I started associating with both able-bodied and disabled men. The movement was fragile, but I was sure about my desire to find a partner who would give me sexual fulfilment. In a way, I was moving to an intimate core in which I started dating men. Some meetings were online, but not all. I recall that ASL (age, sex and location) was the first query. In such chats, I would, did share with them about my visible disability. I would also reiterate that my disability could never be cured. Some would clearly indicate repulsion, but some did venture to meet me personally. However, it was one of my brother's friends who said that disability does not feature into his "mental" landscape. He said, "What if we had an accident after marriage? Would I leave him?" Each day would end with long calls to each other. I was attracted to and flirted with him. I became a needy woman who started loving him deeply. However, there were many blocks in my mind. My internalisation of morality would demand a commitment in the form of marriage. In a patriarchal society, a husband is the only person who should be gifted with virginity. If the woman breaks away from the normative, moral panic is always created that can be associated with corruption and sin. It took me three years, but I realised that commitment was not possible. This cul-de-sac created serious questions in my mind. Should I or shouldn't I? Should I fulfil my desire, notwithstanding my resistance against panic, morality and virginity that I learnt from patriarchy? My resistance brought some pleasure about my womanhood, but a few months were enough for my so-called "able-bodied friend" to break the relationship. I know that break-ups can happen to everyone, but for a disabled woman, the relationship is tied to vulnerability. Also, his insistence was always to live under cover. In such a scenario, desire took a beating as I experienced excruciating pain, but I did explore love for a man, even though it might have only been an illusion. Notwithstanding my depression, my daring attempt clarified my desire and pleasure in sexuality, though it is still fraught with uncertainty.

## Mapping the terrain of the cultural context

Disabled sexuality raises critical questions of what is at stake in the cultural imaginary in India that requires such a closing down of possibilities for sexuality. While it is difficult to trace sexuality in cultural history in this paper, I am picking up some strands that can connect to my understanding of disability and sexuality. I think there have been many variations in different historical phases. Vatsyayana's classic work "Kama sutra" (i.e., Aphorisms of love), written somewhere between the first and sixth centuries, includes the three pillars of the Hindu religion (Doniger and Kakar 2002), "Dharma," "Artha" and "Kama" representing religious duty, worldly welfare and sensual aspects of life, respectively. Further, in addition to Kama sutra, the erotic carvings of Khajuraho temples have become well known for their sensual figurines. For me, the cultural imaginary of the temples is that they signify a remembrance of human life, including sex. Human figures seem to engage in the mithunas[4] or sexual union, a combination of Shiva[5] and Shakti.[6] I really have not seen the temples, because of access issues, but disability is missing in the union.

One strand that is very interesting for me is the work of poets of the Bhakti movement with a connotation of spirituality and the erasure of sexuality in India. They expressed a desire for God that crosses through and refigures the body. A well-known scholar Ramanujan (1992, p. 55) suggests that "the lines between male and female are continuously crossed and re-crossed" in the lives of the Bhakti saints. The deflating of body and sexuality seems to desexualise the body by concentrating exactly on it, by celebrating its autonomy, by dismantling the codes and conventions that "sex" the body. Ckakravarti suggests that, among women bhakti saints and their demystifying of body and sexuality (Chakravarti 1989a; 50–52), there is a mention of women saints like Mahadeviyakka, lalla Ded of Kashmir, Andal, Avvaiyar and Karaikalammaiyar, who desexualised their bodies by focusing on them and celebrated their autonomy by demolishing the patriarchal codes and customs that "sex the body" (ibid). However, for disabled people, such an understanding is not a choice, as it is for advocates of the bhakti movement, for whom desexualisation is a *choice*, a spiritual orientation in which de-sexualisation is the path to salvation.

Kakar, a practising psychoanalyst, discusses (2007, p. 64, cited in Roy 2019) a fantasy central image: that of *jodi* (pair) This fantasy of constituting a "couple" within the extended family is a prominent desire that runs through women's lives and which exists in actual life. Kakar (1996) contends that a wife's greatest aspiration, during her entire adulthood, is to create a "two-person universe" with her husband, where each finally "recognises" the other. In contrast to much of popular western fiction, where women explore the depths of erotic passion or are swept off their feet by a masterful man, the Indian "romantic" yearning is a much quieter affair, which, when unsatisfied, shrivels the emotional life of many women. For the disabled, however, the unconscious desire for the "couple" is very much there. The *mithuna* couple in temple sculptures mentioned above reflects the highest manifestation of the *jodi* as *ardhanarishwara*—the Lord that is half woman, a visualisation of the *Jodi* as a single two-person entity. This is the cultural ideal, which makes a Hindu invoke Sita-Ram and not Site and Ram, Radhakrishna and not Radha and Krishna. To confess to an unfortunate marriage is not only to reconcile one's self to the loss of a cherished personal goal but also to betray a powerful cultural ideal. The persistence and importance of the *Jodi* for the disabled person's sense of identity helps us grasp the need for marriage, which might persist despite the attainment of economic independence. With the advent of economic liberalisation in India in 1991, which has had a lasting impact on the freedom of the mass media in India, sexuality has become more visible. This has brought about important shifts in social mindsets, regarding sexuality.

## Predicament, resistance and change

The split between the personal and political is always critical for disability. It is quite paradoxical, as it is quite evident that the sexual lives of disabled men and women have been dissected in public forums. Furthermore, equal opportunities in education, employment and other areas of life could not have been possible without evolving collective or group efforts. Thus, the introduction of private issues into public spheres was not difficult. Yet, paradoxically, the personal lives of disabled women and men continued to be excluded from the sexuality discourse in several ways. The disabled need to establish authorship of being not only human but also sexual beings. The need to prove and assert desire in normative society is significant. Theoretically, Overboe (1999) argues, there must be an acceptance that the embodied self emerges in discourse and through the "lived, living and sensuous body." For disabled women specifically, there is a reduction in life choices, which has an impact on their sense of self-worth and this, in turn, affects their notions of sexuality. There is always a fear of being ostracised if their sexual desires are discovered. As R. Kaila (personal communication, 14 April 2017) puts it,

Even if I am ready to break away from the codes that have been imposed on me, how [do] I go about searching for a partner? My dad's younger brother once suggested that he might be able to 'help' me, but you think I can complain about him to my parents? I am doomed either way as, for them, the need to feel good about oneself cannot be associated with me. I feel that they have always devalued me so much that I have no clue [where] to look for love and acceptance.

Similarly, S. Agrawal (personal communication, 20 March 2016), a 20-year-old with cerebral palsy, stated,

Unlike my non-disabled peers, it is very difficult for me to connect with other young people, who can be potential partners. To make them understand that I, too, have sexual desires is an uphill task, as the non-disabled appear very close-minded to our sexuality. To say that I have cerebral palsy and I enjoy having sex is immoral. The onlooker's gaze itself creates panic in the psyche: What if they can read my mind?

Shildrick (2000) points out that disabled sexualities are seen as "dangerous," leading to their denial and, as a result, regulatory obstacles are placed in the way by those who interact with them (e.g., parents, care providers, educators, health professionals). These obstacles act to hinder disabled people from achieving sexual development and engagement in sexual activities. For instance, R. Juneja, a 24-year-old young woman with visible mobility impairment, constructs society as the villain, saying,

[It] treats me as not human. If I am not married, I am not entitled to a sexual gratification; this consciousness is excruciatingly painful as I am human too. Our bodies might not be beautiful, but the inner desires are the same; however, [the] need to look for relationships [is] always undercover. Though I am happy as my desires are fulfilled, the loneliness of my life does not go away. Further, if my family finds that I have crossed my boundaries, even my gratification is always fraught with turmoil.

*(personal communication, 29 March 2016)*

It is interesting that when a disabled woman, A. Kapur (personal communication, 6 June 2016) goes to a restaurant, the gaze of the waiter and the manager become exceedingly careful. Meeting a man results in the manager making incursions into their private conversations. "As the man went to the toilet, the manager came in and quietly said, 'Don't worry. You are not out of my sight. In fact, I have warned my waiter, too.' Notwithstanding the care bestowed on me, the privacy of having a dreamy evening was ruined" (personal communication 2014).

According to S. Agrawal (personal communication, 20 March 2016), one of the unique challenges that disabled people face in having their sexuality legitimised in the public sphere is that non-disabled people "do not grow up thinking of disabled people as sexual beings," unlike disabled people. Ashley X is such an example. She was [a] six-year-old female with severe static encephalopathy, who was treated with high-dose oestrogen therapy that would freeze growth, a hysterectomy to prevent periods and removal of her breast buds, with the result that, whatever the intention, she will remain permanently childlike and asexual. As Kittay (2011, p. 221) stated, "Despite all the criticisms, the parents actively defend and promote AT ('the Ashley Treatment') as a way to care for other 'pillow angels,' as they call Ashley."

Basu's (2018) article "Can we ever have a conversation about desire without centring my disability?" brings to focus how men look at women with disabilities. She believes that her life is not centred on her disability, that there is certainly more to her than her disability. She writes,

As a person with disabilities, I don't exist despite them, nor do I exist only because of them. I am writing a story about my life in which I am at the centre, and my disabilities are sketching bits of the background. It is a story of bold moves. It is a story of gentle swaying. It is mine.

Although her wheelchair desexualises her, it also unshackles her femininity without any preconception that is painful. Foucault (1988, p. 5) says that "it is through the inter subjective demands of relationships that individuals 'discover,' in desire, the truth of their being, be it natural or fallen."

Notwithstanding the apprehensions, disabled women mentioned above are constructing novel ways of considering subjects of disability and sexuality as craving and desirable beings. Similarly, Menon (cited in Breckenridge's blog 2017), a queer feminist activist, maintains that people in mental institutions are primarily seen as ill bodies that need to be rescued so that they can live a "normal" life. But this idea of normal doesn't encompass sexual wellbeing. This is because sexuality is considered a luxury, not a necessity. She also interrogates sexual access in the context of incarceration. Menon asked,

Do we mean access to someone else or to the self? If fantasising and masturbating were considered, would it be okay to think of the people they meet the most, their doctor or care staff? This would be considered out of bounds, so even the fantasies of those who are incarcerated are regulated! I often think about how one is to do this work, since even the understanding of who the person is varies greatly across class, education, sexual and gender identities and expressions, caste, experiences of abuse and more.

In 2011, Chib, a disability rights activist and author who has cerebral palsy, wrote in her blog

The words 'sex' and 'disability' don't go together. Can disabled people have sex? Tauba tauba! (My oh My): A topic best not mentioned. Even though I have been brought up in a westernised, liberated family and social strata—the topic has rarely been brought up with me. Most people think that, if they start the conversation, they will hurt my feelings. Why does the topic sex frighten everyone when it comes to disabled people? Yet, the truth is that I am a woman who hasn't been sexually touched. Does that mean I am unfulfilled, unloved?

However, the consolation is that she has many other privileges and blessings that she has been richly endowed with. Her gratitude for care and material benefits seem to at least partially compensate for the absence of sexual relationships. Similarly, Kurian (2018) noted that a young guy experimenting with an app called Tinder, stated,

Although many of my Tinder matches left our meetings undefined, and a few wanted a platonic relationship, some of my matches did call me their date. I was cool with all of this. After all, all of them had had the choice to swipe left and had not exercised it. Many asked me thoughtful questions, and gave me new perspectives on disability, and life. Tacitly, they lifted me from a kind of slumber I'd been in and asked me to be myself.

The fact that disability per se is never devoid of sexual desire is indicated by the blog narrative of Manjari Indurkar (2018). She expressed extreme apprehension about her disability and noted,

[The fact that] persons with multiple disabilities are capable of expressing desire and acting on their sexuality and needs, with or without assistance, is the prima facie argument for the rights of persons with disabilities (PWD) in experiencing and expressing their sexuality.

Therefore, there is an aliveness in the need to interrogate the gaze of the patriarchal and normative society beyond the binary of able/abject understanding. For instance, she is extremely apprehensive about her illness, that brings in her the images of sexuality as shame. As she expresses,

I am worried that my weight makes me unfuckable. I know in my heart that none of that is true. But because I have anxiety, there is no one truth for me. There are just voices. I am worried you are tolerating me. In reality, I am afraid of showing too much skin. I am afraid you might not like the boil scars on my breasts, or my skin pigmentation. I am worried you'll hate my stretch mark. You'll realise I can't eat much of anything because I am ill all the time, unlike the perfect woman who can eat all she wants and not worry about gaining weight or getting sick. I am not that woman. I am worried you will run the minute you see me naked, or worse, that you will give me pity sex. I am worried that you will not want to spend your life with someone who comes with an expiry date that might arrive a little too soon.

The narratives mentioned above create a critical question with reference to opportunities for sexual agency, self-determination and autonomy, and have been marked by essentialist discourse that equates disabled sexualities as "other." Despite the heterogeneity of disability, disabled people are all subject to the government principle of compulsory able-bodiedness and compulsory heterosexuality (McRuer 2002, p 89). As indicated in earlier writings, It is only by challenging prevailing socio-cultural values and the binaries of normal and abnormal, that disabled people can resist normative constructions of them as dependent, asexual or deformed, and begin to forge new identities. We need to contest the notion that biology is a given destiny and, and, identity is always fixed. Though the task may appear formidable, constructed meanings, and not irrefutable facts, will provide the catalyst for change (Ghai, 2015, p. 159).

It is important to reflect on Grosz (1994), who maintains that the historical construal of the male body as "bounded" and therefore "normal" in western societies is central to the legitimation of certain strategies of violence and oppression against those bodies deemed as dangerously "Others," such as the female body and the disabled body. Within the culture in India, patriarchy ("rule of the father") is similar to that described by Grosz, where social systems are such where power is concentrated in the hands of men (Father, Son, Husband). In this type of system, men hold authority over wives, daughters and mothers. Here, if the woman is disabled, she is subordinate and weak. The intrinsic "horror" associated with disability is further reflected in the socio-cultural understanding that continues to obstruct, or barely mention, the issue of disability. Davis (1995, p. 5) wrote that the

disabled body is a nightmare for the fashionable discourse of theory because that discourse has been limited by the very predilection of the dominant, ableist culture. The body is seen as a site of jouissance, a native ground of pleasure, a scene of excess that defies reason,

leaving the disabled body at the margins of the margins. In relation to this, the question of how and in what ways the disabled body in India can be attracted and desired and a site of sensuality and gratification that facilitates a bodily sense of (a)sexuality has not been addressed. For instance, systems of inequality based on caste, race, ethnicity and gender seem to rely on dichotomies, such as "Us" versus "Them," "Self" versus "Other," or "One" versus "Other," meaning not only difference and opposition but also superiority and inferiority. In such dichotomies, the primary term corresponds to the subject, while the opposite ("Other") is reduced to the object, which leads to denying the "Other" personhood or even humanity. From the point of view of sexuality, all binaries, in psychological parlance, operate in the same way as splitting and projection. Thus, the centre expels its anxieties, ambiguities and irrationalities onto the inferior term, filling it with the converse of its own identity. The other, in its very strangeness, simply mirrors and represents what is deeply familiar to the centre but projected outside of it. It is this process of marginality that produces the resentment, enmity and repugnance for the one who is sensed as the other. Framing the argument in this form mandates a justification for inclusion of disability in the categorisation discourse. However, sexuality of a disabled body provokes fears and anxieties about "able body" mortality, and very easily renders itself as the "other."

As Crowley (2012) claims, we are never outside dangerous terrain in the arena of disability, and the very utterance of the term "disability" heralds the theoretical, conceptual and experiential storm.

> The storm is not diminished when we attend to and immerse ourselves further into the worlds of queer disability—a disparate world that brings the deep unevenness of the mundane, the celebratory and the perversities of sex and sexualities to another.
>
> *(pp. 258–259)*

For the disabled people, it is really not the bodily difference which counts, but it also requires self confidence that can resist the differences, stated Anu Dhawan (personal communication, 5 July 2014), a 27-year-old visually impaired woman, "more than being perfect, being sexual demands a confidence in yourself. How can I develop that confidence without a good education and job? At least if I am earning well, someone might decide to marry me."

The material conditions thus curtail the full expression of sexuality. However, I am definitely not suggesting that, if advocacy efforts and policy development can provide universal access, expression of sexuality would be ensured, as sexuality is further embedded in the cultural and moral issues. The theoretical divide between what is categorised as a need and what is categorised as a desire is critical as it distinguishes between those claims that a society is committed to equity. There is always a fear of being ostracised if discovered. What is more painful is that the "tragedy" of disability for a woman has been, not as the assumed effect on sexual activity, but on her ability to become a mother.

## Sex education

In India and elsewhere, sex education is a contested terrain both for non-disabled as well as disabled children and adolescents. Disabled people are often desexualised and infantilised, and the so-called experts on sexuality don't think that disabled people would need sex education. As a young 18-year-old woman, Binneta's (personal communication, 5 January 2013) anguish is articulated when she stated, "There is a general feeling that I am not human and socially the education I have a right to [*sic*]," she says, adding that she has encountered my peers who do not know the basic facts about sex and are hence vulnerable to abuse. Angela, a hearing-impaired woman, stated, "The only sex education I have ever received was from my peer group." Although her grandmother did tell her about menstruation, no one ever discussed the issue of womanhood with her, the only precaution being "*that no one should touch me.*" However, sex education modules do not contain issues about young disabled people.

> In any case, without anyone consenting to marry me, what is the advantage of knowing about sex? Society still thinks that people like me are not entitled to a sex life. I find this very frustrating as my body is just like any other woman's inside. Imagine how you would feel if playing tennis, and playing it well, was considered a major component of happiness and a major sign of maturity, but no one told you how to play, you never saw anybody else play, and everything you ever read implied that normal and healthy people just somehow 'know' how to play and really enjoy playing the very first time they try!
>
> (*personal communication, 12 September 2016*)

As an advocate, I started thinking about sex education as associated with pleasure and not merely addressing issues of vulnerability of sexual violence that disabled people are subjected to. According to Khanna (cited in Tarshi 2018), Director of the Centre for Blind Women, National Association for the Blind, "Adult blind women have a burning issue of sexuality to be addressed as they feel hesitant in expressing their right to sexual freedom and pleasure." These women need to be informed about "the birds and the bees" in a constructive and positive dispassionate way that can empower them to make the right choices and to be given the vocabulary with which to articulate their desires.

It is difficult to place the onus on any one agency in particular, for excluding disabled people from in-depth sex education. Is it the government, is it schools, be they mainstream or special, is it disability organisations and campaigners, or is it families and care providers? The truth is that all parties bear responsibility. Khetarpal (2018), president of the non-profit organisation Cross the Hurdles, recently launched a self-paced online course on comprehensive sexuality education for teens and adults with disabilities, parents of children with disabilities and special educators. In her online blog she notes, "In India, disabled people are regarded as non-sexual and this has also led to the neglect of sexual health," she points out. During her extensive interactions with disabled women, Khetarpal (2018) observed that they have negligible knowledge of sexual health and hygiene. "I found there was a lot of suppression, something which is also the case for people without disabilities. I took a year to prepare the online course and put it together after studying the best practices."

Goyal (2017), on behalf of the Mumbai-based non-profit organisation Point of View, explored women with disabilities. The stories of men with or without disabilities taking sexual advantage of women with disabilities are very common. As Goyal (2017) stated, "When I ask young, visually impaired girls whether they have faced sexual harassment, they remain silent. When pressed, they say that being touched inappropriately while seeking help to cross the street is an everyday experience." These women either end up repressing their sexual desire or permit unsuitable sexual or relationship advances. This author further mentions,

> Sexuality is not on the mainstream disability agenda and the language of positive sexuality, choice and pleasure is a dream yet to be realised. For girls and women with disabilities, sexuality is still about the triple Cs: controversy, control and coercion, but not really the Cs of choice and consent.

However, such initiatives are available in metropolitan cities. Sex education should be provided in schools and colleges. As Anil, a person with cerebral palsy, stated during an interview (personal communication, 6 August 2018) stated; "I think people do not rate sexuality as important because I am a wheelchair user; that is itself self-explanatory." Thus, both disabled men and women get oppressed in an ableist society, which results in them living a single and isolated life against their will. It requires courage to challenge sexual normativity, particularly by those perceived as "weak." The fact that they can be vulnerable should indicate the need for teaching those with disabilities to have appropriate and safe sexual relationships. However, we often find that programmes on HIV, as well as STDs, do not adequately cover the issues of disability. Sex education can play a role in assisting disabled people assert their sexuality. Most disabled people grow up without receiving any form of sex education at school or home. This is assuming that they are not actively discouraged from the idea of sex altogether. It is difficult to place the onus on any one agency, in particular, for excluding disabled people from in-depth sex education. Is it the state, is it schools themselves, be they mainstream or special, is it disability organisations and campaigners, or is it families and care providers? The truth is that all parties bear responsibility.

Education about sexual transmitted infections is another important area. Education about sexuality cannot become a reality unless the attitudes and perception of non-disabled people is worked upon. We need to be unlocking the disabled body and creating avenues for a healthy sexuality, which is not constructed only within marriage. I have come across many examples of professionals, such as doctors, gynaecologists and other service providers, who do not provide relevant information when requested and required. In general, they have a very unfriendly and negative approach towards sexuality and related queries. In India, one has to be married to ask questions related to your sexual and reproductive health. The attitude with which you are given any information is always preceded with the question, why do you need this information at all? My contention is that precise information must be made available. Education is a key area as a lot of parents are anxious about their disabled child/adult being sexually active. The parents think that it's better for them not to know about it.

## Disability and sexuality: Some positive changes

The realisation is that, in an intimate relationship, a wheelchair or a white cane is not all that the disabled person brings to the relationship. S/he will also have their personal interpretations, idiosyncrasies, happy and sad stories. Goyal (2014) shares her life experiences in which she is open about the societal prejudices that exist around disability. However, during her work at Point of View, a non-profit organisation, she noted,

> I realised that I, myself, had internalised some of those prejudices. I always had a sense that if I ever entered a relationship with a non-disabled man, the relationship would be somewhat unequal because of my disability. But after meeting so many women with disabilities, and seeing how they deal with their lives, this idea began to change.

Goyal (2014) brings an extremely inventive idea of dating. She was on a date with a visually impaired man who was holding her hand and said to her, "Nice nail paint, but you could have used a coloured one." And she gasped and asked, "How the hell did you know?" It was true: she was wearing a transparent coat of nail polish. He responded by telling her it was possible to distinguish the two by feeling the density. If the paint felt thicker, it was coloured. Just like Nidhi (Goyal), I was amazed at this small moment in a new romance, that showed just how wonderfully creative dating can be. Stories such as Goyal's (2014) suggest that, although disabled people are socialised into a form of desexualised subjectivity, the voices of disabled people point to resistance by recognising their sexuality.

## Conclusion

Disabled men and women in India have many mediators in their lives, and there is insignificant familial and state-level support for and encouragement of their sexual rights. Disabled women face unique issues given their cultural roles as Indian women with disabilities. Sexuality for disabled people in India has come to be understood as intimate but troublesome, covert but satisfying, and violent but loving where our "inescapability" is read and reflected upon. While battling with discrimination, stigma and multiple power structures, the above experiences reverberate in the lives of disabled men and women with diverse disabilities. Sexuality for disabled people in India is both oppression as well as an agency that can overcome the victimhood created by society. Although it is not an easy endeavour, hope that the empowerment of all disabled people to resist the "silence" of the personal, which will be translated to the political, brings optimism.

## Notes

1 http://www.infochangeindia.org/agenda-issues/sexual-rights-in-india/5630-regulation-of-disabled-womens-sexuality.html accessed on 30th December, 2019.
2 This is a conversation, which the designer had with my father in 1999 on July 24th in Vijay Furnishing in Jhande wala, a place in New Delhi. The designer, a stranger, was oblivious to my presence and the impact on my psyche. He truly believed that he was helping my "unfortunate" father There are many such communications with fellow disabled women and men. This was a part of my research in NMML, for which a report was submitted.
3 Object relations theory is derived from psychoanalytic theory, that highlights interpersonal relations, primarily in the family and especially between mother and child. An object in his theory relates to persons, parts of persons, or symbols of one of these. Relations refer to interpersonal relations and suggest the residues of past relationships that affect a person in the present. Object relations theorists are interested in inner images of the self and other, and how they manifest themselves in interpersonal situations.
4 Drawing a distinction between depictions of mithunas (amorous couples) and maithuna (coitus), Desai states that the earliest maithuna scenes in temple art belong to the 6th–7th century, a period "when the Tantras came to be accepted by the literate class" (p. 206).
5 The temples depict the many different manifestations of Shakti and Shiva, the female and male divine principles.
6 Shakti is the female manifestation.

## References

Addlakha, R. (2007) Gender, subjectivity and sexual identity: How young disabled people conceptualise the body, sex and marriage in urban India. *Occasional Paper Series 46*. New Delhi: Centre for Women's Development Studies.

Basu, S. (2018) Can we ever have a conversation about desire without centering my disability? *Point of view*. Viewed on 13.02.20. https://medium.com/skin-stories/by-soumita-basu-64274f2bcd62

Breckenridge, J. (2017) People with mental illnesses who are incarcerated face a near-total denial of their sexual rights. Viewed on 13.02.20. http://blog.sexualityanddisability.org/2017/03/people-mentalillnesses-incarcerated-face-neartotal-denial-sexual-rights/

Chakravarti, U. (1989) The world of the Bhakti in South Indian traditions: The body and the beyond. *Manus Tenth Anniversary Issue* 50, pp. 51–52.

Chib, M. (2011) I'm single because of my body. *Mumbai Mirror*. Viewed on 12.02.20. http://timesofindia.indiatimes.com/life-style/relationships/lovesex/Im-single-because-of-mybody/articleshow/10051306

Cohen, S. (1972) *Folk devils and moral panics*. London: MacGibbon and Kee.

Crowley, V. (2012) A rhizomatics of hearing: Becoming deaf in the workplace and other affective spaces of hearing. In A. Hickey-Moody & V. Crowley (eds) *Disability matters: Pedagogy, media and affect*. New York: Routledge, pp. 138–153.

Davis, L. (1995) *Enforcing normalcy: Disability, deafness and the body*. London: Verso.

Doniger, W., & Sudhir, K. (trans) (2002) *The Kamasutra of Vatsyayana*. London and New York: Oxford World Classics.

Finger, A. (1992) Forbidden fruit. *New Internationalist* 233, pp. 8–10.

Foucault, M. (1979) *The history of sexuality volume 1: An introduction*. London: Allen Lane.

Foucault, M. (1988) An aesthetics of existence. In L. Kritzman (ed.) *Foucault, M.: Politics, philosophy, culture: Interviews and other writings, 1977–1984*, transl. A. Sheridan, et al. London: Routledge, pp. 47–56.

Garland-Thomson, R. (1997) *Extraordinary bodies: Figuring physical disability in American culture and literature*. New York: Columbia University Press.

Ghai, Anita. (2002) Disabled women: An excluded agenda of Indian feminism. *Hypatia* 16(4), pp. 34–52.

Ghai, A. (2003) *(Dis)Embodied form: Issues of disabled women*. New Delhi: Shakti Books.

Ghai, A. (2015) *Rethinking disability in India*. New Delhi: Routledge.

Ghai, A. (ed). (2018) *Disability in South Asia: knowledge and experience*. Delhi: Sage Publications.

Goyal, N. (2014) Why should disability spell the end of romance? Viewed on 13.2.20. https://in.news.yahoo.com/why-should-disability-spell-the-end-of-romance-055837779.html

Goyal, N. (2017) Sexual rights: Off-limits for Indian women with disabilities? Viewed on 11.02.14. https://www.opendemocracy.net/en/5050/sexual-rights-disability-indian-women/

Grosz, E. (1994) *Volatile bodies: Towards a corporeal feminism*. London: Routledge.

Hanisch, Carol. (2006) The Personal is Political. Available online: http://www.carolhanisch.org/CHwritings/PIP.html (accessed on 29 June 2018).

Indurkar, M. (2018) Love in the time of depession and anxiety. Viewed on 9.02.20. https://www.dailyo.in/lifestyle/love-depression-mental-health-anxiety-sexuality-disability/story/1/22856.html

Kakar, S. (1996) Emergence of Jodi. India Today, December 31st, pp. 105–106.

Kakar, S. (2007) cited in Eshmi Lahiri Roy, 2019. The urban Hindu arranged marriage in contemporary Indian society. In W. Sweetman & A. Malik (eds) *Hinduism in India: Modern and contemporary movements*. Delhi, India: Sage Publishing, p. 29.

Khanna, Shalini, cited in Tarshi (2018) Sexuality and Disability in the Indian Context, working paper. https://www.tarshi.net/index.asp?pid=

Khetarpal, Abha. (2018) Welcome initiatives that shine a light on need for sexuality education for disabled people. https://newzhook.com/story/20708/

Kittay, E.F. (2011) Forever small: The strange case of Ashley X. *Hypatia* 26(3), pp. 610–631.

Kurian, T. (2018, January 9) What it means to be on Tinder as a person with an identifiable disability. Viewed on 22.04.19. https://medium.com/skin-stories/what-it-means-to-be-on-tinder-as-a-person-with-an-identifiable-disability-b3ad50d4a677

McRuer, R. (2002) Compulsory able-bodiedness and queer/disabled existence. In S.L. Snyder, B. Brueggemann, & R. Garland-Thomson (eds) *Disability studies: Enabling the humanities*. New York: Modern Language Association, pp. 88–99.

Menon, Nivedita. (2007) Sexualities Kali/Women Unlimited.

Nair, Janaki, & John, Mary E. (eds). (2000) *A question of silence: the sexual economies of modern India*. London: Zed Books.

Nisha (2006) Regulation of disabled women's sexuality. http://www.infochangeindia.org/agenda-issues/sexual-rights-in-india/5630-regulation-of-disabled-womens-sexuality

Overboe, J. (1999) Difference in itself: Validating disabled people's lived experience. *Body & Society* 5(4), pp. 17–49.

Ramanujan, A.K. (1992) Talking to God in the Mother Tongue. India International Centre.

Shakespeare, T. (1999) The sexual politics of disabled masculinity. *Sexuality and Disability* 17, pp. 53–64.

Shildrick, M. (2000) The body which is not one: Dealing with differences. In M. Featherstone (ed.) *Body modification*. London: SAGE, pp. 77–92.

Wilkerson, A. (2002) Disability, sex radicalism, and political agency. *NWSA Journal* 14(3), pp. 33–57.

Winnicott, D.W. (1991) *Playing and reality*. London: Routledge.

# 25

# DISABILITY AND *ASEXUALITY?*

*Karen Cuthbert*

## Introduction

Not all disabled people are sexual. Some are asexual. The term "asexual" may make some recoil in horror given that this has been (and still is) a label coercively applied to disabled bodies. Within disability studies and disability activism, asexuality is regarded as a harmful myth which has been used to justify exclusionary and violent practices. Asexuality, it is often argued, is empirically and demonstrably untrue—disabled people are simply *not* asexual.

However, asexuality is also increasingly being understood as a sexual orientation or sexual subjectivity in its own right, sitting alongside heterosexuality, bisexuality, homosexuality, pansexuality, queer etc. In basic terms, it refers to a feeling of not being sexually attracted to anyone. Like other sexualities, asexuality is something that a person can actively identify with/as, if it is a term that feels right to them (rather than something that is ascribed by others).

This chapter explores what happens when these two very different understandings of asexuality come together in the person who is disabled—but who also self-identifies as asexual. First of all, I give an overview of how asexuality has emerged as a form of self-identity and, with it, as a field of academic study. I then give some more context regarding the "asexualisation" of disabled bodies, and the ways in which this has been rejected and resisted. However, this disavowal of asexuality by disability activists and scholars, combined with a tendency of asexual activists to draw on discourses of "healthy" and "normal" bodies, creates a kind of intersectional black hole (or "mutual negation") into which disabled asexual subjectivities are lost. Drawing on empirical qualitative research with disabled asexual people, I discuss some of the contours of living in and with this gap. I conclude this chapter by providing some recommendations for how those working within disability studies and in disability activism might adopt a more asexual-inclusive praxis.

## The emergence of asexuality as an identity and field of study

The term asexual is common within the biological sciences to refer to self-reproducing organisms, but there has also been some limited historical use of the term to refer to human sexuality. In the mid and latter parts of the 20th century, human sexuality researchers sometimes included asexuality in their typologies of sexual orientation. Kinsey used the term "X" to describe those who could not be placed on his heterosexual-homosexual continuum, due to them having "no socio-sexual contacts or reactions" (Kinsey et al. 1948, p. 656). Later sexologists would explicitly use the term asexual, such as Storms (1980), who conceptualised asexuality as scoring low on both homo-erotic and hetero-erotic scales, whereas Masters et al. (1988) used asexual to describe (what they saw as) a pathological subset of homosexuality. However, in an excellent genealogy of this asexual "pre-history," Przybylo (2013, p. 227) writes that, in these studies, asexuality tended to function merely as a kind of descriptive placeholder, of which there was a distinct "disinterest in exploring its definitions, parameters and implications."

The sense of asexual as a *self-identity* (as opposed to a label applied by sexologists) began to crystallise and gain traction around the millennium (although undoubtedly, people had been using the term to describe themselves much earlier than this). The Asexuality Visibility and Education Network (AVEN) website and forums was formed in 2001, and this moment is often marked in the literature as the genesis of the modern day understanding of asexuality (Renninger 2015, p. 1515), although there had also been some internet groups and mailing lists before this (Titman, 2017). Although there were some definitional disagreements early on (Titman 2017), in time asexuality came to be defined by AVEN as feeling *little or no sexual attraction towards others* (AVEN n.d.). Also emphasised in the definition is 1) the importance of self-identity (no-one else can tell you you're asexual), 2) the non-pathological nature of asexuality (asexuality is not something to be cured) and 3) asexuality as different from celibacy or abstinence (it is not conceptualised as a *choice*, nor as a description of behaviour). This has now become the widespread and generally agreed-upon definition of asexuality, both popularly and academically. However, there is also huge variation and diversity underneath the rubric of asexuality.[1]

Academic research soon began to reflect this emerging sense of asexuality, although the study that is often credited as inaugurating asexuality research—Anthony Bogaert's 2004 article in the *Journal of Sex Research*—was more akin to earlier sexological work in that it labelled survey respondents as asexual based on their responses to questions asking about sexual desire and sexual activity. However, work that recognised asexuality as self-identity soon followed (e.g., Prause and Graham 2007). Academic research on asexuality has subsequently been conducted from a variety of disciplines. Psychologists and sexologists have been engaged in a positivistic pursuit of "knowledge" of asexuality, determined to define exactly what asexuality *is*, how it comes about, and the biological and psychological characteristics of asexual people (e.g., Yule et al. 2014). Demographers have looked at the prevalence of asexuality in various populations (e.g., Höglund et al. 2014). Social psychologists and sociologists have been interested less in capturing some kind of objective existence of asexuality, and more in the lived experiences of asexual people themselves, exploring identities, relationships and community (e.g., Carrigan 2011; Dawson et al. 2016; Scherrer 2008). Counselling and social work academics have looked at how practitioners can be more asexual-inclusive (e.g., Steelman and Hertlein 2016). Scholars from within the humanities have explored asexuality textually and historiographically (Hanson 2014; Przybylo and Cooper 2014). There has also been literature on the queer, feminist, and otherwise liberatory *potential* of asexuality (Cerankowski and Milks 2014; Gressgård 2013). Whereas still a marginal (and often margina*lised*) area of academic interest, to date asexuality has been the subject of a journal special issue (Carrigan et al. 2013), of an edited collection (Cerankowski and Milks 2014), of a National Women's Studies Association (NWSA) interest group and of several panels at NWSA annual conferences (e.g., NWSA 2015; NWSA 2016), of a full-length undergraduate course (Tran 2018), and of a number of dissertations and theses deposited in institutional repositories (e.g., OATD.org returns 90 matches for the search term "asexuality," the vast majority of which refers to asexuality in the sense of self-identity).

# The asexualisation (and re-sexualisation) of disabled bodies

If we understand asexuality as self-identity, then it makes sense that there are disabled asexual people, just as there are disabled straight people, or disabled bisexual people, or disabled people of every other sexual subjectivity. However, the intersection of disability with asexuality has a particular significance. Disabled people have long been subject to what Shuttleworth (2012 p. 62) calls the "cultural imposition of asexuality." This has many facets. It refers to the construction of disabled people as sexually *undesirable* as well as the assumption that disabled people are *incapable* of feeling sexual desire or being sexually active, since sexuality (or at least how it is discursively constructed) is predicated on the normative able body (Shildrick 2007). The "cultural imposition of asexuality" also refers to the moral mandate that disabled people *should not* be sexually active. This latter facet reflects a long eugenicist history where the re-productivity of disabled bodies are seen as a threat to the (able)body politic (Pfeiffer 1994), although such prohibitions tend to be framed nowadays in terms of a concern for disabled people as being particularly vulnerable, or in terms of the putative threat that disabled parents will pose to their children (Feely 2016, p. 726–728). As such, disabled people's sexuality has been denied and repressed in various violent, coercive and insidious ways (Campbell 2017).

Given this, disability activists and scholars have understandably focused on contesting the idea that disabled people are asexual. Asexuality is frequently described as a "myth" that can be easily dispelled by research findings that point to the fact that disabled people are active and desiring sexual subjects (e.g., East and Orchard 2014; Milligan and Neufeldt 2001; O'Toole and Bregante 1992; Peta et al. 2017; Shakespeare and Richardson 2018). It is sometimes acknowledged that disabled people might not be interested in sex, but this is often accounted for in terms of the internalisation of a pervasive societal dis/ableism. For example, Esmail et al. (2010, p. 1151) state: "The impact of this external stigma on the individual creates a downward spiral of asexuality and evidence suggests that individuals living with a disability may internalise such notions of asexuality." Note the use of the term "downward spiral": asexuality is envisaged as a kind of self-fulfilling prophecy, a pit into which one can fall and may never escape from. Another example can be found in one of the most influential books to have been published on disability and sexuality—Shakespeare, Gillespie-Sells, and Davies' *The Sexual Politics of Disability*:

> damaging emotional and psychological barriers that prevent disabled people from becoming fully functioning human beings, with healthy sexual identities and active, life-enhancing sex lives: these include certain attitudes and assumptions held by oppressed people that create a situation of self-denial or self-harm.
>
> *(1996, p. 40)*

Here, sexuality is seen as fundamental to being a fully functioning human being. This speaks to the framing of sexuality as a natural and universal human drive (Kim 2010), which can also be seen in some of the language used by disability activists and organisations. For example, *Enhance the UK*, a UK-based educational and training charity, formed to "change society's views on disability," states:

> Did you know—disabled people have sex and they like it? Yep. Shocking though it may seem, disabled people are just like, well, everyone, really. Sex is important. We all want to be loved, we all crave a bit of intimacy from time to time and, like it or not, disabled people are sexual beings. Take sex out of the equation and we will quickly start to feel insecure and unattractive.
>
> *(Enhance the UK, n.d.)*

Furthermore, the disabled body has also been theorised as a site for expanded and enhanced erotics: Not only is the disabled body *sexual,* but it also represents the possibility of understanding and experiencing sex and intimacy in new ways, which is of benefit to everyone (Siebers 2012).

## Mutual negation

However, these efforts by scholars, activists and organisations to refute the "charge" of asexuality may also have the unintended effect of erasing disabled asexual people, who <u>are</u> asexual. Kim (2011, p. 483-484) argues: "When asexuality is considered only as disempowering, sexuality quickly becomes a normalising mechanism that rides on its power to construct and alienate asexual bodies." Not only does this mean that disabled people who self-identify as asexual might be marginalised or made invisible, but "compulsory sexuality" (i.e., that everyone is and should be sexual) is also perpetuated in these discourses, which itself has a regulatory and disciplinary power (Gupta 2015). Kim argues that the denial of asexuality feeds into a form of "sexual liberalism" aimed at empowering disabled people to enjoy and participate in sexual culture(s), "without first problematising the power dynamics operating within that culture" (2011, p. 490).

But disabled asexual people may also find themselves marginalised by asexuality discourse. There is a legacy of low sexual desire being pathologised as a condition in and of itself (for example, "hypoactive sexual desire disorder" and "sexual interest/arousal disorder" are diagnosable conditions in the current edition of Diagnostic and Statistical Manual of Mental Disorders). Low sexual desire/interest is also regarded as symptomatic of various illnesses, diseases and disorders, and, more recently, as an iatrogenic side effect of certain medications (Bennett 2014). This understanding of asexuality-as-pathology is also rife in popular culture. For example, Pamela Anderson, of Baywatch fame and now occasional sex advice columnist, said the following to a reader whose boyfriend was possibly asexual:

> I heard that this is an epidemic. Or maybe it's an evolution in the age of technology and germ phobias. Does he watch a lot of explicit pornographers or video games? Does he feel numb? Is he sure of his sexuality? Too much masturbation or fantasies about cyber film stars or video games like Fortnite seem to be an addiction.
>
> *(Dazed, 2018)*

Here, asexuality is framed as a kind of social contagion—an "epidemic" caused by degenerate habits of excess and immoderation, or as a manifestation of neuroticism around cleanliness and contamination. Given how asexuality is frequently presented in this way, asexual activists have often strategically emphasised their own healthiness and otherwise "normal" bodies in order to counter accusations that their asexuality is a problem to be fixed, rather than a legitimate sexual subjectivity (Kim 2010). For example, in interviews to mainstream media, asexual people might state "there's nothing wrong with me, I'm just asexual" (Rothwell 2015), or affirm that "I'm physically capable of it, but mentally I'm just not engaged" (Bell 2018). In the documentary (A) sexual (2011), the "healthiness" of asexuality is similarly emphasised. In one scene, during an interview with Lori Brotto, a psychologist who has authored a number of papers on asexuality, text of various "disorders" appear on screen. These include "post-traumatic stress," "psychotic features," "anxiety," "health problems," "depression," "anger control," "sexual distress," "alexithymia," "suicidal thinking" and "alienation." As Brotto discusses the fact that her research has found no connection between asexuality and any of these things, the words disappear. The implication here is that asexuality is being given a "clean bill of health," which works as an implicit claim for asexuality and asexual individuals to be taken seriously (Sheehan 2015, p. 81). Of course, asexuality itself is not a pathology, but this discursive distancing of asexuality from anything other than "normal," "capable," or "healthy" falls into what Kim (2010, p. 160) calls the "moral and ableist binary of good body and bad body."

Kim (2014) and Gupta (2014) both use the term "mutual negation" to describe this dual process of erasure happening within asexuality narratives and disability narratives. The narratives come together to create a kind of "intersectional black hole" which engulfs disabled–asexual subjectivities. In the next section, I draw upon my own qualitative research to discuss what living in this intersection might mean in practice in the lives of disabled asexual people themselves.[2]

## Living the intersection

The data presented in the following sections come from eleven semi-structured interviews. All the interviews took place online (a pragmatic decision, given the relatively small pool of potential participants), and participants were given a choice of medium through which to take part: Email (asynchronous, text-based), instant messaging (synchronous, text-based) or Skype (synchronous, verbal). These options were intended to make the research somewhat more accessible, albeit with an understanding that they still privileged certain forms of communication and embodiment.

Participants were recruited from online asexuality spaces, and were physically located in the US, the UK, Australia and France. Perhaps as a result of this recruitment strategy (as opposed to recruiting from contexts where disability is foregrounded, for example), the study suffered from the same demographic limitations of much previous asexuality research, where participants were overwhelmingly white, young and female. As a researcher, I myself was also a young white woman, with complicated and shifting relationships to both asexuality and disability that made for multiple nodes of dis/identification.

My analytical approach was heavily influenced by Holstein and Gubrium (2011), who suggest that researchers be attentive to the meaning-making process (how participants and researchers create narratives, the resources drawn upon in order to do so, the subject positions claimed, etc.) but without losing sight of the meanings themselves. This approach avoids the "unrepentant naturalism of documenting the world of everyday life as if it were fully objective and obdurate," but also the "indifference to the lived realities of experience" that can be discerned in some forms of discourse analysis (Holstein and Gubrium 2011, p. 352).

## *Embodied identities*

Most of the participants in my research had an understanding of their own embodied identities which they did not see reflected in either disability or asexuality narratives. Rather than distancing their asexuality from disability, most participants talked about how their asexuality was connected in some way to their disability and/or impairment. Dawn (aged 31) spoke about how her lack of sexual interest and sexual attraction to others could possibly be symptomatic of her myalgic encephalomyelitis (ME)/chronic fatigue syndrome (CFS), which remains a condition about which very little is known. Camille (aged 28), who had a number of neurological conditions, felt her asexuality was linked to her having a "differently structured brain." And Jo (aged 28) and Erin (aged 29) both explicitly talked about impairment and how living in their particular bodies may have played a role in shaping their identities as asexual. Erin, who had a genetic connective tissue condition, said:

Learning about my particular condition and that a part of it, joint hypermobility, has a lot to do with a lack of proprioception and weird sensory issues, well, sometimes I wonder if I have a simple disconnect between my mind and my body, and that is why I don't feel like my body wants to engage with anyone sexually.

Here, Erin muses about whether her difficulties in sensing her own body and its placement in the world has affected how she feels about sex and sexuality. She speaks of her *body* not wanting to engage with anyone sexually—for her, asexuality is something which is embodied and has arisen from her phenomenological experience of being in the world. Jo, who had a condition that caused non-malignant but chronically painful bone tumours, also spoke of a "disconnect":

I do have some kind of a vague theory as to where it [asexuality] all comes from, that being in constant pain from a young age that maybe I didn't have the same connection to my body that other people do and maybe that had something to do with how I view bodies or physical interaction in general.

In both Erin and Jo's accounts, sexuality is implicitly framed as something which requires a kind of mind-body integration. For both of these women, impairment has paradoxically meant both being painfully present in their bodies, whilst at the same time living outside of their bodies in order to manage their pain, which has led to an experience of disconnect. Erin and Jo thus see their asexuality as a complex biopsychosocial process. These stories go against the grain of prevailing asexuality narratives, where asexuality is not "caused" by anything (which, it is assumed, would delegitimise asexuality). Participants were very aware of this. Jo said:

I think there can be a sticking point with the asexual community at large. They are frustrated with always having asexuality be something [that is] wrong with them... people are always presuming there is some kind of defect that makes a person asexual. So, there is a bit of friction there.

And Ryan (aged 24) said:

[T]he view that asexuality is not related to physiological, hormonal or psychological problems is very deeply ingrained in the asexual community. One of the most common dismissals we get from non-asexuals...is that it [asexuality] is caused by some sort of deformity or deficiency. Given our struggle to get people to accept that asexuality exists and is a valid sexual orientation, it [considering that disability and asexuality to be linked] is very much not welcome.

In this context, participants' embodied self-understandings are imagined as a political risk to the goal of asexual acceptance and inclusion. But some participants went on to say how this didn't need to be the case. They challenged the idea that asexuality was valid only if it had no "cause." Jo said, "It's legitimate, it's however you feel, it doesn't matter how that came about...it wouldn't make it any less legitimate if it had a cause." And Kate said, "I don't think a link between asexuality and autism would make asexuality not valid. Autistic asexuals would still be asexual, it just wouldn't be for the same reasons as other asexuals."

At the same time, their stories also go against how asexuality is understood in disability narratives. Participants' embodied experience not only contests the idea that asexuality is a myth, but also that asexuality is something inherently undesirable. Even for Erin and Jo whose asexuality might be understood as a kind of "impairment effect" (Thomas, 1999), they had no desire to eradicate or change it, since it was understood as part of their embodied experience of the world.

## Emotional labour and self-management

Although the disabled asexual people I spoke to resisted dominant asexuality and disability narratives and instead advanced more nuanced and complex understandings, it still remained the case that being a disabled asexual person required a certain kind of emotional labour and self-management. They were not immune from the effects of mutual negation. Some participants spoke about how they worried about "confirming the stereotype"— that is, all disabled people are asexual, and that asexuality is a disorder or condition. This worry manifested itself particularly in relation to being amongst other asexuals at meet ups or events, as, whereas their asexuality was invisible, some had visible disabilities which they felt created a conspicuous presence. For example, Dawn worried about how her using a wheelchair might look in a group of asexual people, whereas Kate had to very carefully manage and monitor her behaviour when she marched in the asexual contingent at Pride:

> I worry about having to justify myself. The other weekend, at Pride, I found myself worrying that I might be making asexuality look bad if I wasn't appearing sufficiently neurotypical.

She spoke of the effort it entailed to pass as neurotypical. Kate was a committed asexuality and disability activist, and, although she was very critical of the idea of *having to* hide her autism (or her asexuality), she felt that there was a certain burden on her as a visible representative of her communities. Being seen as an autistic asexual at a very public event like Pride might mean that an onlooker comes to the conclusion that "asexuality is a medical symptom" or "asexuality is an excuse for socially awkward people who can't get dates." She went on to acknowledge:

> Obviously, if anyone used our group demographics to make negative assumptions based on stereotypes, that would be their problem. But that knowledge didn't stop the issue from being in the back of my mind.

It is interesting to compare Kate's experience to that of Ruth, one of the participants of Shakespeare et al. (1996). Ruth, a disabled lesbian, says that her decision to participate in a Pride march was motivated by a desire to be visible *as* a disabled lesbian to show the world that disabled LGBT persons exist. "I am not sure if I would have gone…if I was not disabled…It's [about] knowing that somebody will see that there are disabled people there" (1996, p. 164). Ruth did not conceive that her being visibly disabled would make the LGBT community "look bad," or confirm any stereotypes. We might understand this in terms of the LGBT community (even back in 1996) as one which has largely been successful in the demedicalisation of homosexuality. In this context, disability may pose less of a risk to the ontological security of the LGBT community. Kate's account, on the other hand, perhaps shows a keen awareness of the "precarity" of asexuality's existence. Whereas this kind of self-consciousness and concern with representation may be shared by non-disabled asexual persons, in Kate's case, she felt a particular personal responsibility due to her positioning at the intersection of asexuality and disability to micro-manage her behaviour and appearance in public.

## Disability negating asexuality?

Some participants spoke about how, paradoxically, they felt that their identity as asexual was negated by others precisely *because of* their disability or impairment. This makes sense when we understand what is being meant by the term "asexual." Asexual as ascribed onto disabled bodies is also an ascription of passivity, of dependence, of inertness. Conversely, asexual as self-identity involves a process of self-reflection, questioning, and laying claim to a label that reflects one's felt subjectivity, all which contravene prevailing ableist norms about how we understand disabled people. Kate speaks about how she is presumed innocent and childlike due to being an autistic woman (and therefore is *ascribed* an asexual identity), but this is something which the process of *identifying as asexual* contradicts ("I realised I didn't feel sexually attracted to others, questioned, had sex and found it wasn't a turn-on, questioned again and eventually decided 'asexual' was the closest fitting label"). This is also tied up with how disability becomes a kind of totalising force. As one participant, Bobby (aged 18) said:

> A bit more of a scepticism there, a bit more of—"well you already have this"…for a disabled person, you go "it's a disabled person" and you might think that asexuality just comes along with that or something, or maybe their medication. Yeah, I think there is more scepticism toward a disabled person saying that they were asexual.

Here, Bobby is articulating what has been termed "identity spread" by medical sociologists and disability scholars, where the presence of disability or impairment is seen to overwhelm all other aspects of identity (Shakespeare, 2006). O'Toole (2000, p. 210) wrote about this specifically in the context of sexuality, arguing that: "For disabled people, it is presumed that any sexual expression is an expression of illness or disability." This has particularly been the case with non-heterosexual identities and expressions: For example, Kafer (2003, p. 82) writes about how queerness in disabled women has been dismissed as "disability-related confusion," or as a kind of last resort since they were incapable of finding male partners. In the case of asexuality, even though asexuality is cathected to disabled bodies, asexuality is simultaneously negated as a possible subject position due to the presence of impairment or disability. This is something that is (ironically) even perpetuated by disability researchers themselves. After commenting on the higher "incidences" of asexuality amongst autistic people, the author of a 2018 paper in *Sexuality and Disability* adds the following caveat: "These data should be interpreted with caution, bearing in mind the difficulty of establishing social relations in persons with ASD [autism spectrum disorder]" (Parchomiuk, 2018, p.11). In effect, the possibility of asexual subjectivity is dismissed because of the already-existing presence of an autism diagnosis.

However, disability could also negate asexuality not just in the eyes of others, but also in terms of self-understanding. Steff (aged 22) spoke about how her journey to an asexual identity didn't chime with the narrative of the asexual community. Her feeling of not being sexually attracted to anyone was not perceived as something "wrong," but rather as something *normal* and to be expected of a person with autism. Her mother had wanted to "prepare" her by letting her know that many autistic people do not have or want sexual relationships. As such, Steff interpreted her feelings of non-attraction and non-desire as part of her autism: "I assumed my lack of interest in sex was because of my disability." Thus, understandings of autism, particularly as they intersect with gender (as reflected in Kate's account, above), worked to prevent Steff identifying with asexuality as a sexual orientation, since she was already *supposed* to be asexual. It was only upon googling "intimacy" and "Aspergers" that Steff came upon AVEN, and, after much reading and questioning, began to see her asexuality in terms of a *marked presence* as an identity or subjectivity, rather than a simple lack or absence. Steff went on to say: "If I did not have Asperger's, I think I would have suspected I was asexual a lot sooner."

## Recommendations

Disabled asexual people are thus positioned at an intersection with a long and fraught history. The contemporary activisms of each constitutive "strand" have adopted strategies of distancing and disavowal for understandable reasons but disabled asexual people have been lost in the shuffle. This chapter has brought the voices of some disabled asexual people to the fore and illustrated some of the lived effects of "mutual negation." Participants spoke about developing an understanding of their embodied identities that was at odds with dominant disability and asexuality discourses, of the emotional labour and self-management necessitated by their positioning at this intersection, and of the paradoxical experience where their disability negated their asexuality, both inter-and intra-personally.

Given that this is a volume centred on disability (rather than asexuality), this last section will discuss how disability organisations, academics and activists might become more asexual inclusive (although this is in no way to let asexuality activism off the hook!). The first thing is to recognise that "asexual" has dual meanings. "Asexual" can no longer be used to describe the erasure or suppression of disabled people's sexuality, without also acknowledging that asexual is also a sexual subjectivity and form of self-identity. Liddiard's book (2018) on disabled people and intimacy is a good example of how to do this. On page 2 of her introduction, she clarifies how she uses the term asexual in the book:

> The asexuality I speak of here…differs from the emergent asexual identity category powerfully claimed by those who, for a variety of reasons, do 'not experience sexual attraction'…Instead the asexuality to which I refer is a set of processes purposefully imposed upon disabled bodies and minds through the processes of ableism and disableism…however, this is not to negate the fact that…disabled people can identify with an asexual identity too.

Alternatively (and possibly preferably), one might use a word such as "desexualised." Unlike asexual, "desexualised" implies an active social process of the kind Liddiard hints at above. It perhaps better encapsulates the power dimensions of the process—as in, something is being taken coercively from a person. We might see parallels with the use of "racialisation" in critical race studies, which is a term used to capture the sense of "race" being actively "done" within particular power relations. Either way, a clearer heuristic distinction needs to be made between asexuality as *ascribed* and asexuality as *agentic* (even if this distinction might be "muddier" in practice).

Secondly, it is important that sexuality itself and the assumptions around it (particularly regarding its universality) are interrogated. While we might interrogate different facets of how sexuality is socially organised or regulated, the idea that everyone "has" a sexuality is not always subject to the same critical scrutiny. Whereas early sociologists of sexuality argued against the idea of an innate sexual drive and urged us to see sexuality as "humanity at its most social" (Gagnon and Simon, as quoted in Jackson and Scott, 2010, p. 13), this has not always been taken up in subsequent sexualities scholarship, particularly within certain strands of psychoanalytically informed queer theory (Hammers 2015). We can detect the "sexual assumption" at work within disability studies—not just in terms of affirming the presence of disabled people's sexuality, which we have already seen—but also as woven through theory. For example, Tobin Siebers' (2012) argument that disabled people should be recognised as a sexual minority implies, to my mind, some kind of uninterrogated "sexualisation" of disability. There is also a similar impulse detectable in Anna Mollow's (2012) work where she theorises that sexuality and disability in psychoanalytical terms might be two different names for the same thing: A "self-rupturing force" (p. 287) that involves a "loss of self, of mastery, of integrity, and control" (p. 297). As such, she sees disability as always "erotically charged" (p. 305). Such readings draw on the idea of a kind of essential libidinal energy that work (however inadvertently) to make asexual subjectivities unthinkable. It is important then that disability scholars and activists not just temper statements such as "disabled people are sexual," but it may be necessary to also look critically at theoretical presuppositions.

Thirdly, and finally, it is important that disability scholars and activists recognise the commonalities they might have with asexual people and asexual activists (and vice versa), and the possibilities for alliance and coalition. Gupta (2014) suggests that both groups ultimately have similar goals: "Both groups seek freedom from sexual coercion—whether that coercion is directed at imposing sexuality on asexuals…or preventing people with disabilities from engaging in desired sexual activity") and "both groups also seek to combat the over-medicalisation of human variation" (2014, p. 296). Recognising asexuality need not impinge on the claiming of sexual citizenship rights for disabled people—indeed, as Kim (2011) argues, they must be part of the same project if it is to have any real meaning or effect at all. A right to be sexual must also be the right *not to be sexual*.

## Notes

1 For example, asexuality is often conceptualised as a spectrum with "asexual" at one pole and "sexual" at the other. This leaves room for subjectivities in-between, with terms emerging to convey nuances in strength and frequency of sexual attraction (e.g., grey-A, demisexuality). Within asexuality discourse, romantic attraction is seen as something distinct from sexual attraction, as so many asexual people also define their romantic orientation (e.g., asexual and homo-romantic, asexual and a-romantic etc.). See Decker (2014) for a comprehensive and accessible introduction.
2 Of course, this presumes that disabled asexual people share the same interpretation of disability and asexuality narratives as has been presented in this chapter. This is not always the case—elsewhere, I discuss how some participants had different perspectives on activist strategies, particularly in relation to asexuality (Cuthbert 2017).

## References

(A)sexual (2011) Documentary. Directed by A. Tucker.

AVEN (n.d.) Overview. Available at: https://www.asexuality.org/?q=overview.html [accessed 04/09/18].

Bell P. (2018) I'm asexual, and this is how I show love to my partner. *i News*. Available at: https://inews.co.uk/inews-lifestyle/women/im-asexual-and-this-is-how-i-show-love-to-my-partner/ [accessed 06/09/18].

Bennett J. (2014) Asexuality: A curious parallel. *Rxisk*. Available at: https://rxisk.org/asexuality-a-curious-parallel/ [accessed 19/09/18].

Bogaert A.F. (2004) Asexuality: Prevalence and associated factors in a national probability sample *Journal of Sex Research* 41(3), pp. 279–287.

Campbell M. (2017) Disabilities and sexual expression: A review of the literature. *Sociology Compass* 11(9), pp. 1–19.

Carrigan M. (2011) Theres more to life than sex? Difference and commonality within the asexual community. *Sexualities* 14(4), pp. 462–478.

Carrigan M., Gupta K. and Morrison T.G. (2013) Asexuality special theme issue editorial. *Psychology and Sexuality* 4(2), pp. 111–120.

Carvalho J., Lemos D. and Nobre P.J. (2017) Psychological features and sexual beliefs characterizing self-labeled asexual individuals. *Journal of Sex and Marital Therapy* 43(6), pp. 517–528.

Cerankowski K.J. and Milks M. (2014) *Asexualities: Feminist and queer perspectives*. London: Routledge.

Cuthbert K. (2017) You have to be normal to be abnormal: An empirically grounded exploration of the intersection of asexuality and disability. *Sociology* 51(2), pp. 241–257.

Dawson M., McDonnell L. and Scott S. (2016) Negotiating the boundaries of intimacy: The personal lives of asexual people. *The Sociological Review* 64, pp. 349–365.

Dazed (2018) Pamela Anderson answers your DMs about sex and love. Available at: http://www.dazeddigital.com/life-culture/article/40759/1/pamela-anderson-sex-relationships-reader-questions-dms-interview [accessed 20/09/18].

Decker J.S. (2014) *The invisible orientation*. New York: Carrel Books.

East L.J. and Orchard T.R. (2014) Somebody Else's Job: experiences of sex education among health professionals, parents and adolescents with physical disabilities in Southwestern Ontario. *Sexuality and Disability* 32, pp. 335–350.

Enhance the UK (n.d.) Sexuality & disability. Available at: http://www.enhancetheuk.org/new-sex-disability-page/ [accessed 04/09/18].

Esmail S., Darry K., Walter A. and Knupp H. (2010) Attitudes and perceptions towards disability and sexuality. *Disability and Rehabilitation* 32(14), pp. 1148–1155.

Feely M. (2016) Sexual surveillance and control in a community-based intellectual disability service. *Sexualities* 19(5–6), pp. 725–750.

Gressgård R. (2013) Asexuality: From pathology to identity and beyond. *Psychology and Sexuality* 4(2), pp. 179–192.

Gupta K. (2014) Asexuality and disability: Mutual negation in *Adams v. Rice* and new directions for coalition building. In K.J. Cerankowski and M. Milks (eds.), *Asexualities: Feminist and queer perspectives*. London: Routledge.

Gupta K. (2015) Compulsory sexuality: Evaluating an emerging concept. *Signs: Journal of Women in Culture and Society* 41(1), pp. 131–154.

Hammers C. (2015) The queer logics of sex/desire and the missing discourse of gender. *Sexualities* 18(7), pp. 838–858.

Hanson E.H. (2014) Toward an asexual narrative structure. In K.J. Cerankowski and M. Milks (eds.), *Asexualities: Feminist and queer perspectives*. London: Routledge.

Höglund J., Jern P., Sandnabba N.K. and Santtila P. (2014) Finnish women and men who self-report no sexual attraction in the past 12 months: Prevalence, relationship status, and sexual behavior history. *Archives of Sexual Behavior* 43(5), pp. 879–889.

Holstein J.A. and Gubrium J.F. (2011) The constructionist analytics of interpretive practice. In N.K. Denzin and Y.S. Lincoln (eds.), *The SAGE handbook of qualitative research*. 4th edition. Thousand Oaks, CA: SAGE.

Jackson S. and Scott S. (2010) *Theorizing sexuality*. Maidenhead: Open University Press.

Kim E. (2010) How much sex is healthy? The pleasures of asexuality. In J.M. Metzl and A. Kirkland (eds.), *Against health: How health became the new morality*. New York: New York University Press.

Kim E. (2011) Asexuality in disability narratives. *Sexualities* 14(4), pp. 479–493.

Kim E (2014) Asexualities and disabilities in constructing sexual normalcy. In K.J. Cerankowski and M. Milks (eds.), *Asexualities: Feminist and queer perspectives*. London: Routledge.

Kafer A. (2003) Compulsory bodies: Reflections on heterosexuality and able-bodiedness. *Journal of Women's History* 15(3), pp. 77–89.

Kinsey A.C., Pomeroy WB and Martin C.E. (1948) *Sexual behaviour in the human male*. London: W.B. Saunders.

Liddiard K. (2018) *The intimate lives of disabled people*. London: Routledge

Masters W.H., Johnson V.E. and Kolodny R.C. (1988) *Masters and Johnson on sex and human loving*. Boston: Little, Brown and Company.

Milligan M.S. and Neufeldt A.H. (2001) The myth of asexuality: A survey of social and empirical evidence. *Sexuality and Disability* 19(2), pp. 92–109.

Mollow A. (2012) Is sex disability? Queer theory and the disability drive. In R. McRuer and A. Mollow (eds.), *Sex and disability*. Durham: Duke University Press.

NWSA (2015) NWSA 36th annual conference meeting program. Available at: https://www.nwsa.org/Files/Program%20PDFs/%202015ProgramFINAL.pdf [accessed 04/09/18].

NWSA (2016) NWSA 37th annual conference meeting program. Available at: https://www.nwsa.org/Files/Program%20PDFs/2016-NWSA-Conference-Program-Web.pdf [accessed 04/09/18].

O'Toole C.J. (2000) The view from below: Developing a knowledge base about an unknown population. *Sexuality and Disability* 18(3), pp. 207–224.

O'Toole C.J. and Bregante J.L. (1992) Disabled women: The myth of the asexual female. In S.S. Klein (ed.), *Sex, equity and sexuality in education*. Albany, NY: SUNY Press.

Parchomiuk M. (2018) Sexuality of persons with autism spectrum disorder (ASD). *Sexuality and Disability* 37(2), pp. 259–274.

Peta C., McKenzie J., Kathard H. and Africa A. (2017) "We are not asexual beings": Disabled women in Zimbabwe talk about their active sexuality. *Sexuality Research and Social Policy* 14(4), pp. 410–424.

Pfeiffer, D. (1994) Eugenics and disability discrimination. *Disability & Society* 9, pp. 481–499.

Prause N. and Graham C.A. (2007) Asexuality: Classification and characterization. *Archives of Sexual Behavior* 36(3), pp. 341–356.

Przybylo E. (2013) Producing facts: Empirical asexuality and the scientific study of sex. *Feminism & Psychology* 23(2), pp. 224–242.

Przybylo E. and Cooper D. (2014) Asexual resonances: Tracing a queerly asexual archive. *GLQ: A Journal of Lesbian and Gay Studies* 20(3), pp. 297–318.

Renninger B.J. (2015) "Where I can be myself … where I can speak my mind": Networked counterpublics in a polymedia environment. *New Media & Society* 17(9), pp. 1513–1529.

Rothwell J. (2015) There's nothing wrong me, I'm just asexual. *Stuff*. Available at: https://www.stuff.co.nz/life-style/love-sex/74528610/theres-nothing-wrong-with-me-im-just-asexual [accessed 06/09/18].

Scherrer K. (2008) Coming to an asexual identity: Negotiating identity, negotiating desire. *Sexualities* 11(5), pp. 621–641.

Shakespeare T. (2006) *Disability rights and wrongs*. Oxon: Routledge.

Shakespeare T., Gillepsie-Sells K. and Davies D. (1996) *The sexual politics of disability*. London: Cassell.

Shakespeare T. and Richardson S. (2018) The sexual politics of disability, twenty years on. *Scandinavian Journal of Disability Research* 20(1), pp. 82–91.

Sheehan R. (2015) *A-identity politics: Asexual exceptionalism, precarity and activism*. Masters thesis, George Mason University. Available at: http://mars.gmu.edu/xmlui/bitstream/handle/1920/9758/Sheehan_thesis_2015.pdf [accessed 20/09/18].

Shildrick M. (2007) Dangerous discourses: Anxiety, desire, and disability. *Studies in Gender and Sexuality* 8(3), pp. 221–244.

Shuttleworth R. (2012) Bridging theory and experience: A critical-interpretive ethnography of sexuality and disability. In R. McRuer and A. Mollow (eds.), *Sex and disability*. Durham: Duke University Press.

Siebers T. (2012) *Disability theory*. Ann Arbor: University of Michigan Press.

Steelman S.M. and Hertlein K.M. (2016) Underexplored identities: Attending to asexuality in therapeutic contexts. *Journal of Family Psychotherapy* 27(2), pp. 85–98.

Storms M.D. (1980) Theories of sexual orientation. *Journal of Personality and Social Psychology* 38(5), pp. 783–792.

Thomas C. (1999) *Female forms: Experiencing and understanding disability*. Buckingham: Open University Press.

Titman N. (2017) Asexuality BC (before cake). *Graphic Explorations Blog*. Available at: https://graphicexplanations.info/2017/10/29/asexuality-bc-before-cake/ [accessed 04/09/18].

Tran H. (2018) SFU becomes the first university to offer asexuality studies. *The Peak*. Available at: https://the-peak.ca/2018/02/sfu-becomes-first-university-to-offer-asexuality-studies/ [accessed 04/09/18].

Yule M.A., Brotto L.A. and Gorzalka B.B. (2014) Biological markers of asexuality: Handedness, birth order, and finger length ratios in self-identified asexual men and women. *Archives of Sexual Behavior* 43(2), pp. 299–310.

# 26

# THROUGH A PERSONAL LENS

## A participatory action research project challenging myths of physical disability and sexuality in South Africa

*Poul Rohleder, Xanthe Hunt, Mussa Chiwaula, Leslie Swartz,*
*Stine Hellum Braathen and Mark T. Carew*

### Introduction

Many societies have become significantly more open about sex and sexuality. Sexuality is everywhere and pornography sites are some of the most visited websites on the Internet. Yet, as many readers of this handbook, and as authors in other chapters have articulated, sex and sexuality are not as equally accessible to everyone as they might be. Persons with disabilities have historically experienced sexual oppression and exclusion (Shakespeare 2000; 2014), and whereas things may be changing in some parts of the world, with increased recognition of the sexual rights of people with disabilities, this is not enjoyed by all people with disabilities in all countries (Rohleder et al. 2018).

It is important in any exploration of "sexuality" to define what is meant by the term "beyond the act of sex or sexual orientation." We draw on the understanding of sexuality as articulated by the World Health Organisation (WHO 2006), which argues that sexuality encompasses not just sex and sexual orientation, but also gender roles, intimacy, reproduction, pleasure and eroticism. Sexuality is fluid and can be expressed in many ways, and encompasses the sexual fantasies, desires and attitudes that we may have, our sexual practices and behaviours and the intimacy and relationships that we have with others. These aspects of sexuality do not just depend on bodily experiences, but also social and cultural influences.

In this chapter, we focus on sexuality and disabilities with reference to a participatory research project that the authors were involved in, which aimed to explore and challenge common myths surrounding the sexuality of people with physical disabilities in South Africa. We briefly describe the project, the methods and the process we embarked on, and the different forms of data that were collected in a participatory, collaborative way. We then discuss key findings that indicate a relatively negative societal attitude towards the sexuality and sexual and reproductive health needs and rights of South Africans with disabilities, and how people with physical disabilities themselves have experienced this exclusion. Finally, we discuss how employing several innovative research methods enabled us to creatively apprehend disabled people's sense of their sexual lives in this local context and offered opportunities to challenge hegemonic attitudes and assumptions.

Why South Africa? Partly, this is because, for the majority of us (the authors), South Africa is where we are from, and where we have historically worked. However, conducting research in this context is important because the large majority of people with disabilities across the world live in low- and middle-income countries (WHO and World Bank 2011). In the most recently published South African census (Statistics South Africa 2014), it was reported that 7.5% of South Africa's population (an estimated 2.87 million people) were reported to have a disability. Yet, recent reviews of the literature on the sexual and reproductive health rights of girls and young women with disabilities (Braathen et al.2017) and sexual and reproductive health issues for people with disabilities in low- and middle-income countries (Carew et al. 2017), indicates a paucity of published research from South Africa, as compared to research from North America and western Europe. Sexuality research in South Africa has historically also been very much concerned with HIV and the response to the epidemic. Thus, we know less about the sexual lives of people with disabilities from the regions where they are most likely to reside.

## Disability and access to sexuality and sexual health in South Africa

South Africa has the biggest HIV epidemic in the world. In the context of this serious public health crisis, the sexual health of people with disabilities, and their risk of HIV infection was highlighted as a significant issue for concern (Groce et al. 2013). Since 2008, the South African national household survey of HIV incidence and prevalence, which is conducted every few years, has included statistical data for people with disabilities. In the most recent iteration conducted in 2012 (Shisana et al. 2014), people with disabilities are reported as a key HIV-risk population group, with an estimated HIV prevalence rate of 16.7%, which is higher than the national average (an estimated 12.2% national prevalence).

These HIV statistics highlight a number of "risk factors" that are well-documented to be associated with sexual oppression for people with disabilities: high incidence of sexual abuse and exploitation; lack of access to sexual health education, care and treatment; and misinformed, even hostile societal attitudes about the sexuality of people with disabilities in many parts of the world (Braathen et al. 2017; Rohleder et al.2018). In South Africa, research has shown that people with disabilities experience high rates of sexual abuse and rape (Maart and Jelsma 2010; Mall and Swartz 2012). Survey studies of HIV and sexual health knowledge and behaviour among samples of people with disabilities in South Africa indicate low levels of knowledge, access to education, information and treatment and high levels of unsafe sex, including sex without a condom (Eide et al. 2011; Rohleder et al. 2010; 2012). These surveys also indicate that women with disabilities are most disadvantaged. Some studies report that, for young people with disabilities, particularly young women, the need for the young person with disability to be accepted and loved made negotiating safe sex difficult (Rohleder 2010; Wazakili et al. 2006). Our research project did not focus specifically on HIV, but these statistics emphasise the importance of increasing the knowledge and awareness of the sexuality and sexual health needs of people with disabilities, which have, for the most part, been ignored.

South Africa is a multi-cultural society, where disability may be understood and responded to in various ways. As with other traditional African cultures (Groce 1999), traditional African cultures in South Africa may understand disability as divine punishment from ancestors or spirits, or a curse from God (Hanass-Hancock 2009). However, with a dominant influence of colonial perspectives from the Global North (Chappell 2019), disability is also regularly understood as attributed to medical causes or viewed as a physical tragedy. As is the case in many parts of the world (Rohleder et al. 2018), reasons why people with disabilities in South Africa have experienced sexual oppression and barriers to expressing their sexuality and access to sexual and reproductive health include the societal attitudes, myths and misconceptions held about them: where they may be assumed to be asexual (regardless of whether they identify as asexual or not); that they may have diminished sexual desires, needs and capabilities; or even that they are to be feared, as uninhibited and sexually dangerous, with their sexuality needing to be controlled (Chappell 2016; Hannass-Hancock 2009; McKenzie 2013; Rohleder and Swartz 2009). South African young people with disabilities also report growing up receiving messages of disability stigma and being perceived as undesirable partners (Chappell 2014; McKenzie and Swartz 2011). As a result, people with disabilities in South Africa may be excluded from living fully sexual lives and accessing sexual and reproductive health care.

Much of the knowledge about negative societal attitudes towards disability and sexuality in South Africa is anecdotal or from small qualitative studies. In order to address this research gap, we wanted to explore what societal attitudes prevail in the South African context and how people with disabilities experience their sexuality.

## Disability and sexuality: Stories from South Africa

Given the background discussed briefly above, and our involvement in research in disability and sexual health, we embarked on a research project to explore and challenge the societal myths of the sexuality of people with physical disabilities in South Africa. The focus was on physical disabilities for two reasons: Firstly, the funding available limited the scope of what we could do, and it was felt that we could be more effective in focusing on a particular population, rather than trying to be overly inclusive. Secondly, we adopted a method of data collection that involved visual representation (discussed further below), with the aim of using this personal, visual representation as counter-representations to societal misrepresentations (Garland-Thomson 2009). We thus thought that focusing on visual, physical disabilities would facilitate these ambitions, within the limited resources available to us. The project was supported by funding from the International Foundation of Applied Disability Research and the National Research Foundation of South Africa (supporting a PhD scholarship student on this project).

## About us

The core research team, namely the authors of this chapter, comprised colleagues who often work together, as well as new colleagues. We are an inter-disciplinary team, with backgrounds in psychology, anthropology, disability studies, journalism and international development. Five were academics and researchers, and one a non-academic, a chief executive officer of a regional disabled people's organisation, but with experience of involvement in research. There were senior academics, mid-career and early-career academics, and a PhD student (since qualified). The team comprised of four men and two women, with some who identify as disabled and others as nondisabled, having different ethnic, cultural and religious backgrounds, and different sexualities. As the core research team, we were the primary developers of the project, conceptualising and managing its core activities, and leading on data collection, analysis and dissemination. Whereas the core research team involved the participation of a disability organisation, a core principle and feature of our project was the active participation of the research participants themselves (people with physical disabilities), who took part in some of the decision making, set the agenda for interviews, and were involved in designing and producing project outputs for a public audience. We did not want to do research for the sake of research, but rather to embark on a project that would lead to social change, and that included the participation of what are usually regarded as the research "subjects." This is in line with the principles of a community, participatory psychology approach (Freire 1972; Kagan et al.2017)

The research participants with whom we worked included seven men and six women with physical disabilities, recruited through known networks and through disability organisations in South Africa. All identified as having a physical disability, with eleven making use of a wheelchair. Three had congenital disabilities, two had acquired their disability in childhood and the remaining eight had acquired their disability as young adults. Participants were from different ethnic and social backgrounds, with differing levels of education (although most had high levels of education). Although we were hoping for a diversity of sexual orientations, all participants who volunteered identified as heterosexual, the result of this being that the project and its findings have a heteronormative bias. There is a very noticeable lack of research looking at the intersection of lesbian, gay, bisexual and trans identities with disability, with what studies there are coming mostly from the USA and the UK (Chappell 2019; Duke 2011;).

## About the project

The project comprised two complementary studies, and a public dissemination plan. We provide a brief overview below. For a more detailed description of what we did, how we did it, and our reflections on the process, see Rohleder et al. (in press).

### *Study 1: A survey of societal attitudes*

The core research team, together with the thirteen participants, worked together in a research workshop to conceptualise and design a survey questionnaire to investigate attitudes and understanding of the general South African population. While there was a need to ensure research reliability and validity by including previously existing measures on certain topics (such as attractiveness or dating), we discussed and agreed on the areas on which to ask questions, the different ways of asking questions (open-ended or scale measures), and ways of reaching potential respondents to the survey. As a collective, it was felt that, whereas we primarily wanted to investigate the attitudes of non-disabled people towards the sexuality of people with disabilities, it was also important to include people with disabilities as respondents to the survey, to explore what attitudes might exist amongst people with disabilities. It is beyond the scope of this chapter to discuss the merits and limitations of "measuring" disability in a survey questionnaire, as this depends on the model for understanding disability. As a group, we agreed to include the questions developed by the Washington Group on Disability Statistics (http://www.washingtongroup-disability.com/), recognising that these questions were developed by an international task group that included the participation of disabled peoples' organisations, and are currently considered the best method for generating reliable disability data. In addition to the questions on disability, the survey questionnaire included rating-scale items and open-ended questions related to thoughts about Attractiveness; Dating; Sex; Sexual Orientation; and the Sexual and Reproductive Health Rights and Needs (as compared with non-disabled people). The questionnaire was designed as both an online and a paper-and-pen survey and was trialled by the participants themselves prior to finalising the survey and opening it up for general responses.

The survey ran from April to August 2016 and was available in four of the most commonly spoken languages in South Africa: Afrikaans, English, isiXhosa, and isiZulu. The online survey was advertised through social media, the online webpages of two prominent newspapers, the student email list of two large South African universities, and through our own colleagues and social networks. To ensure that the survey also reached people without internet access, or who were from more deprived socio-economic backgrounds, pen-and-paper survey responses were collected by three field data collectors in a large low-socio-economic area of one of the main cities in South Africa.

A total of 1990 people responded to the survey (see Hunt et al. 2017). Of these, 1865 reported being non-disabled. There were slightly more female respondents (57%) than male respondents (43%). The majority of respondents (75%) identified as heterosexual. Ages ranged from 18 to 59 with a mean age of 24.4 years. Half of the respondents had completed high school (51%), with 23% having at least one degree qualification. In South Africa, during the years of Apartheid, the population was classified and divided into racial groups: "black" (African), "white," "coloured" (mixed-race), and "Indian." These categories are social constructions and contested but are nevertheless frequently used in contemporary South Africa as a broad indication of representation. Among the survey respondents, 46% identified as "black," 40% as "white," 8% as "coloured," 4% as "Indian" and 2% as "other." The sample of respondents tended to be less representative of the general South African population, with regards to race and education.

## Study 2: Photovoice qualitative study

The project also collected the personal stories and experiences of the thirteen people with physical disabilities with whom we worked. The design of this study component adopted the use of photovoice, an arts-based participatory research method which has been used to collect narratives of personal subjective experiences, often for sensitive topics, or with marginalised voices (Catalani and Minkler 2010; Wang and Burris 1997). In this method, participants are invited to take photographs that represent or are symbolic of the topic or experience being explored. Participants are then invited to bring a self-chosen selection of their photographs to an individual interview, where the photographs are then used as the prompts from which to elicit personal narratives. In this way, the interview agenda, albeit focused on a specific topic, is set by the participant rather than the interviewer (Vaughan 2015).

As mentioned briefly earlier, we felt that the use of photovoice was an important method to use for an issue where representation matters. People with disabilities are frequently subject to the gaze, lens and representation of non-disabled others, who construct them as "other" (Garland-Thomson 2009). By inviting participants to create their own self-representations, it not only facilitates the exploration of their personal experience but also provides a personal lens for which to challenge dominant representations.

Photovoice usually involves the production of photographs, but we also invited participants to choose to create other forms of representation, or indeed to not use this method at all. Of the thirteen participants who were interviewed, nine took and shared photographs; one made drawings; and three participants chose to just talk. One participant also shared poems and text alongside her photographs. The photographs and parts of the accompanying narratives were then used to be shown on the project webpage (www.disabilityandsexualityproject.com) and created as posters which were exhibited at a stakeholders' conference at the end of the project. The photographs, poems and drawings, and personal narratives are also being used for the creation of a book, co-written by academics of the research team and some of the participants. The book is written for both academic and non-academic audiences, featuring mostly personal narratives, with some theoretical material to accompany it. It is currently being completed, for publication as an open-access book in 2021.

In addition to the photovoice interviews, four of the participants (two men and two women) also consented to be interviewed on film, and these interviews were used to produce a short documentary film. This was co-written and produced with the four participants, and a fifth participant, who narrated the film. The film is available to view on the project website (a subtitled version is also available on the website, for viewers who are hard of hearing).

We have published findings from this project in a number of academic journal articles. We provide a summary of the key findings below. We include only one photograph in this chapter. Readers can view other photographs on the website, and in other publications (Hunt et al. 2018a; Hunt et al. 2019, and the forthcoming book).

## Stories from society

Mostly anecdotal evidence and some empirical evidence has suggested that people with disabilities are assumed to be asexual or having diminished sexual interests, needs and capacity (e.g., Milligan and Neufeldt 2001). The survey we conducted is the first attempt we know of to collect empirical evidence from South Africa as to society-wide attitudes and understanding of the sexuality and, in particular, the perceived sexual and reproductive health rights of people with disabilities.

We found some clear differences in the assumptions and views held towards people with disabilities when compared to those held towards non-disabled people. The survey contained questions about respondents' beliefs about the sexuality and sexual rights of non-disabled people and people with physical disabilities. The responses to these questions indicate that, on average, people with disabilities are viewed as having less sexual rights and less sexual and reproductive health care needs than non-disabled people (Hunt et al. 2017).

Whereas, on the whole, there were some positive feelings expressed as to the sexuality of people with disabilities, we found that more negative feelings were expressed by non-disabled respondents when the question was about more intimate contact (going on a date or being in a relationship) with a person with a physical disability, than being a friend or having a person with disability marry in to the family. Respondents to a dating beliefs scale on the survey indicated some feelings of ambivalence, uncertainty and discomfort. Moreover, non-disabled respondents who expressed higher levels of prejudice towards dating people with physical disabilities were less likely to view them as having sexual rights or sexual healthcare needs (Carew et al. in press).

Respondents' views were further explored in a story completion task, where respondents were provided with a vignette where a man or woman with disabilities (gender matched to the respondent's gender and sexual orientation) asks a non-disabled man/woman they just met at a party for a future date. The respondent is then invited to complete the scenario in terms of what the non-disabled person might feel and say. While story completion tasks such as this do not ask respondents to report on what they have previously done, only to imagine what a character might do or feel, it nevertheless allows access to the respondents' way of thinking about the topic (Kitzinger and Powell 1995), and thus is an indication of the respondent's potential views and attitudes. There were some positive views expressed about dating a person with disabilities, with many suggesting they would say yes to a date. However, our analysis of the open responses suggested that many non-disabled people tend to desexualise persons with disabilities. Some suggest that going on a date may be out of pity, while others express some anxiety, even fear when thinking about dating or having a potential relationship with someone with a disability (Hunt et al. 2018b). As an example, one respondent replied that the non-disabled character "goes for one date out of sympathy for him but unfortunately does not return his calls afterwards," indicating that the motivation to go on the date was out of pity, but the intention not to engage further seems a pre-existing thought; it would only be one date. Another respondent wrote: "this might be a deal-breaker in terms of disabilities, as in if she is paralysed from the waist down and does want someone sexually active then he wouldn't want to date her, but at the least they would be friends," indicating that the perceived inability to have sex is a barrier to having a more intimate relationship (see these and other quotes in Hunt et al. 2018b).

The survey questionnaire, also included a free association item, asking respondents to state three traits that come to mind when thinking of women who have physical disabilities, and three which come to mind when thinking of men who have physical disabilities. This question has been used by Nario-Redmond (2010) in previous research on attitudes towards women and men with disabilities in the USA. The terms and phrases used in responding to this question may reflect socially constructed stereotypes (Dixon 2017) held by the respondents. Our analysis of responses (Hunt et al., 2018c) indicate that, similar to the findings of Nario-Redmond (2010), prominent overall stereotypes for women and men with disabilities were the same: perceived as withdrawn and dependent, stereotypes of the "SuperCrip," being a "nice guy/girl," perceived as angry/irritable and aloof, and lacking sexuality or being sexually undesirable. This suggests a diminished and desexualised perception of people with disabilities. Although the stereotype of "SuperCrip" may be understood as positive, it has been argued as epitomising the imperative to be able-bodied and "un-disabled" (McRuer 2010; Watermeyer 2013).

In summary, while many of the responses to the survey reflected some positive attitudes, understanding and views about the sexuality of people with disabilities, there were notable ambivalences and anxieties, with an overall tendency to view people with disabilities as less sexual, less desirable, and as having less sexual and sexual health needs. It needs to be noted, however, that the survey respondent sample was not entirely representative of the South African population, representing some bias towards higher levels of education, and thus influenced by a westernised educational system. With this social context in mind, we then turn to the experiences of persons with physical disabilities themselves.

## Stories from people with disabilities

At the time that the thirteen individual participants were interviewed, four were in a relationship or married, three were separated or divorced, and six were single. They ranged in age from 25 to 60.

All participants spoke of how they were often regarded as asexual, or had their sexuality questioned by non-disabled family and peers (Rohleder et al. 2018). Those who had a congenital disability or acquired their disability as a child spoke of growing up with messages from family and friends that sexuality was something that was not meant for them. For example, one of the women spoke about how "for the family it's like, I'll never get married [...] for them it was just a friend. No idea about the wishes or dreams of this young woman to one day get married, because it's impossible." Similarly, another woman said: "everybody just took it for granted, because I couldn't have children, I wouldn't marry." Similar experiences are reported in other South African studies, where gendered construction about womanhood in a patriarchal traditional culture centres on being a wife, having children and being a mother (Hanass-Hancock 2009; McKenzie 2013). As women with disabilities are assumed to be unable to have sex and thus unable to have children, they may be viewed as not being "suitable" for marriage.

Other participants spoke of the general perception by others that, because of their disability, they are unable to have sex and are thus asexual. Frequent comments made across interviews allude to this. For example:

> this is what people do to you when you are disabled—they mean well, but they don't approach you.
>
> *(Nedah)*[1]

people don't expect me to be in a relationship or be able to do anything because I'm in a wheelchair.

*(John)*

Most of the ladies like to tease me and they say, come, I want to marry you. And then they say, if I marry you, how are you going to satisfy me sex-wise?

*(Edward)*

As the above example quotes suggest, both the men and women interviewed were often perceived as being unable to have sex, and thus unable to sexually satisfy a potential partner, making them undesirable partners. Edward, for example, reports on how he gets teased and mocked. Nedah and John suggest a more active avoidance on the part of others to consider them as potential partners.

Attitudes about disability and sexuality may also be mixed with cultural beliefs about disability, particularly for participants from Xhosa or Zulu cultures. The understanding that disability may be caused by a curse or divine punishment may result in people with disabilities being actively feared. In our study, a Xhosa woman refers to some of the negative cultural meanings that disability can evoke;

in our community, in the black community[2], they are not aware of disability. Like if your child has got a disability, it's got that myth as if you are cursed or you are bad luck, or something isn't right with your family.

*(Pride)*

A Xhosa man, speaks more in-depth about cultural beliefs of disability that constructed him as a feared "other" in his community:

In our society, when you are sitting in a wheelchair, people think that you are bewitched. Now if you are bewitched, no one wants to come close to you. No one wants to be your friend. No one wants to be your family. No one wants to be close to you at all because you are going to curse them. The curse that you are under is going to affect them. So, they don't want to come close to you. They don't want your help. Some believe that, if they come across you and you say hello and you carry on with the conversation a little bit, you are planting that curse onto them. So, if the situation is like that, then there is no way that you will have a friend in a society like that, not a girlfriend. A girlfriend is difficult to get while you are still walking, how if you are cursed?

*(Sipho)*

Participants reflected on how societal attitudes about bodies and attractiveness, impacted on their sense of self in terms of being disabled and feminine or masculine. For women, femininity was very much linked to ideals of being a wife, a lover and a mother, which, as discussed above, are held as hegemonic constructions of womanhood, but these identities were perceived as being foreclosed to them. They spoke about feeling that many men do not see them as sexual beings, because the men think that they will not be able to have children and be good wives due to their disability. Two of the women in our study did have children. One of the women struggled for many years to get pregnant, finally getting pregnant with the help of in vitro fertilisation (IVF). Female participants also spoke about their bodies, and the impact that this had on feeling attractive and feminine:

I also realised that, by the world's majority standards, that I wasn't beautiful and I didn't have a good body. I wasn't attractive, and, for any teenager, any young person, it's hard. I'm delighted that it gets easier as one grows older, and that there is so much else. Now it's easier, when I do see an inkling of something sensuous, I feel better equipped to respond. But in terms of how did it make me feel, well, I felt ugly. I felt rejected and that I missed out a lot. I felt angry, which is probably the overriding emotion that I felt.

*(Mary)*

For the men, many spoke about societal views on disability challenging notions of masculinity in relation to strength, independence and being a provider. This was particularly important for traditional African patriarchal cultures, where men are perceived as needing to take up these positions. Typical comments to this effect were, "is she going to be able to perceive this person as a complete male? Or, is it like a…I don't know…a diluted or minimised version of a man?" (Bubele).

Some participants with paralysis spoke about experiencing some difficulties having sex, experiencing a lack of feeling in their genitals, and, for men, difficulty achieving an erection (Hunt et al. 2018a). For men, the difficulty in achieving an erection, and being disabled, impacted heavily on their sense of masculinity. Four of the men interviewed specifically reported using medication, typically Viagra, but this was not always financially accessible. Their sense of (not) being able to sexually satisfy their female lover weighed heavily on them:

a physically disabled person is more likely to end up in a short-term, like I've mentioned, a short-term relationship due to their challenges and intimacy. Why I'm saying intimacy phase, an intimacy phase from my perspective, because like now, I've got to use equipment or I have to use some medication to give me that stamina, that feeling back, that energy back, to be intimate with a partner. I mean, it's part of life. Sometimes, for example, I can't afford to get that specific treatment, then I'm stranded, which makes my partner stranded, and that same strandedness will make the partner look for someone else to fulfil her wants and needs.

*(Fazil)*

Literature has indicated that constructions of masculinity are often associated with notions of "action" and "potency," and so disability can often be experienced by men as a kind of emasculation (Nolan 2013; Robertson & Smith 2014): "You know, at first I couldn't get an erection. The nurses would usually tease me about it. So, it felt like my manhood was taken away from me" (Tazz).

However, participants' experiences of sexuality and sex were not all negative or constructed as lacking. Some of the participants, particularly those who had acquired a disability, spoke about having found ways to enjoy sexual mutuality with their partners, and to feel erotic sensations and experience themselves as sexual again (Hunt et al. 2018a; Rohleder et al. 2018). In fact, acquiring a disability meant having to renegotiate a sense of themselves as sexual. One of the participants, Kate, had a series of sensual photographs taken of herself, as part of her journey towards reclaiming a positive sense of herself as a woman and a sexual being (Figure 26.1). She said:

I can't hide my flaws and imperfections like other people can. Mine are there for the world to see. In the process of rediscovering my sexuality, I have learned to use what I have to seduce and entice. The silent battles I have fought of self-acceptance and vali-

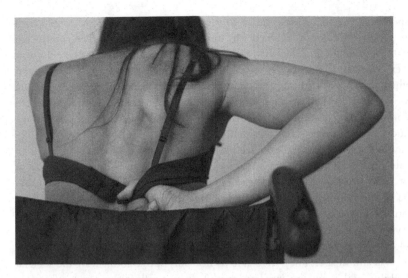

*Figure 26.1* Kate.

dation has left me with the realisation that I no longer have to hide the naked beautiful truth of who I am…a woman in every essence of the word.

(Kate)

This renegotiation of sexuality involved exploring new ways of experiencing eroticism, and mutuality that may not necessarily involve genital penetration. For example, Kate described trying new ways of exploring and enjoying sex:

Well, because I felt so very comfortable with [name], and right from the start I felt completely comfortable with him. I think that, number one, that is a big bonus in any relationship and for me that was a big plus to it. So, physically, he would learn. I mean, we kind of explored because it was new to me and it was new to him as well. So we just had to learn what worked and what didn't work sexually, you know, different positions, what he liked, what I liked, and we were open to… and not very much, but a little bit…to be able to explore.

(Kate)

Similarly, Timothy described finding new erotic spots on his body:

And also, to be sensitive to what you feel. I've had one or two girls that really understood me, that would say, 'Okay, Timothy, what do you enjoy?' And when your body starts adjusting to your disability and that type of thing, it changes, and your sensation changes and all those things. Like if you kiss me on my neck or on my ear, I go nuts!

(Timothy)

## Final reflections

Our findings from the survey data and the interviews with people with physical disabilities indicate that, in South Africa, people with physical disabilities may often be perceived as asexual, and as undesirable potential partners. This is similar to the reported experiences of people with disabilities in other parts of the world (Rohleder et al.2018). Some of these perceptions are informed by cultural understanding of disability, masculinity and femininity, which are not too dissimilar from other contexts. Using data from multiple informant groups, perspectives and using multiple methodologies gives us a broad, comprehensive view of the myths and stigma related to disability and sexuality. The survey used rating scale measures, open-ended questions and story completion tasks, to capture in-depth different ways that attitudes may be expressed. These societal attitudes are reflected in the reported experiences of people with physical disabilities themselves, where participants reported feeling stigmatised and excluded.

Our project, while aiming to investigate dominant societal attitudes and an understanding of disability and society, also attempted to provide the opportunity for people with physical disabilities to challenge and "shoot back at" dominant representations of disabilities and make their own self representations. In the process of taking photographs and reflecting on their experiences, participants were able not only to describe their experience of exclusion and the myths and assumptions held by others, but also to challenge them and to represent themselves employing alternative narratives and images. They were able to not only speak about how the societal constructions of disability and asexuality (predominantly) do not reflect their own subjective experiences of themselves as sexual beings, but also to make visual (photo and video) representations of this so as to challenge these assumptions.

Disability and sexuality are typically researched in terms of investigating sexual functioning and sexual and reproductive health, with comparatively little attention paid to the lived experience of disability and sexuality. To capture this, creative methods allow for participants to set their own agenda about what is meaningful to them and what best captures that experience.

This is a small project, but the use of creative methods has allowed us to have a relatively far reach. Our website has been visited by people from over 50 different countries. Whereas our focus here is on South Africa, the issues highlighted, and the experiences of the people we worked closely with are by no means unique to this context. Indeed, as we have presented this work to different audiences, in different countries, the relevance for other contexts is made clear to us in discussions and feedback that the issues presented generate. Readers may have themselves heard similar stories before.

## Notes

1 Pseudonyms have been used.
2 In this quote, Pride refers to the "black community" in speaking to a white interviewer. It should be noted that there is no single "black community" as there are various African cultures within South Africa.

## References

Braathen, S.H., Rohleder, P., & Azalde, G. (2017) *Sexual and reproductive health and rights of girls with disabilities*. A report written for the United Nations Office of the Special Rapporteur on the Rights of People with Disabilities. Available from: https://www.sintef.no/en/projects/literature-review-sexual-and-reporductive-health-a/

Carew, M., Braathen, S., Swartz, L., Hunt, X., & Rohleder, P. (2017) The sexual lives of people with disabilities within low- and middle-income countries: A scoping study of studies published in English. *Global Health Action* 10(1), p. 1337342.

Carew, M.T., Braathen, S.H., Hunt, X., Swartz,L., & Rohleder, P. (in press) Predictors of negative beliefs toward the sexual rights and perceived sexual healthcare needs of people with physical disabilities in South Africa. *Disability & Rehabilitation*.

Catalani, C., & Minkler, M. (2010) Photovoice: A review of the literature in health and public health. *Health Education & Behavior* 37(3), pp. 424–451.

Chappell, P. (2014) How Zulu-speaking youth with physical and visual disabilities understand love and relationships in constructing their sexual identities. *Culture, Health & Sexuality* 16(9), pp. 1156–1168.

Chappell, P. (2016) Secret languages of sex: Disabled youths experiences of sexual and HIV communication with their parents/caregivers in KwaZulu-Natal, South Africa. *Sex Education* 16(4), pp. 405–417.

Chappell, P. (2019) Situating disabled sexual voices in the global south. In P. Chappell & M. De Beer (Eds.), *Diverse voices of disabled sexualities in the global South*. London: Palgrave Macmillan, pp. 1–25.

Dixon, J. (2017) Thinking ill of others without sufficient warrant? Transcending the accuracy–inaccuracy dualism in prejudice and stereotyping research. *British Journal of Social Psychology* 56(1), pp. 4–27.

Duke, T.S. (2011) Lesbian, gay, bisexual, and transgender youth with disabilities: A meta-synthesis. *Journal of LGBT Youth* 8(1), pp. 1–52.

Eide, A.H., Schür, C., Ranchod, C., Rohleder, P., Swartz, L., & Schneider, M. (2011) 'Disabled persons' knowledge of HIV prevention and access to health care prevention services in South Africa. *AIDS Care* 23(12), pp. 1595–1601.

Freire, P. (1972) *Pedagogy of the oppressed*. Harmondsworth: Penguin.

Garland-Thomson, R. (2009) *Staring: How we look*. Oxford: Oxford University Press.

Groce, N.E. (1999) Disability in cross-cultural perspective: Rethinking disability. *The Lancet* 354(9180), pp. 756–757.

Groce, N.E., Rohleder, P., Eide, A.H., MacLachlan, M., Mall, S., & Swartz, L. (2013) HIV issues and people with disabilities: A review and agenda for research. *Social Science & Medicine* 77, pp. 31–40.

Hanass-Hancock, J. (2009) Interweaving conceptualizations of gender and disability in the context of vulnerability to HIV/AIDS in KwaZulu-Natal, South Africa. *Sexuality & Disability* 27, pp. 35–47.

Hunt, X., Braathen, S., Swartz, L., Carew, M., & Rohleder, P. (2018a) Intimacy, intercourse and adjustments: Experiences of sexual life of a group of people with physical disabilities in South Africa. *Journal of Health Psychology* 23(2), pp. 289–305.

Hunt, X., Carew, M., Braathen, S.H., Swartz, L., Chiwaula, M., & Rohleder, P. (2017) The sexual and reproductive rights and benefit derived from sexual and reproductive health services of people with physical disabilities in South Africa: Beliefs of non-disabled people, *Reproductive Health Matters* 25(50), pp. 66–79.

Hunt, X., Swartz, L., Braathen, S.H., Carew, M., Chiwaula, M., & Rohleder, P. (2019) Shooting back and (re)framing: Challenging dominant representations of the disabled body in the global South. In P. Chappell & M. De Beer (Eds.), *Diverse voices of disabled sexualities in the global South*. London: Palgrave Macmillan, pp. 33–53.

Hunt, X., Swartz, L., Carew, M., Braathen, S., Chiwaula, M., & Rohleder, P. (2018b) Dating persons with physical disabilities: The perceptions of South Africans without disabilities. *Culture, Health & Sexuality* 20(2), pp. 141–155.

Hunt, X., Swartz, L., Rohleder, P., Carew, M., & Braathen, S. (2018c) Withdrawn, strong, kind, but de-gendered: Non-disabled South Africans stereotypes concerning persons with physical disabilities. *Disability & Society* 33(10), pp. 1579–1600.

Kagan, C., Burton, M., & Siddiquee, A. (2017) Action research. In C. Willig & W. Stainton-Rogers (Eds.), *Handbook of qualitative research methods in psychology*. 2nd edition. London: SAGE, pp. 55–73.

Kitzinger, C. and Powell, D. (1995) Engendering infidelity: Essentialist and social constructionist readings of a story completion task. *Feminism & Psychology* 5(3), pp. 345–372.

Maart, S., & Jelsma, J. (2010) The sexual behaviour of physically disabled adolescents. *Disability & Rehabilitation* 32(6), pp. 438–443.

Mall, S., & Swartz, L. (2012) Parents anxieties about the risk of HIV/Aids for their deaf and hard of hearing adolescents in South Africa: A qualitative study. *Journal of Health Psychology* 17(5), pp. 764–773.

McKenzie, J.A. (2013) Disabled people in rural South Africa talk about sexuality. *Culture, Health & Sexuality* 15(3), pp. 372–386.

McKenzie, J.A., & Swartz, L. (2011) The shaping of sexuality in children with disabilities: A Q methodological study. *Sexuality & Disability* 29(4), pp. 363–376.

McRuer, R. (2010) Compulsory able-bodiedness and queer/disabled existence. In L. Davis (Ed.), *The disability studies reader*. London: Routledge, pp. 383–392.

Milligan, M.S., & Neufeldt, A.H. (2001) The myth of asexuality: A survey of social and empirical evidence. *Sexuality & Disability* 19(2), pp. 91–109.

Nario-Redmond, M.R. (2010) Cultural stereotypes of disabled and non-disabled men and women: Consensus for global category representations and diagnostic domains. *British Journal of Social Psychology* 49(3), pp. 471–488.

Nolan, M. (2013). Masculinity lost: A systematic review of qualitative research on men with spinal cord injury. *Spinal Cord* 51, pp. 588–595.

Rohleder, P. (2010) "They dont know how to defend themselves": Talk about disability and HIV risk in South Africa. *Disability & Rehabilitation* 32(10), pp. 855–863.

Robertson, S., & Smith, B. (2014). Men, masculinities and disability. In J. Swain, S. French, C. Barnes, & C. Thomas (Eds.), *Disabling barriers – Enabling environments* (3rd Edition) (78–84). London: Sage Publications.

Rohleder, P., Braathen, S.H., & Carew, M. (2018) *Disability and sexual health: A Critical exploration of key issues.* London: Routledge

Rohleder, P., Braathen, S.H., Carew, M., Chiwaula, M., Hunt, X., & Swartz, L. (in press) Creative collaboration on a disability and sexuality participatory action research project: A reflective diary account. *Qualitative Research in Psychology*.

Rohleder, P., Braathen, S.H., Hunt, X., Carew, M., & Swartz, L. (2018) Sexuality erased, questioned, and explored: The experiences of South Africans with physical disabilities. *Psychology & Sexuality* 9(4), pp. 369–379.

Rohleder, P., Eide, A.H., Swartz, L., Ranchod, C., Schneider, M., & Schür, C. (2012) Gender differences in HIV knowledge and unsafe sexual behaviour among disabled people in South Africa. *Disability & Rehabilitation* 34(7), pp. 605–610.

Rohleder, P., & Swartz, L. (2009) Providing sex education to persons with learning disabilities in the era of HIV/AIDS: Tensions between discourses of human rights and restriction. *Journal of Health Psychology* 14(4), pp. 601–610.

Rohleder, P., Swartz, L., Schneider, M., Groce, N., & Eide, A.H. (2010) HIV/AIDS and disability organizations in South Africa. *AIDS Care* 22(2), pp. 221–227.

Shakespeare, T. (2000) Disabled sexuality: Toward rights and recognition. *Sexuality & Disability* 18(3), pp. 159–166.

Shakespeare, T. (2014) *Disability rights and wrongs revisited.* 2nd edn. Oxon: Routledge, Taylor & Francis Ltd.

Shisana, O., Rehle, T., Simbayi, L.C., Zuma, K., Jooste, S., Zungu, N., Labadarios, D., Onoya, D., et al. (2014) *South African national HIV prevalence, incidence and behaviour survey, 2012.* Cape Town: HSRC Press.

Statistics South Africa (2014) *Census 2011: Profile of persons with disabilities in South Africa.* Pretoria: Statistics South Africa. Available from: http://www.statssa.gov.za/publications/Report-03-01-59/Report-03-01-592011.pdf

Vaughan, C. (2015) Participatory research. In P. Rohleder & A. Lyons (Eds.), *Qualitative research in clinical and health psychology.* Basingstoke: Palgrave Macmillan, pp. 257–276.

Wang, C., & Burris, M. (1997) Photovoice: Concept, methodology, and use for participatory needs assessment. *Health Education & Behavior* 24, pp. 369–387.

Watermeyer, B. (2013) *Towards a contextual psychology of disablism.* London: Routledge.

Wazakili, M., Mpofu, R., & Devlieger, P. (2006). Experiences and perceptions of sexuality and HIV/AIDS among young people with physical disabilities in a South African township: A Case study. *Sexuality & Disability* 24, pp. 77–88.

World Health Organization (WHO) (2006) *Defining sexual health: Report of a technical consultation on sexual health, 28–31 January 2002, Geneva.* Geneva: World Health Organisation.

World Health Organisation & The World Bank (2011) *World report on disability.* Geneva: WHO and The World Bank.

# 27

# "THAT'S MY STORY"

## Transforming sexuality education by, for and with people with intellectual disabilities

*Amie O'Shea and Patsie Frawley*

### Introduction

I'm 40 now. A long time ago, I told them that I'm homosexual, I'm gay. The whole house knows I'm gay. I don't tell the staff at the day centre.

*("Johnno's story," SL&RR 2019, p. 163)*

Johnno's story is one in a growing collection of over 35 stories that have been co-developed in Australia over the past two decades (Frawley et al. 2019; Johnson et al. 2001). These stories have been at the centre of two programmes developed to promote the sexuality rights of people with intellectual disabilities. The first of these programmes, Living Safer Sexual Lives (Frawley et al. 2003), was the result of a research programme that used inclusive research approaches to gather the stories of 25 people, aged 19–57, with intellectual disabilities, and to use these stories in a training programme for people with an intellectual disability, for staff of disability services and families, using only these stories as the core learning materials. This research also informed relationships and sexuality policy in disability services and engaged with people with an intellectual disability to advocate for social opportunities to enable the development of relationships. Kelley Johnson, a leading Australian disability academic and researcher was the driver behind this work, the first of its kind in Australia to place people with an intellectual disability at the centre of research, programme development and advocacy about sexuality.

Building on this work, the Sexual Lives and Respectful Relationships (SL&RR) programme was developed as part of a focus on prevention of violence and abuse in Australia (Frawley et al. 2011). This programme used the same stories, co-developing a sexual rights programme that would be co-facilitated by people with an intellectual disability as peer educators in partnership with community professionals (Frawley and O'Shea, 2020; Frawley et al. 2012; Frawley and Bigby 2015). The use of people's stories at the centre of sexuality education with people with an intellectual disability has been recognised as a productive way of promoting the voices and experiences of people with an intellectual disability as authoritative and instructive about sexuality in their lives and in sexuality education (Barger et al. 2009). Underpinning these two bodies of work has been a commitment to hearing the voices of people with an intellectual disability in sexuality education that acknowledges their "expertise by experience."

As Atkinson and Walmsley note, people with intellectual disabilities "are the ultimate lost voices in terms of autobiographical records [that document their lives]" (1999, p. 203). Prompted by this observation, there has been a rise in the use of life story, oral history and narrative approaches in inclusive research with people with intellectual disability. These approaches have been central to the development of self-authored knowledge about the lived experiences of people with intellectual disability and bring a great deal to our understanding and to this field of research. However, to paraphrase *Animal Farm* (Orwell 1945), perhaps not all stories are created equal. And not all voices are heard. The library of stories drawn on in this chapter are, in a similar way to Atkinson & Walmsley's (1999) description, "lost," underutilised in shaping knowledge about sexual rights and disability. For example, in Kirsty Liddiard's overview of researching disability and sexuality, she notes that Shakespeare et al.'s 1996 book, *The Sexual Politics of Disability*, "for the first time voiced disabled people's own sexual stories" (Liddiard 2017, p. 8). This chapter will illuminate some of the sexual stories gathered and used with the goal of deepening the opportunities for understanding and including the voices of people with intellectual disability in discussions about their sexuality and relationships.

Since the early 2000s, the body of work referred to in this chapter has aimed to bring the voices of people with intellectual disability into research about their lived experiences of sexuality through their stories. It began with a project led by Kelley Johnson, which first gathered the stories of people with intellectual disability, with a focus on sexuality (Johnson et al. 2000), then later developed into a training programme for staff, families and people with intellectual disability (Frawley et al. 2003), then adapted to a peer-led violence and abuse prevention programme (Frawley et al. 2017). Central to this work, and other research informed by it, is the practice of research which prioritises the lived experience of people with intellectual disability and its subsequent use to inform research and practice. People with intellectual disability have taken on roles as researchers and storytellers (Frawley et al. 2003; Johnson et al. 2001), project workers and programme developers (Frawley et al. 2017; Frawley et al. 2019; O'Shea and Stokoe 2018), research participants speaking for themselves (Frawley and Wilson 2016), resource developers (Family Planning NSW, 2015) and peer educators (Frawley and Bigby 2014).

Michael Gill (2015) highlights the almost entire absence of work which acknowledges the capacity of people with an intellectual disability "to contribute to the development of sex education materials that reflect their experiences and desires" (2015, p. 52). We argue, then, that this contribution is transformative, not only through their stories but through the roles that have seen people with intellectual disability enter a space from where previously they have been excluded, based on spurious arguments about "capacity" to know, understand and act in relation to their sexuality and relationship experiences (Foley 2017).

Unlike the work on ontological awareness and multiple identities found outside intellectual disability, we know that, in intellectual disability literature, there is a strong focus on the assessment of knowledge and its subsequent development—a process which notably fails to involve people with intellectual disability themselves (Schaafsma et al. 2017). Recent work in crip theory has been charged with limited usefulness in producing understandings of agency, stigma and visibility for people so-labelled (Löfgren-Mårtenson 2013). Stacy Clifford Simplican, too, has reflected that:

> Telling 'fun and feisty' stories about disabled people's sexuality and sense of community unsettles tired and ableist stereotypes, but it also risks ignoring the ways in which disabled people gain a sense of community by maintaining prejudice against other disabled people.
>
> *(Simplican 2017, p. 115)*

This chapter presents and reflects on the body of work encompassed in the Sexual Lives and Respectful Relationships programme in Australia. This programme and the research around it has been the focus of a research team of people with intellectual disabilities and the authors for over a decade, across two academic institutions and several funded research projects. We draw on this body of work to highlight the relevance of the sexuality and relationship lived experiences of people with an intellectual disability in developing a theoretical, socio-political and contemporary understanding of sexuality and disability research and education practice, with people with intellectual disability.

## Background

Relationship and sexuality education for people with intellectual disability has rarely involved people with intellectual disability in its development or delivery (Frawley and O'Shea 2020; Schaafsma et al. 2017). This is not surprising, given the way that sexuality has been framed for people with intellectual disability, positioning them variously (and confusingly) as asexual, hypersexual, deviant and lacking capacity. Again, we draw on Gill's description of the *extraordinary sexuality* that has been produced for people with intellectual disability, one that is characterised by the absence of reproduction, desire, self-authored sexual identity or desire for intimacy (2015, p. 5). The resulting protectionist regime that surrounds their sexuality prevents people with an intellectual disability from agentic roles in their social and sexual lives (Foley 2017, p. 6).

Ideas of "capacity" particularly dominate discussions about the sexuality of people with intellectual disability, despite the opaqueness of our understanding of this term and the problems we have identified above. Stacey Clifford-Simplican, in her 2015 book, *The Capacity Contract*, argues strongly for the need to reframe capacity as the measurement of what people with an intellectual disability can aspire to and achieve broadly in their lives. Deliberations underpinning the formulation of the Convention of the Rights of Persons with Disabilities about sexual rights, and the question of codifying these in the Convention, highlight the urgency of this reconsideration. Schaaf (2011) has described how a degree of moral panic underpinned these deliberations, largely based on restrictive ideas of capacity and of religious and cultural concerns. Subsequently, Jaramillo-Ruiz described how the deliberations produced a "silence on affirmative sexual and reproductive rights, [which] reinforces prejudices that equate disability with incompetence, incapacity, impotence and asexuality" (Jaramillo-Ruiz 2017, p. 94).

To assert then, as we do in our work and practice over the past decade, that people with an intellectual disability can and should be sexuality educators and have a key place in the development of relationship, sexuality and education programmes is perhaps a progressive, or even radical act. However, increasingly, the inclusion of people with an intellectual disability in this work is being seen as not only possible, but vital to its outcomes (Barger et al. 2009; McCann et al. 2016).

## Sexual Lives and Respectful Relationships

The Sexual Lives and Respectful Relationships (SL&RR) programme engages people with an intellectual disability as storytellers, peer educators and participants. It uses a library of stories gathered through narrative research with people with an intellectual disability to focus discussions on shared experiences of sexuality and relationships. The programme is delivered over four sessions, usually spread over four weeks. Firmly rooted within a rights framework that recognises the capacity and agency of people with intellectual disability to be sexual and to determine their sexual lives, the SL&RR programme is based on and developed from stories developed using a narrative research approach. Participants are facilitated to discuss these stories, using a set of key messages derived from the stories and co-developed by people with intellectual disabilities. Key messages provide the opportunity to further explore sexual rights, sexual health and intimacy, supported by carefully selected and co-developed activities which augment discussions about rights, intimacy in relationships, sexual health and sexual diversity.

The Sexual Lives and Respectful Relationships team confidently claim, "we don't teach anything" (see, for example, Frawley and O'Shea 2020). Instead, we highlight a practice which recognises several key factors: that people with intellectual disability have limited access to effective and meaningful information and education on sexuality, sexual health and relationships; that people with intellectual disability are routinely and inaccurately positioned as passive, unknowing objects awaiting (and requiring) education; and that education and responses must be "specialised" to disability, or recognised in some other way as exceptional to this group.

The programme approach is broadly modelled on adult learning principles and draws on the ideas of liberationist education philosopher Paulo Friere and his work in *The Pedagogy of the Oppressed*. Friere's work, as we use it, makes space for those most marginalised from education to be the educators, thereby liberating and democratising education. Freire argued that "one cannot expect positive results from an educational or political action programme, which fails to respect the particular view of the world held by the people" (1970, p. 95). In SL&RR this idea is operationalised through peer education and expanded through the use of the stories told by people with intellectual disabilities.

## Stories of sexuality and relationships: The voices of people with an intellectual disability

Hearing from people with an intellectual disability about sexuality and relationships in their lives came from earlier life-story work of academic colleagues, such as Dorothy Atkinson and Jan Walmsley. These researchers did important work with people whose voices were otherwise lost to the public and academic records, in particular in relation to their experiences of institutionalisation (Atkinson and Walmsley 1999). The approach taken in the stories represented in this chapter is also shaped by a well-grounded body of research that has developed collections of life stories of people with intellectual disabilities (Atkinson et al. 2000; Meininger 2010; Mitchell et al. 2006; Traustadóttir and Johnson 2000). Kelley Johnson was an early adopter of this work in Australia and led the initial Living Safer Sexual Lives project (Frawley et al. 2003; Johnson et al. 2001; 2002). With ethical approval from the La Trobe University Human Research Ethics Committee, this was the first of its kind in Australia to document the sexuality and relationship experiences of people with intellectual disabilities in a story co-written with an academic researcher using an inclusive narrative approach. The researchers described their findings:

[T]hat people with intellectual disabilities lead diverse sexual lives and that their desires and many of their experiences reflect those of other adults. The stories were complex, emotional, passionate and joyful. They were accounts of adults struggling with issues around sexual expression and relationships.

(Johnson et al. 2001, p. 1)

The Living Safer Sexual Lives research participants were storytellers who shared their experiences of sexual assault, violence and abuse at the hands of people they knew—family, carers, and people they did not know—people on trains, taxi drivers. They told of rushed sex in public places, aspiring to marriage and children, being parents, whose children were involuntarily removed, and claiming LGB identities. Many of these elements are common to several storytellers, along with a description of the constant surveillance by and at the hands of carers, family and others, who mediated and managed their sexual lives. We present excerpts from some of those stories, using pseudonyms originally allocated by the researchers and presented in the training manual, which was produced from the research:

### Hannah's story: "It's hard to get privacy"

Kevin is my boyfriend. He came here about six months ago and we fell in love. We wanted to share a room. The staff sometimes say things that we don't like. When they were saying it wasn't all right to share a room, we'd leave that particular night and go off on our own and then come back the next day. In the end, they decided that we could have the room together.

*(Frawley et al. 2003, p. 179)*

The authors identified six key themes in their analysis of the key themes emerging from the stories. This reflected on the struggles "with issues of love, passion and desire" (Johnson et al. 2001, p. 58), such as those told by Hannah (above) and her partner Kevin, whose story excerpt is featured below.

### Kevin's story: "I'll spend my life with her"

Her name is Hannah. We loved each other *that much*, and we wanted to have sex. [...] I said, 'Do you want to have sex?' and she said 'Yes.' I asked her if I could do things to her and she said 'Yes.' The first time I slept with her, I went to bed with her, and I took my clothes off and I put my penis into her. Yeah. It was quite good. I enjoyed it.

*(Frawley et al. 2003, p. 184)*

The theme of rights and sexuality connected across many of the stories—the right to information, the right to relationships and importantly, the right not to be sexually abused.

### Molly's story: "My dream is to get married"

I went on Depo because I kept forgetting to take the pill and I accidentally fell pregnant once before to a friend. A one-night stand I had. My mum and dad and sister didn't want me to have the baby, because I've got a disability and they reckoned I couldn't look after it. My sister booked me into having an abortion. And the day before, I had a miscarriage. Mum said to me, 'Don't you ever do that again, because if you did you'd be disowned from this family.'

*(Frawley et al. 2003, p. 170)*

### Angela's story: "I like to go on trains"

I like to go on trains...I like to look out the windows and see things. A man touched me on the train, and it was against the law wasn't it? He came up to me and said to touch him. I didn't want to, but he made me. It happened two times and I told the teachers. […] Nothing happened. The second time was this year…then I went up to the police and I had to make a statement, a few statements…I made the police catch the man.

*(Frawley et al. 2003, p. 218)*

Many of the stories reflected the hidden nature of people's sexual lives, particularly so when their sexuality did not fit the prototypical heterosexual, monogamous mould.

### Hussein's story: "Don't tell me parents, it's a secret"

He took his jacket off, his shoes and socks off, his watch, undid his belt, pulled down his pants and take them off, took his jumper off. We started touchin' each other at the same time. I pulled his undies down, he pulled my ones down. We got carried away. He couldn't stop and he kept goin'. We kept on going and he came about four times. I did too. He came on me and I came on him, about four times and he did it to me. I didn't have any condoms. He had a few in his pocket. And we stopped about 1:15pm to get dressed to get going.

*(Frawley et al. 2003, p. 189)*

People with intellectual disability worked with the authors to review the stories and produce what they felt were the key messages to come from each story. The key messages are the primary tool for guiding the discussions, led by peer educators and programme partners in programme delivery.

### Hearing from more hidden voices: LGBTQ people with intellectual disability

The collection of stories has been added to over time, with new stories gathered in a project with LGBTQ people with intellectual disabilities (O'Shea and Rakidzic 2018), and an adaptation of the programme for people with Acquired Brain Injuries (O'Shea et al. 2018). At this point we were able to draw on methodological advances in the co-development of stories (O'Shea 2016) and in inclusive research broadly (Milner and Frawley 2019). In their unpacking of the epistemological examinations of the processes of inclusive researching, and constructing research as a space of subjectified invitation, Milner and Frawley assert:

If we are serious about our democratizing intent not becoming just another form of discursively organized colonisation, we argue that a similar investment in understanding (and responding) to the subject positions of our partners in the research performance ought to become the foundation principle of relationally reconfigured research methods.

*(Milner and Frawley 2019, p. 11)*

Using a narrative research approach, but with a clear intent to position the voices of people with an intellectual disability as authoritative and instructive, framed the way the stories were developed. The process of gathering the stories with LGBTIQ people with an intellectual disability involved repeated opportunities to meet with storytellers, in a location of their choice, and to hear what they had to say, as they wanted to say it. Unlike the previous Sexual Lives and Respectful Relationships stories, these were gathered with a specific purpose, to be shared with others in the programme and to be stories of LGBTIQ people with intellectual disabilities. We share some excerpts here:

### Carol's story: "To be the way I want"

I am a lesbian. I knew from about 15, 16. I knew when I was living with my sister but I never told her. I was terrified she would kick me out. I didn't even know a single person that was gay. So, I didn't know if it was real or not. Maybe it was wrong to be gay. I was scared and I thought they would say, 'Oh, you can't be lesbian. You have to be straight. You have to be not telling anybody. You have to keep it to yourself. You can't go out.' I worried that a lot of things would happen and stop my freedom, if I got caught. So, I keep myself hidden. I thought I might get beaten up or something. Get teased. Even harder with a disability.

*(Frawley et al. 2019, p. 172)*

### Soraya's story: "Not the nothing girl"

In my country, brain injury brings shame for the family. They have to stay at home, they lock the door and don't let them out. It is shame. […] [then] my brain burst inside, and I woke up six weeks later. It was a brain haemorrhage and, when I woke up, I was like a child again, slowly learning words and walking. That reminded me of being a child the first time, and I remembered it all over again. I told my Mum what had happened, and she said, 'Oh well, he has died now.' She didn't even say sorry.

*(O'Shea et al. 2018)*

### Emmet's story: "Less lost than I used to be"

I have an intimate relationship, but we don't really have a sexual relationship. That's my partner's choice, she is not comfortable with being sexual. It's something we have struggled with because, to me, it's such an important way of connecting and being close. […] She said she didn't care if I wanted to have sex with other people, but I don't, I don't want to be closer to other people. I'm comfortable with my sexuality, and my masculinity, and I don't feel that I need to prove it.

*(O'Shea et al. 2018)*

Individually and collectively these stories speak directly of the experiences of people with an intellectual disability as fellow humans trying to negotiate the most human of things —intimacy, friendships, sexual expression and sexual identity. Although they often sound familiar to many, they ring loudly and true to people with an intellectual disability who hear these stories through the programmes referred to here. In particular, they make clear that the barriers that stand in the way of the quest to be sexual, to be safe, and to be the sexual person you want to be for people with an intellectual disability are primarily barriers created by the sexual ableism that operates to diminish their sexuality (Gill 2015).

Storytellers saw a copy of the current programme manual and heard from the researcher about how their stories would be used. The stories were to be alive, read and heard firstly by their peers—not, as is perhaps more usual, by academics or professional staff with journal access or conference attendees. This arguably created a more intimate dialogue, from one person to their peers through two intermediaries—the non-disabled researcher and the peer educators who would facilitate its discussion.

## Deepening understanding: Expanding the SL&RR programme

The SL&RR programme uses a peer education model, where people with intellectual disability are trained to deliver the programme in local sites to their peers (Frawley and O'Shea 2020). Programme partners (from local organisations in areas such as health, sexual health, disability advocacy) work alongside peer educators in planning and organising the programme (Frawley and Bigby 2015). In this way, the voices of people with intellectual disability in works about them and their sexual rights can be extended beyond the stories and into the delivery of programmes (Frawley et al. 2012).

The newest program developed with people with Acquired Brain Injury gathered and collaboratively produced stories with people with ABI. In the same way as the existing SL&RR programme, peer educators (in this case people with acquired brain injury), programme partners and a research advisory group, including people with brain injury, helped to shape the key messages and inform the additional activities to be used alongside the stories in the programme. Although outside the scope of this chapter, we note here that the storyteller contributions demonstrate the way that people can and do have an important place in sexuality education and development of information that is "for them".

Another recent expansion of the SL&RR model included a LGBTQA+ focused pilot. For the remainder of this chapter, we reflect on what has been learned through the stories collected in these expansions, and how they can contribute to and enhance our understanding of sexuality for people with disability.

### *LGBTQ people with intellectual disability*

We learned from the LGBT storytellers what it meant for them to consider safety in their lives and relationships. Their stories shared many of the elements identified in the previous stories—searching for a relationship, and to be recognised by others as a sexual being. Like other storytellers, they reflected on the power, often held by others in their life, around their opportunities to express and experience themselves as sexual. However, this had an additional layer, as storytellers considered the impact of potentially homophobic, biphobic or transphobic attitudes. It is worth noting that these stories were collected shortly after the conclusion of the marriage equality postal survey in Australia, during which there was a high degree of public debate, the long-term impact of which on the LGBT community has yet to be measured. This added layer was visible in their stories as they reflected on the deeper elements of inclusion and acceptance relevant to their lives as LGBT people with intellectual disability. One example is the theme of safety, as seen in this excerpt from Carol:

> I go to church, I try to go every week. It gives me peace—peace with life, and with God. [...] I choose who to tell because some Christians don't like people being gay. Sometimes it's hard because I want to be open and be me, but if I be too open, they might ask, 'please leave, we don't want your type here.' Then it's not safe to go back there and I would not have a place to worship.
>
> *(Frawley et al. 2019, p. 173)*

Like Carol, Johnno made decisions about who and when to share his identity with others:

> I don't want to tell everyone I'm gay. If I tell them, they want to know why I'm gay, I
> don't want them to ask about that. I don't know why I'm gay. Or they might call me a
> 'queer' and I don't like that. It happens at my day service and I told the staff, but they
> didn't do anything. Say 'queer' and bash me up.
>
> *(Frawley et al. 2019, p. 165)*

Both Carol and Johnno considered their physical safety to a greater degree than that reported by previous storytellers. This was primarily, though not exclusively, located with strangers—concerns about being verbally or physically abused on the street. The first author's research notes indicated that Johnno chose not to enter the park where they had agreed to meet, waiting on the street until she arrived, and they walked in together. It was not without resistance, though, as Johnno concluded his story:

> people call me 'poofter' and things, and it's sad but I just tell them 'I don't care, you can
> call me anything you like' and then it would be: 'See you later, alligator. Don't forget
> your toilet paper.' They want to destroy us, but I don't give a shit what they say.
>
> *(Frawley et al. 2019, p. 165)*

## Conclusion

Our goal for this chapter was to share some of what we have learned and heard from the people with intellectual disability. We are optimistic that it will illuminate some of the rightful and valuable contributions which have been made to work, which is about them, and help to argue for their rights to respectful and satisfying intimate relationships. As we stated at the beginning of this chapter, we hope their contributions will promote opportunities for learning and inclusion to inform future directions around disability and sexuality.

We suggest that the Sexual Lives and Respectful Relationships programme presents one answer to the point made so strongly by others (Gill 2012; 2015; Schaafsma et al. 2015) that people with intellectual disability must be involved in work about their sexuality and relationships. We acknowledge the multilevel inclusion which can happen when people with intellectual disability are included in multiple ways, in this case, as storytellers, colleagues and advisors, peer educators and trainers. We know there is much more to be done, to take a greater intersectional analysis and to continue to capture the diverse stories and experiences to be told by people with intellectual disability.

## References

Atkinson, D., Mccarthy, M., Walmsley, J., Cooper, M., Rolph, S., Aspis, S., Barette, P., Coventry, M., & Ferris, G. (eds.) (2000) *Good times, bad times: Women with learning difficulties telling their stories.* Kidderminster: BILD.

Atkinson, D., & Walmsley, J. (1999) Using autobiographical approaches with people with learning difficulties. *Disability & Society* 14, pp. 203–216.

Barger, E., Wacker, J., Macy, R., & Parish, S. (2009) Sexual assault prevention for women with intellectual disabilities: A critical review of the evidence. *Intellectual and Developmental Disabilities* 47, pp. 249–262.

Family Planning NSW (2015) Outing disability [Online]. Accessed 26/7/19.

Foley, S. (2017) *Intellectual disability and the right to a sexual life: A continuation of the autonomy/paternalism debate.* London: Routledge.

Frawley, P., Barrett, C., & Dyson, S. (2012) *Real people, core business. Living safer sexual lives: Respectful relationships*. Report on the development and implementation of a peer-led violence and abuse prevention program for people with an intellectual disability. Melbourne, Victoria: La Trobe University. http://dro.deakin.edu.au/view/DU:30073078

Frawley, P., & Bigby, C. (2014) "I'm in their shoes": Experiences of peer educators in sexuality and relationship education. *Journal of Intellectual and Developmental Disability* 39, pp. 167–176.

Frawley, P., & Bigby, C. (2015) "Molly is just like me": Peer education and life stories in sexuality programmes in Australia. In: Chapman, R., Ledger, S., Townson, L., & Docherty, D. (eds.) *Sexuality and relationships in the lives of people with intellectual disabilities: Standing in my shoes*. London: Jessica Kingsley.

Frawley, P., Johnson, K., Hiller, L., & Harrison, L. (2003) *Living safer sexual lives: A training and resource pack for people with learning disabilities and those who support them*. Brighton: Pavilion Publishing Ltd.

Frawley, P. & O'Shea, A. (2020) 'Nothing about us without us': sex education by and for people with intellectual disability in Australia. *Sex Education*, 20(4), pp. 413–424. doi:10.1080/14681811.2019.1668759

Frawley, P., & O'Shea, A. (2019b) Peer education: A platform for sexuality rights advocacy for women with intellectual disabilities. In: Soldatic, K., & Johnson, K. (eds.) *Global perspectives on disability activism and advocacy*. United Kingdom: Routledge.

Frawley, P., O'Shea, A., SL&RR Review Team, & Wellington, M. (2019) *Sexual lives & respectful relationships training manual*. Geelong: Deakin University.

Frawley, P., O'Shea, A., Stokoe, L., Cini, V., Davie, R., & Wellington, M. (2017) *Sexual lives and respectful relationships training manual*. Geelong: Deakin University.

Frawley, P., Slattery, J., Stokoe, L., Houghton, D., & O'shea, A. (2011) *Living safer sexual lives: Respectful relationships. Peer educator and co-facilitator manual*. Melbourne: Australian Research Centre in Sex, Health & Society, La Trobe University.

Frawley, P., & Wilson, N.J. (2016) Young people with intellectual disability talking about sexuality education and information. *Sexuality and Disability* 34, pp. 469–484.

Freire, P. (1970) *Pedagogy of the oppressed*. London: Penguin Random House.

Gill, M. (2012) Sex can wait, masturbate: The politics of masturbation training. *Sexualities* 15, pp. 472–493.

Gill, M. (2015) *Already doing it: Intellectual disability and sexual agency*. Minneapolis: University of Minnesota Press.

Hamilton, C., & Atkinson, D. (2009) "A story to tell": Learning from the life-stories of older people with intellectual disabilities in Ireland. *British Journal of Learning Disabilities* 37, pp. 316–322.

Jaramillo Ruiz, F. (2017) The committee on the rights of persons with disabilities and its take on sexuality. *Reproductive Health Matters* 25, pp. 92–103.

Johnson, K., Hillier, L., Harrison, L., & Frawley, P. (2000) *People with intellectual disabilities: Living safer sexual lives*. Melbourne: Australian Research Centre in Sex, Health and Society, La Trobe University

Johnson, K., Hillier, L., Harrison, L., & Frawley, P. (2001) *Living safer sexual lives: Final report*. Melbourne: Australian Research Centre in Sex, Health and Society, La Trobe University.

Johnson, K., Strong, R., Hillier, L., & Pitts, M. (2002) *Screened out! Women with disabilities and cervical screening*. Carlton, Victoria: Cancer Council Victoria.

Liddiard, K. (2017) *The intimate lives of disabled people*. London: Routledge.

Löfgren-Mårtenson, L. (2013) "Hip to be crip?" About crip theory, sexuality and people with intellectual disabilities. *Sexuality & Disability* 31, pp. 413–424.

Mccann, E., Lee, R., & Brown, M. (2016) The experiences and support needs of people with intellectual disabilities who identify as LGBT: A review of the literature. *Research in Developmental Disabilities* 57, pp. 39–53.

Meininger, H.P. (2010) Connecting stories: A narrative approach of social inclusion of persons with intellectual disability. *Alter* 4, pp. 190–202.

Milner, P., & Frawley, P. (2019) From 'On,' to 'With' to 'By:' People with a learning disability creating a space for the third wave of inclusive research. *Qualitative Research* 19(4), pp. 382–398.

Mitchell, D., Traustadóttir, R., Chapman, R., Townson, L., Ingham, N., & Ledger, S. (eds.) (2006) *Exploring experiences of advocacy by people with learning disabilities: Testimonies of resistance*. London: Jessica Kingsley.

Orwell, G. (1945) *Animal farm*. New York: Harcourt, Brace and Company.

O'Shea, A. (2016) *"Just a rare girl": Gender and disability in life stories produced with young women with disabilities*. PhD, La Trobe University.

O'Shea, A., & Rakidzic, S. (2018) Sexual lives & respectful relationships for LGBTIQ people with intellectual disability. In: ASID national conference. Gold Coast, Queensland.

O'Shea, A., & Stokoe, L. (2018) Sexual lives & respectful relationships: A peer education sexuality and relationships program by and for people with intellectual disability. In: *National virtual disability conference.*

O'Shea, A., Stokoe, L., Wellington, M., & Frawley, P. (2018) *Sexual lives & respectful relationships: Brain injury training manual.* Geelong: Deakin University.

Schaaf, M. (2011) Negotiating sexuality in the convention on the rights of persons with disabilities. *Sur – International Journal of Human Rights* 8, p. 113.

Schaafsma, D., Kok, G., Stoffelen, J., & Curfs, L. (2017) People with intellectual disabilities talk about sexuality: Implications for the development of sex education. *Sexuality and Disability* 35, pp. 21–38.

Schaafsma, D., Kok, G., Stoffelen, J.M., & Curfs, L.M. (2015) Identifying effective methods for teaching sex education to individuals with intellectual disabilities: A systematic review. *Journal of Sex Research* 52, pp. 412–432.

Simplican, S.C. (2015) *The capacity contract: Intellectual disability and the question of citizenship.* Minneapolis: University of Minnesota Press.

Simplican, S.C. (2017) Feminist disability studies as methodology: Life-writing and the abled/disabled binary. *Feminist Review* 115(1), pp. 46–60.

Traustadóttir, R., & Johnson, K. (2000) *Women with intellectual disabilities: Finding a place in the world.* London: Jessica Kingsley.

# PART VI

# Accommodation, support and sexual well-being

# 28

# SEXUAL WELLNESS FOR OLDER ADULTS WITH A DISABILITY

## A life course perspective

*Maggie Syme, Stacy Reger, Christina Pierpaoli Parker*
*and Sally J. Hodges*

## Introduction

Access to rights as a sexual citizen have long been privileged, excluding many from sexual rights theoretically conferred to all persons (World Association for Sexual Health 2014). This exclusion has a double jeopardy impact on being older and a person with a disability (PWD), and represents the dual impact of ageism (i.e., age discrimination) and ableism (i.e., discrimination against people with disabilities). Older PWDs are often cast as asexual and/or sexual deviants, as they do not fit the societal standard for sex being for the young, fit and beautiful (Esmail et al. 2010; Hillman 2011). Activists and scholars have begun to draw attention to these biased views (Barrett and Hinchliff 2018; Simpson et al. 2018) and instead ask that we regard sexual citizenship as an ageless and ability-encompassing concept.

## Setting the stage

This chapter will examine sexual experiences of older PWDs, utilising common definitions and available research to set the stage. Sexuality involves experiences, expressions, desires, beliefs and attitudes, values, roles and relationships (World Health Organization (WHO) 2018). It is shaped by the "interaction of biological, psychological, social, economic, political, cultural, ethical, legal, historical, religious and spiritual factors" (p. 5). Sexual well-being, then, is defined as the fluctuation of physical, emotional, mental and social well-being in relation to sexuality (WHO 2018).

There is very little research literature devoted to the sexuality of older PWDs. What is known about older adults and PWDs illustrates a diversity of sexual expression, enjoyment and experiences. For example, current scholarship on sexual experiences of older adults (OAs) suggests that types of behaviours remain diverse (Waite et al. 2009), levels of engagement remain fairly consistent across adulthood until the mid- to late 70s (Lindau et al. 2007), and sexual scripts often shift focus from primarily intercourse to include more intimate behaviours (DeLamater 2012). More importantly, sexual wellness remains key to quality of life and healthy relationships for many OAs across the lifespan (Fisher 2010). Sexual wellness involves a multitude of experiences for OAs, including sexual and intimate behaviours as well as abstinence and asexual experiences (Syme et al. 2018). A key part of wellness (or lack thereof) for OAs is freedom, in that they are free to define their sexual identity and experiences as they wish, without coercion and/or discrimination from other individuals, groups or society (Fileborn et al. 2015; Simpson et al. 2018). Herein lies a primary link between OAs and PWDs, in that their freedom—or sexual citizenship—is often challenged, or wholly dismissed.

The sexual experiences of PWDs are well canvassed throughout the accompanying series; they evidence similar conclusions in that disabled persons have varied sexual experiences, behaviours, desires, needs and often experience elevated functional and social/environmental challenges to their sexuality across their lifespan (Mona et al. 2014). For instance, PWDs report fewer opportunities to be sexual, have lower sexual self-esteem and encounter more barriers to sexual and reproductive healthcare than individuals without a disability (Nosek et al. 2001; Rowen et al. 2015). Furthermore, similar to OAs, their sexualities are often medicalised and devoid of a pleasure narrative, that is critical to sexual wellness (Mona et al. 2014). The intersections of these two identities—older and disabled—will likely result in unique strengths, environmental constraints/supports, and dis/advantages across their lifespan, that will affect the sexual trajectories of older PWDs.

## Purpose and framework

A sex-positive approach will be utilised to expand on the concepts and developmental experience of sexual citizenship as an older disabled person. This goes beyond being *pro sex* to encompass diversity in expression and behaviours, empowerment, choice, desire and enjoyment of sex (Williams et al. 2015). Sexual citizenship in later life must be viewed from a life course perspective (Elder 1994), as the accumulation of dis/advantages across a lifetime of environmental constraints and supports result in sexual identities and experiences for older, disabled persons in the moment. In short, how and who we have been as a sexual citizen affects our current sexual life and identity. Within these sexual developmental trajectories, the biopsychosocial and cultural aspects of sexual experiences and identities will be canvassed. Additionally, resources for empowering older disabled persons to actualise their sexual citizenship, within their own sexual trajectories, will be presented.

## Conceptualising older PWD sexual citizenship

The sexuality of older PWDs can be conceptualized from a life-course perspective, which is a socio-developmental theory of the co-occurring influences of time, place, history and social structures on individuals across domains (Elder 1994). Thus, the sexual trajectory of an older disabled person is the result of "the lifelong accumulation of advantageous and disadvantageous experiences, and their adoption and rejection of sexual scripts, within specific socio-historic contexts" (Carpenter et al. 2010, p. 157). It fits well with the social model of disability, which emphasises the importance of the dis/advantages inherent in the social environment and how those shape diverse experiences (Dunn and Andrews 2015). These phenomena cannot be understood outside of the context of ageism and ableism, as they are prime suspects for the sexual marginalisation of older disabled persons.

### *The culture of ageism*

Ageism refers to negative stereotypes, prejudicial beliefs and discriminatory behaviours toward older persons, widely accepted across cultures. It is unique in some ways, in that demographics like sex or race are generally lifetime-lived experiences, but age—and ageism—is defined by a specific transitional period in later life, and is one of the few (if not only) socially sanctioned prejudices (Gendron et al. 2016). Assumptions, jokes, interpersonal communication (e.g., Elderspeak) and overly careful behaviours are all part of "othering," that leads to prejudice towards older persons. Ageist attitudes are so engrained in western culture that, many times, ageist beliefs are assumed to be facts (Gendron et al. 2016). Though it is aimed at older ages, individuals experience age-related prejudice across their life course, that is often insidious, yet powerful (Nelson 2016), evidence of which is observable at even young ages. It has the potential to severely impact cognitive, physical and overall health of individuals as they age. The Stereotype Embodiment Theory (SET) explains how age stereotypes infiltrate our psychological self across our life and manifest as real-life changes in our physical and social well-being (Levy 2009). There is evidence across domains for the impact of negative internalised age stereotypes, including decreased longevity and increased risk of dementia (Nelson 2016).

The culture of ageism and associated negative stereotypes are especially palpable for sexuality, manifesting as dichotomies of sexual older beings, such as the "asexual oldie" versus the "dirty old man" or the "cougar" (Hillman 2011). These are often tied to socio-historical standards of beauty and vitality that are youth-oriented, illustrating the advantageous sexual scripts of the young and the contrary for older persons. For example, in a qualitative study of older men and women in the US (Syme et al. 2018), sexual well-being was filtered through the experience of being older, in that sexual wellness trajectories depended on age salience and were often highly impacted by gendered scripts for older women. The transition of menopause was highly impactful for these older women, and many evidenced internalised ageism and sexist beliefs about older female sexuality (e.g., asexual oldie, sex only for reproduction). Similar age-salient aspects of well-being were evidenced in a national study of OA sexual wellness in the UK (Tetley et al. 2016). Being young is the standard for being sexy, desired and attractive, which is a highly influential script in the sex lives of many OAs.

## The culture of ableism and intersectionality

The medical model has historically defined disability, which situates disability in the biological and operates on a deficit-based, disease viewpoint. Newer, and more culturally informed models of disability understand it as a form of identity and individual difference, similar to race, gender and sexual identity (Dunn and Andrews 2015). Disability, then, is situated in person-environment interactions, physical limitations are opportunities to respond with accommodating resources, and disability is understood within the lived socio-political context (Olkin 1999). Similar to OAs, society has a stereotyped, negative bias against disabled persons—or ableism. This stems from historical beliefs rooted in the medical and moral models of disability that suggest it is a moral failing and a disease, rendering the person both deviant and not "normal" (Dunn and Andrews 2015). Ableism can be defined as "denoting an attitude that devalues or differentiates disability through the valuation of able-bodiedness equated to normalcy" (Campbell 2008a, p. 2). The experiences of discrimination and consistent societal messages, that disability is a stigmatised identity, accumulate across the life course, and the result may be internalised ableism, or a negative bias against the disabled self.

How might this affect sexual citizenship and trajectories? Disability studies on body and sexual esteem (Taleporos and McCabe 2002) and intimate relationships (Esmail et al. 2010) suggest that internalised stigma about disability will have an effect on how the PWD views his/her ability to be attractive, have sexual opportunities and successfully navigate sexual experiences, but also if they "ought" to do so. Studies suggest this may be more challenging for women with a disability (WWD) (e.g., Basson 1998), including reproductive/sexual healthcare being privileged for able-bodied women in many contexts (Bremer et al. 2010). Another potential result of internalised ableism is distancing or rejecting disability as an identity. Scholars have likened this to internalised racism and explained that it causes an over-identification with the dominant, acceptable culture—able bodied, in this case—at the expense of rejecting your identity as disabled (Campbell 2008b). This has many potential negative implications for sexual wellness, including isolation from a support network and community, that can be instrumental in wellness, and potentially rejecting resources that are "not for me."

For older men and women with disabilities, sexual expression varies depending on the unique nature of the disability, and the individual's attitudes and values around sex. In line with SET theory (Levy 2009), those with visible disabilities may be more subject to internalised biases due to the increased saliency of disability in everyday life. SET and other stereotype threat theories would suggest that the more the characteristic (e.g., disability) emerges into awareness, the more impact it may have on the individual's psychological and social functioning (Steele and Aronson 1995). In addition, acquired, as compared to congenital, disability may have a significant effect on a sexual trajectory, representing a critical turning point. Though, as an older PWD, the time to adjust/adapt to sexual challenges is greater than for a younger PWD, a phenomenon which has been shown to be beneficial to outcomes such as satisfaction with life for PWDs (Bogart 2014).

An older PWD is at double jeopardy for stigma and, thus, for internalising messages about being both disabled and older. Sexual citizenship is limited via self-fulfilling prophecies. The internal messages become "I do not deserve/need/want" or "I cannot" engage in sexual experiences and therefore they act in accordance, a process often operating outside of awareness (Levy 2009). Both ableism and ageism must be openly addressed, and sex-positive messages must be promoted that empower an older PWD to pursue sexual citizenship on his/her terms.

## Environmental supports and constraints

Significant institutional and systemic barriers prevent optimal education, advocacy and care for older disabled persons, thus limiting sexual citizenship. First, there are general barriers PWDs and older adults face in accessing quality healthcare, dependent on country of residence and its respective healthcare system. Barriers may include cost, adequate insurance coverage, transportation, complexity of health concerns spanning multiple medical specialties and accessible offices and facilities. There are also specific challenges to accessing sexual healthcare, many of which are related to sexual stigma and the culture of communication between providers and patients around sex. Healthcare providers often feel embarrassed or are not prepared to have conversations about sexuality with OAs (Hinchliff and Gott 2011), and have similar discomfort in discussing sexual health with PWDs (Rubin, 2005). In addition, older PWDs may feel uncomfortable initiating conversations about sexual health (Hinchliff and Gott 2011) and often prefer that their providers initiate such conversations, though they report feeling that their providers are not interested (Colton 2008). Providers across professions may harbour negative attitudes about older PWDs and sex, that influence whether and how they broach the topic of sexuality as part of routine care (Colton 2008; Rubin 2005). This is particularly concerning, given that some OAs and PWDs do engage in risky sexual behaviours (e.g., unprotected sex), might underestimate that risk and are unlikely to communicate this to their providers (Bauer et al. 2016; Rubin 2005; Syme 2017) For LGBTQ, older, disabled persons, concerns about social stigma and judgement from healthcare providers create an additional hurdle in communication and receipt of adequate sexual healthcare (Furlotte et al. 2016).

A discussion of environmental constraints and supports for practicing sexual citizenship must include institutional living environments, primarily residential care/nursing homes. Of note, the built environment of residential care/nursing homes has been cited as restrictive toward sexual expression (Frankowski and Clarke 2009). The physical set up of most nursing home rooms across western nations is dual—if not multiple—occupancy, lacking privacy from consistent monitoring and open-door policies and bed and room set-ups are not well equipped for intimate behaviours (Simpson et al. 2018). There are few, if any, private spaces accessible to residents.

In addition, several psychological and social constraints are inherent in nursing home environments. Primary among these is the perceived and real dependency of some nursing home residents on those in power positions. Residents are often stripped of autonomy and left to rely on those "in charge" of their rights and decisions, such as family members, staff and administrators (Doll 2016). Unfortunately, administration, providers, and family, who are often in positions of power, often harbour negative/prohibitive attitudes toward sexuality, ageing and disability that are driven by lack of awareness and ageist/ableist beliefs about the asexual and/or deviant nature of sex for older PWDs. This is often compounded when considering sexual and gender minority status (i.e., LGBTQ older adults). Research suggests that older sexual and gender minorities are more likely to experience discrimination in institutional settings, may re-closet due to fear of mistreatment in absence of protective policies, and may not receive culturally competent care, due to lack of training, knowledge or policies to address social and health disparities (Hinrichs and Vacha-Haase 2010; Porter et al. 2016).

In terms of social constraints, there are often limited opportunities for sexual partners within a nursing home (Bauer et al. 2013). Furthermore, privacy and discretion are of concern to older PWDs living in nursing facilities. Residents express concern that staff are not discreet regarding sexual relationships among residents, and that fellow residents will form negative opinions of those who engage in sexual activity (Bauer et al. 2013). One study found nursing home resident attitudes toward sexual activity among residents tended to be neutral to negative (Tzeng et al. 2009). In addition, residents express worry that those who engage in sexual activity will be the subjects of gossip (Frankowski and Clark 2009).

Issues related to protection and consent may further limit the autonomy of older PWDs in nursing homes. There is an ongoing discussion about the best way to address consent and sexual activity in residential settings for individuals with cognitive or capacity deficits (Bauer et al. 2013; Syme 2017). Because inter/national laws have largely failed to address consent specifically for OAs, residential facilities often have no formal policy. There remains debate and discomfort in ensuring the rights of residents to autonomy, privacy and dignity, regarding sexuality (Wilkins 2015). Studies show that most staff in nursing facilities have fewer concerns when two cognitively intact residents have a sexual relationship (Bauer et al. 2013); however, paternalistic attitudes towards any functional challenges among residents can result in overly restrictive actions and policies (e.g., all-or-nothing capacity determinations), that fail to appreciate the nuance of capacity, dementia and functionality (Wilkins 2015). A balanced approach involves decision making on an individual basis, and planning that includes the resident and partner, family members, medical and mental healthcare providers and an ombudsperson, when appropriate, to determine the best policy for an individual (Syme 2017; Wilkins 2015). This should be based on what is known of the situation, cognitive function, capacity for consent and the resident's values (Bauer et al. 2016).

## Biopsychosocial and cultural aspects of sexual wellness for older PWDs

In addition to the concepts described above, a thorough understanding of sexual wellness trajectories for older adults with disabilities is filtered through a biopsychosocial lens (Marshall 2011).

The following outlines the physiological, emotional, psychological and socio-cultural factors that influence sexual wellbeing for individuals with disabilities across the life course.

## Biological

Older PWDs may have unique physiological attributes that impact sexual health. Ageing often involves health changes that affect sexual function and expression. For men, erectile dysfunction (ED) is a common concern with ageing. A recent review of the sexual dysfunction literature suggested that ED affects 2–40% of men aged 40 to 69, and 50–100% of older men (McCabe et al. 2016). Ageing men report concerns related to decreased sexual desire, decreased penile sensation, and longer plateau and refractory periods, which may be related to lower testosterone levels (Albersen et al. 2012). Ageing women also experience physical changes that influence sexual expression, primarily the decline in oestrogen during menopause, which causes vaginal dryness, atrophy of the vaginal wall, narrowing of the vaginal canal and decreased blood flow (Hillman 2011; McCabe et al. 2016). In women, pelvic floor disorders increase in incidence with age, and older women report concerns about urinary and faecal incontinence, which lower interest in sex due to embarrassment (Ratner et al. 2011). Postmenopausal women also frequently report concerns about sexual wellness and function, including low desire and arousal, decreased intensity or frequency of orgasm, pain with intercourse and self-consciousness about body appearance (Ambler et al. 2012).

Older PWDs may also experience physical symptoms that affect their experience of sexual activity or their ability to engage in certain sexual behaviours. One such physical consideration fatigue which has been described as one of their most debilitating symptoms for some PWDs (Goedendorp et al. 2014). In addition, chronic physical pain and treatments for pain can impede sexual expression for older PWDs, either directly (e.g., pelvic pain), or indirectly, due to pain that causes fatigue, limits mobility, worsens with sexual activity or requires medication that affects desire or energy for sexual activity (Dorado et al. 2018).

Older PWDs may face challenges related to increased incidence of chronic health conditions and their management. Both men and women may experience changes in sexual function related to health changes common in ageing that disrupt blood flow, affect neurovasculature and impact sex hormones. Additionally, medications used to manage chronic health conditions can have side effects that interfere with sexual health (DeLamater 2012). For example, antihypertensives used to manage cardiovascular disease may lead to lowered blood pressure and consequently ED, or hormones used to treat prostate or breast cancer may decrease sexual desire.

## Psychological

Individuals' sexual trajectories are shaped by the accumulation of sexual experiences that have been filtered through our thoughts and feelings and have shaped our views about our sexual self. This happens in the context of an ageist and ableist society that impacts our access to sexual citizenship. The psychological experience of sex is pivotal to the sexual wellness of an older PWD, and should be considered along several dimensions (e.g., cognitive, affective, attitudinal).

Many OAs perceive sex as integral to quality of life, and report that their mental health both impacts, and is impacted by, sexual health (Fisher 2010). Older adults are also more likely than younger people to have conservative beliefs about sex (e.g., sex only as penile-vaginal intercourse, heteronormative, monogamy) (Waite et al. 2009). Another study found that WWD report more negative views of sex than do non-disabled women, though, in the same study, there was no difference in sexual desire and fantasy between the groups (Vansteenwegen et al. 2003).

For ageing PWDs, cognitions around sex based on the heteronormative sexual experiences may prove a hindrance to sexual expression, limiting creativity and fulfilment.

Capacity and cognitive function are of particular interest for older disabled people, as both age and the nature of disability may have increased risk of cognitive limitations. For individuals with cognitive impairments, family members or providers may have questions about the ability to fully comprehend and consent to sexual activity (Wilkins 2015). Individuals with intellectual disability (ID) have historically faced discrimination and rights violations with regard to sexual expression, such as forced sterilisation, invasive birth control methods and gender-segregated living arrangements to prevent sexual activity (Gill 2015). With age, concerns about consent and capacity can be compounded by cognitive decline, and there is a delicate balance between respect for the rights of the individual for self-determination and agency regarding sex, and efforts to protect vulnerable adults who could be victimised as a result of cognitive limitations (Syme 2017).

Self-esteem and body image also influence sexual wellness in older PWD. For both OAs and PWDs, body image concerns are a major component of self-esteem (Bauer et al. 2016), and mainstream culture has historically rejected or ignored older and disabled bodies as sexually undesirable, which can lead to internalised body shame (Hillman 2011; Taleporos and McCabe 2002). In addition, there is some evidence that having a disability may negatively affect sexual self-esteem for some individuals (Moin et al. 2009), a risk that tends to be greater as functional status (i.e., ability to complete activities of daily living (ADLs) and independent activities of daily living (IADLs) ) decreases (McCabe and Taleporos 2003). Interestingly, McCabe and Taleporos (2003) also found that WWD appear to be less negatively affected than their male counterparts, which is a flipped gendered script from that of older women without a physical disability (Syme et al. 2018).

The emotional context of sexual activity shapes experience and sexual wellness. For older PWD, as for all people, a fulfilling sex life includes self-determination, safety and dignity (Marshall 2011; Syme et al. 2018). Mental health concerns and their pharmacological treatments may also impact sexual health and expression. For example, depression may lower interest in or pleasure derived from sex, and some antidepressant medications used to manage mood can also lower libido (DeLamater 2012). For OAs who have lost a partner, grief or guilt about seeking a new partner may interfere with pursuing relationships or sexual activity (DeLamater 2012). When discussing psychological issues in ageing, loneliness must be addressed, though it is not a phenomenon reserved for older age. In a recent longitudinal study in the US, over 40% of those aged 65+ reported experiencing chronic loneliness, which was associated with several physical and functional limitations (Perissintotto et al. 2012). It also affects psychological and social functioning, and can increase your risk of mental health concerns (Cornwell and Waite 2009), inhibit the quality of relationships (Santini et al. 2016), and potentially distance lonely older PWDs from the benefits of partnered sexual and intimate activity.

## Social/relational

Social and situational factors will affect sexual expression in OAs with disabilities. Potential social barriers include limited availability of sexual partners due to various reasons, including death of a partner or gender disparity—older women outnumber older men in the United States, creating a "partner gap"; older men are more likely to be partnered and report more sexual activity than women (Lindau and Gavrilova 2010). Also, PWDs may be overlooked or dismissed as potential sexual partners by others as a result of the myth of disabled asexuality and the stigma of social judgements about desirability (Howland and Rintala 2001). For older PWDs, lack of privacy in the living environment (e.g., a shared family home, a private home with a live-in caregiver, residential care facility) can also limit sexual expression (Mona et al. 2014).

The decrease in number of potential sexual partners may also be the result of self-selection. As we age, we tend to prioritise and engage in meaningful activities and relationships (Charles and Carstensen 2010), which results in fewer but higher-quality relationships. Thus, older-adult PWDs may be more intentional, emotionally regulated and less reactive in their relationships because they cultivate relational climates that have less stress. To illustrate, a qualitative study of the sexual well-being of OAs showed that sexual citizenship for some older men and women included purposeful abstinence, and wellness was at times derived from social intimacy rather than sexual intimacy (Syme et al. 2018).

Psychological and physical safeties are primary concerns for those with less power in sexual relationships, including OAs (Syme et al. 2018) and PWDs (Mona et al. 2014). It can have an impact on wellness in various ways, through coercion and interpersonal violence, or otherwise. For example, older women in heterosexual relationships have reported less self-efficacy with condom negotiation and specifically report feeling less power in relationships to negotiate sexual safety (Durvasala 2014). Power differentials expand beyond interpersonal sexual and intimate relationships as well. Adult children may not view their parents as sexual beings, which may be expressed as unsupportive language or paternalistic actions. People in positions of power—medical providers, long-term care administrators, adult children—have the responsibility to reflect on their own paternalism when considering their role in facilitating the sexual citizenship of older PWDs (Simpson et al. 2018).

## Cultural/environmental

A key socio-cultural consideration for the older PWD is an examination of the cultural structures and resources available to or absent from the sexual trajectory. For instance, it is well known that older generations report less sexual knowledge than younger generations, with historically limited opportunities for sex (Syme et al. 2017). This is likely compounded by the lifelong lack of quality sex education provided to PWDs (McCabe et al. 2000). This may disadvantage an older disabled person in terms of health and risk decisions, as well as self-efficacy in negotiating sexual experiences without practice or adequate information.

Socio-historical events may differentially shape the sexual trajectories of older disabled persons and must also be considered. For instance, many of the current older PWD generations residing in western nations experienced social upheaval in their sexual prime, including civil rights movements, Stonewall (i.e., gay rights movement in the US), women's rights movements, and the impact of the HIV/AIDS epidemic of the 1980s. Current events that may shape the sexual trajectories of the now and future older PWDs include healthcare laws, such as the Affordable Care Act, and human rights laws such as various marriage equality laws enacted globally in the past five to ten years (e.g., Canada, US, Europe, Australia). Also, the recent global #MeToo movement had a significant impact on older women, as many reported re-experiencing sexual traumas from earlier life that they had not yet reconciled, due to several years of burying/ignoring memories (Dvorak 2018). Older PWDs also have experienced advancements in sexual health, such as erectile enhancement drugs, birth control, Plan B, and general medical-technological leaps that provide new avenues for sexual choices. However, sexual citizenship challenges may limit the accessibility of these resources to older PWDs due to patient and provider ignorance, prejudicial attitudes of providers and family, and a host of social and environmental barriers.

In addition to the identities of older and disabled people, there are further intersections to consider that have impacted and will impact the sexual trajectories of older PWDs, such as gender. Men and woman experience different environmental constraints and supports across the life course, resulting in different trajectories. The double standard of ageing has a trickle-down effect for sexual citizenship as it is more acceptable for older men to be sexually active than for older women (Hillman 2011), which is documented in responses of older women related to their sexual well-being (Syme et al. 2018). Other examples include the fact that older women are disproportionately assigned the caregiver role in middle and later life, which affects their livelihood, their own physical and mental health and impacts their intimate and sexual relationships directly and indirectly (Drummond et al. 2013).

Layered stress comes in when factors, such as age, sex, gender, race, sexual orientation, disability, socio-economic status and religion intersect, manifesting in multiple forms of dis/advantage in terms of opportunity for sexual citizenship and freedom from disempowerment in sexual relationships. Older LBGTQ individuals—approximately 2% of the older population—have evidenced numerous physical and sexual health disparities (Fredriksen-Goldsen et al. 2015). In addition, environmental constraints such as discriminatory healthcare settings and nursing home environments, that are unsupportive and/or discriminatory, often result in older residents going "back in the closet" for fear of mistreatment (Furlotte et al. 2016). This revocation of sexual citizenship can result in limited contact with an existing partner or prohibit a resident from pursuing future intimate relationships.

## Supporting sexual citizenship and wellness across the life course

Educating older PWDs and healthcare providers helps address misinformation and misconceptions surrounding ageing, disability and sexuality to promote a stronger dialogue. For provider education, integrating standardised didactics into medical and graduate curricula or providing continuing education opportunities to existing primary care and other providers (e.g., physical therapists) will increase both visibility and acceptability. The impact of brief education is promising, both with OAs (Falvo and Norman 2004) and providers—including physicians, nurses (Bauer et al. 2013) and nursing home staff (DiNapoli et al. 2013).

However, information alone may not suffice. The systemic environments (e.g., medical settings) in which providers and older PWDs coexist often recapitulate narratives of older PWD as passive, dependent, ineffectual sexual agents (Pierpaoli Parker 2018). To address more systemic barriers, providers can start by recognising, inventorying and challenging biases about ageing, sexuality and disability (Levy 2009). Imperative to this process is acknowledging that power providers have to significantly enhance or subvert PWDs' sexual wellness through opening the discussion. Recall that many OAs are more likely to report issues with sexual functioning if and when their healthcare provider prompts them (Ports et al. 2014), further illustrating the influence providers can have on optimising sexual wellness. Providers can also be trained to assess sexual health using tools such as the PLISSIT model (Annon 1976), which provides a concrete, brief strategy for initiating and maintaining the discussion of sexuality— developed for use with OAs.

Empowerment of older PWDs to enhance their sexual health can also be addressed through psychoeducational interventions directed toward the older PWD. For example, the *Senior Sex Education Experience (SEXEE)* study develops, implements and evaluates sex education curricula tailored to the unique and specific needs of older adults (Pierpaoli Parker 2018), while integrating findings into an evidence-based protocol for improving, enhancing, and promoting sexual knowledge, functioning and communication among OAs. Enhancing psychological flexibility around sex is often a major goal, as older PWDs are subject to prevailing (heteronormative) performance-oriented intercourse models of sexuality that emphasize synchronicity, erection, penetration and orgasm, that render as insignificant a significant portion of sexual experiences in later life (McCarthy and Pierpaoli 2015). Embracing the full range of diverse sexual and intimate activities and asynchronous sexual scenarios can relieve older PWDs of sexual anxieties, increase sexual self-efficacy and expand their sexual repertoires (McCarthy and Pierpaoli 2015). Some of these scenarios may include mutual masturbation, oral sex, petting, erotic massage, "outercourse" and deep kissing. Learning to see partners as intimate friends, sex as a team sport and process (vs. performance) and being open to sampling the continuum of erotic scenarios, when sex does not flow to intercourse, can help older PWDs reclaim their sexual citizenship and pursue wellness.

While support interventions for older PWDs should provide education about sexual health and risk, curricula should also empower and promote positive sexuality and emphasise flexibility in conceptualising and expressing sexuality. This includes clear, humble communication, appropriate experimentation with intercourse and non-intercourse scenarios, affirming sensuality and intimacy, attaining sexual competence (e.g., the ability to give and receive pleasure), setting sexual boundaries based on preferences, values and safety (Bockting and Forberg 1998). Teaching dating (e.g., online, in-person), relationship and communication skills will also help PWDs to improve self-efficacy in negotiating safer sex and expressing sexual needs, questions and concerns to partners and providers alike. The eventual development of peer-based sexual wellness programmes (Pierpaoli Parker 2018) may also improve ongoing empowerment and sexual citizenship efforts. These efforts should honour individual agency and differences in the pursuit of sexual wellness, emphasising that one size does not fit all.

## Call to action

This exploration of sexual expression across the life course of older disabled individuals was purposefully approached with a sex-positive framework, prizing choice and empowerment for older PWDs. As discussed, the sexual citizenship of older PWDs continues to be jeopardised, and requires not just exploration, but action. Scholars are tasked with moving knowledge forward in an under-appreciated and, likely, under-funded area, well-served by social action research and other methods that empower diverse voices. Providers can begin with self-reflection on their own ageist and ableist beliefs about sexuality and continue by challenging those with education and training. Older disabled persons should also consider their own internalised ageist and ableist messages about sex that may inhibit their sexual wellness trajectories. Also, co-creation of environments and connections that support sexual wellness for older disabled persons and advocating for sexual citizenship for older PWDs in all professional spaces is a necessity.

## References

Albersen, M., Orabi, H. and Lue, T. (2012) Evaluation and treatment of erectile dysfunction in the aging male: A mini-review. *Gerontology* 58(1), pp. 3–14.

Ambler, D., Bieber, E. and Diamond, M. (2012) Sexual function in elderly women: A review of current literature. *Reviews in Obstetrics & Gynecology* 5, pp. 16–27.

Barrett, C. and Hinchliff, S. (2018) *Addressing the sexual rights of older people: Theory, policy, and practice.* New York: Routledge.

Basson, R. (1998) Sexual health of women with disabilities. *C.M.A.J: Canadian medical association Journal* 159, pp. 359–362.

Bauer, M., Fetherstonhaugh, D., Tarzia, L., Nay, R., Wellman, D. and Beattie, E. (2013) "I always look under the bed for a man." Needs and barriers to the expression of sexuality in residential aged care: The views of residents with and without dementia. *Psychology and Sexuality* 4(3), pp. 296–309.

Bauer, M., Haesler, E. and Fetherstonhaugh, D. (2016) Let's talk about sex: Older people's views on the recognition of sexuality and sexual health in the healthcare setting. *Health Expectations* 19(6), pp. 1237–1250.

Bockting, W. and Forberg, J. (1998) *All gender health: Seminars for Minnesota's transgender community, leaders manual.* Minneapolis, MN: University of Minnesota Medical School.

Bogart, K. (2014) The role of disability self-concept in adaptation to congenital or acquired disability. *Rehabilitation Psychology* 59(1), pp. 107–115.

Bremer, K., Cockburn, L. and Ruth, A. (2010) Reproductive health experiences among women with physical disabilities in the Northwest Region of Cameroon. *International Journal of Gynecology & Obstetrics* 108(3), pp. 211–213.

Campbell, F. (2008a) Exploring internalized ableism using critical race theory. *Disability & Society* 23(2), pp. 151–162.

Campbell, F. (2008b) Refusing able(ness): A preliminary conversation about ableism. *Journal of Media and Culture* 11 [online]. Available at: http://journal.media-culture.org.au/index.php/mcjournal/article/view/46 (Accessed: 15 Jan. 2019).

Carpenter, L. (2010) Gendered sexuality over the life course: A conceptual framework. *Sociological Perspectives* 53(2), pp. 155–177.

Charles, S. and Carstensen, L. (2010) Social and emotional aging. *Annual Review of Psychology* 61(1), pp. 383–409.

Colton, J. (2008) Sex and the elderly: What physicians should know about their older patients. Yale Medicine Thesis Digital Library 317, pp. 1–131.

Cornwell, E. and Waite, L. (2009) Social disconnectedness, perceived isolation, and health among older adults. *Journal of Health and Social Behavior* 50(1), pp. 31–48.

DeLamater, J. (2012) Sexual expression in later life: A review and synthesis. *Journal of Sex Research* 49(2–3), pp. 125–141.

Di Napoli, E., Breland, G. and Allen, R. (2013) Staff knowledge and perceptions of sexuality and dementia of older adults in nursing homes. *Journal of Aging and Health* 25(7), pp. 1087–1105.

Doll, G. (2016) Dementia and consent for sex reconsidered. *NAELA Journal* 12(2), pp. 133–144.

Dorado, K., McDonnell, C., Edwards, R. and Lazaridou, A. (2018) Sexuality and chronic pain. In: Enok, O. and Rolf, J. (eds.) *Understanding sexuality: Perspectives and challenges of the 21st century.* Hauppauge, NY: Nova Scotia Publishers, pp. 79–104.

Drummond, J., Brotman, S., Silverman, M., Sussman, T., Orzeck, P., Barylak, L. and Wallach, I. (2013) The impact of caregiving. *Affilia* 28(4), pp. 415–428.

Dunn, D. and Andrews, E. (2015) Person-first and identity-first language: Developing psychologists cultural competence using disability language *American Psychologist* 70(3), pp. 255–264.

Durvasula, R. (2014). HIV/AIDS in older women: Unique challenges, unmet needs. *Behavioral Medicine*, 40(3), pp. 85–98. doi:10.1080/08964289.2014.893983

Dvorak, P. (2018) "I never told anyone": Christine Blasey Ford has unleashed a torrent of sexual assault stories. *Washington Post* [online]. Available at: https://wapo.st/2R8YVVd (Accessed: 10 Jan. 2019).

Elder, G. (1994) Time, human agency, and social change: Perspectives on the life course. *Social Psychology Quarterly* 57(1), p. 4.

Esmail, S., Darry, K., Walter, A. and Knupp, H. (2010) Attitudes and perceptions towards disability and sexuality. *Disability and Rehabilitation* 32(14), pp. 1148–1155.

Falvo, N. and Norman, S. (2004) Never too old to learn: The impact of an HIV/AIDS education program on older adults knowledge. *Clinical Gerontologist* 27(1–2), pp. 103–117.

Fileborn, B., Thorpe, R., Hawkes, G., Minichiello, V., Pitts, M. and Dune, T. (2015) Sex, desire and pleasure: Considering the experiences of older Australian women. *Sexual and Relationship Therapy* 30(1), pp. 117–130.

Fisher, L. (2010) *Sex, romance, and relationships: AARP survey of midlife and older adults* [online]. Washington, DC: AARP. Available at: https://doi.org/10.26419/res.00063.001 (Accessed: 15 Jan. 2019).

Frankowski, A. and Clark, L. (2009) Sexuality and intimacy in assisted living: Residents perspectives and experiences. *Sexuality Research and Social Policy* 6(4), pp. 25–37.

Fredriksen-Goldsen, K., Kim, H., Shiu, C., Goldsen, J. and Emlet, C. (2015) Successful aging among LGBT older adults: Physical and mental health-related quality of life by age group. *The Gerontologist* 55(1), pp. 154–168.

Furlotte, C., Gladstone, J., Cosby, R. and Fitzgerald, K. (2016) "Could we hold hands?" Older lesbian and gay couples perceptions of long-term care homes and home care. *Canadian Journal on Aging/La Revue canadienne du vieillissement* 35(4), pp. 432–446.

Gill, M. (2015). *Already doing it: Intellectual disability and sexual agency*. Minneapolis, MN: University of Minnesota Press.

Gendron, T., Welleford, E., Inker, J. and White, J. (2016) The language of ageism: Why we need to use words carefully. *The Gerontologist* 56(6), pp. 997–1006.

Goedendorp, M., Knoop, H., Gielissen, M., Verhagen, C. and Bleijenberg, G. (2014) The effects of cognitive behavioral therapy for postcancer fatigue on perceived cognitive disabilities and neuropsychological test performance. *Journal of Pain and Symptom Management* 47(1), pp. 35–44.

Hillman, J. (2011) *Sexuality and aging: Clinical perspectives*. New York: Springer.

Hinchliff, S. and Gott, M. (2011) Seeking medical help for sexual concerns in mid- and later life: A review of the literature. *Journal of Sex Research* 48(2–3), pp. 106–117.

Hinrichs, K. and Vacha-Haase, T. (2010) Staff perceptions of same-gender sexual contacts in long-term care facilities. *Journal of Homosexuality* 57(6), pp. 776–789.

Howland, C. and Rintala, D. (2001) Dating behaviours of women with physical disabilities. *Sexuality and Disability* 19(1), pp. 41–70.

Levy, B. (2009) Stereotype embodiment: A psychosocial approach to aging. *Current Directions in Psychological Science* 18(6), pp. 332–336.

Lindau, S. and Gavrilova, N. (2010) Sex, health, and years of sexually active life gained due to good health: Evidence from two US population based cross sectional surveys of ageing. *BMJ* 340, p. c810.

Lindau, S., Schumm, L., Laumann, E., Levinson, W., OMuircheartaigh, C. and Waite, L. (2007) A study of sexuality and health among older adults in the United States. *New England Journal of Medicine* 357(8), pp. 762–774.

Marshall, B. (2011) The graying of "sexual health": A critical research agenda. *Canadian Review of Sociology – Revue canadienne de sociologie* 48(4), pp. 390–413.

McCabe, M., Cummins, R. and Deeks, A. (2000) Sexuality and quality of life among people with a physical disability. *Sexuality and Disability* 18(2), pp. 115–123.

McCabe, M., Sharlip, I., Lewis, R., Atalla, E., Balon, R., Fisher, A., Laumann, E., Lee, S. and Segraves, R. (2016) Incidence and prevalence of sexual dysfunction in women and men: A consensus statement from the Fourth International Consultation on Sexual Medicine 2015. *Journal of Sexual Medicine* 13(2), pp. 144–152.

McCabe, M. and Taleporos, G. (2003) Sexual esteem, sexual satisfaction, and sexual behavior among people with physical disabilities. *Archives of Sexual Behavior* 32(4), pp. 359–369.

McCarthy, B. and Pierpaoli, C. (2015) Sexual challenges with aging: Integrating the GES approach in an elderly couple. *Journal of Sex & Marital Therapy* 41(1), pp. 72–82.

Moin, V., Duvdevany, I. and Mazor, D. (2009) Sexual identity, body image and life satisfaction among women with and without physical disability. *Sexuality and Disability* 27(2), pp. 83–95.

Mona, L., Syme, M. and Cameron, R. (2014) Sexuality and disability: A disability affirmative approach to sex therapy. In: Binik, Y. and Hall, K. (eds.) *Principles and practices of sex therapy*. New York: Guilford.

Nelson, T. (2016) Promoting healthy aging by confronting ageism. *American Psychologist* 71(4), pp. 276–282.

Nosek, M., Howland, C., Rintala, D., Young, M. and Chanpong, G. (2001) National study of women with physical disabilities: Final report. *Sexuality and Disability* 19(1), pp. 5–40.

Olkin, R. (1999) *What psychotherapists should know about disability*. New York: Guilford.

Perissinotto, C., Stijacic Cenzer, I. and Covinsky, K. (2012) Loneliness in older persons: A predictor of functional decline and death. *Archives of Internal Medicine* 172(14), pp. 1078–1083.

Pierpaoli Parker, C. (2020). The Senior Sex Education Experience (SEXEE) Study: Considerations for the development of an adult sex education pilot intervention (Unpublished doctoral dissertation). The University of Alabama, Tuscaloosa, AL.

Porter, K., Brennan-Ing, M., Chang, S., Dickey, L., Singh, A., Bower, K. and Witten, T. (2016) Providing competent and affirming services for transgender and gender nonconforming older adults. *Clinical Gerontologist* 39(5), pp. 366–388.

Ratner, E., Erekson, E., Minkin, M. and Foran-Tuller, K. (2011) Sexual satisfaction in the elderly female population: A special focus on women with gynecologic pathology. *Maturitas* 70(3), pp. 210–215.

Rowen, T., Stein, S. and Tepper, M. (2015) Sexual healthcare for people with physical disabilities. *Journal of Sexual Medicine* 12(3), pp. 584–589.

Rubin, R. (2005) Communication about sexual problems in male patients with multiple sclerosis. *Nursing Standard* 19(24), pp. 33–37.

Santini, Z., Fiori, K., Feeney, J., Tyrovolas, S., Haro, J. and Koyanagi, A. (2016) Social relationships, loneliness, and mental health among older men and women in Ireland: A prospective community-based study. *Journal of Affective Disorders* 204, pp. 59–69.

Simpson, P., Wilson, C., Brown, L., Dickinson, T. and Horne, M. (2018) We've had our sex life way back: Older care home residents, sexuality and intimacy. *Ageing and Society* 38(7), pp. 1478–1501.

Steele, C. and Aronson, J. (1995) Stereotype threat and the intellectual test performance of African Americans. *Journal of Personality and Social Psychology* 69(5), pp. 797–811.

Syme, M. (2017) Supporting safe sexual and intimate expression among older people in care homes. *Nursing Standard* 31(52), pp. 52–63.

Syme, M., Cohn, T. and Barnack-Tavlaris, J. (2017) A comparison of actual and perceived sexual risk among older adults. *Journal of Sex Research* 54(2), pp. 149–160.

Syme, M., Cohn, T., Stoffregen, S., Kaempfe, H. and Schippers, D. (2018) "At my age": Defining sexual wellness in mid- and later life. *Journal of Sex Research*, Advanced publication online, pp. 1–11.

Taleporos, G. and McCabe, M. (2002) The impact of sexual esteem, body esteem, and sexual satisfaction on psychological well-being in people with physical disability. *Sexuality and Disability* 20(3), pp. 177–183.

Tetley, J., Lee, D., Nazroo, J. and Hinchliff, S. (2016) Let's talk about sex: What do older men and women say about their sexual relations and sexual activities? A qualitative analysis of ELSA Wave 6 data. *Ageing and Society* 38(3), pp. 497–521.

Tzeng, Y., Lin, L., Shyr, Y. and Wen, J. (2009) Sexual behaviour of institutionalised residents with dementia: A qualitative study. *Journal of Clinical Nursing* 18(7), pp. 991–1001.

Vansteenwegen, A., Jans, I. and Revell, A. (2003) Sexual experience of women with a physical disability: A comparative study. *Sexuality and Disability* 21(4), pp. 283–290.

Waite, L., Laumann, E., Das, A. and Schumm, L. (2009) Sexuality: Measures of partnerships, practices, attitudes, and problems in the National Social Life, Health, and Aging Study. *Journals of Gerontology. Series B – Psychological Sciences and Social Sciences* 64B(Supplement 1), pp. i56–i66.

Wilkins, J. (2015) More than capacity: Alternatives for sexual decision making for individuals with dementia. *The Gerontologist* 55(5), pp. 716–723.

Williams, D.J., Thomas, J.N., Prior, E.E. and Walter, W. (2015) Introducing the multidisciplinary framework of positive sexuality. *Journal of Positive Sexuality* 1, pp. 6–11.

World Association for Sexual Health (2014) *Declaration of sexual rights* [online]. Available at: http://www.worldsexology.org/resources/declaration-of-sexual-rights/ (Accessed: 10 Jan. 2019).

World Health Organization (2018) *Defining sexual health* [online]. Available at: https://www.who.int/reproductivehealth/topics/sexual_health/sh_definitions/en/ (Accessed: 15 Jan. 2019).

# 29

# TOWARD SEXUAL AUTONOMY AND WELL-BEING FOR PERSONS WITH UPPER LIMB MOBILITY LIMITATIONS

## The role of masturbation and sex toys

*Ernesto Morales, Geoffrey Edwards, Véronique Gauthier, Frédérique Courtois, Alicia Lamontagne and Antoine Guérette*

## Introduction

Our understanding of human sexuality has undergone a sea change in the past several decades, based largely on the ideas put forward by Michel Foucault (1990, 2012). Before Foucault's seminal work, sexuality was viewed by the medical community as a necessary component in human health, but it was also circumscribed by a range of paradoxical attitudes and ideas in the ways that it manifests itself in daily life. This is nowhere more evident than in the way sexuality among people with disability was determined. Margrit Shildrick writes (2009):

> In the contemporary western world, considerations of sexual pleasures and sexual desire in the lives of disabled people play very little part in lay consciousness, and practically none in the socio-political economy… The problem is that in the context of mainstream values, the conjunction of disability and sexuality troubles the parameters of the social and legal policy that purports both to protect the rights and interests of individuals, and to promote the good of the socio-political order. It is as though the very being of anomalous embodiment mobilises both an overt and an unspoken anxiety, an anxiety that is at its most acute in relation to sexuality. The outcome is the strange paradox evident in western society that alternates between denying that sexual pleasure has any place in the lives of disabled people and fetishising it. Both responses constitute a refusal of sexuality as a regular element in disability experience, an effective silencing that damages not just self-esteem, but—bearing in mind Merleau-Ponty's evaluation of sexuality as a mode of existence—all aspects of the capacity for self-becoming.

As Shildrick and Foucault suggest, the relationship between sexuality and well-being is much broader than that simply subsumed within the standard medical framing of sexual health and functioning. Furthermore, Rosi Braidotti's comments (1994) are also relevant to our subject:

Let us take as our starting point Foucault's analysis of the political economy of truth about sexuality in our culture. The distinction between technologies of reproductive power—scientia sexualis—and the practices of pleasure of the self—ars erotica—thus becomes capital.

Foucault argues that, since the eighteenth century, the bodily material has been situated at the heart of techniques of control and analysis aimed at conceptualising the subject. The term "bodily material" refers to the body as a supplier of forces, energies, whose materiality lends them to being used, manipulated and socially constructed. This critical perspective situates the two main approaches to sexuality in relation to technological aids, what she calls "scientia sexualis" for the medical perspective, and "ars erotica" for a counter framing, closer to Shildrick's idea of the "capacity for self-becoming." These perspectives challenge not only our understanding of sexuality (and by extension, gender) but also of disability.

Based on this reframing, there is a need to shift support for sexuality among people with disability towards an approach more focused on pleasure and the whole person, that is, integrate ars erotica with the existing scientia sexualis preoccupations. Furthermore, this needs to be carried out with sensitive awareness of how bodily material, what Braidotti calls the "organs without bodies" (a play on Deleuze's concept of the "Body without Organs"—Deleuze and Guattari 1987), can sometimes be manipulated within medical contexts without sufficient attention paid to the individual's full process of self-becoming. In this paper, the reframing of sexuality integrating ars erotica in the context of disability will be discussed in terms of self-becoming by focusing in issues related to masturbation practices. In this paper, we address this reframing of what is possible in terms of self-becoming by focusing in issues related to masturbation practices.

For many adults with disabilities, sexual self-stimulation is their only sexual activity and a significant number are limited in their ability to self-stimulate (Dupras 2012). On the other hand, the sexual reality of people with disabilities varies greatly from one person to another (Dupras 2012), a fact reinforced by our own study (Morales et al. 2016b). Their pathologies (e.g., stroke, spinal cord lesion, poliomyelitis, thalidomide, morbid obesity, cerebral palsy) and related issues (e.g., paralysis, amputation, spasms, fatigue, decreased sensitivity in erogenous areas, urinary problems, erectile dysfunctions, vaginal dryness secondary to changes of arousal capacity) will differ for each person (Giami 1987; Sevène 2014). Moreover, people with disabilities have higher rates of sexual dysfunction and have decreased desire, due to their condition and/ or the resulting psychological distress (McCabe et al. 2003), such as depression and anxiety that might or might not be related to their limitations.

Masturbation is a sexual practice that must be defined since it includes a complex set of sexual and social behaviours that are constantly under construction (Tiefer 1998). Studies investigating the biopsychosocial benefits of masturbation have observed that it can improve a person's overall health (Coleman, 2002), promote learning about one's own body (McCormick, 1994), be used as therapy to enhance body image or to reduce certain erectile dysfunctions in men (Coleman 2002; Tiefer 1998), and constitutes a less expensive alternative to sexual practices that present a certain risk (e.g., turning to a sex worker in order to experience pleasure) (Kaestle and Allen 2011). A priori, masturbation seems to be a kind of practice that is equally important to men and women (Action des femmes handicapées 2012). In general, men are presumed to masturbate more and to have a more permissive attitude to sexuality (Petersen and Hyde 2010). Women are said to feel ambivalent emotions about their masturbation habits, such as both satisfaction and shame (Bowman 2014). Culturally, masturbation is more valued in men, and women have less knowledge of their own bodies (Schwartz 2007; Wade et al. 2005), which could explain women's more marked ambivalence about these practices (Hogarth and Ingham 2009). On the other hand, there are an extensive number of products that can be found in sex shops for women only, some unisex, and a more limited number of products for men only, while some products cater to transgender individuals.

Masturbation, with its focus on giving oneself pleasure, as well as on giving free reign to the erotic imagination, embodies, quite literally, the re-engagement with ars erotica, as described by Braidotti. Furthermore, as highlighted by the literature discussed above, masturbation practices among people with disability reveal a much broader palette of practices than is usually assumed within a normative understanding of sexuality (Morales et al. 2016a; 2016b). For example, one man with spinal cord injury in our study had developed a highly eroticised oral practice which took the place of intercourse usually associated with sexual practice with a partner. Whereas some dealt with masturbation as a stress relief process, others turned the process of masturbation into an extended and prolonged sensual encounter with their own body. Furthermore, these different means of eroticising masturbation practice did not always follow expected gender divisions (Morales et al. 2016b). In addition, many of the women we interviewed had experienced sexual abuse (Morales et al. 2016a), and several of these found that masturbation offered a kind of therapy, enhancing their sense of being in control of their sexual lives after having had this control perturbed (Morales et al. 2016a). This highlights how personal history intersects with masturbation practice, further enhancing the heterogeneous function of the latter.

## Research purpose and design

In the following pages, we summarise the studies we carried out, first exploring adult users' own perceptions of their masturbatory practices, and then how we used these assessments to drive the (co)design of masturbation accessories, or sex toys, adapted to the needs of people with upper limb deficits and other mobility limitations. We then draw on these results to re-examine the broader issues this work raises on the way sexual pleasure and well-being are understood in the context of disability and, indeed, in society at large. We framed the work by viewing sex toys as more than objects that people use to increase their sexual pleasure, such as a dildo or a vibrator. In this broader perspective, any object was viewed as potentially a sexual aid to promote pleasure and we were not focusing only on the genital area. However, we had planned to 3D print the solutions from the beginnings of the project, and, due to funding considerations, we discarded many ideas and finally concentrated on stimulating female and male genitalia.

The research we developed focused on adults with disabilities in their upper limbs. We did not include adolescents within our participants in order to avoid extensive delays to obtain the ethics approval of the project, however we are well aware that adolescents with disabilities experience similar issues in their own sexual practices.

## The promotion of pleasure via the use of sex toys

The definition and use of sex toys has changed a good deal since vibrators were introduced in the 1800s. In the 1980s, a real change in the perception of these objects spread via the growing number and acceptance of sex shops. The booming sex industry re-defined them as "self-pleasure" tools (Laqueur 2003).

In North America today, many people still perceive the use of sex toys as something reserved for single, lonely or "desperate" people, whereas, in reality, more and more people are using them during solitary masturbation or in couples simply to enhance their pleasure (Herbenick et al. 2011). We now know that few women achieve vaginal orgasms (approximately 30%) since the vaginal walls have far fewer nerve endings than the head of the clitoris (Koedt 2010; Maines

1999). In general, masturbation is still more taboo for women than for men (Office des personnes handicapées du Québec 2017). Although sex toys are increasingly viewed as "self-pleasure" objects, medical professionals are more likely to prescribe these objects as a treatment for "sexual dysfunction." Even though the definition of the term "sexual dysfunction," which, in the 1970s, was used only for women, has changed and is now used for both men and women (lack of desire and lubrication, pain and dissatisfaction), it is still a medical term and implicitly treats sexuality as a physical disorder, whereas lack of desire or lubrication may be due more to sexual practices that do not mentally and emotionally satisfy women (Maines 1999) or to performance anxiety, especially in men (Donada 2012). The literature on sex toys suggests that, although using these is one of the most effective ways of achieving orgasm, this is still considered to be less acceptable and socially desirable than other methods (e.g., as a couple or manually with a partner) (Marcus 2011). Men still experience strong social pressure to perform sexually and the "cult of the phallus," supporting the myth that penis size is proportional to the partner's pleasure (Donada 2012), whether this be a woman, a gay man or a transgendered individual of any sex. This explains why some men may be intimidated by sex toys or fear that heterosexual women will no longer need men (Maines 1999).

With the growing spread of bricks and mortar sex shops, as well as online markets for sex products, there is now available a wide range of sex toys for women (e.g., vibrators, dildos, Ben Wa balls, and stimulators for the clitoris, G-spot, breasts and anus) and more and more toys for men (e.g., penile sleeves such as the Fleshlight, prostate/anal stimulators, penis rings). Vibration in sex toys may be variable or non-existent, since vibration can be noisy, making it impossible to remain private when a person does not live alone (living with family or roommates) (Marcus 2011). Sex toys with a human-like appearance seem to be more popular with men (toys that look like a vagina for men with a heterosexual orientation or like a penis for men with a homosexual orientation), while women use more sex toys in bright colours and fanciful shapes (Westermann 2013). To increase sales, there is an ever-increasing variety of sex toys meant to be more innovative, creative, realistic, luxurious or high-tech.

Sex toys are most often designed for women these days, partly due to the fact that these are more frequently prescribed by health care professionals to treat sexual dysfunctions (both in couples and alone) and also because of sex toy parties, which are still popular for "girls' nights in" (Albrecht 2012; Herbenick et al. 2011; Reece et al. 2009). Although companies, such as Eros & Cie, are starting to include couples, and therefore men, at these events, the objects presented that are designed for men are often described on the basis of how they can be used to satisfy women (Maines 1999). The aim is to give women the tools to better understand their sexuality so they can then teach men about their preferences and demystify the use of sex toys, which are not reserved for single people (Maines 1999). However, more and more items specifically for men are being created and sold in sex shops and online, and men are increasingly buying and using them. In fact, according to recent studies, approximately half of all Americans (44.8% of men and 52.5% of women) have used sex toys in their sex play (Herbenick et al. 2009; Reece et al. 2009). In November 2015, a survey carried out by Quebec's association of sex shops and erotic videos found that their customer base included as many men as women (Association des Boutiques ET Vidéos érotiques du Québec 2015).

In women without chronic health conditions affecting their sexuality, numerous beneficial effects of using sex toys have been documented, including increased desire, sexual awakening and lubrication, and more orgasms, including multiple orgasms and more intense and rapid orgasms (Davis et al. 1996; Herbenick et al. 2009; Marcus 2011; Shirley 1973). In fact, women who use a vibrator in their sex life generally have both better mental health (e.g., they feel more desirable) and better physical health (e.g., better knowledge of their body) than women who do not (Herbenick et al. 2009). Frequent use of a vibrator can even allow them to perceive anomalies faster (e.g., breast cancer or vulvar cancer) (Herbenick et al. 2009). Few studies have examined the issue in men, but it has been found that the use of sex toys can also enhance their masturbation practices by increasing and maintaining erections (Reece et al. 2009; Widow and Wassersug 2016). Some studies have found that it is not unusual for a male partner in a heterosexual relationship to masturbate with his wife's sex toy (Reece et al. 2009). A study carried out in 2016 also found that men who used sex toys on the prostate could have multiple orgasms without necessarily having an erection or ejaculating (i.e., without using the penis to have an orgasm) (Wibowo and Wassersug 2016). In the authors' view, this offers an interesting solution for men with erectile dysfunction (Wibowo and Wassersug 2016).

## The use of sex toys by individuals with disabilities

As can be inferred, sex toys can be a useful aid for people with physical disabilities involving erectile dysfunction, motor disabilities affecting the arms, chronic pain, decreased sensitivity and involuntary movements, as well as for people whose medication affects their mobility (RFSU 2010). There are sex toys on the market that were not created specifically for people with this kind of disability but whose design makes them easy to use (e.g., Sybian (Feelgood2 Ltd. 2016), Pulse II Solo (Abco Research Associates 2016), Vibrator Gripping Aid (The Active Hands Co. Ltd. 2016). In fact, some websites, such as *Come as You Are* (1997–2016), present sex toys that can be inserted (e.g., love egg, We-Vibe, vaginal or anal dildo) or placed on the genitals (e.g., butterfly vibrator for women, Fleshlight for men) and operated with a remote control (with a cord or cordless). Others provide suction cups that can be attached to the wall (e.g., dildo or Fleshlight attached to the wall); they can have different kinds of handles (e.g., dildo with an oval handle), whereas others have more intense vibration for people who want a stronger sensation or have decreased sensitivity (e.g., original Magic Wand similar to Ferticare-Acuvibe). This website also suggests the use of pillows to assist in positioning the body (Come as You Are 2016). In fact, there are cushions specifically designed to facilitate positioning of the body, such as the Liberator (2016). Some guides cover the topic of sexuality in people with disabilities and may recommend sex toys whose design can facilitate masturbation (Alberta Health Services 2013; Bailar-Heath et al. 2010; Kaufman et al. 2003).

Unfortunately, there are not many design solutions for adaptable sex toys on the market. Design is probably the discipline that excels in exercising discrimination, otherwise the world would be significantly more accessible. Regardless of the efforts made in terms of universal design and accessibility, the area that has not been explored is the design of technical aids for sexuality or sex-aids. The unprecedented boom of sexual toys and the porn industry in the past three decades has not yet taken into consideration the needs of adults with disabilities. There have been some attempts by companies such as mypleasure.com and sportsheets.com, although the distribution of their products remains limited.

Furthermore, many people with disabilities experience problems if their personal spaces are not adapted, restricting their ability to engage in sexual activity in comfort or with any measure of privacy (McCabe et al. 2003). Concerning the use of sex toys, people with disabilities must often employ care, since the insertion of an object into a bodily orifice may carry a risk of injury if movements cannot be controlled (e.g., spasms or cramps in the arms or hands) (Bailar-Heath 2010). Furthermore, with regard to hygiene, it is important to always wash sex toys properly after use. This task can be difficult, even impossible, for people with physical disabilities, so they may be obliged either to ask for help from someone else (e.g., a family member, friend or caregiver) (Bailar-Heath 2010) or stop using them. The use of sex toys as a therapeutic tool is recommended only when they are adapted to the individual's condition (Jannini et al. 2012). The numerous difficulties that people with motor disabilities may encounter greatly impair the pleasure and relaxation that masturbation could give them (Dupras 2012), especially for people with upper limb disabilities. The lack of appropriate sex education, added to the lack of privacy and accessibility in the area of sexuality (e.g., adapted sex toys), and society's failure to acknowledge the sexual needs of people with disabilities, can only hinder their autonomy in this regard (Masson 2013). According to Wilkerson (2001), sexual autonomy is a fundamental right for all people, and ignoring this fact contributes to the oppression of people with disabilities.

As highlighted by this review of sex toys and the ways they are used in both the broader culture and in relation to people with disabilities, it can be seen that sexuality is still largely treated as a medical condition and not as a source of pleasure and well-being for people with disabilities. With the wider public moving away from viewing masturbation as reprehensible, and with a widespread acceptance of sex toys as a source of pleasure and well-being, there is a need to not only include people with disabilities within this discussion, but also to ensure that these same sex toys can be used safely and readily. In the next section, we present efforts to carry out such a programme.

## The assessment of masturbatory practices and their use in support of the co-design of adapted sex toys

The *Centre interdisciplinaire de recherche en réadaptation et intégration sociale* (CIRRIS; Interdisciplinary Centre for Research on Rehabilitation and Social Integration) in Québec City, Canada, contributes to the development and knowledge transfer in the field of rehabilitation and social integration. This mission is made possible via research activities studying both personal (impairments and disabilities) and environmental factors (barriers and facilitators) that influence the social participation of persons with disabilities. This research is focused on interdisciplinary and intersectoral approaches that allow the study of complex issues with an integration of biomedical and social research. With the exception of Dr Courtois, all the authors of this chapter are researchers or research professionals at the CIRRIS. The present chapter results from funding by the Fonds de recherche du Québec—Société et Culture granting programme for "innovative projects," where the main objectives were to explore the experiences of men and women with upper limb disabilities in relation to masturbation and present a design process for sex toys that would meet their needs.

In order to develop this particular issue, a qualitative methodology was developed that included meetings involving participants with motor disabilities and a co-design focus group with professionals. The sample included 17 participants aged 18 years and over with motor disabilities of the upper limbs, able to take part in verbal exchanges (expression and comprehension) and with no cognitive impairments, and eight professionals (one caregiver, four occupational therapists, and three sexologists).

### Session 1: Individual semi-structured interviews with participants with motor disabilities

The first meeting with each participant was a two-hour, semi-structured interview to identify their individual experiences and the problems they had experienced in their sex lives. The interviews took place at either the participant's home or at CIRRIS and were recorded. At the first interview, the project was explained to the participants. The principal researcher or a research professional conducted the interview, sometimes together or with a graduate student, depending on the participant's preference as to the sex of the interviewer. The interview included open-ended questions focusing on participants' experiences, challenges, barriers, opportunities and facilitators during masturbation. The interview guide also included general inquiries about date of birth, gender, sexual orientation, education level, annual income, pathology and type and model of technical aid used. At the end of the first session, participants were asked to think about possible design solutions for one or more sex aids that would help them to engage in sexual self-stimulation. They were provided with a notebook, pencil and eraser and encouraged to write down and/or draw their projected solutions in the notebook in preparation for the next session. None of these individuals needed alternative methods of writing or sketching.

## Session 2: Individual co-design interviews with participants with motor disabilities

One week after the first interview, the principal researcher or a student, sometimes accompanied by a research professional (depending on the participant's preference) conducted the second co-design interview, which lasted two hours and was also recorded. The co-design sessions drew on participants' experiences to explore possible design solutions aimed at improving accessibility and inclusion (Sanders and Stappers 2008). This research was based on a user-centred design approach (Sanders and William, 2001), which incorporates participants' views and sees them as "partners," who contribute their expertise in the early phases of the design process (Sanoff 2007). Co-design methodologies have been successfully used in many different domains and with diverse populations (Francis et al. 2009; Frauenberger et al. 2011; Morales et al. 2012). For our study, during the co-design sessions, the participants were asked to describe suggestions for the design of alternative solutions for sex aids. The participants described their own ideas to the principal researcher, who simultaneously drew the idea on a large piece of paper. Input was provided by the researchers in order to give concrete form to the ideas and translate them into design proposals. This process generated new ideas and a graphic record (Morales et al. 2012), which were collected by the researcher.

## Session 3: Focus group with professionals

An expert focus group, involving caregivers, occupational therapists and sexologists, was conducted by the first author (an architect able to create drawings for validation). In assembling the members of the focus group, a special effort was made to find persons with different professions related to disability, sexuality and problems experienced by people with motor disabilities. We presented the ideas generated earlier by the participants (i.e., both their experiences and the graphic data). This allowed the stakeholders to critique, enrich and validate the design solutions proposed by the participants. Questions were asked about the feasibility and safety of each design and other possible modifications. New design solutions were generated based on these discussions. The main objective of this session was to assess and improve the design proposals from a practical, adaptive and clinical point of view.

## Data analysis

All three sessions were studied via content analysis techniques (Mucchielli 2009; Ryan and Bernard 2000) to determine participants' contributions. Several questions were added during the interviews to adapt to the respondents' discourse. All sessions were recorded and then transcribed. Using appropriate software, two members of the research team independently read through all the transcripts and noted the emerging themes, using an open coding procedure in the case of the first session. More detailed explanations were then requested and the feasibility, difficulties and problem-solving characteristics were identified in a text developed in tandem with the graphic data (drawings). Once this process was complete, tables and charts were developed to organise the data. The graphic results of the focus group were then compared with the main concepts defined in session 1 with user participants (Miles and Huberman 2003). This process enabled us to validate or challenge the design solutions adopted.

## Results of the study

Even though experiences with sex education and masturbation in men and women with disabilities often present more differences than similarities (Action des femmes handicapées 2012), masturbation is very important for both sexes and, despite their different anatomy, they have similar problems related to their disabilities. The main topics covered in the interviews are discussed here in relation to the literature concerning sex education, masturbation in people with disabilities, both in adolescence and in adulthood, and difficulties they may encounter related to masturbation.

Sex education for Quebecers aged over 30 was strongly influenced by the church (Roman Catholic religion), which was predominant in Quebec until the 1970s. As observed in the literature, most of the male and female participants received no information on the topic because it was a taboo subject in their families or the institutions they attended (schools and rehabilitation centres) (Browne and Russell 2005), largely because of the dominant influence of Roman Catholicism (Desjardins 1997). The lack of education on the subject seems to have had more of an impact on women, however, since the female participants mentioned that they generally had their first sexual experiences later, in late adolescence or early adulthood, and found it more difficult to get to know how their bodies functioned in order to achieve orgasm. The lack of knowledge most women had of their own bodies (Schwartz 2007; Wade et al. 2005) might explain why the women in the study found it very difficult to explain the masturbation techniques they used: they were unable to name the body parts they stimulated (e.g., clitoris, G spot, vaginal wall). Furthermore, during adolescence, teens with disabilities, and especially girls, were overprotected by their parents and caregivers, who did not encourage, emphasise or normalise their relationship with sexuality; the result was a lack of education, which was detrimental to their sexual satisfaction (Vaughn et al. 2015). Despite the lack of knowledge, men on the other hand, had no difficulty accurately describing their techniques and the body parts they stimulated.

Regarding the participants' masturbation practices, both men and women expressed physiological needs, and not just the men, as many earlier studies had suggested (Bajos and Bozon 2008). Just like the men, the women pointed out that masturbation could meet their need to reduce stress and fulfil their physical needs (sexual tension). Furthermore, contrary to the social discourse regarding motivations to engage in sexual activity, the male participants did not mention the concept of performance. However, in accordance with the prevailing discourse in today's society, the men who were interviewed said that they were motivated by sexual desire and the need to reduce stress, while the women participants mentioned that they were motivated by love, sensuality and the need for affection (Boivin 2014). In addition, the men were more likely to use sexuality as a means of mitigating negative psychological states, while negative emotions such as suffering might diminish the women participants' sexual desire (Carvalho and Novre 2011; Hill and Preston 1996). As for the environment, most participants said that they masturbated in their bedrooms and several mentioned that they found this difficult because their living space was not well adapted and did not allow them to do so optimally (McCabe et al. 2003). Both men and women highlighted privacy concerns that discouraged them from masturbating in places such as the bathroom, because of the presence of a personal care attendant.

The women in this study were more likely to use sex toys for masturbating than the men. Nevertheless, as observed in the literature, the use of such objects appeared to be problematic for both sexes (Bailar-Heath et al. 2010). Participants mentioned certain risks such as the possibility of injuring themselves while using sex toys that are not adapted to their motor skills (e.g., spasms or cramps in the arms) (Bailar-Heath et al. 2010). They also mentioned problems cleaning these items, the impression that the toys were incompatible with their body, buttons that were too hard to press, overly complex instructions, and excessively high purchase prices (given that it is impossible to test effectiveness before buying one). All participants revealed an interest in using sex toys as therapeutic tools to facilitate masturbation if these tools were adapted to their condition (Jannini et al. 2012). As observed in the literature, the men in the sample had a more permissive attitude to their sexuality (Petersen and Hyde 2010), and some of them had used the services of sex workers to meet their needs. The women, on the other hand, were ambivalent about their masturbation practices, feeling both satisfaction and shame (Bowman 2014). The fact that sexual activity is generally more valued in men in our society, whereas women often lack knowledge in this domain (Schwartz 2007; Wade et al. 2005), could explain why the latter felt more ambivalent about masturbation (Hogarth and Ingham 2009) and sex workers. In addition, the fact that sex workers' services are not legalised in our society or are otherwise controversial, makes some people reluctant to use them. As for the biopsychosocial benefits of masturbation, this study also observed that this activity can improve health (Coleman 2002), foster bodily learning (McCormick 1994) and enhance body image (Christensen 1995).

## *Developing prototypes*

*Rhinoceros 3D* software was used to create designs for all the proposed objects. This software allows one to quickly and accurately model curved objects in two and, primarily, three dimensions. The main advantage of using this software in our project was its compatibility with the 3D printer available at CIRRIS. As a result, it was easy to work with "polysurfaces" (mathematical representations of curves and surfaces) and convert a model into a "mesh" (triangulation of the model, allowing it to be exported in different formats) compatible with the 3D printer software. Prototypes of sex toys were then made from acrylonitrile butadiene styrene (ABS) plastic, heated to a high temperature, layered down on the printer's platform, and then cooled. The prototypes were then placed in a bath of solution to remove all the residual parts that were used for support during printing. Once this process was completed, some of the models were covered with silicone of a specific elasticity and density in a mould that had also been 3D printed. Note that all the materials used were regulated, flexible and non-allergenic, so that the prototypes were safe and pleasant to use.

The first prototype unisex sex toy we developed was a nipple stimulator. This is an object with numerous silicone nubs that stimulate the nipples, with the help of a vibrator. The idea was that this system would be integrated into a garment to increase users' privacy. However, the research team decided to stop working on this prototype because it ran into major difficulties in creating certain parts out of silicone.

As for prototypes for women, two prototype sex toys that can stimulate three female erogenous zones were selected for development. The first was a simple dildo in standard male dimensions, covered with transparent silicone. This prototype could potentially be placed on the end of a wand (extension) to artificially lengthen a user's arms. It was designed to reach the

G-spot while allowing for vaginal stimulation. The second object stimulated the third erogenous zone (the clitoris) and could be used at the same time as the dildo. The prototype, which was C-shaped, could be placed with one side on the clitoris and the other on the vaginal wall at the level of the G-spot. This object produced a vibration on the external clitoris that could also be felt throughout the internal clitoral area and controlled remotely.

However, due to limited options of materials for the 3D printer model available at the research centre, along with complications in the development of the moulds, we decided to adapt an existing dildo model and adapt the suction cup to fit a wand extension that we created (see Figure 29.1).

Among prototypes developed for men, the development of the prototype prostate stimulator was halted for safety reasons; although the research team tried to work on this important aspect, participants' safety could not be assured. Regarding stimulation of the penis, a prototype selected for development was a wraparound pump. This prototype consisted of a rectangular piece of silicone with rings that allowed it to be rolled around to create a tube similar to a Fleshlight. Thus, the man's penis can be placed on the rectangular part and the rings make it possible to roll the silicone around it. However, the inflatable feature of the sex toy posed several difficulties for the fabrication and we decided to replace this idea with a smooth silicone surface made with a 3D-printed mould for the interior part of the sex toy. This prototype toy could be attached to one end of a wand, enabling men with motor disabilities to use it even if they do not have the motor skills to reach their genitals with their hands (Figure 29.2). All of these characteristics allow a man with erectile dysfunction to use this item more easily than toys now available on the market (he can easily insert his penis into the toy even if he does not yet have an erection).

The prototype sex toys mentioned above were developed with the option of attaching a wand (extension) that would make masturbation easier for users with a motor disability who are unable to reach their genitals with their hands. All parts of the wand can be screwed together; they are easy to take apart and are interchangeable (Figures 29.3). The upper end of the wand was configured so that different-shaped handles could be installed. The first handle was similar to a joystick and could be grasped by the middle or the top. This first type of handle also allowed a strip of Velcro to be attached so the user could have better control. It was designed to deal with the problems experienced by people with tetraplegia, spasticity of the arms, arthritis, pain in the

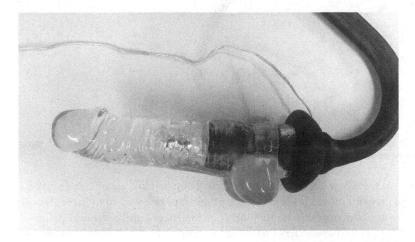

*Figure 29.1* Prototype of adaptation to suction cup of existing dildo models.

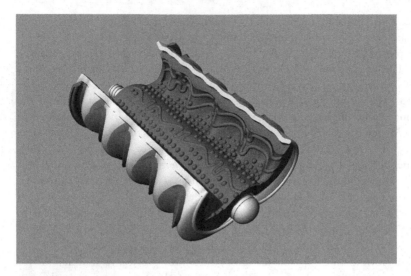

*Figure 29.2*  Prototype penis stimulator.

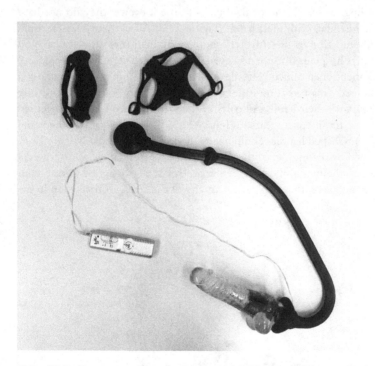

*Figure 29.3*  Prototype of wand extension adapted to suction cup of existing dildo models and three different types of handle.

hands or wrists or loss of sensitivity. The second kind of handle looked like a star with four symmetrical arms; like the previous model, it allowed strips of Velcro to be attached to make it more versatile for users, who could operate it with their arms or forearms. This handle was designed for people with quadriplegia (if they can flex their elbows), pain in the hands and/or wrists, loss of sensitivity, amputation of the upper limbs or shortened upper limbs. The third kind of handle was a simple sphere the approximate size of a tennis ball, which fitted into the palm of the hand. This kind of handle was primarily for people who have a certain amount of motor ability in their hands, enabling them to grasp spherical objects.

These prototypes were printed as described, and then incorporated into a validation experiment, the results of which are still being assessed. There were several technical issues that limited our ability to fully assess their utility. For example, some of the extensions which we developed, broke. Furthermore, it was easier to manufacture robust designs for women than the more complex designs required for men. Nonetheless, some users reported success and satisfaction with the products. Ultimately, of course, difficulties of this kind could be expected. 3D printing technology is still very new, so that finding and using robust materials for applications, such as this one, is still a challenge.

In summary, we identified several devices and/or adapted extensions that could render sex toys more easily accessible to people with disabilities, especially those with limited upper limb mobility. A subset of these devices was developed, and some preliminary evaluations obtained. Although technical issues remain a challenge, the process as a whole has been extremely informative. Furthermore, in addition to calling into question or seeking to enhance a whole set of masturbatory practices, the study itself raised important issues. Even carrying out a study of this sort within a research centre posed its own challenges. Indeed, discussions within the research team addressed sexuality within an open-ended framework, including preoccupation with sexual orientation, gender identity and, indeed, the relation between private sexual practices and public discourse about sexuality, since each of us have our own, private, sexual practices.

## Rethinking sexual pleasure and well-being in the context of disability

Whereas it is clear that the sexual revolution still has a way to go in the context of disability, it is now understood, within the context of health care and the promotion of health, that sexuality is not only a right, but also a source of well-being, for people with disabilities as well as the wider public. That said, it is also clear from the above discussion that the framing of sexuality in terms of pleasure is still problematic for the health community, perhaps particularly in the context of people with disabilities, partly as a result of broader public perceptions of disability as identified in the opening paragraphs of this chapter. Indeed, we have been discussing the context of people with upper limb mobility limitations in particular, but the same arguments apply to other populations, including people with intellectual disabilities, mental health issues, and those with autism, where some have argued that encouragement of sexual expression be limited. The use of sex toys as a source of pleasure poses similar difficulties to at least some health professionals. The tendency towards prescription of sex toys primarily to address sexual dysfunction is an example of such a reductive understanding of sexual practices and their benefits. Furthermore, the commercial context under which sex toys are promoted, and the links between this and the porn industry, contribute to increasing a sense of unease. The marketplace, in addition, exercises its own form of censorship. And yet, as our studies highlight, a focus on designing sex toys that enhance pleasure for people who experience a wide range of disabling environments may not only open up new possibilities of sexual experience but may also contribute to broadening commercial markets. Indeed, while several of our prototypes resemble existing products, others incorporate new ideas and approaches which might well find application to a broad group of users, not only those who self-identify as living with disability.

In addition, it can be argued that the issue of sexuality within the context of disability opens up broader questions about the nature of what it is to be human. This issue is part of the discussion as raised by Foucault (1990; 2012) and followed through by Shildrick (2009) and Braidotti

(1994). For, while our bodies may not be social constructs per se, all our understandings of the body are socially constructed. The context of disability itself situates this argument. The modern understanding of disability is that this is not a "property" of anyone's body—in a sense, no such entity as a "disabled person" exists. The social model of disability views disability as arising when a given body is not well supported by the environment (Shakespeare and Watson 2002). So, a person is "disabled" by the environment. Furthermore, and this is also important, different functional limitations will lead to different forms of disability. So, saddling all these different people with a common label ("disabled") is doubly problematic—this is, in itself, a way of denying the reality of the experience of disability, which is anything but uniform.

## Conclusion

Our study highlighted the wide range of sexual masturbation practices undertaken by "people living in disabling situations" in order to work around their limitations, and access pleasurable experiences. Sexuality, in this context, is characterised by its heterogeneity. But once one has opened up our conception of sexuality in this way, why stop there? Sexual orientation and gender identities (and gender fluidity) are similarly much more varied than the traditional, normative models would have us believe. These areas of the human experience are interconnected. Once one moves beyond a perspective which tries to categorise sexuality into dysfunctional or normative (scientia sexualis) and adopts an approach which recognises the variability of the human form and our modes of finding pleasure (ars erotica), there is, in all likelihood, no going back to a narrower perspective.

## References

Abco Research Associates (2016) Sybian Store: Sybian. Available from: https://sybian.com/store/ [Accessed 8th August 2019].

Albrecht, L. (2012) *Home sex toy parties: A non-traditional, uniquely situated venue of sexuality education for women.* Edmonton: University of Alberta Press.

Bailar-Heath, M.B., & Hough, S. (2010) Sexual device manual for persons with disability. *Sexuality and Disability* 28(1), pp. 61–62.

Bajos, N., & Bozon, M. (2008) *Enquête sur la sexualité en France.* Paris: La Découverte.

Boivin, J. (2014) *L'association entre la dépendance, l'autocritique, les motivations sexuelles et la satisfaction sexuelle.* Montréal: Université du Québec à Montréal Press.

Bowman, C.P. (2014) Women's masturbation experiences of sexual empowerment in a primarily sex-positive sample. *Psychology of Women Quarterly* 38(3), pp. 363–378.

Braidotti, R. (1994) *Nomadic subjects: Embodiment and sexual difference in contemporary feminist theory.* New York: Columbia University Press.

Browne, J., & Russell, S. (2005) My home, your workplace: People with physical disability negotiate their sexual health without crossing professional boundaries. *Disability & Society* 20(4), pp. 375–388.

Carvalho, J., & Nobre, P. (2011) Différences de genre et désir sexuel. Comment les facteurs émotionnels et relationnels déterminent-ils le désir sexuel selon le genre? *Sexologies Journal* 20(4), pp. 235–240.

Coleman, E. (2002) Masturbation as a means of achieving sexual health. *Journal of Psychology & Human Sexuality* 14(2/3), pp. 5–16.

Come as You Are Co-operative (2016) Sexuality and disability. Available from: http://www.comeasyouare.com [Accessed 8th August 2019].

Davis, C., Blank, J., Lin, H.Y., & Bonillas, C. (1996) Characteristics of vibrator use among women. *Journal of Sex Research* 33, pp. 215–320.

Deleuze, G., & Guattari, F. (1987). In B. Massumi (Trans.), *A thousand plateaus: Capitalism & schizophrenia.* Minneapolis: University of Minnesota Press.

Desjardins, G. (1997) Une mémoire hantée: L'histoire de la sexualité au Québec. *Cap-aux-Diamants: La revue d'histoire du Québec* 49, pp. 10–14.

Donada, G. (2012) Sexe: Hommes, l'angoisse de la performance. Available from: http://www.psycholo-gies.com/Couple/Sexualite/Desir/Articles-et-Dossiers/Sexe-hommes-l-angoisse-de-la-performance [Accessed 8th August 2019].

Dupras, A. (2012) Handicap et sexualité: Quelles solutions à la misère sexuelle? *ALTER* 6(1), pp. 13–23.

Feelgood2 Ltd (2016) Humpus; single or together, always a pleasure! Available from: http://www.humpus.com/ [Accessed 8th August 2019].

Foucault, M. (1990) *The history of sexuality: An introduction*, volume 1. New York: Vintage.

Foucault, M. (2012) *The history of sexuality: The use of pleasure*, volume 2. New York: Vintage.

Francis, P., Balbo, S., & Firth, L. (2009) Towards co-design with users who have autism spectrum disorders. *Universal Access in the Information Society* 8(3), pp. 123–135.

Frauenberger, C., Good, J., & Keay-Bright, W. (2011) Designing technology for children with special needs: Bridging perspectives through participatory design. *Journal on CoDesign* 7(1), pp. 1–28.

Giami, A. (1987) Coping with the sexuality of the disabled: A comparison of the physically disabled and the mentally retarded. *International Journal of Rehabilitation Research* 10(1), pp. 41–48.

Herbenick, D., Reece, M., Sanders, S.A., Dodge, B., Ghassemi, A., & Fortenberry, J.D. (2009) Prevalence and characteristics of vibrator use by women in the United States: Results from a nationally representative study. *J Sex Med* 6, pp. 1857–1866.

Herbenick, D., Schick, V., Reece, M., Sanders, S., Dodge, B., & Fortenberry, J.D. (2011) The Female Genital Self-Image Scale (FGSIS): Results from a nationally representative probability sample of women in the United States. *Journal of Sexual Medicine* 8, pp. 158–166.

Hill, C.A., & Preston, L.K. (1996) Individual differences in the experience of sexual motivation: Theory and measurement of dispositional sexual motives. *Journal of Sex Research* 33(1), pp. 27–45.

Hogarth, H., & Ingham, R. (2009) Masturbation among young women and associations with sexual health: An exploratory study. *Journal of Sex Research* 46(6), pp. 558–567.

Jannini, E.A., Limoncin, E., Ciocca, G., Buehler, S., & Krychman, M. (2012) Ethical aspects of sexual medicine. Internet, vibrators, and other sex aids: Toys or therapeutic instruments? International Society for Sexual Medicine 9(12), pp. 2994–3001.

Kaestle, C.E., & Allen, K.R. (2011) The role of masturbation in healthy sexual development: Perceptions of young adults. *Archives of Sexual Behavior* 40(5), pp. 983–994.

Kaufman, M., Silverberg, C., & Odette, F. (2003) *The ultimate guide to sex and disability*. Berkeley: Cleis Press Inc.

Koedt, A. (2010) Le mythe de l'orgasme vaginal. *Nouvelles Questions Feministes* 29(3), pp. 14–22.

Laqueur, T. (2003) *Solitary sex: A cultural history of masturbation*. New York: Zone Books.

Liberator (2016) Liberator. Available from: https://www.theliberator.co.uk [Accessed 8th August 2019].

Maines, R. (1999) *The technology of orgasm: 'Hysteria', the vibrator, and women's sexual satisfaction*. Baltimore: John Hopkin's University Press.

Marcus, B.S. (2011) Changes in a woman's sexual experience and expectations following the introduction of electric vibrator assistance. *Journal of Sexual Medicine* 8(12), pp. 3398–3406.

Masson, D. (2013) Femme et handicap. *Recherche fémine* 26(1), pp. 111–129.

McCabe, M.P., Taleporos, G., & Dip, G. (2003) Sexual esteem, sexual satisfaction, and sexual behaviour among people with physical disability. *Archives of Sexual Behavior* 32(4), pp. 359–369.

McCormick, N.B. (1994) *Sexual salvation: Affirming women's sexual rights and pleasures*. Praeger: Westport.

Miles, M.B., & Huberman, A.M. (2003) *Analyse des données qualitatives*. 2nd edition. Brussels: De Boeck Supérieur.

Morales, E., Gauthier, V., Edwards, G., & Courtois, F. (2016a) Women with disabilities' perception of sexuality, sexual abuse and masturbation. *Sexuality and Disability* 34(3), pp. 303–314.

Morales, E., Gauthier, V., Edwards, G., & Courtois, F. (2016b) Masturbation practices of men and women with upper limb motor disabilities. *Sexuality and Disability* 34(3), pp. 417–431.

Morales, E., Rousseau, J., & Passini, N. (2012) Using a co-design methodology for research on environmental gerontology. Journal of Gerontology and Geriatric Research 106(1), pp. 1–10.

Mucchielli, A. (2009) *Dictionnaire des méthodes qualitatives en sciences humaines*. 2nd edition. Paris: Armand Colin.

Office des personnes handicapeés du Québec (2017) Les personnes avec incapacités au Québec; Prévalence et caractéristiques de l'incapacité. Enquête canadienne sur l'incapacité de 2012, Drummondville: Gouvernement du Québec.

Petersen, J.L., & Hyde, J.S. (2010) A meta-analytic review of research on gender differences in sexuality, 1993–2007. Psychological Bulletin 136(1), pp. 21–38.

Reece, M., Herbenick, D., Sanders, S.A., Dodge, B., Ghassemi, A., & Fortenberry, J.D. (2009) Prevalence and characteristics of vibrator use by men in the United States. *Journal of Sexual Medicine* 6(7), pp. 1867–1874.

Ryan, G., & Bernard, R. (2000) Data management and analysis methods. In: Denzin, N., & Lincoln, Y. (eds.), *Handbook of qualitative research*, 2nd edition. Thousand Oaks: SAGE, pp. 769–802.

Sanders, E.B.N., & Stappers, P.J. (2008) Co-creation and the new landscapes of design. *Journal of CoDesign* 4(1), pp. 5–18.

Sanders, E.B.N., & William, C.T. (2001) Harnessing people's creativity: Ideation and expression through visual communication. In: Langford, J., & McDonagh-Philip, D. (eds.), *Focus groups: Supporting effective product development*. London: Taylor & Francis, pp. 145–156.

Sanoff, H. (2007) Special issue on participatory design. *Design Studies* 28(3), pp. 213–215.

Schwartz, P. (2007) The social construction of heterosexuality. In: Kimmel, M.S. (ed.), *The sexual self: The construction of sexual scripts*. Nashville: Vanderbilt University Press, pp. 80–92.

Sevène, A. (2014) Fonction sexuelle et handicap physique. *La Presse Médical* 43(10), pp. 1116–1119.

Shakespeare, T., & Watson, N. (2002) The social model of disability: An outdated ideology? *Research in Social Science and Disability* 2, pp. 9–28.

Sherfey, M.J. (1973) *The nature and evolution of female sexuality*. New York: Random House.

Shildrick, M. (2009) *Dangerous discourses of disability, subjectivity and sexuality*. New York: Palgrave Macmillan.

The Active Hands Co Ltd (2016) Vibrator gripping aid Eupore. Available from: http://www.eastin.eu/en-ie/searches/products/ detail/database-dlf%20data/id-0107181 [Accessed 8th August 2019].

Tiefer, L. (1998) Masturbation: Beyond caution, complacency and contradiction. Journal of Sex and Marital Therapy 13(1), pp. 9–14.

Vaughn, M., Silver, K., Murphy, S., Ashbaugh, R., & Hoffman, A. (2015) Women with disabilities discuss sexuality in San Francisco focus groups. *Sexuality and Disability* 33(1), pp. 19–46.

Wade, L.D., Kremer, E.C., & Brown, J. (2005) The incidental orgasm: The presence of clitoral knowledge and the absence of orgasm for women. *Journal of Women's Health* 42(1), pp. 117–138.

Westermann, E. (2013) *La commercialisation d'objets érotiques: Représentations, discours et communication dans un marché en croissance. Le cas des présentations à domicile*. Montréal: Université du Québec à Montréal Presse.

Wibowo, E., & Wassersug, R.J. (2016) Multiple orgasms in men—What we know so far. *Sexual Medicine Reviews* 4(2), pp. 136–148.

Wilkerson, A. (2001) Disability, sex radicalism, and political agency. In: Hall, K.Q. (ed.), *Feminist disability studies*. Bloomington: Indiana University Press, pp. 193–217.

# 30

# PAID SEXUAL SERVICES FOR PEOPLE WITH DISABILITY

## Exploring the range of modalities offered throughout the world

*Rachel Wotton*

## Introduction

Discourse pertaining to the rights of people with disability are most often embedded in issues concerning increased inclusivity within the workforce, accessible parking and buildings, meaningful participation within societal activities and access to affordable technology, support aids, accommodation and transportation. Increased visibility and representation of people with disability within the film industry, television programmes, advertisements, performances and exhibitions have also contributed towards a growing awareness of the rights of people with disability (Wotton 2016b). Acknowledgement and recognition of people with disability as sexual beings has become the "last frontier." The quest for connectivity and basic human affection for people with disability has often been marred by the politics of body autonomy, religious and moral stances and physical and legal barriers. Additionally, within social media and public commentary, there is a constant debate regarding whether sexual expression should be regarded as a human right or merely a human desire (Heidari 2015; Miller et al. 2015).

This chapter explores the different sexual and socially structured frameworks some people with disability utilise to explore their sexual expression, outside of the usual dating and relationship paradigms. It examines the professional capacities of sex workers, sexual surrogates, sexual assistants and masturbation services, identifying their current geographical locations and reviewing the nuances within each modality. The legal frameworks in which each operates will also be discussed, as well as identifying their similarities and differences.

Information and observations included within this chapter are derived from the author's lived experience of being a sex worker for over 26 years, from the accumulative research which informed her Master's dissertation, *Sex Workers Who Provide Services to Clients with Disability in New South Wales, Australia* (Wotton 2016b), knowledge gained during her Churchill Fellowship (Wotton 2017), professional networking with academics and individuals, participation at a range of conferences and ongoing involvement with the organisation Touching Base Inc., which she co-founded in 2000. The integration of these personal and professional experiences has assisted in framing the different sexual service modalities on offer for people with disability, from a complex and holistic perspective. This chapter is an extension of the author's research in this area (Wotton and Isbister 2007; 2017)

## The right to sexual expression and pleasure

Regardless of increased visibility in the media, multimedia platforms and on TV, the sexual rights of people with disability is still an incredibly contentious issue. As Tepper (2000a) noted, "the intersection of sexuality and disability is often associated with victimisation, abuse and purity." People with disability are still often regarded as being either hyper-sexual (McCabe 1999; Taylor Gomez 2012) or completely asexual (Esmail 2010). Both these fallacies are shrouded in fear, categorising the person as either being someone "out of control," whose deviant behaviour needs to be controlled and managed, <u>or</u> someone who is both socially and sexually vulnerable. These misguided beliefs lead to further disempowerment of people with disability. This has included a lack of opportunity to receive any sex education, as well as forced sexual segregation, physical and sexual confinement and even marital prohibition (Jungels and Bender 2015; Sakellariou 2006; Taylor Gomez 2012; Tepper 2000). Other barriers identified, precluding people from enjoying full sexual expression, include a lack of accessible and meaningful resources around sex, sexuality, choices and options; a lack of clear policies in disability organisations to guide workers and support staff; an unwillingness from staff to assist in transportation, communication or purchasing even basic equipment, such as lubricant or sex toys; a fear of ridicule from service providers, support staff or carers; pervasive religious beliefs of others or their service providers; disapproval from family and friends; poverty; infantilism; isolation and a lack of peer support (Magnan and Reynolds 2006; O'Dea et al.2012; Shuttleworth and Taleporos 2016; Silverberg and Odette 2011). These restrictive environments strip people of their dignity, bodily autonomy and their own agency and yet, as George Taleporos once stated, *disability does not abolish the need and potential for sexual expression* (2001, p. 156).

Whereas the majority of people with disability around the world enjoy being sexually active with no need for specialised or specific assistance, some people with disability have chosen to utilise a number of different paid sexual services. These include booking a sex worker, a sexual surrogate, a sexual assistant or receiving a un/paid masturbation service. The remainder of this chapter will explore these differing modalities, outlining the description of the services, the geographical location of where services are offered, the pricing structure of each and the aims and objectives of the service. Additional information covered includes examining the motivation of the client to seek such services, the availability and eligibility of both the service provider and the client, as well as outlining specific organisations who may facilitate training for those who provide the sexual services.

## Sex work

Terminology used to describe the sex industry and those who participate in it, have varied with time and societal norms. The term *sex worker* was coined by sex worker and activist, Carol Leigh (aka Scarlet Harlot), in 1978. "Sex work" and "sex worker" are now listed in both the Merriam Webster Dictionary and the Oxford English Dictionary. There are a myriad of definitions utilised within academia (Allman and Ditmore 2016), all mirroring, to differing degrees, the one that Weitzer uses, being "the exchange of sexual services, performances, or products for material compensation. It includes activities of direct physical contact between buyers and sellers ... as well as indirect sexual stimulation" (Weitzer 2010).

Recent online media, short films, documentaries and feature films have all paid tribute to clients with disability choosing to see a sex worker. These include *The Price of Intimacy: The Time*

*I Hired a Sex Worker* (Gurza 2016), *For One Night Only* (BBC One 2007), *Hasta la Vista (Come as You Are)* (Clercq & Philpot 2011), *Short on Short* (Swartz 2015), *Scarlet Road* (Paradigm Pictures 2011; Wallace 2013), TEDx talk: *Open Your Mind to What Goes on Behind Closed Doors* (Wotton 2016a), *The Too Hard Basket* (Blades, 2009), *I Have Cerebral Palsy and I Enjoy Having Sex* (Wright, 2014), *The Intouchables* (Nakache and Toledano 2011), *L'Assistante* (Guiraud and Jaulin 2012) and *The Gift* (Harvey and Harvey 2013).

## Services offered

The most commonly-held assumption is that sex work is *only* about heterosexual vaginal intercourse. This is categorically wrong. Not only are sex workers and their clients of all genders, but the services provided by sex workers to their clients vary as much as one's imagination can reach. A recent study in New South Wales, Australia (Wotton 2016b) identified over 20 different types of services that sex workers provided to their clients with disability. This included hand relief (masturbation), oral sex, vaginal sex, anal sex, fantasies, roleplays, nude body slides, nude massage, striptease, Bondage, Discipline, Sadism and Masochism (BDSM), strap-on services (pegging), use of vibrators and Spanish ("tit fucking"). Additional comments made mention of personal intimacy, cuddling, conversation and company which, when assessed overall, reflect the broad range of desires and sexual activity enjoyed by the general population within Australia (Richters et al. 2014; de Visser et al. 2014). This data also supports other empirical literature, where sex workers also identified providing many therapeutic and emotional aspects to their clients, outside of the usual sexual parameters (Bates and Berg 2014; Hartley 2000; Sanders 2006).

## Geographical location

Sex work has often been proclaimed as the "oldest profession," with sex workers being revered in some ancient cultures as temple or "sacred prostitutes" (Califia 2002). Regardless of the legal frameworks and criminal codes, the sex industry today operates in nearly every country in one way or another. The provision of sexual services can occur in a wide variety of locations and situations, including massage parlours, brothels, some strip clubs, hotels and motels, private homes, apartments, safe houses, in cars or trucks, online (through web-cam and porn sites), in retirement homes, nursing homes, in disability supported accommodation and any other location where the sex worker and the client have negotiated. One research study identified at least 25 different types of sex work, based on varied locational settings, where this kind of transaction can occur (Harcourt and Donovan 2005).

## Pricing structure

Sex work services vary enormously, depending upon a wide variety of factors. These include the location for service delivery, type and length of service, the individual or establishment's pricing structure, whether there is travel involved, whether additional room rental costs are incurred, and whether speciality items are requested to be bought for the appointment and need to be reimbursed.

## Service aims and objectives and client motivation

The primary objective of sex workers is to provide mutually consenting services to their clients, with each worker choosing what services they wish to provide as well as which ones they don't offer. There are many roles sex workers may play in the lives of their clients with disability. As well as seeking sexual enjoyment, on par with the general client population, clients with disability have identified a number of other significant motivations for seeing a sex worker. These include: learning about their sexual capacities after a significant injury or illness, gaining confidence and social skills before embarking on the dating scene, alleviating loneliness, increasing their experience, knowledge and acceptance about their own bodies, enjoying the opportunity to give and receive touch in a sensual way and to lose their virginity (Kulick and Rydström 2015; Liddiard 2014; McGath 2016; TLC Trust; Touching Base Inc. 2016a; Wotton and Isbister 2010). Furthermore, sex workers can become safer sex educators, creating a safe and respectful environment for their clients to explore their sexual expression and to increase their confidence, knowledge and skill-set to utilise in potential future dating experiences (Bates and Berg 2014; Hartley 2000; Sanders 2006). Finally, some people with disability (e.g., those born with shortened or no arms, those who have had multiple amputations and others who may have significant mobility and/or dexterity impairments) are sometimes unable to masturbate, so utilise the services of sex workers in order to gain equitable access to this form of sexual fulfilment (Heckendorf 2013).

## Availability and eligibility of service providers and clients

The frequency of appointments can vary greatly. Clients may choose to book the same sex worker regularly, sometimes over many years, while others may opt to see a variety of sex workers and pursue a range of services. Clients may choose to see sex workers only in between relationships, as a one-off event, for special occasions (e.g., birthday or Christmas), later in life when their life partner has passed away or after an accident or injury, when their physical abilities and confidence have changed. Sex workers may work part or full time within this occupation, with many qualified and working in other areas of the workforce. Some sex work advertising directories now have the option to find "disability-friendly" sex workers. While some sex workers actively advertise as being "specialised" in providing services to clients with disability, they generally all have extensive experience in seeing clients from all walks of life and of all abilities, *not* just those with disability. Clients can self-refer or also have the assistance of a third party (e.g., carer, professional support person, parent, sibling, friend) to help co-ordinate the appointment.

## Organisations who may facilitate training and referrals between service providers and clients

There are a few internationally recognised organisations who support sex workers and clients with disability. The two that are most prominent are Touching Base Inc. (Australia) and TLC-Trust (UK), with both organisations having their genesis in 2000. These organisations were created to better support the ongoing needs of both marginalised communities, to advocate for labour rights and sexual rights of sex workers and people with disability, and to facilitate easier communication and referral pathways between the two groups. While Touching Base Inc. has provided their Professional Disability Awareness Training (PDAT) to sex workers since 2001 (Touching Base Inc. 2016b), a number of sex worker organisations have also provided training, workshops and information days for local sex workers to increase their knowledge and confidence in providing services to clients with disability. They include Kassandra in Germany (Mittler 2013), Bliss without Risk, Czech Republic (Fraňková 2015) and the Sex Workers Education and Advocacy Taskforce (SWEAT)/Sisonke in South Africa (Dockney 2013)

## Sexual surrogacy

Masters and Johnson developed sexual surrogacy in America in the early 1960s as a new type of sexual therapy (Runciman 1975). Other terms most commonly used now include "surrogate partner therapy" and "surrogate partnership" to describe the therapy, and "sex surrogate," "sexual surrogate" or "partner surrogate" to describe the person who has physical, intimate contact with the client. Masters and Johnson opened their first sex therapy clinic in 1964 and continued to operate until 1994. This therapy, introduced within their second book, *Human Sexual Inadequacy* (1970), was deemed revolutionary at the time, and it remains a contentious therapy option, even today.

Sexual surrogacy recently enjoyed mainstream media attention since the release of the Hollywood film, *The Sessions* (2012). Based on the true story of Mark O'Brien, a man who contracted polio as a child and spent most of his life in an iron lung due to complications from the disease, he hired a sexual surrogate to help him lose his virginity and enjoy sexual contact with a woman. Written observations from both O'Brien and the sexual surrogate, Cheryl Green, have been published, outlining their experiences working with each other and were used to develop the screenplay (Cohen-Greene and Garano 2012; O'Brien 1990; Saratogian News 2012). With *The Sessions* being based in the USA, where sex work is predominately illegal, seeing something like this on the "big screen" was an eye-opening moment for people, especially as it was screened at the Sundance Film Festival. Additionally, with highly respected actors, such as Helen Hunt and William H Macy, taking on pivotal roles, those working in the surrogacy field hoped to gain increased credibility and professional respect in what they do, drawing away from the perception that it was a form of sex work.

## *Services offered*

Services are provided, according to what the therapist believes will best assist with the development of the client's physical and emotional personal growth. Specially trained sexual surrogates are employed to follow these directions and provide hands-on, tactile services to the client. This can include sexual and non-sexual, clothed or naked services, but neither the client nor the surrogate goes beyond the directives outlined by the therapist. While intercourse may eventually occur during a session, it is not the final or only activity that is focused on. Feedback from both the surrogate and the client are given to the therapist, who then, in conjunction with further sessions with the client, will formulate what the surrogate should focus on next time they meet their client. Overall, this therapy is structured to be both short- and long-term goal-orientated, with the client always moving forward in their development and learning curve. Essentially, each session will address new and different activities and issues, with constant revision of previous learnt skills. Although this triangulated model of therapy, with therapist, client and surrogate working together, is the best-known framework, there are also surrogates who, after their training—or drawing upon training from similar modalities—work more independently, advertising for clients themselves and structure the sessions according to what they professionally believe would be best for their client.

## *Geographical location*

Sexual surrogacy predominantly operates in the USA and Israel, but has also been practiced to varying degrees in the UK, Australia and different parts of Europe (Colley 2018; Freckelton 2013). Whereas still operating in a quasi-legal "grey zone" within the USA (often being conflated with sex work, which is illegal everywhere, except for a handful of brothels operating in certain counties within the state of Nevada), in Israel it enjoys the full support of the Government. This was primarily due to Governmental guilt in sending so many abled-bodied men to war and having a massive increase in permanently injured soldiers returning home.

Sexual surrogacy was initially designed to be provided at a specific therapy centre and, though this still occurs in Israel, other countries, such as the USA, often allow for surrogates and clients to meet elsewhere. This could include the client's home or the workplace of the sexual surrogate.

## Pricing structure

The sexual surrogate is paid a fixed amount each session, regardless of what activities occur. When working in a triangulated model, the rate is determined by the organisation or the specific therapist, *not* the sexual surrogate (Ridley 2012). Only if the surrogate were working and advertising independently would they set their own pricing structure.

## Service aims and objectives and client motivation

Clients first work with therapists to deal with specific problems, anxieties or issues they wish to address. If they need specific "hands-on" practice to achieve their learning goals, then the use of sexual surrogate services would then be introduced into the therapy schedule. Vena Blanchard, president of the International Professional Surrogates Association (IPSA), describes it as "a therapeutic treatment that combines psychotherapy with experiential learning … it's a programme designed for people who struggle with anxiety, panic, and past trauma—things that can distort a person's experience in the moment" (Savage 2018). Overall, surrogacy aims to create a safe learning environment that simulates real-life experiences that the client has been too afraid or nervous to pursue.

## Availability and eligibility of both the service provider and the client

Unlike sex work, a surrogacy client doesn't normally get to individually choose the surrogate with whom they will be working but is assigned one, based on a range of demographics and availability (Rosenbaum et al. 2014). The clients often work with a therapist before being introduced to a surrogate. Clients who simply want to experience sexual and compassionate touch or to lose their virginity straight away are generally ineligible. Clients normally need to travel to a specific location, so anyone whose disability precludes them from doing this will be automatically excluded. In fact, Israeli representatives at the 2014 1st Global Conference: Sexualities and Disabilities, in Lisbon, Portugal were very clear during a Q&A that they would not send a surrogate out to the home of a person with disability, even if that potential client were unable to access their specific therapy rooms.

## Organisations who may facilitate training

The International Professional Surrogacy Association (IPSA 2018a) is based in San Diego, America, and operates the best-known well-established training course designed specifically for sexual surrogates. Whereas surrogacy sessions generally remain a private interaction, one educational documentary, *Beruf: Berührerin*, encompasses the discussions and work of three sexual surrogates in Austria, Germany and Switzerland. Due to the sensitive nature of the work, this film is not available online (Dworschak and Müller 2012). In 2018, IPSA also released *Surrogate Partners:*

*Intimate Profiles*, also not permitted to be uploaded online. This 23-minute documentary features interviews with therapists, clients and surrogates, as well as footage of surrogates and clients doing experiential activities that form the basis of their Surrogate Partner Therapy (IPSA 2018b).

In Israel, the Dr Ronit Aloni Clinic is the only major centre providing sexual surrogacy and has been operating since the late 1980s, when Dr Aloni brought her formal surrogacy training back from the USA (Kolirin 2017). This centre provides training for their surrogates and receives referrals, from the sex therapy clinic at Tel Aviv's Ichilov Hospital, as well as from other sex therapists working from other clinics and hospitals across Israel.

## Sexual assistants

Sexual assistants appear to be sexual surrogates who are trained to provide services only to clients with disability, but the client does not go through parallel therapy sessions with a therapist while participating in appointments with a sexual assistant. EPSEAS, the European Platform Sexual Assistance, describes their role as:

> Sexual assistance is supporting adults with disabilities in the whole spectrum of their sexuality. It could be to help them to learn or improve their skills when it comes to interpersonal relationships, intimacy and intimate and/or sexual relationships. Each person is unique, as is their sexuality. Each relationship between a Sexual Assistant and the beneficiary is unique and made of unique circumstances. Sexual assistance is determined as much by the particular needs imposed by disability as by the sexual experience itself. These two aspects are systematically present in each Sexual Assistance.
>
> *(EPSEAS European Platform Sexual Assistance, n.b)*

### *Services offered*

Services offered are driven from what the clients feel they need and desire, although the sexual assistants also drive the direction of the appointments. Speaking with a number of sexual assistance organisations in 2017, it appears that different organisations have different structures. For example, in Austria, the sexual assistants only provided services up to and including masturbation, and, if the client was "ready and able" to experience other sexual experiences, then they would refer them to sex workers. Other organisations said that it was totally up to the sexual assistant as to what level of sexual contact they were willing to provide.

### *Geographical location*

Sexual assistance services and organisations have emerged predominantly in Europe. The European Platform Sexual Assistance (EPSEAS) is a platform of non-profit organizations who are active in providing sexual assistance for people with disabilities and the elderly. Their partner organisations are located in France, Belgium, Switzerland, Spain, Italy and the Czech Republic.

### *Pricing structure*

The sexual assistant is paid a fixed amount each session, regardless of what activities occur. The rate is generally determined by the organisation, not the sexual assistant, unless the sexual assistant has decided to advertise and work independently.

### *Service aims and objectives and client motivation*

Sexual assistance services are provided to those who need a level of specific assistance in learning about how to enjoy sexual expression, where their disability currently impedes their ability to move forward through the usual channels of dating and forming a relationship in society. This can include embarrassment of their bodies and their disability, lack of sexual and social experience, fear of rejection in dating forums, lack of confidence and the fear of approaching sex workers due to societal discrimination and stigma around who is involved in the sex industry (i.e., fear of being arrested, being abused or having money stolen from them).

### *Availability and eligibility of both the service provider and the client*

The only clients that sexual assistants work with are those with disability or older adults. Clients are often interviewed/assessed for their eligibility by the organisation to determine what their needs are and then paired with the most appropriate sexual assistant. Some sexual assistants eventually have their own website and clients contact them directly for appointments. While this mirrors how sex workers operate, they are clear that they provide this type of therapy/support *only* to clients with disability.

Sexual assistants are also assessed for their eligibility. In one organisation, they need to be interviewed by a psychologist for an hour before being accepted to work with clients. Other organisations require sexual assistants to pay for and complete intense specialist training courses and graduate successfully before being allowed to work with clients. One of the trainers outlined the initial requirements used by several French-speaking organisations within France, Belgium and Switzerland (SEHP—Sexualité et Handicaps Pluriel, Corps Solidaires and CH(s)OSE),

> We ask that candidates be over 25 years old, have physical and mental health compatible with the sexual assistance activity, have financial revenue equivalent to a part-time job, a clean judicial slate and have spoken with their partner about their wish to become a sexual assistant.
>
> *(Warembourg 2016)*

While there are definitely a number of former or current sex workers who are also qualified as sexual assistants, the majority of them are not. Some sexual assistants go on to create and manage their own websites, while others may receive referrals directly from an organisation or from advertising on a specific website established to give direct information and contact details for sexual assistants (Freya 2019).

### *Specific organisations that facilitate training and referrals between service providers and the clients*

The European Platform Sexual Assistants (EPSEAS) network has members operating in Belgium, France, Spain, Italy, Switzerland, the Netherlands and the Czech Republic. InSeBe, based in Zurich, Switzerland has been one of the training organisations partnering with others throughout Europe to provide concise but thorough sexual assistant courses. Some courses, like the one CH(s)OSE facilitates (Corps Solidaires; Credi 2017) requires over 120 contact hours for participants, spread out over a year.

## Masturbation services

There are a couple of organisations and businesses that have been set up to "lend a help-ing hand" and offer paid or volunteer masturbation services to people with disability. These include Handisex (Denmark) (2012), White Hands (Japan) (2008) and Hand Angels (Taiwan). Another lesser- known Japanese organisation called Sexual Volunteers (SV) Sakura no Kai describes itself as an organization "that helps people with disabilities masturbate (ji'i)."(Nakamura 2014).

## *Services offered*

Only masturbation services are provided to the client and the touch is one-way only. In Japan, the services are carried out using a glove and the person masturbating the client remains fully clothed. Remarkably, Nakamura mentioned that, with the Sexual Volunteers organisation, "care services can be provided in the handicap-accessible bathroom of the train station if the client doesn't have a private space at home" (2012). In Denmark, the person is *only* assisted to mastur-bate via the use of sex toys, and the worker/assistant only assists with masturbation by putting their hand over the client's hand to help the client simulate the action other people are able to do independently. This organisation provides a range of other services where they can provide assistance with couples needing positioning, a referral to sensual massage service (though their website is a bit ambiguous as to what is actually included during that service (Persson 2014)), sexual guidance, sexual training and sexual education. In Taiwan, the service providers have spoken about sometimes being naked with the client, with media articles indicating that each service is always 90 minutes in duration.

## *Geographical location*

HandiSex is based in Denmark, White Hands and Sexual Volunteers are in Japan and Hand Angels is located in Taiwan.

## *Pricing structure*

For those in Denmark and Japan, the service is a set price which the organisation determines. Hand Angels in Taiwan only offered volunteers, so they masturbated the clients for free, as a type of benevolent "community service."

## *Service aims and objectives and client motivation*

The organisations facilitate individuals to provide sexual relief only to clients with disability. Media previously carried out by Hand Angels mentioned that "these volunteers use only their hands for second-base kind of stuff—hugging, caressing and kissing on the face are fine, but anything penetrative (fingering, oral sex, vaginal sex, and anal sex) is not" (Moura and Jie Zou 2015). Clients want to experience sexual expression and to engage these services either because they do not have the capacity to organise a sex worker or may not have a supportive care worker they trust to take them to a sex worker. The clients may not be aware that sex workers would be willing and able to provide a range of services (for example, COSWAS, a Taiwanese sex worker organisation, has done referrals to sex workers since 2010 (Kuo 2016)) or they may have previous negative experiences with either dating or with sex workers. Also, clients may not have sufficient funds to pay for the services of a sex worker.

### Availability and eligibility of both the service provider and the client

In Japan, these services are only available to men with disability. The Japanese organisation stated, "White Hands is a welfare organisation that caters to the sexual needs of men with severe physical disabilities." When asked about whether they provided services to women, they replied "We haven't pursued this because we haven't received any requests from them. At White Hands, the suggestions and criticisms we usually receive is in regards to the fact that we have no care services for women with disabilities. Our response to them though, is what kind of service would be appropriate?"

HandiSex, in Denmark, provides assisted masturbation services to both men and women but *only* if they are unable to masturbate:

> If you are unable to use your hands, the option of using sex toys to masturbate is present. Assisted masturbation differs from prostitution in that the assistant does not participate in the act itself, but rather facilitates the possibility for bodily pleasure.
>
> *(HandiSex 2012)*

The Taiwanese Hand Angels provided services to both men and women but, whereas the sessions ran for 90 minutes, the preparation time could be up to six months. One media article also indicated that some male applicants, requesting a female worker, were on the waiting list for up to two years before being provided with a service (Henderson, 2017). Services were offered to any person "recognised by the government as having a serious physical impairment but can't be mentally disabled. Once they're cleared, the service is totally free" (Moura and Jie Zou 2015). Successful applicants were entitled to three hand jobs in a lifetime, accompanied by caresses and kisses (strictly face only) (McHugh 2015).

### Specific organisations who may facilitate training

One of the founders of Hand Angels explained in an interview that every volunteer needed to pass a two-month training programme in order to know and understand disabled people's bodies and their thoughts (Queer Taiwan 2018). There is no mention of specific training given to staff on the HandiSex website. While the author has no specific references outlining the training for White Hands staff, it must be noted that the language barrier precluded the author from having detailed access to further information at this time.

### Similarities among service modalities

The most important similarity among all four modalities is that only mutually consenting activities are pursued and that both client and service provider are of legal age of consent for sexual activities within each country. Additionally, while there may be third party assistance in establishing the booking/session, it is only ever the service provider and the client who will interact in a sexual, hands-on way. Excluding the Taiwanese Angel Hands volunteers, all service providers are paid. Overall, all service provision is clearly negotiated, with boundaries, payment, services offered, time and date of the appointment and location of service delivery all clearly established prior to the delivery of services. With each, there is a consistent aim to always create a safe and supportive environment to allow the client to explore aspects of their sexual expression without fear of judgement or ridicule. Excluding the masturbation services, the former three modalities all offer an educational component that can include learning about positive communication methods, non-verbal cues, hands-on experience, body anatomy, practicing how to positively engage with another person during a sexual interaction and practical tips around safer sex practices and positioning. All modalities have been shown to increase the confidence and well-being of the client. For some people with disability, participation in any of these services can also address a physiological desire and need, known as skin hunger (Findlay 2012).

Another similarity among all modalities is that training—especially formally recognised and accredited training—is extremely limited and not universal in the scope of their educational content. While certain leading organisations have emerged with their own detailed workshop and training packages, a common criticism is that, aside from sexual surrogacy in Israel, none are supported by Government nor recognised within formal academic fields.

## Differences among service modalities

There are a range of significant differences between each modality. They include autonomy of choice, types of services available, the financial/pricing structure, frequency of service delivery, the process of securing an appointment, the geographical location and availability of services and the confidentiality protocols established for the services.

### *Autonomy of choice*

While sex work is based on a two-way working relationship with the services offered and provided by direct negotiation between the sex worker and the client (or clearly identified at the sex services premises as to what is available), the other sexual service options generally all have third party influences determining the scope and direction of the services provided each time. Sex work sessions are initiated at the client's discretion, with the client having full control over who they engage. They can choose to see the same sex worker each time or change locations and person/s according to their personal desires and situation. With the other modalities, the service provider is generally allocated to the client by a third person and they will only ever see that same person. This is often done by trying to match their desires and needs with who is available but, overall, unless the client self-refers to a sexual surrogate or assistant via their direct website (only available in limited locations) then their autonomy of choice is removed.

### *Types of services available*

For masturbation service provision, there is only one type of service available, with the aim to explicitly elicit an orgasm and provide sexual relief. Clients of sexual surrogacy and sexual assistants generally map out specific short- and long-term goals at the start of their therapy to work through a series of objectives, tasks and activities with a gradual progression towards explicit sexual contact. Orgasm and intercourse are not the primary or initial activities experienced and, in fact, may not be part of the client's goals at all. Sex work services can encompass any or all of these, depending on the rationale for the booking, with the kind of sexual interaction varying dramatically or remaining the same each time, depending on what the client wishes to experience and what the sex worker is comfortable providing. Within sex work, the difference is also that appointments are made explicitly at the client's request and matches their individual sexual desires, needs, finances and availability, instead of being limited by a rigid framework dictated by a third party.

## Financial structure

The financial aspects of the transaction are different with each type of service. Sex work has a wide variety of charges, depending on the services provided, the length of the booking, the location of the booking, whether it's with a private, independent worker or in a sex services premises (i.e., in a brothel or massage parlour); differing standard prices apply within certain regional areas or if additional transportation and travel costs are incurred. The other modalities (excluding the volunteer services provided by Hand Angels in Taiwan) all have set prices that are pre-determined by the organisations and not the service provider. These fixed prices all appear to be universally lower than sex industry prices in each location.

## Frequency of service delivery

The frequency and variety of services received by the client are different for each modality. The sex work client can choose to see the same sex worker multiple times, sometimes spanning months, years or decades, or choose to see multiple sex workers during their lifetime. At odds with this, the volunteer Taiwanese masturbation services have put a "cap" on only three services provided to a client during their lifetime. It is unknown how often the other masturbation services can be requested by a client. Distinct from sex work, clients of both surrogacy and sexual assistant services must work consistently with only one provider and commit to a finite number of sessions. At completion of this therapy arc, the professional working relationship between the client and the provider is terminated.

## Securing an appointment

Although some clients with disability certainly must work through many barriers before having the opportunity to book a sex worker, an appointment can occur nearly immediately once these issues have been addressed, based purely upon the availability of the sex worker and the client. With the other modalities, clients need to go through a process of assessment, therapy or interview with third parties—either a therapist and/or a Government agency—before interactions can begin between the service provider and themselves. They may also be put on a waiting list before this can occur.

## Geographical location and availability of services

Sex work occurs almost universally around the world. Surrogacy, sexual assistance and masturbation services are only offered in specific and limited geographical locations. The use of the latter three terms (surrogacy, sexual assistance and masturbation services) and available service options can sometimes be attributed to different legal parameters in each country and the ongoing stigma associated with the sex industry. Distancing themselves from the historically nuanced term of "prostitute," people involved try to focus on the more benevolent aspects of their work, framing it as a type of therapy and support, one which both society and the legal justice system can more easily embrace. As Kulick and Rydstrom (2015) noted though:

> this might help to make paying for sex more palatable to some people with disabilities or to caregivers of disabled people who are frightened or repelled by the idea of prostitution. But a claim to provide more empathy, care and concern than is offered

by women who work in "classic prostitution" should also perhaps be heard as a way of staking a class (and probably a race) distinction.

*(p. 186)*

## Confidentiality protocols established for the services

While confidentiality is an integral aspect of all services, as third parties are generally involved in sexual surrogacy, sexual assistants and the masturbation service, their confidentially protocols are *explicitly* discussed and agreed to prior to commencement of service provision. This is different to sex work where confidentiality between the two parties is *implicit*, due to the time-honoured principles of discretion and respect of privacy between sex workers and clients. That said, by request of the client, referrals and recommendations from one sex worker to another can often occur, either in the same location, interstate or even from one country to another (Phamodi 2017). The fear of different laws, language barriers, or the exhaustion of having to explain their disability and how it affects them to potentially many new people can create additional barriers for clients with disability to continue enjoying sexual expression, so this peer-to-peer professional "hand over" can be of great benefit to the client.

## Conclusion

There are many factors that contribute to the complexity of sexual services for people with disability. They include prescriptive policies; an omission of appropriate education, support and advocacy; the absence of good legal frameworks; a lack of awareness and attention to the sexual rights of people with disability; and no dedicated funding for specific organisations to develop meaningful and appropriate resources. These types of issues, along with moral and ethical debates, present additional challenges faced by people with disability, often precluding them from enjoying sexual experiences and enjoyment. While it should always remain an individual decision to seek out sexual services, or not, ultimately the fundamental issue is about choice and respecting the person's autonomy, with no difference in this equation as a result of a person's disability. While it's imperative to emphasise that this should never be the only "go-to" option for people with disability, it should not be dismissed either. Rather, it should be regarded as just one of many options people with disability should be able to choose from, as with anyone else in society.

Dearing and Isbister (2014) examined sexuality, disability and the right of self-determination by stating:

> the dignity of risk includes the right of all adults to make their own choices and preferences about their health, care and lifestyle, even if others—including healthcare professionals or other support providers—believe those choices will endanger a person's health or longevity, or otherwise disapprove. In the context of sexual rights of people with a disability, the dignity of risk supports the right of people with disability to a personal sphere of sexuality free from arbitrary or unlawful interference by third parties.
>
> *(p. 26)*

Building awareness around these current and emerging practices can significantly assist people with disability and other third parties to differentiate between the different types of sexual services that may be available in their location. It is anticipated that this information can support positive legislative changes to allow for these options to be legally recognised occupations, in all countries, in the future. This information can also inform different disability and ageing sectors, that are starting to review and discuss policies, procedures and training needs to best support their clients' needs and sexual expression. "Equal rights for all" can never be proclaimed until the sexual rights of people with disability are adequately addressed and supported. Paid sexual services must be included as a choice under these sexual rights.

# References

Allman, D., & Ditmore, M.H. (2016) Introduction to the culture, health & sexuality virtual special issue on sex, sexuality and sex work. *Culture, Health & Sexuality*, pp. 1–8. doi:10.1080/13691058.2016.1180855.

Bates, J., & Berg, R. (2014) Sex workers as safe sex advocates: Sex workers protect both themselves and the wider community from HIV. *AIDS Education and Prevention* 26(3), pp. 191–201. doi:10.1521/aeap.2014.26.3.191.

BBC ONE. (2007). For one night only [Film]. Retrieved from http://www.astaphilpot.co.uk/#!bbc/cee5

Blades, J. (2009) The too hard basket [Audio File]. Retrieved from http://www.abc.net.au/rn/360/stories/2009/2762394.htm

Califia, P. (2002) Whoring in utopia. In A. Soble & N. Pwer (Eds.), *The philosophy of sex: Contemporary readings*. Plymouth, UK: Rowman & Littlefield.

Clercq, P.D., & Philpot, A. (2011) Come as you are [Film]. Retrieved from http://www.imdb.com/title/tt1753887/

Cohen-Greene, C., & Garano, L. (2012) *An intimate life: Sex, love, and my journey as a surrogate partner*. A memoir: Catapult.

Colley, D. (2018) My life as a sex surrogate: A different side to the industry. Retrieved from https://www.news.com.au/lifestyle/relationships/sex/my-life-as-a-sex-surrogate-a-different-side-to-the-industry/news-story/c09db1982d09dcc9692d8b25a07ccdb3

Corps Solidaires. Certified training for sensual accompanying and sexual assistance for people with disabilities. Retrieved from http://corps-solidaires.ch/formation/

Credi, B. (2017) Seven graduates in sex assistance for the disabled. Retrieved from https://www.west-info.eu/seven-graduates-in-sex-assistance-for-the-disabled/

Dearing, C., & Isbister, S. (2014). The Right To Say Yes: Upholding The Dignity Of Sex Workers And Their Clients With Disability'. *Newsletter of the International Bar Association's Human Rights Law Working Group, a Subcommittee of the Rule of Law Action Group,, v. 1, no. 2, PART II*, 25–30.

de Visser, R.O., Richters, J., Rissel, C., Badcock, P.B., Simpson, J.M., Smith, A.M.A., & Grulich, A.E. (2014) Change and stasis in sexual health and relationships: Comparisons between the First and Second Australian Studies of Health and Relationships. *Sexual Health* 11(5), p. 505. doi:10.1071/SH14112.

Dockney, J. (2013) South Africa: Sex work and disability—A crucial need seldom spoken about. Retrieved from https://allafrica.com/stories/201307241210.html

Dworschak, C., & Müller, A. (2012) Beruf: Berührerin (Occupation: Berührerin) [Video File]. Retrieved from https://www.dorftv.at/video/7318 / https://www.film.at/beruf-beruehrerin

EPSEAS European Platform Sexual Assistants (n.b.) What is sexual assistance? Retrieved from http://www.jpgeas.eu/en/page/181

Esmail, S., Darry, K., Walter, A., & Knupp, H. (2010) Attitudes and perceptions towards disability and sexuality. *Disability & Rehabilitation* 32(14), pp. 1148–1155. doi:10.3109/09638280903419277.

Findlay, C. (2012) Interview with sex worker Rachel Wotton—Skin Hunger part 2. Retrieved from https://carlyfindlay.com.au/2012/05/22/interview-with-sex-worker-rachel-wotton-skin-hunger-part-2/

Fraňková, R. (Producer) (2015) Czech project offers services for disabled. Retrieved from https://www.radio.cz/en/section/curraffrs/czech-project-offers-sex-services-for-disabled

Freckelton, I. (2013) Sexual surrogate partner therapy: Legal and ethical issues. *Psychiatry Psychology and Law* 20(5), pp. 643–659. doi:10.1080/13218719.2013.831725.

Freya (2019) About sexual assistance. Retrieved from https://www.freya.live/cs/sexualni-asistence/o-sexualni-asistence

Guiraud, D., & Jaulin, A.-C. (2012) L'Assistante [Film]. Retrieved from http://www.adastra-films.com/en/films/films-produced/163-l-assistante

Gurza, A. (2016) Price of intimacy: The time i hired a sex worker. Retrieved from http://www.out.com

HandiSex (2012) For people with disabilities. Retrieved from https://handisex.dk/?lang=en#bruger

Harcourt, C., & Donovan, B. (2005). The many faces of sex work. *Sexually Transmitted Infections* 81(3), pp. 201–206.

Hartley, N. (2000) Bodhisattvas amongst us: Compassionate sex workers. In M.B. Sycamore (Ed.), *Tricks and treats: Sex workers write about their clients*. London: Routledge, pp. 71–74.

Harvey, L., & Harvey, S. (2013) The gift [Short Film]. Retrieved from https://vimeo.com/64540956

Heckendorf, D. (2013) David Heckendorf: Submission No. 634, Senate Inquiry into the National Disability Insurance Scheme Bill 2012 [10]. Retrieved from http://www.aph.gov.au/DocumentStore.ashx?id=01cac46b-7140-45ac-aa46-d2dc1f1bca8f

Heidari, S. (2015) Sexual rights and bodily integrity as human rights. *Reproductive Health Matters* 23(46), pp. 1–6. doi:10.1016/j.rhm.2015.12.001.

Henderson, C. (2017) Hand Angels: The Taiwanese movement helping disabled people to masturbate. Retrieved from https://vt.co/news/weird/hand-angels-taiwanese-movement-helping-disabled-people-masturbate/

International Professional Surrogacy Association (IPSA) (2018a) About IPSA—Statement of IPSA purpose and mission. Retrieved from http://www.surrogatetherapy.org/

International Professional Surrogacy Association (IPSA) (2018b) Surrogate partners: Intimate profiles DVD. Retrieved from http://www.surrogatetherapy.org/surrogate-partners-intimate-profiles/

Jungels, A.M., & Bender, A.A. (2015) Missing intersections: Contemporary examinations of sexuality and disability. In *Handbook of the sociology of sexualities*. New York: Springer, pp. 169–180.

Kolirin, L. (2017) Israeli therapist offers surrogate sex partners. Retrieved from https://www.thejc.com/news/israel/israeli-therapist-offers-surrogate-sex-partners-1.438667

Kulick, D., & Rydström, J. (2015) *Loneliness and its opposite: Sex, disability, and the ethics of engagement*. Durham: Duke University Press.

Kuo, P. (2016) COSWAS. Retrieved from http://shadainternational.com/all-impairments/sexual-services/

Lewin, B. (2012) The Sessions [Film]. Retrieved from http://www.imdb.com/title/tt1866249/

Liddiard, K. (2014) "I never felt like she was just doing it for the money": Disabled menws intimate (gendered) realities of purchasing sexual pleasure and intimacy. *Sexualities* 17(7), pp. 837–855.

Magnan, M.A., & Reynolds, K. (2006) Barriers to addressing patient sexuality concerns across five areas of specialization. *Clinical Nurse Specialist* 20(6), pp. 285–292.

Masters, W.H., & Johnson, V.E. (1970) *Human sexual inadequacy*, Vol. 1. Boston: Little, Brown.

McCabe, M.P. (1999) Sexual knowledge, experience and feelings among people with disability. *Sexuality and Disability* 17(2), pp. 157–170.

McGath, M. (2016) "We bring happiness into their lives"—Meet the sex workers providing services for clients with disabilities. Retrieved from https://www.independent.ie/life/health-wellbeing/health-features/we-bring-happiness-into-their-lives-meet-the-sex-workers-providing-services-for-clients-with-disabilities-34984671.html

McHugh, J. (2015) Taiwanese charity giving out hand-jobs to the physically disabled. *Cosmopolitan*. Retrieved from https://www.cosmopolitan.com/uk/reports/a33516/taiwanese-charity-giving-out-hand-jobs-physically-disabled/

Miller, A.M., Kismödi, E., Cottingham, J., & Gruskin, S. (2015) Sexual rights as human rights: A guide to authoritative sources and principles for applying human rights to sexuality and sexual health. *Reproductive Health Matters* 23(46), pp. 16–30. doi:10.1016/j.rhm.2015.11.007.

Mittler, D. (2013) In Germany, sex workers get special training to serve the disabled. Retrieved from http://www.worldcrunch.com/culture-society/in-germany-sex-workers-get-special-training-to-serve-the-disabled/handicapped-sex-germany-prostitution-kassandra/c3s11876/

Moura, N., & Jie Zou, Y. (2015) These volunteers give hand jobs to the severely disabled. Retrieved from https://www.vice.com/en_us/article/av4m8p/hand-angel-hand-jobs-taiwan-748

Nakache, O., & Toledano, E. (2011) The intouchables [Film]. Retrieved from http://www.imdb.com/title/tt1675434/

Nakamura, K. (2012) *Transl/disability: Disability, queer sexualities, and transsexuality from a comparative ethnographic perspective*. Paper presented at the forum 'Shōgai, kuia, shitizunshippu'(Disability, queer, citizenship), held at the Center for Barrier-Free Education at the University of Tokyo.

Nakamura, K. (2014) Barrier-free brothels: Sex volunteers, prostitutes, and people with disabilities. In S. Kawano, G. Roberts, & S. Long (Eds.), *Capturing contemporary Japan: Differentiation and uncertainty*. Hawai'i: University of Hawai'i Press, pp. 202–220.

O'Brien, M. (1990) On seeing a sex surrogate. Retrieved from http://thesunmagazine.org/issues/174/on_seeing_a_sex_surrogate

O'Dea, S.M., Shuttleworth, R.P., & Wedgwood, N. (2012) Disability, doctors and sexuality: Do healthcare providers influence the sexual wellbeing of people living with a neuromuscular disorder? *Sexuality and Disability* 30(2), pp. 171–185.

Paradigm Pictures (2011) Scarlet road: A sex worker's journey [documentary]. Retrieved from http://www.scarletroad.com.au

Persson, A. (2014) What is sensuality massage? Retrieved from http://www.sensualitetsmassage.dk/om-massagen/

Phamodi, S. (2017) A sex worker, a disabled client and an email that changed everything [Audio Interview with Article]. Retrieved from https://www.takebackthetech.net/blog/sex-worker-disabled-client-and-email-changed-everything

Queer Taiwan (2018) 'Queer Taiwan' interviews the sexual services organization Hand Angels. Retrieved from https://lalatai.com/en/tvmovie/news/queer-taiwan-interviews-sexual-services-organization-hand-angels

Richters, J., de Visser, R.O., Badcock, P.B., Smith, A.M., Rissel, C., Simpson, J.M., & Grulich, A.E. (2014) Masturbation, paying for sex, and other sexual activities: The Second Australian Study of Health and Relationships. *Sexual Health* 11(5), p. 461. doi:10.1071/SH14116.

Ridley, J. (2012) Secrets of the sex surrogates. New York Post. Retrieved from https://nypost.com/2012/10/25/secrets-of-the-sex-surrogates/

Rosenbaum, T., Aloni, R., & Heruti, R. (2014) Surrogate partner therapy: Ethical considerations in sexual medicine. *Journal of Sexual Medicine* 11(2), pp. 321–329. doi:10.1111/jsm.12402.

Runciman, A. (1975) Sexual therapy of Masters and Johnson. *The Counseling Psychologist* 5(1), pp. 22–30. doi:10.1177/001100007500500107.

Sakellariou, D. (2006) If not the disability, then what? Barriers to reclaiming sexuality following spinal cord injury. *Sexuality and Disability* 24(2), pp. 101–111.

Sanders, T. (2006) Female sex workers as health educators with men who buy sex: Utilising narratives of rationalisations. *Social Science & Medicine* 62(10), pp. 2434–2444.

Saratogian News (2012) Mark O'Brien, the real person behind Hollywood's. *The Sessions*. Retrieved from https://www.saratogian.com/news/mark-o-brien-the-real-person-behind-hollywood-s-the/article_d2acec3e-eceb-5be0-8708-c2557d3e1290.html

Savage, D. (2018) Is there such a thing as sexual surragacy? Is it legit? Retrieved from https://www.chicagoreader.com/chicago/dan-savage-love-sexual-surrogate-partner-masters-johnson/Content?oid=56093929

Shuttleworth, R., & Taleporos, G. (2016) *Facilitated sex for adults with disability: Sexual right and disability accommodation*. Paper presented at the sexual dysfunction conference, Auckland, New Zealand. Retrieved from http://sdc.conferenceworks.com.au/presentations/facilitated-sex-for-adults-with-disability-sexual-right-and-disability-accommodation/

Silverberg, C., & Odette, F. (2011) *Sexuality and access project : Survey summary*. Retrieved from Canada http://www.ryerson.ca/content/dam/crncc/knowledge/relatedreports/sexualityintimacy/Sexuality%20and%20Access%20Survey%20Summary.pdf

Swartz, P. (2015) Short on short [Video File]. Retrieved from https://www.youtube.com/watch?v=xR-QCUCsXPM&feature=youtu.be&list=PLfPWpm2llg4sluDQR4qX_a4iUwZbvVQ9v

Taleporos, G. (2001) Sexuality and physical disability. In *Sexual positions: An Australian view*. Melbourne: Hill of Content Publishing, pp. 155–166.

Taylor Gomez, M. (2012) The s words: Sexuality, sensuality, sexual expression and people with intellectual disability. *Sexuality and Disability* 30(2), pp. 237–245. doi:10.1007/s11195-011-9250-4.

Tepper, M.S. (2000) Sexuality and disability: The missing discourse of pleasure. *Sexuality and Disability* 18(4), pp. 283–290.

TLC Trust. (2008) Testimonials from service users. Retrieved from http://www.tlc-trust.org.uk/page56.html

Touching Base Inc. (2016a) Client stories. Retrieved from http://www.touchingbase.org/clients-stories

Touching Base Inc. (2016b) Professional disability awareness training (PDAT) for sex workers. Retrieved from http://www.touchingbase.org/workshops-and-training/pdat

Wallace, B. (2013) Scarlet road documentary (review). Retrieved from http://www.pla.qld.gov.au/Resources/PLA/reportsPublications/newsletters/documents/InTouch77Apr2013.pdf

Warembourg, S. (2016) Sensual and sexual assistants in Europe. Retrieved from http://shadainternational.com/all-impairments/sexual-services/

Weitzer, R. (2010) *Sex for sale: Prostitution, pornography, and the sex industry*, Vol. 2. New York: Routledge.

White Hands (2008) Sexual rights for all!! Retrieved from http://www.whitehands.jp/e.html

Wotton, R. (2016a) Open your mind to what goes on behind closed doors [Video File]. Retrieved from http://tedxtalks.ted.com/video/Open-Your-Mind-to-What-Goes-on

Wotton, R. (2016b) *Sex workers who provide services to clients with disability in New South Wales, Australia* (Masters of Philosophy). University of Sydney. Retrieved from https://ses.library.usyd.edu.au/handle/2123/16875

Wotton, R. (2017) To review training programs for sex workers providing services to clients with disability—UK, Germany, Austria, Switzerland, Czech Republic, Denmark. Retrieved from http://www.churchilltrust.com.au/fellows/detail/4190/%20Rachel+Wotton

Wotton, R., & Isbister, S. (2007). *Sexual Surrogacy and Sex Work: similarities and differences.* Paper presented at the 1st World Congress in Sexual Health: XVIII World Congress of WAS, Sydney, Australia. http://www.sexualhealthvisual.com/Video_by_Rachel_Wotton_on_Sexual_Surrogacy_And_Sex_Work_Similarities_And_Differences.html

Wotton, R., & Isbister, S. (2010) A sex worker perspective on working with clients with a disability and the development of Touching Base Inc. In R. Shuttleworth & T. Sanders (Eds.), *Sex & disability: Politics, identity and access.* Leeds: Disability Press, pp. 155–178.

Wotton, R., & Isbister, S. (2017). Similarities and differences between sexual surrogacy, sex work and facilitated sexual assistance. *The Journal of Sexual Medicine* 14(5), pp. e252–e253. doi:10.1016/j.jsxm.2017.04.249

Wright, C. (2014) I have cerebral palsy and I enjoy having sex [Video File]. Retrieved from http://www.sbs.com.au/

# 31

# PROMOTING SEXUAL WELL-BEING FOR WOMEN WITH DISABILITIES THROUGH FAMILY-CENTRED INTEGRATED BEHAVIOURAL HEALTHCARE

*Colleen Clemency Cordes and Christine Borst*

## Introduction

Approximately one billion people worldwide live with disability (World Health Organization [WHO] 2011), making this population the largest minority group in the world (WHO 2011). In the United States, variations in definitions of disability, methodological approaches for identifying disability, and cultural factors that may influence individuals' self-report of disability have resulted in estimations of the prevalence of women with disability in the US. (WWD) ranging from 12% (US Census Bureau 2012, as cited by the Center for Research on Women with Disabilities [CROWD] n.d.), to approximately 25% (Centers for Disease Control and Prevention 2015). Rates of disability increase across the lifespan, with more than 35% of individuals ages 65 years and older living with a disability (Kraus 2017), in part due to the morbidity associated with multiple chronic conditions. For people with disabilities (PWD), sexual health, defined as a "state of physical, emotional, mental and social well-being in relation to sexuality...a positive and respectful approach to sexuality and sexual relationships, as well as the possibility of having pleasurable and safe sexual experiences free of coercion, discrimination and violence" (WHO 2002, p. 5), is an often central, but overlooked, aspect of their identity and life experience. In order to ensure appropriate whole-person care, behavioural health providers (BHPs) working with women with disabilities (WWD) are called to provide disability culturally competent sexual healthcare (Mona et al. 2017). In order to do so, BHPs must be cognisant of the unique historical, sociocultural, biological and interpersonal factors that influence sexual health among WWD.

Over the past fifty years, conceptualisations of disability have evolved from moral models, wherein disability was associated with a moral failing on the part of the individual living with disability and their families (Andrews 2017), to biopsychosocial models that acknowledge disability as a complex interaction of medical, social, environmental and political factors (WHO 2011). Perhaps among the most dominant conceptualisations of disability over the decades has been that of the medical model, which views disability as residing within the individual, and individuals are classified by their disability (e.g. "the spinal cord injury," "the amputee" or "the multiple sclerosis patient"). This view of disability may result in a desire to "fix" or "cure" one's disease or condition, and fails to acknowledge social, environmental and political factors that influence one's overall well-being. Comparatively, the psychosocial model emphasizes the role of diverse factors in enabling or restricting participation in activities (Andrews 2017), and recognises the dynamic nature of impairment and activity limitations across individuals, environments and situations. A healthcare provider's conceptualisation of disability invariably influences their approach to assessment and intervention, and BHPs working in integrated healthcare settings—wherein BHPs are working alongside medical providers as part of a care team—are encouraged to challenge conceptions of disability that simplistically view WWD solely on the basis of diagnosis, and to openly consider how biopsychosocial factors may influence WWD's experiences of disability.

## Integrated care

There has been a shift in healthcare delivery over the past several decades, evolving from a dissected view of patient care, to a whole-person, biopsychosocial approach (Corso et al. 2016). The collaboration between behavioural and physical health care providers, widely known as "integrated care," or "integration," has gained momentum in recent years as a means of delivering this comprehensive health care (Heath et al. 2013). Integrated care is widely conceptualised by six levels of collaboration/integration that lie on a continuum, and fall under three main categories: coordinated, co-located, and integrated care (Heath et al. 2013). At the far end of the continuum is Level 1: Minimal Collaboration, in which behavioural health and medical providers work separately, and rarely communicate unless specific information is required for a shared patient. At the other end of the continuum falls the highest level of integration, Level 6: Full Collaboration in a Transformed/Merged Practice, where BHPs and medical providers work side by side, functioning as a single unit, set out to treat the whole person. As noted by Marlowe and colleagues (2014), the function of BHPs in medical settings is different from mental health practitioners in traditional settings:

> The professional practice of onsite BHPs differs from that of traditionally trained mental health providers (Strosahl 2001) in terms of the nature of the health concerns that prompt referrals (i.e., not just for behavioural health assessment, diagnosis and treatment), the timeframe available to work with patients (i.e., 15–30 minute therapy encounters), and the dissemination of information among providers, patients and BHPs (via exam room consults, curbside consults [impromptu conversations between healthcare providers] and a shared electronic health record [EHR]).
>
> *(p. 79)*

As the integration movement has gained momentum in healthcare, experts have identified another essential piece of the whole-person care puzzle: a systemic, family-centred approach to care (Hodgson et al. 2014). There is extensive evidence in the literature supporting the impact of relationships on health, and vice versa (Borst et al. 2018), highlighting the importance of utilising a family systems lens, wherein providers "focus on the interrelatedness of family members and their interactions rather than on the individual family members" (Hecker et al. 2003, p. 57), and how it influences health and functioning. For example, Kiecolt-Glaser and colleagues (2005) found that non-disabled couples with high relational hostility experiencing a chronic health condition healed at 60% of the rate of couples with low hostility. High levels of family stress can also increase the impact of a chronic health condition by decreasing functioning and quality of life (Parkerson et al. 1995). The definition of family varies among research studies (e.g., family of origin, family of choice, both); however, providers are encouraged to focus less on identifying a universal definition of family and more on who each *individual patient* identifies as their family. It is this relational piece of the family dynamic that is crucial to healthcare delivery (Hodgson et al. 2014).

Combining a family systems lens with the BPS model allows both medical providers and BHPs to fully conceptualise the interactive relationship between individual, family and health (Borst et al. 2018). While providers from all disciplines are being called to approach health with a relational lens (DeGruy and McDaniel 2013; Talen and Valeras 2013), integrating BHPs with specialised training in systems-focused behavioural health integration can ensure that comprehensive patient needs are being met. This is particularly relevant when addressing sexual health concerns in an integrated care environment, where diverse inter- and intrapersonal experiences interact to influence overall sexual well-being. For example, although exploring *current* relational dynamics is a vital part of addressing sexual health, it does not provide the complete picture. Previous relationships, including family of origin experiences, can have a long-lasting impact on sexual well-being (Atwood 2006). This, coupled with the tendency of medical providers—and families of children and adolescents with disabilities—to be pessimistic about the potential for fulfilling sexual and relational lives in adulthood (Berman et al. 1999), may result in BHPs needing to more comprehensively and intentionally address sexuality and sexual needs to ensure the well-being of WWD. Additionally, consideration of unique factors that might influence a WWD's experience of sexual health and well-being is essential to the delivery of culturally competent sexual healthcare.

## Disability and culture

In order to provide culturally competent healthcare services, BHPs are encouraged to examine the intersectionality of cultural identities for WWD. A cultural group in and of itself, disability culture is defined by a set of shared experiences and values, including a willingness to accept help, an appreciation of the absurd (Gill 1995), social justice and a need to give back to others in the community (Forbert-Pratt 2018). The degree to which WWD identify with the disability culture varies, and as such, providers are encouraged to use person-first ("a woman with a disability") along with identity-first ("a disabled woman") language as appropriate, given a woman's embracement of disability as central to her identity (Dunn and Andrews 2015). Additionally, one should consider the intersectionality of identities, as Hays' (1996) ADDRESSING model promotes the identification of multiple factors that dynamically influence one's identity and life experiences: *A*ge and generational influences, *D*evelopmental or acquired *D*isability, *R*eligion, *E*thnicity, *S*ocio-economic status, *S*exual orientation, *I*ndigenous heritage, *N*ation of origin, and *G*ender. For WWD, as with all individuals, the salience of one aspect of their identity may ebb and flow within the context of dynamic situations; these experiences, in turn, may influence one's view of themselves as a sexual being and influence their sexual self-esteem. Consider for example, how one's nation of origin, and the socio-political views of disability and sexuality therein, might influence their view of WWD's sexuality, as depicted below.

> Maria is a 25-year-old heterosexual cisgender female who came to the United States from Columbia with her family when she was three years old. She has severe rheumatoid arthritis which requires the use of a wheelchair. While she strongly identifies with the disability community, she recognizes that her experiences as a WWD are different

from those of her friends from earlier generations that grew up prior to passing of the civil rights American Disabilities Act, or those of her family of origin who are living without disabilities, and her family members still living in Columbia. She additionally experiences tensions in her views of herself as a sexual being, influenced in part by her parents' religious and cultural beliefs about sexuality, and their opinions about masturbation and sexual pleasure outside of marriage.

## WWD and sexuality

While BHPs working to address sexual health and well-being must consider a woman's intersecting identities, it is critical to note that disability is unlike other cultural identities in how it influences the unique experiences of WWD. As noted by Mona et al. (2017), "disability per se is unique from other cultural experiences and identities in that it represents a difference in the body, inclusive of non-normative physical, sensory, cognitive and/or psychological representation" (p. 1005). The embodiment of these differences—whether apparent or non-apparent—invariably influence a WWD's sexual self-esteem and sexual expression, which are strongly associated with overall well-being, sexual satisfaction and quality of life, particularly among WWD (Taleporos and McCabe 2002). Providers working with WWDs must challenge ableist—and often heteronormative—assumptions of "normative" sexual behaviour, which often focus on penile-vaginal intercourse and orgasm as the "goal" of sexual interactions, and instead promote strategies that allow for diverse means of sexual expression. For BHPs working with WWDs, it is critical that they assess for, and be aware of, issues related to body image and sexual self-esteem and inequities in access to partnered relationships (Nosek et al. 2001) as well as explore WWDs adaptive approaches to sexuality and sexual activities in order to maximise pleasure and intimacy.

> Consider, again, the case of Maria. Maria presents to her primary care provider (PCP) as part of a well-woman check. As is routine in her PCP's office, a BHP also engages Maria in an assessment of her biopsychosocial needs. During this assessment, when inquiring about sexual activity, Maria indicates that she has recently began a new relationship, and she is interested in birth control. She also notes some anxieties about engaging in sexual activity with her new boyfriend, who has never been intimate with a woman who utilizes a wheelchair. In exploring Maria's anxieties and encouraging open conversation between Maria and her boyfriend about her concerns, the BHP encourages Maria to consider strategies for encouraging sexual pleasure, including erotic touch, incorporating transferring from the wheelchair into the bed as part of the sexual act, and identifying positions to minimize pain during sexual activity, while focusing on the goal of mutual pleasure in the sexual encounter. The BHP then expresses a willingness to follow up with Maria and her boyfriend as needed to further discuss and explore opportunities for sexual intimacy in their relationship, and/or refer Maria to a sexologist.

BHPs working in team-based medical settings are uniquely poised to both engage with and advocate for the sexual well-being of WWD like Maria. Whereas the initial assessment of behavioural health needs was a routine practice in this particular primary care office, it revealed a need for the provision of reproductive health care, as well as an exploration of adaptive strategies to promote sexual health and intimacy. In a large US national study of WWD, 31% of women reported having been denied reproductive healthcare by a physician due to their disability (Nosek et al. 2001); a BHP may thus need to take an educative and advocacy role with the integrated care team to address any assumptions related to the sexuality of WWD that ensure equitable access to preventive healthcare, while approaching care from a holistic perspective that takes into account the biopsychosocial needs of WWD, rather than distilling their concerns into a single diagnostic category.

## Disability culturally competent sexual healthcare

The Association of Reproductive Health Professionals (ARHP 2010) outlined sexual health clinical competencies that encompassed medical knowledge, understanding of relevant psycho-social issues and clinical skills as they pertain to sexual health and well-being. Acknowledging that "sexual function does not exist in a vacuum—it is influenced by the relationship, fatigue, stress and other sociocultural factors" (p. 1), BHPs must engage in ongoing professional develop-ment to ensure that they have the requisite knowledge, skills and abilities to work with WWD in relation to sexual health and well-being in integrated environments. Although BHPs are called to provide competent services across a broad domain of concerns in the primary care environ-ment, few receive explicit training in disability and sexuality to achieve these goals for WWD; even fewer receive additional training on addressing these complexities within the context of the family system. To this end, Mona et al. (2017) developed the Disability and Sexuality Health Care Competency Model (DASH-CM) in order to encourage self-awareness and skill develop-ment in the delivery of sexual health care to persons with disabilities (PWD). While an in-depth discussion of DASH-CM is beyond the scope of this chapter, at its core, DASH-CM notes the importance of developing competency within the following domains:

1. Disability and sexuality critical awareness: Understanding and examination of the indi-vidual, social and political context in which disability and sexuality are experienced.
2. Disability and sexuality knowledge: Understanding of disability and chronic health condi-tions in the context of one's identity and sexual experiences.
3. Disability and sexuality skill development: Developing disability-affirmative skills through training, attention to implicit bias and practice.
4. Disability and sexuality practice and application: Utilisation of evidence-based practices, related to the assessment and treatment of sexual health among WWD, and attending to the need to promote the accessibility of clinical interactions and sexual encounters, while recognising diverse opportunities for sexual expression and pleasure (Mona et al. 2017). As BHPs work as part of a healthcare team in the delivery of sexual health care, they are encouraged to not only examine their own beliefs and biases about sexuality and sexual health for WWD, but to engage interdisciplinary professionals and the families of WWD (as appropriate) in similar processes.

## Critical clinical issues influencing sexual well-being for WWD

Sexual functioning has long been considered to be multifaceted: "Sexuality is best understood as a multicausal, multidimensional phenomenon with psychological, biological, relational and cultural components" (McCarthy and Thestrup 2008, p. 592). Not only is sexual health influ-enced by general health, but it is impacted by both *individual* emotional well-being as well as *interpersonal* relational stress (Stayton 1996). In order to adequately address sexual well-being, providers must assess for psychological, situational and relational factors in addition to identifying any biological concerns (Atwood 2006). They must similarly consider the role of, and engage with, both the integrated and interprofessional healthcare team and the WWD's family system (as desired). The interplay between these dynamic systems is complex, and BHPs must be prepared to navigate a variety of common concerns related to sexual well-being among WWD.

## Abuse

Persons with disabilities experience higher rates of emotional, physical, sexual and intimate partner abuse than their non-disabled peers (Krnjacki et al. 2016), with one study reporting 56% of WWD experiencing a lifetime history of physical or sexual abuse, inclusive of a caregiver's refusal to provide access to essential resources, such as a wheelchair or medication (Milberger et al. 2003), or abuse in the context of personal assistance (e.g., unnecessary force, touching or roughness during activities of daily living such as bathing, dressing or transferring to a wheelchair or bed; Salwen et al. 2016). Among these WWD who have experienced abuse, 70% reported experiencing abuse at the hands of an intimate partner, and WWD are more likely than women without disabilities to experience rape, sexual violence, physical violence, psychological aggression and control over reproductive or sexual health at the hands of an intimate partner (Breiding and Armour 2015) than women without disabilities. Since many WWD who experience abuse may not seek help for abuse, but present to a primary care or medical environment for other assistance, BHPs must be sensitive to their unique vulnerabilities and screen accordingly for risk factors for partner abuse, including a history of WWD or partner depression, substance use, history of abuse, marital dissatisfaction, fear of one's partner, or the partner's refusal to provide care (Salwen et al. 2016). This is particularly important for the BHP-provided family-based sexual health services, as establishing a patient's safety and providing the patient with the appropriate resources to seek help if desired is a critical first step. Although a comprehensive discussion of screening and assessment for abuse and intimate partner violence among WWD is beyond the scope of this chapter, readers are directed to Salwen et al. (2016) for recommended strategies for risk factor screening, including tools which are brief and readily implementable in a medical environment, as well as questions that should be incorporated into a thorough assessment. Additionally, BHPs are implored to ensure that resources provided to WWD who are requesting assistance in leaving an abusive relationship are accessible, as only 50% of women who do seek help report having a positive experience with the resources provided to them (Milberger et al. 2003).

## Depression

Primary care medical settings have become the "de facto" mental health system in the United States, and as such, a BHP working in an integrated care setting is likely to encounter individuals with depression, regardless of disability status. Persons living with disabilities have higher rates of depression than their non-disabled peers, with WWD experiencing higher rates of depression than men with disabilities (Noh et al. 2016); depression may additionally be complicated by societal and attitudinal barriers in WWD's lived experiences. Given the bidirectional nature of the relationship between depression and sexual dysfunction (Clayton et al. 2015), a BHP may need to address underlying mood disorders when working with WWD to promote sexual health and well-being. In integrated environments, the Patient Health Questionnaire (PHQ-9) is frequently utilised as a brief screen for depression. For WWD with a positive screen, BHPs are encouraged to acknowledge and inquire about how a WWD's experience of depression may be influencing her sexual relationships and functioning. While BHPs need to be aware of the higher rates of depression among WWD and their impact on sexual health, it is additionally critical to not assume that a WWD's depression is a direct result of their disability. BHPs are encouraged to utilise a family systems framework, coupled with an understanding of the social determinants of health that disproportionally influence a WWD's mental health status, such as lack of access to healthcare, employment, etc.—when enquiring about depression among WWD, as individual depression can be influenced by family functioning (Keitner et al. 1987), and may further influence a woman's sexual intimacy and well-being with her partner(s).

## *Reproductive Health*

More than four million (6%) American parents are living with a disability; WWD experience a variety of barriers to reproductive healthcare, including, but not limited to, inaccessible facilities, lack of provider knowledge about the reproductive needs of WWD, provider bias, limited contraceptive options (Becker et al. 1997), as well as resistance and discrimination regarding a WWD's ability to conceive, carry a foetus to term and physically and emotionally care for a child (Clemency Cordes et al. 2017). WWD are less likely to receive prenatal care in their first trimester and are more likely to report greater levels of maternal stress associated with medical complications during their pregnancy than their non-disabled peers, and this may, in part, be due to negative experiences with healthcare providers and higher rates of intimate partner violence for these women (Mitra et al. 2015). For WWD who desire to pursue motherhood, additional barriers and biases exist in the form of laws that allow disability to be grounds for termination of parental rights and increased rates of interaction with child welfare agencies (NCD 2012), despite evidence that WWD exhibit higher levels of resilience and creativity than parents without disabilities (Prilleltensky 2004). Additionally, children of WWD exhibit typical development and enhanced levels of functioning than those raised by non-disabled women (NCD 2012). As such, BHPs must be prepared to educate members of the integrated medical team regarding WWD and parenting issues, attending to adaptive parenting strategies and the need for universal parental rights.

## Assessment and treatment of sexual health and well-being

As noted above, few BHPs are explicitly trained in the assessment and treatment of sexual health and well-being for WWD; however, most, if not all, BHPs working in an integrated healthcare environment will be tasked to address this dynamic relationship at some point in their careers. The American Psychological Association (2012) has highlighted the importance of collaboration when working with WWD; this collaboration extends beyond the individual relationship between a BHP and a WWD to encompass collaboration with the WWD's healthcare team and family, as relevant and desired. For BHPs working as part of a medical team, disability-competent care requires that the provider attends to content and communication style, as WWD often experience microaggressions in their interactions with healthcare providers (Mona et al. 2017). Relational and attitudinal microaggressions are defined as the intentional (or unintentional) messages conveyed by non-disabled individuals that are negative, demeaning or even hostile about disability (Coble-Temple and Bell 2013). With regards to sexual health and well-being, microaggressions can range from a provider's belief that WWD are asexual or disinterested in childbearing, to overtly dismissing a WWD's desire to promote sexual intimacy with a current or future partner, to failing to inquire about sexual health and well-being during an initial assessment when a provider would otherwise assess for these domains among persons without disabilities (Clemency Cordes et al. 2016). As rapport and trust between a WWD and her care team are critical, BHPs should monitor their own potential for engaging in microaggressions, as well as educating other providers on the importance of open and non-judgmental communication in both personal and professional interactions with WWD (Mona et al. 2017).

## General assessment

Individuals presenting for care in the primary care setting have indicated that providers should initiate conversations in relation to sexual health and well-being (Ryan et al. 2018); however, more than 40% do not routinely ask about sexual health (Wimberly et al. 2006). In order to promote sexual health and well-being among WWD, it is critical that BHPs enquire about sexual health at the initial visit and subsequently, as relevant. For BHPs in the fast-paced integrated primary care setting, a comprehensive sexual health assessment for a WWD may be beyond the scope of services that are able to be provided. As a result, BHPs are encouraged to leverage screening tools that are already commonplace in primary care to screen for common mental health concerns that might contribute to diminished sexual health and well-being, such as the PHQ-9 questionnaire for depression, or the Alcohol Use Disorders Identification Test (AUDIT) for alcohol misuse, while noting that these measures were not originally validated for use with WWD, and should therefore be interpreted with caution (Clemency Cordes et al. 2016). BHPs may similarly encourage medical providers to initiate conversations related to sexual health and well-being with WWD who may be presenting for care related to management of a chronic health conditions associated with diminished sexual well-being, such as diabetes or chronic pain, during team meetings.

## Brief sexual health assessment

Figuero-Haas (2012) recommends that providers begin conversations about sexual health and well-being with open-ended questions, such as "Do you have any questions or concerns related to sex?" Providers are reminded that executing this type of questioning is a skill and requires practice in order to elicit open and honest responses. While the actual protocol regarding which provider asks the questions (i.e., the medical provider or the BHP) can vary from clinic to clinic, the medical provider and the BHP should work together to ensure both providers feel confident openly discussing sexual health.

Given the time constraints in primary care, Hunter and colleagues (2009) provide sample assessment questions for rapid identification of sexual dysfunction, including "Have you ever experienced an orgasm?" "Were there changes in your health, relationships or other areas when the problems began?" and "Do you experience any pain with intercourse?" (p. 209). Such focused questions should be adapted in order for the BHPs to attend to the unique experiences of WWD, and adapt their assessments accordingly. For example, many WWD may not experience physiological orgasms; as a result, an assessment focus may instead be on the WWD's experiences of sexual pleasure, or positions that that enhance intimacy and pleasure, rather than a prior history of climax. This brief approach to assessment may subsequently be followed by paper-pencil questionnaires such as The Sexual Quality of Life—Female (SQOL-F), which was developed with the intent of assessing the role of sexual dysfunction on the quality of life in women, and was validated with women with a spinal cord injury (Symonds et al. 2005). Statements to be assessed include, "When I think about my sexual life, it is an enjoyable part of my life overall," and "I worry about the future of my sexual life" (Symonds et al. 2005, p. 393). A list of additional brief sexual health assessment options for use in primary care can be seen in Table 31.1. BHPs should be prepared for adaptive strategies for the delivery of traditional

*Table 31.1* Female sexual dysfunction assessment tools* (adapted from Figuero-Haas (2012), p. 157)

| Tool | Assessment area |
| --- | --- |
| Female Sexual Function Index<br>http://www.fsfiquestionnaire.com/ | Desire, arousal, orgasm and pain |
| Profile of Female Sexual Function<br>http://www.ncbi.nlm.nih.gov/pubmed/14660291 | Desire in postmenopausal women |
| Brief Profile of Female Sexual Function<br>http://www.menopausematters.co.uk/quizbpfsf.php | Self-screener for low desire in<br>    surgically postmenopausal women |
| Sexual Interest and Desire Inventory—Female<br>http://www.thecshe.com/pdf/Sexual%20Interest%20and<br>    %20Desire%20Inventory-%20Female%20(SIDI-F).pdf | Severity of symptoms in women with<br>    hypoactive sexual desire disorder |
| Female Sexual Distress Scale—Revised<br>http://www.obgynalliance.com/files/fsd/FSDS-R_<br>    Pocketcard.pdf | Distress |
| Sexual Quality of Life—Female<br>http://www.prolutssh.com/fdescrip.html | Quality of life in women with sexual<br>    dysfunction |
| Golombok Rust Inventory of Sexual Satisfaction (GRISS) | Quality of sexual relationship |
| Decreased Sexual Desire Screener<br>http://www.obgynalliance.com/files/fsd/DSDS_Pocketcard.pdf | Brief diagnostic tool for hypoactive<br>    sexual desire disorder |

*Please note, many of these tools are focused on sexual dysfunction (per the medical model). Not all of them have been validated for use with women with disability.

paper-pencil questionnaires, such as delivering assessments verbally or via alternative means to promote accessibility.

Although a thorough sexual history-taking is commonplace in couples sex therapy (McCarthy and Thestrup 2008), it is rare to be allotted the necessary time in medical settings to conduct a comprehensive interview, let alone a comprehensive interview with the patient *and* her partner (if applicable). It is notable that WWD report greater dissatisfaction with sexual relationships and dating opportunities than women without disabilities (Nosek et al. 2001); this may be particularly true for those with early-onset disabilities, who had limited access to sexual health education and/or awareness of sexual techniques to accommodate disability (Olkin 2016). As a result, providers can be purposeful in their questioning, being mindful of the impact of family of origin experiences on sexual wellness along with the systemic influence that a partner or significant other may have on the WWD's sexual health. It is vital that providers ask permission from the WWD prior to discussing sexual wellness in front of others present in the exam room (e.g., caregiver, spouse).

## Treatment

As noted above, medical models of disability have focused their attention on *dysfunction* and/or *disease*; when applied to sexual health and well-being, this has resulted in a focus on eliminating sexual dysfunction, rather than the promotion of sexual satisfaction and well-being. Like all healthcare providers, BHPs may make ableist assumptions of what constitutes sexual pleasure and satisfaction; providers are encouraged to challenge such assumptions in order to assist in WWD taking agency of their sexual lives (Clemency Cordes et al. 2017). Although comprehensively addressing sexual health and well-being may be outside the scope of a BHP functioning in an integrated primary care clinic, BHPs should be familiar with treatment approaches in order to best counsel WWD regarding their options and to facilitate appropriate referrals.

In order to ensure delivery of culturally-appropriate services, BHPs are encouraged to adopt the meta framework of disability affirmative therapy (DAT; Olkin 2016). Approaching disability from a biopsychosocial lens, DAT encourages BHPs (and all providers) to partner with WWD in the treatment process in order to reduce barriers, challenge assumptions and create socially and physically enabling environments. For BHPs working in integrated care environments, this may include educating other members of the team regarding strategies for encouraging sexual well-being and pleasure, advocating for systems-level changes to enhance inclusivity (e.g., providing longer healthcare visits for a WWD to accommodate issues related to transferring to/ from a wheelchair), and addressing microaggressions that WWD may otherwise experience from members of the healthcare team.

For BHPs who often have limited exposure to explicit training in sexology, the Ex-PLISSIT model (Taylor and Davis 2007) for assessment and treatment of sexual health, which expands on Annon's (1976) model, may serve as one strategy for discussing sexuality and sexual expression. The original PLISSIT model utilized four levels of intervention: Permission, Limited Information, Specific Suggestions, Intensive Therapy; comparatively, Ex-PLISSIT emphasizes that permission-giving is central across all four levels of intervention, and that BHPs should always be attuned to their own level of competency, thereby referring (with permission) to providers or community resources to comprehensively address the needs of WWD (Taylor and Davis 2007). For example, a BHP might engage a WWD in a brief assessment of their sexual concerns as noted above, and, with permission, provide both psychoeducation, regarding safe sexual practices, and provide specific recommendations to the WWD and her partner (if relevant), related to enhancing one's sexual scripts so as to not solely focus on penile-vaginal intercourse for WWD in heterosexual relationships. However, when noting that the WWD's history of sexual abuse is influencing relational dynamics and sexual functioning, the BHP would again request permission to refer the WWD to qualified providers, who could deliver evidence-based practices (e.g., cognitive behavioural therapy, disability affirmative therapy) to comprehensively address the WWD's needs.

It is notable that the function and role of the BHP in the comprehensive promotion of sexual health and well-being among WWD may, in part, be dictated by systems-level variations between clinics. For example, in Level 6 integration, as described above, which is also referred to as the Primary Care Behavioural Health (PCBH) model, the on-site BHP works alongside the medical team, addressing not only traditional "mental health" concerns (e.g., depression, anxiety), but also anything behavioural that may influence health (e.g., adherence to medication, lifestyle changes) in brief (15–30 minutes) consultations (Clemency Cordes et al. 2018). In this model of integration, the BHP is well-suited to routinely inquire about sexual well-being as a part of comprehensive wellness, and deliver brief interventions on an intermittent basis, but may refer out to community-based specialists when more intensive sex therapy is needed.

In contrast, in a Level 3 co-located clinic, although there *is* a BHP onsite, the role of the BHP can be quite different than in the PCBH model (Clemency Cordes et al. 2018). The BHP may do some brief consultations in exam rooms, but much of their time is spent in a designated therapy space, providing traditional (50-minute) sessions addressing patient mental health concerns. Although certainly capable of addressing issues such as sexual well-being among WWD, the BHP in a co-located setting may not be as visible in the clinic or as available to discuss anything beyond mental health/comorbid medical conditions as a BHP in a PCBH setting (unless identified as an issue by the PCP). However, if and when a WWD is referred to a BHP at this level of integration, they may be able to more comprehensively engage WWD and their partners in evidence-based approaches such as family therapy, cognitive behavioural therapy, sensate focus and other strategies, that can be delivered on a more regular, long-term basis.

## Conclusion

Behavioural health providers working in integrated care are uniquely positioned to assist WWD in the promotion of sexual health and well-being, given their role as part of a medical team. However, few BHPs working in this environment have received explicit training on the unique lived experiences of WWD and how this influences relational and sexual dynamics between WWD and their partners. As such, BHPs are encouraged to engage in ongoing training related to the sexual health needs of this population, as well as self-reflection regarding their own beliefs and biases to ensure they are prepared to deliver culturally competent sexual healthcare. In doing so, BHPs are in a unique position to not only influence sexual health outcomes, but to also enhance the WWD's lived experience through their validation of their humanity, and their core sense of self as a sexual being.

## References

American Psychological Association (2012) *Guidelines for the assessment and intervention with persons with disabilities*, viewed 28 June 2019. https://www.apa.org/pi/disability/resources/assessment-disabilities.aspx

Andrews, E.E. (2017) Disability models. In M.A. Budd, S. Hough, S.T. Wegener, and W. Steirs (eds.), *Practical psychology in medical rehabilitation*. New York, NY: Springer.

Annon, J.S. (1976) 8—The PLISSIT Model: A proposed conceptual scheme for the behavioral treatment of sexual problems. In J. Fischer and H.L. Gochros (eds.), *Handbook of behavior therapy with sexual problems; volume 1: General procedures*. New York, NY: Elsevier.

Association of Reproductive Health Professionals (2010) *Clinician competencies for sexual health*, viewed 28 June 2019. https://www.arhp.org/clinical-fact-sheets/clinician-competencies-for-sexual-health

Atwood, J.D. (2006) Sexual dysfunctions and sex therapy. In L.L. Hecker and J.L. Wetchler (eds.), *An introduction to marriage and family therapy*. New York: Haworth Press.

Becker, H., Stuifbergen, A., and Tinkle, M. (1997) Reproductive health care experiences of women with physical disabilities: A qualitative study. *Archives of Physical Medicine and Rehabilitation* 78(12) Supplement 5, pp. S26–S33.

Berman, H., Harris, D., Enright, R., Gilpin, M., Cathers, T., and Bukovy, G. (1999) Sexuality and the adolescent with a physical disability: Understandings and misunderstandings. *Issues in Comprehensive Pediatric Nursing* 22, pp. 183–196.

Borst, C., Martin, M., and Macchi, C.R. (2018) Health problems in couple and family therapy. In J. Lebow, A. Chambers, and D. Breunlin (eds.), *Encyclopedia of couple and family therapy*. New York, NY: Springer.

Breiding, M.J., and Armour, B.S. (2015) The association between disability and intimate partner violence in the United States. *Annals of Epidemiology* 25, pp. 455–457.

Center for Disease Control and Prevention (2015) *Key findings: Prevalence of disability and disability types among adults, United States—2013*, viewed 28 June 2019. https://www.cdc.gov/ncbddd/disabilityand health/features/key-findings-community-prevalence.html

Center for Research on Women with Disabilities (CROWD) (n.d.) *Demographics*, viewed 28 June 2019. https://www.bcm.edu/research/centers/research-on-women-with-disabilities/general-info/demographics.

Clayton, A.H., Croft, H.A., and Handiwala, L. (2015) Antidepressants and sexual dysfunction: Mechanisms and clinical implications. *Journal of Postgraduate Medicine* 126(2), pp. 91–99.

Clemency Cordes, C., Cameron, R.P., Eisen, E., Coble-Temple, A., and Mona, L.R. (2017) Leveraging integrated health services to promote behavioral health among women with disabilities. In K.A. Kendall-Tackett and L.M. Ruglass (eds.), *Women's mental health across the lifespan: Challenges, vulnerabilities, and strengths*. New York, NY: Routledge.

Clemency Cordes, C., Reiter, J., and Manson, L. (2018) Integrated behavioral healthcare models. In C.R. Macchi and R. Kessler (eds.), *Training to deliver integrated care: Skills aimed at the future of healthcare*. New York, NY: Springer.

Clemency Cordes, C., Saxon, L.C., and Mona, L.R. (2016) Women veterans with disabilities: An integrated care perspective. In S. Miles-Cohen & C. Signore (eds.), *Eliminating inequities for women with disabilities: An agenda for health and wellness.* Washington, DC: American Psychological Association.

Coble-Temple, A., and Bell, A. (2013) Nothing about us without us: A qualitative investigation of ableist microaggressions through the voices of people with visible disabilities. PhD thesis, John F. Kennedy University.

Corso, K.A., Hunter, C.L., Dahl, O., Kallenberg, G.A., and Manson, L. (2016) *Integrating behavioral health into the medical home: A rapid implementation guide.* Phoenix, MD: Greenbranch Publishing.

DeGruy, F., and McDaniel, S. (2013) The landscape of integrated behavioral health care Initiatives. In M. Talen and A. Valeras (eds.), *Integrated behavioral health in primary care: Evaluating the evidence, identifying the essentials.* New York, NY: Springer.

Dunn, D.A., and Andrews, E.E. (2015) Person-first and identity-first language: Developing psychologists' cultural competence using disability language. *American Psychologist* 70(3), pp. 255–264.

Figueroa-Haas, C. (2012) Screening for female sexual dysfunction in primary care settings. *The Journal for Nurse Practitioners* 8(2), pp. 156–157.

Forbert-Pratt, A.J. (2018) (Re)defining disability culture: Perspectives from the American Disabilities Act generation. *Culture & Psychology*, pp. 1–16.

Gill, C.J. (1995) Research on urban planning and architecture for disabled persons in Iran establishing design criteria. *Independent Living Institute*, viewed 29 June 2019. http://www.independentliving.org/docs3/gill1995.html

Hays, P.A. (1996) Addressing the complexities of culture and gender in counseling. *Journal of Counseling and Development* 74, pp. 332–338.

Heath, B., Wise Romero, P., and Reynolds, K.A. (2013) *Review and proposed standard framework for levels of integrated healthcare.* Washington, DC: SAMHSA-HRSA Center for Integrated Health Solutions.

Hecker, L.L., Mims, G.A., and Boughner, S.R. (2003) General systems theory, cybernetics, and family therapy. In L.L. Hecker and J.L. Wetcher (eds.), *An introduction to marriage and family therapy.* Binghampton, NY: Haworth Clinical Practice Press.

Hodgson, J., Lamson, A., Mendenhall, T., and Tyndall, L. (2014) Introduction to medical family therapy: Advanced applications. In J. Hodgson, A. Lamson, T. Mendenhall, and D.R. Crane (eds.), *Medical family therapy.* New York: Springer.

Hunter, C.L., Goodie, J.L., Oordt, M.S., and Dobmeyer, A.C. (2009) *Integrated behavioral health in primary care: Step-by-step guidance for assessment and intervention.* Washington, DC: American Psychological Association.

Kiecolt-Glaser, J.K., Loving, T.J., Stowell, J.R., Malarkey, W.B., Lemeshow, S., Dickenson, S.L., and Glaser, R. (2005) Hostile marital interactions, proinflammatory cytokine production, and wound healing. *Archives of General Psychiatry* 62, pp. 1377–1384.

Kienter, G.I., Miller, I.W., Epstein, N.B., Bishop, D.S., and Fruzzetti, A.E. (1987) Family functioning and the course of major depression. *Comprehensive Psychiatry* 28, pp. 54–64.

Kraus, L. (2017) *2016 disability statistics annual report.* Durham, NH: University of New Hampshire.

Krnjacki, L., Emerson, E., Llewellyn, G., and Kavanagh, A.M. (2016) Prevalence and risk of violence against people with and without disabilities: Findings from an Australian population-based study. *Australian and New Zealand Journal of Public Health* 40, pp. 16–21.

Marlowe, D., Hodgson, J., Lamson, A., White, M., and Irons, T. (2014) Medical family therapy in integrated primary care: An interactional framework. In J. Hodgson, A. Lamson, T. Mendenhall, and D.R. Crane (eds.), *Medical family therapy.* New York: Springer.

McCarthy, B.W., and Thestrup, M. (2008) Couple therapy and the treatment of sexual Dysfunction. In A. Gurman (ed.), *Clinical handbook of couple therapy* (4th ed.). New York: Guilford Press.

Milberger, S., Israel, N., LeRoy, B., Martin, A., Potter, L., and Latchak-Schuester, P. (2003) Violence against women with physical disabilities. *Violence and Victims* 18, pp. 581–591.

Mitra, M., Clements, K.M., Zhang, J., Iezzoni, L.I., Smeltzer, S.C., and Long-Bellil, L.M. (2015) Maternal characteristics, pregnancy complications, and adverse birth outcomes among women with disabilities. *Medical Care* 53, pp. 1027–1032.

Mona, L.R., Cameron, R.P., and Clemency Cordes, C. (2017) Disability culturally competent sexual healthcare. *American Psychologist* 72, pp. 1000–1010.

National Council on Disability (2012) *Rocking the cradle: Ensuring the rights of parents with disabilities and their children.* Washington, DC: NCD.

Noh, J.W., Kwon, Y.D., Park, J., Oh, I.H., and Kim, J. (2016) Relationship between physical disability and depression by gender: A panel regression model. *PLoS One* 11(11), p. e0166238.

Nosek, M.A., Howland, C., Rintala, D.H., Young, M.E., and Chanpong, G.F. (2001) National study of women with physical disabilities: Final report. *Sexuality and Disability* 19, pp. 5–39.

Olkin, R. (2016) Disability affirmative therapy. In I. Marini and M.A. Stebnicki (eds.), *The professional counselor's desk reference* (2nd ed). New York, NY: Springer.

Parkerson, G.R., Broadhead, W.E., and Tse, C.K.J. (1995) Perceived family stress as a predictor of health related outcomes. *Archives of Family Medicine* 4(3), p. 253.

Prilleltensky, O. (2004) My child is not my carer: Mothers with physical disabilities and the well-being of children. *Disability and Society* 19(3), p. 210.

Ryan, K.L., Arbuckle-Bernstein, V., Smith, G., and Phillips, J. (2018) Let's talk about sex: A survey of patients' preferences when addressing sexual health concerns in a family medicine residency program office. *PRiMER* 2, p. 23.

Salwen, J.K., Gray, A., and Mona, L.R. (2016) Personal assistance, disability, and intimate partner violence: A guide for healthcare providers. *Rehabilitation Psychology* 61, pp. 417–429.

Stayton, W.R. (1996) Sexual and gender identity disorders in relational perspective. In F.W. Kaslow (ed.), *Handbook of relational diagnosis and dysfunctional family patterns*. Hoboken, NJ: John Wiley & Sons.

Strosahl, K. (2001) The integration of primary care and behavioral health: Type II changes in the era of managed care. In N.A. Cummings, W. O'Donohue, S.C. Hayes, and V. Follette (eds.), *Integrated behavioral heatlhcare: Positioning mental health practice with medical/surgical practice*. San Diego, CA: Academic Press.

Symonds, T., Boolell, M., and Quirk, F. (2005) Development of a questionnaire on sexual quality of life in women. *Journal of Sex & Marital Therapy* 31, pp. 385–397.

Talen, M.R., and Valeras, A.B. (2013) Introduction and overview of integrated behavioral health in primary care. In M.R. Talen and A.B. Valeras (eds.), *Integrated behavioral health in primary care: Evaluating the evidence, identifying the essentials*. New York, NY: Springer.

Taleporos, G., and McCabe, M.P. (2002) The impact of sexual esteem, body esteem, and sexual satisfaction on psychological well-being in people with physical disability. *Sexuality and Disability* 20, pp. 177–183.

Taylor, B., and Davis, S. (2007) The extended PLISSIT model for addressing the sexual well-being of individuals with an acquired disability or chronic illness. *Sexuality and Disability* 25, pp. 135–139.

Wimberly, Y.H., Hogben, M., Moore-Ruffin, J., Moore, S.E., and Fry-Johnson, Y. (2006) Sexual history-taking among primary care physicians. *Journal of the National Medical Association* 98, pp. 1924–1929.

World Health Organization (2002) *Defining sexual health: Report of a technical consultation on sexual health 28–31 January 2002, Geneva*, viewed 28 June 2019. http://www.who.int.ezproxy1.lib.asu.edu/reproductivehealth/topics/gender_rights/defining_sexual_health.pdf

World Health Organization and World Bank Group (2011) *Summary: World report on Disability*. Malta: World Health Association.

# 32
# OCCUPATIONAL THERAPY'S ROLE EMPOWERING SEXUALITY FOR INDIVIDUALS WITH DISABILITY

*Kathryn Ellis and Dikaios Sakellariou*

## Introduction

Sexuality is a part of many people's lives, whether they live with disability or not. Affirming engagement in chosen sexual experiences and forms of expression can contribute to an individual's quality of life (Diamond and Huebner 2012; Kleinstäuber 2017; Lindae et al. 2007; Viejo et al. 2015). The importance of expressing and experiencing one's self as a sexual being does not dissipate with disability. The study by Esmail et al. (2001) showed sexuality to be the most serious concern in post-disability relationships. Unfortunately, pervasive socio-cultural norms related to sexuality can often lead to a discourse of asexuality, whereby people with disabilities are seen by others, and sometimes by themselves, as asexual, leading to a silencing around their sexual needs. This presumption of asexuality is not the only misrecognition people with disabilities face; disability is also associated with hypersexuality or sexual deviancy. Socio-cultural norms, reserving sexuality for young, White, healthy, heterosexual men of reproductive age, perpetuate the neglect of the sexual rights of people with disabilities, among many other groups (Collins et al. 2017; McGrath and Sakellariou 2015; Tepper 2000). These norms also decrease access to accurate information and competent service providers for people with disabilities. The occupational therapy profession's dedication to holistic interventions to enhance independence and satisfaction with engagement in all the activities people do, situates the profession in a perfect position to enable sexuality for all.

Our aim in this chapter is to explore occupational therapy's engagement with sexuality assessment and interventions for people with disabilities. First, sexuality will be framed as an integral part of everyday life, or, in occupational therapy parlance, "an activity of daily living." Activities of daily living are defined as, "activities that are oriented towards taking care of one's own body," (American Occupational Therapy Association 2014, p. 620). Occupational therapists consider these activities to be "fundamental to living in a social world" (Christiansen and Hammecker 2001, p. 156). Then, we will provide a validation for occupational therapy's involvement and an overview of the tensions that often lead to the invisibility of sexuality in occupational therapy educational curricula and practice. We will then proceed to present a framework to support the role occupational therapy could, and sometimes already does, play in enabling sexuality. This chapter will conclude by offering specific suggestions for occupational therapy education on how to leverage a distinct occupational therapy skill-set to address sexuality.

## Importance of sexuality for people with disabilities

Sexual engagements and experiences can have either enriching or detrimental influences on individuals' lived experiences. Sexuality can be affirming, facilitating the development of self-advocacy skills, promoting psychological adjustment, reducing social anxiety and enhancing self-concept and quality of life (Diamond and Huebner 2012; Kashdan et al. 2014; Smith et al. 2011; Viejo et al. 2015). Socio-cultural norms often give the greatest acceptance to the sexual expression and relevancy of healthy, young, White, heterosexual, men of reproductive age (Collins et al. 2017; McGrath and Sakellariou 2015; Tepper 2000). Additionally, the occupational therapy profession is largely based on White middle-class western values, which highlight productivity as paramount alongside a delegitimisation or distrust of anything hedonistic and pleasure-focused, or considered as such (McGrath and Sakellariou 2015). These norms support an adherence to the privacy, shame and guilt around sexual activity and expression. These socio-cultural norms pose a challenge for people's sexual identity even for those given cultural *permission*. Compounding this tension, sexuality can also be a site of exercising power, control and abuse through disrespect, coercion, manipulation, harassment, bullying and sexual assault (Deering et al. 2014; Diamond and Huebner 2012; Espelage et al. 2015; Papp et al. 2017).

People with disabilities are often viewed as lacking desirability and ability as sexual partners. While they often rely on healthcare professionals for information regarding sexuality, this information is not always forthcoming. This reluctance to deal with sexuality is highly problematic; while people with disabilities are sexual beings and require sexual health education, they may be less likely to receive it. For example, adolescent women with disabilities are more likely than adolescent women without disabilities to become pregnant, experience sexual assault and forgo the use of contraceptive means, yet they are less likely to receive sexual health education than their non-disabled peers (Linton and Adams Rueda 2014).

It is important to note that disability is not an exclusive identity and does not exist in a vacuum; individuals who identify as disabled may also identify with other groups. This is known as intersectionality, which considers the experience of possessing multiple identities and the compounding privilege or oppression arising at the intersection of those identities (Cole 2009). For example, we know that long-term health, quality of life and social participation are impacted for people who identify as a racial or sexual minority and live with a chronic illness or disability (Dispenza et al. 2016), leading to challenges with employment, stigma, internalisation of heterosexism, lack of social support and anticipated discrimination, which compromise wellbeing (Dispenza et al. 2016; Williams et al. 2017).

## Reasons for occupational therapy's reluctance to address sexuality

Occupational therapists often perpetuate the silence and delegitimisation surrounding disability and sexual activity, despite it being identified as an activity of daily living (American Occupational Therapy Association 2014). Occupational therapy literature has long documented a lack of preparedness to address sexuality in clinical practice (Conine et al. 1979; Lohman et al. 2017). Despite the importance of sexual activity for people with disabilities and the holistic nature of the occupational therapy profession, occupational therapists often hesitate to discuss sexuality with their clients (Dyer and das Nair 2013; Hattjar et al. 2008; Hyland and McGrath 2013; Jones et al. 2005; McGrath and Sakellariou 2015). Although occupational therapists address other intimate and culturally sensitive activities of daily living, such as toileting, feeding and dressing, addressing sexuality appears to pose a significant challenge and they often avoid it.

Several of the reasons for occupational therapists' reluctance to address sexuality with people with disabilities have been well documented in the literature, and include socio-cultural norms, institutional limitations and lack of competency and comfort (Dyer and Nair, 2013; Hattjar et al.2008; McGrath and Lynch 2014; McGrath and Sakellariou 2015). Occupational therapists, probably influenced by the cultural privacy of sexual discussion, report a fear of causing offence to service users, concerns for disapproval from superiors and awkwardness as reasons for not being sexuality-inclusive in their interventions. Also, the prioritisation of sexual activity for pro-creation means that sexual activity outside of reproductive goals can be considered subordinate to other occupations such as employment, education and household chores, which constitute rehabilitation *needs* (McGrath and Sakellariou 2015).

Institutional entities should understand the role they could play in deconstructing the restrictions of socio-cultural norms. Occupational therapy curricula related to sexuality for people with disabilities are sparse and inadequate (Conine et al.1979; Couldrick, 1999; Eglseder and Webb, 2017; Jones et al. 2005; Lohman et al.2017; Payne et al.1988). It is important to develop sexuality education activities, which provide more time and opportunities for self-reflection of personal sexuality beliefs, values, attitudes, and biases. Also, institutions that provide services for people with disabilities should consider a top-down approach to counter the perception of disapproval surrounding sexual activity. Hosting training events, including sexuality in their mission and vision statements and conducting awareness campaigns can bolster the likelihood that a clinician will address the topic of sexuality (Gianotten et al.2006).

Lastly, an individual clinician's perception of lack of competency or comfort to adequately address sexuality with people with disabilities, and the perceived lack of institutional guidance or support, can also often restrict occupational therapists' preparedness to address sexuality in practice. Occupational therapists often report that their academic programme did not prepare them to address the sexuality needs of their clients and they would not be comfortable to do so (Jones et al. 2005). The current Accreditation Council for Occupational Therapy Education (ACOTE) Standards in the USA do not include assessment or intervention standards specific to sexuality for entry-level occupational therapy or occupational therapy assistant education, perpetuating the disconnect between service users' needs, the scope of practice and academic translation of clinical skills (ACOTE 2018).

Occupational therapy students, clinicians and educators have given several reasons why they do not engage more with assessing the sexuality of people with disabilities. Some reasons given by our students and colleagues include the following:

> We were told by our professors that sex is important to our clients, but that's all they said.

> The accreditation for our hospital dinged us for not addressing sexuality. Our leadership got together and determined OT [occupational therapy] would address it. They laughed and said, 'don't waste too much time on it.'

> As an OT professor I understand why it is important for sexuality to be included in our curriculum. I posed this to my programme director and she informed me that since sexuality was not connected to an education standard that I was not to include it.

# Occupational therapy and people with disabilities

The occupational therapy profession is grounded in a belief that all people need to engage in personally and socially meaningful occupations. Occupational therapists often acknowledge the relevance of sexuality to service users' lives (Hattjar et al. 2008; Lohman et al. 2017; Sakellariou and Algado 2006). Often, occupational therapy discourse and policy documents include sexual activity and intimate relationship management within the scope of occupations which occupational therapists should address (American Occupational Therapy Association [AOTA] 2014).

The way sexuality is addressed in occupational therapy is very important. Occupational therapists are trained to be aware of socio-cultural norms, individuals' abilities and lived experiences. Occupational therapists are educated in activity analysis, which is a process used to deconstruct the required components of engagement in an occupation, revealing any friction points or challenges. This includes identifying the cognitive, emotional and physical requirements of the activity, as well as the requirements of the environment to support the activity. Occupational therapists have a distinct ability to identify activity components, determine which are enabling and which are disabling, and proceed to address those disabling components through modification, adaptation, advocacy and remediation. Integral to an activity analysis and resultant intervention is client-centred care, which positions the client's experiences, desires and goals as the paramount factor that guides intervention. Whereas occupational therapists are often challenged to engage in self-reflection regarding socio-cultural norms and complex activity analysis within occupational therapy curricula, these hallmark assignments infrequently venture into the realm of sexuality, leaving professionals feeling unable to address sexuality topics, despite having the primed skills.

To improve outcomes for people with disabilities and their partners, occupational therapists can advocate for mutual responsibility for sexual pleasure and education on understanding how disability can impact sexuality. Occupational therapists can teach effective verbal and non-verbal communication between partners about their sexual needs, modification and accommodation strategies, encouragement of attitude and behaviour changes to include a broader expectation of sexual activity and suggest prescription/provision of sexual aids (Esmail et al. 2001).

## *Enabling sexuality for all: Occupational therapy interventions*

Traditionally, occupational therapy sexuality-related interventions have focused on modifications and adaptations to mitigate physical impairment. However, interventions ought to also consider the issues pertaining to consent, ownership, respecting the diversity of sexual and gender-related identities, sexual behaviours and personal responsibility. Interventions might include discussions related to limit-setting, abstaining from sexual activity, risk management, safety planning, relaxation techniques, self-esteem, self-expression, identity and unlearning harmful messages and beliefs. People can present with a variety of abilities, preferences and past experiences, which are important during the intervention process.

To address these topics, it is essential to work closely with and listen to service users, to adopt a non-judgemental attitude and to minimise assumptions. Occupational therapists should conduct a self-reflection of sexual attitudes, values, beliefs and bias and consider sharing these self-reflections with their professional peers. Only by building self-awareness and considering sexuality as part of people's everyday life can occupational therapists correct misinformation, reconcile implicit or explicit bias, and negotiate differences between themselves and clients. Self-reflection is critical to awareness and understanding of other's assumptions, beliefs and experiences and this process enhances therapeutic use of self, i.e. an intentional relationship and preparedness to address sexuality and cultural sensitivity (Whitney and Fox 2017).

## Occupational therapy and sexuality assessment and interventions

Below, we present evidence of some occupational therapy engagements with sexuality. Peer-reviewed literature is sparse, so this is a limited and not an exhaustive list. Sakellariou (2006) studied men's perspectives about sexuality post spinal cord injury (SCI) and found that the men were limited by perceived societal opinions that they were asexual, feelings of infantilisation and dependency, social disapproval and anxiety. He argued that it is necessary to develop comprehensive sex education programmes and also to ensure that appropriate personal assistance is available. Similarly, Fritz and Lysack (2015) explored the experiences of 20 women with SCI, focusing on their sexual and reproductive health. These women self-identified reproduction and sexual confidence as important components of their rehabilitation and additionally noted their occupational therapy intervention did not include reproduction or sexuality. The authors recommended that future occupational therapy intervention and education should include a broader concept of sexuality to include sexual intimacy, education on erogenous zones and suggestions to "foster the woman's sexual self-esteem" as important for sexual wellness (2015, p. 8).

Adolescents with and without disability are potential clients who may benefit from receiving occupation-focused education about sexuality and dating. In a qualitative study of high school sexual health education for adolescents with disabilities, occupational therapists identified three themes: sexuality is different for each student; parents and teachers "do not know what to do"; and occupational therapy practitioners may be well suited to address these needs (Krantz et al. 2016, p. 4). Gontijo et al. (2016) provided a potential model for adolescent intervention. They implemented an occupational therapy group series for adolescents without disabilities, aged 13–17 in Brazil. The group structure included games to learn about sexual and reproductive health and resulted in a positive correlation between sexuality knowledge and group attendance. Education topics included developmental changes, sexually transmitted diseases and their prevention, reflections about consequences of pregnancy, contraception, how "gender perceptions are constructed and valued during adolescence and how these can influence this public's occupational performance and sexual and reproductive health" (p. 27). This study can be adapted to fit the specific needs of people with specific impairments, as well as be adapted to be inclusive of considerations such as body image, consent, communication of physical assistance needs, boundaries and sexual preferences.

Engel-Yeger et al. (2015) provide evidence for the potential role of occupational therapy practitioners assisting people with post-traumatic stress syndrome (PTSS) to engage in intimate social participation. PTSS is a result of exposure to a traumatic event whereby "the person experienced, witnessed or was confronted with an event or events that involved actual or threatened death or serious injury, or threat to the physical integrity of self or others" and "the person's response involved intense fear, helplessness or horror" (American Psychiatric Association 2000). PTSS can result in depression and anxiety diagnoses, which are often referred to as invisible disabilities. Investigating the relationship between the sensory profile of individuals with PTSS and their attitudes towards intimacy, Engel-Yeger et al. (2015) identified correlations between scoring "more/much more than most people" on the Adult Sensory Profile and fear of intimacy on the Fear of Close Personal Relationship Questionnaire, with individuals with PTSS scoring higher on sensory processing tendencies than the control group. The anxiety resulting from the emotional burden from sensory processing disorders can have an impact on the social requirements for intimate relationships and is an area which occupational therapists can help mitigate with sensory processing, emotional regulation and intimacy interventions (Engel-Yeger et al. 2015). Navigating social environments and interactions and engaging in intimate relationships are often required preceding activities to sexual activities. Occupational therapists can enable sexual activity with this population by addressing the noted barriers to social and intimate activities.

There are some intervention strategies that can be very useful for sexuality-focused interventions. Positive visualisation of sexual activity, establishing a sexual role model to emulate, positive reframing, positive body image awareness, masturbation, practicing assertive communication and grading activities required to meet a potential partner are suggestions by which occupational therapy practitioners can assist their clients with improving their sexual self-esteem. Additionally, intervention can explicitly allow the individual to explore their sexual self-view and concept and recognise the often-damaging effect of societal norms. For example, people with disabilities may be viewed as asexual or oversexualised, possibly leading to a dislike of their bodies or the attention they receive. Such external stigma can skew internal sexual self-view development or feelings of attractiveness. Occupational therapy focuses on guiding individuals to take ownership over intentionally shaping and embracing their sexual self-view unapologetically by utilising non-judgemental and empowering attitudes towards their clients. The process can look like this: collaborate with service users so they can (1) identify their core values, (2) build self-awareness of harmful messaging and replace that messaging with positive cognitive reframing, (3) consider lifestyles and outward image that affirms and/or excites them, (4) establish steps to asserting the image and engaging in sexual activity, (5) conduct safety planning for asserting that image or engaging in sexual activity, if needed, and (6) assert the chosen image and engage in or pursue chosen sexual activities.

There are a few examples of occupational therapists who have specifically explored sexuality for people with disabilities. Ellis and Dennison (2015) published a guide which follows an activity analysis format for veterans engaging in sexual activity and intimacy. They highlight common combat-related injuries, including PTSS and urogenital injuries, and offer holistic suggestions to enhance performance and satisfaction with sexual activity. Naphtali et al. (2009) provide a resource for people who are interested in learning about specific assistive devices to enhance sexual performance. Both resources offer suggestions for sexual activity and sexual relationships with self and/or others, such as communication scripts, strategies to enhance sexual desire and modifications, adaptations and accommodations for sexual positioning.

## Occupational empowerment framework and rights-based approach

The concept of empowerment, as presented by Hammell (2016), and a rights-based approach could be used by occupational therapists to conceptualise the broad possibilities of sexuality interventions and advocacy relevant for people with disabilities. Hammell uses the World Bank Empowerment Framework, which defines empowerment as "the process of increasing the assets and capabilities of individuals or groups to make purposive choices and to transform those choices into desired actions and outcomes" (World Bank 2014). The Empowerment Framework includes four components, which can be used as guides for client intervention and a call to action for a practice change in occupational therapy.

### Increasing assets

First, occupational therapists should consider how they can encourage utilisation and strengthening of the assets already possessed. For the client, this might be increasing the range of motion in order to attain a certain sexual position or guiding improved self-awareness to promote stress management skills for an individual who identifies with a sexual minority group and experiences discrimination and bullying. Occupational therapy practitioners can acknowledge inherent assets of the profession, including activity analysis, holistic understanding of human occupation and client-centred care as assets enabling therapists to address sexuality topics with people with disabilities.

## Increasing capabilities

Increasing capabilities relates to how people are enabled within their communities to potentialise themselves with the available assets they have. Occupational therapists promote and advocate for safe and inclusive environments for their clients, as well as working with clients to mitigate the challenges of the environment. For example, an occupational therapist can work with a person who identifies as transgender on a safety plan for accessing unfriendly environments. Furthermore, occupational therapists should reflect on their unique understanding of how environment impacts occupation and influence policy to enable safe community access for sexual and gender minorities. Similarly, it is important to develop opportunities for occupational therapists to learn and practice their skills related to sexuality intervention for people with disabilities.

## Transforming choices into action

"Transforming choices into action" refers to understanding what the goal of the client is and the specific barriers to bridging the goal with attainment. It is important for the occupational therapist to understand the social structures that limit sexual fulfilment and expression for clients and allow the reluctance of the profession to address this aspect of the human experience. For example, an occupational therapist who is unfamiliar with sexual bondage may become uncomfortable or judgemental if the client's goal is to reengage in that activity. The occupational therapist can uphold this component by researching the activity components of sexual bondage, respecting the client's choices and collaborating with the client on this engagement goal. Transforming choices into action highlights the importance of humans' freedom to choose and do.

## Accomplishing desired actions and outcomes

Client-centred care places the client at the centre of the therapeutic decision-making process and is paramount to successful intervention. Bright et al. encourage occupational therapy professionals to ask themselves, regarding their clients, "who are they and what do they need?" versus "what's wrong with them and how do I help them?" (Bright et al. 2012, p. 1002, cited in Hammell 2016). Asking this question helps to prevent the clinician's personal biases and experiences from limiting the therapeutic relationship. These questions are particularly important for sexuality interventions. Considering the minimal social discussion related to the wide variety of preferences and activities that are experienced by humans, individuals can often think, "normal is what I do, abnormal is what I don't do," as it relates to sexual expression and engagement. For this reason, self-reflection of values, attitudes, beliefs, and biases is important to maintain focus on the client's goals and how they wish to achieve them.

## Curriculum focused on unlearning harmful and misleading messages

The barriers to occupational therapy clinicians' readiness to address sexuality with people with disabilities is multi-factorial. Centring attention on occupational therapy education is a possible solution at the institutional level, which also enhances clinician competency and comfort, and addresses socio-cultural norms. Occupational therapy academic programmes should include opportunities for students to learn inclusive and accurate information regarding human sexuality and sexual development. Occupational therapy clinicians who have a medically accurate and inclusive understanding of sexuality are better equipped to address some of the challenges people with disabilities can have, related to sexuality, procreation and sexual relationships.

It is important to include disability in the curricula of all healthcare specialisations, addressing the subject in a coherent and integrated way across a variety of educational experiences, rather than in one-off sessions (Shakespeare and Kleine 2013). Effective pedagogy includes opportunities for students to consider the messages they learned related to sexuality and consider whether they might act as biases during therapy (Bruess and Schroeder 2013; Green et al. 2015; Hattjar 2017; Stayton 1998; Whitney and Fox 2017). This process might lead students to consider whether their own attitudes, values or beliefs are harmful or helpful to their own concept of sexuality, sexual expression and sexual experiences. For example, the simple message that it is not only okay, but constructive to discuss sexual activity preferences with sexual partners might make participants aware of permission to do this in their personal life. Verbally communicating sexual activity preference, requests and limits could perhaps directly counter their own previous beliefs. Educators can acknowledge the importance of this process and create a supportive and safe environment in which students can reflect.

Situated or experiential learning is a key factor for sexuality education (Eglseder and Webb 2017; Green et al. 2015). Small-group discussion, role playing and interviewing people with disabilities are among the activities that can provide students the opportunity to practice professional discourse and language. Engaging in dialogue counters the tendency to "not talk about it," and normalises sex-related conversation. Lastly, exposure to sexually explicit material, inclusive of diverse sexual activities and lifestyles, sexual orientations and gender identities is critical to habituating the normalcy of sexual expression and experiences (Green et al. 2015; Pillai et al. 2015; Stayton 1998).

Shakespeare et al. (2009) argue that to achieve best outcomes, education on disability needs to include people with disabilities, who can help students understand the lived experience of living with disability. For example, educators could consider including as learning materials, empowering and sexually explicit pictures of people with disabilities, and hosting people with disabilities to share their experiences. Educators could also consider inviting people who identify with specific sex-related communities, such as the Kink community, to educate students.

## Conclusion

In this chapter, we explored occupational therapy's current and potential role in addressing sexuality with people with disabilities. Occupational therapy clinicians should be aware that healthcare professionals are often not prepared or willing to address the sexuality-related issues of people with disabilities, thus leading to sexuality being neglected as an activity of daily living. The reluctance to address sexuality is a multi-level complex problem, involving socio-cultural norms and institutional limitations. Ensuring that occupational therapy curricula are inclusive of evidence-based strategies for sexuality topics can enhance widespread clinical adoption of clinical practices which view people with disabilities truly holistically, acknowledging them as sexual beings. Occupational therapists are aware of the paramount impact of satisfying engagement and performance, and are trained in the therapeutic use of self, cultural competency and activity analysis. These core occupational therapy beliefs and skills position them perfectly to address sexuality in clinical practice.

# References

American Council of Occupational Therapy Education (2018, August) *ACOTE standards*. Available at: https ://www.aota.org/~/media/Corporate/Files/EducationCareers/Accredit/StandardsReview/2018-ACOTE-Standards-Interpretive-Guide.pdf

American Occupational Therapy Association (2014) Occupational therapy practice framework: Domain and process. *American Journal of Occupational Therapy* 68(Suppl. 1), pp. S1–S48.

American Psychiatric Association (2000) *Diagnostic and statistical manual of mental disorders*, 4th ed. Washington, DC: American Psychiatric Association.

Bright, F.A.S., Boland, P., Rutherford, S.J., Kayes, N.M., and McPherson, K.M. (2012) Implementing a client-centred approach in rehabilitation: An autoethnography. *Disability and Rehabilitation* 34, pp. 997–1004. https://doi.org/10.3109/09638228.2011.629712

Bruess, C., and Schroeder, E. (2013) *Sexuality education: Practice and theory*, 5th ed. Burlington, MA: Jones & Bartlett Learning.

Christiansen, C.H., and Hammecker, C.L. (2001) Self care. In Bonder, B., and Wagner, M. (eds.) *Functional performance in older adults*. Philadelphia: F.A. Davis, pp. 155–175.

Cole, E.R. (2009) Intersectionality and research in psychology. *American Psychologist* 64, pp. 170–180. https://doi.org/10.1037/a0014564

Collins, R., Strasburger, V., Brown, J., Donnerstein, E., Lenhart, A., and Ward, M. (2017) Sexual media and childhood well-being and health. *Pediatrics* 140(5), pp. S162–S166. https://doi.org/10. 1542/ peds. 2016- 1758X

Conine, T., Christie, G., Hammond, G., and Smith, M. (1979) An assessment of occupational therapists' roles and attitudes toward sexual rehabilitation of the disabled. *American Journal of Occupational Therapy* 33(8), pp. 515–519.

Couldrick, L. (1999) Sexual issues within occupational therapy, part 2, implications for education and practice. *British Journal of Occupational Therapy* 62(1), pp. 26–30. https://doi.org/10.1177/03080226990 6200107

Deering, K.N., Amin, A., Shoveller, J., Nesbitt, A., Garcia-Moreno, C., Duff, P., and Shannon, K. (2014) A systematic review of the correlates of violence against sex workers. *American Journal of Public Health* 104(5), pp. e42–e54. https://doi.org/10.2105/AJPH.2014.301909

Diamond, L., and Huebner, D. (2012) Is good sex good for you? Rethinking sexuality and health. *Social and Personality Psychology Compass* 6(1), pp. 54–69. https://doi.org/10.1111/j.1751-9004.2011.00408.x

Dispenza, F., Viehl, C., Sewell, M.H., Burke, M.A., and Gaudet, M.M. (2016) A model of affirmative inter-sectional rehabilitation counseling with sexual minorities: A grounded theory study. *Rehabilitation Counseling Bulletin* 59(3), pp. 143–157.

Dyer, K., and das Nair, R. (2013) Why don't healthcare professionals talk about sex? A systematic review of recent qualitative studies conducted in the United Kingdom. *Journal of Sexual Medicine* 10(11), pp. 2658–2670. https://doi.org/10.1111/j.1743-6109.2012.02856.x

Eglseder, K., and Demchick, B. (2017) Sexuality and spinal cord injury: The lived experiences of intimate partners. *OTJR: Occupation, Participation and Health* 37(3), pp. 125–131. https://doi. org/10.1177/1539449217701394

Ellis, K., and Dennison, C. (2015) *Sex and intimacy for wounded veterans: A guide to embracing change*. Los Angeles, CA: Sager Group.

Engel-Yeger, B., Palgy-Levin, D., and Lev-Wiesel, R. (2015) Predicting fears of intimacy among individuals with post-traumatic stress symptoms by their sensory profile. *British Journal of Occupational Therapy* 78(1), pp. 51–57. https://doi.org/10.1177/0308022614557628

Esmail, S., Esmail, Y., and Munro, B (2001) Sexuality and disability: The role of health care professionals in providing options and alternatives for couples. *Sexuality and Disability* 19(4), pp. 267–282. https://doi. org/10.1023/A:1017905425599

Espelage, D.L., Basile, K.C., Rue, L.D.L., and Hamburger, M.E. (2015) Longitudinal associations among bullying, homophobic teasing, and sexual violence perpetration among middle school students. *Journal of Interpersonal Violence* 30(14), pp. 2541–2561. https://doi.org/10.1177/0886260514553113

Fritz, H., Dillaway, H., and Lysack, C. (2015) "Don't think paralysis takes away your woman-hood": Sexual intimacy after spinal cord injury. *American Journal of Occupational Therapy* 69(2), pp. 6902260030p1–6902260030p10. https://doi.org/10.5014/ajot.2015.015040

Gianotten, W., Bender, J., Post, M., and Höing, M. (2006) Training in sexology for medical and paramedical professionals: A model for the rehabilitation setting. *Sexual & Relationship Therapy* 21(3), pp. 303–317. https://doi.org/10.1080/14681990600754559

Gontijo, D., de Sena e Vasconcelos, A., Monteiro, R., Facundes, V., de Fátima Cordeiro Trajano, M., & de Lima, L. (2016). Occupational therapy and sexual and reproductive health promotion in adolescence: A case study. *Occupational Therapy International* 23(1), pp. 19–28. https://doi.org/10.1002/oti.1399

Green, E.R., Hamarman, A.M., and McKee, R.W. (2015) Online sexuality education pedagogy: Translating five in-person teaching methods to online learning environments. *Sex Education* 15(1), pp. 19–30. https://doi.org/10.1080/14681811.2014.942033

Hammell, K. (2016) Empowerment and occupation: A new perspective. *Canadian Journal of Occupational Therapy* 83(5), pp. 281–287. https://doi.org/10.1177/0008417416652910

Hattjar, B. (2017) Addressing sexual activity: A structured method for assisting clients with intimacy. *OT Practice* 22(19), pp. 8–12.

Hattjar, B., Parker, J., and Lappa, C. (2008) Addressing sexuality with adult clients with chronic disabilities: Occupational therapy's role. *OT Practice* 13(11), p. CE-2p.

Hyland, A., and Mc Grath, M. (2013) Sexuality and occupational therapy in Ireland—A case of ambivalence? *Disability and Rehabilitation* 35(1), pp. 73–80. https://doi.org/10.3109/09638288.2012.688920

Jones, M., Weerakoon, P., and Pynor, R. (2005) Survey of occupational therapy students' attitudes towards sexual issues in clinical practice. *Occupational Therapy International* 12(2), pp. 95–106.

Kashdan, T., Adams, L., Farmer, A., Ferssizidis, P., McKnight, P., and Nezlek, J. (2014) Sexual healing: Daily diary investigation of the benefits of intimate and pleasurable sexual activity in socially anxious adults. *Archives of Sexual Behavior* 43(7), pp. 1417–1429. https://doi.org/10.1007/s10508-013-0171-4

Kleinstäuber, M. (2017) Factors associated with sexual health and well-being in older adulthood. *Current Opinion in Psychiatry* 30(5), pp. 358–368. https://doi.org/10.1097/YCO.0000000000000354

Krantz, G., Tolan, V., Pontarelli, K., and Cahill, S. (2016) What do adolescents with developmental disabilities learn about sexuality and dating? A potential role for occupational therapy. *Open Journal of Occupational Therapy* 4(2), pp. 1–17. https://doi.org/10.15453/2168-6408.1208

Lindau, S., Schumm, L., Laumann, E., Levinson, W., O'Muircheartaigh, C., and Waite, L. (2007) A study of sexuality and health among older adults in the United States. *New England Journal of Medicine* 357(8), pp. 762–775. https://doi.org/10.1056/NEJMoa067423

Linton, K.F., and Adams Rueda, H. (2014) Experiences with pregnancy of adolescents with disabilities from the perspectives of the school social workers who serve them. *Health & Social Work* 39(2), pp. 92–100. https://doi.org/10.1093/hsw/hlu010

Lohman, H., Kobrin, A., and Chang, W.P. (2017) Exploring the activity of daily living of sexual activity: A survey in occupational therapy education. *Open Journal of Occupational Therapy (OJOT)* 5(2), pp. 1–11. https://doi.org/10.15453/2168-6408.1289

McGrath, M., and Lynch, E. (2014) Occupational therapists' perspectives on addressing sexual concerns of older adults in the context of rehabilitation. *Disability & Rehabilitation* 36(8), pp. 651–657. https://doi.org/10.3109/09638288.2013.805823

McGrath, M., and Sakellariou, D. (2015) Why has so little progress been made in the practice of occupational therapy in relation to sexuality? *American Journal of Occupational Therapy* 70(1), pp. 7001360010p1–7001360010p5. https://doi.org/10.5014/ajot.2016.017707

Naphtali, K., MacHattie, E., and Elliott, S. (2009) *PleasureAble: Sexual device manual for persons with disabilities.* Vancouver, BC: Disabilities Health Research Network.

Papp, L., Erchull, M., Liss, M., Waaland-Kreutzer, L., and Godfrey, H. (2017) Slut-shaming on Facebook: Do social class or clothing affect perceived acceptability? *Gender Issues* 34, pp. 240–257. https://doi.org/10.1007/s12147-016-9180-7

Payne, M., Greer, D., and Corbin, D. (1988) Sexual functioning as a topic in occupational therapy training: A survey of program. *American Journal of Occupational Therapy* 42(4), pp. 227–230.

Pillai-Friedman, S., Pollitt, J.L., and Castaldo, A. (2015) Becoming kink-aware—A necessity for sexuality professionals. *Sexual and Relationship Therapy* 30(2), pp. 196–210. https://doi.org/10.1080/14681994.2014.975681

Sakellariou, D. (2006) If not the disability, then what? Barriers to reclaiming sexuality. *Sexuality and Disability* 24(2), pp. 101–111.

Sakellariou, D., and Algado, S. (2006) Sexuality and disability: A case of occupational injustice. *British Journal of Occupational Therapy* 69(2), pp. 69–76.

Shakespeare, T., Iezzoni, L.I., and Groce, N. (2009) Disability and the training of health professionals. *The Lancet* 374(9704), pp. 1815–1816.

Shakespeare, T., and Kleine, I. (2013) Educating health professionals about disability: A review of interventions. *Health and Social Care Education* 2(2), pp. 20–37.

Smith, A., Lyons, A., Ferris, J., Richters, J., Pitts, M., Shelley, J., and Simpson, J.M. (2011) Sexual and relationship satisfaction among heterosexual men and women: The importance of desired frequency of sex. *Journal of Sex & Marital Therapy* 37(2), pp. 104–115. https://doi.org/10.1080/0092623X.2011.560531

Stayton, W. (1998). A curriculum for training professionals in human sexuality using the Sexual Attitude Restructuring model. *Journal of Sex Education and Therapy* 23(1), pp. 26–32.

Tepper, M. (2000) Sexuality and disability: The missing discourse of pleasure. *Sexuality & Disability* 18(4), pp. 283–290.

Viejo, C., Ortega-Ruiz, R., and Sánchez, V. (2015) Adolescent love and well-being: The role of dating relationships for psychological adjustment. *Journal of Youth Studies* 18(9), pp. 1219–1236. https://doi.org/10.1080/13676261.2015.1039967

Whitney, R.V., and Fox, W.W. (2017) Using reflective learning opportunities to reveal and transform knowledge, attitudes, beliefs, and skills related to the occupation of sexual engagement impaired by disability. *Open Journal of Occupational Therapy* 5(2), pp. 1–12. https://doi.org/10.15453/2168-6408.1246

Williams, S.L., Mann, A.K., and Fredrick, E.G. (2017) Proximal minority stress, psychosocial resources, and health in sexual minorities. *Journal of Social Issues* 73(3), pp. 529–544. https://doi.org/10.1111/josi.12230

World Bank (2014) *Poverty net: What is empowerment?* Available at: http://web.worldbank.org

# 33

# DISABILITY AND SOCIAL WORK

## Partnerships to promote sexual well-being

*Sally Lee*

## Introduction

This chapter explores ways in which social work practitioners can work with disabled people to support their sexual expression, esteem and confidence. Sex is a sensitive topic and an aspect of social work practice which faces multi-layered barriers, ranging from social taboos around sex and disability, to personal values, culture and experience (Lee et al. 2020). In this chapter, I argue that it is through recognising the centrality of sexual well-being to general well-being (rather than it being "the icing on the cake") that practitioners can be empowered to use their interpersonal and advocacy skills to confront the sexual disenfranchisement often experienced by disabled people (Lee et al. 2018). The chapter offers a structured approach to practice focused on sexual well-being, which is underpinned by the concept of sexual citizenship and involves knowledge of legislation, use of sex counselling models and specific aspects of sexual well-being as a framework by which practitioners can explore concerns people may experience and thus overcome these barriers.

Concern for general well-being is at the heart of modern social work in the UK, underpinning adult social care legislation (for example the Care Act 2014). In the UK, the Department of Health (DoH) (2014) defines well-being as feeling good and functioning well. This comprises firstly, subjective well-being, including life satisfaction, positive emotions and meaning, and secondly, objective well-being, based on assumptions about human needs and rights (DoH 2014). Sexual esteem, positive body image and sexual satisfaction are strong predictors of subjective well-being (Taleporos and McCabe 2002) and Shakespeare (2000), citing Finger (1992), argues that sexuality and sexual relationships are often the source of disabled people's deepest oppression, impacting on general well-being. As with general well-being, sexual well-being is both a subjective and objective experience, involving the whole person: body, mind, emotions and self in connection with others (Lee et al. 2020). Emphasising general well-being as the route to enable social work practitioners to engage with sexual well-being is legitimised by the International Federation of Social Work (IFSW 2014) definition of social work which emphasises "empowerment, social justice, human rights, collective responsibility and respect for diversities to engage people and structures to address life challenges and enhance well-being." The IFSW definition highlights the multi-dimensional nature of well-being, ranging from the lived experience of individuals to the structural inequalities resulting in the oppression of communities. In relation to sexual well-being, this means both the felt or subjective experience of sexual esteem and the objective experience of sexual citizenship, where the individual is recognised as a sexual being with the associated sexual rights and responsibilities.

International interest in the concept of well-being is also of interest to social policy makers: for example, in the UK, social policy and legislation are underpinned by the commitment to advance well-being (including the Care Act 2014). For social work practice with adults, the focus on well-being means working with individuals to investigate and address both the subjective and objective aspects. The emphasis on well-being is a good fit for social work practice which spans the subjective and objective domains of people's lives, working with people as they interact with the external world but experience this world "from the inside" (Galvin and Todres 2013). Social workers' skill in engaging with sensitive topics that touch on people's public and private experiences means they are well placed to discuss sexual well-being with clients. This accords with research that indicates sexual well-being is experienced in both public and private ways (Lee et al. 2018; 2020).

However, social workers may experience barriers to engagement with sexual well-being ranging from social taboos around sex and disability, to personal values, cultural norms and uncertainty about the law (Lee et al. 2020). In addition, concern for sexual well-being has been largely overlooked in social work practice and education policy, meaning practitioners are uncertain about what support they can or should offer (Bywaters and Jones 2007; Lee et al. 2018; Myers and Milner 2007; Sloane 2014). Sex tends to be viewed as a private matter and people using social services are sexually disenfranchised when services fail to acknowledge their sexual identities. When sex and sexuality are addressed in social work practice and policy, the focus tends to be on abusive, exploitative and other negative experiences of sexual activity or expression, rather than on sexual well-being as an aspect of physical and mental well-being (Myers and Milner 2007). The focus on sexual harm ignores the significance of sexual identity, expression and pleasure to human well-being, as well as the rights of sexual citizenship. This chapter explores approaches to social work practice which enable practitioners to engage with issues around sexual well-being through person-centred, humanised practice focused on well-being. Such humanised approaches are essential to effective practice because the sexual self is part of human self-hood and relationships, both of which lie at the heart of social work (Schaub et al. 2017).

It is important to state at the outset of the chapter that it is not assumed that everyone who uses social services has concerns regarding their sexual well-being. Rather, this chapter argues that sexual well-being is an aspect of general well-being, and social workers need to be able, and willing, to work with individuals to acknowledge and provide support with any sexual well-being concerns as part of their concern for general well-being (Lee and Fenge 2016). This chapter emerges from reflection on my social work career, revealing how positive my move to a team focused on supporting physically disabled people was for me, personally and professionally. Working with a politically engaged group of disabled people enabled a sense of real partnership to characterise the social work relationships we developed together. Approaching these relationships with the understanding that I was a resource from which people could draw meant that the relationships were directed by the individuals, as far as possible, and created openness to issues which really mattered to the people concerned. The doctoral and post-doctoral research underpinning this chapter emerged from this partnership approach and was initiated by a question posed by a young disabled person regarding their desire for sexual experiences and, ultimately, a family of their own. Looking for ways to support this person revealed the dearth of information and services around disability and sexuality, and led to a personal appreciation that social workers need to work with issues that are meaningful to the individual, rather than priorities assumed in policy. The original doctoral research, undertaken whilst I was in social work practice, has led to further participatory research projects also using qualitative methodologies to gain insight into the lived experience of disability and sexual well-being. Ethical approval for the research was gained from the Research Governance Review Group at Bournemouth University and all participants in the research project were able to give informed consent. Further details can be found in Lee and Fenge (2016), Lee et al. (2018; 2020). References made throughout the chapter to my own professional experiences and participatory research with disabled people and social work practitioners provide "real world" insights which illustrate and breathe life into key points. All contributors have been given pseudonyms, and this chapter features quotes from Eli, Daryl, Jackie, Morgan and Kay, who provide insights from living with impairment, and quotes from Kim, Ash and Jay, practitioners.

## Social work with disabled people

Western social work with disabled people has been criticised for being oppressive and controlling, effectively dehumanising people by treating them as problems to be managed rather than individuals with agency (Oliver et al. 2012). The dominance of the medical, individual and personal tragedy models of disability has led professionals to define disabled people's needs and to control services (Oliver et al. 2012): "doing to" rather than "doing with" disabled people. In the UK, the individual model of practice grew from the early roots of social work in religious and charitable organisations and developed into case work approaches, where practice focused on the individual as both the location of the "problem" and the site of intervention, with social work becoming part of the management of "social problems," including disability (Oliver et al. 2012). Currently, in the UK, most social workers are employed by local government agencies and undertake interventions determined by the legislative framework but interpreted through professional values (Davies 2012). Social work policy and practice are therefore directly influenced by political decision making taking place within the wider social context. Alternative, more radical, models of social work have developed in other countries (for example, China, India and South Africa; Ornellas et al. 2018), such as community engagement and collective social action, in which social workers are employed by non-governmental agencies and therefore less influenced by political ideology.

More recent changes to social work theory, policy and practice can be directly linked to the international impact of rights-based social movements, including the disability movement. These organisations and campaigns for social change have led to significant progress in promoting social equality (for example, in the UK, the implementation of legislation, such as the Human Rights Act 1998 and the Equality Act 2010, which makes discrimination on the basis of protected characteristics, including disability, illegal). However, austerity policies implemented in response to the global financial crash of 2007/8 have negatively impacted on the social inclusion of marginalised groups, including disabled people, profoundly affecting the well-being of citizens across the globe. For example, in 2017, the United Nations Committee on the Rights of Persons with Disabilities reviewed the UK's progress with the United Nations Convention on the Rights of Persons with Disabilities (CRPD 2006) and found that progress had slowed or even reversed. As such, to be effective in improving well-being social work with disabled people requires practitioners and organisations to commit to working in partnership with disabled people to remove disabling barriers (Oliver et al. 2012), embedding the social model of disability (Oliver 1983) into practice. The social model underpins the human rights model of disability and has profoundly influenced modern social work practice which:

- Focuses on the dignity of disabled people, views disabled people as part of human diversity, and acknowledges that disabled people have human rights.
- Recognises that people are disabled not by their impairment, but by barriers in the society and environment they live in, such as negative attitudes and inaccessible public spaces (Equality and Human Rights Commission 2017, p. 5).

Such partnerships build on collective strengths rather than individualistic practice, developing equal relationships and collaboration to address oppression.

In relation to sexual well-being, the social model enables insights into disabling barriers and discourses associated with disabled people's sexual expression. For example, the physical barriers disabled people experience included accessing social spaces, such as bars and clubs where relationships might begin, and discourse around sex, which is dominated by prescribed notions of beauty and desirability. For the young person who triggered my research, the lack of opportunity to sexually express themselves was a source of emotional distress and frustration. They described this lack as reinforcing their awareness that they had lost out in their personal development by not engaging in the "usual" activities and relationships of someone their age; this further underlined their sense of difference from others. An important part of the work we undertook was to identify and access opportunities for social interactions with peers in a diverse range of settings where they could "try out" communication skills.

Progressive social work with disabled people requires social work educators, leaders and practitioners to embrace the spirit of rights-based, person-centred, collaborative approaches to social work. Key to progressive practice is recognising users of services (or their representatives) as their own experts and re-positioning disabled people as active participants in decision making and service design (Hunter and Titterton 2012). An important way of achieving this progressive goal is through the inclusion of learning about the social origins of disability discrimination and oppression in pre- and post-qualifying social work education to enable practitioners to address the negative assumptions about impairment and dependency (Oliver et al. 2012). To support disabled people with sexual expression, social workers need to be mindful of oppressive practices and understand disabled people's experience of sexual well-being in the context of oppression. As described by Morgan (participant), social workers should: "Come from a heritage of non-repression, with interpersonal skills to be able to gain information, non-verbally, as well as to determine where their comfort level is in terms of the conversation and to be able to provide educational material."

Modernised social work, underpinned by theories of personalisation, provides opportunities for practice to be humanising, prioritising concern for well-being. Humanisation theory explores the world "lived from the inside" investigating the meaning of well-being and suffering (Galvin and Todres 2013). Humanised social work practice acknowledges dehumanising processes and seeks to counter these through care founded on genuine empowerment. Reframing my social work practice in terms of concern for general well-being enabled me to articulate the principles of humanisation, part of which was recognising sexual well-being as a key aspect of human experience, but it also highlighted the barriers to practitioner engagement with sexual well-being.

## Barriers to practitioners engaging with sexual well-being

Social work struggles to engage with issues around sexuality in its widest sense (Schaub et al. 2017). Kim and Jay (practitioners) explain the reasons that they avoid discussion around sexual well-being include personal feelings, such as embarrassment:

You know, those are the sorts of things that we know we probably should try and talk about, but wouldn't because it's, probably, it's just embarrassing, I think.

I think I just try and avoid it probably. And hope they talk to someone else, that they've got friends to talk to.

But it can be just as difficult for disabled people to bring up the topic:

*Daryl:* On their assessment, you know where they go through and tick all the boxes? "Do you need help with this?" "Do you need help with that?" Maybe if that was included on that, or perhaps they could bring up the subject themselves because then it wouldn't be relying on the disabled person to say "Hey can we talk about my sex life?"

*Eli:* I think the social worker should say "And are you okay with your personal life?" I think if a social worker did lead me towards that, I would be able to pick up on that and move forward with it, whereas bringing the subject up yourself is kind of a big deal, I think, going back to because people have been treated so long as non-sexual and for a long time I was, but actually I am normal and I do have my own personal life, it's a big deal for me to say vocally to someone who I've probably not met before.

Lack of awareness from practitioners also creates a barrier:

*Ash:* I just don't know how I would start that conversation.

*Jay:* I need to know what I can offer or what is available that I am going to provide that person based on the information they give me, and then I think that would give…that would be key for me.

*Kim:* If my clients, you know: "Well I'd like to have sex but I'm struggling with it." I just wouldn't know what to do or where to go for the information. I've got no…no, not even websites or…I wouldn't even…and then I would be worried about googling it.

Cultural norms and religious traditions can also create barriers, for example taboos around sex and disability, or the negative discourse which assumes that sexual expression is not important or relevant for disabled people. Jay explains: "It doesn't cross my mind, because I probably subconsciously think, well, they're not going to be having sex. They're not going to want to have sex or they're not going to be because they're disabled."

Structural barriers, such as economic systems devaluing disabled people (Blackburn et al. 2015; Dyer and Das Nair 2013; Lee et al. 2018; Shuttleworth 2010) can also prevent practitioner engagement. Eli describes the negative impact of experiencing frequent changes of one's allocated social worker as limiting her ability to develop a trusting relationship, where she could imagine discussing her more intimate concerns. Frequent staff and case turnover are aspects of service efficiency policies which prioritise resources over people.

I can't build a relationship with my social worker where I can get to the point of telling them I have these intimate issues or anything like that, because I don't know the person. I might know them for a few weeks and then they are gone again. I then have to go back through everything again, how my condition affects me, everything to do with my care again, and that is seriously depressing.

(Eli)

However, barriers cannot be an excuse for inaction, as practice which refuses to address the issues that matter to people is oppressive and dehumanising (Galvin and Todres 2013). Recognition that sexual well-being is part of general well-being means that practitioners need to use their professional skills to enable individuals to express their concerns, which could occur through them raising the issue of personal relationships.

## Overcoming barriers to sexual well-being-engaged practice

### Understanding the law

Some people need support or help with their sexual expression. Domestic law determines what is legal in each country, so it is essential that readers investigate the law related to their own context and circumstances. Social workers must balance the promotion of autonomy with the need to protect individuals at risk of harm (British Association of Social Workers (BASW) 2012). Having such a dual agenda can lead practitioners and organisations to prioritise safeguarding and risk-averse practice rather than empowering people to engage in the full range of relationships (de Than 2015). Participants in my research identified lack of awareness of the law as a key barrier to practitioners supporting individuals with sexual expression:

> One of the big barriers is ignorance of the law so they [practitioners] are very concerned they are breaking the law and in actual fact in most cases it's the opposite but once I'd found that information and the support worker I had at the time was confident where she stood, she did actually help me to meet up with a sex worker, once she was confident that she was on the right side of the law, that she was doing a good thing, not a bad thing, she then helped.
>
> *(Daryl)*

Practitioners need to be aware that Article 8 of the European Convention of Human Rights (ECHR) (1953) (incorporated into domestic UK law by the Human Rights Act 1998) concerns the right to respect for private and family life, and includes quality of life, personal and sexual autonomy (decision making), confidentiality, privacy, dignity and the forming and maintaining of personal relationships. Articles 9, 10 and 11 also concern rights in relation to expression, including rights to information, education and potentially support (de Than 2015). These rights extend to people who lack capacity. Article 19 of the United Nations Convention on the Rights of Persons with Disabilities (2006), entitled *Living independently and being included in the community*, requires governments to take action to facilitate disabled people's full inclusion and participation in the community. Everyone has the right to sexual expression and to relationships of their choosing, and any restrictions, for example on the grounds of protection from harm, need robust reasons and must be proportionate (de Than 2015). Social workers, organisations and social service users must weigh up the right to protection and autonomy to prevent over-protection and to avoid breaking human rights law.

### Citizenship and social justice

Professional codes of social work practice, along with the global definition, make clear that social workers are responsible for promoting disabled people's citizenship and inclusion, and addressing social injustice (BASW 2012; IFSW 2014). Citizenship involves having rights and responsibilities; the concept of sexual citizenship (Weeks 1998) promotes rights specifically connected to sexual well-being. The sexual citizenship of physically disabled people challenges exclusion on the basis of disability and is grounded in principles of human sexual rights (Kanguade 2010). This involves three requirements:

- Demand for control over bodies, feelings and relationships.
- Demand for access, representations, relationships and public spaces.
- Demand for choice about identities, lifestyles and experience (Kanguade 2010, p.210–211).

Acknowledging links between sexual well-being and sexual citizenship, social inclusion and the experience of social justice enables social workers to engage in sexual well-being- focused practice through alignment with professional values and responsibilities (Lee and Fenge 2016), including advocacy. Advocacy is core to social work practice and is at the heart of the IFSW definition of global social work. It means supporting personal agency so individuals can define and create their own lives and is a vital aspect of empowerment and safeguarding (Hafford-Letchfield and Carr 2017). Advocacy is linked to sexual citizenship as it concerns working to improve people's lives by advocating for access to information, services, financial support or opportunities for sexual expression.

Well-being is also a matter of social justice and includes the experience of oppression and how societal benefits, such as access to services, are distributed (Bell 2017). Disabled people's access to sexual health information and services, including sex education, has been unequal (Liddiard 2018), suggesting that such services are not considered to be relevant to disabled people. Disabled people have the right to full participation in every aspect of life, and social justice-informed social work practice should contribute to social change through alignment with disabled people to develop mutually supportive social structures which create well-being and build resilience (Ferguson and Lavalette 2013; Quarmby 2011). Alignment can be demonstrated by focusing on personal issues which matter to individuals, such as sexual well-being, as well as on social justice issues, such as the fair distribution of resources.

Recognising diversity requires that social workers respond positively to the differences which impact people's lives (Bell 2017). Recognition of diversity includes what constitutes sexual well-being for someone, not only in terms of their sexuality, but also their sexual expression. Some individuals need help from carers, social workers, occupational therapists and other professionals with respect to sexual well-being. This help ranges from professionals being thoughtful about seating positions so that consenting adults, who want to express affection, such as being able to hold hands or kiss, are able to do so with minimal intervention. Other individuals may need support with accessing or using equipment; for example, Jackie, a research participant, described how carers supported him into a position where he could masturbate privately. Westernised notions of independence, framed in terms of self-sufficiency, marginalises people who require assistance, and underplays the interdependence which characterises human lives. Independence means different things to different people (see Buckley and Lee 2019, for an account of the lived experience of living independently). Oliver (1990, p. 94) argued: "Dependency is created amongst disabled people, not because of the effects of functional limitations on their capacities for self-care, but because their lives are shaped by a variety of economic, political and social forces which produce it." The sexual disenfranchisement of disabled people is an example of the impact of disabling forces on people's lives

## Using models of sex counselling in social work practice

The absence of policy and guidance on sexual well-being-focused social work practice suggests that sex is a concern that is not relevant to social work, or that it requires specialist expertise. Owens (2015) observes that, over many years of running a sex and disability helpline, multiple queries have been answered by listening and providing information, rather than by employing complex interventions. Models used in sex therapy, such as PLISSIT, are helpful in guiding work concerning sexual well-being and offer guidance on levels of expertise needed for different levels of support. PLISSIT is an acronym for Permission giving, whereby the therapist opens the topic up to discussion enabling the client to articulate their concern; Limited Information giving provides non-specific information which may address general concerns; Specific Suggestions and Intensive Therapy (Dunk-West 2007) requires higher-level skills and addresses in detail more specific concerns. These four levels of sex counselling and therapy encourage professionals to engage with service users' sexual concerns commensurate with the nature of the concern and their level of experience.

The first two levels of the PLISSIT model involve social workers being "perceived as someone to whom clients can talk to about issues related to sexuality" (Dunk-West, 2007, p. 5). The worker recognises that talking about sexuality is relevant to everyday life and is a legitimate topic within social work encounters. The second level (Limited Information) of the PLISSIT model requires knowledge about sexual issues, human physiology and sexual values to be able to offer limited information. Social workers should ensure that they use reliable sources of information and that service users know how to access information when they wish (Blackburn et al. 2015). The third and fourth levels of the PLISSIT model, Specific Suggestions and Intensive Therapy, require in-depth knowledge and training, which are not part of the general social work role. In these cases, referral to a specialist should be the course of action.

The Recognition Model (Couldrick et al. 2010), designed for use in general health and social care practice, details five stages of interaction requiring progressively higher levels of skill and knowledge. The first stage recognises the service user as a sexual being, taking account of the diversity of human sexuality and expression across cultures and the life course; the second stage is sensitive permission giving, being open and able to engage with what is meaningful to the other; the third stage focuses on exploration of the sexual concern, with stage four addressing issues which fit within the team's expertise and boundaries. Stage five is onward referral when appropriate (Couldrick et al. 2010). Social workers are not sexual therapists, but they need to become humanising, competent sexual well-being collaborators when required. The PLISSIT and Recognition Models provide useful frameworks in which to engage with sexual well-being-focused work.

Throughout this chapter, I have argued that that social work concerned with human well-being requires practitioners to explore with individuals what is meaningful to them, including (if appropriate) sexual well-being. Thinking about specific aspects of sexual well-being develops this argument further by highlighting how practitioners can incorporate concern for sexual well-being into their practice.

## Approaching sexual well-being through specific aspects of well-being

### Personal dignity

Person-centred and humanising approaches to social work practice enable practitioners to focus on human well-being through collaborative exploration of factors impacting on well-being, as succinctly expressed by Kim (practitioner): "Just getting a bit of a feel about who she is, how she perceives the world, her ambitions, what she wants to achieve." Humanisation adds depth to collaborative exploration, enabling social workers to use empathy to think about factors that enhance feelings of being human. To support personal dignity, social workers need an understanding of what it means to the individual and the barriers they experience, but building genuine relationships can take time:

> I felt quite embarrassed about it all as well. And I didn't really know where to start with it. But yeah, I think, it just took a long time and I needed that year to build enough of a relationship to make sure we were sort of meeting all of her needs, not just the, you know, the basics like personal care, accommodation, food, nutrition— those sorts of things.
>
> *(Ash)*

Dignity can be defined as "being treated as if you matter" (DoH 2018). Without a sense of self-worth, the expectations of treatment from others will be low (for example, feeling that I do not deserve better) (Mayers et al. 2003). The significance of sex and desirability to a sense of self-worth is evidenced by research (for example, Grossman et al. 2003; Lee et al. 2018; 2020; Taleporos and McCabe 2002), a point echoed by Daryl who described his reluctance to use online dating sites, expecting to be rejected and found undesirable, as there are: "so many sweeties in the sweetie shop." Loss of confidence negatively impacts on a sense of dignity, potentially leading to passivity, where life events happen to a person rather than through their active exercise of agency. Promoting sexual self-esteem, which aids confidence, becomes an imperative in preventative-focused social work. Enabling the individual to feel sexually competent, attractive and acceptable to others by acknowledging the person's sexual identity and sexual expression builds self-esteem and confidence (Mayers et al. 2003).

Social workers have to be aware of the negative narrative of disability and how it impacts on self-worth and psychological well-being, creating a more hostile world for disabled people (Quarmby 2011; Simcock and Castle 2016), and to counter the narrative around disability in their practice if they are to understand a disabled person's situation. Western cultures promote ideological notions of independence and self-sufficiency as being integral elements of dignity and self-worth, but such notions are detrimental to the well-being of individuals who do not conform to these constructs (Oliver et al. 2012). Social workers can be active participants in changing such negative narratives of disability by being knowledgeable about the issues impacting on disabled people's lives, such as gender issues, cultural notions of beauty, loss of agency, risk aversion and human rights contraventions (de Than 2015). In addition, social workers can promote alternative views of disability such as the Affirmative Model (Swain and French 2000), which promote the positive social identities and perspective of disability. The Affirmative Model challenges the notion of disability as tragedy and that the only possible response to impairment is a negative one, and questions the assumption that disabled people want to be "normal," "independent" or to "adjust" (Swain and French 2000).

To work in humanising ways, social workers must always promote personal agency by keeping the person's voice central. Agency is linked to human dignity so "when agency is taken away, one's sense of personhood is diminished" (Galvin and Todres 2013, p. 12). Empowering people to assert their agency, including defining their personal meaning of dignity, is crucial. Agency is dependent on being able to make informed decisions through the provision of accurate information. The participants in my research expressed frustration at the limited information regarding sexual well-being and the barriers to accessing information in accessible formats.

Social workers need the time to research and seek information in appropriate formats to meet individual needs.

## Psychological and emotional well-being

Psychological well-being is linked to physical and sexual esteem (McCabe and Taleporos 2003; Taleporos and McCabe 2002) and "sexual expression is a natural and important component of self-concept, emotional well-being and quality of life" (Milligan and Neufeldt 2001, p. 92). To improve well-being, it is necessary to improve body image and sexual esteem. Taleporos and McCabe (2002, p. 182) argue that "sexual esteem, body esteem and sexual satisfaction were strong predictors of self-esteem in people with physical disability… if they felt good about their body and were sexually satisfied, they were less likely to feel depressed." Social workers need awareness of the connection between well-being and sexual self-esteem and how to alleviate any negative effects by exploring meaningful ways of developing self-esteem, through positive role models, connecting people to networks of peer support, and investigating opportunities where a person can express their sexuality (for example, by going on a date). Social workers can help build sexual self-esteem by recognising individuals as sexual beings and making sexual well-being an aspect of holistic practice, as described by Jay (practitioner):

> They need to talk about sexuality openly, so they don't feel stigmatised talking about it. So that's the starting point, to talk about it, and to be able to incorporate it into the conversation and, particularly with a young person, they may not be aware that sexual well-being is part of an overall well-being, so there has to be an educational aspect to it.

Sex has an increasingly important role to play in modern society in the construction of the self, and sexual intimacy has become an essential part of self-constitution (Dunk-West 2007). Because humans are sense-makers, we look for significance and make explanatory stories about ourselves and our world (Galvin and Todres 2013). Sense-making is done within a social context, with external messages becoming integrated into the sense-making narrative. Positive role models are important to our narratives, creating psycho-emotional well-being and positive self-development. When sense-making is negative or distorted, people become dislocated and a sense of meaninglessness or hopelessness can prevail, resulting in low well-being. Person-centred humanised approaches enable sense making, reconnecting people to their life journey through exploration of their past and anticipated future. Continuity of the self is a significant factor in well-being (Galvin and Todres 2013), so social workers need to be mindful of enabling continuity for the person through periods of transition, including in the event of impairment or change. For example, aiding people to connect their "old" self to the "new" self, finding aspects which have remained the same, whilst valuing the new perspectives impairment has brought. Interventions using biography and personal narrative can be helpful in enabling a person to make sense of their experience; for example, I worked with someone following the diagnosis of a neurological condition and, as part of exploring his changing situation, we thought about ways his sexual expression had changed, making connections to his "old, "current" and "future" selves, using a narrative approach. By using the Affirmative Model, social work can help disabled people counteract negative narratives of disability and positively construct personal identity (French and Swain 2008).

## Protection from abuse and neglect

Living free from abuse is a human right and is key to well-being (DoH 2018; Human Rights Act 1998). Safeguarding is core to social work; however, concerns about the safety of service users and their exposure to risk can create barriers to addressing sexual well-being in those with whom we work (de Than 2015). Empowerment is vital to effective safeguarding (Hafford-Letchfield and Carr 2017), including empowering people to be fully involved in determining risk and protection from abuse and neglect. Such empowerment means engaging with service users about matters which improve their quality of life, including sexual well-being, and may concern sexual relationships or activities where, even though there is risk of harm, the person wishes to continue those activities or relationships. To approach safeguarding co-operatively, social workers require skills of openness and honesty to understand the risks so that the person can make informed decisions (Lawson 2017). Addressing sexual well-being in social work practice means helping service users, carers and advocates to obtain the relevant information about sex and sexuality, so that they can make informed choices to better safeguard themselves (Hollomotz 2010). The enduring narrative of disability leading to sexual disenfranchisement is a form of harm which denies full humanity and can affect the ability to establish sexual relationships (Owens 2015; Shakespeare et al. 1996). For instance, Jackie describes how overprotection, due to fears of his exposure to risk, denies him the privacy he craves. A humanised approach requires social workers to explore what sexual well-being means with service users, in order to build sexual self-confidence and plan effective interventions.

By promoting the concept of sexual citizenship by, for example, providing information or sex education, social workers work toward social inclusion and enabling people to protect themselves (Grossman et al 2003; Shakespeare 2006; Shuttleworth 2010; de Than 2015), as succinctly expressed by Kim (practitioner): "Why not just educate everyone so they can have better protection, and everyone can be safe?" However, because sex education for disabled people has been inadequate (Owens 2015; Shakespeare 1996; 2006) social workers cannot assume that disabled people have the necessary information to enable their sexual well-being. Raising awareness is crucial to protection from and prevention of abuse and neglect (DoH 2018), and social workers have a role in informing, empowering and advocating for those at risk of harm (Lawson 2017). For example, Jay (a practitioner) suggests:

> find[ing] out do they need advice, do they need a YouTube video to kind of look at or do they need equipment? I guess everybody is different and it's having that list of different things available. You might get somebody who is really computer literate, for example, so you suggest some websites.

## Relationships

The quality of an individual's social relationships is one of the strongest indicators and predictors of well-being (Helliwell et al. 2015; Markey et al. 2007). Social networks provide protection from harm as isolation creates risk through lowered self-esteem and removal from supportive relationships (Campaign against Loneliness 2019). Good-quality relationships are beneficial to mental and physical health (Galvin and Todres 2013). But the nature of meaningful relationships depends on personal perceptions; for example, for some people a relationship does not have to involve direct contact to be meaningful (on-line friendships). Social workers can emphasise and strengthen bonds by collaboratively exploring service users' networks and enabling the development of social links to reduce isolation. Work addressing the social structural causes of social isolation at a strategic level is also necessary to improve disabled people's access to opportunities of togetherness (Grossman et al. 2003; Shakespeare 2006; Shuttleworth 2000) and social workers need to align themselves with disability organisations to effect structural change (for example, campaigning for accessible public transport). Relationships should be a focus of care plans, as evidence demonstrates how social participation reduces morbidity and promotes well-being (Handley et al. 2015).

Impairment can change relationships, disrupting established roles of exchange and reciprocity and may lead to relationship re-evaluation and even separation (McCabe and Taleporos 2003). Adjustment to different roles and expectations may be necessary and the ability to discuss sexual difficulties is helpful, but can be difficult, as expressed by Daryl:

> It is very hard if you're asking someone to help you with the shower, for example, to actually have a sexual life with them. Where are the boundaries? I honestly think that it is an impossible situation, which is why I decided to reply (to consent to participate). To have someone who is a carer, who is also a husband...I think it puts them in an insidious situation, both the person, the cared for and the carer.

Social workers undertaking well-being-focused work can use humanising approaches to explore "insider perspectives" and assist individuals to re-evaluate in a constructive way, facilitating communication, which aids adjustment to change (Esmail et al. 2007; Galvin and Todres 2013; Grossman et al. 2003). The dominant discourse of the ideal (nuclear) family can lead to a negative self-perception by those outside of such relationships and, to promote social inclusion and equality, such normative ideas should be challenged by social workers.

## Conclusion

This chapter has demonstrated that sexual well-being is a key factor in general well-being and an area of practice where social workers can put their professional and personal values into action. The chapter has suggested that specific well-being domains provide a framework to help explore concerns people may experience, and sex counselling models offer a structured approach to sexual well-being-engaged social work practice. Personalised approaches to care and support, framed within the theory of humanisation, enable social workers to collaboratively explore life lived from the inside to identify what matters to people. To be effective, social workers need to engage with issues which may be challenging or sensitive and acknowledge the barriers to engaging in aspects of life such as sexual well-being. Social work encounters people in their public and private domains, and practitioners are often skilful in creating relationships of trust where sensitive, private issues can be discussed. The international interest in human well-being provides practitioners with opportunities to work with people, to think widely about factors which contribute to or detract from their well-being. By making the links between sexual and general well-being explicit, modernised social work can incorporate concern for sexual well-being into holistic practice. This chapter has argued that the key skills involved in sexual well-being-engaged work centre on concern for human well-being and I end the chapter with a comment from Eli, one of the research participants, which sums up why such work is so important:

> The biggest thing I want to say is how important it is for me personally, you can't strip life to the basics all the time, that's not a life to me, that's just surviving.

## References

Bell, J. (2017) Values and ethics. In Lishman, J., Yuill, C., Brannan, J. and Gibson, A. (eds.) *Social work an introduction*. London: SAGE, pp. 3–18.

Blackburn, M., Chambers, L., Earle, S. and Raeburn, D. (2015) *Talking about sex, sexuality and relationships: Guidance and standards*. Milton Keynes: Open University.

British Association of Social Workers (2012) *The Code of Ethics for Social Work* (2012) Birmingham: British Association of Social Workers. Available at: https://www.basw.co.uk/resources/basw-code-ethics-social-work (Accessed: 10.10.2019).

Buckley, S. and Lee, S. (2019) I was assessed under the Care Act to enable me to live independently. In Hughes, M. (ed.) *A guide to statutory social work interventions: The lived experience*. London: Palgrave Macmillan.

Bywater, J. and Jones, R. (2007) *Sexuality and social work*. Exeter: Learning Matters.

Campaign to End Loneliness (2019) *Threat to health*. Available at: https://www.campaigntoendloneliness.org/threat-to-health/ (Accessed: 10.10.2019).

Care Act 2014. Available at: http://www.legislation.gov.uk/ukpga/2014/23/contents/enacted (Accessed: 10.10.2019).

Convention on the Rights of Persons with Disabilities (CRPD) (2006) Available at: https://www.un.org/development/desa/disabilities/convention-on-the-rights-of-persons-with-disabilities.html (Accessed: 10.10.2019).

Couldrick, L., Sadlo, G. and Cross, V. (2010) Proposing a new sexual health model of practice for use by physical disability teams: The recognition model. *International Journal of Therapy and Rehabilitation* 17(6), pp. 290–299.

Davies, M. (2012) *Social work with adults*. London: Palgrave Macmillan.

de Than, C. (2015) Sex, disability and human rights. In Owens, T. (ed.) *Supporting disabled people with their sexual lives*. London: Jessica Kingsley, pp. 86–103.

Department of Health (2014) *Well-being why it matters to health policy*. Available at: https://www.gov.uk/government/uploads/system/uploads/attachment_data/file/277566/Narrative__January_2014_.pdf (Accessed: 10.10.2019).

Department of Health (updated 2018) *Care and support statutory guidance issued under the Care Act 2014* [online]. Available at: http://www.gov.uk/guidance/care-and-support-statutory-guidance (Accessed: 10.10.2019).

Dunk-West, P. (2007) Everyday sexuality and social work: Locating sexuality in professional practice and education. *Social Work and Society* 5(2), pp. 135–152.

Dyer, K. and das Nair, R. (2013) Why don't healthcare professionals talk about sex? A systematic review of recent qualitative studies conducted in the United Kingdom. *Journal of Sexual Medicine* 10, pp. 2658–2670.

*Equality Act 2010*. Available at: http://www.legislation.gov.uk/ukpga/2010/15/contents (Accessed: 10.10.2019).

Equality and Human Rights Commission (2017) *How well is the UK performing on disability rights? The UN's recommendations for the UK*. Available at: https://www.equalityhumanrights.com/en/publication-download/how-well-uk-performing-disability-rights (Accessed: 10.10.2019).

Esmail, S., Munro, B. and Gibson, N. (2007) Couple's experience with multiple sclerosis in the context of their sexual relationship. *Sexuality and Disability* 25, pp. 163–177.

European Court of Human Rights (1953) European convention on human rights. Available at: https://www.echr.coe.int/Documents/Convention_ENG.pdf (Accessed: 10.10.2019).

Ferguson, I. and Lavalette, M. (2013) Critical and radical social work: An introduction. *Critical and Radical Social Work* 1(1), pp. 3–14.

Finger, A (1992) Forbidden fruit. *New Internationalist* 233, pp. 8–10.

French, S. and Swain, J. (2008) *Understanding disability: A guide for health professionals*. Edinburgh: Churchill Livingstone.

Galvin, K. and Todres, L. (2013) *Caring and well-being: A lifeworld approach*. London: Routledge.

Grossman, B., Shuttleworth, P. and Prinz, P. (2003) Locating sexuality in disability experience. *Sexuality Research & Social Policy* 1(2), pp. 91–96.

Hafford-Letchfield, T. and Carr, S. (2017) Promoting safeguarding: Self-determination, involvement and engagement in adult safeguarding. In Cooper, A. and White, E (eds.) *Safeguarding adults under the Care Act 2014: Understanding good practice*. London: Jessica Kingsley Publishers, pp. 91–109.

Handley, S., Joy, I., Hestbaek, C. and Marjoribanks, D. (2015) *The best medicine? The importance of relationships for health and well-being*. Doncaster: Relate and New Philanthropy Capital.

Helliwell, J., Layard, R. and Sachs, J. (2015) *World happiness report 2015*. Available at: https://worldhappiness.report/ed/2015/ (Accessed: 10.10.2019).

Hollomotz, A. (2010) Sexual "vulnerability" of people with learning difficulties: A self-fulfilling prophecy. In Shuttleworth, R. and Sanders, T. (eds.) *Sex & disability: Politics, identity and access.* Leeds: Disability Press, pp. 21–40.

*Human Rights Act 1998.* Available at: https://www.legislation.gov.uk/ukpga/1998/42/contents (Accessed: 10.10.2019).

Hunter, S. and Titterton, M. (2012) Conceptual foundations and theory building in personalisation. In Davies, M. (ed.) *Social work with adults.* London: Palgrave Macmillan, pp. 48–63.

International Federation of Social Work (2014) *Definition of social work.* Available at: https://www.ifsw.org/what-is-social-work/global-definition-of-social-work/ (Accessed: 10.10.2019).

Kanguade, G. (2010) Advancing sexual health of persons with disabilities through sexual rights: The challenge. In Shuttleworth, R. and Sanders, T. (eds.) *Sex and disability: Politics, identity and access.* Leeds: Disability Press, pp. 197–215.

Lawson, J. (2017) The "Making Safeguarding Personal" approach to practice. In Cooper, A. and White, E. (eds.) *Safeguarding adults under the Care Act 2014: Understanding good practice.* London: Jessica Kingsley Publishers, pp. 20–39.

Lee, S. and Fenge, L. (2016) Sexual well-being and physical disability. *British Journal of Social Work* 46(8), pp. 2263–2281.

Lee, S., Fenge, L. and Collins, B. (2018) Promoting sexual well-being in social work education and practice. *Social Work Education* 37(3), pp. 315–327.

Lee, S., Fenge, L. and Collins, B. (2020) Disabled people's voices on sexual well-being. *Disability and Society* 35(2), pp. 303–325.

Liddiard, K. (2018) *The intimate lives of disabled people.* London: Routledge.

Markey, C., Markey, P. and Fishman Gray, H. (2007) Romantic relationships and health: An examination of individuals' perceptions of their romantic partners' influences on their health. *Sex Roles* 57, pp. 435–445.

Mayers, K., Heller, D. and Heller, J. (2003) Damaged sexual self-esteem: A kind of disability. *Sexuality and Disability* 21(4), pp. 269–282.

McCabe, M. and Taleporos, G. (2003) Sexual esteem, sexual satisfaction and sexual behaviour among people with physical disability. *Archives of Sexual Behavior* 32(4), pp. 359–369.

Milligan, M. and Neufeldt, A. (2001) The myth of asexuality: A survey of social and empiricalevidence. *Sexuality and Disability* 19(2), pp. 91–109.

Myers, S. and Milner, J. (2007) *Sexual issues in social work.* Bristol: BASW.

Oliver, M. (1983) *Social work with disabled people.* Basingstoke: Palgrave MacMillan.

Oliver, M. (1990) *The politics of disablement.* Basingstoke: Macmillan.

Oliver, M., Sapey, B. and Thomas, P. (2012) *Social work with disabled people.* Basingstoke: Palgrave MacMillan.

Ornellas, A., Spolander, G., Engelbrecht, L., Sicora, A., Pervova, I., Martínez-Román, M., Law, A,' Shajahan, P.K., das Dores Guerreiro, M., Casanova, J., Garcia, M., Acar, H., Martin, L. and Strydom, M. (2018) Mapping social work across 10 countries: Structure, intervention, identity and challenges. *International Social Work* 62(4), pp. 1183–1197.

Owens, T. (2015) *Supporting disabled people with their sexual lives.* London: Jessica Kingsley.

Quarmby, C. (2011) *Scapegoat: Why we are failing disabled people.* London: Portobello Books.

Schaub, J., Willis, P. and Dunk-West, P. (2017) Accounting for self, sex and sexuality in UK social workers' knowledge base: Findings from an exploratory study. *British Journal of Social Work* 47, pp. 427–446.

Shakespeare, T. (2000) Disabled sexuality: Toward rights and recognition. *Sexuality and Disability* 18, pp. 159–166.

Shakespeare, T. (2006) *Disability rights and wrongs.* London: Routledge.

Shakespeare, T., Davies, D. and Gillespie-Sells, G. (1996) *The sexual politics of disability.* London: Cassell.

Shuttleworth, R. (2000) The search for sexual intimacy for men with cerebral palsy. *Sexuality and Disability* 18(4), pp. 263–282.

Shuttleworth, R. (2010) Towards an inclusive sexuality and disability research agenda. In Shuttleworth, R. and Sanders, T. (eds.) *Sex and disability: Politics, identity and access.* Leeds: Disability Press, pp. 1–20.

Simcock, P. and Castle, R. (2016) *Social work and disability.* Cambridge: Polity Press.

Sloane, H. (2014) Tales of a reluctant sex radical: Barriers to teaching the importance of pleasure for wellbeing. *Sexuality and Disability* 32, pp. 453–467.

Swain, J. and French, S. (2000) Towards an affirmation model of disability. *Disability & Society* 15(4), pp. 569–582.

Taleporos, G. and McCabe, M. (2002) The impact of sexual esteem, body esteem and sexual satisfaction on psychological well-being in people with physical disability. *Sexuality and Disability* 20(3), pp. 283–290.

Weeks, J. (1998) The sexual citizen. *Theory, Culture and Society* 15(4), pp. 35–52.

# 34

# INTERSECTIONS OF DISABILITY, SEXUALITY, AND SPIRITUALITY WITHIN PSYCHOLOGICAL TREATMENT OF PEOPLE WITH DISABILITIES

*Sarah S. Brindle and Samantha Sharp*

## Introduction

Religion and spirituality have become more prominent in psychological research and practice and may have unique considerations for people with disabilities, in terms of conceptualisations of disability within various belief systems, as well as accessibility issues in engaging in religious practice. For those individuals with disabilities seeking psychological treatment around sexual intimacy, sexual expression, and close relationship functioning, many can benefit from consideration of this important variable if they identify as religious or spiritual. Identifying disability-affirming sexuality resources can be challenging for people with disabilities due to stigma around sexuality and disability (Mona et al. 2014), especially within a context of particular religious or spiritual beliefs. However, finding ways to access positive sexual expression can also lead to improvements in mental health outcomes and overall wellness (Sanchez-Fuentes et al. 2014). This chapter will explore the intersections among disability, sexuality, and spirituality with the goal of increasing understanding of their interrelatedness and potential impact on psychological health.

Definitions of disability and religion/spirituality (R/S) will first be addressed to create a foundation on which to illustrate how sexuality and intimacy intersect with this important life identity and experience. A review of the demographics of R/S will be presented, followed by examination of the reasons that mental health providers have traditionally excluded R/S in clinical work and reasons that this variable may be helpful to address. A discussion of the various potential psychological and physical health outcomes associated with R/S practice will follow. Specific religious traditions and their approaches to the human experience of disability will be reviewed, accompanied by examination of those traditions' views and approaches to sexual expression and sexual relationships. Lastly, the chapter will offer brief practical suggestions regarding clinical assessment of these diverse issues.

## Definitions

Disability is a complex life experience. For the purposes of this paper, disability will be discussed as a culturally identifying variable, although there is wide variability among people with disabilities themselves and non-disabled people in terms of preferences regarding person-first language ("people with disabilities") versus identity-first language ("disabled people") (Dunn and Andrews 2015). Many people with disabilities identify strongly with and take pride in the cultural experience of disability and prefer identity-first language, whereas others choose person-first language, which historically was encouraged as a way to address the dehumanising social stigma around disability (Andrews et al. 2019). This preference is highly variable and dependent on numerous factors. Disability itself—aside from identity language—is defined as "a physical or mental impairment that substantially limits one or more major life activity" (ADA 2019) within the United States, but this definition does not fully embrace the cultural identity aspects of the lived disability experience. It is also noted that some individuals do not identify as having a disability per se but may present for psychological treatment in the context of a chronic health condition or other, physical/mental state that negatively impacts life activities.

Within the research field of the psychology of R/S there has been a great deal of discussion regarding definitions and conceptualisations of these two terms, adding much complexity—and some debate—as to what these terms really mean. Pargament (2007) in his approach to spiritually integrated therapy conceptualises religiousness and spirituality together as "a set of pathways that people follow in search of the sacred" (p. 196). The word "religion" itself has origins in Latin, coming from the root word "religio," meaning a bond between humans and some power greater than humanity (Hill et al. 2000). Spirituality, as a concept separate from religion, has been defined as an internal experience towards the sacred that an individual nurtures on a personal basis and that is "motivated by interest in meaning, purpose and significance" (Johnstone et al. 2007, p. 1154). Religion, in contrast, has been defined as a larger organised institutional system meant to *facilitate* spirituality, including traditions, practices, doctrines, beliefs and moral codes (Exline et al. 2014; Pargament et al. 2013). It is also generally accepted in the field of the psychology of R/S that an individual can be religious without being spiritual or spiritual without being religious (Pargament 1999). In more recent socio-cultural contexts—aside from the academic discipline of the psychology of R/S—there has been an evolution of the terms, such that religion has increasing been understood as a more static institutional religious experience, with spirituality meaning an elevated dynamic experience that is considered more in vogue culturally (Pargament 1999). These findings suggest that using both terms in assessment should prove useful in identifying those individuals who may benefit from the integration of R/S in clinical work.

## Demographics

Religion and spirituality have historically been important components in socio-cultural landscapes throughout the world, although fluctuations in that importance can occur over time and within certain age groups. Worldwide, only 16.4% of the world's population are thought to be unaffiliated with any particular religious faith, with Christians (31.4%), Muslims (23.2%), Hindus (15%), and Buddhists (7.1%) representing the major organised religious affiliations (Pew Research Center 2010). Recent research on adult Americans (Pew Research Center 2014) indicates that 53% of those surveyed stated that religion was *very important* to them and 24% stated that it was *somewhat important*. In order of importance level, the following groups had more than

50% of respondents indicating that religion was very important them: Jehovah's Witness (90%), Historically Black Protestant (85%), Mormon (84%), Evangelical Protestant (79%), Muslim (64%), Catholic (58%), Mainline Protestant (53%) and Orthodox Christian (52%). Whereas there are virtually no reliable statistics on the affiliations of R/S for people with disabilities in particular, it is recognised that people with disabilities represent the largest minority group in the world, comprising approximately 15% of the global population (World Health Organization 2011). As such, people with disabilities represent a large cultural group that may benefit from culturally tailored healthcare services. It is also noted that some individuals with disabilities may carry "dual-minority" status, such as those who identify as sexual minorities (lesbian, gay, bisexual), or gender minorities (e.g., transgender), or ethnic minorities (Olkin 1999 p. 20).

## Psychology and religion

For a myriad of reasons, R/S has only recently received attention as a variable to include in assessment and treatment within psychological practice, although the psychology of R/S has been studied as a basic science for a much longer time within the history of the field. Psychologists have, in the past, avoided addressing religion at all due to its association with superstition or magic and the fear that this could delegitimise psychology as a professional field (Wulff 1997). It is noted also that, as a profession, psychologists tend to identify less as being religious or spiritual themselves, with surveys of psychologists finding much lower religiosity than the general population (Delaney et al. 2007; McMinn et al. 2009). As such, psychologists may unintentionally underestimate the significance of this aspect of people's lives (Pargament et al. 2013). Rehabilitation psychologists—those clinicians who work primarily with individuals with disabilities—who are not religious themselves may see R/S as an inappropriate method for coping with new physical disability or simply inappropriate to address (Johnstone et al. 2007). Others feel that R/S, like politics, may be too personal of a topic to include in assessment or therapeutic enquiry and the question might be experienced as an intrusion for many patients (Trieschmann 2001). Within the medical profession, efforts by physicians to include R/S in treatment have, in some cases, been met with criticism by faith leaders for being out of their scope of practice (Lawrence 2002; Sloan et al. 2000). In a practical sense, many professionals may simply feel unprepared to address these issues due to lack of training in many professional programmes and textbooks (Pargament et al. 2013; Plante 2016).

Despite these various negative sentiments and cautions to using R/S in clinically oriented psychology work, Pargament et al. (2013) acknowledged the more recent shift from a now-strong research base in R/S to gradually seeing clinicians applying this positive research in psychology practice. On a more basic level, one of the most important reasons for including R/S in clinical work is that approximately 84% of the world's population, including people with disabilities, identify religion and/or spirituality as being important in their lives (Pew Research Center 2010). Some research has also noted that people who identify as religious or spiritual may choose to engage more frequently in religious practices after they experience an acquired disability or as they are living with a progressive disabling condition (Chen and Boore 2008; Haley, Koenig and Bruchett 2001; Idler and Kasl 1997).

## Religion/spirituality and health outcomes
### Traditional religious practices

The potential health benefits of R/S have been noted in many studies, including less prevalent drug and alcohol abuse (Gorsuch 1995); more effective coping with negative life events

(Pargament 1997; Poloma and Pendleton 1989); and higher resilience (Connor and Davidson 2003). In terms of physical health, one systematic review of 724 quantitative studies on the relationship between religion and health found that 66% of those studies noted a positive relationship between religion and health (Koenig et al. 2001). The specific practice of prayer has been positively associated with pain coping and post-traumatic growth, being linked to providing a sense of meaning/purpose, increasing a sense of control, fostering relaxation, and serving as a means of giving and receiving support (Harris et al. 2010; Wachholtz et al. 2007). Specific research on individuals undergoing rehabilitation for traumatic injuries or other physical medical conditions has found a positive relationship between positive spiritual experiences and better physical and mental health outcomes (Johnstone and Yoon 2009). These authors also note value in addressing R/S in those individuals with acquired disabilities (e.g., traumatic brain injury, stroke, spinal cord injury), who will probably live for a long time with their disability and may find R/S helpful in their adjustment and re-evaluation of life's meaning over time.

## Non-traditional religious practices

Specific less traditional spiritual practices, including yoga and meditation, have also found support in the healthcare literature. Most western yoga instruction is centred on physical postures, breathing, and meditation—and this type of yoga practice has received a great deal of research support for multiple health problems. Ross et al. (2013) note positive associations for individuals with cancer, diabetes, cardiovascular disease, and metabolic syndrome, as well as depression and anxiety. Others have offered personal narrative accounts detailing the use of yoga in their own lives as a path to healing. Sanford (2006) speaks about his journey after becoming paralysed at age 13, then discovering yoga as a young adult and finding a new mind-body connection that he went on to share as a yoga teacher and mentor for people with disabilities. Meditation has a rich tradition itself as a separate spiritual practice and, like yoga, is meant to be a system to alleviate suffering, with much empirical support suggesting its effectiveness in managing pain, depression, and anxiety (Bergemann et al. 2013). Mindfulness-based meditation in particular has been studied quite extensively and is widely used in mental health settings (Davidson et al. 2003).

## Cultural competence

Congruent with psychologists' ethical responsibility to provide culturally competent and culturally informed care, addressing R/S is required. According to the American Psychological Association (APA), cultural differences include religion and are specifically mentioned in the code of ethics. Thus, not recognising this cultural lens within psychological practice would violate specific ethical standards (APA 2002; Plante 2016). This is similar to psychologists' conceptualisation of disability as a cultural variable that must be recognised when working with people with disabilities (APA 2005; Hays 2009). Pargament and colleagues (2013) call it "more than good sense" (p. 6) to include R/S in clinical practice, given that such a large portion of the population sees the world though a "sacred lens" (p. 6). Johnstone (2007) also notes that ignoring these important cultural variables can negatively affect an individual's experience in psychotherapy, such that they may be less open to certain interventions or even discontinue therapy prematurely, not having their "total being" understood. A report from the American Association of Pastoral Counselors (2000) found that this effect may be even more pronounced in individuals with more conservative or devout religious identities, who may be more concerned that clinicians won't take their beliefs seriously. In fact, in this same report, 75% of respondents indicated that it was important to seek counselling from professionals who would integrate their religious beliefs and values into counselling.

## Religious/spiritual traditions and their views on disability and sexuality

Turning our focus from the research on the health benefits and outcomes associated with R/S to the more experiential aspects of disability and R/S, Nosek and Hughes (2001) propose that spirituality and disability are closely related in that both represent journeys of discovery, of developing a clearer sense of self. Trieschmann (2001) suggests that "the illness or disability experience becomes an opportunity for the individual to get in touch with the needs of his or her true Self, i.e., to use this opportunity to manifest the true essence of oneself that exists regardless of physical function." Others have discussed the experience of people with acquired disability and the process of spiritual development with incorporation of living with disability. Tilka and Brindle (2018) note that some individuals develop a deeper connection with God in living with disability and find that their re-evaluation of core values (as part of rehabilitation and healing) leads them to a stronger religious faith—or back to a faith that they had previously neglected. Chen and Boore (2008) also describe a process of spiritual re-evaluation and—for some—a strengthening of religiosity, following an injury.

Within the field of Christian theology and practice, several authors have explored ideas related to disability—both positive and negative—within church teachings. Wilder (2016), in her book exploring disability inclusion in Christian faith communities, discusses many biblical references to disability that may or may not be interpreted as liberating by church leaders, and argues that any spiritual analysis of disability must recognise both the high prevalence of disability in our world as well as its significant stigmatisation. She presents disability theology as a theology that "affirms the value of people with disabilities, critiques the social and religious practices that diminish their well-being, and argues for justice for people with disabilities in the church" (p. 8). She also examines several problematic ideas within the Christian tradition. One of these ideas (also noted by Eisland 1994) is that people with disabilities are understood to be either "divinely blessed or damned" (p. 13), meaning either existing as a more sacred example of humanity for *other* people's religious growth and benefit or that their disability is the result of punishment for a sin. Related to these notions is the belief that disability is due to a lack of faith or insufficient faith (Reynolds 2016). Another more general problematic idea is that the experience of people with disabilities (as represented in the bible and in sermons) is appropriate to exploit in providing religious education for non-disabled people, rather than recognising people with disabilities as equal members of a faith community, and as such, deserving of equal participation and inclusion in all aspects of the community—not just for the edification or pity of non-disabled members. She further proposes a model of liberating theology that recognises that "[God's] justice comes not from reshaping the bodies of people with disabilities, but from reshaping the social and even physical world around them to provide access and welcome" (p. 20). Finally, Black (1996) examines biblical narratives of people with disabilities being cured by God or Jesus, making an important distinction between *cure* and *healing*. She proposes that healing can have multiple positive meanings, but that preachers too often interpret healing in the bible as an individual being cured of their disease, implying that a cured body is always more desirable than a disabled one.

Within Christian theology, there are varied perspectives as to the function and purpose of human sexuality; however, there are some specific elements that hold true across denominations. Most Christian theologians establish the role of sex and sexuality as beginning with the Creation story, wherein God created man, and from him, woman, in His own image, complementary to one another to serve as stewards of the Earth (Joo 2015). Their union, according to Christian theology, serves manifold functions: satisfying man's need for companionship, generating happiness, and creating complementary but distinct conceptualisations of masculinity and femininity (Joo 2015; Wilson 2017). Marriage also serves as the confines in which sexual acts can be engaged in in accordance with the intentions of the Divine, per scripture, as human sexuality serves as a motivator for men and women to engage in an intimate and unified knowledge of one another (Joo 2015; Wilson 2017). Amongst Christian theologians there is ongoing debate as to the role of Lesbian, Gay, and Bisexual (LGB) Christians and the permissibility of their sexual practices from Church leadership (Childs 2013; Daughrity 2015). Amongst various denominations, there is marked variance with regards to the acceptance of LGB individuals, sex acts outside of marriage, and sex for other than procreative purposes (Childs 2013). With regards to the sexual practices of Christians with disabilities, little is written on this topic.

Iozzio (2016) noted similar themes to Christian disability perspectives in her examination of the Catholic Church and views on disability. Central to Church teachings is the sanctity of all life, as the *imago Dei* concept reflects that all human life is created in God's image. Iozzio notes that, while this belief protects the most vulnerable (e.g., unborn children, children with disabilities, the elderly), it also places people with disabilities in a more diminutive position or deserving of pity; Catholics are tasked to "do to the least of our brothers and sisters" as they do to Christ in providing service and protection (p. 125), which, of course, does not place people with disabilities in an equal status in the faith community, including them in the definition of "least." Wilder (2016) also makes reference to the *imago Dei*, but calls on the faithful to acknowledge that they themselves fail to recognise the image of God in many of their fellow human beings and that it is the social response to people with disabilities that represents the real sin that creates stigmatisation and marginalisation.

Catholic conceptualisations of sexuality are predicated on the Holy Trinity and creation of man: that God created man in His image with specific purposes—including sexual pleasure—in mind (United States Conference of Catholic Bishops, n.d.). The Catholic perspective on the sanctity of life, as referenced above, is telegraphed from the belief that man is in the image of the divine, from which the parameters around sexual acts also stem (United States Conference of Catholic Bishops, n.d.). Catholic doctrine emphasises the specifics of the Creation story as paramount in understanding God's desires for human sexuality; the bodies of men and women are, from the Catholic perspective, designed for unique and specific sexual purposes (Trujillo and Sgreccia, 1995; United States Conference of Catholic Bishops, n.d).

In his seminal text, *Man and Woman He Created Them: A Theology of the Body* (1986), Pope John Paul II unequivocally states that the ultimate purpose of humanity is the unification of the Church (humankind) with Christ, this spiritual marriage being represented in the earthly marriages of Catholics. Predicated upon this notion is that sex and human sexuality serve the purpose of literally and divinely unifying man and woman (John Paul II 1986). Equally, because marriage—and the implied sexual activity within—are considered sacraments, Catholic doctrine also stresses the importance of celibacy, abstinence, and chastity for those who are not married within the Church (Trujillo and Sgreccia 1995). Cardinal López Trujillo spoke to the need for people who, through self-discernment, decide that marriage—and particularly the conception and/or rearing of children—is not their preference, be familiarised with the value of celibacy (Trujillo and Sgreccia 1995). He essentially recommended that any Catholic who, due to disability, mental health needs, or any other number of personal reasons, feels that the Catholic conceptualisation of marriage is not reasonable, practical, or desirable to them, has a role of service that they are not only wanted to fill within their Church community, but expected to. What this translates to is a belief system and community practice that places marriage and conception as the ideal but makes space for people who decide marriage is not for them to still have a position of value, respect, and service within their faith community.

Given the theological and doctrinal belief amongst Catholics of the explicit heterosexual and bioessentialist purpose of the human body, LGB identities are not affirmed in most western Catholic churches. Catholic doctrine, such as the writings mentioned above, takes an explicitly pro-heterosexual stance, and stresses the need for celibacy amongst those for whom traditional male-female marriage is not possible (Trujillo and Sgreccia 1995; United States Conference of Catholic Bishops n.d). However, there are movements within the Church which speak to the need for the Church to embrace and integrate LGB Catholics. With regards to transgender individuals, the Catholic Church has no explicit doctrine or policy that speaks to their status within the Church; however, as spelled out above, the equation of birth anatomy and gender is prolific within Catholic doctrine (Human Rights Campaign 2018). The experiences of LGB and transgender Catholics vary widely between parishes and Catholic leadership, with many western parishes engaging in LGBT+ outreach and support programming, countered with frequent non-affirmative practices (Human Rights Campaign 2018).

Other religious traditions carry their own history, biases and challenges in conceptualising and including the experience of disability. These elements are often reinforced by family and friends, not only by religious figures in an individual's life, which can be more influential on self-worth (Nosek and Hughes 2001). The Hindu tradition does involve some positive representation of disability, however, its emphasis on fatalism and karma (morally good actions leading to morally superior states and vice versa) have represented cultural obstacles for people with disabilities and have also prevented people with disabilities from participating in some religious practices (Donahue 2016). Bhatt (1963) states that in ancient India "it was believed that the disabled were reaping what they had sowed in lives bygone and any attempt to ameliorate their lot would, therefore, interfere with this divine justice" (p. 96). She further states that, in modern India, people with disabilities have not found a strong voice compared to western countries and still resign themselves to karmic fate in their conceptualisation of their condition.

Uniform contemporary beliefs about sexuality amongst Hindus are difficult to enumerate, in large part due to the dynamic and variable definitions of what specifically Hinduism is, as well as the utilisation of multiple sacred texts. Hinduism in practice is varied, dialectical at times, and unique to practitioners and faith leaders (Soherwordi 2011).

What this means for sexual practices amongst Hindus is that, simply put, it varies. Doniger, in her text *On Hinduism* (2014), speaks to the tension between traditions within Hinduism with regards to sexuality; sex is simultaneously praised as "inherently sacred," with the human body celebrated for its fertility and erotic capabilities within some religious texts, while simultaneously the sex act is viewed as anxiety provoking, dangerous, and potentially rife with violence. She also asserts that the *Kama sutra*, a famed instructional erotic Hindu text, is geared toward diverse audiences, implied by the inclusion of procreative and non-procreative sex acts, such as oral sex, as well as encouragement to engage in same-sex sexual practices, which are discouraged only amongst the religious elite. Within the *Vedas*, there is also reference to the "third sex," individuals for whom sex is not procreative, due to same-sex preferences or impotence, and these individuals are central to some Hindu traditions (Human Rights Campaign 2018b). Harris (2016) notes that Buddhist teachings generally emphasise a universality that can be considered inclusive for people with disabilities, although little attention to disability per se is given in ancient texts. The concept of karma is used here as well, stemming from harmful action or simply positive or negative *intention*, which Harris notes can make spiritual growth more accessible for people with disabilities, as intention can be present when physical action is not possible. What is most non-affirming for people with disabilities is the focus on suffering (and removal of suffering) and a generally negative view of sickness, which negates the idea that some aspects of disability can be positive.

Sexuality and sexual practices amongst Buddhist practitioners are less delineated than other religious traditions detailed above, in part because of the lack of one specific religious text pertaining to the topic. José Ignacio Cabezón speaks to the disparity between the understanding and practices of Buddhists in the West with regards to sex, and the proscriptions of sexual behaviours and sexuality detailed in Buddhist scholastic literature (Cabezón 2017). Anecdotally, he describes encountering perspectives espoused by western Buddhists touting the lack of prohibition within Buddhism towards sexual practices; in truth, Buddhist texts are incredibly specific about acceptable sexual practices, with guidelines regarding partners, the body parts permissible for inclusion in sex acts, as well as the amount of allowable orgasms for men per night. This speaks to the disparity between practitioners and religious doctrine within Buddhism that also impacts the LGB population. Cabezón (2017) describes a 1997 interaction with the Dalai Lama and LGB Buddhists during which the Dalai Lama affirmed the protections of LGB Buddhists against discrimination on the basis of sexual orientation; however, per the Dalai Lama, due to historical texts prohibiting sex between men, these relationships are not in accordance with Buddhist teachings. In contrast, sexual acts between women are not prohibited. At present, Cabezón and other contemporary Buddhist scholars are seeking to create a singular gender identity-inclusive text to speak to the role of sex and sexuality within Buddhism.

The Islamic faith is examined from a disability perspective by Rispler-Chaim (2016), who notes that the tradition has a rich history of tolerance and compassion, although there is no Arabic word that encompasses people with disabilities as a whole (rather, there is reference separately to blind, deaf or lame individuals). He notes that most of the texts referring to those with disabilities relate to the ability/inability of individuals to perform various religious duties, with the law being "tolerant, accepting, accommodating and forgiving" (p. 179) in exemptions. These exemptions might relate to prayer (which involves physical movement), hajj (pilgrimage to Mecca), or fasting during Ramadan. There is also some indication that Islamic societies demonstrated awareness of human rights (including rights for people with disabilities) well before more organised international work in this area, with a focus on inclusion in society and religious practice (Baderin 2009).

Within Islam, marriage and sex are intimately linked. The Prophet Mohammed recognised in his teachings that marriage and shared sexual pleasure between heterosexual spouses are in congruence with God's will (Dialmy 2010; Gutiérrez 2012). Eroticism within the confines of heterosexual marriage is encouraged, however, the Prophet is specific about appropriate sexual positions, which should be limited to male on top—typically referred to as the "missionary" position (Dialmy 2010; Gutiérrez 2012). It is the prevailing belief that female sexuality, if uncontrolled, could lead to "social chaos," thus husbands are instructed to attend to the sexual satisfaction, via orgasm, of their wife in order to maintain both fidelity and greater social order (Dialmy 2010).

In conversations around disability, there is an ongoing debate amongst Muslim theological scholars, disability advocates, and people with disabilities as to how disabled individuals may engage in marriage within Islam (Al-Aoufi et al. 2012). Some Islamic theologians argue that disabled individuals have a right to be married and have their own families, as this is considered a divine imperative, coupled with specific recognition of the sexual needs of people with disabilities which the institution of marriage may facilitate (Al-Aoufi et al. 2012). With regards to homosexuality, there is variance amongst the Islamic doctrines; the major doctrines of *Sunna* and *Shi'a* dictate that homosexuality must be punished, whereas minor doctrines, such as *Zahirism* and *Rafida* indicate that LGB individuals should not. In contemporary practice, in some Muslim-majority countries LGB individuals are "imprisoned, whipped or sentenced to death" (Dialmy 2010). In contrast, transgender individuals experience recognition and acceptance under Islamic law, due to the belief of Islamic leadership that individuals are born transgender while they choose their sexual preferences; however, community level experiences vary and frequently include social rejection (Human Rights Campaign 2018).

Lastly, the Jewish faith also proposes that God created all humanity in the divine image and that—as such—all humans have inherent worth. However, as Belser (2016) notes, there has been active debate about the role of people with disabilities in Jewish life and law, particularly since the religious practices of Judaism are just as (or more) important than assertions of belief and faith, and many practices (and sacred areas of temples) have not been traditionally accessible for disabled Jews. That being said, there are many historical examples of leaders in the faith with disabilities and the Deaf Jewish community has pioneered the development of Deaf Jewish culture, which has grown significantly over time (Burch 2004).

Amongst adherents to Judaism, the prevailing belief regarding the purpose of human (hetero) sexuality is the fulfilment of a divine imperative, first established in the Garden of Eden: that God created woman to be man's sexual mate—as opposed to merely his helper (Levine 2009). Per Jewish theology, woman was created from man's rib, so that sex drive is considered an urge of reunification between two bodies once comprised of one. Jewish teachings stress that sexual abstinence is an "unnatural" state and that sexual gratification is not exclusively for procreation, but also for pleasure (Levine 2009; Nelson 1999). Notable also under Jewish law is that sexual pleasure is considered a right specifically for women, with the ethical onus upon men being to consider the enjoyment of their female partner in sexual acts (Biale 1984; Nelson 1999). Contemporary Jewish perspectives also allow for sexual pleasure to be derived from varied sexual positions, aside from exclusively male partner on top, and the kissing of body parts aside from merely the mouth, provided it does not become habitual (Nelson 1999). This perspective on the purpose and goals of human sexuality within Judaism are specifically limited to the confines of heterosexual marriage, speaking to the overall tension within Judaism between divinely imperative sexually pleasurable behaviours and the proscribed situations in which they are permitted (Nelson 1999).

Contemporary experiences amongst LGB Jews vary with regards to their inclusion in faith communities and expressions of faith, such as marriage, and are largely dependent upon the sect of Judaism with which the individual identifies—Reform, Conservative, Reconstructionist, or Orthodox. Within all forms of contemporary Orthodoxy, same-sex marriage and sexual activity are prohibited, although some rabbis and Orthodox communities are welcoming of LGB Jews, as outlined by Human Rights Campaign (2018c). Amongst the other three sects of contemporary Judaism, there is an institutional commitment to inclusion of LGB Jews that falls along a spectrum from full and affirming to markedly less so, with variance occurring at a rabbinic level. The experiences of transgender Jews mirror those of LGB Jews, with LGB- affirmative sects engaging in trans-affirmative practices, and more conservative sects engaging in varying levels of strict bioessentialism (Human Rights Campaign 2018a, 2018d, 2018e).

## Disability and the physical practice of religion/spirituality

On a much more practical level, people with disabilities of any faith often face multiple environmental and architectural barriers in simply putting their religious or spiritual beliefs into practice. These barriers may be even more noticeable and upsetting than deeper theological conceptualisations. Challenges with respect to accessible transportation to religious services and basic access into churches, temples, or mosques and all parts of those buildings (including altar and social/community areas) can not only be frustrating, but also convey an unwelcoming message—both for religious practitioners and people with disabilities who might have interest in serving their faith community in any leadership role (e.g., preaching, reading scripture, cantoring). Eisland (1994) notes that some churches take communion to disabled parishioners in their seats if they are not physically able to receive the sacrament at the altar. For her, she reports, this experience was an exclusionary one: "Hence, receiving the Eucharist was transformed for me from a corporate to a solitary experience; from a sacralization of Christ's broken body to a stigmatisation of my disabled body" (p. 112). Prayer itself—in its most traditional physical form in the Christian faith, with hands folded and kneeling on one's knees—is not physically possible for many. Daily prayer for Muslims involves more complex bodily movements, with a combination of standing, kneeling and prostrating (Rispler-Chaim 2016). Of course, prayer is much more than a physical gesture, but this aspect of it may represent a significant loss for many people who practice their faith more traditionally, even if they are "exempt" from this part of the practice. A simple search for the word "knees" on a database of Christian church hymns returns a total of 647 church hymns, with text referencing being on "bended knee" or "falling on my knees" as a method of worship that is not accessible for people with mobility impairments (Hymnary.org 2019). Other issues related to the unpredictability of chronic pain, the need for intermittent time lying in bed or resting, and complex needs in terms of personal care or bowel and bladder management may further frustrate efforts to participate in organised religious practices or community. In the more spiritual-type practices of activities, such as yoga and meditation, people with disabilities may feel excluded from the more stereotypical versions of these activities, observing others engaging in complicated physical yoga poses that require strength and balance or listening to meditation exercises involving a calm voice inviting the practitioner to walk down a peaceful winding staircase or feel warm sand between their toes. Many of these spiritual practices are based on ancient teachings with no cultural precedent to include people with disabilities until recently.

## Disability and sexuality

Many barriers exist that prevent people with disabilities access to sexual healthcare information and development of positive sexual expression in their lives. Often people with disabilities are considered asexual or without sexual interest or ability, due to general stigma around the disability experience (Clemency Cordes et al. 2016). Body image concerns, lack of sexual health information in formative years (compared to non-disabled peers), and social isolation due to accessibility issues can all contribute to lack of sexual self-esteem in this population, yet adequate attention to sexual health and functioning in healthcare settings for people with disabilities can contribute significantly to adaptive coping and well-being (Mona et al. 2017).

## Assessment and clinical considerations

Screening for sexuality and intimacy concerns should be included in all initial assessments and should be prefaced by a statement normalising sexual health as an important component of overall health. Hatzichristou et al. (2004) suggests a single basic screening question, such as "What questions or concerns might you have with your sexual functioning or sexual intimacy" or a question about changes in this area of their functioning. This initial question may not lead to further discussion of sexual issues but may serve to normalise this topic for the individual raising this issue in future sessions. If more extensive assessment is indicated—particularly for individuals who present specifically for treatment of sexuality issues—Mona et al.

(2010) provide a comprehensive sexual health assessment model that takes into account various diversity issues that may be relevant for this population.

Initial screening around R/S should also be included in all initial assessment of individuals and couples, both for the reasons outlined in the introduction to this chapter, as well as to simply open the door to this topic as something that the therapist is willing to incorporate in therapy. A commonly recommended spiritual screening assessment, proposed by Pargament (2007), consists of four yes/no questions, followed by further prompts if the individual answers affirmatively. They are the following: 1. Do you consider yourself a religious or spiritual person? If so, in what way? 2. Are you affiliated with a religious or spiritual denomination or community? If so, which one? 3. Has your problem affected you religiously or spiritually? If so, in what way? 4. Has your religion or spirituality been involved in the way that you have coped with your problem? If so, in what way? In working with individuals with disabilities, one could also add the following disability-focused screening question, regardless of which type of R/S assessment is being used: Has living with your disability affected your religious beliefs or practices? If so, in what way?

As outlined above, several issues may be relevant to explore in assessment for those who indicate that R/S is of great importance to them. These include: arranged marriage/consent in marriage, traditional/acceptable gender roles and roles within sexual practices, beliefs about intimacy outside the primary relationship or marriage, acceptable/permissible sexual practices, beliefs on masturbation, and beliefs about sexual orientation and gender minorities. Hall and Graham (2014) note that religion tends to be less influential on sexual practices and beliefs in many western countries and Russia, compared to the Arab and Persian countries of the Middle East. More conservative religious groups in general, such as Evangelical Christians, may also have more traditional beliefs in these areas (Eliason et al. 2017). Regardless, it is important to be sensitive to variations in religious prohibitions or proscriptions, especially since there can be significant differences in sexual attitudes among those who identify with the same religious group (Ahrold et al. 2011). These potential religious prohibitions around sexual behaviour are also separate from clients' actual or lived-out behaviour and may be relevant in addressing discrepancies that may be causing R/S struggle or other negative emotions, such as guilt (Woo et al. 2012). In addition, for sexual minority individuals, direct inquiry regarding R/S struggles around sexual identity can be helpful in guiding spiritually positive interventions (Fontenot 2013).[1]

## Conclusion

Understanding the experience of spirituality in people with disabilities as it relates to sexuality issues and overall well-being can be challenging and complex, but, when recognised, can lead to better physical and mental health outcomes. Many individual variables are considered when incorporating R/S in this type of clinical work and careful assessment is integral to this process. Scholars and healthcare providers are encouraged to seek additional education and mentorship to gain greater understanding, and clinical training programmes should consider this information in their planning of curriculum around cultural competency. Future directions for research enquiry could involve targeting more specific disability groups (e.g., those with vision or hearing impairments, those with developmental disabilities) in order to learn more about their lived experiences with R/S and sexuality. As Kersting (2003) notes, addressing R/S can bring "a beautiful aspect of the human experience into the therapy room" (p. 40) and—within the context of more intimate sexuality issues—can be extremely helpful to individuals in having their overlapping identities fully recognised.

## Note

1 Those providers who seek additional educational training on incorporating R/S in assessment and treatment intervention are referred to Johnstone et al. (2007), Miller (1999) and Plante (2016) for specific resources.

## References

ADA National Network (2019) What is the definition of disability under the ADA? [online]. Retrieved from https://adata.org/faq/what-definition-disability-under-ada [Accessed 11 Sep. 2019].

Ahrold, T.K., Farmer, M.A., Trapnell, P.D., & Meston, C.M. (2011) The relationship among sexual attitudes, sexual fantasy, and religiosity. *Archives of Sexual Behavior* 40, pp. 619–630.

Al-Aoufi, H., Al-Zyoud, N., & Shahminan, N. (2012) Islam and the cultural conceptualisation of disability. *International Journal of Adolescence and Youth* 17(4), pp. 205–219.

American Association of Pastoral Counselors (2000) Samaritan Institute report. Washington, DC: Greenberg Quinlan Research.

American Psychological Association (2002) Ethical principles of psychologists and code of conduct. *American Psychologist* 57, pp. 1060–1073.

American Psychological Association (2005) *Report of the presidential task force on evidence-based practice.* Retrieved from http://www.apa.org/practice/ebpreport.pdf

Andrews, E.E., Forber-Pratt, A.J., Mona, L.R., Lund, E.M., Pilarski, C.R., & Balter, R. (2019) #SaytheWord: A disability culture commentary on the erasure of disability. *Rehabilitation Psychology* 64(2), pp. 111–118.

Baderin, M.A. (2009) *International human rights and Islamic law.* Oxford: Oxford University Press.

Belser, J.W. (2016) Brides and blemishes: Queering women's disability in rabbinic marriage law. *Journal of the American Academy of Religion,* 84(2), pp. 401–429.

Bergeman, E.R., Siegel, M.W., Belzer, M.G., Siefel, D.J., & Feuille, M. (2013) Mindful awareness, spirituality, and psychotherapy. In K.I. Pargament, J.J. Exline, & J.W. Jones (Eds.), *APA handbook of psychology, religion, and spirituality Vol. 2: An applied psychology of religion and spirituality.* Washington, DC: American Psychological Association, pp. 207–222.

Bhatt, U. (1963) *The physically handicapped in India (A growing national problem).* Bombay: Popular Book Depot.

Biale, R. (1984) *Women and Jewish Law, An exploration of women's issues in Halakhic sources.* New York: Schocken Books.

Black, K. (1996) *A healing homiletic: Preaching and disability.* Nashville, TN: Abington Publishing.

Burch, S. (2004) *Signs of resistance: American deaf cultural history, 1900 to World War II.* New York: New York University Press.

Cabezón, J.I. (2017) Revisiting the traditional Buddhist views on sex and sexuality. In *Lions roar: Buddhist wisdom for our time.* Retrieved from https://www.lionsroar.com/rethinking-buddhism-and-sex-2/

Chen, H., & Boore, J.R.P. (2008) Living with a spinal cord injury: A grounded theory approach. *Journal of Clinical Nursing* 17(5a), pp. 116–124.

Childs, J. (2013) Human Sexuality. In G. Forell (Ed.), *Christian social teachings: A reader in Christian social ethics from the bible to the present.* 2nd ed. Minneapolis: Augsburg Fortress, pp. 377–409. doi:10.2307/j.ctt22nm868.42.

Clemency Cordes, C., Cameron, R.P., Mona, L.R., Syme, M.L., & Coble-Temple, A. (2016) Perspectives on disability within integrated healthcare. In L. Suzuki, M. Casas, C. Alexander, & M. Jackson (Eds.), *Handbook of multicultural counselling.* 4th ed. Thousand Oaks, CA: SAGE, pp. 401–410.

Connor, K.M., & Davidson, J.R.T. (2003) Development of a new resilience scale: The Connor-Davidson resilience scale (CD-RISC). *Depression and Anxiety* 18, pp. 76–82.

Daughrity, D. (2015) Marriage and sexuality. In *To whom does Christianity belong? Critical issues in World Christianity.* Minneapolis: Augsburg Fortress, pp. 217–236. doi:10.2307/j.ctt13www91.16.

Davidson, R.J., Kabat-Zinn, J., Schumacher, J., Rosenkranz, M., Muller, D., Santorelli, S., & Sheridan, J. (2003) Alterations in brain and immune function produced by mindfulness meditation. *Psychosomatic Medicine* 65, pp. 564–570.

Dialmy, A. (2010) Sexuality and Islam. *European Journal of Contraception and Reproductive Health Care* 15(3), pp. 160–168. doi:10.3109/13625181003793339.

Donahue, A. (2016) Hinduism and disability. In D.Y. Schumm & M. Stoltzfus (Eds.), *Disability and world religions.* Waco, TX: Baylor University Press, pp. 1–23.

Doniger, W. (2014) *On Hinduism.* New York: Oxford University Press.

Delaney, H.D., Miller, W.R., & Bisono, A.M. (2007) Religiosity and spirituality among psychologists: A survey of clinician members of the American Psychological Association. *Professional Psychology: Research and Practice* 38, pp. 538–546.

Dunn, D.S., & Andrews, E.E. (2015) Person-first and identity-first language: Developing psychologists cultural competence using disability language. *American Psychologist* 70, pp. 255–264.

Eisland, N.L. (1994) *The disabled god: Toward a liberatory theology of disability.* Nashville, TN: Arbington Press.

Eliason, K.D., Hall, M.E.L., Anderson, T., & Willingham, M. (2017) Where gender and religion meet: Differentiating gender role ideology and religious beliefs about gender. *Journal of Psychology and Christianity* 36(1), pp. 3–15.

Exline, J.J., Pargament, K.I., Grubbs, J.B., & Yali, A.M. (2014) The religious and spiritual struggles scale: Development and initial validation. *Psychology of Religion and Spirituality* 6(3), pp. 208–222.

Fontenot, E. (2013) Unlikely congregations: Gay and lesbian persons of faith in contemporary U.S. culture. In K.I. Pargament, J.J. Exline, & J.W. Jones (Eds.), *APA handbook of psychology, religion, and spirituality. Vol. 1: Context, theory, and research.* Washington, DC: American Psychological Association, pp. 617–633.

Gorsuch, R.L. (1995) Religious aspects of substance abuse and recovery. *Journal of Social Issues* 51(2), pp. 65–84.

Gutiérrez, R.A. (2012) Islam and sexuality. *Social Identities: Journal for the Study of Race, Nation and Culture* 18(2), pp. 155–159. doi:10.1080/13504630.2012.652840.

Haley, K.C., Koenig, H.G., & Bruchett, B.M. (2001) Relationship between private religious activity and physical functioning in older adults. *Journal of Religion and Health* 40, pp. 305–312.

Hall, K.S.K., & Graham, C.A. (2014) Culturally sensitive sex therapy: The need for shared meanings in the treatment of sexual problems. In Y.M. Binik & K.S.K. Hall (Eds.), *Principles and practice of sex therapy.* New York: Guilford Press, pp. 334–358.

Harris, J., Erbes, C., Engdahl, B., Tedeschi, R., Olsdon, R., Winskowski, A.M.M., & McMahill, J. (2010) Coping functions of prayer and posttraumatic growth. *International Journal for the Psychology of Religion* 20, pp. 26–38.

Harris, S.E. (2016) Buddhism and disability. In D.Y. Schumm & M. Stoltzfus (Eds.), *Disability and world religions.* Waco, TX: Baylor University Press, pp. 25–45.

Hatzischristou, D., Rosen, R., Broderick, G., Clayton, A., Cuzin, B., & Derogatis, L. (2004) Clinical evaluation and management strategy for sexual dysfunction in men and women. *Journal of Sexual Medicine* 1(1), pp. 49–57.

Hays, P. (2009) Integrating evidence-based practice, cognitive-behavior therapy, and multicultural therapy: Ten steps for culturally competent practice. *Professional Psychology: Research and Practice* 40, pp. 354–360.

Hill, P.C., Pargament, K.I., Hood, R.W., McCullough, M.E., Swyers, J.P., Larson, D.B., & Zinnbauer, B.J. (2000) Conceptualizing religion and spirituality: Points of commonality, points of departure. *Journal for the Theory of Social Behaviour* 30(1), pp. 51–77.

Human Rights Campaign (2018a) Stances of faiths on LGBTQ issues: Conservative Judaism. Retrieved from https://www.hrc.org/resources/stances-of-faiths-on-lgbt-issues-conservative-judaism

Human Rights Campaign (2018b) Stances of faiths on LGBTQ issues: Hinduism. Retrieved from https://www.hrc.org/resources/stances-of-faiths-on-lgbt-issues-hinduism

Human Rights Campaign (2018c) Stances of faiths on LGBTQ issues: Orthodox Judaism. Retrieved from https://www.hrc.org/resources/stances-of-faiths-on-lgbt-issues-orthodox-judaism

Human Rights Campaign (2018d) Stances of faiths on LGBTQ issues: Reconstructionist Judaism. Retrieved from https://www.hrc.org/resources/stances-of-faiths-on-lgbt-issues-reconstructionist-judaism

Human Rights Campaign (2018e) Stances of faiths on LGBTQ issues: Journal of Reform Judaism. Retrieved from https://www.hrc.org/resources/stances-of-faiths-on-lgbt-issues-reform-judaism

Hymnary.org (2019) Text results: Search term "knees." Retrieved from https://hymnary.org/search?qu=all%3Aknees%20in%3Atext

Idler, E.L., & Kasl, S.V. (1997) Religion among disabled and nondisabled persons: II. Attendance at religious services as a predictor or the course of disability. *Journals of Gerontology: Series B* 52B(6), pp. S306–S316.

Iozzio, M.J. (2016) Catholicism and disability. In D.Y. Schumm & M. Stoltzfus (Eds.), *Disability and world religions.* Waco, TX: Baylor University Press, pp. 115–135.

Johnstone, B., Glass, B.A., & Oliver, R.E. (2007) Religion and disability: Clinical, research and training considerations for rehabilitation professionals. *Disability and Rehabilitation* 29(15), pp. 1153–1163.

Johnstone, M., & Yoon, D.P. (2009) Relationships between the brief multidimensional measure of religiousness/spirituality and health outcomes for a heterogenous rehabilitation population. *Rehabilitation Psychology* 54(4), pp. 422–431.

Joo, C.G. (2015) Marriage and sexuality in terms of Christian theological education. *Procedia: Social and Behavioral Sciences* 174, pp. 3940–3947. doi:10.1016/j.sbspro.2015.01.1137.

Kersting, K. (2003) Religion and spirituality in the treatment room. *American Psychology Association Monitor* 34(11), p. 40.

Koenig, H.G., McCullough, M., & Larson, D.B. (2001) *Handbook of religion and health.* New York: Oxford University Press.

Lawrence, R.J. (2002) The witches brew of spirituality and medicine. *Annals of Behavioral Medicine* 24(1), pp. 74–76.

Levine, E. (2009) Validating human eroticism. In *Marital relations in ancient Judaism.* Wiesbaden: Harrassowitz Verlag, pp. 140–151. Retrieved from http://www.jstor.org.lib.pepperdine.edu/stable/j.ctvc2rm5c.11

McMinn, M.R., Hathaway, W.L., Woods, S.W., & Snow, K.N. (2009) What American Psychological Association leaders have to say about Psychology of Religion and Spirituality. *Psychology of Religion and Spirituality* 1, pp. 3–13.

Miller, W.R. (1999) Diversity training in spiritual and religious issues. In W.R. Miller (Ed.), *Integrating spirituality into treatment.* Washington, DC: American Psychological Association, pp. 253–263.

Mona, L.R., Cameron, R.P., & Clemency Cordes, C. (2017) Disability culturally competent sexual healthcare. *American Psychologist* 72(9), pp. 1000–1010.

Mona, L.R., Goldwaser, G., Syme, M., Cameron, R.P., Clemency, C., Miller, A.R., Lemos, L., & Ballan, M.S. (2010) Assessment and conceptualization of sexuality among older adults. In P.A. Lichtenberg (Ed.), *Handbook of assessment in clinical gerontology.* 2nd edn. New York: Elsevier, pp. 331–356.

Mona, L.R., Syme, M.L., Cameron, R.P., Clemency Cordes, C., Fraley, S.S., Baggett, L.R., & Roma, V.G. (2014) Sexuality and disability: A disability-affirmative approach to sex therapy. In Y.M. Binik & K. Hall (Eds.), *Principles and practices of sex therapy.* 5th edn. New York: Guilford Press, pp. 457–481.

Nelson, W. (1999) Sexuality in Judaism. Retrieved from http://www.mesacc.edu/~thoqh49081/StudentPapers/JewishSexuality.html

Nosek, M.A., & Hughes, R.B. (2001) Psychospiritual aspects of sense of self in women with physical disabilities. *Journal of Rehabilitation* 67(1), pp. 20–25.

Olkin, R. (1999) *What psychotherapists should know about disability.* New York: Guilford Press.

Pargament, K.I. (1997) *The psychology of religion and coping: Theory, research, practice.* New York: Guilford Press.

Pargament, K.I. (1999) The psychology of religion and spirituality? Yes and no. *The International Journal for the Psychology of Religion* 9(1), pp. 3–16.

Pargament, K.I. (2007) *Spiritually integrated psychotherapy: Understanding and addressing the sacred.* New York: Guilford Press.

Pargament, K.I., Mahoney, A., Exline, J.J., Jones, J.W., & Shafranske, E.P. (2013) Envisioning an integrative paradigm for the psychology of religion and spirituality. In K.I. Pargament, J.J. Exline, & J.W. Jones (Eds.), *APA handbook of psychology, religion, and spirituality: Vol. 1, Context, theory, and research.* Washington, DC: American Psychological Association, pp. 1–19.

Pargament, K.I., Mahoney, A., Shafranske, E.P., Exline, J.J., & Jones, J.W. (2013) From research to practice: Toward an applied psychology of religion and spirituality. In K.I. Pargament (Ed.), *APA handbook of psychology, religion, and spirituality: Vol. 2. An applied psychology of religion and spirituality.* Washington, DC: American Psychological Association, pp. 3–22.

Paul II, J. (1986) *Man and woman he created them: A theology of the body.* Boston, MA: Pauline Books & Media. ISBN 0-8198-7421-3.

Pew Research Centre (2010) Pew-Templeton global religious futures project. Retrieved from http://www.globalreligiousfutures.org/explorer/custom#/?subtopic=15&chartType=pie&data_type=percentage&destination=from&year=2010&religious_affiliation=all&countries=Worldwide&gender=all&age_group=all

Pew Research Centre (2014) *2014 religious landscape study (RLS-II).* Retrieved from http://www.pewforum.org/religious-landscape-study/importance-of-religion-in-ones-life/

Plante, T.G. (2016) Principles of incorporating spirituality into professional clinical practice. *Practice Innovations* 1(4), pp. 276–281.

Poloma, M.M., & Pendleton, B.F. (1989. Exploring types of prayer and quality of life research: A research note. *Review of Religious Research* 31, pp. 46–53.

Reynolds, T. (2016) Protestant Christianity and disability. In D.Y. Schumm & M. Stoltzfus (Eds.), *Disability and world religions.* Waco, TX: Baylor University Press, pp. 115–135.

Rispler-Chaim, V. (2016) Islam and disability. In D.Y. Schumm & M. Stoltzfus (Eds.), *Disability and world religions.* Waco, TX: Baylor University Press, pp. 167–187.

Ross, A., Friedmann, E., Bevans, M., & Thomas, S. (2013) National survey of yoga practitioners: Mental and physical health benefits. *Complementary Therapies in Medicine* 21, pp. 313–323.

Sanchez-Fuentes, M., Santos-Iglesias, P., & Sierra, J.C. (2014) A systematic review of sexual satisfaction. *International Journal of Clinical and Health Psychology* 14, pp. 67–75.

Sanford, M. (2006) *Waking: A memoir of trauma and transcendence*. U.S.: Rodale.

Sloan, R.P., Bagiella, E., VandeCreek, L., Hover, M., Casalone, C., Hisrch, T.J., Hasan, Y., & Poulos, P. (2000) Should physicians prescribe religious activities? *New England Journal of Medicine* 342(25), pp. 1913–1916.

Soherwordi, S.H. (2011) Hinduism: A western construction or an influence? *South Asian Studies: A Research Journal of South Asian Studies* 26(1), pp. 203–214. Retrieved from https://pdfs.semanticscholar.org/40da/e6ac33e7a5373d5a5ecaedeef8870329789b.pdf

Tilka, N., & Brindle, S.S. (2018) Understanding spiritual well-being among veterans with spinal cord injury. Unpublished raw data.

Trieschmann, R.B. (2001) Spirituality and energy medicine. *Journal of Rehabilitation* Jan/Feb/March, pp. 26–32.

Trujillo, A.L., & Sgreccia, E. (1995) The truth and meaning of human sexuality: Guidelines for education within the family. Retrieved from http://www.vatican.va/roman_curia/pontifical_councils/family/documents/rc_pc_family_doc_08121995_human-sexuality_en.html

United States Conference of Catholic Bishops (n.d.) Love and sexuality. Retrieved from http://www.usccb.org/beliefs-and-teachings/what-we-believe/love-and-sexuality/index.cfm

Wachholtz, A.B., Pearce, M.J., & Koenig, H. (2007) Exploring the relationship between spirituality, coping, and pain. *Journal of Behavioral Medicine* 30, pp. 311–318.

Wilder, C. (2016) *Disability, faith, and the church: Inclusion and accommodation in contemporary congregations*. Santa Barbara, CA: Praeger.

Wilson, T.A. (2017) *Mere sexuality: Rediscovering the Christian vision of sexuality*. Retrieved from https://ebookcentral-proquest-com.lib.pepperdine.edu

Woo, J.S.T., Morshedian, N., Brotto, L.A., & Gorzalka, B.B. (2012) Sex guilt mediates the relationship between religiosity and sexual desire in East Asian and Euro-Canadian college-aged women. *Archives of Sexual Behavior* 41, pp. 1485–1495.

World Health Organization (2011) *World report on disabilities*. Retrieved from https://www.who.int/disabilities/world_report/2011/report/en/

Wulff, D.M. (1997) *Psychology of religion: Classic and contemporary*. 2nd ed. New York: Wiley.

# INDEX

Page numbers in **bold** denote tables, in *italic* denote figures

Printed in the United States
By Bookmasters